The Collected Works of Eugene Paul Wigner

Part A · Volume IV

Springer-Verlag Berlin Heidelberg GmbH

Photo: 1983

Eugene Paul Wigner

# The Collected Works of Eugene Paul Wigner

## Part A

### *The Scientific Papers*

Editor: Arthur Wightman

Annotated by
Nandor Balazs  Herman Feshbach  Brian Judd  Walter Kohn
George Mackey  Jagdish Mehra  Abner Shimony  Alvin Weinberg
Arthur Wightman

## Part B

### *Historical, Philosophical, and Socio-Political Papers*

Editor: Jagdish Mehra

Annotated by
Conrad Chester  Gérard Emch  Jagdish Mehra

# The Collected Works of Eugene Paul Wigner

## Part A

### *The Scientific Papers*

*Volume I*
Part I: Eugene Paul Wigner – A Biographical Sketch
Part II: Applied Group Theory 1926–1935
Part III: The Mathematical Papers

*Volume II*
Nuclear Physics

*Volume III*
Part I: Particles and Fields
Part II: Foundations of Quantum Mechanics

*Volume IV*
Part I: Physical Chemistry
Part II: Solid State Physics

*Volume V*
Nuclear Energy:
Part I: Eugene Wigner and Nuclear Energy
Part II: Memoir of the Uranium Project
Part III: Articles, Reports, and Memoranda on Nuclear Energy
Part IV: The Wigner Patents

## Part B

### *Historical, Philosophical, and Socio-Political Papers*

*Volume VI*
Philosophical Reflections and Syntheses

*Volume VII*
Historical and Biographical Reflections and Syntheses

*Volume VIII*
Socio-Political Reflections and Civil Defense

# The Collected Works of

# Eugene Paul Wigner

Part A

## *The Scientific Papers*

Volume IV

Part I: Physical Chemistry
Annotated by Nandor Balazs

Part II: Solid State Physics
Annotated by Walter Kohn

Edited by Arthur S. Wightman

Springer

Arthur S. Wightman

Department of Physics, Princeton University
Joseph Henry Laboratories, Jadwin Hall
Princeton, NJ 08544, USA

Nandor Balazs

Department of Physics
State University of New York at Stony Brook
Stony Brook, NY 11794, USA

Walter Kohn

Department of Physics
University of California at Santa Barbara
Santa Barbara, CA 93106, USA

The photograph of E. P. Wigner on page II is reproduced from the frontispiece of the publication
*How Far Are We from Gauge Forces?* Proceedings of the Twenty-First Course of the International School of Subnuclear
School Physics, held August 3–14, 1983, in Erice, Trapani, Sicily, Italy. Plenum Press, New York 1985.

Library of Congress Cataloging-in-Publication Data
(Revised for A, v.1, pt.1-3)
Wigner, Eugene Paul, 1902– The collected works of Eugene Paul Wigner.
Includes bibliographical references.
Contents: pt. A. The scientific papers – v. 1, pt. 1. A bibliographical sketch/by Jagdish Mehra.
pt. 2. Applied group theory 1926–35 annotated/by Brian Judd.
pt. 3. Mathematical papers annotated/by George Mackey –
– v. 5. Nuclear energy. 1. Mathematical physics. I. Weinberg, Alvin Martin, 1915–. II. Title
QC19.3.W54   1992   530   92-38376
ISBN 978-3-642-63824-4     ISBN 978-3-642-59033-7 (eBook)
DOI 10.1007/978-3-642-59033-7

Typesetting of the annotations, the translated contributions (from Hungarian) on pp. 63ff. and pp. 121ff., and the reset
contribution on pp. 333ff.: Springer T$_E$X in-house system
SPIN 10080947          55/3143 – 5 4 3 2 1 0 – Printed on acid-free paper

# Editors' Preface

The papers have been divided, necessarily somewhat arbitrarily, into two parts

Part A: The Scientific Papers
Part B: Historical, Philosophical, and Socio-Political Papers

Within each part, the papers have been divided by subject, and within each subject printed chronologically. With some exceptions, every scientific paper is reprinted in its original form. One class of exceptions consists of papers that are simply translations into Hungarian from German or English; they are omitted, but listed in the bibliographies. Scientific papers originally in Hungarian have been translated into English. Some of the papers of Volume V/Part III, Articles, Reports, and Memoranda on Nuclear Energy, have been reset and the figures redrawn. The originals were declassified reports, some in nearly illegible shape. Some reports and patents in Volume V/Part III and Part IV are listed by title only. In contrast to the scientific papers where the coverage is essentially complete, in Part B, a selection has been made. We believe it is representative of Wigner's far ranging concerns. The five books in which Wigner was involved as author, co-author, or lecturer are not reprinted in The Collected Works, but are noted in the annotations and bibliographies.

After the publication of Volume I, V, and VI of The Collected Works, Eugene Wigner died (1 January 1995). The heirs of his estate have donated his scientific papers and correspondence to Princeton University where they are available to researchers at

The Eugene P. Wigner Papers
Manuscripts Division
Department of Rare Books and Special Collections
Princeton University Libraries

Springer-Verlag and the co-editors of The Collected Works intend to give to Princeton University for inclusion in this collection, whatever scientific papers are left over after the publication of The Collected Works is complete.

*Jagdish Mehra*
*Arthur S. Wightman*

# Contents

# PART II

## Solid State Physiscs

# PART I

## Physical Chemistry

# Wigner on Physical Chemistry

Annotation by Nandor Balazs

## I. Studies in Chemical Reactions

Though young Eugene Wigner's original interests did lie in the direction of physics, he nevertheless started his university studies reading chemical engineering. He himself describes the reasons for this choice as follows [1]. "Father once asked me, 'Tell me, son, what do you want to be?' After thinking briefly I replied, 'Daddy, if I may be quite honest, I would like to become a scientist, preferably a physicist.' Father had expected this answer, and said, 'Tell me, how many positions are there in this country for a physicist?' I thought, and, exaggerating a little, answered, 'Four'. Father, in turn, 'And do you think that you are sufficiently clever to get one of the four?' Thus, we then decided that if I go to the university I should become a chemical engineer." After a year at the Technical University in Budapest (1920–1921) he continued his studies at the Technical University, Berlin. Upon his arrival to Berlin his friendly and professional ties with Michael Polanyi increased. When, in 1923, Polanyi became a departmental head in the Kaiser Wilhelm Institute of Physical and Electrochemistry he also returned to his earlier interest in chemical reactions. Already before, during 1920 Polanyi came to the conclusion that the theories then in existence cannot be correct; and he even questioned whether the laws of physics known at that time were adequate to construct a correct theory of chemical reactions. He communicated these concerns to his close friend, Eugene Wigner, who had a very high opinion of Polanyi. It seemed thus natural that Wigner decided to write his doctoral dissertation on the problem of chemical reactions.

While working on his Thesis he also collaborated with others in widely different fields. His first published paper (1923, with A. Szegvari [2]) deals with the concentration dependence of the dielectric constant of colloidal vanadium pentoxide, using a simple model to describe the polarizability of a colloid particle. The next one, [3], with H. Mark, is a beautiful and detailed analysis in X ray crystallography, determining the crystal structure of $\alpha$ sulfur. This is an impressive achievement, since the unit cell contains 128 atoms!

During the next year, 1925, he received his degree from the Technische Hochschule in Berlin. Wigner's copies of the Thesis did not survive. One in Budapest was destroyed in 1945 during the turmoil of the Soviet occupation, the other copy in Berlin was burned in the bombing.

The times were auspicious to consider the theory of chemical reactions. 1925 was the year Heisenberg, and Dirac constructed the foundation of quantum mechanics, and the following year brought Schrödinger's celebrated series of papers on wave mechanics. And Polanyi was indeed correct: it required new laws of physics to achieve progress in the proper understanding of chemical reactions.

Already Boltzmann noticed that the theory of chemical reactions falls naturally into distinct parts [4]. One is the description and analysis of the mechanism which leads to a reaction among the colliding constituents, or to the dissociation of a constituent; the other part is the statistical mechanical description of the collection of particles which undergo the reaction. The reaction analysis consists of a) describing the elementary reactions in terms of the known quantal properties of the constituents; b) combining these elementary reactions in order to express in terms of them the actual reactions. These steps absolutely required the advent of quantum mechanics, since even the notion of valence forces were incomprehensible without it. However, most reactions envisaged took place in dilute gases and at normal temperatures; thus in the statistical mechanical description the usual paraphernalia of *classical* statistical mechanics could still be used, (these being in fact the semiclassical approximation of the more correct quantum mechanical results).

On and off Wigner was concerned with the theory of chemical reactions for about fifteen years, from 1925 to 1939, writing eleven articles during this period. These are the following: [5], [6] with M. Polanyi; [7] with H. Pelzer; [8]; [9]; [10] with L. Farkas; [11] with H. Eyring; [12]; [13] with J. O. Hirschfelder; [14], [15]. One can rightly claim that these papers laid the foundation of the modern theory of chemical reactions as we know it.

Paper [5], written together with Polanyi, is based in part on his Berlin thesis, and a brief account of it is also given by Wigner in Polanyi's Obituary, written by him, ([16], p. 421). In it Wigner and Polanyi anticipated some of the most basic results of the yet to be invented quantum mechanics, to wit, that *the energy spread $\Delta\varepsilon$ is related to the life time $\Delta\tau$ through $\Delta\varepsilon\Delta\tau = h$, and that the conservation of angular momentum is only valid in the formation of molecules for values of the angular momenta that are multiples of $h/2\pi$.* (Wigner once said jocularly ([1], p. 7) "that he has been accused by several people inventing here Heisenberg's uncertainty relations". This seems to be also the result to which Wigner's criptic remark refers in [17] that the time-energy uncertainty relation was known before Heisenberg's.) Other parts of the paper are not yet quantal at all; among others, it studies monomolecular reactions by considering the probability of the internal energy fluctuation of a molecule exceeding a given threshold. This naturally leads immediately to Arrhenius factors in the decay rates.

With the arrival of quantum theory further advances could be made. The basic step forward was provided by F. London [18] who clearly recognised that in a chemical reaction two different possibilities can arise. 1) The quantum states of the electrons do not change, but the chemical constitution is altered;

this situation is conventionally called the adiabatic case. 2) Both the quantum states of the electrons and the chemical constitution undergoes a change; the diabatic case. (In the discussion of the adiabatic case London makes full use of the so called Born-Oppenheimer approximation, before the Born-Oppenheimer paper!) The adiabatic situation is the usual one; in the hands of Wigner, and others it has led to the development of the now standard Transition State Theory, abbreviated often as TST. (The non-adiabatic theory was first studied also by F. London [19]. However, already in 1929 Wigner, with von Neumann, draws attention to the difficulties which can arise if adiabatic energy surfaces meet; [20].) The first appearance of TST is in [7]; it is then applied to reactions involving ortho- and para-hydrogen, and deuterium; [7], [8], [9], [10]. Paper [11] is a popular account of the TST, while paper [14] is an extensive survey of the developments up to 1937. It still can serve as a superb introduction to the theory.

The basic idea is very simple. In a dilute gas we may consider only the atoms actually partaking in an elementary reaction to form the quantal system under consideration. In general, the nuclei of the atoms move slowly compared to the electrons. This allows one first to neglect the kinetic energy of the nuclei altogether, and compute the lowest energy state of this system, considering the nuclei as fixed. The adiabatic assumption now states that the nuclei will actually move as if this energy were the potential energy of the nuclear motion. Since the nuclei are heavy, their motion can be described classically in the configuration space of the nuclei by the motion of a configuration point on this energy surface. This potential energy landscape is usually quite complicated. However, for many reactions, there will be a valley connecting the initial and final configurations, with a saddle point in the valley. The saddle point configuration is considered to be the transition state, and the value of the energy at the saddle point is the activation energy of the reaction. The transit of the configuration point across the saddle point describes the reaction, and it usually does so using *classical* dynamics. This picture may not be always valid. Firstly, the adiabatic hypothesis may fail. Secondly, it may be necessary to use the whole *phase* space (and not only the *configuration* space!) to describe the dynamical situations which correspond to the transition state, (as for example in association reactions in three body collisions). Finally, the presence of quantum effects may modify this picture. Wigner returned several times to the general discussion of these limitations. (See, for example the review [14], and the discussion after [15], including J. Franck's remarks.)

More specifically, in [13] he addresses himself to the limitations which arise using a classical passage over the quantal energy surface, and then to the use of some corrections to salvage the classical picture. There are two points considered. Firstly, the classical definition of the transition state requires that the energy landscape be sufficiently *flat* across the saddle point; secondly, even then the classical passage across the saddle point must be corrected to take into account the quantal *reflections*, and the *tunnelling* that may occur. In [13] Wigner estimates the condition on the flatness of the barrier, and the size of the quantal reflections so as to allow this theory to be still applicable. The possibility of *chemical reactions through tunnelling* is already mentioned in [6, p. 441] this being the first mention in the literature of a "sub-barrier reaction"! In a remarkable note, [8], all the quantum corrections to the flux across the barrier are computed using the just invented Wigner function [21]. These corrections are then utilised in the description of ortho-to-para-hydrogen conversion. Ironically, the corrections make the agreement worse!

There are, however, reactions for which the TST in its present form cannot be sensibly applied since the description of the reaction requires the *whole phase-space* of the atoms and not only their *configuration* space. Such a reaction is the association of two atoms in a three body collision. Wigner observed that this is more a problem of statistical mechanics, than that of quantum mechanics, [15]. In the phase-space of the constituents of the three body collision a critical hypersurface can be defined which separates the bound states of the associated atoms from the free states, while the third body resides at infinity. The reaction rate constant is then expressible in terms of the average crossing rate across this *phase-space surface* in the required direction.

These papers had considerable direct, and indirect effects on the scientific community.

The *direct* effect was mainly on the chemical constituency, since Wigner's work essentially mapped out the fruitful directions for any future progress. It provided a convenient structural division of the problem into statistical mechanical and quantal parts; it assessed the approximations and shortcomings inherent in the division itself, and in the approximate calculations used in each part. Even today, very little needs to be changed in the excellent review [14], published in 1938, to provide an introduction to the modern theory of reaction rates. In addition to the general theory he also gave explicit calculations for *actual* experiments; see, for example his review article [10] (with L. Farkas) on the rates of elementary reactions in light and heavy hydrogen. The disparities between the actual and computed results were also put to constructive use to refine different models. (See, for example [15].) The basic insights produced in the papers have even been utilised in recent work on chaos, as in the study of fluxes through gaps in cantori [22]. The more recent developments in the theory of chemical reactions occur either in the statistical mechanical treatment, or in the quantal treatment. For the former see the review article [23]; for the quantal development see [25], [26], [27].

The *indirect* effect, particularly in nuclear physics was even greater. The late thirties saw the rise of nuclear physics and the study of fission. *All* the basic results of the theory of chemical reactions, as the use of adiabatic energy surfaces, the compound nucleus as a transition state, the possible tunnelling through the transition state, diabatic effects, ets., etc. could immediately be applied here. (See, for example the basic paper on nuclear fission, N. Bohr and J.A. Wheeler, [28], and a more recent survey [29].) In fact, the theory constructed in the 1925 paper [5], is also valid for nuclear association reactions. Similarly, nuclear fission resembles monomolecular chemical reactions. Thus the theory of chemical reactions was a natural early training ground for Wigner when turning to nuclear physics; and was so used by many others as well.

# II. Studies in Statistical Mechanics

While in quantum theory Wigner's work encompassed both the foundations and the applications, in statistical mechanics he took a more pragmatic view. There are no publications on the foundations, as ergodic theory, approach to thermal equilibrium, origins of irreversibility, etc.; he used statistical mechanics as a given tool to obtain answers to actual problems in physics. (There is one important exception, the invention and use of Wigner's Function. Because of its importance this will be discussed in the adjoining Section III.)

His efforts fall naturally in two major parts, classical, and quantal.

Classical statistical mechanics is used practically throughout in the theory of chemical reactions. As discussed there, he clearly distinguished between the *quantal* description of the atoms partaking in the elementary reaction, and the *statistical mechanical* description of the reaction-process itself. The latter was basically a classical one, corrected for quantum effects. For example, such a quantum notion as a chemical reaction is specified as either a passage of a point in configuration space across a saddle point [7], or across a hypersurface in phase-space [14]. Essentially all statistical mechanical concepts were given in terms of the canonical or microcanonical description.

The full power of classical phase-space methods were exploited in [30] where a general discussion and derivation of the Onsager-reciprocal relations are given. Onsager's original papers were hard to grasp. There, Onsager does essentially two things, 1) he normalises the thermodynamical fluxes and forces in a precise way; 2) next, he shows that if irreversible processes are described as linear relations between the thermodynamical fluxes and forces *so normalised,* then the coefficients appearing in these linear relations obey simple symmetry properties, the Onsager relations. The proof makes use of two assumptions; a) that the decay of a fluctuation proceeds the same way as an irreversible process; b) that the microscopic motion of the underlying dynamical system is time reversible. In [30] Wigner explores the relation between the *macroscopic* description and the *microscopic,* phase-space description of a thermodynamical system, and shows that the principle of detailed balancing holds in the macroscopic description as well if it holds for the microscopic one. Then he replaces the usual linear *differential* relations describing the irreversible process by linear *integral* relations, and shows that by virtue of the *macroscopic* detailed balancing the kernel of the integral equation is such that the Onsager relations immediately follow.

{Let $\vec{Q}$ denote all the macroscopic coordinates needed in the macroscopic description. Then let $P(\vec{Q})$ be the normalised probability density of the $\vec{Q}$ values, and $T_t(\vec{Q} \rightarrow \vec{Q}')$ the transition probability from $\vec{Q}$ to $\vec{Q}'$ in time $t$. Then the linear equations of motion are written as $\int d\vec{Q}' T_t(\vec{Q} \rightarrow \vec{Q}')Q_i' =$

$\sum \varepsilon(t)_i^k Q_k$. The Onsager relations hold if $\varepsilon$ is a symmetric matrix. Wigner now shows that as a consequence of macroscopic detailed balancing the operator $P(\vec{Q})^{1/2} T_t(\vec{Q} \to \vec{Q}') P(\vec{Q}')^{-1/2}$ is a symmetric operator in a real Hilbert space of quadratically integrable functions, and this implies the Onsager relations. Since $\int d\vec{Q}' T_t(\vec{Q} \to \vec{Q}') = 1$, this operator has as its eigenfunctions $P(\vec{Q})^{1/2}$ with eigenvalue unity, and the integral equations show that the submanifold spanned by $Q_i P(\vec{Q})^{1/2}$ is invariant under the action of this operator.}

It is well known that the two body distribution functions of equilibrium statistical mechanics can be approximately computed from an integral equation. In [31] Wigner and Ufford consider points on a circle interacting pairwise with the repulsive potential $v(x) = -kT \ln \sin^2 \pi x/L$, where $x$ is the separation between two points. They then compute both the exact pair distribution function and the solution of the integral equation. The solution of the integral equation differs from the exact solution.

The literature of the integral equation method has since grown tremendously [32], and other exact results are also known [33].

Wigner's work on the theory of random matrices (that was utilised here) has also other links with statistical mechanics. These will be discussed independently.

## III. Studies on Wigner's Function

Soon after the invention of quantum mechanics it became natural to study the quantum corrections arising in equilibrium statistical mechanics. Elementary arguments show that they will be more and more important as the temperature of the system is lowered. Consequently the experimental efforts increased to measure at low temperatures the two standard experimentally accessible functions of thermodynamics, the specific heat of the system as a function of the temperature and pressure, and the pressure as a function of the density and temperature.

The question then arises as to the quantum corrections to these two functions, or better, to the thermodynamical potential of the system. However, this question is only part of a more general, and more fundamental query.

*On what mathematical object, by what mathematical method can one display the existence of the classical limit; and what expansion scheme, in terms of what parameters will systematically provide the correction terms to the classical results?*

For several reasons the answer is by no means obvious. 1) The most efficacious display of classical dynamics is given in the *phase-space*, a manifold which has no natural equivalent in quantum theory, and thus would appear only *in* the classical limit, but not *before*, or *during* the limiting *process*. 2) The intuitively most natural, and mathematically most accessible quantum objects (as wave functions, matrix elements, transition probabilities etc.) usually refer to *particular representations*, while the experimentally observed corrections are clearly representation independent. 3) While it is obvious that the deviation from the classical results will be more and more negligible as Planck's constant tends to zero, the proper choice of the *dimensionless small parameter*, or parameters, is by no means clear.

In order to answer these points a particularly attractive approach was conceived by Wigner, [21], for non-relativistic quantum theory in the absence of spin. His subsequent papers on this topic are [8], [14], [34], [35], and [36] with R.F. O'Connell, [37] with R.F. O'Connell, [38], [39] with M. Hillary, R.F. O'Connell and M.O. Scully, [40] with R.F. O'Connell, [41] with Y.S. Kim, [42] with Y.S. Kim, [43] with Y.S. Kim ([38] demonstrates the immense popularity of the original memoir. It is from Citation Classic, and informs one that between 1961 (the year Citation Index started) and December 1983 over 655 publications referred to this article.)

(Identical suggestions to Wigner's were put forward by Dirac in 1930 [44], and Heisenberg in 1931 [45]. These were ad hoc proposals without exploiting the full scope of the idea.)

Wigner considers the density operator and associates with this *operator* a *function* of real parameters. (The association is invertible and one to one. For Hermitian operators the resulting function is real.) The parameters appearing in this function can be so chosen that *in the classical limit* they are identical to a *particular* set of classical canonical variables, i.e. to Cartesian rectangulars for the coordinates, with canonically conjugate momenta; also, in this limit the associated function becomes identical to the classical phase-space density in this particular set of variables. In the physics literature this function is called Wigner's function, while in the mathematical literature it is named the Weyl symbol of the density operator. (See the later discussion on pseudodifferential operators). It turns out that even outside the classical limit these parameters may still be thought of as some mock coordinates and momenta, and Wigner's function as some sort of mock phase-space density. This comes about since in the present scheme the computation of the correct *quantal expectation values* follows the same formal operations as the evaluation of standard, *classical statistical mechanical averages*, for any value of $\hbar$, (thus *not* only in the classical limit).

The association proposed can be expressed as

$$A_{\mathrm{W}}(p,q) \approx \mathrm{Trace}\left\{A \Delta_{\mathrm{W}}(p,q)\right\},$$

with its inverse

$$A(p_{\mathrm{op}}, q_{\mathrm{op}}) \approx \int dp\,dq\, A_{\mathrm{W}}(p,q) \Delta_{\mathrm{W}}(p,q).$$

Here $p$ and $q$ are the mock momentum and coordinate parameters, $A_{\mathrm{W}}(p,q)$ the function associated with the operator A. The singular operator $\Delta_{\mathrm{W}}(p,q)$ mediates the association; (I do not indicate the dependence on the operators, but only on the parameters.) It is a sort of hybrid Dirac $\delta$ whose argument is part operator, part $c$ number, of the type $\delta(p_{op} - p, q_{op} - q)$, but suitably symmetrised in the operators. It is given by

$$\Delta_{W}(p, q) \approx \int du dv \, \exp[iu(p_{op} - p) + iv(q_{op} - q)]/\hbar \; .$$

(The coeffficients in both formulae depend upon the normalisation used.)

The expectation value of A in the state described by the density operator $\rho$ is

$$\text{Trace } A\rho = \int dp \, dq \, A_{W}(p, q)\rho_{W}(p, q) \; .$$

(Wigner's original paper expresses his function differently. The convenient $\Delta_{W}$ operator was introduced by de Groot; see his Montreal lectures, or the review by Balazs and Jennings, both quoted later.)

Thus, the exact quantal expectation value can be written as in classical statistical mechanics, since the right hand side of this expression is formally the same as used there. (The similarity is only formal, since $\rho_{W}(p, q)$, the analogue of the classical probability density, will in general not be positive as the classical probability density would be. In fact, there is an equivalence only between *expectation values* and *averages over associated functions using $\rho_{W}$ as the weight*. Outside the classical limit the mock parameters $p$ and $q$ and their functions on the mock phase-space only "resemble" their classical equivalents.)

Wigner did not exhibit this relation in its full generality. In particular, he did not discuss (because he did not need it) that *all* the functions appearing in the mock phase-space average must be generated by this association, and not only the function associated with the density operator.

Later, in 1949, Moyal [46] pointed out that Wigner's construction is identical to the particular association proposed by H. Weyl in 1927 [47] between classical dynamical quantities and operators. Weyl used this association *in reverse*, as a general prescription to construct operators associated with classical dynamical quantities. The so constructed operators correspond to a special ordering of the $p$ and $q$ operators, to the so called Weyl ordering.

Wigner finds the asymptotic expansion of Wigner's function by showing a) that if the density operator satisfies its quantal equation of motion, Wigner's function satisfies an integral-differential equation; b) that in a formal expansion of the integral in a power series of $\hbar$, the $\hbar$ independent term leads to Liouville's equation, while the others contain formally the quantum corrections; c) that the solution of the full equation can be computed explicitly as a formal power series in $\hbar$. The conditions on the nature of the convergence is left open. A necessary condition is that the forces of interaction should be slowly variable. Wigner also points out that the expression as it stands does not take into account the exclusion principle.

Much of the later work, both by Wigner, and by others was in the following directions.

a)  General reviews
b)  Properties of the mock phase space density

c) Questions of uniqueness of the association
d) Relativistic versions
e) Versions applicable if a spin variable is also present
f) Applications in physics
g) Mathematical studies of the problem of associations.

Wigner has contributed much to the study of these points as well. The first practical applications were in [8], where quantum corrections to chemical reaction rates are studied; and in [21], where the leading quantum corrections to the free energy are derived in general, (but without taking into account the effects of statistics.) The latter were then used to evaluate the quantum corrections for a low temperature He gas, and compared with the experimental results. These results were never published. They were too sensitive to the details of the, then, poorly determined interatomic forces, and were thus inconclusive. (Actually, the desire for an explanation of these particular experimental results gave Wigner the impetus to invent Wigner's function.)

A *general* Lorentz invariant theory of Wigner's function is lacking. This is not surprising, since, a) there is no general, satisfactory, *classical* Lorentz invariant formalism of interacting particles either; b) the form of the association used by Wigner is deeply linked to Galilean transformations. For *special cases*, however, it is possible to extend the theory in a Lorentz invariant manner using light-cone, or null coordinates [40], [41]; this was applied to the description of light waves [40], [41]; and to the description of a collection of non-relativistic harmonic oscillators [42]. The latter, in turn, can be utilised to express in this manner, for a particular model the bound state of hadrons composed of quarks, and to calculate the nucleon form factor in some approximation [42]. (For Wigner's general opinions on the problem of Lorentz invariance in this description, see his observations at the end of [35]. For other applications, see also the book by de Groot et al. on relativistic kinetic theory, quoted below).

The incorporation of statistics and spin into this association is more delicate. This is natural, since neither the spin variable, nor the Bose-, or the Fermi-statistics have any classical equivalent; thus, it is not to be expected that a formalism tailored to resemble the classical one should describe these modifications in simple terms. [40] contributes to this problem. Other work on these matters by different authors are recorded in the reviews listed below.

There are several general review articles available on the work of Wigner and others. The most recent such reviews are by S.R. de Groot [48], S.R. de Groot, W.A. van Leeuwen and C.G. van Weert [49], P. Carruthers and F. Zachariasen [50], N.L. Balazs and B. Jennings [51], M. Hillary, R.F. O'Connell, M. Scully and E.P. Wigner [39]; F.J. Narcowich [52]. There is a considerable overlap among the general discussions given in the different reviews, although some of them have more special aims as well. Carruthers et al. concentrate on scattering problems; Balazs et al. on the variety of different possible associations, and their relation to the transformation properties of the different mock phase-spaces engendered by them; de Groot et al. stress the applications in problems of relativistic kinetic theory.

Until the 1960's the mathematical community took scant notice of these ideas, when they were independently rediscovered and used in the theory of ordinary and partial differential equations, and in the asymptotic analysis of their solutions.

The theory of linear differential equations with constant coefficients is well understood. What happens, however, if the coefficients are no longer constant, but, say, *slowly* variable? Formally, the equation with *constant* coefficients can be written as an operator function acting on a function where the operator contains $p_{op}$ *alone*. If the coefficients depend on the variable $x$ as well the operator can be now thought of as depending on $q_{op}$ and $p_{op}$ (usually in an antinormal ordering, $p_{op}$ acting first). Then, if the $x$ dependence is weak, we can take into account this "weak" dependence, by considering $p_{op}$ and $q_{op}$ as "nearly" commuting, resembling thus a quantal description near the classical limit. Hence, it will be useful to recast the solution scheme in such a way that the "classical" solution should form a natural *starting* point. We are back, in the problem-setting of quantum theory, to the problem of the classical limit. Thus, the considerations which lead to Wigner's function will become useful in this context as well.

Wigner's function, and other similar functions are now reintroduced for this purpose. In the mathematical literature they are designated as *symbols* associated with operators. It is easy to associate a function with an operator, since the *matrix elements* of the operator in a representation, the *kernel of an integral operator*, can be thought of as the associated function. These objects, however, have shortcomings. The integral kernels in the simplest $q$ or $p$ representation are not only highly singular, but in general it is not easy to see immediately their functional relation to the classical dynamical quantities to which they correspond. The introduction of these symbols remedies these defects by requiring them to have the following convenient properties. a) They should be singularity free; b) if the original operator is an operator function of only the $p_{op}$, or only of $q_{op}$, the associated symbol should be the same function of the parameter $p$ or of the parameter $q$ as the operator function was of the corresponding operator; c) if the operator is a general operator function of $p_{op}$ and $q_{op}$, the symbol should have a well defined asymptotic expansion in $\hbar$, with the leading term being the same function of the parameters $p$ and $q$ as the operator function was of $p_{op}$ and $q_{op}$; thus the leading term should be the classical dynamical quantity; d) the, in general, singular integral kernel representing the operator in the $q$ representation should be constructible from this, non-singular symbol and a well defined, universal singular contribution. *Thus the symbol should have a classical significance, have an asymptotic expansion, and should express the non-singular part of the kernel in a universal manner.* The precise formulation of these conditions, the demonstration of the existence of such objects, and their actual construction is the task of the theory of pseudo-differential operators.

Let $q_{op}$, resp. $p_{op}$ have the eigenvalues $x'$, resp. $k'$, with eigenfunctions $|x'\rangle$, resp. and $|k'\rangle$. Then the action of an operator $A$ on a general ket $|\rangle$

can be expressed in the $q$ representation as an integral operator with the kernel $\langle x''|A|x''\rangle$. The simplest symbol of the operator $A$ is given by $\sigma_A(q, p) = \langle x'|A|k'\rangle/\langle x'|k'\rangle$, evaluated for $x' = q$, $k' = p$. The kernel is then expressed as

$$\langle x'|A|x''\rangle = \int dk' \sigma_A(x', k')\langle x'|k'\rangle\langle k'|x''\rangle,$$

which gives the standard transformation formula if the above expression for $\sigma_A$ is resubstituted. This particular definition of the symbol gives a *preferential* role to the $q$ representation, or to the anti-normal ordering of operators, while the adjoint symbol gives a preferential role to the $p$ representation and to normal ordering. It has been extensively used by Kirkwood. (See the discussion of the Kirkwood association in the review [51].)

An alternative symbol can be introduced which treats on equal footing all representations engendered by the eigenvectors of a general linear combination of $p_{\text{op}}$ and $q_{\text{op}}$, and which gives a preferential role to operators ordered according to Weyl's prescription. These turn out to be the so called Weyl symbols, with *Wigner's function being the Weyl symbol of the density operator.* The kernel can be simply expressed as

$$\langle x'|A|x''\rangle = \int dpdq\, A_W(p, q)\langle x'|\Delta_W(p, q)|x''\rangle.$$

(Thus we see that the $\Delta_W$ operator introduced to produce Wigner's function serves also to express the singular part of the kernel, while the Weyl symbol is used for the non-singular factor. A similar $\Delta$ operator can also be introduced to specify the singularity of the kernel when using the $\sigma_A$ symbol. In general, the singular part of the kernel is always extracted by the introduction of a suitable singular $\Delta$ operator. The coefficient function is the associated symbol. This, then, insures that for each class of symbols the singular part of the kernel is extracted by the same singular operator.)

There is a large literature on pseudo-differential operators. See, for example, the magisterial work of L. Hörmander [53], with extensive references. (Where, however, Wigner is not even mentioned!), or the excellent short review by the same author [54].

Paper [55] with T. Teichmann deals with the scattering of an electromagnetic field, while paper [56] describes problems in the multiple scattering of particles. The two papers are independent of each other but both show a masterful understanding of the mathematical tools required.

In [55] a perfectly conducting, simply connected cavity is considered with holes. The tangential components of the electric field is to vanish on the perfect conductor, but not necessarily in the holes. What is a *complete* set of characteristic fields in terms of which any cavity field can be expanded? In the absence of holes the complete set is given by the normal modes of the cavity, with the tangential components of the electric field vanishing on the whole boundary. Then, the associated normal frequencies are *all different from zero*. The presence of holes, however, *allows zero frequency solutions*, associated with possible static and irrotational electric and magnetic fields. For standard boundary conditions only the static, irrotational magnetic field, or its potential is needed to complete the normal modes. This potential is a harmonic function which *couples at the boundary* with the normal modes. The paper then gives a constructive method to solve this complicated boundary value problem, and to find approximations to it.

The study [56] applies the ideas and methods of *group theory* to problems *in multiple scattering*. Here, the prototype of an elementary act is a free flight, followed by a scattering, whose properties are given probabilistically. There are, however, many different possibilities. In general, an elementary act changes the state of the system. Let the elementary acts form a group, and suppose that, a) all states can be reached from a standard state through the operation of the group; b) the probability that an elementary act changes the state is invariant under the group. Then the evolution equation of the probability distribution function in terms of the transition probabilities, the Kolmogorov equation, can be written as an integral equation over the group. Two cases can be now distinguished. 1) the group has only *one* element which transforms the standard state into a particular given state; in this case the number of parameters which characterise the state is the same as the number of parameters of the group, and the different states of the particle can be labeled by the corresponding group element. 2) If the *number of group parameters exceeds that of the state parameters* there will be a subgroup whose elements do not change the standard state. Now the left cosets of this subgroup correspond to different states of the particle. Wigner shows that in *both* cases the *same* equation holds, but for case 2 additional conditions must be imposed both on the distribution function and on the transition probability.

The evolution equation reads as

$$f_{n+1}(t) = \int ds\, f_n(s) P(s^{-1}t)\,,$$

where $f_n(s)ds$ is the probability that the state $s$, labeling group element $s$, (within the volume element $ds$ in the invariant group space) has been reached in $n$ steps, and $P(s^{-1}t)dt$ is the transition probability that in one step the element $t$ within $dt$ is reached starting from $s$. ($P$ depends only on this combination due to *the invariance of $P$ under the group*. For example, in the usual free flight problem this invariance asserts that the change of state depends only on the free flight connecting the two states, but not on the starting state as well; $s^{-1}t$ is the separation between $s$ and $t$.)

The great advantage of this abstract form – the equation appearing as a group integral – is the following. The group integral of the product of the convolute and the matrix of the representation is the product of two matrices. From this it immediately follows, multiplying the evolution equation by $D(t)$, that

$$\int dt f_{n+1}(t) D(t) = \int ds f_n(s) D(s) \Pi\,,$$

where $\Pi = \int ds P(s) D(s)$, and $D(s)$ is a representation of $s$. This equation can then be trivially solved. For example, it corresponds to solving convolutions in terms of Fourier transforms, if the elementary act corresponds to a free flight in an arbitrary direction. There the transition probability was invariant under translations, and a Fourier transform appeared, the exponentials in the Fourier transformation being "really" the representation of the translation group. Here one sees that what is needed in general is the *group representation of the group under which the transition probability is invariant*.

In fact the equation expressing $f_{n+1}$ in terms of $f_n$ can be used as the prototype equation to define the convolution for the two functions $f_n$ and $P$, both defined over a group and one of them being invariant under the action of the group. It is by now clear that this idea can be exploited in different directions. It forms the basis of harmonic analysis over groups; it can be used to invent random walks and central limit theorems for random variables invariant with respect to groups of transformations. See for example the magnificent review by G.W. Mackey [57], or the book by A. Terras [58].

# V. Studies in Conventional Quantum Theory

This is an annotation to several papers dealing with some *non-group theoretical aspects* of quantum theory. These are [59], [20] with von Neumann, [60] with von Neumann, [61], [62], [63]. Save the last one, all belong to the early period of Wigner's work. Three of them, [59], [61], and [62] were commissioned (essentially) by Rudolf Ortvay, who was (until the end of 1944) the head of the Institute of Physics at the Peter Pazmany Scientific University in Budapest, and the Editor of the Hungarian Journal of Mathematics and Physics. They were eventually presented by Ortvay in two sessions of the Hungarian Academy of Sciences. One can find some information on these matters in [64] which is a reprint of Wigner's letters to Ortvay. In a letter dated May 20, 1932 there is a brief outline of [61], a paper which deals with the nature of "heavy" (or Chadwick's) neutrons, and "light" (or Pauli's) neutrons, the latter being now called neutrinos. (There are also further allusions to two other possible papers, conceivably to be published in the Hungarian journal. One of these would deal with time reversal, the other with the magnetic moment of stationary states. They were apparently not published.) On June 25 of the same year Wigner answers some observations of Ortvay on the neutron, neutrino problem, although in the mean time (June 13) [61] was read at the Hungarian Academy of Sciences.

In the paper itself Wigner discusses the question whether the light neutrons are to be the same as the heavy neutrons, or the possibility that the two are different; the light neutrons' mass being that of an electron, ("for symmetry reasons", as Wigner says.) He further argues that possibly in $\beta$-decay a light neutron and electron leave always together, and that a nucleus always contains an equal number of electrons and light neutrons. This would allow the correct statistics for a nucleus, even if electrons are present in the nucleus, as was supposed at that time. He adds an interesting note in proof, arguing against Heisenberg's supposition (1932) that a neutron is a bound state of an electron proton system. If this were so, then the lifetime of an $H$ atom would be finite due to a transition from a *2p* state of $H$ to a neutron state. Such *2p* states do appear if the $H$ is in a chemical compound, and Wigner estimates the lifetime of such a compound to be of the order of magnitude of a month.

A letter dated Oct. 14, 1934 is a covering letter to [62] which describes what is now known as the Wigner-Brillouin perturbation theory. A letter dated Dec. 3 of the same year communicates to Ortvay Wigner's realisation that Brillouin already published a formal series which corresponds to his result. However, as he points out, Brillouin obtains his results in a different way, and does not discuss the most important property of the perturbation theory, to wit, that it approximates the energy eigenvalues from *above*. In the published paper reference is made to Brillouin's series. (In fact, both Brillouin and Wigner missed an earlier paper by Lennard-Jones [65] where the same perturbation theory has been proposed and utilized!)

To summarise, I quote here from the Hungarian letter. "The present study aims to find an approximation scheme which approximates the ground state energy from above, similarly to a variational method where each approximation is a mean value of the operator $H + V$ with respect to a normalised wave function. If we evaluate the expectation value of $H + V$ with respect to a trial function {4a}, the problem amounts to minimising the expression {4b}. If in this expression we omit from the numerator the double sum, the minimum is reached when the {trial parameter} $\alpha$ has the value given in {5a}. Reinserting this value in the *complete* expression {4b} gives then the implicit equation {5b} for the determination of this value. This solution, which corresponds to the third approximation of the Rayleigh-Schrödinger equation, always gives a value too large. The higher approximations can now be obtained by replacing in the double sum, which is bilinear in $\alpha$, one of the $\alpha$ factors with the other one as unknown."... "The approximate expressions ... can be easily generalised for $n = \infty$, giving a formal solution of the eigenvalue equation. In particular cases one can easily show that the infinite series converge to the true solution, and it is clear that the present method converges in many situations when the original method fails." (The labeling of the equations via the {} braces are by me.)

Paper [61] is a pedagogical paper showing that for composite systems Bose-Einstein, or Fermi-Dirac statistics is applicable depending whether the number of constituents are even or odd. It also contains an extremely up to date (1928) discussion on Heisenberg's new (1928) observation that the absence of even, or odd lines in the bandspectra of molecules signal the symmetric or antisymmetric states of the nuclei; thus they can be used to specify the statistics of nuclei using optical observations. (There seems to be an accidental error in the conclusion of the Hungarian text (p. 582), where the words even and odd are interchanged. The German summary also contains an error; $^{14}$N is not a fermion.)

Papers [20] and [60], both written with v. Neumann and published back to back, raise two important problems in quantum theory. [20] deals with the problem of possible degeneracies in adiabatic perturbations. Let the Hamiltonian of a system depend on a set of real parameters and consider in a parameter space a curve describing an orbit of the parameters. To each point on the orbit a spectrum of the Hamiltonian (with the associated eigenfunctions) can be computed. What can one say about the behaviour of the spectrum? In particular, under what conditions can the spectrum pass through degeneracies while the parameters advance along the path? If one considers the *energy levels* as a function of the parameters, the points where this does occur are called by M.V. Berry diabolical points, since the intersecting energy levels have the local appearance of a diabolo (and can be devilish to treat). The paper shows that this cannot happen in general unless the parameter space is at least three dimensional. (For real Hamiltonians two dimensions are already sufficient, although this is not explicitly discussed in the paper.) The utilisation of these results became essential in chemistry and in nuclear physics where the intersection and non-intersection of energy levels in adiabatic perturbations are crucial for many phenomena. (See, for example the report by M.V. Berry [66], or the collection of papers in [67].)

The notion of investigating the properties of the *spectrum* while the parameters orbit in the parameter space turned out to be extremely useful. Remarkably, the next step, the systematic study of the *eigenfunctions* as a function of the parameters was done only much later and led, in the hands of others, in particular M.V. Berry, to beautiful and deep results, as, for example to the notion of the geometric phase (when the orbits in the parameter space are closed). (See the collection [67]).

Paper [60] studies two examples of eigenvalue problems with two particular potentials, which can give rise to discrete eigenvalues embedded in the continuum. The Hamiltonians studied describe three dimensional separable motions with a spherically symmetrical potential. The physics of each example is the following. In the first case the existence of this anomalous eigenvalue corresponds classically to a motion which for the given energy reaches infinity during a finite time, rebounds from there and performs thus oscillations. (This is the consequence of a potential which for large $r$ behaves as $-r^4$.) In the other example the potential is oscillatory and falls of slowly at $\infty$. Now the positive eigenvalue has an eigenfunction that is generated by "resonance" oscillations on the peaks and valleys of the potential. The oscillatory potential problem has been put in a more general mathematical context and discussed by B. Simon, who has also corrected a mistake in [60], and has displayed a class of oscillatory potentials for which there is one, and only one such positive eigenvalue. (See [68].)

The behaviour of the Hamiltonian operator at infinity can give rise to difficulties in perturbation theory as well. Paper [63] considers the Coulomb potential in the hydrogen atom Hamiltonian as a perturbation in a spherical box of radius $R$, using conventional Rayleigh-Schrödinger perturbation theory (with Dirichlet or Neumann boundary conditions at $R$), where the charge functions as the perturbation parameter. The first order correction vanishes, and the second order correction leads, in the limit $R \to \infty$, to the correct *form* for the energy but with a numerical coefficient which is too small, irrespectively of the boundary condition selected. The remarkable feature here is the occurrence of the combination of a *finite* result with its being *incorrect*. The incorrectness is a consequence of a non uniform convergence in the parameters $e^2$ and $1/R$; while the finiteness is due to the inverse *first* power dependence of the potential.

# References

[1] E. Wigner: Laws of Nature and Initial Conditions. Fizikai Szemle XXXIII, *19*, 1–11 (1983)

[2] A. Szegvari and E. Wigner: Über elektrische Erscheinungen bei Stäbchensolen. Koll.-Zeitschr. *33*, 218–222 (1923); Vol. IV, Part I

[3] H. Mark and E. Wigner: Die Gitterstruktur des Rhombischen Schwefels. Z. Phys. Chem. *111*, 398–414 (1924); Vol. IV, Part I

[4] L. Boltzmann: Lectures on Gas Theory (translated by Stephen G. Brush). University of California Press 1964, pp. 376–411 (Original publication date 1895)

[5] M. Polanyi and E. Wigner: Bildung und Zerfall von Molekülen. Z. Phys. *33*, 429–434 (1925); Vol. IV, Part I

[6] M. Polanyi and E. Wigner: Über die Interferenz von Eigenschwingungen als Ursache von Energieschwankungen und chemischer Umsetzungen. Z. Phys. Chem. A (Haber-Band) *43*, 439–452 (1928); Vol. IV, Part I

[7] H. Pelzer and E. Wigner: Über die Geschwindigkeitskonstanten von Austauschreaktionen. Z. Phys. Chem. B *15*, 445–471 (1932); Vol. IV, Part I

[8] E. Wigner: Über das Überschreiten von Potentialschwellen bei chemischen Reaktionen. Z. Phys. Chem. B *19*, 203–216 (1933); Vol. IV, Part I

[9] E. Wigner: Über die paramagnetische Umwandlung von Para-Orthowasserstoff. Z. Phys. Chem. B *23*, 28–32 (1933); Vol. IV, Part I

[10] L. Farkas and E. Wigner: Calculation of the Rate of Elementary Reactions of Light and Heavy Hydrogen. Trans. Faraday Soc. *32*, 708–723 (1936); Vol. IV, Part 1

[11] E. Wigner and Henry Eyring: On the Rate of Chemical Reactions. Scientific Monthly *44*, 564–567 (1937); Vol. IV, Part I

[12] E. Wigner: Calculation of the Rate of Elementary Association Reactions. J. Chem. Phys *5*, 720–725, (1937); Vol. IV, Part I

[13] J. O. Hirschfelder and E. Wigner: Some Quantum-Mechanical Considerations in the Theory of Reactions Involving an Activation Energy. J. Chem. Phys. *7*, 616–628 (1939); Vol. IV, Part I

[14] E. Wigner: The Transition State Method. Trans. Faraday Soc. *34*, 29–41 (1938); Vol. IV, Part I

[15] E. P. Wigner: Some Remarks on the Theory of Reaction Rates. J. Chem. Phys. *7*, 646–652 (1939) (including discussion); Vol. IV, Part I

[16] E. P. Wigner and R. A. Hodgkin: Obituary: Michael Polanyi. Biographical Memoirs of the Royal Society *23*, 413–448 (1977); Vol. VII

[17] E. P. Wigner: On the Time-Energy Uncertainty Relation, Chap. 14 in: Aspects of Quantum Theory (Cambridge University Press 1972), pp. 237–247; Vol. IV, Part I

[18] F. London: Über den Mechanismus der homöopolaren Bindung, pp. 104–113 in: Probleme der modernen Physik, herausgegeben von P. Debye Arnold Sommerfeld zum 60. Geburtstag gewidmet, S. Hirzel, Leipzig 1928

[19] F. London: Zur Theorie nicht adiabatisch verlaufender chemischer Prozesse. Z. Physik *74*, 143–174 (1932)

[20] J. von Neumann and E. Wigner: Über das Verhalten von Eigenwerten bei adiabatischen Prozessen. Phys. Zeitschr. *30*, 467–470 (1929); Vol. I, Part III

[21] E. Wigner: On the Quantum Corrections for Thermodynamic Equilibrium. Phys. Rev. *40*, 749–759 (1932); Vol. IV, Part I

[22] R.S. McKay, J.D. Meiss, I.C. Percival: Transport in Hamiltonian Systems. Physica *13*D, 55–81 (see p. 57) (1984)

[23] P. Haenggi, P. Talkner, B. Borkovec: Reaction-rate theory: fifty years after Kramers. Rev. Mod. Phys. *62*, 251–341 (1990)

[24] R.D. Levine: Quantum Mechanics of Molecular Rate Processes. Clarendon Press, Oxford 1969

[25] R.D. Levine, R.B. Bernstein: Molecular reaction dynamics and chemical reactivity. Oxford University Press, 1987

[26] Theory of chemical reaction dynamics, Ed. M. Baer, CRC Press, Boca Raton, FL

[27] M.S. Child: Semiclassical Mechanics with Molecular Applications, Clarendon Press, Oxford, 1991

[28] N. Bohr and J.A. Wheeler: The Mechanism of Nuclear Fission. Phys. Rev. *56*, 426–450 (1939)

[29] M. Brack, J. Damgaard, A.S. Jensen, H.-C. Pauli, V.M. Strutinsky and C.Y. Wong: Funny Hills: The Shell-Correction Approach to Nuclear Shell Effects and Its Application to the Fission Process. Rev. Mod. Phys. *44*, 320–505 (1972)

[30] E.P. Wigner: Derivations of Onsager's Reciprocal Relations. J. Chem. Phys. *22*, 1912–1915 (1954); Vol. IV, Part I

[31] C.W. Ufford and E. Wigner: On the Calculation of the Distribution Function. Phys. Rev. *61*, 524–527 (1942); Vol. IV, Part I

[32] J.-P. Hansen and I.R. McDonald: Theory of Simple Liquids. Academic Press N.Y. 1986

[33] E.H. Lieb and D.C. Mattis: Mathematical Physics in One Dimension: Academic Press N.Y. 1966

[34] E.P. Wigner: Quantum Mechanical Distribution Functions. Revisited in: Perspectives in Quantum Theory; Essays in Honor of Alfred Landé; Eds. Wolfgang Yourgau and Alwyn van der Merwe, MIT Press Cambridge MA 1971, pp. 25–36; Vol. IV, Part I

[35] E.P. Wigner: The General Properties of the Distribution Function and Remarks on its Weakness, in: The Physics of Phase Space (Eds. YS. Kim and W.W. Zachary) Lecture Notes in Physics *278*, Springer, 1987; Vol. IV, Part I

[36] R.F. O'Connell and E.P. Wigner: Quantum Mechanical Distribution Functions: Conditions for Uniqueness. Phys. Ltr. *83A*, 145–148 (1981); Vol. IV, Part I

[37] R.F. O'Connell and E.P. Wigner: Some Properties of a Non-Negative Distribution Function. Phys. Ltr. *85A*, 121–126 (1981); Vol. IV, Part I

[38] E.P. Wigner: On the Quantum Correction for Thermodynamic Equilibrium. This Weeks Citation Classic, December 19, 1983

[39] M. Hillary, R.F. O'Connell, M.O. Scully and E.P. Wigner: Distribution Functions in Physics: Fundamentals Phys. Rep. *106*, 121–167 (1984); Vol. IV, Part I

[40] R.F. O'Connell and E.P. Wigner: Manifestations of Bose and Fermi statistics on the quantum distribution function of spin 0 and 1/2 particles. Phys. Rev. A *30*, 2613–2618 (1984); Vol. IV, Part I

[41] Y.S. Kim and E.P. Wigner: Covariant Phase-Space Representations for Localized Light Waves. Phys. Rev. A *36*, 1293–1297 (1987); Vol. III, Part I

[42] Y.S. Kim and E.P. Wigner: Covariant Phase-Space Distributions for Harmonic Oscillators. Phys. Rev. A *38*, 1159–1167 (1985); Vol. IV, Part I

[43] Y.S. Kim and E.P. Wigner: Covariant Phase-Space Representation and Overlapping Distributions. Phys. Rev. A *39*, 2829–2834 (1989); Vol. IV, Part I

[44] P.A.M. Dirac: Note on Exchange Phenomena in the Thomas Atom. Proc. Cam. Phil. Soc. *26*, 376–385 (1930)

[45] W. Heisenberg: Über die inkohärente Streuung von Röntgenstrahlen. Phys. Ztschr. *32*, 737–740 (1931)

[46] J.E. Moyal: Quantum Mechanics as a Statistical Theory. Proc. Camb. Phil. Soc. *45*, 99–124 (1949)

[47] H. Weyl: Quantenmechanik und Gruppentheorie. Zeitschr. Phys. *46*, 1–46 (1927)

[48] S.R. de Groot: La transformation de Weyl et la fonction de Wigner. Les Presses Universitaire de Montreal, Montreal, 1974

[49] S.R. de Groot, W.A. van Leeuwen and C.G. van Weert: Relativistic Kinetic Theory. North Holland, Amsterdam, 1980

[50] P. Carruthers and F. Zachariasen: Quantum Collision Theory with Phase-Space Distributions. Rev. Mod. Phys. *55*, 245–285 (1983)

[51] N.L. Balazs and B.K. Jennings: Wigner's Function and Other Distribution Functions in Mock Phase-Spaces. Phys. Rep. *104*, 347–391 (1984)

[52] F.J. Narcowich: Wigner Distribution Functions, Seminars in Mathematical Physics, No. 1, Texas A&M University, College Station TX, 1986

[53] L. Hörmander: The Analysis of linear Partial Differential Operators, Vol. 3, Chapter 18. Springer-Verlag, Berlin 1985

[54] L. Hörmander: On the Existence and the Regularity of Solutions of Linear Pseudodifferential Equations. L'Enseignement Mathematique Ser. 2, *17*, 99–163 (1971)

[55] T. Teichmann and E.P. Wigner: Electromagnetic Field Expansions in Loss-free Cavities Excited through Holes. J. Appl. Phys. *24*, 262–267 (1953); Vol. IV, Part I

[56] E.P. Wigner: The Problem of Multiple Scattering. Phys. Rev. *94*, 17–25 (1954); Vol. IV, Part I

[57] G.W. Mackey: Harmonic Analysis as the Exploitation of Symmetry. Bull. Amer. Math. Soc. *3*, 543–698 (1980)

[58] A. Terras: Harmonic Analysis on Symmetric Space and Applications. Vol. I, Springer-Verlag, New York 1985

[59] E. Wigner: Összetett rendszerek statisztikája az uj quantummechanika szerint (The Statistics of Composite Systems according to the New Quantum Mechanics). Magyar Tud. Akad. Mat. Ért. *46*, 576–582 (1929); Vol. IV, Part I

[60] J. v. Neumann and E. Wigner: Über merkwürdige diskrete Eigenwerte: Phys. Zeitschr. *30*, 465–467 (1929); Vol. I, Part III

[61] E. Wigner: Adalékok a neutron elméletéhez (Contributions to the Theory of Neutrons). Magyar Tud. Akad. Mat. Ért. *49*, 142–146 (1932); Vol. IV, Part I. Editor's note: What Wigner here calls light neutron is now called neutrino.

[62] E. Wigner: On a Modification of the Rayleigh-Schrödinger Perturbation Theory. Magyar Tud. Akad. Mat. Ért. *53*, 475–482 (1935); Vol. IV, Part I

[63] E.P. Wigner: Application of the Rayleigh-Schrödinger Perturbation Theory to the Hydrogen Atom. Phys. Rev. *94*, 77–78 (1954); Vol. I, Part III

[64] Wigner Jenö levelei Ortvay Rudolfhoz (Letters of E. Wigner to R. Ortvay) Fizikai Szemle *22*, 45–48 (1972)

[65] J.E. Lennard-Jones: Perturbation Problems in Quantum Mechanics. Proc. Roy. Soc. A *129*, 598–615 (1930)

[66] M.V. Berry: In: Chaotic Behaviour in Quantum Systems, Ed. G. Casati, Plenum Press N.Y., 1983, pp. 123–140

[67] Geometric Phases in Physics, Eds. A. Shapere and F. Wilczek, World Scientific, Singapore 1989

[68] M. Reed and B. Simon: Methods of Modern Mathematical Physics. Academic Press, New York, Vol. IV, pp. 223, 352 (1978); Vol. III, p. 155 (1979)

# Über elektrische Erscheinungen bei Stäbchensolen

A. Szegvari und E. P. Wigner

Kolloid-Zeitschrift *33*, 218–222 (1923)

(Eingegangen am 23. Juli 1923.)

1. Stäbchensole zeigen im Gegensatz zu anderen viele merkwürdige Eigenschaften; so z. B. die elektrischen und magnetischen Richtungseffekte, und vielleicht als auffallendste Erscheinung: die hohe induktive Leitfähigkeit, die zur Messung von ungewöhnlich hohen Dielektrizitätskonstanten führt. Wie J. Errera[1] zeigte, ist die „Dielektrizitätskonstante" eines Vanadinpentoxydsols sehr stark abhängig von seiner Konzentration, seinem Alter und sogar von der Temperatur, ferner von der Spannung und Frequenz des bei der Messung angewendeten Wechselstromes. Unter Umständen konnte er Dielektrizitätskonstanten messen, die fünfmal (und noch mehr) größer waren als die des Wassers.

Der zunächstliegende Gedanke zur Erklärung dieses Verhaltens könnte die Annahme sein, daß die Teilchen des Vanadinpentoxydsols, welche bekanntlich[2] ausgeprägten Stäbchencharakter besitzen, als elektrische Dipole aufzufassen sind[3]. Diese Betrachtungsweise ist aber kaum angängig. Dies geht erstens aus dem Verhalten der Vanadinpentoxydstäbchen im elektrischen Wechselfeld unter dem Ultramikroskop hervor[4]; es zeigte sich nämlich, daß die Teilchen im elektrischen Wechselfeld je nach der Feldrichtung sich hin und her bewegten ohne umzuklappen; wären es Dipole, so dürften

---

[1] J. Errera, Koll.-Zeitschr. 31, 59 (1922).

[2] Siehe u. a. Diesselhorst, Freundlich u. Leonhardt, Elster-Geitel-Festschrift 1915, 453.

[3] J. Errera, Koll.-Zeitchr. 32, 373 (1923).

[4] H. R. Kruyt, Koll.-Zeitchr. 19, 161 (1916).

sie sich, da es sich im allgemeinen um homogene Felder handelt, h ö c h s t e n s d r e h e n. Und zweitens kann keine Ursache angegeben werden, die ein polares Verhalten der zwei Stäbchenenden bewirkte.

2. Es möge hier bemerkt werden, daß die Teilchen einen t r i p o l a r t i g e n Charakter besitzen.

Die Oberfläche eines Kolloides verhält sich ja wie eine elektrische Doppelschicht, und zwar sitzt die innere Belegung derselben ziemlich fest an der Teilchenoberfläche, während die äußere als Raumladung vorzustellen ist, die aus Ionen besteht, die durch die innere Belegung elektrostatisch angezogen sind. Demgemäß besitzt die äußere Doppelschicht gegenüber der inneren eine große Beweglichkeit.

Nehmen wir nun an, die innere Belegung entspräche einem an der g a n z e n Oberfläche des Teilchens k o n s t a n t e n Potentialunterschiede gegenüber dem Medium. Da nun diese Teilchen aber n i c h t k u g e l f ö r m i g sind, kann hierbei die O b e r f l ä c h e n d i c h t e n i c h t k o n s t a n t sein. Nehmen wir z. B. an, die Stäbchen wären angenähert Rotationsellipsoide, so müßte sich die Oberflächendichte der Spitze zu derjenigen der Seite verhalten wie die Stäbchenlänge zum Stäbchendurchmesser: sie wäre also a n d e r S p i t z e v i e l g r ö ß e r. Da aber die elektrische Ladung hier an Ionen gebunden ist und demnach einer größeren Ladungsdichte auch eine g r ö ß e r e I o n e n k o n z e n t r a t i o n entspricht, müßte bei der vorherigen Annahme (konstantes Potential) auch die I o n e n k o n z e n t r a t i o n der ä u ß e r e n I o n e n h ü l l e a n d e r S p i t z e g r ö ß e r sein als gegenüber den mittleren Partien. Nun trachten sich aber Konzentrationsunterschiede bei Ionen infolge des osmotischen Druckes a u s z u g l e i c h e n; dies muß auch tatsächlich stattfinden. Dieser Konzentrationsausgleich dauert nun so lange, bis die hierdurch verursachten n e u e n P o t e n t i a l u n t e r s c h i e d e v o n E n d e g e g e n M i t t e den osmotischen Kräften das Gleichgewicht halten (Fig. 1).

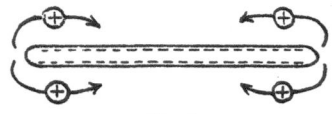

Fig. 1

Es entsteht demnach ein Tripol (an den zwei Enden ein gleiches, aber von der Mitte abweichendes Potential); dies müßte man viel-leicht bei der Erklärung einiger Erscheinungen der Stäbchensole mit in Betracht ziehen. (Auf das vorher erwähnte anomale induktive Verhalten kann dies aber keinen besonderen Einfluß ausüben.)

Die genaue Berechnung des Effektes ist nicht ausführbar, da die Ausgangsdifferentialgleichungen nicht zu integrieren sind.

3. Der letzterwähnten induktiven Erscheinung könnte man durch folgende Betrachtung gerecht werden.

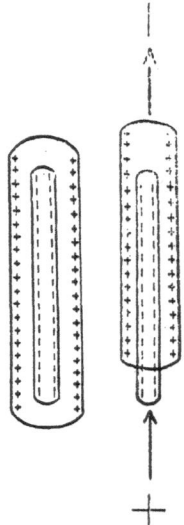

Fig. 2

Wir denken uns der Einfachheit halber die äußere Belegung als eine das Teilchen auf eine gewisse Distanz umgebende Hülle. Wirkt nun ein elektrisches Feld auf dieses Gebilde ein, so wird es trachten, die H ü l l e u n d d e n K e r n i n e n t g e g e n g e s e t z t e r R i c h t u n g z u v e r s c h i e b e n (Fig. 2). Dieser Verschiebung wirkt aber die elektrostatische Anziehung von Kern und Hülle entgegen. Während dieser Verschiebung muß also Arbeit gegenüber elektrostatischen Kräften geleistet werden, diese wird auch vom angelegten Felde geleistet. Wenn sich nun die Feldveränderung umkehrt, wird diese entstandene potentielle Energie wieder in Feldenergie umgewandelt. Dieses Verhalten entspricht aber einer K a p a z i t ä t s v e r g r ö ß e r u n g, also einer Vergrößerung der Dielektrizitätskonstante des Gesamtsystems.

Es möchte noch hinzugefügt werden, daß hierbei eine R i c h t k r a f t auftritt, die die Teilchen (sofern sie nicht von vornherein gerichtet waren) parallel zu den Kraftlinien zu richten sucht (Fig. 3), aber auf bereits gerichtete Teilchen überhaupt kein Drehmoment ausübt, was die eingangs erwähnten kataphoretischen Beobachtungen erklärt.

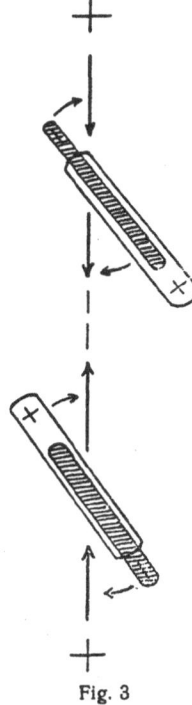

Fig. 3

Die Berechnung des Effektes gestaltet sich am einfachsten wie folgt:

Wir nehmen an, die vorher erwähnte Einstellung der Teilchen parallel zu den Kraftlinien wäre bereits vollständig. Weiter setzen wir voraus, daß die Feldveränderungen so langsam sind, daß die Wirkung der Hülle auf den Kern dem äußeren Feld ständig das Gleichgewicht hält, unbeeinflußt von der Viskosität des Mediums und der Trägheit der Teilchen.

Bestimmen wir zunächst den Einfluß des elektrischen Feldes auf ein Teilchen von der Länge 1 und der Dicke 2 r. Wir setzen des weiteren voraus, daß die durch die Scheitelwerte der Feldstärke verursachte Verschiebung x relativ zur Gesamtlänge des Teilchens gering ist. Die Oberflächendichte sei $\sigma$, die elektrische Feldstärke im Wasser E. Aus Fig. 4 ist er-

sichtlich, daß man nur die Wirkung derjenigen Teile auf das elektrische Feld berücksichtigen muß, bei welchem infolge der relativen Verschiebung die Ladung freigelegt wird.

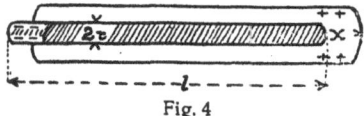

Fig. 4

Die Wirkung des elektrischen Feldes auf sein freigelegtes Ende ist nun entgegengesetzt gleich der Wirkung der anderen Spitze auf das Ende:

$$\frac{(2 r \pi x \sigma)^2}{l^2 D} = E \cdot 2 r \pi x \sigma \qquad (1)$$

$2 r \pi x \sigma$ ist die bei der relativen Verschiebung x „freigelegte" Ladung; D bedeutet die Dielektrizitätskonstante des Zwischenmediums (Wasser).

Aus Gleichung (1) folgt für die „freie" Endladung

$$2 r \pi x \sigma = E l^2 D;$$

multiplizieren wir diesen Wert mit l, so bekommen wir das Moment m des Teilchens:

$$m = l \cdot 2 r \pi x \sigma = E l^3 D \qquad (2)$$

Führen wir jetzt das Verhältnis h von Länge zu Dicke ein; es sei also $h = \frac{l}{r}$. Ist ferner $\varrho$ die Dichte des Stäbchens, $\mu$ seine Masse, so wird:

$$\mu = r^2 \pi l \varrho = \frac{l^3}{h^2} \pi \varrho.$$

Führen wir diesen Wert (für $l^3$) in Gleichung (2) ein:

$$m = \frac{E D h^2 \mu}{\pi \varrho}.$$

In einem gewissen Gesamtvolumen v des Sols ist das von den Teilchen herrührende Gesamtmoment

$$\Sigma m = \Sigma \frac{E D h^2 \mu}{\pi \varrho}.$$

Bei dem rechts stehenden Ausdruck muß nur über $\mu$ summiert werden; $\Sigma \mu$ ist aber gleich der im Volumen v gesamt gelösten Substanz. Beträgt ihre Konzentration c, so ist $\Sigma \mu = c v$, demnach

$$\Sigma m = \frac{E D h^2}{\pi \varrho} \cdot \Sigma \mu = \frac{E D h^2 c v}{\pi \varrho} \qquad (3)$$

Nun ist bekanntlich die Dielektrizitätskonstante K gleich

$$K = 1 + 4 \pi \varepsilon, \qquad (4)$$

wo $\varepsilon$ die Elektrisierungskonstante bedeutet. Für ein Sol ist z. B.:

$$\varepsilon = \frac{\Sigma m + M}{v\,E} \qquad (5)$$

wo $\dfrac{\Sigma m}{v\,E}$ die von den Teilchen herrührende, $\dfrac{M}{v\,E}$ die von den in Betracht kommenden $H_2O$-Molekülen herrührende Elektrisierung bedeutet (M ist die Momentensumme dieser Wassermoleküle). Zur Berechnung der letzteren setzen wir in Gleichung (5) $\Sigma m = O$ (reines Wasser); wird der hierbei sich ergebende Wert von $\varepsilon$ in Gleichung (4) eingesetzt, so folgt, wenn man in Betracht zieht, daß jetzt $K = D$ ist:

$$\frac{M}{v\,E} = \frac{D-1}{4\,\pi} \qquad (6)$$

Setzen wir nun in Gleichung (5) für $\dfrac{\Sigma m}{v\,E}$ den Wert aus Gleichung (3), für $\dfrac{M}{v\,E}$ den Wert aus Gleichung (6) ein, so wird:

$$\varepsilon = \frac{D\,h^2\,c}{\pi\,\varrho} + \frac{D-1}{4\,\pi}.$$

Dieser Wert in Gleichung (4) eingesetzt ergibt endlich für die Dielektrizitätskonstante des Gesamtsystems:

$$K = D \left(1 + 4\,h^2 \frac{c}{\varrho}\right). \qquad (7)$$

Die Formel (7) besagt demnach, daß die für den Effekt ausschlaggebende Größe h, die „relative Länge" der Teilchen, ist. Für annähernd kugelförmige Teilchen, wo h nahe gleich 1 ist, beträgt er höchstens einige Prozent der Dielektrizitätskonstante des Mediums. Ferner ist die Dielektrizitätskonstante von der Oberflächendichte, also vom Entladungszustande des Sols, ferner vom Dispersitätsgrad (in erster Annäherung) u n a b h ä n g i g.

Die aus der Gleichung (7) berechneten Werte sind der Größenordnung nach richtig. Man kann nämlich ultramikroskopisch durch Auszählung feststellen, daß z. B. in einem Fall bei einem älteren Vanadinpentoxydsol die Teilchen, die länger als 15 $\mu$ waren, einen großen Anteil des gesamten Soles ausmachten[5]. Da die Dicke der Vanadinpentoxydnädelchen. nicht auflösbar war (keine Trennung der Interferenzstreifen beider Ränder), kann für die o b e r e Grenze der Dicke 0,5 $\mu$ genommen werden; dieser Wert ergibt für h ungefähr 30, daher ergibt

sich für ein einprozentiges Sol ($c = 10^{-2}$, $\varrho = 3,4$) eine Dielektrizitätskonstante von 960.

Nach der Formel (7) ist dieser Effekt bei jedem Stäbchensol vorhanden, wenn die gemachten Voraussetzungen erfüllt sind, daß also unter anderem die äußere Belegung als eine zusammenhängende Raumionenladung betrachtet werden kann; bei perlschnurartigen Gebilden, wo die Stäbchen durch „gerichtete Koagulation" entstanden sind, wird dies wohl nicht immer der Fall sein. Sind ferner die Schwingungen sehr rasch, so sind zwei weitere Vereinfachungen nicht mehr angängig, daß nämlich von der Trägheit der Teilchen und der Viskosität des Mediums abgesehen werden kann. Würde man auch diese in Betracht ziehen, so würde die Formel (7) eine Zeitgröße enthalten, also z. B. eine Funktion der Schwingungszahl des elektrischen Feldes sein.

Die Anwendung der Formel (7) gestattet auch Schlüsse in anderer Richtung zu ziehen. Es ist nämlich eine vielleicht noch offene Frage, ob die Teilchen des Vanadinpentoxydsols durch Kristallisation oder durch gerichtete Koagulation entstanden sind. J. E r r e r a[6] fand nun, daß die Dielektrizitätskonstante der Vanadinpentoxydsole mit dem Altern stark zunimmt. Dies bedeutet also, daß hierbei h stark wachsen muß; denn von einer sonstigen Dispersitätsgradveränderung ist ja die Dielektrizitätskonstante (in erster Annäherung) unabhängig. Ein Wachsen von h bedeutet aber, daß die Teilchen in der Länge im Verhältnis d a u e r n d rascher, also „beschleunigt", wachsen als in der Dicke. Bei einem Kristallisationsvorgange würde man annehmen, daß die Wachstumsgeschwindigkeit einer jeden Kristallfläche normal zur Oberfläche konstant bleibt, dies würde also für das V e r - h ä l t n i s von Wachstumsgeschwindigkeiten in zwei verschiedenen Richtungen ebenfalls gelten. Der früher erwähnte Befund, nach dem sich also h beim Altern ändert, würde demnach so zu deuten sein, daß das Wachstum der Vanadinpentoxydteilchen n i c h t durch einen g e - w ö h n l i c h e n K r i s t a l l i s a t i o n s v o r g a n g bewirkt wird.

Endlich sei bemerkt, daß die Erscheinung eine typisch kolloidchemische bzw. kapillarphysikalische ist, da sie ja ihren Grund in der elektrischen Doppelschicht hat, also in der H e t e r o g e n i t ä t bzw. Zweiphasigkeit des Systems. Wie die Kataphorese, Elektroendosmose usw. die elektrischen G l e i c h s t r o m -

---

[5] Die Sole sind nämlich polydispers, zum Teil sogar amikroskopisch, und es soll ja ein mittleres h geschätzt werden.

[6] J. E r r e r a, Koll.-Zeitschr. 31, 63 (1922).

222

erscheinungen der Kapillarphysik sind, ist das induktive Verhalten der Stäbchensole eine den Gleichstromerscheinungen dem Wesen nach vollständig analoge Wechselstromerscheinung dieser Disziplin [7]).

Zusammenfassung.

1. Die Teilchen der Stäbchensole müssen sich tripolartig verhalten.

2. Das von Errera beobachtete anomale induktive Verhalten der Stäbchensole läßt sich

qualitativ wie quantitativ durch die elektrische Doppelschicht begründen.

3. Die Dielektrizitätskonstante solcher Systeme ist, wenn man von anderen Einflüssen absieht, proportional mit dem Quadrat der „relativen Länge" $\left( \left( \frac{\text{Länge}}{\text{Dicke}} \right)^2 \right)$ der Teilchen.

4. Die Anwendung der sich ergebenden Formel auf das Verhalten von Vanadinpentoxydsolen beim Altern läßt folgern, daß das Anwachsen der Vanadinpentoxydstäbchen nicht durch eine gewöhnliche Kristallisation erfolgt.

Wir möchten nicht verfehlen, Herrn Professor Freundlich für viele wertvolle Ratschläge und Hinweise bestens zu danken.

[7]) Eine weitere Bestätigung der Gleichung 7 ist die von Errera (Koll.-Zeitschr. 31, 62 [1922]) gemessene Abhängigkeit der Dielektrizitätskonstante von der Konzentration des Sols. Er fand, daß $(K-D)$ proportional zur Konzentration ist, ganz wie es Gleichung 7 verlangt.

*Aus dem Kaiser-Wilhelm-Institut für physikalische Chemie und Elektrochemie, Berlin-Dahlem.*

# Die Gitterstruktur des rhombischen Schwefels

H. Mark und E. P. Wigner

Zeitschrift für physikalische Chemie *111*, 398–414 (1924)

(Eingegangen am 8. 5. 24.)

## I. Einleitung.

Wenn man einen Blick auf die Tabelle der Elemente wirft, deren Struktur bisher bestimmt worden ist, so sieht man, dass hauptsächlich die hochsymmetrischen Strukturen bekannt sind, da zu ihrer Bestimmung eine geringere Zahl experimenteller Feststellungen bereits hinreicht, während die Bestimmung niedrigsymmetrischer Strukturen grössere Schwierigkeiten bereitet. Bei kubischen (tetragonalen, hexagonalen) Elementarkörpern hat man nämlich nur eine bzw. zwei Achsenlängen zu bestimmen, während z. B. im rhombischen System, deren drei festgelegt werden müssen. Darauf ist es wohl zurückzuführen, dass bisher — abgesehen von Kohlenstoff und Silicium — nur Metallstrukturen gut erforscht sind.

Nun sind aber eben die Elemente aus der 5., 6. und 7. Reihe des periodischen Systems, also hauptsächlich $P$, $S$, $J$ solche, deren Strukturen den Chemiker am meisten interessieren. Erstens weiss man hier aus rein chemischen Ursachen (Dampfdichte, osmotischer Druck in Lösungen und Verwandtes), dass die Atome zu Molekülen vereinigt sind und es ist sicher nicht ohne Interesse zu sehen, ob und wie dies in der räumlichen Anordnung der Beugungszentra zum Ausdruck kommt. Zweitens sind hier mehrere allotrope Modifikationen gefunden worden, so dass sich in diesen Fällen die Möglichkeit des Studiums eines Umwandlungspunktes durch Vergleich der beiden Gitter ergibt[1]). Die genaue Untersuchung der beim Umwandlungspunkt obwaltenden Verhältnisse könnte nämlich neues Material zur Beurteilung der Bindung der Atome im Gitter an den Tag fördern.

Unsere Wahl fiel von allen hier in Betracht kommenden Elementen deshalb auf den $\alpha$-Schwefel, weil davon grosse und schön ausgebildete Kristalle vorlagen, die für die von uns verwendete Methode notwendig waren. Auch weiss man ziemlich sicher, dass der Schwefel in der 8., also rhombisch-bipyramidalen, Kristallklasse kristallisiert[2]). Sir W. H. Bragg[3]) hat auch die Identitätsperioden des Schwefels schon im Jahre 1914 zu bestimmen versucht und von K. Weissenberg und dem einen von uns[4]) waren bereits an kleinen Kriställchen eine Reihe von Drehdiagrammen aufgenommen worden, welche die Braggschen Achsenlängen bestätigten, der Annahme einer Basisflächenzentrierung dieses Elementarkörpers jedoch widersprachen. Da bereits in diesen Aufnahmen einige (allerdings sehr schwache) Punkte auftraten, die mit den gefundenen Achsen ($a = 5.3\ \text{Å}$, $b = 6.4\ \text{Å}$, $c = 12.2\ \text{Å}$) im Widerspruch standen, schien es uns notwendig, an grossen, schön ausgebildeten Kristallen die Untersuchung fortzusetzen.

## II. Das verwendete Material und die verwendete Methode.

Für die Untersuchung standen uns drei schön ausgebildete Schwefelkristalle zur Verfügung; zwei hiervon waren natürlich und stammten aus Girgenti[5]), während der dritte ein künstlich aus $CS_2$ gezüchtetes Exemplar war.

Ausserdem verfügten wir zur Herstellung von Drehdiagrammen über eine grosse Zahl gut ausgebildeter kleiner Kriställchen von 1 bis 2 mm Grösse.

An den grossen Einkristallen fanden sich die in der Tabelle 1 enthaltenen Kanten und Ebenen wohlausgebildet.

---

[1]) Vgl. die von Westgren am Eisen durchgeführte Untersuchung. Zeitschr. f. physik. Chemie **102** (1922).

[2]) Herrn Prof. Dr. A. Johnsen verdanken wir den Hinweis, dass auch die rhombisch-bisphenoidische Klasse in Betracht zu ziehen sei.

[3]) W. H. Bragg, Proc. Roy. Soc. London (A) **89**, 575 (1914).

[4]) Zeitschr. f. Elektrochemie (1923).

[5]) Den einen verdanken wir der Güte des Herrn Prof. F. Becke in Wien, den anderen haben uns die Herren Prof. A. Johnsen und C. Belowski in liebenswürdigster Weise zur Verfügung gestellt.

400                        H. Mark und E. Wigner

Tabelle 1.

| Kanten | Ebenen |
|--------|--------|
| [010] | (001) |
| [110] | (010) |
| [101] | (011) |
| [011] | (111) |
|       | (113) |

Da es zur Vermessung des Elementarkörpers darauf ankam, alle drei Achsenlängen aus Drehdiagrammen direkt zu bestimmen und zur Kontrolle, sowie zur sicheren Feststellung des Bravais-Gitters auch die drei Flächendiagonalen und die Raumdiagonale direkt nach der Polanyischen Gleichung zu vermessen, mussten Schichtliniendiagramme um [100], [010], [001], [110], [101], [011], [111] angefertigt werden. Vier dieser Richtungen waren nach Tabelle 1 als Kanten ausgebildet, also leicht zu justieren, von den anderen dreien liess sich [001] ebenfalls sehr leicht einstellen, da die darauf senkrechte Ebene (001) gut ausgebildet war. [100] haben wir dadurch eingestellt, dass die in (001) auf [010] senkrechte Richtung aufgesucht wurde; während die Einstellung von [111] etwas schwieriger zu bewerkstelligen war.

Ausser der Bestimmung dieser Identitätsperioden wollten wir aber auch sämtliche wichtigen Ebenen nach der Braggschen Gleichung zur Reflexion bringen, um über den Intensitätsabfall der verschiedenen Ordnungen sowie über das Fehlen von Interferenzen dieser wichtigsten Netzebenen sicheren Aufschluss zu erhalten. Es wurden also sämtliche in Tabelle 1 aufgeführten Ebenen direkt zur Reflexion gebracht; das als Wachstumsfläche niemals auftretende vordere Pinakoid (100) wurde durch Anschleifen freigelegt.

In der Literatur finden sich über das Gitter des rhombischen Schwefels die beiden bereits zitierten Angaben, welche die Schwierigkeiten erkennen liessen, die sich einer sicheren Bestimmung in den Weg stellen würden. Es sind dies grosse Perioden als Kanten des Elementarkörpers, niedrige Symmetrie und das „schlechte Reflexionsvermögen" des Schwefelatoms.

Die von Bragg angegebenen Perioden sind: $a = 5.3$, $b = 6.4$, $c = 12.2$ Å, über das Reflexionsvermögen findet sich bei Bragg folgende Angabe[1]: „die vom Schwefel gelieferten Spektra sind alle schwach, besonders im Falle der (101) Ebene und in keinem Falle sind Spektra von höherer Ordnung als der ersten aufgefunden worden."

---

[1] W. H. Bragg, Proc. Roy. Soc. London (A) **89**, 575 (**1914**).

Die Umgehung dieser Schwierigkeiten gelingt dadurch, dass man das schlechte Reflexionsvermögen durch ausgiebige Beleuchtung (grosse Intensität des Primärstrahles, lange Beleuchtungsdauer) kompensiert.

Grosse Kantenlängen eines Elementarkörpers bewirken bei gegebener Strahlung, dass die niedrig indizierten Interferenzen sich in der Nähe des Durchstosspunktes anhäufen und besonders bei niedriger Symmetrie sehr bald eine Vieldeutigkeit in der Zuordnung von reflektierenden Ebenen zu den gemessenen Linien auftritt. Um dies zu verhindern, haben wir eine Kamera mit sehr grossem Durchmesser ($d = 30$ cm) gewählt und als monochromatisches Licht im Gegensatz zu Bragg, welcher Molybdän ($\lambda = 0.712$) verwendete, Kupfer ($\lambda = 1.54$)

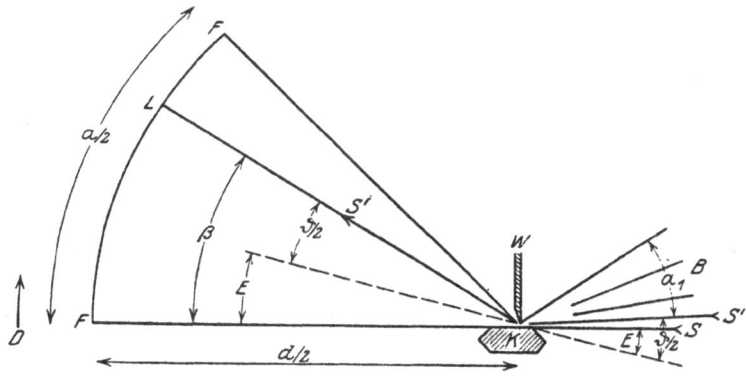

Fig. 1.

gebraucht. Durch Anwendung der Schneidenmethode gelang es fernerhin, scharfe Linien (vgl. etwa die Fig. 5 um [110]) zu erzielen, deren Vermessung grosse Genauigkeit gewährte. Ausserdem wurde der Kristall immer nur um einen bestimmten Winkel und nicht um 360° gedreht, so dass von vornherein nur wenige, ganz bestimmte Ebenen zur Reflexion gelangten. Hierdurch wurde erreicht, dass in keinem Falle bezüglich der Indizierung eine Zweideutigkeit möglich war. Die verwendete Kamera, welche der Seemannschen nachgebildet war, und die Rechenmethode seien im folgenden kurz beschrieben.

Ein Kristall $K$ (Fig. 1) wurde mit einer bestimmten Ebene, der „eingestellten Ebene", gegen eine Platte $W$ gelehnt und mit dem konvergenten Bündel $B$ von monochromatischen Röntgenstrahlen beleuchtet. Als Blende dient die Wolframplatte $W$, die gegen das Kristall

402                          H. Mark und E. Wigner

gerichtet ist. Der Film $F$ war in der Entfernung von $\dfrac{d}{2} = 15$ cm in einer Zylinderfläche von der Höhe von etwa 10 cm ausgespannt. Die Spitze der Blende $W$ war in etwa $^1/_{10}$ mm Entfernung von der Oberfläche des Kristalls, so dass die Linien etwa diese Breite hatten.

Der Öffnungswinkel des Röntgenstrahlenbüschels sei $\alpha_1$, des Films $\alpha_2$. Die reflektierende Netzebene schliesse mit der Richtung $SF$ den Winkel $\varepsilon$ ein. Ist dann $\dfrac{\vartheta}{2}$ der Glanzwinkel dieser Ebene, so erscheint das Bild ihrer Linie $L$ unter dem Winkel $\beta$, wo

$$\beta = \varepsilon + \frac{\vartheta}{2} \tag{1}$$

ist. Dafür dass es überhaupt erscheine, ist (wenn die Linie am Äquator liegt) notwendig und hinreichend, dass

$$0 < \frac{\vartheta}{2} - \varepsilon < \alpha_1 \tag{2}$$

$$0 < \frac{\vartheta}{2} + \varepsilon < \alpha_2 \tag{3}$$

wie aus der Fig. 1 ersichtlich. Wenn (2) nicht zutreffen würde, würde die Ebene überhaupt nicht zur Reflexion gelangen; würde (3) nicht zutreffen, so käme das Bild nicht auf den Film.

Bei unseren Aufnahmen war $\alpha_1 \sim \alpha_2 \sim 45°$. Für diesen Fall ist der für uns in Betracht kommende Bereich in Fig. 2 schraffiert gezeichnet. Wir ersehen daraus, dass die Ebene höchstens den Winkel von $22^1/_2°$ mit der „eingestellten Ebene" einschliessen darf, der Glanzwinkel dem absoluten Wert nach grösser sein muss, als der Winkel $\varepsilon$ zwischen den beiden Ebenen und dass die Summe der Absolutwerte der beiden Winkel $\dfrac{\vartheta}{2} + \varepsilon$ kleiner sein muss als 45°. Diese Bedingungen erwirkten, dass die Linien unzweideutig indizierbar waren.

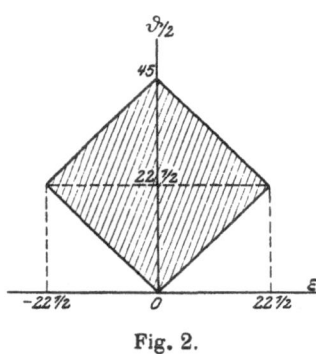

Fig. 2.

Das „konvergente Strahlenbündel $B$" wurde so erzeugt, dass wir die ganze in Fig. 1 abgebildete Vorrichtung um die Achse $O$ um $\alpha_1$ in der Richtung $D$ durch ein Uhrwerk drehen liessen, während der Strahl in Wirklichkeit immer nur in der Richtung $S$ (mit einem Öffnungswinkel von etwa 5°) kam.

Wenn etwa $(uvw)$ die „eingestellte Ebene" war, und etwa $[hkl]$ die Drehrichtung darstellte, so wurden zunächst zur Indizierung des Äquators die Winkel $\varepsilon_1 \ldots \ldots \varepsilon_i$ zwischen den Ebenen $(rst)$, für welche $hr + ks + lt = 0$, und der Ebene $(uvw)$ berechnet nach der Formel

$$\cos \varepsilon_{rst} = \frac{1}{J_{uvw}\,J_{rst}} \left( \frac{ur}{a^2} + \frac{vs}{b^2} + \frac{wt}{c^2} \right), \tag{4}$$

wobei

$$J_{uvw} = \sqrt{\frac{u^2}{a^2} + \frac{v^2}{b^2} + \frac{w^2}{c^2}} \tag{5}$$

$$J_{rst} = \sqrt{\frac{r^2}{a^2} + \frac{s^2}{b^2} + \frac{t^2}{c^2}} \tag{6}$$

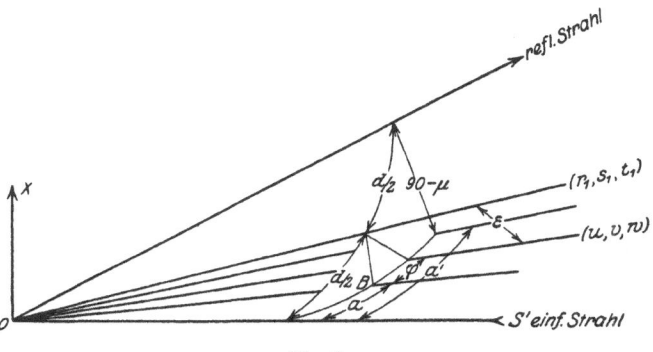

Fig. 3.

ist. Diese Formeln gelten nur für einen rhombischen Elementarkörper mit den Achsen $a$, $b$ und $c$. Dann wurden aus ihrer Gesamtheit diejenigen ausgewählt, welche im Verein mit $\vartheta_{rst}$ die Bedingungen (2) und (3) erfüllten. $\vartheta_{rst}$ wurde hierbei nach der quadratischen Form

$$\sin^2 \frac{\vartheta_{rst}}{2} = \frac{\lambda^2}{4} \left( \frac{r^2}{a^2} + \frac{s^2}{b^2} + \frac{t^2}{c^2} \right) \tag{7}$$

berechnet. Diese Methode hat also die Kenntnis der Achsen $a$, $b$ und $c$ zur Voraussetzung. Da die Kantenlängen des Elementarkörpers vorher in unabhängiger Weise durch die Polanyische Gleichung

$$J = \frac{n\lambda}{\cos \mu} \tag{8}$$

(wobei $J$ die Identitätsperiode, also in diesem Falle die Kantenlänge, $\lambda = 1{\cdot}54$ die Wellenlänge des Lichtes, $90° - \mu$ den Winkelabstand vom Äquator und $n$ die Ordnungszahl der Schichtlinien bedeuten) bestimmt worden war, war sie anwendbar.

Für die Indizierung der Schichtlinie (Fig. 3, dabei ist die Drehachse, mit $OX$ gekennzeichnet, vertikal angenommen) wurden wiederum zunächst die Winkel $\varepsilon$ bestimmt, welche die auf der ersten Schichtlinie auftretenden Ebenen $(r_1 s_1 t_1)$, für welche $h r_1 + k s_1 + l t_1 = 1$, mit $(u v w)$ bilden. Die Winkel, welche die Normalen der Ebenen $(r_1 s_1 t_1)$ mit der Drehrichtung einschliessen, sind

$$\cos(90° - c) = \frac{1}{J_{r_1 s_1 t_1}} \cdot \frac{h r_1 + k s_1 + l t_1}{\sqrt{h^2 a^2 + k^2 b^2 + l^2 c^2}}. \tag{9}$$

Der Winkel zwischen dem einfallenden Strahl und der Projektion $B$ der Ebenennormalen in die Horizontale ist

$$\cos a = \frac{\sin \frac{\vartheta}{2}}{\cos c}. \tag{10}$$

Die Winkel, welche der einfallende Strahl $S$ mit den Projektionen der reflektierten Strahlen in die Horizontale einschliesst, sind:

$$\cos a' = \frac{\sin \vartheta}{\sin \mu}. \tag{11}$$

Die Winkel zwischen der Normale der „eingestellten Ebene" und der Projektion der Normale der reflektierenden Ebene in die Horizontale ist:

$$\cos \varphi = \frac{\cos \varepsilon}{\cos c}. \tag{12}$$

Aus diesen berechnet sich der horizontale Winkelabstand der Linie vom Durchstosspunkt am Film

$$\varsigma = 90° - a' + a \pm \varphi. \tag{13}$$

Der vertikale Abstand ist durch die Schichtlinienbeziehung gegeben.

## III. Bestimmung des Elementarkörpers und der Translationsgruppe.

Die Strukturbestimmung erfolgte in drei Schritten. Zunächst wurden durch Drehdiagramme um die Hauptachsen die Identitätsperioden in diesen Richtungen, d. h. die Grösse des Elementarkörpers festgestellt.

In den nachfolgenden Tabellen sind die den verschiedenen Schichtlinien entsprechenden doppelten Schichtlinienabstände unter $2e$, die ctg des Öffnungswinkels des Schichtlinienkegels unter $\frac{e}{r}$, der cos dieses Winkels unter $\cos \mu$ und die daraus berechneten Identitätsperioden unter $J$ aufgetragen.

Die Gitterstruktur des rhombischen Schwefels.    405

Das Diagramm der Tabelle 2 ist in Fig. 4 abgebildet.

Tabelle 2.
Drehdiagramm um [001] (Fig. 4).

| Nr. | $2e$ | $\dfrac{e}{r}$ | $\cos \mu$ | $J$ | Mittel |
|---|---|---|---|---|---|
| I | 9.5 | 0.0633 | 0.0633 | 24.48 | |
| II | 19.5 | 0.1300 | 0.127 | 24.62 | |
| III | 29.5 | 0.1968 | 0.192 | 24.40 | |
| IV | 40.0 | 0.2670 | 0.259 | 24.60 | $24.56_2$ |
| V | 49.5 | 0.330 | 0.313 | 24.63 | |
| VI | 61.0 | 0.406 | 0.375 | 24.65 | |

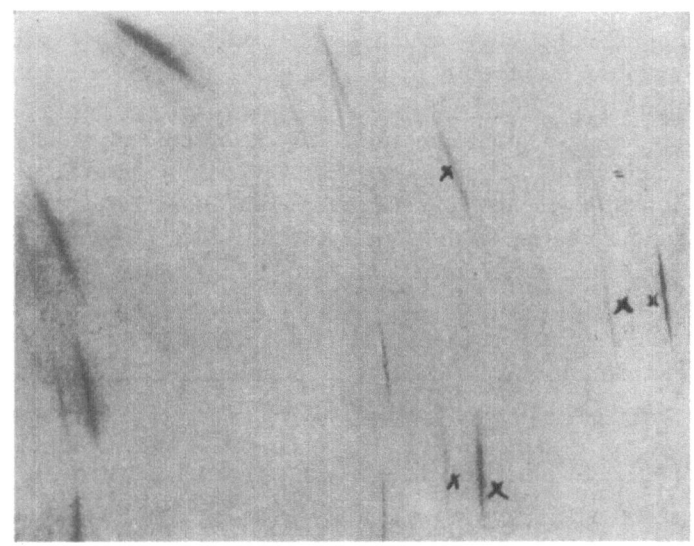

Fig. 4.
Drehdiagramm um die $c$-Achse mit $Cu$-Strahlung aufgenommen.

Tabelle 3.
Drehdiagramm um [010].

| Nr. | $2e$ | $\dfrac{e}{r}$ | $\cos \mu$ | $J$ | Mittel |
|---|---|---|---|---|---|
| I | 18.0 | 0.120 | 0.119 | 12.92 | |
| II | 37.0 | 0.246 | 0.238 | 12.88 | 12.87 |
| III | 57.0 | 0.380 | 0.360 | 12.80 | |

406                                    H. Mark und E. Wigner

Tabelle 4.
Drehdiagramm um [100].

| Nr. | $2e$ | $\dfrac{e}{r}$ | $\cos\mu$ | $J$ | Mittel |
|-----|------|------|------|------|--------|
| I   | 22·0 | 0·147 | 0·144 | 10·62 |        |
| II  | 45·0 | 0·300 | 0·288 | 10·62 | 10·61  |
| III | 66·5 | 0·443 | 0·400 | 10·60 |        |

Das röntgenographisch gefundene Achsenverhältnis ist somit
0·820:1:1·900; es stimmt mit dem goniometrischen 0·8109:1:1·9005[1])
hinreichend überein. Wie man sieht, sind die Achsen genau doppelt
so gross als die von Bragg gefundenen, was seine experimentelle Begründung dadurch erfährt, dass auf unseren Diagrammen an den Stellen
der ungeraden Schichtlinien sicher Linien nachweisbar waren. Einige
solche Reflexionen sieht man z. B. in Fig. 4, sie sind dort durch ein
$x$ hervorgehoben.

Die einzelnen Schichtlinienabstände konnten nicht leicht vermessen werden, da keine Punkte, sondern Linien entstanden. Es musste
also der Abstand zwischen den Mittelpunkten dieser Linien abgeschätzt
werden. Zur grösseren Sicherheit wurden noch andere Diagramme
um dieselben Richtungen aufgenommen, bei welchen aus dem einfallenden Strahl durch einen horizontalen Schlitz ein niedriger Bereich
ausgeblendet, so dass die Interferenzen beinahe punktförmig und die
Schichtlinienabstände gut vermessbar wurden. Auch durch Drehdiagramme mit kleinen Kriställchen wurden diese Identitätsperioden
überprüft.

Aus der Dichte des Schwefels 2·037[2]) ergab sich, dass im Elementarkörper

$$\frac{10 \cdot 61 \cdot 12 \cdot 87 \cdot 24 \cdot 56 \cdot 10^{-24} \cdot 6 \cdot 06 \cdot 10^{23}}{32 \cdot 07},$$

das ist etwa 129 Atome, enthalten sind. Diese Zahl ist später dadurch
verbessert worden, dass die (001)-Ebene — wie wir sehen werden —
erst in 8. Ordnung auftritt, was einer achtfachen Unterteilung der
$c$-Achse entspricht. Die Anzahl der Teilchen im Elementarkörper muss
also durch 8 teilbar sein, was die Zahl 128 festlegt.

Nach derselben Methode wurden durch Drehung um [011], [101],
[110] die Identitätsabstände in der Richtung der Flächendiagonalen

---

[1]) P. v. Groth, Chem. Kristallogr. **1**, 26.
[2]) P. v. Groth, loc. cit.

Die Gitterstruktur des rhombischen Schwefels.    **407**

bestimmt. Es ergab sich, dass der Elementarkörper allseitig flächen-zentriert ist, da die bezüglichen Abstände halb so gross sind, als die aus den Achsen berechneten. Ein Diagramm um [110], welches mit $MoK_\alpha$-Strahlung aufgenommen wurde, ist in Tabelle 5, ein mit $Cu$-Strahlung aufgenommenes in Tabelle 6 wiedergegeben.

## Tabelle 5.
### Drehdiagramm um 110.

| Nr. | $2e$ | $\dfrac{e}{r}$ | $\cos\mu$ | $J$ | Mittel |
|---|---|---|---|---|---|
| I | 4·8 | 0·0839 | 0·0839 | 8·32 | |
| II | 9·7 | 0·170 | 0·168 | 8·30 | 8·35 |
| III | 14·4 | 0·252 | 0·245 | 8·44 | |
| IV | 19·0 | 0·332 | 0·315 | 8·34 | |

In ähnlicher Weise wurde mit Kupferstrahlung für die Identitäts-perioden in der Richtung [101] 13·22, in der Richtung [011] 13·93 gefunden, während die aus den Achsen berechneten Längen 26·60 bzw. 27·7 sind. (Für [110] findet man aus den Achsen 16·38.)

Eine Drehung um die Raumdiagonale [111] bewies, — wie ja vor-auszusehen war —, dass der Elementarkörper nicht raumzentriert ist; (gefunden 29·25, berechnet 29·6).

Nun wurden die Linien in sämtlichen Diagrammen identifiziert; zunächst auf dem Äquator. In den folgenden Tabellen sind bei beob. bzw. ber. die beobachteten bzw. berechneten Werte der Winkelent-fernungen vom Durchstosspunkt eingetragen. Der Durchstosspunkt wurde von einer Linie aus genau bestimmt, die dann in Klammern ein-gesetzt ist. Wie auf S. 402, Formel (1) abgeleitet, ist der Winkel-abstand am Äquator

$$\beta = \varepsilon + \frac{\vartheta}{2}, \tag{1}$$

wo $\dfrac{\vartheta}{2}$ einfach den halben Glanzwinkel und $\varepsilon$ den Winkel zwischen der Ebene, die zur Reflexion kam, und der „eingestellten Ebene" bedeutet.

Der Winkel $\dfrac{\vartheta}{2}$ wurde nach der bekannten quadratischen Form be-stimmt und zwar

$$\sin^2\frac{\vartheta}{2} = 0.005255\, h^2 + 0.003585\, k^2 + 0.000986\, l^2.$$

408                    H. Mark und E. Wigner

## Tabelle 6.
Äquator bei der Drehachse [110].    Eingestellte Ebene (001).    (Fig. 5).

| Indizierung | $(008)_\alpha$ | $(0016)_\alpha$ | $(008)_\beta$ | $(0016)_\beta$ | $(1\,1\,11)_\alpha$ |
|---|---|---|---|---|---|
| beob.<br>ber. | (14·5)<br>(14·5) | 30·2<br>30·2 | 13·1<br>13·1 | 27·0<br>26·9 | 5·5<br>5·5 |

| Indizierung | $(1\,1\,11)_\beta$ | $(1\,1\,13)_\alpha$ | $(1\,1\,15)_\alpha$ | $(2\,2\,20)_\beta$ |
|---|---|---|---|---|
| beob.<br>ber. | 3·5<br>3·5 | 11·7<br>11·7 | 17·4<br>17·4 | 19·3<br>19·4 |

Fig. 5 a.

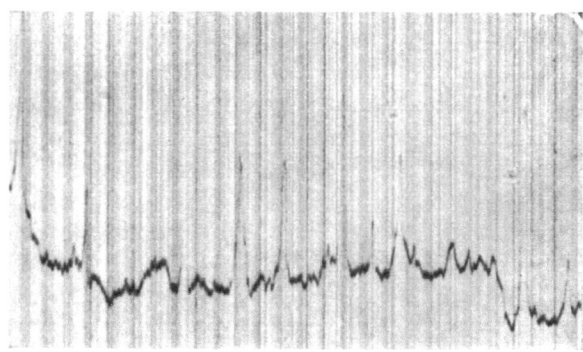

Fig. 5 b.

Drehdiagramm um die [110]-Richtung mit *Cu*-Strahlung.    Eingestellte Ebene war (001).

Eine zur sicheren Vermessung und quantitativeu Intensitätsverwertung angefertigte Photometrierung dieses Diagrammes befindet sich unter 5 b; sie ist im bezug auf das Originaldiagramm seitenverkehrt.

Die Gitterstruktur des rhombischen Schwefels.                  409

## Tabelle 7.
Äquator bei der Drehung um [110].   Eingestellte Ebene (111).  (Fig. 6.)

| Indizierung | $(111)_\alpha$ | $(222)_\alpha$ | $(333)_\alpha$ | $(444)_\alpha$ | $(555)_\alpha$ |
|---|---|---|---|---|---|
| beob. | 5·7 | 11·5 | 17·55 | 23·7 | 29·7 |
| ber. | 5·7 | 11·5 | 17·4 | 23·5 | 29·8 |

| Indizierung | $(553)_\alpha$ | $(331)_\alpha$ | $(442)_\alpha$ | $(111)_\beta$ | $(222)_\beta$ |
|---|---|---|---|---|---|
| beob. | 21·9 | 4·55 | 13·7 | 5·15 | 10·4 |
| ber. | 21·7 | 4·5 | 13·5 | 5·1 | 10·3 |

| Indizierung | $(333)_\beta$ | $(331)_\beta$ | $(442)_\beta$ |
|---|---|---|---|
| beob. | 15·75 | 2·85 | 11·0 |
| ber. | 15·6 | 2·8 | 11·2 |

Fig. 6 a.

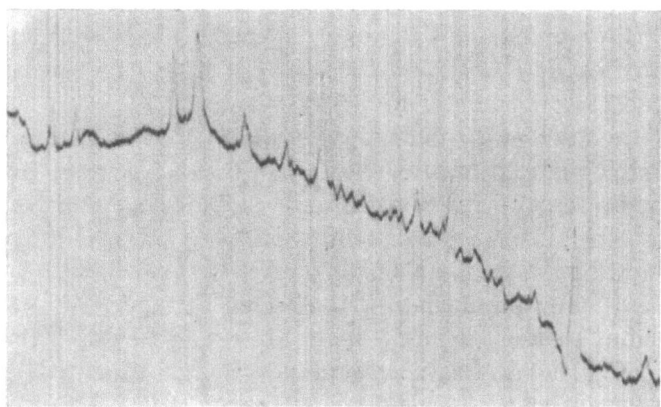

Fig. 6 b.
Drehdiagramm um die [110]-Richtung mit *Cu*-Strahlung.  Eingestellte Ebene war (111).

Die Photometrierung befindet sich unter 6 b.  Bei dieser Aufnahme konnte der Nullpunkt — unabhängig von den Linien — aus der *Br*- und *Ag*-Absorptionskante bestimmt werden.

Des weiteren sei die Auswertung einer Schichtlinie mitgeteilt.

410                    H. Mark und E. Wigner

Tabelle 8.
Schichtlinie bei der Drehachse [110]. Eingestellte Ebene (001). (Fig. 6).

| Indizierung | (119) | (119) | (1 1 11) | (1 1 13) | (1 1 15) |
|---|---|---|---|---|---|
| beob. | 11·45 | 18·55 | 16·4 | 25·4 | 29·4 |
| ber. | 11·4 | 18·6 | 16·2 | 25·4 | 29·4 |

| Indizierung | (028) | (0 2 10) | (2 0 10) |
|---|---|---|---|
| beob. | 6·6 | 31·95 | 8 |
| ber. | 6·5 | 32·05 | 8·2 |

Die $\beta$-Linien sind hier nicht angeführt, da sie von den $\alpha$-Linien durch ihre etwas tiefere Lage leicht zu unterscheiden sind. Die Lage des Nullpunktes war vom Äquator her bekannt.

Auf dieselbe Weise wurden sämtliche in den 7 Drehdiagrammen auftretenden Linien indiziert.

## IV. Bestimmung der Raumgruppe.

Der Statistik der auftretenden Netzebenen entnehmen wir zunächst für die Raumgruppenbestimmung folgende Tatsachen:

    1. Bipyramiden mit lauter ungeraden Indizes besitzen normales $R$; (111), (335), (1 1 11), (1 1 13) dagegen treten (112), (211) usw. nicht auf.

    2. Die Röntgenperioden des vorderen und seitlichen Pinakoids sind viertelprimitiv; (001) tritt erst in achter Ordnung auf.

Die rhombisch bipyramidale Klasse enthält 28 Raumgruppen, von denen $V_h^{25} - V_h^{28}$ durch die Feststellung 1. streng ausgeschlossen werden, denn es sind Ebenen beobachtet worden {(111) usw.}, welche in diesen Gruppen nicht auftreten dürften. Die übrigen Gruppen $V_h^1 - V_h^{24}$ bleiben zur Diskussion stehen. Nimmt man aber den in der Tabelle 5 und in dem nachfolgenden Absatz enthaltenen Befund hinzu, dass die Identitätsperioden auf den Flächendiagonalen die halbe Länge der Hypotenusen besitzen, so gelangt man zu einem allseitig flächenzentrierten Elementarkörper. Dieses Ergebnis steht mit der Feststellung 1. in völliger Übereinstimmung, denn die experimentell gefundenen Auslöschungen sind gerade diejenigen, welche von den beiden allseitig flächenzentrierten Raumgruppen gefordert werden. Es sind nunmehr $V_h^{23}$ und $V_h^{24}$ übrig. Zwischen diesen liesse sich streng unterscheiden, wenn man ein Pinakoid in zweiter Ordnung beobachtet hätte. Dies

Die Gitterstruktur des rhombischen Schwefels.                    **411**

ist aber nicht der Fall; vielmehr sind zwei Pinakoide erst in vierter, das dritte in achter Ordnung vorhanden. Auf Grund dieser Erscheinung kann man zwar zu der Aussage kommen, dass die Gruppe $V_h^{24}$ die gefundenen Verhältnisse am besten wiedergibt; es ist aber nicht möglich $V_h^{23}$ streng auszuschliessen, weil die Auslöschung von (200), (020) und (002) eine durch die Freiheitsgrade der Struktur bedingte sein könnte. Auch wäre es möglich, dass bei besonders langen Expositionen doch Andeutungen dieser Ebenen auftreten könnten. Da wir aber auf verschiedenen Diagrammen von sehr hoch indizierten Ebenen [z. B. (2, 2, 20)] die $\alpha$-, $\alpha'$-, $\beta$- und $\gamma$-Linien noch sicher feststellen konnten, so ist die Auslöschung jedenfalls eine sehr weitgehende und es sind keinerlei experimentelle Anhaltspunkte für eine gegenteilige Auffassung vorhanden. **Demgemäss wollen wir bei der Lokalisierung der Punktlagen nur die Gruppe $V_h^{24}$ diskutieren.**

Unter der Annahme, dass Schwefel bisphenoidisch kristallisiert, ergibt sich nur die allseitig flächenzentrierte Gruppe $V^7$.

In $V_h^{24}$ sind nun 128 $S$-Atome zu lokalisieren. Der einfachste Fall ist der, dass vier verschiedene $S$-Atome da sind, die sich in allgemeinen Punktlagen befinden. Wenn es mehr als vier verschiedenartige $S$-Atome gibt, kommt noch eine Reihe von Möglichkeiten hinzu, bei welchen sowohl in allgemeinen als auch in speziellen Lagen Atome zu lokalisieren sind. Wir wollen im folgenden nur den Fall von vier ungleichwertigen $S$-Atomen diskutieren.

Es sind also vier konstituierende Punkte $S_1$, $S_2$, $S_3$ und $S_4$ zu lokalisieren. Die Koordinaten für eine allgemeine Punktlage sind[1]):

$$[m, n, p] \quad [m+\tfrac{1}{2}, n+\tfrac{1}{2}, p] \quad [m+\tfrac{1}{2}, n, p+\tfrac{1}{2}] \quad [m, n+\tfrac{1}{2}, p+\tfrac{1}{2}]$$

$$[\overline{m}, n, p] \quad [\overline{m}+\tfrac{1}{2}, n+\tfrac{1}{2}, p] \quad [\overline{m}+\tfrac{1}{2}, \overline{n}, p+\tfrac{1}{2}] \quad [\overline{m}, \overline{n}+\tfrac{1}{2}, p+\tfrac{1}{2}]$$

$$[\overline{m}, n, \overline{p}] \quad [\overline{m}+\tfrac{1}{2}, n+\tfrac{1}{2}, \overline{p}] \quad [\overline{m}+\tfrac{1}{2}, n, \overline{p}+\tfrac{1}{2}] \quad [\overline{m}, n+\tfrac{1}{2}, \overline{p}+\tfrac{1}{2}]$$

$$[m, n, \overline{p}] \quad [m+\tfrac{1}{2}, n+\tfrac{1}{2}, \overline{p}] \quad [m+\tfrac{1}{2}, \overline{n}, \overline{p}+\tfrac{1}{2}] \quad [m, \overline{n}+\tfrac{1}{2}, \overline{p}+\tfrac{1}{2}]$$

$$[\overline{m}+\tfrac{1}{4}, \overline{n}+\tfrac{1}{4}, \overline{p}+\tfrac{1}{4}] \qquad [m+\tfrac{1}{4}, n+\tfrac{1}{4}, \overline{p}+\tfrac{1}{4}]$$

$$[m+\tfrac{1}{4}, \overline{n}+\tfrac{1}{4}, p+\tfrac{1}{4}] \qquad [\overline{m}+\tfrac{1}{4}, n+\tfrac{1}{4}, p+\tfrac{1}{4}]$$

$$[\overline{m}+\tfrac{3}{4}, \overline{n}+\tfrac{3}{4}, \overline{p}+\tfrac{1}{4}] \qquad [m+\tfrac{3}{4}, n+\tfrac{3}{4}, \overline{p}+\tfrac{1}{4}]$$

$$[m+\tfrac{3}{4}, \overline{n}+\tfrac{3}{4}, p+\tfrac{1}{4}] \qquad [\overline{m}+\tfrac{3}{4}, n+\tfrac{3}{4}, p+\tfrac{1}{4}]$$

$$[\overline{m}+\tfrac{3}{4}, \overline{n}+\tfrac{1}{4}, \overline{p}+\tfrac{3}{4}] \qquad [m+\tfrac{3}{4}, n+\tfrac{1}{4}, \overline{p}+\tfrac{3}{4}]$$

$$[m+\tfrac{3}{4}, \overline{n}+\tfrac{1}{4}, p+\tfrac{3}{4}] \qquad [\overline{m}+\tfrac{3}{4}, n+\tfrac{1}{4}, p+\tfrac{3}{4}]$$

$$[\overline{m}+\tfrac{1}{4}, \overline{n}+\tfrac{3}{4}, \overline{p}+\tfrac{3}{4}] \qquad [m+\tfrac{1}{4}, n+\tfrac{3}{4}, \overline{p}+\tfrac{3}{4}]$$

$$[m+\tfrac{1}{4}, \overline{n}+\tfrac{3}{4}, p+\tfrac{3}{4}] \qquad [\overline{m}+\tfrac{1}{4}, n+\tfrac{3}{4}, p+\tfrac{3}{4}]$$

---

[1]) P. Niggli, Geometr. Kristallogr. d. Diskontinuums, Leipzig 1919, S. 215.

Die Struktur besitzt also die Freiheitsgrade

$$m_1, \; n_1, \; p_1, \; m_2, \; n_2, \; p_2, \; m_3, \; n_3, \; p_3, \; m_4, \; n_4, \; p_4,$$

und

$$\frac{S_2}{S_1}, \quad \frac{S_3}{S_1}, \quad \frac{S_4}{S_1}$$

die Verhältnisse der Reflexionsvermögen. Diese Verhältnisse kann man mit grosser Wahrscheinlichkeit gleich Eins setzen, da einerseits heteropolare Bindungen zwischen $S$-Atomen wohl nicht in Frage kommen und andererseits Auslöschungen beobachtet wurden, welche durch die Raumgruppe allein nicht bedingt sind, so dass rationale Unterteilungen durch Ebenen vorkommen müssen, welche sich bei gleicher Atomzahl pro Å² nur dadurch unterscheiden, dass die einen Ebenen z. B. $S_1$-, die anderen $S_2$-Atome enthalten; wären die Reflexionsvermögen verschieden, so müsste durch ihre Verschiedenheit die irrationale Unterteilung zufällig gerade kompensiert werden. Nun wurden aber diese Auslöschungen auch noch mit verschiedenen Wellengängen festgestellt, so dass nicht nur die Reflexionsvermögen selbst, sondern auch ihre Abhängigkeit von der Wellenlänge so geartet sein müsste, dass in jedem der Fälle gerade Kompensation des Parameters eintritt. Da dies aber sehr unwahrscheinlich ist, ist es wohl berechtigt, diese Verhältnisse gleich 1 zu setzen. Es bleiben somit noch 12 Parameter zu bestimmen, wofür der Strukturfaktor herangezogen werden muss. Da eine gleichzeitige Bestimmung sämtlicher Unbekannten ausserordentlich kompliziert wird, sei diese zunächst nur für den $p$-Parameter durchgeführt. Die quantitative Vermessung der Intensitäten von $(0\,0\,4)$, $(0\,0\,8)$, $(0\,0\,12)$, $(0\,0\,16)$ und $(0\,0\,24)$ bei $Sr$-Strahlung ($\lambda_{k_u} = 0.87$ Å, $\lambda_{k_\beta} = 0.78$ Å) und $Cu$-Strahlung ($\lambda_{k_\alpha} = 1.54$, $\lambda_{k_\beta} = 1.39$) zeigt Tabelle 9. Der Strukturfaktor für $(0\,0\,4)$ z. B. lautet:

$$F_{004} = S_1 \{ 4e^{8\pi i p_1} + 4e^{-8\pi i p_1} + 4e^{8\pi i (p_1 + \frac{1}{4})} + 4e^{-8\pi i (p - \frac{1}{4})}$$
$$+ 4e^{8\pi i (p_1 + \frac{1}{4})} + 4e^{-8\pi i (p_1 - \frac{1}{4})} + 4e^{8\pi i (p_1 + \frac{3}{4})}$$
$$+ 4e^{-8\pi i (p_1 - \frac{3}{4})} \} + S_2 \{ 4e^{8\pi i p_2} + \cdots \} + S_3 \{ \cdots \}$$
$$+ S_4 \{ \cdots \}$$
$$= 64 \, (S_1 \cos 8\pi i p_1 + S_2 \cos 8\pi i p_2 + S_3 \cos 8\pi p i_3$$
$$+ S_4 \cos 8\pi i p_4).$$

Wenn man

$$S_1 = S_2 = S_3 = S_4 = S$$

setzt, erhält man hieraus

$$F_{004} = 64\,S (\cos 8\pi i p_1 + \cos 8\pi i p_2 + \cos 8\pi i p_3 + \cos 8\pi i p_4).$$

Die Gitterstruktur des rhombischen Schwefels.      413

## Tabelle 9.

| Indizierung | Intensität mit $Cu$-Strahlung | Intensität mit $Sr$-Strahlung |
|:---:|:---:|:---:|
| (0 0 4) | 0 | 0 |
| (0 0 8) | 15 | 12 |
| (0 0 12) | 0 | 0 |
| (0 0 16) | 28 | 30 |
| (0 0 20) | | 0 |
| (0 0 24) | | 8 |
| (0 0 28) | | 0 |

Aus der ersten und dritten dieser Intensitäten folgen bereits die fünfte und siebente, wenn man — wie wir es getan haben — annimmt, dass die Reflexionsvermögen der ungleichwertigen $S$-Atome gleich sind. Berechnet man aus den vier ersten Gleichungen $p_1$ bis $p_4$, so kommt man zu folgenden Werten:

$$p_1 \sim 0.017,$$

$$p_2 \sim 0.142,$$

$$p_3 \sim 0.047,$$

$$p_4 \sim 0.172.$$

Leider stehen keine weiteren Reflexionen zur Verfügung, um die Zuverlässigkeit der $p$-Werte zu prüfen; sie können daher nur als ein ganz versuchsweiser Ansatz gewertet werden. Denn abgesehen davon, dass keine Ebene als Kontrolle benutzt werden konnte, muss auch darauf hingewiesen werden, dass bei der verwendeten experimentellen Methode die theoretischen Grundlagen der für die Intensität gebrauchten Formel keineswegs einwandfrei sind, da weder die Bohrsche Absorption noch die durch die Reflexion bedingte darin zum Ausdruck kommen. Wir sind damit beschäftigt, durch Verwendung härterer Strahlung mehr Ebenen zur Reflexion zu bringen und dann unter Berücksichtigung der Absorptionsverhältnisse aus Drehdiagrammen um [100] und [010] auch die Werte von $m$ und $n$ gesondert zu bestimmen.

Um über die Möglichkeit, in dem gefundenen Gitter abgeschlossene Gruppen erkennen zu können, etwas zu erfahren, wollen wir noch kurz das Reis-Weissenbergsche Zusammengehörigkeitskriterium darauf anwenden. Hierbei ergibt sich, dass sich höchstens Gruppen von 16 $S$-Atomen zusammenfassen lassen. Das Gitter, welches die Schwerpunkte dieser Gruppen bilden, besteht aus zwei um $\frac{1}{4}\frac{1}{4}\frac{1}{4}$ gegeneinander verschobenen flächenzentrierten Komplexen; ist also das „rhombische Diamantgitter".

414     H. Mark und E. Wigner, Die Gitterstruktur des rhombischen Schwefels.

## Zusammenfassung.

Rhombischer Schwefel kristallisiert rhombisch bipyramidal mit den Achsen:

$$a = 10.61 \, \text{Å},$$
$$b = 12.87 \, \text{Å},$$
$$c = 24.56 \, \text{Å},$$

in der Raumgruppe $V_h^{24}$; der allseitig flächenzentrierte Elementarkörper enthält 128 Atome. In diesem Gitter lässt sich eine Gruppe von höchstens 16 $S$-Atomen geometrisch zusammenfassen, deren Schwerpunkte ein rhombisches Diamantgitter bilden.

# Bildung und Zerfall von Molekülen

M. Polanyi und E. P. Wigner

Zeitschrift für Physik *33*, 429–434 (1925)

(Eingegangen am 27. Juni 1925.)

Diskussion der Annahme spontanen Zerfalles und einfacher Bildung gequantelter Systeme auf Grund der Annahme endlich breiter Quantenniveaus.

1. In ihrer Arbeit über das obige Thema gehen Born und Franck[1]) davon aus, daß weder eine Bildung eines Moleküls durch einfache Assoziation, noch der zugehörige Gegenprozeß, nämlich der spontane Zerfall eines Moleküls, vor sich gehen kann. Als Begründung wird angeführt, daß es unendlich unwahrscheinlich ist, daß die Relativenergie der zusammenstoßenden Gebilde gerade einen quantenhaft zulässigen Energiewert des zu bildenden Moleküls beträgt.

Ein abweichendes Verhalten nehmen Born und Franck an, wenn das eine Gebilde zu einem festen Körper ausartet. An einem solchen kann ein Atom oder ein Molekül ohne weiteres haften bleiben, weil hier schon eine kontinuierliche Mannigfaltigkeit von Energiewerten zulässig sein soll.

Die vorliegende Untersuchung befaßt sich im Gegensatz zu Born und Franck damit, die Molekülbildung durch einfache Assoziation und den spontanen Zerfall derselben zu diskutieren. Wir gehen dabei von der Vorstellung aus, daß die Quantenzustände eine endliche Breite haben[2]) und es daher eine endliche — wenn auch zuweilen geringe — Wahrscheinlichkeit dafür gibt, daß zusammenstoßende Moleküle mit quantenhaft zulässiger Energie zu einem neuen Molekül zusammentreten.

2. Das Vorkommen solcher Vorgänge sehen wir bereits durch das vorhin erwähnte Haften von Molekülen an festen Wänden als erwiesen an. Wir stützen uns ferner darauf, daß kürzlich von Auger[3]) ein Beispiel für den spontanen Zerfall eines gequantelten Systems geliefert worden ist. Er beobachtete, daß ein Atom, dem durch Röntgenstrahlen ein $K$-Elektron entrissen wurde, ein zweites Elektron emittiert. Das

---

[1]) Ann. d. Phys. **76**, 225, 1925.

[2]) Vgl. z. B. N. Bohr, ZS. f. Phys. **13**, 117, 1923 (S. 150 ff.).

[3]) C. R. **180**, 65, 1925. Vgl. auch L. Meitner, ZS. f. Phys. **17**, 16, 1923.

Elektron stammt aus der *L*-Schale, die Energie wird durch Übergang eines zweiten *L*-Elektrons in die *K*-Schale geliefert. Aus älteren Versuchen von Barkla über die Ausbeute der Röntgenfluoreszenz schließt Bothe[1]), daß bei $Br_2$, wo die genauesten Messungen vorliegen, die Hälfte der ionisierten Atome die *K*-Linie emittieren, die andere Hälfte dagegen auf die von Auger angegebene Weise zerfällt. Es zeigt sich also, daß die Wahrscheinlichkeit des Zerfalles hier gleich der Ausstrahlungswahrscheinlichkeit ist.

Wir wollen nun zeigen, auf welche Weise die Elementarprozesse bei einem spontanen Zerfall anzunehmen sind, um ein der Thermodynamik entsprechendes Gleichgewicht zu gewährleisten. Als Beispiel möge der vorhin erwähnte Fall des im *K*-Ring ionisierten Atoms dienen. Für die Konzentrationen $c_1$ der doppelt geladenen Ionen, $c_2$ der Elektronen und $c$ der einfach geladenen Ionen gilt die thermodynamische Beziehung

$$\frac{c_1 c_2}{c} = e^{\frac{Q}{kT}} \frac{(2\pi\mu kT)^{3/2}}{h^3}, \tag{1}$$

worin $Q$ die Ionisierungsarbeit und

$$\mu = \frac{m_1 m_2}{m_1 + m_2}.$$

bedeutet. ($m_1$ Masse des doppelt geladenen Ions, $m_2$ Masse des Elektrons.)

Pro Sekunde sollen $c/\tau$ einfache Ionen in ein doppelt geladenes und ein Elektron zerfallen.

Anderseits soll bei einem „Zusammentreffen" von einem Elektron und doppelten Ion eine Rekombination erfolgen, falls die Energie der Relativbewegung in den quantenhaft erlaubten Energiebereich von der Breite $\varDelta\varepsilon$ fällt. Ein Zusammentreffen soll zustande kommen, wenn der Bahnenabstand kleiner als $d$ ist. Die Anzahl der Rekombinationen ist hiernach

$$c_1 c_2 \frac{\sqrt{Q}}{(kT)^{3/2}} \frac{2}{\sqrt{\pi}} e^{-\frac{Q}{kT}} \varDelta\varepsilon \sqrt{\frac{2Q}{\mu}}.$$

Und als Bedingung des stationären Zustandes hat man

$$\frac{c}{\tau} = \frac{c_1 c_2}{(kT)^{3/2}} \sqrt{Q} \frac{2}{\sqrt{\pi}} e^{-\frac{Q}{kT}} \varDelta\varepsilon \sqrt{\frac{2Q}{\mu}} d^2 \pi.$$

Mit (1) kombiniert ergibt dies

$$\tau d^2 = \frac{h^3}{8\pi^2 Q \mu \varDelta\varepsilon}. \tag{2}$$

---

[1]) Phys. ZS. **26**, 410, 1925.

Bildung und Zerfall von Molekülen.    431

Für $\tau \varDelta \varepsilon$ setzen wir nun in Übereinstimmung mit der Theorie der Linienbreite[1]) $h$. Dann folgt

$$d \sqrt{2\,Q\,\mu} = \frac{h}{2\,\pi}. \tag{3}$$

Anderseits ist für eine Rekombination, bei der der Bahnabstand $\delta\ (< d)$ ist, der gegenseitige Drehimpuls gleich $\delta \sqrt{2\,Q\,\mu}$. Gleichung (3) besagt also, daß dieser Drehimpuls kleiner sein muß als $h/2\,\pi$.

Ein ähnliches Bild wollen wir in der Folge auch der Molekülbildung zugrunde legen.

Es muß hervorgehoben werden, daß hierdurch der Drehimpulssatz bei Rekombinationen aufgehoben wird. Die Quantentheorie erlaubt für den Drehimpuls des entstandenen Gebildes nur ganzzahlige Vielfache von $h/2\,\pi$. Die vorgegebenen, nicht gequantelten Drehimpulse der sich vereinigenden Körper müssen daher auf das nächstliegende Vielfache aufgefüllt werden. Unter dieser Annahme ist die Gültigkeit der Reaktionsisochore in allen Fällen gewährleistet, worauf wir uns in der Folge stützen werden.

3. Wir betrachten zwei Atome $A$ und $B$, die ein Molekül $AB$ bilden können. Wenn $A$ und $B$ mit der Relativenergie $\varepsilon$ sich vereinigen, so wird das entstandene Molekül die innere Energie $\varepsilon + Q$ ($Q =$ Dissoziationswärme) haben. Die Konzentration $c'$ solcher $AB$-Moleküle, die eine höhere innere Energie als $Q$ haben, im Gleichgewicht mit Atomen von der Konzentration $c_1$ und $c_2$ ist durch die Gleichung gegeben

$$\frac{c_1 c_2}{c'} = \frac{(2\,\pi\,\mu\,k\,T)^{3/2}}{h^3}\,\frac{h^2}{8\,\pi^2\,J\,k\,T}\,\frac{k\,\Theta}{Q}.$$

Dabei sind die Schwingungen mit der charakteristischen Temperatur $\Theta$ als harmonisch, das Trägheitsmoment $J$ als unveränderlich angenommen[2]). Bei Annahme einer Lebensdauer $\tau$ zerfallen pro Sekunde

$$\frac{c'}{\tau} = \frac{c_1 c_2}{\tau}\,\frac{2^{3/2}\,\sqrt{\pi}\,h\,J\,Q}{k^{3/2}\,\sqrt{T}\,\Theta\,\mu^{3/2}} \tag{4}$$

Moleküle $AB$. Im Gleichgewicht werden ebenso viele gebildet, also ist die Rekombinationsgeschwindigkeit durch die rechte Seite der Gleichung

---

[1]) Vgl. N. Bohr, l. c.

[2]) Diese Annahmen bieten sicher nur eine grobe Annäherung, doch sollen sie hier verwendet werden, weil eine Grundlage für ihre Berichtigung zurzeit fehlt.

432                         M. Polanyi und E. Wigner,

gegeben.    Die Geschwindigkeitskonstante $\varkappa$ dieser Elementarreaktion
hat demnach den Wert

$$\varkappa = \frac{2\sqrt{2\pi}\,h\,J\,Q}{k^{3/2}\sqrt{T}\,\Theta\,\mu^{3/2}}\,\frac{1}{\tau}. \tag{5}$$

Dieselbe Formel gewinnt man gemäß Punkt 2 auch, wenn man annimmt, daß die Rekombination zu einem Molekül mit der inneren
Energie $\varepsilon$ und dem Drehimpuls $\dfrac{nh}{2\pi}$ erfolgt, sobald die Atome die Relativenergie zwischen $\varepsilon - Q$ und $\varepsilon + \varDelta\varepsilon - Q$ und den Drehimpuls zwischen
$(n-1)\,\dfrac{h}{2\pi}$ und $\dfrac{nh}{2\pi}$ besitzen.    Dabei hängt $\varDelta\varepsilon$ durch die Gleichung
$\varDelta\varepsilon\tau = h$ mit $\tau$ zusammen.

Den Vorgang der Dissoziation müßte man sich auf Grund des verwendeten Bildes in zwei Schritten vorstellen.  1. Bildung eines Moleküls,
dessen innere Energie größer als $Q$ ist (durch Zusammenstoß oder
Strahlungsabsorption).    2. Spontaner Zerfall dieses Moleküls.

Der zweite Prozeß wird nur dann erfolgen, wenn das energiereiche
Molekül nicht vorzeitig seine Energie wieder verliert.    Umgekehrt
verläuft die Molekülbildung so, daß die zunächst entstehenden energie-
reichen $AB$-Moleküle durch weitere Zusammenstöße oder Ausstrahlung
stabilisiert werden.    Die Wahrscheinlichkeit für die Bildung energie-
reicher $AB$-Moleküle haben wir bereits angegeben.    Von den $x$ vorhandenen werden, wenn wir von der Strahlung absehen, $xS$ in der
Zeiteinheit stabilisiert, wo $S$ die vom Gesamtdruck abhängige Stoßzahl ist.
Anderseits zerfallen pro Sekunde $x/\tau$.    Im stationären Zustande ist

$$c_1 c_2 \varkappa = \frac{x}{\tau} + x\,S.$$

Die Reaktionsgeschwindigkeit wird durch die Anzahl der pro Zeiteinheit
stabilisierten Moleküle gegeben.    Daher ist die Geschwindigkeitskonstante
der Gesamtreaktion $K$

$$K = \frac{\varkappa\,S}{\dfrac{1}{\tau} + S}. \tag{6}$$

Bei kleinen Drucken geht diese Gleichung (da $\dfrac{1}{\tau} > S$ wird) über in

$$K = \varkappa\,S\,\tau = \frac{2\sqrt{2\pi}\,h\,J\,Q}{k^{3/2}\sqrt{T}\,\Theta\,\mu^{3/2}}\,S. \tag{6a}$$

Bei hohen Drucken wird $\left( \text{da } \dfrac{1}{\tau} < S \right)$

$$K = \varkappa = \frac{2\sqrt{2\pi}\,h\,J\,Q}{k^{3/2}\sqrt{T}\,\Theta\,\mu^{3/2}}\,\frac{1}{\tau}. \qquad (6\,\text{b})$$

Die Größenordnung der Lebensdauer $\tau$ bestimmen wir aus den Versuchen von B o d e n s t e i n und L ü t k e m e y e r[1]) über die Geschwindigkeit der Reaktion $2\,\text{Br} \to \text{Br}_2$. Wir führen dementsprechend ein

$$K = 10^{11,345}\,\text{ccm/Mol. sec} = 3,65 \cdot 10^{-13}\,\text{ccm/Atom sec},$$

ferner $\log J = -37,27$[2]) und $\Theta = 510$[3]) und erhalten $\tau = 3,5 \cdot 10^{-9}$ Es steht hiermit in Übereinstimmung, daß die Reaktion zwischen den Drucken von 0,3 und 1 Atm. druckunabhängig verlief, da $\dfrac{1}{\tau S}$ hier etwa 0,3 bis 0,1 beträgt.

Aus den Messungen der Lebensdauer des aktiven Wasserstoffs von W o o d[4]) und B o n h o e f f e r[5]) folgt für die Reaktion $2\,\text{H} \to \text{H}_2$ $\log K = -15$. Aus Gleichung (6 a) berechnet sich für $\text{H}_2$ (mit $\log J = -40,7$ $\Theta = 5000$)[6]) für kleine Drucke (0,5 mm) $\log \varkappa = -16,3$. Dies ist kleiner als der gemessene Wert, was auf Wandreaktion beruhen kann.

Handelt es sich um die Entstehung heteropolarer Verbindungen, so muß die Ausstrahlung mit berücksichtigt werden. Die Wahrscheinlichkeit dafür, daß das Molekül $AB$ den zur Stabilisierung nötigen Strahlungsverlust von $\varepsilon$ erleidet, sei $1/\tau'$ pro Zeiteinheit.

Dann gilt (wenn wir nun die Stabilisierung durch Stöße vernachlässigen)

$$K = \frac{\tau}{\tau + \tau'},\; \varkappa = \frac{2\sqrt{2\pi}\,h\,J\,Q}{k^{3/2}\sqrt{T}\,\Theta\,\mu^{3/2}}\,\frac{1}{\tau + \tau'}. \qquad (7)$$

Eine Stabilisierung durch Strahlung ist anzunehmen bei der Bildung von $\text{Na}\,\text{J}$ aus den Atomen, die H. B e u t l e r mit einem von uns[7])

---

[1]) ZS. f. phys. Chem. **114**, 208, 1924.
[2]) Berechnet aus der chemischen Konstante des $\text{Br}_2$-Moleküls $i = 2,546$ (L ü d e und S u h r m a n n, ZS. f. Phys. **29**, 71, 1924).
[3]) E u c k e n und F r i e d, ZS. f. Phys. **29**, 36, 1924.
[4]) Phil. Mag. (6) **42**, 729, 1921; (6) **44**, 538, 1922. Proc. Roy. Soc. (A) **97**, 455, 1921; **102**, 1, 1922.
[5]) ZS. f. phys. Chem. **119**, 199, 1924.
[6]) E u c k e n und F r i e d, l. c.
[7]) H. B e u t l e r und M. P o l a n y i, Die Naturwiss. 1925.

**434**    M. Polanyi und E. Wigner, Bildung und Zerfall von Molekülen.

studiert hat.    Die Versuche können gedeutet werden mit einer Lebensdauer $\tau$ von $3 \cdot 10^{-9}$ und $\tau' \curvearrowright \tau$.

4.  Neben der einfachen Rekombination kann die Molekülbildung
auch durch Dreierstöße vor sich gehen.  Die Geschwindigkeit einer solchen
Reaktion ist, wenn man die Dauer eines Stoßes zu $10^{-12}$ sec annimmt [1]),

$$c_1 c_2 \frac{S^2}{c_{\text{gesamt}}} \cdot 10^{-12}.$$

Wie ein Vergleich mit Gleichung (6) zeigt, bleibt die Geschwindigkeit
der Dreierstoßreaktion fast immer unterhalb derjenigen des oben dis-
kutierten Reaktionsweges.  Natürlich ist hierin die wesentliche Annahme
enthalten, daß die Querschnitte gastheoretisch berechenbar sind.

---

[1]) K. F. Herzfeld, ZS. f. Phys. 8, 132, 1922.

# Über die Interferenz von Eigenschwingungen als Ursache von Energieschwankungen und chemischer Umsetzungen

M. Polanyi und E. P. Wigner

Zeitschrift für physikalische Chemie A (Haber-Band) *43*, 439–452 (1928)

(Eingegangen am 10. 10. 28.)

## Inhaltsangabe.

Nach alten Erfahrungen enthält die Konstante der Geschwindigkeit monomolekularer Reaktionen in Lösungen und, wie sich neuerdings zeigt, auch in Gasen, neben einem exponentiellen Faktor, eine Grösse, die stets von etwa der gleichen Grössenordnung $10^{14}$ ist. Um dies zu erklären stellen wir uns vor, dass die Energie im Molekül in elastischen Wellen hin und her schwankt und die Umsetzung dann eintritt, wenn sich zufällig durch Interferenz der Wellen die Amplitude an einer bestimmten Bindung über einen kritischen Betrag erhöht. Das ergibt für die Geschwindigkeitskonstante $\nu e^{-\frac{Q}{RT}}$, worin $\nu$ die Frequenz der Atomschwingung ist, also die richtige Grössenordnung hat.

## 1. Problemstellung.

Unsere Arbeit ist darauf gerichtet, eine Vorstellung über den Mechanismus monomolekular verlaufender Reaktionen (Dissoziation und Umwandlung) grösserer Moleküle zu gewinnen. Die monomolekularen Gasreaktionen sind in letzter Zeit vielfach diskutiert worden, doch gehen die Arbeiten (die später noch berührt werden sollen) auf den Mechanismus innenmolekularer Vorgänge nicht ein, sondern benutzen zu deren Kennzeichnung eine Konstante, die mit keiner anderen Erscheinung in Zusammenhang gebracht wird. Unsere Vorstellung soll zur Deutung dieser Konstante führen. Sie geht davon aus, dass das System aus einer Anzahl von Atomen besteht, die durch quasielastische Kräfte an Gleichgewichtslagen gebunden sind. Die möglichen Bewegungen solcher Systeme lassen sich aus Eigenschwingungen zusammensetzen, die durch Interferenz (Schwebung) zuweilen an einzelnen Stellen eine erhebliche Amplitude erzeugen können. Wenn man z. B. vier in einer Geraden angeordnete Atome hat, so hat das System drei Eigenschwingungen. Wenn nur eine dieser Eigen-

440                           M. Polanyi und E. Wigner

schwingungen angeregt ist, so ist die Amplitude aller Atome zeitlich
konstant und von gleicher Grössenordnung. Sind dagegen etwa alle
drei Eigenschwingungen erregt, so werden die Schwebungen zur Folge
haben, dass jede einzelne Bindung zeitweise den grössten Teil der
Schwingungsenergie aufnimmt. Hierbei wird eine Umwandlung eintreten können, indem eine Konfiguration entsteht, die beim Abfluss
der Energie nicht stets in den Ausgangszustand zurückkehrt[1]), sondern mit etwa gleicher Wahrscheinlichkeit in eine neue Gleichgewichtslage übergeht. Auch eine Dissoziation kann auf diese Weise eintreten,
wenn die kritische Elongation die Reichweite der molekularen Kräfte
erreicht. Von der Erörterung sind gemäss der obigen Vorstellung Bildung und Zerfall zweiatomiger Moleküle im Gasraum von vornherein
ausgeschlossen. Dieser Vorgang war Gegenstand einer früheren Notiz
der Verfasser[2]). Er wurde in Analogie zum Augerprozess gebracht.
Kurz nachdem K. F. BONHOEFFER und L. FARKAS[3]) nachgewiesen
hatten, dass der photochemische $NH_3$-Zerfall einen dem Augerprozess
analogen Vorgang darstellt, ist diese Frage mit Hilfe der neuen Quantenmechanik von R. DE L. KRONIG[4]) angeschnitten und als typisch
quantenmechanischer Effekt gekennzeichnet worden. Die Geschwindigkeit von Vorgängen, die durch diesen Effekt bedingt sind, ist
wesentlich geringer als die der im Text betrachteten Reaktionen.

Eine Vernachlässigung ist die Annahme quasielastischer Bindungen, womit die Streuung der elastischen Wellen und damit die
Überlagerung der gewöhnlichen Wärmeleitung[5]) wegfällt. Auch ist
die Zuständigkeit unserer Ableitung voraussetzungsgemäss auf Umwandlungen beschränkt, die mit keinem Elektronensprung verbun-
den sind.

Eine weitere Einschränkung ist durch die Verwendung der klassischen Vorstellungen an Stelle der quantenmechanischen bedingt. Da

---

[1]) Es war bisher üblich, den Reaktionsvorgang in zwei verschiedenartige Teile
zu zerlegen: die „Aktivierung" und die „Umwandlung". Die „Aktivierung" soll
durch intramolekulare Energieübertragung beliebig schnell erfolgen, die „Umwandlung" dagegen spontan gemäss einer gewissen, die Reaktionsgeschwindigkeit wesentlich bestimmenden Lebensdauer des aktivierten Zustands. Diese Auffassung
lehnen wir ab, weil zufolge der Umkehrbarkeit der Elementarprozesse jener Vorgang, der für die eine Reaktionsrichtung „Aktivierung" ist, für die Gegenreaktion
„Umwandlung" sein muss (unsere eigene Vorstellung entspricht dieser Forderung).
[2]) Z. Physik **33**, 429. 1925.     [3]) K. F. BONHOEFFER und L. FARKAS, Z. physikal.
Chem. **134**, 337. 1928.     [4]) R. DE L. KRONIG, Z. Physik **50**, 347. 1928.     [5]) Vgl.
P. DEBYE, Vortrag über die kinetische Theorie der Materie. Leipzig 1914.

es sich hier jedoch um Zustände mit hohen Quantenzahlen handelt, wird auch die strengere Behandlung mit Hilfe der Quantenmechanik kein abweichendes Ergebnis liefern[1]). Allerdings bringt es die Quantenmechanik mit sich, dass Umwandlungen, die gemäss unserer Vorstellung als möglich erscheinen könnten, es doch nicht sind, weil ausser den klassischen Integralen auch noch weitere Integrale der Bewegungsgleichungen existieren, die die Wahrscheinlichkeit gewisser Übergänge, z. B. von einem Singulett- in einen Triplettzustand auf ein sehr geringes Mass reduzieren[2]).

Ausser dem hier beschriebenen Reaktionsmechanismus eröffnet die Quantenmechanik noch eine ganz andere Möglichkeit für den Umwandlungsvorgang. Es kann das nichtmechanische Überspringen der die beiden Modifikationen trennenden Energiebuckel eintreten, ähnlich, wie dies bei radioaktiven Prozessen der Fall ist[3]). Ein solcher Prozess hätte jedoch einige sonderbare Merkmale, von denen es zunächst nicht wahrscheinlich erscheint, dass sie auf die bekannten Reaktionen zutreffen. Insbesondere müsste die Reaktionsgeschwindigkeit beim absoluten Nullpunkt noch einen endlichen Wert behalten und mit der Temperatur nicht gemäss $Ae^{-\frac{Q}{RT}}$, sondern viel langsamer ansteigen.

---

[1]) Im Falle quasielastischer Bindungen kann man — einer Arbeit von Schrödinger folgend — (Die Naturw. 14, 664. 1926) — den Übergang zu hohen Quantenzahlen auch im einzelnen durchführen.

Klassisch lässt sich die Bewegung als Superposition von Eigenschwingungen auffassen. Führen wir anstelle der Koordinaten der einzelnen Teilchen »Normalkoordinaten« ein, d. h. solche, die bei allen Eigenschwingungen, abgesehen von einer konstant bleiben, bei dieser Eigenschwingung dagegen eine reine Sinusbewegung ausführen, so bewegt sich der Bildpunkt des Systems im Konfigurationsraum nach der Gleichung

$$q_k = a_k \sin (\omega_k + \varphi_k)\, t. \qquad (\alpha)$$

Quantenmechanisch bedeutet das Einführen von Normalkoordinaten die Separation der Variablen. Für jede Normalkoordinate gilt die Gleichung eines reinen, harmonischen eindimensionalen Oszillators. In der erwähnten Arbeit von E. Schrödinger ist dann explizite gezeigt, dass die Koordinate $q_k$ sich im wesentlichen nach der Gleichung ($\alpha$) bewegt.

Aber auch für nichtquasielastische Kräfte muss die Quantenmechanik bei hohen Quantenzahlen mit der klassischen Theorie äquivalente Resultate liefern, wenn es auch schwerer sein dürfte, den Beweis im einzelnen durchzuführen.

[2]) E. Wigner, Göttinger Nachr. 375, 1927. Nach freundlicher Mitteilung von Herrn F. London ist er mit der Erörterung dieser Seite der Frage beschäftigt.

[3]) G. Gamow, Z. Physik 51, 204. 1928.

## 2. Energieschwankungen innerhalb eines festgefügten Systems.

Wir denken uns ein Atomaggregat, das aus $n$ ungefähr gleich schweren Atomen besteht, die von etwa gleich grossen Kräften zusammengehalten werden. Die rascheste Eigenschwingung habe die Periode $\frac{1}{\nu}$, die Gesamtenergie des Systems sei $E$. Wir fragen wie lange es im Mittel dauern wird, bis an einer hervorgehobenen Bindung zwischen zwei Atomen eine Energie auftritt, deren Betrag $\varepsilon$ überschreitet. Hierzu fassen wir zunächst die statistische Wahrscheinlichkeit dieses Zustandes ins Auge. Diese ist[1]):

$$W = \left(1 - \frac{\varepsilon}{E}\right)^{3n-1}. \tag{1}$$

Der Ausdruck geht für grosse Werte von $n$ über in die BOLTZMANNsche Formel für Stoffe mit der spezifischen Wärme $k$

$$W = e^{-\frac{\varepsilon}{kT}}, \tag{1a}$$

wenn man $\frac{E}{3n} = kT$ setzt.

Andererseits gewinnt man eine Abschätzung der mittleren Dauer $\tau$ eines solchen Zustandes, wenn man die Fortschreitungsgeschwindigkeit einer Störung gemäss der Schallgeschwindigkeit $v$ ansetzt:

$$\tau = \frac{d}{v}, \tag{2}$$

wo $d$ den Abstand der Nachbaratome bedeutet. Für $v$ ergibt sich aus der Wellenlänge der raschesten Schwingung, welche $d$ beträgt, gemäss $\nu\lambda = v$

$$v = \nu d \tag{3}$$

und hieraus

$$\tau = \frac{1}{\nu}. \tag{3a}$$

In Verbindung mit (1) bzw. (1a) erhalten wir für die gesuchte Zeit $Z$, innerhalb deren es im Mittel einmal vorkommt, dass die hervorgehobene Bindung die Energie $\varepsilon$ hat

$$Z = \frac{1}{\nu}\left(1 - \frac{\varepsilon}{E}\right)^{-3n+1}, \tag{4}$$

$$Z = \frac{1}{\nu} e^{+\frac{\varepsilon}{kT}}. \tag{4a}$$

Diese Überlegung ist insofern ungenau, als darin angenommen wird, dass die Amplituden benachbarter Atome voneinander stati-

---

[1]) H. A. LORENTZ, Theories statistiques en thermodynamique. Leipzig et Berlin 1916.

stisch unabhängig sind. Doch führt folgender Weg unter Vermeidung dieser Annahme zu einem im wesentlichen gleichlautenden Ergebnis.

Betrachten wir eine Atomreihe von $n$-Atomen, in dem jedes Atom der Einfachheit halber dieselbe Masse $m$ hat. Die Entfernung des $k$-ten Atoms von der Gleichgewichtslage sei $x_k$, die Kraft, die das $k$-te Atom auf das $k-1$-te ausübt, sei $\alpha(x_{k-1}-x_k)$. Dann sind die Bewegungsgleichungen

$$m\ddot{x}_k = \alpha(x_{k+1}-x_k) + \alpha(x_{k-1}-x_k).$$

Die Kreisfrequenzen $\omega$ der Eigenschwingungen sind die Lösungen der algebraischen Gleichung $n$-ten Grades

$$\begin{vmatrix} -m\omega^2+\alpha & -\alpha & 0 & 0 & 0 \\ -\alpha & -m\omega^2+\alpha & -\alpha & 0 & 0 \\ 0 & -\alpha & -m\omega^2+2\alpha & -\alpha & 0 \\ 0 & 0 & -\alpha & -m\omega^2+2\alpha & -\alpha \\ 0 & 0 & 0 & -\alpha & -m\omega^2+\alpha \end{vmatrix} = 0. \quad (5)$$

Führen wir hier für $\frac{m\omega^2}{\alpha}$ die neue unbekannte $y$ ein, so lautet (5)

$$\begin{vmatrix} 1-y & -1 & 0 & 0 & 0 \\ -1 & 2-y & -1 & 0 & 0 \\ 0 & -1 & 2-y & -1 & 0 \\ 0 & 0 & -1 & 2-y & -1 \\ 0 & 0 & 0 & -1 & 1-y \end{vmatrix} = 0. \quad (5a)$$

Die Determinante (5a) hat $n$ Zeilen. Für kleine $n$ kann man die Lösungen von (5a) explizite berechnen, für grosse $n$ noch leicht (im wesentlichen mit Hilfe der DESCARTESschen Regel) die Anzahl der Lösungen in verschiedenen Intervallen angeben[1]). Da alle $y$ zwischen 0 und 4 liegen, können wir die folgenden Intervalle von $y$ betrachten: 0 bis 1, 1 bis 2, 2 bis 3, 3 bis 4. Das Resultat ist in Fig. 1 zusammengefasst. Wir

Fig. 1.

sehen, dass die Eigenfrequenzen ungefähr gleichmässig auf das ganze Intervall verteilt sind. So lange die Massen der einzelnen Atome und

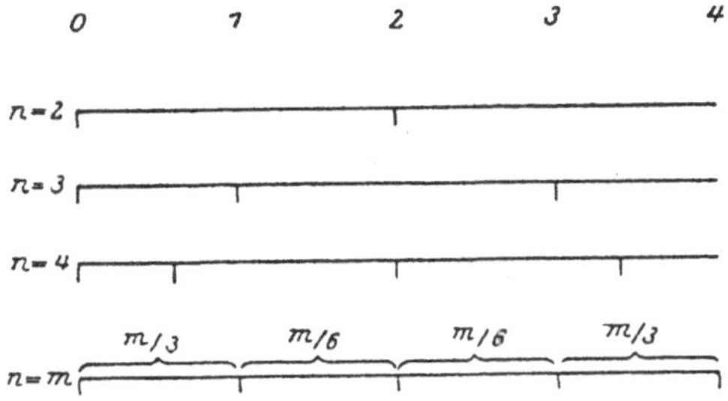

[1]) Für das langwellige Gebiet siehe auch P. DEBYE, Ann. Phys. 39, 789. 1912. Vgl. auch M. BORN, Atomtheorie des festen Zustandes. Berlin 1923. Daselbst weitere Literatur.

444                         M. Polanyi und E. Wigner

die zwischen ihnen liegenden Kräfte von der gleichen Grössenordnung sind, wird sich hieran nichts ändern. Den Fall, dass eine Bindung wesentlich schwächer ist, als die übrigen, werden wir weiter unten kurz berücksichtigen. Die Entfernung von zwei benachbarten Atomen $d$ als Funktion der Zeit $t$ wird die Gestalt

$$d = d_0 + x_k - x_{k-1} = d_0 + a_1 \cos \omega_1 (t + \varphi_1) + a_2 \cos \omega_2 (t + \varphi_2) + \cdots\cdots$$
$$+ a_n \cos \omega_n (t + \varphi_n)$$

haben. Führen wir hierfür die Kreisfrequenz der raschesten Schwingung $2\pi\nu$ ein, so können wir im Intervall $-\dfrac{2}{\nu} < t < \dfrac{2}{\nu}$ mit hinreichender Genauigkeit

$$d = d_0 + c \cos \frac{2\pi\nu}{8} (t + \varphi) + c' \cos \frac{3\cdot 2\pi\nu}{8} (t + \varphi') + c'' \cos \frac{5\cdot 2\pi\nu}{8} (t + \varphi'')$$

$$+ c''' \cos \frac{7\cdot 2\pi\nu}{8} (t + \varphi''') \tag{6}$$

schreiben. Ein hoher Wert für $d-d_0$ zur Zeit $t=0$ wird wegen der gleichmässigen Verteilung der Eigenfrequenzen auf das Intervall in den allermeisten Fällen so zustande kommen, dass die Koeffizienten $c, c', c'', c'''$ schon selber sehr hohe und zwar ungefähr gleich hohe Werte haben. Ausserdem müssen die $\varphi$ alle sehr klein, mit anderen Worten, die vier Schwingungen von (6) in Phase sein. Wenn $d$ besonders hoch ist, so dass $\dfrac{a(d-d_0)^2}{2} > \varepsilon$ ist, wird im Mittel

$$\frac{a(d-d_0)^2}{2} - \varepsilon \sim kT$$

sein. Noch grössere Überschreitungen werden sehr selten sein. Man überzeugt sich mit Hilfe von (6) leicht, dass der Zustand $\dfrac{a(d-d_0)^2}{2} > \varepsilon$ im Mittel, während einer Zeit von der Grössenordnung $\dfrac{1}{\nu}\sqrt{\dfrac{2kT}{\varepsilon}}$ anhalten wird. Da seine statistische Wahrscheinlichkeit $\sqrt{\dfrac{kT}{\varepsilon}}\, e^{-\frac{\varepsilon}{kT}}$ ist, wird er pro sec $\nu e^{-\frac{\varepsilon}{kT}}$-mal erreicht werden.

Bei genauerer rechnerischer Durchführung dieser Überlegung ergab sich $1\cdot 15 \cdot \nu \cdot e^{-\frac{\varepsilon}{kT}}$ an Stelle von $\nu \cdot e^{-\frac{\varepsilon}{kT}}$. Hierbei ist die in Fig. 1 gegebene Verteilung der Eigenschwingungen angenommen, sowie dass $\varepsilon$ gross gegen $kT$ ist.

Das Ergebnis gilt den Voraussetzungen gemäss nur, wenn die Umwandlung durch die Elongation längs einer einzigen Koordinate also etwa durch Längung einer einzelnen Bindung vor sich geht. Kann eine Reaktion durch Elongation eines Atoms (oder Radikals) nach allen drei Raumrichtungen erfolgen („dreidimensionale Umwandlung"), wie dies besonders bei einem mehrfach gebundenen Atom der Fall sein kann, so wird die Reaktionsgeschwindigkeit um den Faktor $\frac{2\,\varepsilon}{kT}$ vergrössert. Für dreidimensionale Umwandlungen ergibt sich hiermit die Reaktionsgeschwindigkeitskonstante zu

$$\approx \frac{2\,\nu\varepsilon}{kT}\, e^{-\frac{\varepsilon}{kT}}.\ \text{[1]}$$

Bei tiefen Temperaturen sind die Schwingungen nicht sämtlich angeregt. Da hierbei die rascheste Schwingung, die noch angeregt wird, eine Frequenz $\frac{kT}{h}$ hat, tritt diese Grösse an Stelle von $\nu$.

Bisher wurde der Fall betrachtet, dass die Festigkeit von allen Bindungen von gleicher Grösse ist. Es besteht aber auch das Bedürfnis, sich über Fälle zu orientieren, wo das nicht zutrifft, z. B. ein adsorbiertes Molekül an einer Wand, sowie der Fall zweier Moleküle, die durch Zusammenstoss im Gasraum miteinander in Berührung getreten sind. In solchen Systemen ist eine Bindung wesentlich schwächer als die anderen und zwar von der Grössenordnung der Kohäsionskräfte, das ist etwa zehnmal kleiner als die interatomaren Bindungen.

Es ist klar, dass das Vorhandensein einer schwachen Bindung zwischen zwei schwingenden mehratomigen Gebilden einen geringeren Energiefluss ermöglicht, als wenn die gleiche Stelle stark gebunden wäre. Ist keine Kraft vorhanden, verschwindet ja der Energiefluss.

### 3. Monomolekulare Reaktionen in Gasen und Lösungen.

Wir wollen hier die Umwandlung bzw. den Zerfall von Molekülen ins Auge fassen, die gross genug sind, um mit hinreichender Genauigkeit der Gleichung (4a) zu entsprechen. Das bedeutet, dass die Zufuhr von Energie an das Molekül als geschwindigkeitsbestimmender Faktor zurücktritt und es lediglich auf die inneren Vorgänge ankommt. Dies

---

[1] Für den 2-dimensionalen Fall gilt entsprechend $\approx \dfrac{2\,\nu\,\varepsilon^{1/2}}{(kT)^{1/2}}\, e^{-\frac{\varepsilon}{kT}}$.

ist dann der Fall, wenn das Molekül gross genug ist, um stets die
Energie $\varepsilon$ in sich zu tragen und auch dann, wenn es von einer so grossen
Zahl äusserer Impulse getroffen wird, dass hierdurch die nötige Energie-
zufuhr gewährleistet ist.

Die Geschwindigkeitskonstante $\varkappa$ solcher monomolekular verlaufender Reaktionen soll gemäss (4a)

$$\varkappa = \nu c^{-\frac{Q}{RT}} \tag{7}$$

sein, worin $Q$ die Aktivierungswärme bedeutet.

Für dreidimensionale Umwandlungen geht dies, wie oben ausgeführt wurde, über in

$$\varkappa = \frac{2\nu Q}{RT} e^{-\frac{Q}{RT}}.$$

Der „temperaturunabhängige Faktor" der monomolekularen
Reaktionen ist also gleich der Atomfrequenz, bei dreidimensionalen
Reaktionen noch mit $\frac{\varepsilon}{kT}$ multipliziert. Die Grössenordnung der Atom-
frequenz ist $5 \cdot 10^{12} - 5 \cdot 10^{13}\,\text{sec}^{-1}$, somit wäre

$$\varkappa = 5 \cdot 10^{12} e^{-\frac{Q}{RT}} \quad \text{bis} \quad 5 \cdot 10^{13} e^{-\frac{Q}{RT}}.$$

Welche Zahl aber bis zum 20fachen Wert ansteigen kann, wenn die
Reaktion durch Verschieben in mehreren Richtungen erfolgen kann.

Dieses Ergebnis steht in Übereinstimmung mit der Erfahrung,
dass der temperaturunabhängige Faktor der Reaktionsgeschwindigkeitskonstante monomolekularer Reaktionen, stets nahezu den gleichen in der Grössenordnung $10^{13}$ bis $10^{14}$ gelegenen Wert aufweist.
Diese Regel kommt bereits in der von S. Dushmann[1]) aufgestellten
Gleichung

$$\varkappa = \nu^* e^{-\frac{h\nu^*}{kT}}$$

zum Ausdruck, da die Abweichungen zwischen $\nu^* = \dfrac{Q}{Nh}$ und unserem
$\nu \left(\text{bzw. } \dfrac{2\nu Q}{RT}\right)$ innerhalb der Streuung der Werte gelegen sind.

Einen Überblick über die temperaturunabhängigen Faktoren monomolekularer Reaktionen gibt die Fig. 2. In die Figur sind aufgenommen
die in Lösung verlaufenden Reaktionen, die von J. A. Christiansen[2])

¹) S. Dushmann, J. Amer. Chem. Soc. 43, 397. 1921.    ²) J. A. Christiansen,
Z. physikal. Chem. 104, 451. 1923.

zusammengestellt worden sind, sowie die Werte für folgende Gas-reaktionen: $2N_2O_5 = 2N_2O_4 + O_2$[1]); Zersetzung von $CH_3COCH_3$[2]); $CH_3N_2CH_3 = N_2 + C_2H_6$[3]); Zersetzung von $C_3H_7N_2Cl_3H_7$[4]); $CH_3OCH_3 = CH_3 + H_2 + CO$[5]); Zersetzung von $C_2H_5OC_2H_5$[6]); Zersetzung von $CH_3CH_2CHO$[7]); Racemisierung von Pinen[8]).

Die Streuung kann zum Teil durch die Unsicherheit der Messung (namentlich der Aktivierungswärme) verursacht sein, doch lässt die Näherung, in der wir unsere Betrachtung durchgeführt haben, auch die Möglichkeit echter Abweichungen zu.

Bei den ausnahmsweise stark nach oben herausfallenden Werten ist die Möglichkeit zu berücksichtigen, dass die Faktoren durch Ketten-

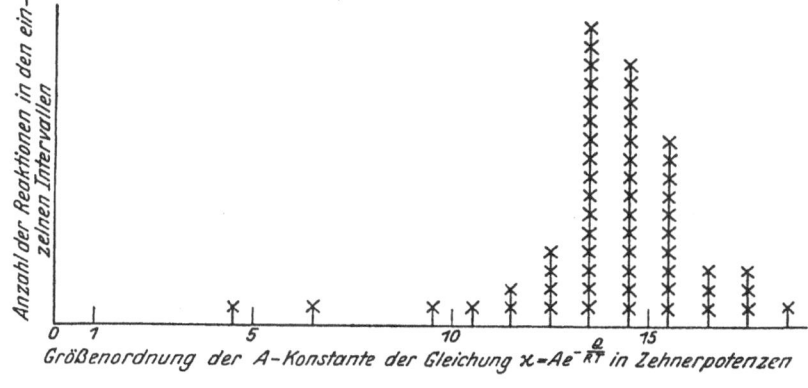

Fig. 2.

reaktionen vergrössert werden können. Für Abweichungen nach unten könnte, falls sich diese als reell herausstellen, eine Erklärung durch Verfeinerung der Theorie gefunden werden.

Bevor wir uns weiteren Anwendungen der Gleichung (7) zuwenden, sei auch noch der Reaktionsverlauf in Gasen bei niedrigen Drucken kurz betrachtet. Im Grenzfalle kleiner Drucke ist stets die Zufuhr-geschwindigkeit der Energie für die Reaktionsgeschwindigkeit mass-

---

[1]) DANIELS und JOHNSTON, J. Amer. Chem. Soc. 43, 53. 1921; G. SPRENGER, Z. physikal. Chem. 136, 49. 1928.   [2]) HIRSCH und HINSHELWOOD, Proc. Royal Soc., London A 111, 245.   [3]) RAMSPERGER, J. Amer. Chem. Soc. 49, 912, 1495. 1927.   [4]) RAMSPERGER, J. Amer. Chem. Soc. 50, 617. 1928.   [5]) HIRSCH und ASKEY, Proc. Royal Soc., London A 115, 215. 1927.   [6]) HIRSCH, Proc. Royal Soc., London A 114, 221. 1926.   [7]) HIRSCH und TOMPSON, Proc. Royal Soc., London A 113, 221. 1926.   [8]) SMITH, J. Amer. Chem. Soc. 49, 43. 1927.

448                        M. Polanyi und E. Wigner

gebend. Wenn bei jedem Stoss voller Energieaustausch zwischen den zusammenstossenden Molekülen stattfinden würde, so wäre die Wahrscheinlichkeit für die Übertragung einer Energie, die grösser als $\varepsilon = \frac{Q}{N}$ ist

$$\frac{1}{(s-1)!}\left(\frac{\varepsilon}{kT}\right)^{s-1} e^{-\frac{\varepsilon}{kT}}$$

und damit die Reaktionsgeschwindigkeitskonstante $\varkappa$

$$\varkappa = S\,\frac{1}{(s-1)!}\left(\frac{\varepsilon}{kT}\right)^{s-1} e^{-\frac{\varepsilon}{kT}}\,{}^1),$$

worin $S$ die Stosszahl bedeutet.

Diese Gleichung ist nun mit Rücksicht darauf zu modifizieren, dass der Energieaustausch, wie in Abschnitt 2 erwähnt, durch die schwache Bindung, welche zwei aneinander stossende Moleküle verbindet, nur in beschränktem Masse vor sich geht, was zu einer Verminderung von $\varkappa$ führt.

### 4. Verdampfungsgeschwindigkeit.

Eine weitere Anwendung können unsere Vorstellungen bei der Verdampfungsgeschwindigkeit finden. Wir wollen diesen Fall jedoch nur als Beispiel für unsere Vorstellungen behandeln, da die Geschwindigkeit des Prozesses selbst schon durch frühere Betrachtungen[2]

---

[1] M. POLANYI, Z. Physik 1, 337. 1920. Mit Hilfe dieser Gleichung hat E. WIGNER (Diss., Technische Hochschule, Berlin 1925) gezeigt, dass die Geschwindigkeit der bis dahin bekannten monomolekularen Gasreaktionen durch Zufuhr von Energie auf dem Wege der Zusammenstösse erklärt werden kann. Dasselbe untersuchen G. N. LEWIS und D. F. SMITH, J. Amer. Chem. Soc. 47, 1514. 1925. Von J. A. CHRISTIANSEN, Proc. Camb. Phil. Soc. 23, 438. 1926 und HINSHELWOOD, Proc. Royal Soc., London 113, 230. 1926 wurde diese Frage weiter bearbeitet und gefördert. Den Übergang vom Grenzgebiet kleiner Drucke zu dem monomolekularen Verlauf, der bei hohen Drucken eintritt, haben RICE und RAMSPERGER (J. Amer. Chem. Soc. 49, 1617. 1927), sowie auch I. S. KASSEL (J. physic. Chem. 32, 225. 1928) neu betrachtet, indem sie die Notwendigkeit des Übergangs der Aktivierungswärme auf einen bestimmten Freiheitsgrad zugrunde legten. An dem Ergebnis ihrer Arbeit wird durch unsere Ausführungen nichts geändert.

[2] B. BAULE, Ann. Phys. 44, 145. 1914; J. LANGMUIR, Physical Review 8, 149, 1916. Siehe die zusammenfassende Darstellung bei K. F. HERZFELD, l. c, S. 229 ff. (Die von HERZFELD an anderer Stelle des Buches gegebene direkte Berechnung der Verdampfungsgeschwindigkeit ist wegen stillschweigenden Annahmen, die sie enthält, nicht als strenge Ableitung zu betrachten.)

Über die Interferenz von Eigenschwingungen usw.    449

mit Hilfe des Gleichgewichts und der Rückreaktion berechnet worden ist. Unsere Behandlung ist nur insoweit befriedigender, als sie die Geschwindigkeit der Verdampfung auch direkt abzulesen gestattet.

Da das Wegfliegen eines Atoms vom Kristall nach allen drei Raumrichtungen erfolgen kann, handelt es sich hier um eine „dreidimensionale Umwandlung". Die Wahrscheinlichkeit $\varphi$ dafür, dass ein bestimmtes Atom in der nächsten Sekunde verdampft, ergibt sich somit gemäss (7a)

$$\varphi = 2\nu \frac{\lambda}{RT} e^{-\frac{\lambda}{RT}},\tag{9}$$

worin $\lambda$ die Verdampfungswärme bedeutet.

Eine Bestätigung dieser Gleichung finden wir darin, dass sie in Verbindung mit der Dampfdruckgleichung für hohe Temperaturen[1]) ($c$ Konzentration des gesättigten Dampfes)

$$c = \left(\frac{2\pi m}{kT}\right)^{3/2} \nu^3 e^{-\frac{\lambda}{kT}}$$

zur Erfahrungstatsache führt, dass ein Atom (Molekül), das auf sein Kondensat trifft, dort haften bleibt: Bezeichnen wir nämlich die Wahrscheinlichkeit hierfür mit $w$, so ergibt sich ($d$ = Gitterkonstante)

$$\sqrt{\frac{1}{2\pi}} c \sqrt{\frac{kT}{m}} w = \frac{2\nu}{d^2} \frac{\lambda}{kT} e^{\frac{\lambda}{kT}},$$

woraus sich $w$ zu

$$w = \frac{1}{\pi} \frac{\lambda}{m\nu^2 d^2} \approx 1\tag{10}$$

berechnet wie man durch Einsetzen der Werte von $\lambda$, $m$, $\nu$ findet

Bei tieferen Temperaturen wird unser Bild etwas komplizierter, es zeigt sich nämlich, dass, wenn $kT \ll h\nu$ wird, der Akkomodationskoeffizient ausgesprochen abfallen muss.

Auch durch direkte Anwendung unserer Grundvorstellungen kommt man zum gleichen Ergebnis. Wenn ein Molekül nämlich auf sein Kondensat trifft, so muss es dort — um wieder wegfliegen zu können — wenigstens eine halbe Schwingung ausführen. Während dieser Zeit wird es aber schon mit erheblicher Wahrscheinlichkeit so viel Energie ($\sim kT$) verlieren, dass es haften bleibt. Es ergibt sich

---

[1]) Siehe z. B. K. F. HERZFELD, Kinetische Theorie der Wärme. S. 223. 1925.

450                           M. Polanyi und E. Wigner

also, die Wahrscheinlichkeit $w$ für die Kondensation auftreffender
Moleküle zu $w \sim 1$[1]).

Der Vollständigkeit halber wollen wir noch bemerken, dass diese
letztere Überlegung nur dann gilt, wenn das Atom eine empfindliche
Kugel seines Kondensats trifft, d. h. auf eine solche Stelle fällt, wo
die potentielle Energie kleiner als die Verdampfungswärme ist. Dass
diese empfindlichen Kugeln die Oberfläche eines Kristalls praktisch
vollkommen bedecken, kommt in (10) zum Ausdruck.

### 5. Wandreaktion.

Wir haben oben die Geschwindigkeit monomolekularer Reaktionen im Gasraume und in Lösung betrachtet. Es ist anzunehmen,
dass ein Molekül, das an einer Wand reagiert, in gleicher Weise zu
betrachten ist, da auch hier die Bedingung ausreichender und daher
die Reaktionsgeschwindigkeit nicht beschränkender Energiezufuhr
vorliegt. Eine Abschätzung der Geschwindigkeit von Wandreaktionen
ergibt sich auf dieser Grundlage wie folgt:

Die Anzahl $N_a$ der an der Oberfläche $\omega$ adsorbierten Moleküle
bei der Konzentration $c$ im Gasraume ergibt sich aus der Adsorp-
tionswärme $W$, wenn wir Dicke $d$ der Wirkungssphäre der Adsorp-
tionskräfte zu $10^{-8}$ cm annehmen:

$$N_a = \omega c d e^{\frac{W}{RT}} = \omega . 10^{-8} c e^{\frac{W}{RT}} .$$

Der Anteil der pro Sekunde umgewandelten Moleküle ist gemäss (7)

$$\frac{1}{N_a} \frac{dN_a}{dt} = \nu e^{-\frac{Q}{RT}} ,$$

wo $Q$ die Aktivierungswärme im adsorbierten Zustande bedeutet.

----

[1]) Vgl. auch die zitierte Arbeit von B. Baule. Man könnte daran denken,
den Kondensationsvorgang vom Standpunkt der gewöhnlichen Wärmeleitung zu
behandeln, indem man versucht, den Wärmefluss zu berechnen, der während des
Aufstossens und Rückprallens stattfindet. Dieser Weg ist jedoch grundsätzlich zu
verwerfen, da die Wärmeleitungsgleichungen nur in Bereichen gültig sind, die gross
sind gegenüber der Debyeschen „mittleren Weglänge" der Wärmeleitung. (Es
kommt in dem hier behandelten Falle noch hinzu, dass der Temperaturgradient
extrem gross ist, so dass auch in dieser Hinsicht die Voraussetzungen der Wärmeleitungsgleichungen nicht erfüllt sind.) In der Tat erhält man nach dieser Rechnungs-
weise für $1 - w$ eine sehr kleine Zahl, was der Erfahrung nicht entspricht, da $1 - w$
(wie $w$ selbst) in der Grössenordnung von Eins gelegen ist (vgl. P. Harteck, Z.
physikal. Chem. 134, 1. 1928).

Hieraus ist

$$\frac{dc}{dt} = c\,\omega\,dv\,e^{-\frac{Q}{RT}} \approx \omega c\,10^6\,e^{-\frac{Q}{RT}}\,e^{\frac{W}{RT}}.$$

Die Anzahl $N^*$ der auf die Fläche $W$ treffenden Moleküle ist

$$N^* = \omega\,\sqrt{\frac{RT}{2\,\pi M}}\,c \sim 10^4\,\omega c$$

($T = 600$, Molekulargewicht $= 60$), demnach

$$\frac{dc}{dt} \sim 100\,N^*\,e^{-\frac{Q-W}{RT}}.$$

Der temperaturunabhängige Faktor ergibt sich demnach im allgemeinen nicht gleich der Anzahl der auf die Wand treffenden Moleküle, sondern wird meist erheblich grösser sein, als diese.

Die Messungen von Schwab und Pietsch[1]) über die Zersetzung von Methan ergeben für den temperaturunabhängigen Faktor den Wert $10^{4\cdot5}\cdot N^*$. Auch die Messungen der Oxydation von Wolfram durch Sauerstoff, die von Langmuir[2]) ausgeführt wurden, sowie noch unveröffentlichte Messungen von L. Marton und M. Polanyi[3]) über die Einwirkung von $H_2O$ auf Wolfram zeigen diesen Effekt; die temperaturunabhängigen Faktoren betragen hier $10^{2\cdot5}\cdot N^*$ bzw. $10^3\cdot N^*$. Es scheint uns somit, dass diese Erscheinung allgemein verbreitet ist. Neben der grundsätzlichen Erklärung, die wir oben angegeben haben (die natürlich noch weitgehend verfeinert werden müsste) ist noch zu berücksichtigen, dass (wie schon Schwab und Pietsch erwähnen) die Oberfläche durch Zerklüftung vergrössert sein kann, wodurch (bei grossen freien Weglängen) die adsorbierte Menge wächst, ohne dass die berechnete Anzahl der Wandstösse sich ändert.

## Schlussbemerkung.

Innerhalb eines durch feste Bindungen zusammengehaltenen Systems von Atomen führt die Interferenz der Eigenschwingungen zu Schwankungen der Amplitude der einzelnen Atome, die folgender Art sind: Die Energie eines Atoms ändert sich jeweils im Mittel während einer Zeit $\frac{1}{\nu}$ in gleichem Sinne. Während dieser Zeit beträgt die Ände-

---

[1]) Schwab und Pietsch, Z. Elektrochem. 32, 430. 1926.     [2]) J. Langmuir, J. Amer. Chem. Soc. 35, 105. 1913.     [3]) L. Marton und M. Polanyi, Aus dem Forschungslaboratorium der Vereinigten Glühlampenfabriken Ujpest.

452 M. Polanyi und E. Wigner, Über die Interferenz von Eigenschwingungen usw.

rung der Energie im Mittel etwa $kT$. Die Häufigkeit, mit der die Energie auf den Betrag $\varepsilon$ $(\varepsilon \gg \varkappa T)$ anwächst, ist $\sim \nu e^{-\frac{\varepsilon}{kT}} \sec^{-1}$. Die Atomfrequenz hat also die Bedeutung einer Wechselzahl der Schwingungsenergie.

Wenn eine chemische Reaktion oder sonstige Umwandlung durch Überschreitung einer kritischen Elongation eines bestimmten Atoms erreicht wird, indem ein Energieberg überschritten wird (Umwandlung), oder das Atom auf ein höheres Energieplateau hinaufgeschafft wird (Dissoziation, Verdampfung), so berechnet sich demnach die Häufigkeit eines solchen Vorganges zu $\nu e^{-\frac{\varepsilon}{kT}}$.

Das Ergebnis liefert die Erklärung der alten Erfahrung, dass der temperaturunabhängige Faktor der Geschwindigkeitskonstante monomolekularer Reaktionen in der Grössenordnung von $10^{14}$ gelegen ist. Auch die Verdampfungsgeschwindigkeit kommt im Einklang mit der Erfahrung heraus. Es erklärt sich ferner, wieso der temperaturunabhängige Faktor bei Wandreaktionen vielfach grösser herauskommt, als die Anzahl der gegen die Wand stossenden Moleküle.

Wenn das System eine schwache Bindung enthält (Moleküle im Zusammenstoss, Moleküle an der Wand), so ist an dieser Stelle die Energieübertragung schwächer, als über eine feste Bindung.

Unsere Behandlung des Gegenstandes war nicht darauf gerichtet, dass die Geschwindigkeitsgleichungen streng richtige Werte für das Gleichgewicht liefern. Das Ziel war, für praktisch wichtige Fälle die Geschwindigkeitskonstante abzuschätzen.

# The Statistics of Composite Systems According to the New Quantum Mechanics

E. P. Wigner

(Original in Hungarian with German abstract). Összetett Rendszerek Statisztikája az Új Quantum-Mechanika Szerint. Magyar Tudományos Akadémia Matematikai és Természettudományi Értesitöje *45*, 576–582 (1929). Translated from Hungarian by Akos Sebestyén

1. Only one kind of statistics applies in classical mechanics. In fact the behavior of a system is completely (i. e., also microscopically) determined if the equations of motion are known. The latter does not influence the macroscopic behavior of the system since according to the quasi-ergodic hypothesis, it will get arbitrarily close to any state of the given energy after a long enough time, and statistical mechanics deals with averages when the system goes through all such states many times. We obtain the probability of the macroscopic state of the system according to this classical, or, as it is called, Boltzmann statistics if we calculate the number of ways its various parts can be placed in the phase-space cells compatible with this state. In counting these distributions we must distinguish two states in which the same number of atoms are placed in the same cells but they differ from each other through an interchange of atoms. In combinatorics this rule is expressed by considering the atoms as distinguishable, and we must calculate the number of possible distributions of the atoms in the cells.

All these can be derived from classical mechanics and the quasi-ergodic hypothesis recalled before, in the simplest way with the help of Liouville's theorem. (See e.g. "Introduction to the Corpuscular Theory of Matter" by R. Ortvay, (1927) Budapest.)

If we consider the combinatorial statement in classical statistics: "The atoms have to be distributed in the cells but if atoms in two different cells in one distribution are interchanged compared to another, then these two distributions must be considered different even if atoms of identical chemical nature are interchanged (for instance in case of two atoms of hydrogen)", then it seems to be obvious to relinquish the latter condition and to consider atoms of the same kind indistinguishable. Bose, who originated this idea has shown that if we apply this statistics to a gas consisting of light quanta we obtain directly Planck's radiation law.

As an example we shall apply these statistics for two atoms and two cells, an even more simple case. We denote the two cells by two squares and the two atoms in Boltzmann's case by the numbers 1. and 2., and in Bose's case by two points.

---

* *Editor's Note:* There are two slips at the end of the paper and in the abstract. Contrary to what is stated there, nitrogen ($^{14}$N) nuclei satisfy Bose statistics as, more generally, do all nuclei that contain an even number of particles.

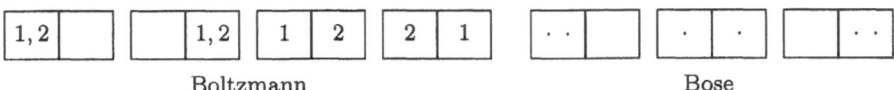

Boltzmann                              Bose

We see, that in Boltzmann's case the probability that the two atoms are in different cells is equal to the probability that they are in the same cell. According to Bose the latter is twice as large as the former. This is very suitable to characterize Bose's statistics qualitatively.

Shortly after Bose's investigations, Einstein applied Bose statistics to an ideal gas. He found interesting phenomena of degeneracy which, following him, have also been described by many others and which we can understand if we suppose that the probability for two atoms to reside in the same cell of phase-space is larger than what is claimed by classical statistics.

Soon after Einstein's paper, Fermi stressed that, in contrast, our experience concerning the electron structure of atoms does not suggest a larger probability for placing two atoms in the same cell than in classical statistics. On the contrary, Pauli's principle establishes (on the basis of experience gained from spectra) that no state can exist in which a cell of phase-space contains more than one particle. Fermi took 0 for the probability of such states while the apriori probabilities of those states are equal where there is at most one particle in the cell. This is called Fermi statistics.

2. As we see, setting up other statistics than Boltzmann's (the Bose and Fermi types) was completely arbitrary. Quantum mechanics however established to a certain extent the motivation for these two new statistics.

We have noted before that in addition to mechanics the quasi-ergodic hypothesis is also needed to formulate Boltzmann statistics.

Heisenberg [1] and Dirac [2] have independently noticed that this hypothesis does not apply in quantum mechanics. They have shown that for instance in the case of two electrons there exist two groups of states, the symmetric and antisymmetric states which do not approach each other. The symmetric states are characterized by wave functions

$$\psi(r_1, r_2, \ldots, r_n)$$

(where $r_1, r_2, \ldots, r_n$ are the coordinates of the particles) symmetric in the arguments; that is

$$\psi(r_1, r_2, \ldots, r_n) = \psi(r_{\alpha_1}, r_{\alpha_2}, \ldots, r_{\alpha_n}) \tag{1}$$

holds for any permutation $\alpha_1, \alpha_2, \ldots, \alpha_n$ of the numbers $1, 2, \ldots, n$. For wave functions of antisymmetric states (1) holds only for even permutations $\alpha_1, \alpha_2, \ldots, \alpha_n$, otherwise

$$\psi(r_1, r_2, \ldots, r_n) = -\psi(r_{\alpha_1}, r_{\alpha_2}, \ldots, r_{\alpha_n}). \tag{1a}$$

Now a symmetric state always remains symmetric at all times and similarly an antisymmetric state remains always antisymmetric. This follows from the time development of $\psi$ which is described by a linear differential equation being symmetric in

$$r_1, r_2, \ldots, r_n.$$

There exist other states as well, besides the symmetric and antisymmetric states (in the case of two particles only these two exist; but for three particles there are three, and for four particles there are five kinds of conceivable states). However, these two cases are the simplest and the most important, and we need not consider the others, since Dirac has shown that Bose statistics applies for particles in symmetric states and Fermi statistics has to be used for those in antisymmetric states. Accordingly, quanta of light are always in symmetric states and electrons are always in antisymmetric states. The latter applies also for protons.

In what follows, we shall see that for composite bodies the first or the second statistics applies depending whether it consists of an even or an odd number of particles. To determine this we have to treat nuclei as being built up of protons and electrons.

3. Consider first two hydrogen atoms in their normal states confined to a box. Let the coordinates of the nuclei be $R_1$ and $R_2$ and those of the electrons $r_1$ and $r_2$; we denote by $L$ the wave function of the normal state of the H-atom (the first Laguerre-function) and the possible stationary wave functions of the box, that is the de Broglie's functions, by $B_k$. Let $M$ and $m$ be the masses of the nucleus and the electron respectively. The following four states have the same energy:

$$\psi_1 = B_k \left( \frac{MR_1 + mr_1}{M + m} \right) L(R_1 - r_1) B_{k'} \left( \frac{MR_2 + mr_2}{M + m} \right) L(R_2 - r_2),$$

$$\psi_2 = B_k \left( \frac{MR_1 + mr_2}{M + m} \right) L(R_1 - r_2) B_{k'} \left( \frac{MR_2 + mr_1}{M + m} \right) L(R_2 - r_1),$$

$$\psi_3 = B_k \left( \frac{MR_2 + mr_1}{M + m} \right) L(R_2 - r_1) B_{k'} \left( \frac{MR_1 + mr_2}{M + mr_2} \right) L(R_1 - r_2), \quad (2)$$

$$\psi_4 = B_k \left( \frac{MR_2 + mr_2}{M + m} \right) L(R_2 - r_2) B_{k'} \left( \frac{MR_1 + mr_1}{M + m} \right) L(R_1 - r_1).$$

Out of these functions, we can construct only one function that is antisymmetric in both of the pairs $R_1, R_2$ and $r_1, r_2$. The states realizable in nature correspond to such functions as

$$\psi_{kk'} = \psi_1 - \psi_3 - (\psi_2 - \psi_4) = \psi_{k'k}. \quad (3)$$

As we see a single state corresponds to any combination $kk'$ irrespective of its order; there is also a state for $k = k'$; we get exactly the same number of states as required by Bose's statistics. If we write

$$\frac{MR + mr}{M + m} \approx R \tag{4}$$

corresponding to the large nuclear mass compared to the the electron we get

$$\psi_{kk'} = [B_k(R_1)B_{k'}(R_2) + B_k(R_2)B_{k'}(R_1)]\, L(R_1 - r_1)L(R_2 - r_2)$$
$$- [B_k(R_1)B_{k'}(R_2) + B_k(R_2)B_{k'}(R_1)]\, L(R_2 - r_1)L(R_1 - r_2)\,. \tag{5}$$

The expression in square brackets is characteristic for Bose statistics. We have just seen that the number of the stationary states of a gas consisting of two H-atoms in their normal states is given by Bose statistics; (5) shows that also the statistics of their position corresponds to Bose statistics, at least if we consider domains large compared with the radius of the H-atom. In this case $\psi_{kk'}$ differs considerably from zero only if either $R_1 - r_1$ and $R_2 - r_2$ or $R_1 - r_2$ and $R_2 - r_1$ is very small. In both cases only one of the two terms is dominant and the factor

$$[B_k(R_1)B_{k'}(R_2) + B_k(R_2)B_{k'}(R_1)]$$

gives Bose statistics for the position, (e. g. the maximal probability of the state $R_1 \approx R_2$.)

4. If the number of atoms is more than two, let us say $n$, we have two write the only antisymmetric combination in place of (5)

$$\psi_{k_1}\dots k_n = \sum_{\sigma}\prod_{x} B_{k_{x(\sigma)}} \sum_{\tau}\prod_{\lambda} L(R_\lambda - r_{\lambda(\tau)})\varepsilon_\tau$$

$$= \sum_{\sigma}\prod_{x} B_{k_{x(\sigma)}}(R_k)\left| \left( \begin{array}{ccc} L(R_1 - r_1) & \dots & L(R_n - r_1) \\ L(R_1 - r_k) & \dots & L(R_n - r_n) \end{array} \right) \right|, \tag{5a}$$

where $\sigma$ and $\tau$ denote the permutations $1\,(\sigma)$, $2\,(\sigma)$, $\dots$, $n\,(\sigma)$ and $1\,(\tau)$, $2\,(\tau)$, $\dots$, $n\,(\tau)$ respectively of the numbers $1, 2, \dots, n$. The summarizations over $\sigma$ and $\tau$ must be extended for all the $n!$ permutations, $\varepsilon_\tau = +1$ or $\varepsilon_\tau = -1$ depending whether $\tau$ is an even or an odd permutation. Now we can derive the requirements of Bose statistics from (5a) similarly to the case (5), rigorously for stationary states, but also for the statistics of the position provided we consider domains large compared to the radius of the atom.

For the case of more complicated atoms or molecules we must proceed similarly and we obtain within the same limitations Bose or Fermi statistics depending whether the body is built up from an even or odd number of particles.

In the preceding considerations we did not take into account either the magnetic moments of the protons and of the electrons or the possibility that the normal state may be degenerate in so far as it has an "internal" quantum number that is greater than 0. Our results are exactly valid for the states $^1S$ only. In other cases we have to remedy this by counting all the cells of the phase space as many times as the degree of degeneracy of the state considered, and then apply Bose's or Fermi's statistics. We must proceed in this way even in the case of electrons since their lowest states are doubly degenerate on account of their magnetic moments. (See e. g. W. Pauli jr.: Zs. f. Phys. *41*, 81, 1927.)

5. In the paper just mentioned W. Pauli jr. studies the behavior of the electron gas on the basis of Fermi's statistics. A. Sommerfeld [3] gave with great success further applications of this statistics.

This application is not so easy for compound bodies. In gases with temperatures so low and densities so high that a considerable deviation from Boltzmann's statistics can be expected the atoms are nearer to each other than allowed by the restrictions of the preceding sections which made our results applicable, or gas condensation may take place.

The statistics of atomic nuclei can be established far more certainly than the statistics of entire atoms. Heisenberg first called attention to the absence of even or odd lines in the band spectra of molecules which can be traced back to nuclei that can be in only antisymmetric or else in only symmetric states. It can be shown generally (E. Wigner and E. E. Witmer: Zs. f. Phys. *51*, 859, 1928, loc. cit. in what follows) that the absence of lines can be expected only for such molecules consisting of identical atoms ($H_2$, $He_2$, $N_2$, $O_2$ etc.) which have no "electric moment". There are four kinds of such states which we denote by $O_+$, $O'_+$ and $O_-$, $O'_-$. In loc. cit. it is also established that in case of $O_+$ and $O'_+$ the even lines are symmetric while, on the contrary in case of $O_-$ and $O'_-$ the even lines are antisymmetric. If only symmetric states exist in the nuclei and they have no angular momentum then the antisymmetric states are absent and vice versa, if they have angular momenta the antisymmetric states yield far weaker lines.

Experiments show for the four molecules mentioned before that the odd lines are always strong, the even ones are completely absent for $He_2$ and $O_2$. Consequently, the nuclei of O and He have no angular momentum at all while the angular momentum of H and N are $\frac{1}{2}\frac{h}{2\pi}$ and $\frac{h}{2\pi}$ respectively (see R. de L. Kronig: Die Naturwissenschaften *16*, 225, 1928.) In order to decide whether the nuclei are in symmetric or antisymmetric states we must know whether the spin angular momentum of the electron is $O_-$ or $O_+$. It is not quite elementary to determine this. However, the basis of loc. cit., starting from certain criteria (especially on the basis of band convergence) it is nevertheless possible. Based on this it can be concluded that the normal state of $H_2$ and $N_2$ is $^2O_+$, for $O_2$ it is $^3O_-$, and for $He_2$ it is probably $^1O_-$.

This leads us to the conclusion that the following rule is also correct for nuclei: bodies built up of an even number of particles follow Fermi's statistics.

## References

[1] Zs. f. Phys. *38*, 411 (1926)
[2] Proc. Roy. Soc. *112*, 661 (1926)
[3] Die Naturwissenschaften *16*, 372 (1928), papers in Zs. f. Phys. in 1928

(From the session of Division III of the Hungarian Scientific Academy, 8. Oct. 1928.)

# STATISTIK ZUSAMMENGESETZTER SYSTEME NACH DER NEUEREN QUANTEN-MECHANIK.

### Von E. WIGNER.

Es wird die Frage untersucht, durch welche Statistik (BOLTZ-MANN, FERMI oder BOSE) man die Entartungserscheinungen, statistischen Gewichte usw. von zusammengesetzten Systemen (Atomen oder Molekülen) beschreiben kann. Natürlich wird diese Beschreibung nur solange gelten, als man die Systeme als Massenpunkte ansehen darf.

Zur Entscheidung der Frage wird die Wellenfunkzion eines Gases, bestehend aus den betreffenden Atomen, bezüglicherweise Molekülen, aufgeschrieben ($Gl$ (5) für ein Gas aus zwei Wasserstoffatomen, $Gl$ (5a) für den allgemeinen Fall. Die grossen $R$ sind die Schwerpunktskoordinaten, die kleinen $r$ die inneren Koordinaten des Atoms). Aus dieser Wellenfunktion wir dabgelesen, dass das ganze System FERMI-scher oder BOSE-scher Statistik genügt, je nachdem die Anzahl der das System aufbauenden FERMI-schen Elementarteilchen ungerade oder gerade ist.

Für Atomkerne ist die Bedingung, dass man sie als Massenpunkte ansehen kann, bei optischen Versuchen allenfalls erfüllt. Die Bandenspektren zeigen in der Tat eine Bestätigung der Theorie. Das Spektrum von $H_2$ (Grundzustand $^1\Sigma_+$) zeigt, dass der $H$-Kern FERMI-scher, das Spektrum von $N_2^+$ (Grundzustand $^2\Sigma_+$), dass der $N$-Kern ebenfalls FERMI-scher, das Spektrum von $O_2$ (Grundzustand $^3\Sigma_-$), dass der $O$-Kern BOSE-scher Statistik genügt.

---

(Aus der Sitzung der III. Klasse der Ungarischen Akademie der Wissenschaften vom 8. Okt. 1928.)

# Über die Geschwindigkeitskonstante von Austauschreaktionen

H. Pelzer und E. P. Wigner

Zeitschrift für physikalische Chemie B *15*, 445–471 (1932)

(Eingegangen am 19. 11. 31.)

Es wird zunächst versucht, die Annahme von LONDON zu begründen, dass die chemischen Austauschreaktionen adiabatisch verlaufen. Die wesentliche Bedingung hierfür scheint es zu sein, dass die unterste Energiefläche energetisch weit getrennt von den übrigen liegen soll. Ferner wird für die Umsetzung von Parawasserstoff in normalem Wasserstoff, für die EYRING und POLANYI die Aktivierungswärme berechnet haben, die absolute Grösse der Geschwindigkeitskonstante bestimmt.

1. Wir kennen drei verschiedene Typen von chemischen Reaktionen:

I. Elementare Addition und spontaner Zerfall (Prädissoziation):

$$\left.\begin{array}{l} H + H \rightleftarrows H_2 \\ AlH \rightleftarrows Al + H. \end{array}\right\} \tag{I}$$

Dass solche Reaktionen wegen des gleichzeitigen Bestehens von Energie- und Impulssatz nicht ohne Schwierigkeiten vor sich gehen können, haben zuerst M. BORN und J. FRANCK bemerkt[1]).

Die Reaktion (I) kann zweierlei Ursachen haben. Im ersteren Falle[2]) kann man von einem Molekül ausgehen, das keine Elektronenanregung, nur Rotationsenergie und Schwingungsenergie hat. Man kann nämlich bei einem rotierenden zweiatomigen Molekül die Zentrifugalkräfte durch ein „Zentrifugalpotential" $\frac{M^2}{mr^2}$ ($M$ = Drehimpuls, $m$ = relative Masse der beiden Atome, $r$ = ihr Abstand) ersetzen. Wenn dieses Potential zusammen mit dem Anziehungspotential der beiden Atome bei grösseren Kernabständen ein Maximum hat, so sind klassisch solche Bewegungen des Moleküls möglich, deren Energie kleiner als dieses Maximum, aber doch grösser als die Dissoziationsenergie ist, und die trotz dieses letzten Umstands stabil sind. Nach der Quantenmechanik sind aber auch diese Bewegungszustände nicht vollkommen stabil, sie zerfallen infolge des „Tunneleffekts", weil ein Potential-

---

[1]) M. BORN und J. FRANCK, Z. Physik **31**, 411. 1925.    [2]) J. OLDENBERG, Z. Physik **56**, 563. 1929.

446                          H. Pelzer und E. Wigner

berg in der Quantenmechanik, selbst wenn die Energie zu seiner
Überschreitung nicht hinreicht, kein absolutes Hindernis ist und
„unmechanisch" übersprungen werden kann. Das einfachste Beispiel
hierfür bildet der radioaktive Zerfall[1]).

Die zweite Reaktionsart nach dem Schema (I) ist die Prädissozia-
tion[2]). Es handelt sich hierbei um den Zerfall eines Moleküls mit
Elektronenanregung, oder in der Rückreaktion um die Bildung eines
Moleküls mit Elektronenanregung aus den Atomen.

Die absolute Grösse der Geschwindigkeitskonstante der Reaktionen nach dem Schema (I) hängt eng mit der Linienbreite zusammen
und lässt sich mit ihrer Hilfe absolut angeben[3]).

II. Den zweiten Reaktionstyp bilden die sogenannten doppelten
Umsetzungen:

$$\left.\begin{array}{l} Na + ClH \rightleftarrows NaCl + H \\ H + Cl_2 \rightleftarrows HCl + Cl. \end{array}\right\} \qquad \text{(II)}$$

Wenn sich ein $Na$-Atom einem $HCl$-Molekül nähert, wird es von
diesem zunächst abgestossen. Hat es aber eine genügende kinetische
Energie, um diese Abstossung zu überwinden, so lockert es gleichzeitig die $Cl$–$H$-Bindung. Bei einem gewissen kritischen Abstand hört
dann die Abstossung zwischen dem $Na$ und $Cl$ auf, während die bis-
herige Anziehung zwischen dem $Cl$ und $H$ in Abstossung übergeht.
Daher wird das $H$ abgestossen, das $Na$ und $Cl$ werden dagegen ein
neues Molekül bilden. Wesentlich ist für die Diskussion dieses Reak-
tionsverlaufs, dass die potentielle Energie nicht die Gestalt

$$f(r_{12}) + g(r_{13}) + h(r_{23})$$

hat, vielmehr die Kräfte zwischen dem $H$ und $Cl$ durch die Anwesen-
heit des $Na$ auch ihrerseits beeinflusst werden.

Der vorbeschriebene Mechanismus wurde von F. LONDON[4]) ent-
deckt und von EYRING und POLANYI[5]) genauer besprochen und zur
Berechnung von Aktivierungswärmen verwandt. Charakteristisch für
den ganzen Mechanismus ist, dass er adiabatisch, d. h. ohne Elektronensprung verläuft. Die vorliegende Note wird sich ausschliesslich
mit dieser Reaktionsart beschäftigen und wird erstens Bedingungen

[1]) G. GAMOW, Z. Physik 51, 204. 1928. E. U. CONDON und R. W. GURNEY,
Nature 122, 439. 1928.     [2]) R. DE L. KRONIG, Z. Physik 62, 300. 1930. Siehe daselbst auch weitere Literatur.     [3]) M. POLANYI und E. WIGNER, Z. Physik 33,
429. 1925. O. K. RICE, Physic. Rev. 33, 748. 1929. 35, 1538. 1930.     [4]) F. LON-
DON, SOMMERFELD - Festschrift, S. 104. S. Hirzel, Leipzig 1928.     [5]) H. EYRING
und M. POLANYI, Z. physikal. Ch. (B) 12, 279. 1931.

anzugeben versuchen, unter welchen man annehmen kann, dass die Reaktion adiabatisch verläuft. Zweitens soll versucht werden für einen Fall, in dem diese Bedingungen erfüllt sind, die absolute Grösse der Geschwindigkeitskonstante zu bestimmen.

In die Gruppe (II) können wir auch die sogenannten Energie-übertragungen[1])

$$Hg^* + Ba \leftrightarrows Hg + Ba^* \tag{IIa}$$

und die Ladungsübertragungen[2])

$$Hg^- + Na \rightleftarrows Hg + Na^- \tag{IIb}$$

einreihen, die allerdings keineswegs adiabatisch verlaufen.

Auch die Assoziation durch Dreierstoss

$$\left.\begin{array}{l} Na + Na + He \rightleftarrows Na_2 + He \\ H + Cl + He \rightleftarrows HCl + He \end{array}\right\} \tag{IIc}$$

gehört hierher, über die wir allerdings theoretisch sehr wenig wissen. Praktisch überwiegt sie gewöhnlich an Geschwindigkeit die Reaktion (I), die dasselbe Resultat hätte.

Auch diese Reaktion kann adiabatisch verlaufen. Betrachten wir z.B. die Reaktion $H + H + H \rightleftarrows H_2 + H$, und nehmen wir als einfachsten Fall an, dass die drei Atome zunächst auf einer Geraden liegen, und dass auch die Geschwindigkeiten so beschaffen sind, dass die gerade Konfiguration erhalten bleibt. Dann kann man die Bewegung durch die Abstände $c$ und $b$ der seitlichen Atome vom mittleren Atom beschreiben. Die zeitliche Änderung von $c$ und $b$ wird, wie in dem von EYRING und POLANYI näher untersuchten Fall $H_2 + H$ durch die zeitliche Änderung der Koordination $c$ und $b$ einer Kugel gegeben, die auf der in Fig. 1 gegebenen Energiefläche abrollt. Der Unterschied gegenüber dem von EYRING und POLANYI untersuchten Fall besteht nur darin, dass man hierbei nicht von einem der beiden Molekültäler (I) und (II), sondern vom Hochplateau ausgehen muss, das die Mitte der Figur einnimmt. Erteilt man der Kugel auf diesem Hochplateau etwa die in der Figur eingezeichnete Geschwindigkeit, so wird die Kugel längs der eingezeichneten Linie rollen und aus dem Molekültal (II) nicht mehr herauskommen, sondern darin unter Schwingungen in Richtung wachsender $c$ fortschreiten. Dies entspricht der Bildung

---

[1]) H. KALLMANN u, F. LONDON, Z. physikal. Ch. (B) **2**, 207. 1929.     [2]) H. KALL-MANN und B. ROSEN, Z. Physik **61**, 61. 1930.

448                        H. Pelzer und E. Wigner

eines Moleküls $H_2$, das also auf diesem Wege adiabatisch und ohne
Aktivierungswärme aus den Atomen entstehen kann.

III. Der dritte Typ ist schliesslich die Dissoziation durch Interferenz der Eigenschwingungen, etwa:

$$N_2O \rightarrow N_2 + O. \qquad \text{(III)}$$

Der Reaktionsmechanismus besteht darin[1]), dass die verschiedenen thermisch angeregten Eigenschwingungen, die die Bindungen
bald dehnen, bald zusammendrücken, nach einer hinreichend langen
Zeit an einer Bindung in Phase kommen und diese dann infolge ihres
Zusammenwirkens sprengen. Über die Aktivierungswärme solcher
Reaktionen können wir im voraus fast gar nichts aussagen. Wenn diese
bekannt ist, kann man die absolute Grösse der Geschwindigkeitskonstante roh abschätzen[2]). Interessant ist auch die Abhängigkeit der
Reaktionsgeschwindigkeit von der Konzentration[3]). Die Geschwindigkeit dieser Reaktion übersteigt die von (I) bei höheratomigen Molekülen. Bei zweiatomigen Molekülen ist dagegen der Reaktionsmechanismus (III) ausgeschlossen, da bei diesen nur eine Eigenschwingung
existiert und eine Interferenz von Eigenschwingungen daher nicht möglich ist.

2. Wir wollen hier zuerst die Frage untersuchen, unter welchen
Umständen eine Reaktion vom Typus (II) adiabatisch verlaufen wird.
Die Vorstellung, die den nachstehenden Betrachtungen zugrunde liegt,
ist dabei folgende:

[1]) M. Polanyi und E. Wigner, Z. physikal. Ch. (A) **139**, 439. 1928.
[2]) H. Kummerow und M. Volmer (Z. physikal. Ch. (B) **9**, 141. 1930) haben die
Reaktion (III) quantitativ untersucht und finden eine 100- bis 1000mal kleinere
Geschwindigkeitskonstante, als man nach der erwähnten Rechnung erwarten würde.
Dies dürfte aber nicht so sehr auf der Ungenauigkeit der Rechnung beruhen, als
vielmehr darauf, dass sich während der Reaktion (III) das Multiplettsystem des
Gesamtsystems ändert: $N_2$ ist ein Molekül im Singulettzustand, der Grundzustand
von $O$ ist ein Triplettzustand. Daher ist das System, das durch die rechte Seite
von (III) dargestellt ist, im Triplettzustand, während $N_2O$ ein Singulettmolekül ist.
Die Reaktion $N_2O \rightarrow N_2 + O$ ist also eigentlich „verboten", und diese Tatsache
kann für den Faktor von etwa 0·001 verantwortlich gemacht werden. Natürlich
würde man ohnehin keine gute Übereinstimmung in diesem Falle zwischen Rechnung und Experiment erwarten, da es sich um die Zersetzung eines sehr wenig-
atomigen Moleküls handelt, während die Rechnung für hochatomige Moleküle an-
gestellt worden ist. Doch dürfte dies keinen Faktor von der Grössenordnung 0·001
bedingen.        [3]) Vgl. etwa die Diskussion bei O. K. Rice und H. Ramsperger, J.
Am. chem. Soc. **46**, 1617. 1927. L. S. Kassel, J. physical Chem. **32**, 225. 1928.

Über die Geschwindigkeitskonstante von Austauschreaktionen.    **449**

Wir können uns die SCHRÖDINGER-Gleichung für das ganze System aufgeschrieben denken

$$-\frac{h}{2\pi i}\frac{\partial\Psi(X,x,t)}{\partial t} = -\frac{h^2}{8\pi^2 M}\sum_k\frac{\partial^2\Psi}{\partial X_k^2} - \frac{h^2}{8\pi^2 m}\sum_k\frac{\partial^2\Psi}{\partial x_k^2} + V(X,x)\Psi. \quad (1)$$

Hierbei sollen die $x_k$ die Koordinaten der Elektronen, die $X_k$ die Koordinaten der Kerne bedeuten, die Massen der Kerne wurden der Einfachheit halber gleich angenommen. Lassen wir für einen Augenblick den Operator der kinetischen Energie der Kerne weg, so können wir nach dem Vorgang von BORN und OPPENHEIMER[1]) im übriggebliebenen Operator $H_1$ die Kernkoordinaten $X$ als Parameter behandeln und die Eigenwertgleichung

$$E_\varkappa(X)\psi_\varkappa(X,x) = H_1\psi_\varkappa(X,x) = -\frac{h^2}{8\pi^2 m}\sum_k\frac{\partial^2\psi_\varkappa}{\partial x_k^2} + V(X,x)\psi_\varkappa \quad (2)$$

für jeden Wert der Parameter $X_k$ gelöst denken. Es werden natürlich sowohl die Eigenwerte $E_\varkappa(X)$, wie auch die Eigenfunktionen von den Werten der Parameter $X_k$ abhängen. Die Orthogonalitäts- und Normalisierungsbedingungen lauten[2])

$$\int\psi_\varkappa(X,x)^2\,dx = 1; \quad \int\psi_\varkappa(X,x)\psi_\lambda(X,x)\,dx = 0 \text{ für } \varkappa \neq \lambda \quad (3)$$

oder nach $X_k$ differentiiert

$$\left.\begin{array}{c} \displaystyle\int\psi_\varkappa\frac{\partial\psi_\varkappa}{\partial X_k}\,dx = 0; \quad \int\psi_\varkappa\frac{\partial\psi_\lambda}{\partial X_k}\,dx = -\int\psi_\lambda\frac{\partial\psi_\varkappa}{\partial X_k}; \\[2ex] \displaystyle\int\psi_\varkappa\frac{\partial^2\psi_\varkappa}{\partial X_k^2} + \left(\frac{\partial\psi_\varkappa}{\partial X_k}\right)^2\,dx = 0, \end{array}\right\} \quad (3\,\mathrm{a})$$

wo beidesmal nur über die Elektronenkoordinaten zu integrieren ist. Die $E_\varkappa(X)$ bilden im Konfigurationsraum der Kerne allein Flächen, die wir als **Energieflächen**[3]) bezeichnen werden. Jede Energiefläche entspricht einem Zustand der Elektronen, der stationär wäre, wenn die Kerne wirklich an ihre Lagen festgenagelt wären. Jeder Punkt einer Energiefläche entspricht einer Konfiguration der Kerne.

---

[1]) M. BORN und I. R. OPPENHEIMER, Ann. Physik 84, 457. 1927. Vgl. auch C. ZENER, Physic. Rev. 38, 277. 1931. Unsere Überlegung beruht ebenso wie die von BORN und OPPENHEIMER darauf, dass die Kernmassen gross im Verhältnis zu den Elektronenmassen sind.    [2]) Die $\psi_\varkappa$ können und sollen als reell vorausgesetzt werden.    [3]) Eine solche Energiefläche, nämlich die unterste Energiefläche für den Fall dreier $H$-Atome ist in Fig. 1 gegeben. Sie wurde der Arbeit von POLANYI und EYRING entnommen.

Man sagt, dass die Bewegung des Systems adiabatisch etwa auf der ersten Energiefläche abläuft, wenn man so rechnen darf, wie wenn die Kerne unter dem Einfluss eines Potentials $E_1(X)$ stehen würden

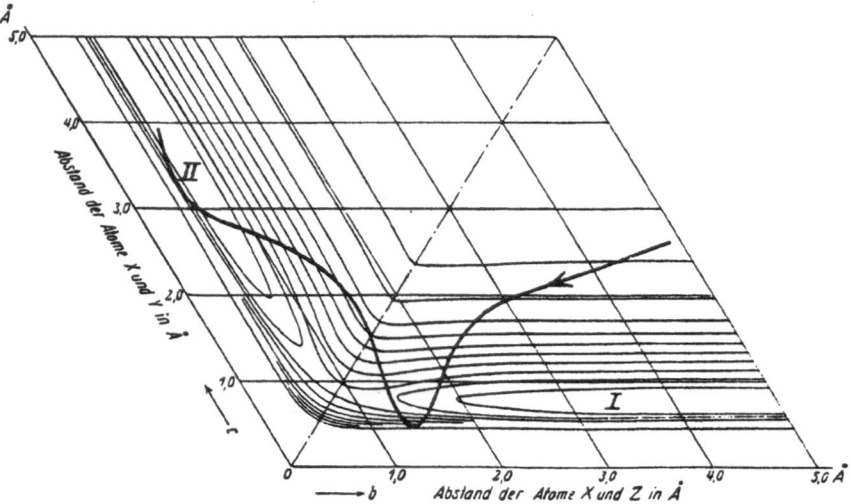

Fig. 1. Potentialgebirge dreier geradlinig angeordneter Wasserstoffatome. Die eingezeichneten Linien sind Niveaulinien. Nach EYRING und POLANYI in schiefwinkligen Koordinaten gezeichnet, um die Dynamik des Systems richtig wiederzugeben. Die mit einem Pfeil versehene Linie stellt eine mögliche Bewegung des Bildpunktes dar.

und sich so langsam bewegen würden, dass die Elektroneneigenfunktion der Bewegung immer folgen kann, d. h. immer $\psi_1(X, x)$ bleibt. Um dies auszudrücken, zerlegen wir $\Psi$

$$\Psi(X, x, t) = \sum_{\varkappa} \chi_\varkappa(X, t)\, \psi_\varkappa(X, x)\,. \tag{4}$$

Setzen wir dies in (1) ein, so können wir durch Multiplikation mit $\psi_\varkappa(X, x)$ und Integration über die Elektronenkoordinaten die Bewegungsgleichung für die $\chi_\varkappa$ in bekannter Weise ableiten

$$
\begin{aligned}
-\frac{h}{2\pi i}\frac{\partial \chi_\varkappa(X, t)}{\partial t} &= -\frac{h^2}{8\pi^2 M}\sum_k \frac{\partial^2 \chi_\varkappa}{\partial X_k^2} + E_\varkappa(X)\chi_\varkappa \\
&- \sum_\lambda \sum_k \frac{h^2}{8\pi^2 M}\left[2\int \frac{\partial \psi_\lambda}{\partial X_k}\psi_\varkappa\, dx \frac{\partial \chi_\lambda}{\partial X_k} + \int \frac{\partial^2 \psi_\lambda}{\partial X_k^2}\psi_\varkappa\, dx\, \chi_\lambda\right].
\end{aligned}
\tag{5}
$$

Man kann die Wellenfunktion $\Psi$ durch die Wellenfunktionen $\chi_1, \chi_2, \chi_3, \ldots$ ersetzen, die Bewegungen der Kerne allein entsprechen.

Allerdings braucht man nicht nur eine Wellenfunktion dieser Art, sondern unendlich viele.

Wären in (5) die Glieder der zweiten Zeile, die tatsächlich klein sind, nicht vorhanden, so würde sich $\chi_x$ genau so verhalten, als ob es die Wellenfunktion von Massenpunkten wäre, zwischen denen das Potential $E_x(X)$ herrscht. In Wirklichkeit sind aber die zu den verschiedenen Energieflächen gehörenden Wellen $\chi_1, \chi_2, \chi_3, \ldots$ nicht ganz unabhängig voneinander, weil sie sich wegen der Glieder der zweiten Zeile von (5) gegenseitig beeinflussen.

Die Glieder der zweiten Zeile von (5) sind deswegen klein, weil in ihnen die Masse des Kerns im Nenner auftritt. In der Tat ist $\int \frac{\partial \psi_\lambda}{\partial X} \psi_x \, dx$ von der Grössenordnung eines reziproken Atomradius. Daher hat die zweite Zeile die Grössenordnung $\frac{\chi h^2}{8 \pi^2 M}$ /Atomradius (Wellenlänge der Kernwellen) und ist daher der Atomradius/(Wellenlänge der Kernwellen) kleiner als das Glied mit der kinetischen Energie der Kerne. Wir wollen dies aber noch genauer ausführen und zeigen, dass die Glieder der zweiten Zeile von (5) bei der Berechnung der Reaktionsgeschwindigkeit der Umsetzung $H + H_2 \rightleftarrows H_2 + H$ wirklich vernachlässigt werden können.

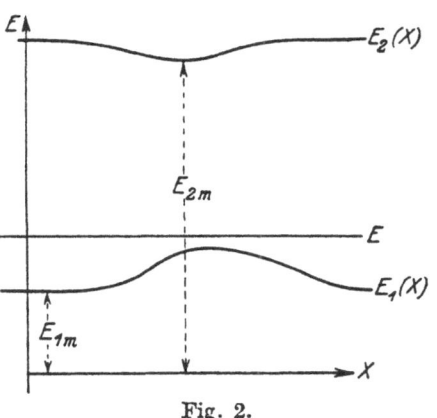

Fig. 2.

3. Es sei die Energie des Systems $E$, und wir wollen die Voraussetzung machen, dass $E$ kleiner ist als selbst der tiefste Punkt $E_{2\,m}$ der zweiten Energiefläche (vgl. Fig. 2). Dies ist für die oben erwähnte Reaktion für die meisten Zusammenstösse, die zur Reaktion führen, erfüllt. Ihre Energie ist nämlich nur um weniges grösser als die Aktivierungswärme, während der tiefste Punkt der nächsten Energiefläche drei getrennten Atomen entspricht und daher wesentlich höher ist. Wenn wir die Glieder in (4), die den höheren Energieflächen entsprechen, zu einem Glied vereinigen

$$\Psi(X, x, t) = \chi_1(X, t)\,\psi_1(X, x) + u(X, x, t), \qquad (4\,a)$$

so lautet $H\Psi = E\Psi$

$$E_1(X)\,\chi_1\,\psi_1 - \frac{h^2}{8\,\pi^2\,M}\sum_k\left(\psi_1\frac{\partial^2\chi_1}{\partial X_k^2} + 2\frac{\partial\psi_1}{\partial X_k}\frac{\partial\chi_1}{\partial X_k} + \chi_1\frac{\partial^2\psi_1}{\partial X_k^2}\right) + Hu$$
$$= E\,\chi_1\,\psi_1 + E\,u \qquad\qquad (6)$$

und wir wollen zeigen, dass die Wahrscheinlichkeit der Anregung aller höheren Energieniveaus zusammen $\iint |u^2|\,dx\,dX$ immerfort (für alle Zeiten) sehr klein ist. Wir multiplizieren (6) mit $u^*$ und integrieren über alle Koordinaten, wegen (3) verschwindet $\int u^*\psi_1\,dx$ schon über die Elektronenkoordinaten integriert. Ausserdem ist

$$\iint u^*\,H\,u\,dx\,dX$$
$$= \frac{h^2}{8\,\pi^2\,M}\sum_k\iint\left|\frac{\partial u}{\partial X_k}\right|^2 dx\,dX + \sum_{\varkappa=2}^{\infty}\iint \chi_\varkappa^*\psi_\varkappa\,E_\varkappa(X)\,\chi_\varkappa\psi_\varkappa\,dx\,dX$$
$$> \frac{h^2}{8\,\pi^2\,M}\sum_k\iint\left|\frac{\partial u}{\partial X_k}\right|^2 dx\,dX + E_{2m}\iint u^2\,dx\,dX,$$

so dass wir

$$\iint u^*\,H\chi_1\,\psi_1\,dx\,dX = -\frac{h^2}{8\,\pi^2\,M}\iint\sum_k\left(2\frac{\partial\chi_1}{\partial X_k}\frac{\partial\psi_1}{\partial X_k}u^* + \chi_1\frac{\partial^2\psi_1}{\partial X_k^2}u^*\right)dx\,dX$$
$$< -\frac{h^2}{8\,\pi^2\,M}\sum_k\iint\left|\frac{\partial u}{\partial X_k}\right|^2 dx\,dX - (E_{2m} - E)\iint u^2|\,dx\,dX \qquad (7)$$

erhalten. Hier kann man das erste Glied rechts weglassen, dadurch wird die rechte Seite sicher nur vergrössert, und es folgt mit Hilfe der BESSELschen Ungleichung

$$\frac{h^2}{8\,\pi^2\,M}\sqrt{\iint u^2|\,dx\,dX}\left(\sqrt{\iint\left|\sum_k 2\frac{\partial\chi_1}{\partial X_k}\frac{\partial\psi_1}{\partial X_k}\right|^2 dx\,dX}\right.$$
$$\left. + \sqrt{\iint\left|\sum_k\chi_1\frac{\partial^2\psi_1}{\partial X_k^2}\right|^2 dx\,dX}\right) > (E_{2m} - E)\iint u^2|\,dx\,dX. \qquad (8)$$

Dies kann man mit $(\iint u^2\,dx\,dX)^{1/2}$ kürzen, und die linke Seite folgendermassen abschätzen: $\int\sum_k\left(\frac{\partial\psi_1}{\partial X_k}\right)^2 dx$ ist eine Funktion der $X$, das für irgendein Wertesystem dieser ein Maximum hat, das wir mit $\gamma_1^2$ bezeichnen wollen. Ähnliches gilt von $\int\left(\sum_k\frac{\partial^2\psi_1}{\partial X_k^2}\right)^2 dx$, doch können wir das zweite Glied in (8), da es klein gegenüber dem ersten ist — $\chi_1$ weist ja wesentlich stärkere Schwankungen auf, als $\psi_1$ — überhaupt weglassen [vgl. auch (*)]. Dann bleibt aus (8):

$$\frac{h^2}{8\,\pi^2 M}\,2\,\gamma_1 \sqrt{\int \sum_k \left|\frac{\partial \chi_1}{\partial X_k}\right|^2 dX} > (E_{2\,m} - E)\sqrt{\iint |u^2\ dx\,d\overline{X}}. \qquad (9)$$

Nun müssen wir $\int \sum_k \left|\frac{\partial \chi_1}{\partial X_k}\right|^2 dX$ abschätzen. Zu diesem Zweck multiplizieren wir das konjugiert Komplexe von (6) mit $\chi_1 \psi_1$ und integrieren wieder. Den kleinsten Wert von $E_1(X)$ bezeichnen wir mit $E_{1\,m}$, das zweite Glied von (6) gibt nach einer partiellen Integration eben das, was wir für (9) brauchen. Wegen (3a) fällt das dritte Glied ganz weg und auch das vierte können wir weglassen, weil es klein dem zweiten gegenüber ist [die Schwankungen von $\psi_1$ sind viel kleiner als die von $\chi_1$, vgl. auch (*)]:

$$\frac{h^2}{8\,\pi^2 M} \int \sum_k \left|\frac{\partial \chi_1}{\partial X_k}\right|^2 dX + \iint \chi_1 \psi_1 H u^* dx\,dX < E - E_{1\,m}. \qquad (10)$$

Statt $\iint \chi_1 \psi_1 H u^* dx\,dX$ können wir auch $\iint u^* H \chi_1 \psi_1 dx\,dX$ schreiben, und dies haben wir schon in (7) berechnet. Das erste Glied der ersten Zeile von (7), das grösser als das zweite ist, ist wegen der BESSEL-schen Ungleichung kleiner als

$$\left.\begin{aligned} \iint u^* H \chi_1 \psi_1 dx\,dX &< \frac{h^2}{4\,\pi^2 M} \sqrt{\iint |u^2\ dx\,dX}\sqrt{\iint \left|\sum_k \frac{\partial \chi_1}{\partial X_k}\frac{\overline{\partial \psi_1}}{\partial X_k}\right|^2 dx\,dX} \\ &< \frac{h^2 \gamma_1}{4\,\pi^2 M} \sqrt{\iint |u^2\ dx\,dX}\sqrt{\int \sum_k \left|\frac{\partial \chi_1}{\partial X_k}\right|^2 dx\,dX} \end{aligned}\right\} (11)$$

und kann daher neben dem ersten Glied von (10) ebenfalls weggelassen werden. Es bedeutet dann (10), dass die kinetische Energie der Kerne kleiner ist als der maximale Überschuss der Gesamtenergie $E$ über die potentielle Energie. Setzt man (10) in (9) ein, so erhält man schliesslich

$$\iint |u^2| dx\,dX < \frac{4\,h^2 \gamma_1^2}{8\pi^2 M(E_{2\,m} - E)}\frac{E - E_{1\,m}}{E_{2\,m} - E}. \qquad (12)$$

Der letzte Faktor ist in der Grössenordnung 1 (nämlich etwa 0`1), wir müssen daher nur den ersten Faktor, d. h. $\gamma_1$ abschätzen.

Wir wollen diese Abschätzung nicht ganz streng ausführen. Da in die Definition von $\gamma_1^2$ ausser $h$, $e$ nur die Elektronenmasse eingeht, kann — abgesehen von einer numerischen Konstante — $\gamma_1$ nur $\frac{1}{a_0}$ sein, wenn $a_0$ der Radius der ersten BOHRschen Bahn ist. Um $\gamma_1$ doch etwas genauer abzuschätzen, kann man für $\psi_1(X, x)$ die erste Näherung einsetzen, die starren Atomen entspricht (die Näherung, die LONDON

H. Pelzer und E. Wigner

immer benutzt). Im Falle dreier Wasserstoffatome kann man dann die Integration leicht ausführen und erhält

$$\int \sum_k \left(\frac{\partial \psi_1(X,x)}{\partial X_k}\right)^2 dx = \frac{3}{a_0^2} \quad \text{und} \quad \int \left(\sum_k \frac{\partial^2 \psi_1}{\partial X_k^2}\right)^2 dx = \frac{11}{a_0^4}. \qquad (*)$$

In diesem Falle ist also schon das Integral (*) unabhängig von den Kernkoordinaten $X$, von den Abständen der drei Wasserstoffatome voneinander. Der Maximalwert der Integrale (*) ist also gleich ihrem Wert. Durch Einsetzen von $\gamma_1 = \dfrac{\sqrt{3}}{a_0}$ in (11) erhält man

$$\iint |u^2| \, dx \, dX < 0{\cdot}7 \cdot 10^{-3}. \qquad (**)$$

Die Wahrscheinlichkeit der angeregten Zustände, der zweiten und aller höheren Energieflächen zusammen, ist immerfort kleiner als diese Zahl. Man kann weiterhin auch sagen, dass die Grössenordnung der Wahrscheinlichkeit einer Anregung wirklich durch (**) gegeben ist, also nicht sehr viel kleiner ist, da die Abschätzungen, die wir vorgenommen haben, nur verhältnismässig kleine numerische Faktoren bedingen können (etwa einen Faktor $^1/_5$).

So günstig, wie bei drei $H$-Atomen liegt die Sache natürlich nur selten, weil die Anregungsspannung nur selten so gross gegen die Aktivierungsenergie ist, wie hier. Deshalb ist ja eine Anregung in diesem Falle scheinbar überhaupt unmöglich und wird erst durch die Wechselwirkungsenergie von Elektronenbewegung und Kernbewegung ermöglicht. Andererseits ist natürlich bei anderen Stoffen $M$ grösser als bei Wasserstoff, was die Wahrscheinlichkeit der angeregten Zustände wieder herabdrückt.

Wir wissen jetzt, dass die Wahrscheinlichkeit der angeregten Zustände klein ist. Dies schliesst natürlich noch nicht aus, dass die angeregten Zustände die Entwicklung des unangeregten Zustands wesentlich beeinflussen. Wir müssen daher noch untersuchen, wie gross der Fehler sein kann, den man macht, wenn man die Änderung von $\chi_1$ anstatt mit (5) mit

$$-\frac{h}{2\pi i}\frac{\partial \chi_1^0}{\partial t} = -\frac{h^2}{8\pi^2 M}\sum_k \frac{\partial^2 \chi_1^0}{\partial X_k^2} + E_1(X)\chi_1^0 - \frac{h^2}{8\pi^2 M}\sum_k \int \frac{\partial^2 \psi_1}{\partial X_k^2}\psi_1 \, dx \cdot \chi_1^0. \quad (5\,a)$$

berechnet. In (5a) wurde zur ersten Zeile von (5) noch ein Glied

$$-\frac{h^2}{8\pi^2 M}\sum_k \int \left(\frac{\partial \psi_1}{\partial X_k}\right)^2 dx$$

hinzugefügt, damit die weggelassenen Glieder $\chi_1$ nicht mehr enthalten sollen (wegen $\int \frac{\partial \psi_1}{\partial X_k} \psi_1 dx = 0$). Das Zusatzglied, das als ein Zusatzpotential aufgefasst werden kann, ist wie (*) zeigt, sehr klein, von der Grössenordnung von $0\cdot02$ Volt.

Berechnet man die Änderung von $\chi_1^0$ mit (5a), die von $\chi_1$ mit (5), d. h. setzt nach einiger Umformung

$$-\frac{h}{2\pi i}\frac{\partial \chi_1}{\partial t} = -\frac{h^2}{8\pi^2 M}\sum_k \frac{\partial^2 \chi_1}{\partial X_k^2} + E_1(X)\chi_1 - \frac{h^2}{8\pi^2 M}\sum_k \int \frac{\partial^2 \psi_1}{\partial X_k^2}\psi_1 dx \chi_1 \left. \right\} \quad (5\,\mathrm{b})$$

$$+\frac{h^2}{8\pi^2 M}\sum_k \int \left(2\frac{\partial u}{\partial X_k}\frac{\partial \psi_1}{\partial X_k} + u\frac{\partial^2 \psi_1}{\partial X_k^2}\right)dx,$$

so wird sich $\chi_1^0$ von $\chi_1$ im Laufe der Zeit entfernen. Aus (5a) und (5b) folgt für die Differenz

$$\frac{\partial}{\partial t}\int |\chi_1^0 - \chi_1|^2 dX =$$

$$-\frac{h}{2\pi M}J\sum_k \iint \left(2\frac{\partial u}{\partial X_k}\frac{\partial \psi_1}{\partial X_k} + u\frac{\partial^2 \psi_1}{\partial X_k^2}\right)(\chi_1^0 - \chi_1)^* dx\, dX, \left. \right\} \quad (12)$$

wo das $J$ bedeutet, dass vom darauffolgenden Ausdruck der Imaginärteil genommen werden soll. Es folgt, wenn man das zweite Glied rechts als klein dem ersten gegenüber weglässt, mit Hilfe der Besselschen Ungleichung:

$$\frac{\partial}{\partial t}\sqrt{\int |\chi_1^0 - \chi_1|^2 dX} < \frac{h\gamma_1}{2\pi M}\sqrt{\iint \sum_k \left|\frac{\partial u}{\partial X_k}\right|^2 dx\, dX}.$$

Nun folgt aus (7), wenn wir diesmal das zweite Glied rechts weglassen und für die linke Seite (11) einsetzen, und noch darin (12) und (10) berücksichtigen:

$$\frac{h^2}{8\pi^2 M}\sum_k \iint \left|\frac{\partial u}{\partial X_k}\right|^2 dx\, dX < \frac{h^2 \gamma_1^2}{2\pi^2 M}\frac{E - E_{1\,m}}{E_{2\,m} - E}. \quad (13)$$

Die Wahrscheinlichkeit, dass $\chi_1^0$ ein von $\chi_1$ verschiedener Zustand sein soll, ist daher kleiner als

$$\iint |\chi_1^0 - \chi_1|^2 dX < \left(\frac{h\gamma_1^2}{\pi M}\right)^2 \frac{E - E_{1\,m}}{E_{2\,m} - E}t^2. \quad (14)$$

Diese Grösse wächst zwar nach hinreichend langer Zeit über alle Grenzen an, es genügt aber, für $t$ die Dauer eines Zusammenstosses

$$\frac{2 a_0}{v} \approx \frac{a_0 \sqrt{M}}{\sqrt{E - E_{1\,m}}}$$

($v$ ist die Geschwindigkeit) einzusetzen, nach Ablauf dieser Zeit werden sich ja die Reaktionsteile voneinander entfernt haben und es wird überhaupt keine Umsetzung mehr stattfinden können. Dass unsere Formel diesen Tatbestand nicht wiedergibt, beruht darauf, dass wir unabhängig von $t$ mit den grösstmöglichen Werten der für die Störung massgebenden Grössen gerechnet haben. Wir können daher [mit Hilfe von (*)] für die Reaktion $H + H_2 \rightleftarrows H_2 + H$

$$\int |\chi_1^0 - \chi_1|^2 dX < \frac{9\,h^2}{\pi^2 M \cdot a_0^2} \frac{1}{E_{2\,m} - E} \sim \frac{1}{10} \qquad (\dagger)$$

schreiben. Dies ist $72\,\dfrac{m}{M}$-mal das Verhältnis einer atomaren Energiegrösse $\dfrac{h^2}{8\,\pi^2 m a_0^2}\, m a_0^2$ zu einer anderen $E_{2\,m} - E$. Diese Zahl wäre für schwerere Kerne noch kleiner und zeigt, dass man die Vorstellung von Energieflächen für die Atome sehr gut verwenden kann.

Die Abschätzung (†) ist im Gegensatz zu der Abschätzung (**) eine sehr schlechte, weil sie gewissermassen annimmt, dass die Störung immer die günstigst mögliche Phase hat. Diese Annahme ist in dem Übergang von (12) zu (14) enthalten. Die Abweichung von $\chi_1^0$ von $\chi_1$ ist in Wirklichkeit wesentlich kleiner, als die Grenze in (14) es erlaubt, weil die Störungen im Laufe der Zeit sich nicht immer stärken, sondern zuweilen auch durch Interferenz schwächen werden. Wir wollen aber hierauf an dieser Stelle nicht eingehen und nur noch bemerken, dass noch ein Effekt in unserer Rechnung nicht ganz sinngemäss berücksichtigt worden ist.

Nachdem sich nämlich die Reaktionspartner getrennt haben, können offenbar keine Elektronenanregungen mehr stattfinden und alle Partner müssen — da die Energie zu ihrer Anregung nicht hinreicht — im Grundzustand sein. Die Glieder in (5), die die Übergänge zwischen den verschiedenen Energieflächen bedeuten, verschwinden aber auch für weit getrennte Reaktionspartner nicht. Dies hat seine Ursache darin, dass die von uns als adiabatisch bezeichneten Bewegungen längs der Energieflächen solchen Bewegungen entsprechen, bei denen die „Mitbewegung des Kerns" für die einzelnen Reaktionspartner nicht berücksichtigt ist. Bei Berücksichtigung der Mitbewegung ist ja die Wellenfunktion $\psi_1(X, x)\chi(MX + mx)$ nicht $\psi_1(X, x)\chi(X)$. Solange die Atome nahe zueinander sind, kann man die Mitbewegung natürlich nicht abtrennen, weil man nicht weiss, welcher Kern von welchen Elektronen mitbewegt wird, und tatsächlich erzeugt die Mitbewegung während dieser Zeit Übergänge zwischen den Energieflächen. Wenn sich die Atome trennen, sollte aber die als adiabatisch bezeichnete Bewegung automatisch in eine solche übergehen, bei der jeder Kern von seinen eigenen Elektronen mitbewegt wird. Doch scheint es nicht ganz einfach zu sein, dies zu erreichen.

Natürlich ist mit dieser Rechnung über die Frage der Sprünge zwischen zwei Energieflächen nur sehr wenig ausgesagt. Erstens gibt es Fälle, in denen der Sprung in eine andere Energiefläche nicht nur ein häufiger Vorgang ist (ein Beispiel hierfür findet sich am Schluss dieser Arbeit), sondern sogar mit Sicherheit auftritt. Die unterste Energiefläche wird ja nur selten so weit getrennt von den übrigen Energieflächen liegen, wie in unserem Beispiel, häufig werden sich die Energieflächen fast überschneiden.

Aber auch wenn unsere Voraussetzungen bezüglich der Lage der Energieflächen zutreffen, würde noch der Fall eine gesonderte Untersuchung erheischen, dass eine Umsetzung auf der untersten Energiefläche sehr unwahrschein-

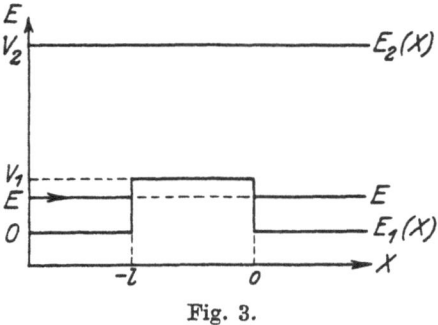

Fig. 3.

lich ist. Dies kann z. B. vorkommen, wenn die Energie $E$ zu klein ist zur Überschreitung des Potentialwalles, der der Reaktion im Wege steht. So ein Fall wird schematisch in Fig. 3 veranschaulicht.

Man könnte denken, dass die Welle, die auf der zweiten Energiefläche ist, ungehindert nach rechts hinüberlaufen kann, und dort wieder auf die untere Fläche hinunterfallen wird. Die Wahrscheinlichkeit einer Umsetzung auf diesem Wege wäre zwar im Sinne der vorangehenden Ausführungen klein, da sie sich aber schon bei niedrigerer Energie abspielen könnte, als die normale Umsetzung, würde sie doch meistens vorherrschen, weil eine Herabdrückung der Aktivierungswärme die Geschwindigkeit der Reaktion viel mehr erhöht, als sie die Einführung eines sterischen Faktors erniedrigt.

Wir haben die Rechnung für die zwei Energieflächen der Fig. 3 unter der Voraussetzung ausgeführt, dass die beiden Integrale (*) konstant sind. Dabei hat sich gezeigt, dass die Welle auf der oberen Energiefläche ebenso zurückgeworfen wird, wie auf der unteren. Dies beruht darauf, dass an den Stellen, wo $\chi_1$

verschwindet, für $\chi_2$ eine Differentialgleichung mit einer Eigenwertkonstante $E$ gilt, die kleiner ist, als alle Eigenwerte der Differentialgleichung (diese sind ja alle grösser als $E_{2\,m}$). Daher existieren zwei Lösungen, deren eine, die exponentiell ansteigende, natürlich ausgeschlossen ist. Die andere fällt aber exponentiell ab und verschwindet praktisch innerhalb 0˙05 Å.

4. Wir wollen jetzt zu unserem eigentlichen Gegenstand übergehen und die absolute Grösse (die sogenannte temperaturunabhängige Konstante) der Reaktion $H + H_2 = H_2 + H$ berechnen.

Anstatt die Änderung von $\chi_1$ mit (5a) zu berechnen, wollen wir klassisch rechnen, weil die klassische Rechnung viel leichter ist. Die klassische Rechnung wird — wegen der Nullpunktsenergie — eine unrichtige Aktivierungswärme geben, der sterische Faktor, für den einige Prozent Abweichungen nichts ausmachen, wird aber wohl nicht sehr falsch herauskommen.

Wir wollen dabei von dem Eyring-Polanyischen Diagramm (Fig. 1) ausgehen. Dieses Diagramm gilt für dieReaktion

$$Y + XZ = YX + Z$$

für den Fall, dass das freie Atom $Y$ in der Richtung der Verbindungslinie der beiden Molekülatome $X$ und $Z$ herankommt, weil dieser Weg die geringste Aktivierungsenergie benötigt. In den Diagrammen sind längs der Koordinatenachse die Abstände $b$ und $c$ der beiden äusseren Atome vom Mittelatom eingetragen und die Niveaulinien gleicher Energie $W$ des Gesamtsystems stellen eine Sattelformation dar, die die beiden Täler des Anfangs- und Endzustands trennt; die Höhe des Sattelpunktes entspricht dabei (wenn man noch die Nullpunktsenergie berücksichtigt, so wie dies in der Eyring-Polanyischen Arbeit näher ausgeführt wird) der Aktivierungsenergie. Sie wurde dort für die drei Reaktionen $H + H_2$ (para)$= H_2 + H$, $H + HBr = H_2 + Br$ und $H + Br_2 = HBr + Br$ nach einem graphischen Verfahren bestimmt, wobei die Niveaulinien $W$ berechnet wurden nach der Formel:

$$W = D - V\overline{\alpha^2 + \beta^2 + \gamma^2 - \alpha\beta - \beta\gamma - \gamma\alpha}. \tag{15}$$

Hierin bedeutet $D$ die um die Nullpunktsenergie vermehrte Dissoziationsenergie des Ausgangszustands; die Wechselwirkungsenergien

$$\alpha\,(b+c); \quad \beta\,(b); \quad \gamma\,(c)$$

wurden nach der Morseschen Gleichung für die Potentialkurve zweiatomiger Moleküle als Funktion der Abstände $b$, $c$, $b+c$ berechnet.

Über die Geschwindigkeitskonstante von Austauschreaktionen.     459

Es könnte gegen die EYRING-POLANYIsche Betrachtungsweise Bedenken verursachen, dass nur Reaktionswege betrachtet werden, bei denen das dritte Atom in der Verbindungslinie der beiden Molekülatome herankommt. Denn dies ist beispielsweise sicher nicht richtig für die Umwandlung von Parawasserstoff in normalen, weil dabei der Drehimpuls sich ändern muss. Ausserdem könnte man befürchten, dass, wenn nur dieser Weg erlaubt wäre, zu selten ein Zusammenstoss eintritt, bei dem Reaktion möglich ist. Es ist mit eine Aufgabe vorliegender Arbeit, sich von der Bedingung freizumachen, dass die drei Atome genau auf einer Geraden liegen müssen.

Wir stellen uns nun also als nächste Frage die nach der Geschwindigkeitskonstante, nachdem in der EYRING-POLANYIschen Arbeit gezeigt wurde, wie die Aktivierungswärme zu berechnen ist. Oder anders ausgedrückt: Wie ändern sich quantitativ die Sattelverhältnisse, wenn man auch Zusammenstösse berücksichtigt, bei denen die drei Atome nicht in einer Geraden liegen?

Der Gedanke, der dieses Problem lösen soll, ist der folgende. Wir denken uns die Wechselwirkungsenergie in der Umgebung des Sattelpunktes entwickelt als Funktion der beiden Abstände $b$ und $c$ und des Winkels $\delta$ zwischen $b$ und $c$. Wenn wir nun auch diese Funktion dreier Variabler uns natürlich nicht als Fläche im dreidimensionalen Raum vorstellen können, so bedeutet sie gleichwohl eine Sattelformation in einem Raum höherer Dimensionszahl und es wird auch bei ihr einen „Grat" geben, der die beiden Talgebiete niedriger Energie voneinander trennt und dessen tiefster Punkt unser Sattelpunkt ist; bloss wird unser Grat keine eindimensionale Linie sein, wie bei den Gebirgen unserer dreidimensionalen Umwelt, sondern er wird ein Gebilde von zwei Dimensionen sein. Aber wir wollen unsere Überlegungen im folgenden so aussprechen, dass man bei der anschaulichen Vorstellung eines Gebirgssattels im gewöhnlichen Raum auch alles Wesentliche sehen kann.

Um die Frage nach der Geschwindigkeitskonstante zu beantworten — das ist die Frage nach der sekundlichen Zahl der Prozesse, bei denen das trennende Sattelgebiet in der Richtung vom Ausgangstal nach dem Endtal überschritten wird —, stellen wir uns zunächst vor, wir hätten im Phasenraum des Bildpunktes Gleichgewicht, bei dem eben soviel Überschreitungen aus dem Ausgangstal in das Endtal vorkommen, wie umgekehrt aus dem Endtal in das Ausgangstal zurück. Da ist es nicht schwer, nach der BOLTZMANNschen Formel

für eine bestimmte Temperatur die Dichte der Bildpunkte in unserem
Sattelgebiet und seiner Umgebung auf den Grathöhen anzugeben.
Ferner können wir sagen, wie gross bei dieser Temperatur die Geschwindigkeiten unserer Bildpunkte sein werden, und zwar interessieren wir uns gerade für diejenige Geschwindigkeitskomponente,
welche auf dem Grat orthogonal steht und aus dem einen Tal ins
andere führt. Wir können sie über den Geschwindigkeitshalbraum,
der aus dem Ausgangstal in das Endtal führt, mitteln, und da wir,
wie gesagt, auch die Dichte der Bildpunkte überall auf dem Grat
kennen, können wir sofort sagen, wieviele Bildpunkte pro Sekunde
bei Gleichgewicht aus dem Ausgangstal in das Endtal hinüberwandern.
Die Dichte der Bildpunkte und damit der Überschreitungen nimmt
nach der BOLTZMANN-Formel sehr rasch (exponentiell) ab, wenn wir
aus dem Sattelpunkt den Grat hinaufklettern, so dass man sie sehr
bequem über das gesamte trennende Gebiet der Grathöhen weg-
integrieren kann, um die Zahl der Überschreitungen festzustellen.
Daher werden wir uns (wegen dieser exponentiellen Abnahme der
Dichte) bei unserer Entwicklung der Energie nach den Abständen in
der Umgebung des Sattelpunktes mit den quadratischen Gliedern begnügen, was die Rechnung sehr erleichtert. Die Zahl der so berechneten Übergangsprozesse beim Gleichgewicht ist aber bekanntlich
dieselbe, wie wenn die Rückreaktion fehlen würde, weil das End-
produkt noch nicht vorhanden ist. Mithin haben wir die Zahl der
Übergangsprozesse berechnet, womit die Frage nach der Geschwindigkeitskonstante beantwortet ist.

Was wir in dieser Weise berechnen, ist streng genommen die
Anzahl der sekundlichen Gratüberschreitungen im Gleichgewicht, während wir uns sehr weit vom Gleichgewicht befinden. Das Tal II ist
ja ganz leer, obwohl es im Gleichgewicht ebenso viele Bildpunkte hat,
wie das Tal I. Natürlich ist die Zahl der Gratüberschreitungen in
Richtung des Tals II unabhängig von der Besetzungsdichte dieses,
und hängt nur von der Besetzungsdichte des Tals I ab, wo wir Gleichgewicht annehmen wollen. Wir rechnen aber so, als ob überhaupt
keine Gratüberschreitungen in umgekehrter Richtung in das Tal I
vorkommen würden. Dies ist, obwohl das Tal II anfangs ganz leer
ist, nicht ganz berechtigt, weil es im allgemeinen möglich ist, dass
ein Bildpunkt, der den Grat eben nach Tal II überschritten hat, von
dort sofort in das Tal I zurückgeworfen wird. Einen solchen Fall
würden wir als Reaktion mitrechnen, weil wir ja nur die Zahl aller

Gratüberschreitungen nach Tal II berechnen können. Ein Blick auf die Fig. 1 lehrt aber, dass solche Reflexionen des Bildpunktes kaum vorkommen können.

Wir haben nun den vorher ausgeführten Gedanken durchzuführen. Wir wollen es an dem Beispiel der Umwandlung des Parawasserstoffs in normalen ausführen, für den A. FARKAS[1]) den Mechanismus

$$H + H_2 \,(\text{para}) = H_2 \,(\text{normal}) + H$$

angegeben hat. Bei der Rechnung wollen wir nur die Wahrscheinlichkeit der Umsetzung eines Moleküls $H_2$ mit einem Atom $H$ berechnen, da die Reaktionsgeschwindigkeit im Falle mehrerer Teilchen doch proportional zur Konzentration ist. Der erste Schritt hierbei ist die Entwicklung der Energie in der Umgebung des Sattels.

Es gilt jetzt nicht mehr $a = b + c$, da wir uns ja gerade von der Voraussetzung, dass die Atome in geradliniger Anordnung reagieren, freimachen wollen. Nach dem Cosinussatz haben wir vielmehr:

$$a^2 = b^2 + c^2 + 2bc \cos \delta.$$

Am Orte des Sattelpunktes sei $b = c = b_0$ und $\delta = 0$. In der Umgebung dieses Punktes haben wir dann die Lage gegeben durch die Grössen

$$b = b_0 + B; \quad c = c_0 + C; \quad \delta, \tag{16}$$

wobei wir nach $B, C, \delta$ entwickeln wollen und uns mit Gliedern zweiter Ordnung begnügen. Dann wird

$$a = b_0 + B + c_0 + C - \frac{b_0 c_0}{2\,(b_0 + c_0)}\,\delta^2.$$

Und in der zu entwickelnden Funktion

$$\left.\begin{aligned} W &= D - V\overline{a^2 + \beta^2 + \gamma^2 - \alpha\beta - \beta\gamma - \gamma\alpha} \\ &= D - V\overline{f(a)^2 + f(b)^2 + f(c)^2 - f(a)f(b) - f(b)f(c) - f(c)f(a)}, \end{aligned}\right\} \tag{17}$$

wobei $f(r)$ die MORSEsche Funktion für das Wasserstoffatom bedeutet:

$$f(r) = D(e^{-2k(r-r_0)} - 2e^{-k(r-r_0)}).$$

$D =$ Dissoziationsenergie[2]) + Nullpunktsenergie, $r_0 = 0{\cdot}75\ \text{Å} =$ normaler Kernabstand, $k = 2{\cdot}09\ \text{Å}^{-1}$ ergibt sich aus der Grösse des Grund-

[1]) A. FARKAS, Z. physikal. Ch. (B) 10, 419. 1930.    [2]) Der in der EYRING-POLANYIschen Arbeit auf S. 286 angegebene Wert 39132 = 111 kcal ist zu ersetzen durch 37000 = 107 kcal, wie es übrigens auch den Rechnungen in dieser Arbeit richtig zugrunde gelegt wurde.

462                          H. Pelzer und E. Wigner

schwingungsquants (in der Tabelle auf S. 286 der Arbeit von EYRING und POLANYI gilt die Angabe $cm^{-1}$ für die letzte Spalte natürlich nicht mehr, $k$ ist in $Å^{-1}$ ausgedrückt).

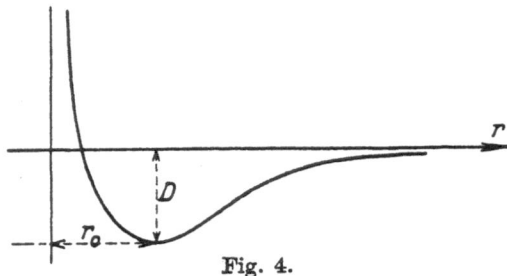

Fig. 4.

Wir haben $W$ um die Stelle $B = C = \delta = 0$ zu entwickeln; da an dieser Stelle ein Sattelpunkt ist, verschwinden dort die ersten Differentialquotienten und unsere Entwicklung beginnt mit den quadratischen Gliedern, mit welchen wir uns auch begnügen:

$$W(b_0 + B, c_0 + C, \delta) = W(b_0, c_0; \delta = 0) + \frac{1}{2}\left[\frac{\partial^2 W}{\partial B^2}B^2 + \frac{\partial^2 W}{\partial C^2}C^2 + \frac{\partial^2 W}{\partial \delta^2}\delta^2\right] + \frac{\partial^2 W}{\partial B \partial C}B \cdot C.$$

(Die gemischten Differentialquotienten nach $\delta$ und einem der Abstände $B$ und $C$ fallen fort.) Ferner ergibt sich in dem von uns betrachteten Falle dreier Wasserstoffatome die weitere Vereinfachung, dass $b_0 = c_0$ ist[1]), und man daher nach dem Differenzieren $\beta = \gamma$ setzen kann, wodurch der Wurzelausdruck den einfachen Wert $\alpha - \beta$ bekommt.

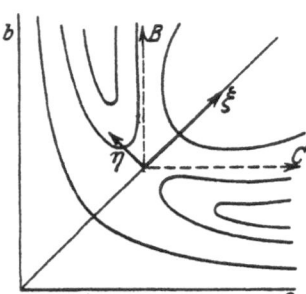

Fig. 5. Transformation von den Grössen $b$, $c$ auf $B$, $C$ und schliesslich auf Hauptachsenkoordinaten $\xi$, $\eta$.

Wir haben nun die Differentialquotienten zu berechnen und sie in (3) einzusetzen. Man erhält so

$$W = W_0 + D[0.65(B^2 + C^2) + 1.49 BC + 0.09\delta^2]. \quad (18)$$

Dass die quadratische Form in der eckigen Klammer bezüglich $B$ und $C$ wirk-

---

[1]) In der Arbeit von EYRING und POLANYI ist der auf S. 290 angegebene Wert $0.95$ Å für den Abstand $b_0 = c_0$ zu ersetzen durch den Wert $0.91$ Å, wie es ja auch aus der dortigen Fig. 7 auf S. 291 zu entnehmen ist.

lich einen Sattel bedeutet und nicht etwa eine Mulde, sieht man, wenn man durch die Substitution

$$B = \frac{(\xi + \eta)}{\sqrt{2}}$$
$$C = \frac{(\xi - \eta)}{\sqrt{2}}$$
$$\left.\right\}\qquad(19)$$

auf Hauptachsen transformiert. Schreiben wir nämlich unsere quadratische Form in (18) allgemein:

$$W = W_0 + l\,(B^2 + C^2) + mBC + u\delta^2, \qquad (18\,\text{a})$$

so wird durch die Substitution (5) daraus:

$$W = W_0 + \left(l + \frac{m}{2}\right)\xi^2 - \left(\frac{m}{2} - l\right)\eta^2 + u\delta^2 = p\xi^2 - q\eta^2 + u\delta^2; \quad (18\,\text{b})$$

da $m = 1{\cdot}49\,D > 2\,l = 1{\cdot}3\,D$, so bedeutet dies tatsächlich Hyperbeln als Niveaulinien und $\xi$ bedeutet die Gratlinie, $\eta$ die Talverbindungslinie.

Schon jetzt können wir den sterischen Faktor abschätzen. Wir sehen aus (18), dass der Koeffizient von $\delta^2$ etwa die Grösse $u = \dfrac{D}{10}$ besitzt. Nähert sich das freie Atom genau in der Richtung der Kernverbindungslinie, so muss es die Aktivierungsenergie $W_0$ besitzen, um reagieren zu können; kommt es aber in einen Winkel $\delta$ gegen die Kernverbindungslinie heran, so muss es also um $\dfrac{D}{10}\delta^2$ mehr Energie besitzen, um erfolgreich zu sein. Nun ist aber allgemein bei zwei quadratischen Freiheitsgraden die mittlere Überschreitung einer beliebig vorgegebenen Energie $W_0 = RT$[1]). Wir fragen nun einfach,

---

[1]) Man sieht das ein, wenn man bedenkt, dass der Schwerpunkt der unter der Kurve $e^{-\frac{\varepsilon}{RT}}$ und der $\varepsilon$-Achse liegenden Fläche immer im Abstand $RT$ von der sie linksseitig begrenzenden Abszisse entfernt liegt (vgl. Fig. 6):

$$y = \frac{\displaystyle\int_{W_0}^{\infty} \varepsilon \cdot \varepsilon^k\, e^{-\frac{\varepsilon}{RT}}}{\displaystyle\int_{W_0}^{\infty} \varepsilon^k\, e^{-\frac{\varepsilon}{RT}}} \approx W_0 + RT + \text{Glieder mit } \frac{(RT)^2}{W_0}\ \text{usw.}$$

464                    H. Pelzer und E. Wigner

welchen Bruchteil der Einheitskugel diese mittlere Überschreitung bedeckt. Dazu muss also sein $\frac{D}{10}\delta^2 = RT$, oder mit $D = 1\cdot07\cdot10^5$ cal; $T \sim 300°$ abs. (entsprechend den Temperaturen bei GEIB und HARTECK)

$$\delta^2 \approx 0\cdot05.$$

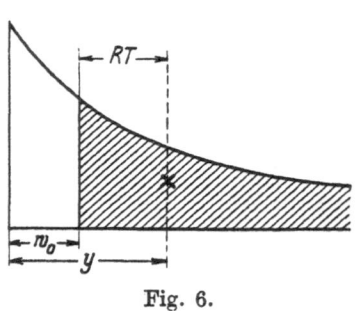
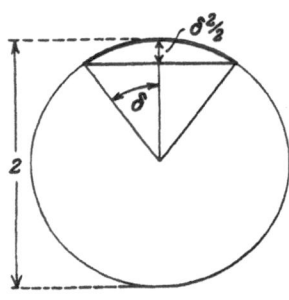

Fig. 6.                                    Fig. 7.

Bis zu diesem Winkel werden also im Mittel erfolgreiche Stösse vorkommen; nun verhält sich die in der Fig. 7 gezeichnete Calotte zur ganzen Kugeloberfläche wie die zugehörige Höhe $\frac{\delta^2}{2}$ zum ganzen Durchmesser 2; sie erfüllt also den Bruchteil $\frac{\delta^2}{4} = \frac{1}{80}$ der ganzen Oberfläche; da das Atom aber auch noch auf der anderen Seite herankommen kann, ist der sterische Faktor doppelt so gross, also $\sim \frac{1}{40}$. Experimentell finden FARKAS[1]) sowie GEIB und HARTECK[2]) etwa das Fünffache.

### Berechnung der Dichte im Phasenraum.

Solange die drei Wasserstoffatome als unabhängig betrachtet werden, und wir ihnen die kartesischen Koordinaten zuteilen:

$$x_1\ y_1\ z_1 \left.\right| \text{ sind im Abstand } b \text{ voneinander}$$
$$\text{Mittelatom:}\quad x_2\ y_2\ z_2 \left.\right\} \quad \text{(ursprüngliches Molekül)}$$
$$\text{ursprünglich freies Atom:}\quad x_3\ y_3\ z_3\,,$$

ist das Volumenelement des Konfigurationsraums einfach gegeben durch

$$dx_1\,dy_1 \ldots dz_3$$

[1]) FARKAS, Z. physikal. Ch., loc. cit.     [2]) GEIB und HARTECK, Z. physikal. Ch., BODENSTEIN-Festband, 849. 1931.

und die Zahl der in ihm enthaltenen Bildpunkte ist ihm proportional. Um aber ·die Wechselwirkungsenergie bequemer berücksichtigen zu können, müssen wir auf neun andere Koordinaten transformieren, und wir wählen als solche

1. Wie oben die drei kartesischen Koordinaten des Mittelatoms 2:

$$\left.\begin{aligned} x_2 &= X \\ y_2 &= Y \\ z_3 &= Z \, . \end{aligned}\right\} \tag{19 a}$$

2. Für das Atom 1, das mit 2 zusammen das Ausgangsmolekül bildet, Polarkoordinaten $b$, $\vartheta$ und $\varphi$ mit Atom 2 als Ursprung:

$$\left.\begin{aligned} x_1 &= X - b \sin \vartheta \cos \varphi \\ y_1 &= Y - b \sin \vartheta \sin \varphi \\ z_1 &= Z - b \cos \vartheta \, . \end{aligned}\right\} \tag{19 b}$$

3. Für das Atom 3 Polarkoordinaten $c$, $\delta$ und $\psi$, ebenfalls mit 2 als Ursprung, aber mit der Molekülachse als Polarachse:

$$\left.\begin{aligned} x_3 &= X + c \sin \delta \cos \psi \cos \vartheta \, \cos \varphi - c \sin \delta \sin \psi \sin \varphi + c \cos \delta \sin \vartheta \cos \varphi \\ y_3 &= Y + c \sin \delta \cos \psi \cos \vartheta \, \sin \varphi + c \sin \delta \sin \psi \cos \varphi + c \cos \delta \sin \vartheta \sin \varphi \\ z_3 &= Z - c \sin \delta \cos \psi \sin \vartheta \qquad\qquad\qquad\qquad\quad + c \cos \delta \cos \vartheta \, . \end{aligned}\right\} \tag{19 c}$$

Die Funktionaldeterminante ergibt sich nach einiger Rechnung zu

$$F = \frac{\delta \, (x_2 y_2 z_2 \, x_1 y_1 z_1 \, x_3 y_3 z_3)}{\delta \, (X \, Y \, Z \, b \, \vartheta \, \varphi \, c \, \vartheta \, \psi)} = b^2 \sin \vartheta \, c^2 \sin \delta \, .$$

Das Volumenelement in den neuen Koordinaten ist also:

$$d \, v = b^2 c^2 \, d b \, d c \, \sin \delta \, d \delta \, \sin \vartheta \, d \vartheta \, d \psi \, d \varphi \, d X \, d Y \, d Z \, .$$

Wir denken uns nun ein Molekül und ein Atom eingeschlossen in einen Kasten vom Volumen $V$. Dann hat der Konfigurationsraum neun Dimensionen: $X$, $Y$, $Z$, $b$, $\vartheta$, $\varphi$, $c$, $\delta$, $\psi$. Um die Wahrscheinlichkeit dafür zu berechnen, dass der Bildpunkt sich eben auf dem Grat befindet, müssen wir den allgemeinen Ausdruck für die Wahrscheinlichkeit:

$$\text{const} \, e^{-\frac{W}{kT}} \, b^2 c^2 \, d b \, d c \, \sin \delta \, d \delta \, \sin \vartheta \, d \vartheta \, d \psi \, d \varphi \, d X \, d Y \, d Z \tag{20}$$

über einen Streifen von der Breite $\varDelta \eta$ längs des Grates integrieren. Die Const bestimmt sich daraus, dass das Integral über den ganzen Konfigurationsraum gleich Eins ist. Da der Ausdruck (20) überall viel kleiner ist, als in der Nähe der Stelle $b = r_0$, was dem normalen Abstand der beiden $H$-Atome im Molekül entspricht, brauchen wir

466                        H. Pelzer und E. Wigner

bei der Bestimmung der Const nur über diese Gegend zu integrieren.
Dabei können wir $W = s(b-r_0)^2$ setzen, wo $s$ die elastische Konstante
des $H_2$-Moleküls ist. Die Integration über $X, Y, Z$ und $c, \delta, \psi$ ergibt $V^2$,
die Integration $\vartheta, \psi$ den Faktor $4\pi$. Wir erhalten daher

$$\text{const } 4\pi V^2 \int_0^\infty e^{-\frac{s(b-r_0)^2}{kT}} b^2 \, db = 1.$$

Hier können wir für $b^2 db = r_0^2 db$ schreiben und erhalten

$$\text{const} = \frac{1}{4\pi V^2} \frac{1}{r_0^2} \sqrt{\frac{s}{\pi k T}}.$$

Dies müssen wir in (20) einsetzen. Um die Integration längs des
Grates ausführen zu können, führen wir nach (19) wieder

$$b - b_0 = B = \frac{(\xi + \eta)}{\sqrt{2}}; \quad C = \frac{(\xi - \eta)}{\sqrt{2}}$$

ein und erhalten für die Anzahl der Bildpunkte auf den beiden Graten
(der eine entspricht der Annäherung des freien Atoms von der Seite
des Atoms 1, der andere einer Annäherung von der Seite des Atoms 2
an das Molekül)

$$n = \frac{\Delta \eta}{2\pi V^2 r_0^2} \sqrt{\frac{s}{k T}} \int e^{-\frac{W}{kT}} b^2 c^2 d\xi \int \sin\delta \, d\delta \int \sin\vartheta \, d\vartheta \, d\psi \, d\varphi \, dX \, dY \, dZ. \quad (21)$$

Hier ist wieder $\Delta \eta$ die Breite des Gebiets, das wir noch zum
Grat rechnen. Für $W$ muss hierbei die Entwicklung (18b) eingesetzt
werden, die in der Nähe der
tiefsten Stelle des Grates gilt.
Die höheren Stellen des Grates
sind ja so unwahrscheinlich,
dass wir sie ausser acht lassen
können, die überwiegende Mehr-
zahl der Bildpunkte des Grates
ist in der Nähe der Stelle
$b = c \approx b_0$, $\delta = 0$, wenn $b_0$ der
Abstand der seitlichen Atome
vom Mittelatom für die Sattel-
konfiguration ist.

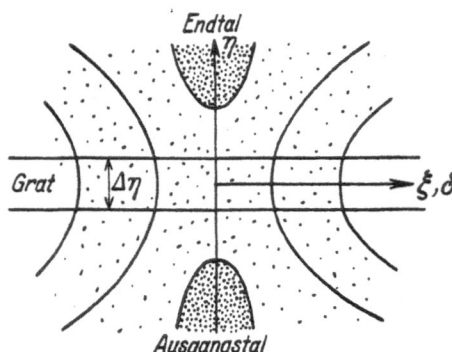

Fig. 8. Die Umgebung des Sattels. Nach
rechts und links, gegen die beiden Grathöhen
zu, nimmt die Besetzung mit Bildpunkten
sehr rasch ab.

Die Integration über $X$,
$Y, Z, \vartheta, \varphi$ gibt $4\pi V$, die
über $\psi$ gibt $2\pi$, für $\sin\delta$

schreiben wir $\delta$ und erhalten so (längs des Grates ist $\eta = 0$) mit Hilfe von (18b)

$$n = \frac{2\pi \cdot 2\pi \, V \cdot \Delta\eta}{\pi \, V^2 \, r_0^2} \sqrt{\frac{s}{\pi \, k T}} \, b_0^4 \, e^{-\frac{\pi_0}{kT}} \int e^{-\frac{(p\,\dot\xi^2 + u\,\delta^2)}{kT}} \, \delta \, d\xi \, d\delta \, .$$

$$n = \frac{4\pi \, \Delta\eta \, b_0^4}{V \, r_0^2} \sqrt{\frac{s}{p}} \frac{kT}{u} \, e^{-\frac{\pi_0}{kT}} \, . \tag{22}$$

Dies ist die Zahl der Bildpunkte auf dem Grat. Ein Bildpunkt, der die Geschwindigkeit $\dot\eta$ senkrecht zum Grat besitzt, braucht die Zeit $\frac{\Delta\eta}{\dot\eta} = \tau$ zum Durchlaufen des Gratgebiets. Ist der Bruchteil der Bildpunkte mit der Geschwindigkeitskomponente senkrecht zum Grat zwischen $\dot\eta$ und $\dot\eta + d\dot\eta$ gleich $w(\dot\eta)\,d\dot\eta$, so passieren aus diesem Geschwindigkeitsbereich pro Sekunde $n\,w(\dot\eta)\frac{d\dot\eta}{\tau}$ den Grat. Die gesamte Reaktionsgeschwindigkeit ist daher

$$\frac{n}{\Delta\eta} \int \dot\eta \, w(\dot\eta) \, d\dot\eta = \frac{4\pi \, b_0^4}{V \, r_0^2} \sqrt{\frac{s}{p}} \frac{(kT)^{3/2}}{u} \cdot \sqrt{\frac{3}{2\pi M}} \cdot e^{-\frac{\pi_0}{kT}} \, . \tag{23}$$

wo für $\int \dot\eta \, w(\dot\eta)\,d\dot\eta$ sein Wert eingesetzt wurde (vgl. den Anhang).

Das ist die relative Zahl der $H_2$-Moleküle, die pro $H$-Atom im Volumen $V$ pro Sekunde umgesetzt werden.

5. Aus den Versuchen von A. FARKAS und GEIB und HARTECK ergibt sich folgende Tabelle für die Geschwindigkeitskonstante $k_3$ in

$$\frac{1}{p} \frac{dp}{dt} = - k_3 c ,$$

wo $p$ für den Augenblick die Konzentration des Parawasserstoffs und $c$ die der Wasserstoffatome in Mol/Liter bedeutet.

| $T$ | $k_3$ | $\ln \frac{k_3}{T^{3/2}}$ | $C$ |
|---|---|---|---|
| 283 | $8\cdot 6 \cdot 10^4$ | $2\cdot 9$ | $2 \cdot 10^6$ |
| 373 | $2\cdot 7 \cdot 10^6$ | $5\cdot 9$ | $5 \cdot 10^6$ |
| 873 | $1\cdot 2 \cdot 10^9$ | $10\cdot 7$ | $2 \cdot 10^6$ |
| 1023 | $1\cdot 9 \cdot 10^9$ | $11\cdot 0$ | $1\cdot 5 \cdot 10^6$ |

Aus den Zahlen der 3. Spalte erhält man für die Aktivierungswärme $W_0$ Werte zwischen 6300 und 7100 cal. Das ist etwas weniger, als GEIB und HARTECK erhalten, weil sie eine Formel verwenden, die $T^{1/2}$ enthält an Stelle unseres $T^{3/2}$, doch sei erwähnt, dass die Aktivierungswärme, wie wir sie berechnen, mehr streut, als nach

468    H. Pelzer und E. Wigner

Geib und Hartecks Berechnungsweise. Mit Hilfe einer Aktivierungswärme von 6600 cal erhält man dann die Werte der letzten Spalte für die Konstante $C$ in der Formel

$$k_3 = C T^{3/2} e^{-\frac{W_0}{RT}}.$$

Aus (23) ergibt sich dagegen

$$C = 6 \cdot 10^{20} \cdot \frac{4\pi b_0^4}{r_0^2} \sqrt{\frac{s}{p}} \frac{k^{3/2}}{u} \sqrt{\frac{3}{4\pi M}} \qquad (23\,\mathrm{a})$$

oder mit $b_0 = 0{\cdot}91 \cdot 10^{-8}$ cm, $r_0 = 0{\cdot}75 \cdot 10^{-8}$ cm, $\dfrac{u}{k} = \dfrac{D}{10R} = 5 \cdot 10^3$,

$M = 1{\cdot}6 \cdot 10^{-24}$ g, $\sqrt{\dfrac{s}{p}} = 1{\cdot}75$

$$C = 2{\cdot}1 \cdot 10^6.$$

Die gute Übereinstimmung zwischen beobachtetem und berechnetem $C$ wird man zweifellos zum grossen Teil einem Zufall zuschreiben, wenn man die vielen Fehlerquellen in Betracht zieht.

Als Fehlerquellen in unserer Rechnung kommen drei Ursachen in Frage. Erstens die Unsicherheit im Energieausdruck (18b) betreffs der Konstanten $b_0$, $u$, $p$. Doch zeigt (23a), dass nur die erste Grösse sehr wesentlich ins Gewicht fällt, und diese ist wohl richtig bestimmt: sie streut bei allen Rechnungen von Eyring und Polanyi nur sehr wenig, obwohl diese Rechnungen unter sehr verschiedenen Annahmen durchgeführt worden sind. Nimmt man an, dass die Aktivierungswärme wirklich so klein ist, wie es die von Farkas angegebene erste Zahl andeutet, so würde die Energiefläche wohl noch flacher sein, als von uns angenommen, und dies würde $C$ in der falschen Richtung ändern. Es scheint daher, dass die grössere Aktivierungswärme wahrscheinlicher ist.

Als zweite Ursache kommt in Frage, dass Bildpunkte, die den Grat schon überschritten haben, wieder in das erste Tal zurückgeworfen werden. Dies wird durch die Fig. 12 von Eyring und Polanyi nahegelegt, in der zwei Grate sind, die von einer Mulde (einer instabilen Verbindung $H_3$ entsprechend) getrennt sind. Ist so eine Mulde vorhanden, so muss jeder Bildpunkt zwei Grate überschreiten, und ein Bildpunkt, der den ersten Grat bereits überschritten hat, kann auf eine ungünstige Stelle des zweiten stossen und von dort in die Mulde zurückgeworfen werden. Er wird dann etwas in der Mulde herumrollen und diese schliesslich durch einen der Grate verlassen. Im ungünstigsten Falle — bei sehr tiefer Temperatur —

würde der Bildpunkt die Mulde durch beide Grate mit gleicher Wahrscheinlichkeit verlassen, und dies würde einen Faktor $^1/_2$ für (23a) bedeuten. Wahrscheinlich ist aber bei 1000° dieser Faktor schon nahezu Eins.

Als dritte Fehlerquelle ist schliesslich in Betracht zu ziehen, dass die Berechnung der Bewegung des Bildpunktes mit der klassischen Theorie ausgeführt worden ist. Das ist wohl im Falle von Wasserstoffatomen nicht berechtigt, wie schon daraus hervorgeht, dass die Nullpunktsenergie von $H_2$, ein Quanteneffekt ($^1/_2 h\nu = 6$ kcal) in der Grössenordnung der ganzen Aktivierungswärme ist. Wie gross allerdings der hierdurch verursachte Fehler ist, lässt sich schwer angeben. Bei allen anderen Reaktionen, die sich nicht unter $H$-Atomen abspielen, wäre er sicher kleiner. Wir hoffen hierauf bei anderer Gelegenheit noch einmal zurückzukommen.

Wir wollen noch unsere Gleichung (23a) für die Geschwindigkeitskonstante vergleichen mit der sich aus der einfachen kinetischen Gastheorie sich ergebenden Stosszahl[1]), die wir mit $k_2^0$ bezeichnen wollen, $r_0' = $ Radius von $H_2$, $r_1' = $ Radius von $H$:

$$k_2^0 = 4\pi \cdot 6\cdot 10^{20} \cdot (r_1' + r_0')^2 \sqrt{\frac{kT}{2\pi}} \sqrt{\frac{3}{2M}}. \tag{24}$$

Es ergibt sich:

$$\frac{CT^{3/2}}{k_2^0} = \frac{\sqrt{2}\,b_0^4}{(r_1' + r_0')^2 r_0^2} \sqrt{\frac{s}{p}} \frac{ekT}{u}. \tag{25}$$

Obwohl man denken sollte, dass aus der LONDONschen Theorie eine sehr niedrige wirksame Stosszahl folgt, weil die Reaktion eine sehr spezielle Konfiguration, nämlich die gerade Konfiguration verlangt, ist dies nicht der Fall. Es tritt zwar ein Faktor $\frac{kT}{u}$ auf, der nach der Überlegung auf S. 463 bis 464 einem sterischen Faktor entspricht und bei niedriger Temperatur besonders klein ist. Dagegen tritt auch $\frac{b_0^4}{(r_1' + r_0')^2 r_0^2}$ auf, wo sich der Zähler auf ein durch die Reaktion gestrecktes Molekül, der Nenner auf ein normales Molekül bezieht. Ausserdem tritt $\sqrt{\frac{s}{p}}$ auf, das grösser als Eins ist, dem Umstand entsprechend, dass die Krümmung der Potentialfläche längs des Grates kleiner als im Normalzustand des Moleküls ist. Daher kommt für

---

[1]) Vgl. K. F. HERZFELD, Kinetische Theorie der Wärme.

470                         H. Pelzer und E. Wigner

die Bildpunkte ein längerer Teil des Grates in Frage, als der Abstand
der Atome im Molekül variieren kann, und dies erhöht die statistische
Wahrscheinlichkeit des Grates. Zudem ist die Aktivierungswärme,
wenn man sie nach der Formel

$$Q = RT^2 \frac{d}{dT} \ln \frac{k_3}{\sqrt{T}}$$

berechnet, um $RT$ höher als $W_0$. Berechnet man nun aus $k_3 = C' e^{-\frac{Q}{RT}}$
zurückzu das $C'$, so wird sie das $e$-fache von $CT^{3/2}$. Daher tritt bei dem
Vergleich des in dieser Weise berechneten „temperaturunabhängigen
Faktors" $C'$ mit der Stosszahl zu (25) noch ein Faktor $e$ hinzu.

Eine halbe Potenz der Temperatur tritt in (23a) [wie auch in (24)]
auf, weil die Anzahl der Zusammenstösse proportional zur Geschwindigkeit der Moleküle ist. Weitere zwei halbe Potenzen rühren daher,
dass für die Reaktion zwei Bedingungen mehr ($b = b_0$, $c = c_0$, $\delta = 0$)
erfüllt sein müssen, als für den Normalzustand ($b = r_0$).

6. Wir haben bisher so gerechnet, wie wenn durch die Umsetzung
normaler $H_2$ entstehen würde, d. h. Ortho- und Parawasserstoff im
Verhältnis der statistischen Gewichte 3 : 1. Dies erscheint angesichts
der Tatsache, dass die vorhandene Energie gross gegen den Abstand
der Ortho- und Parazustände ist, von vornherein sehr wahrscheinlich.
Wir wollen es aber noch genauer begründen.

Stellen wir uns ein Parawasserstoffmolekül vor, so sind die beiden
Kernspins sicher antiparallel und jedes hat mit der Wahrscheinlich-

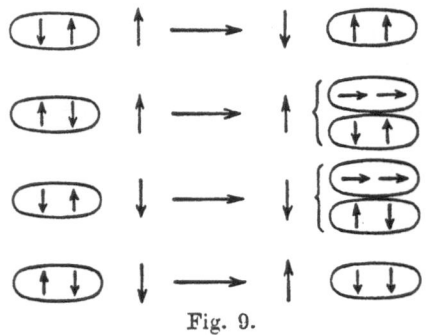

Fig. 9.

keit $^1/_2$ die $+Z$- und mit der Wahrscheinlichkeit $^1/_2$ die $-Z$-Richtung.
Betrachten wir nun einen Zusammenstoss, bei dem der Kernspin
des freien Atoms die $+Z$-Richtung
hat, dies ist bei der Hälfte aller
Zusammenstösse der Fall. Nun
wird sich hieran im Laufe
des Zusammenstosses nichts
ändern, weil die Zeit eines
Kernaustausches zu gross

ist: Er ist in der Grössenordnung der Periode der einquantigen Rotation, während die Zeit des Zusammenstosses der Periode einer sehr
vielquantigen Rotation entspricht. Die Atome, die nach dem Zusammenstoss das Molekül bilden werden, haben also mit der Wahr-

scheinlichkeit $^1/_2$ einen gesamten Kernspin $+1$ in der $Z$-Achse. Tritt dies ein, so ıst das entstandene Molekül sicher im „Kerntriplett"-Zustand, ein Orthowasserstoffmolekül mit der gesamten $Z$-Spinkomponente $^2/_2$. Die Wahrscheinlichkeit, dass bei einem Zusammenstoss dieser Zustand entsteht, ist also $^1/_4$. Da aber die anderen Orthowasserstoffzustände mit derselben Wahrscheinlichkeit entstehen müssen, ist die Gesamtwahrscheinlichkeit für das Entstehen eines Orthowasserstoffmoleküls $^3/_4$. Insbesondere zeigt es sich, dass bei dem Zusammenstoss, bei dem die Spinkomponente des entstehenden Moleküls in der $Z$-Achse Null ist (zweite und dritte Zeile der Fig. 9) mit gleicher Wahrscheinlichkeit ein Ortho- und ein Para-molekül entsteht.

### Anhang.

Den Mittelwert von $\dot\eta$ berechnet man wohl am einfachsten, indem man sich daran erinnert, dass

$$\eta = \frac{(B-C)}{\sqrt 2} = \frac{(b-c)}{\sqrt 2}$$

ist, wo $b$ und $c$ die Abstände des ersten und dritten Atoms vom zweiten Atom sind. Wegen der Geradlinigkeit der Konfiguration ist aber $\dfrac{(b-c)}{2}$ der Abstand des Schwerpunktes der Atome 1 und 3 vom Atom 2, so dass $\dot\eta$ das $\sqrt 2$-fache der Relativgeschwindigkeit des Atoms 2 gegen den Schwerpunkt der Atome 1 und 3 ist. Die Wahrscheinlichkeit einer Relativgeschwindigkeit $\dfrac{\dot\eta}{\sqrt 2}$ ist aber

$$e^{-\frac{1}{kT}\frac{MM'}{2(M+M')}\frac{\dot\eta^2}{2}},$$

wo $M$ die Masse des zweiten Atoms ist und für $M'$ die Summe der Massen der Atome 1 und 3, d. h. $2M$ eingesetzt werden muss. Wir erhalten somit

$$w(\dot\eta) = \text{const} \cdot e^{-\frac{1}{kT}\frac{\dot\eta^2}{6}},$$

und hieraus berechnet sich der Mittelwert von $|\dot\eta|$ zu $\dfrac{\sqrt{6kT}}{\sqrt{\pi M}}$. Dies muss noch mit 2 dividiert werden, weil nur die Hälfte der Bildpunkte die richtige Geschwindigkeitsrichtung hat.

Die Arbeit wurde ausgeführt im Kaiser Wilhelm-Institut für physikalische Chemie in Berlin-Dahlem. Wir möchten es nicht versäumen, Herrn Prof. POLANYI auch an dieser Stelle für viele wertvolle Besprechungen herzlichst zu danken.

# Über das Überschreiten von Potentialschwellen bei chemischen Reaktionen

E. P. Wigner

Zeitschrift für physikalische Chemie B *19*, 203–216 (1932)

(Eingegangen am 30. 7. 32.)

Es wird die Durchlässigkeit einer Potentialschwelle gegenüber einem Atomstrom mit MAXWELLscher Geschwindigkeitsverteilung berechnet für den Fall, dass es erlaubt ist, die Quantenkorrektion nur bis zu Gliedern mit der zweiten Potenz der PLANCKschen Konstante zu berücksichtigen. Das Resultat wird mit ECKARTS exakter Berechnung der Durchlässigkeit einer bestimmten Potentialschwelle verglichen und auf die Umwandlungsgeschwindigkeit von Parawasserstoff in normalen Wasserstoff angewandt.

1. Es wurde schon wiederholt auf die Möglichkeit hingewiesen, dass das unmechanische Überschreiten von Potentialschwellen bei chemischen Reaktionen eine Rolle spielt[1]. Die meisten chemischen Reaktionen bestehen ja darin, dass ein Atom eine Potentialschwelle überschreitet und so eine neue Konfiguration des Systems herbeiführt. Die Höhe der Schwelle äussert sich als Aktivierungswärme.

Es sind hierbei drei Fälle zu unterscheiden: Entweder ist die Schwelle so dick, dass die Atome sie praktisch gar nicht durchdringen können — dies wird gewöhnlich stillschweigend angenommen —, dann verläuft die Reaktion klassisch. Wenn die Schwelle sehr dünn und hoch, die Temperatur sehr niedrig ist, so tragen hauptsächlich die Atome mit geringen Geschwindigkeiten zur Reaktion bei. Es ist wahrscheinlich, dass für keine bis jetzt bekannte Reaktion dieser Fall vorliegt, sie würde sich durch nur geringe Temperaturabhängigkeit der Reaktionsgeschwindigkeit bemerkbar machen. Bei genügend tiefer Temperatur muss zwar dieser Mechanismus immer der ausschlaggebende sein, aber diese Temperatur ist in den allermeisten Fällen so niedrig und die zugehörige Reaktionsgeschwindigkeit so klein, dass sie

---

[1] F. HUND, Z. Physik **43**, 805. 1927. J. R. OPPENHEIMER, Physic. Rev. **31**, 66. 1928. BOURGIN, Pr. Nat. Acad. Washington 15, 357. 1929. R. M. LANGER, Physic. Rev. **34**, 92. 1929. M. BORN und J. FRANCK, Nachr. Götting. Ges. **1930**, 77. S. ROGINSKI und L. ROSENKEWITSCH, Z. physikal. Ch. (B) **10**, 47. 1930. **15**, 103. 1931 M. BORN und V. WEISSKOPF, Z. physikal. Ch. (B) **12**, 206. 1931.

praktisch überhaupt keine Rolle spielt. Der dritte Fall, den wir zu unterscheiden haben, ist schliesslich der, dass die reagierenden Moleküle fast genug Energie haben, um die Schwelle mechanisch überschreiten zu können, dass aber ein endlicher Prozentsatz unter ihnen doch eine etwas niedrigere Energie hat und durch den obersten Teil des Berges unmechanisch hindurchgeht. Dieser Effekt wird nur eine Korrektion für die klassisch berechnete Reaktionsgeschwindigkeit bedeuten, wegen ihrer prinzipiellen Bedeutung soll sie aber im folgenden doch ausführlich behandelt werden.

Schon C. ECKART[1]) hat darauf hingewiesen, dass man in diesem Falle die bekannte WENTZEL-BRILLOUINsche Methode zur Berechnung der Wahrscheinlichkeit der Schwellenüberschreitung nicht benutzen kann. Er stellte jedoch ein Potential auf, für das man die SCHRÖDINGER-Gleichung für jede Energie exakt lösen kann. Wir brauchen viel weniger, nämlich die Durchdringungswahrscheinlichkeit nur für Geschwindigkeiten, die nahezu gleich der Geschwindigkeit sind, bei der das Überschreiten der Schwelle mechanisch eben noch möglich ist. Dagegen wollen wir diese Wahrscheinlichkeit für eine beliebige Potentialschwelle berechnen. Am Schluss wollen wir unser Resultat auf die ECKARTsche Potentialschwelle anwenden und die Resultate vergleichen.

2. Wir denken uns eine Potentialschwelle $V(x)$ und einen Strom von Atomen, der von links (negative $x$) im Temperaturgleichgewicht auf die Schwelle zuströmt und fragen nach der Zahl derjenigen Atome, die reflektiert und die durchgelassen werden. Da wir es mit einem System zu tun haben, das im wesentlichen der klassischen Mechanik gehorchen soll, und bei dem die Quanteneffekte eine nur geringere Rolle spielen sollen, können wir versuchen, mit der Wahrscheinlichkeitsfunktion $P(x, p)$ zu rechnen. Diese Wahrscheinlichkeitsfunktion entsteht aus der statistischen Matrix[2]) $U(x, x')$ des Systems durch die Transformation[3])

$$P(x,p) = \int_{-\infty}^{\infty} U(x+y, x-y)\, e^{\frac{2\,i p y}{h}}\, dy \qquad (1)$$

und ihre Veränderung mit der Zeit ist durch

$$\frac{\partial P(x,p)}{\partial t} = -\frac{p}{m}\frac{\partial P}{\partial x} + \frac{\partial V}{\partial x}\frac{\partial P}{\partial p} - \left(\frac{h}{2}\right)^2 \frac{1}{3!}\frac{\partial^3 V}{\partial x^3}\frac{\partial^3 P}{\partial p^3} + \left(\frac{h}{2}\right)^4 \frac{1}{5!}\frac{\partial^5 V}{\partial x^5}\frac{\partial^5 P}{\partial p^5}\cdots \qquad (2)$$

¹) C. ECKART, Physic. Rev. **35**, 1303. 1930.    ²) L. LANDAU, Z. Physik **45**, 430. 1927. H. WEYL, Z. Physik **46**, 1. 1927. J. v. NEUMANN, Nachr. Götting. Ges. **246**, 245. 1927.    ³) E. WIGNER, Physic. Rev. **40**, 749. 1932.

Über das Überschreiten von Potentialschwellen bei chemischen Reaktionen.    205

gegeben ($h$ ist dabei die PLANCKsche Konstante dividiert durch $2\pi$).
Wir interessieren uns für eine Lösung dieser Gleichung, die stationär
ist, für sehr grosse negative $x$ und
positive $p$ gleich $e^{-\beta p^2/2m}$ wird, für
sehr grosse positive $x$ und nega-
tive $p$ dagegen Null ist. Die erste
Bedingung bedeutet, dass wir es
mit einem stationären Zustand zu
tun haben, die zweite, dass die auf
die Schwelle zuströmenden Atome
im Temperaturgleichgewicht sind
und ihre Temperatur $T = \frac{1}{k\beta}$ ist, die
dritte Bedingung besagt schliess-
lich, dass von rechts keine Atome
auf die Schwelle zuströmen. Für
den klassischen Fall sind die Strö-
mungslinien der Atome im $x, p$-
Raum (Phasenraum) in Fig. 1 ein-
gezeichnet.

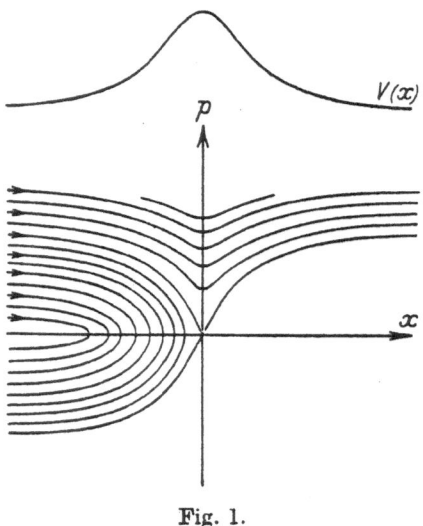

Fig. 1.

Eigentlich interessieren wir uns nicht für das ganze Strömungs-
bild, sondern nur für den Gesamtstrom $\int\limits_0^\infty p P(x,p)\,dp$ für sehr grosse $x$.
Es liegt daher nahe, (2) von 0 bis $\infty$ zu integrieren (die linke Seite
kann gleich Null gesetzt werden)

$$-\frac{1}{m}\frac{\partial}{\partial x}\int\limits_0^\infty p\,P(x,p)\,dp = \sum_{n=0}^\infty \left(\frac{h}{2i}\right)^{2n}\frac{1}{(2n+1)!}\frac{\partial^{2n+1}V(x)}{\partial x^{2n+1}}\frac{\partial^{2n}P(x,0)}{\partial p^{2n}}. \quad (3)$$

Diese Formel erlaubt es uns, den Strom mit Hilfe der Werte der
Wahrscheinlichkeitsfunktion an der Stelle $p = 0$ zu berechnen.

Da ein unmechanisches Überschreiten der Schwelle nur für Geschwindigkeiten in Frage kommt, die ganz in der Nähe der kritischen
Geschwindigkeit $v = \sqrt{\frac{2V_0}{m}}$ liegen ($V_0$ ist die Höhe der Schwelle), wird
die Wahrscheinlichkeitsfunktion, abgesehen von der Umgebung der
Stellen $\frac{p^2}{2m} + V(x) = V_0$ genau so wie im vollständigen thermodynamischen Gleichgewicht aussehen. Für $P(x, p)$ in der Mitte des schraffierten

206                               E. Wigner

Teils der Fig. 1 können wir daher die Werte für das thermodynamische Gleichgewicht einsetzen[1])

$$P(x,p) = e^{-\frac{\beta p^2}{2m}} e^{-\beta V} \left(1 - \frac{h^2\beta^2}{8m}\frac{\partial^2 V}{\partial x^2} + \frac{h^2\beta^3}{24m}\left(\frac{\partial V}{\partial x}\right)^2 + \frac{h^2\beta^3}{24m^2}p^2\frac{\partial^2 V}{\partial x^2} + \cdots\right). \quad (4)$$

Ebenso können wir für den inneren Teil des Gebietes, in dem in Fig. 1 $P(x,p) = 0$ ist, auch jetzt $P(x,p) = 0$ annehmen. Nur für das Trennungsgebiet des schraffierten Teiles vom unschraffierten müssen wir eine etwas genauere Überlegung anstellen, die Bahnen werden dort. etwa so verlaufen, wie es in Fig. 2 angedeutet ist. Dies wird durch die höheren Glieder von (2) bewirkt, die den in der klassischen Wahrscheinlichkeitsfunktion vorhandenen Sprung auszuglätten bestrebt sind.

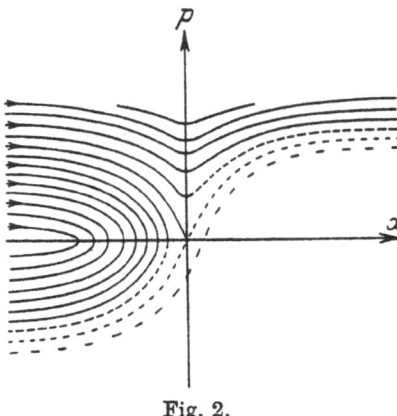

Fig. 2.

Wir wollen zunächst annehmen, dass die Verteilung für das thermodynamische Gleichgewicht (4) innerhalb des ganzen schraffierten Teiles von Fig. 1 gilt, also insbesondere für $p = 0$ bis $x = 0$, und dass die Wahrscheinlichkeitsfunktion ausserhalb des schraffierten Bereichs wirklich Null ist. Berechnen wir den Strom unter dieser Voraussetzung mit Hilfe von (3), so haben wir nur noch in der Umgebung von $x = 0$, $p = 0$ die Gleichung (2) streng zu lösen und in dem Teil von (3), in dem wir die falschen Werte für $P(x,p)$ benutzt haben, diese durch die richtigen Werte zu ersetzen.

3. Setzen wir (4) für $x < 0$ in (3) ein, setzen wir dagegen für $x > 0$ das $P(x,0) = 0$, so erhalten wir bei Vernachlässigung von Gliedern mit höherer als der zweiten Potenz von $h$ für $x \lessgtr 0$

$$\int_0^\infty \frac{p}{m} P(x,p)\,dp = e^{-\beta V(x)}\left(\frac{1}{\beta} - \frac{h^2\beta}{24m}\frac{\partial^2 V(x)}{\partial x^2} + \frac{h^2\beta^2}{24m}\left(\frac{\partial V(x)}{\partial x}\right)^2\right) \quad (5)$$

---

[1]) Die Normierung ist dabei so, dass an den Stellen vollen thermodynamischen Gleichgewichts $\sqrt{\dfrac{2\pi m}{\beta}}$ Atome pro Zentimeter sind, oder, was dasselbe bedeutet, bei $x = -\infty$ sind $\sqrt{\dfrac{\pi m}{2\beta}}$ nach rechts laufende Atome.

Über das Überschreiten von Potentialschwellen bei chemischen Reaktionen.    207

und für $x \gtreqless 0$

$$\int_0^\infty \frac{p}{m} P(x,p)\, dp = \int_0^\infty \frac{p}{m} P(0,p)\, dp = e^{-\beta V_0}\left(\frac{1}{\beta} - \frac{h^2 \beta}{24\,m} V_2\right), \qquad (5\,\mathrm{a})$$

wo $V_2 = \dfrac{\partial^2 V(0)}{\partial x^2}$ wie im folgenden allgemein $V_i = \dfrac{\partial^i V(0)}{\partial x^i}$ sein soll. Bereits (5a) gibt einen guten Näherungswert für die Durchlässigkeit der Potentialschwelle gegenüber einem Atomstrom von der Temperatur $\frac{1}{k\beta}$.

Schreiben wir nämlich (2) für $x \sim 0$ und eine symmetrische Schwelle auf

$$-\frac{p}{m}\frac{\partial P}{\partial x} + \sum_{n=0}^\infty \left(\frac{h}{2i}\right)^{2n} \frac{1}{(2n+1)!}\, V_{2n+2}\, x\, \frac{\partial^{2n+1} P}{\partial p^{2n+1}} = 0 \qquad (6)$$

und behalten wir für den Augenblick nur zwei Glieder der Summe. Dann können wir durch eine Substitution $x = \alpha y$, $p = \alpha' q$ alle numerischen Koeffizienten gleich machen, so dass (6) in

$$-q\frac{\partial P}{\partial y} - y\frac{\partial P}{\partial q} - y\frac{\partial^3 P}{\partial q^3} = 0 \qquad (6\,\mathrm{a})$$

übergeht. Dabei ist $\alpha = \sqrt{\dfrac{h^2 V_4}{24\,m\,V_2^2}}$, und dies wird daher auch die ungefähre Grösse des Übergangsgebietes der Fig. 2 sein, in dem $P$ von $e^{-\beta V_0}$ auf Null herabfällt. In einem Gebiet dieser Grössenordnung haben wir daher in (3) ein falsches $P$ verwendet. Der für den Strom hierdurch verursachte Fehler ist von der Grössenordnung $e^{-\beta V_0} \dfrac{h^2 V_4}{24\, V_2\, m}$. Obwohl nun dies auch mit der zweiten Potenz von $h$ geht, ist es doch viel kleiner als das andere Korrektionsglied, das in (5a) steht, das Verhältnis der beiden ist von der Grössenordnung $\dfrac{V_4}{\beta V_2^2} \sim \dfrac{1}{V_0 \beta}$. Das ist aber für alle praktisch in Frage kommenden Temperaturen sehr klein.

4. Im folgenden sei das zuletzt besprochene Korrektionsglied eingehender betrachtet, d. h. sein genauer Koeffizient berechnet[1]). Wir beschränken uns dabei auf eine symmetrische Potentialschwelle. Es muss das Übergangsgebiet in $P(x,p)$ in der Umgebung der Stelle $x = 0$, $p = 0$ genauer untersucht werden.

Die Überlegung am Schlusse des vorigen Abschnitts gibt uns die Grössenordnung von $\dfrac{\partial}{\partial p}$ in diesem Bereich, sie ist $\dfrac{1}{\alpha'} = \sqrt{-\dfrac{24\, V_2}{h_2\, V_4}}$. Daher

---

[1]) Da das Glied, dessen Koeffizient in diesem Punkt berechnet werden soll, in den praktisch wichtigen Fällen sehr klein ist, ist die Kenntnis von 4 für das spätere nicht nötig.

208                                    E. Wigner

ist in dem fraglichen Gebiet jedes Glied von (2) von gleicher Grössenordnung, und wir dürfen die Reihe (6) nicht abbrechen, wie wir das
in der provisorischen Betrachtung in (6a) getan haben. Wir müssen
eine Lösung von (6) finden, die ausserhalb des Übergangsgebietes
(eines Gebietes von der Grösse $a = \sqrt{\dfrac{h^2 V_4}{24\,m\,V_2^2}}$ in der $x$-Richtung und
der Grösse $a' = \sqrt{-\dfrac{h^2 V_4}{24\,V_2}}$ in der $p$-Richtung) sich so verhält, dass es
mit wachsendem Abstand vom Nullpunkt zu $e^{-\beta V_0}$ (eigentlich zu
$e^{-\beta V_0}(1 + h^2\ldots)$, aber die Glieder mit $h^2$ können wir in diesem Korrektionsglied vernachlässigen) geht, wenn wir uns auf einer Geraden vom
Nullpunkt wegbewegen, deren Richtung in das schraffierte Gebiet
der Fig. 1 weist, die also einen Winkel zwischen $\varphi$ und $\varphi + \pi$ mit der
$x$-Achse einschliesst, wo tg $\varphi = \sqrt{-m\,V_2}$ ist. Entfernt man sich dagegen
auf einer Geraden vom Nullpunkt, die einen Winkel zwischen $\varphi - \pi$
und $\varphi$ mit der $x$-Achse einschliesst, so soll die Lösung zu Null gehen.

An Stelle dieser Lösung können wir auch eine Lösung betrachten,
die im ersten Winkelbereich zu $\dfrac{1}{2}\,e^{-\beta V_0}$, im zweiten zu $\dfrac{1}{2}\,e^{-\beta V_0}$ geht.
Wenn wir zu dieser Lösung noch die triviale Lösung $e^{-\beta V_0}$ von (6)
addieren, erhalten wir die Lösung, die die vorerwähnten Grenzbedingungen befriedigt.

Nun ist (6) linear sowohl in $x$, wie auch in $p$. Durch eine LAPLACE-
Transformation in beiden Argumenten

$$P(x,p) = \iint\limits_{-\infty}^{\infty} Q(y,q)\, e^{-i(xy - pq)}\, dy\, dq \tag{7}$$

können wir sie daher zu einer partiellen Differentialgleichung ersten
Grades machen (die in sehr naher Beziehung zur gewöhnlichen SCHRÖ-
DINGER-Gleichung steht)

$$\left.\begin{aligned}
&\frac{y}{m}\frac{\partial Q}{\partial q} - \frac{2}{h}\sum_{n=0}^{\infty}\frac{V_{2n+2}}{(2n+1)!}\left(\frac{hq}{2}\right)^{2n+1}\frac{\partial Q}{\partial y} = 0 \\
&\frac{y}{m}\frac{\partial Q}{\partial q} - \frac{2}{h}\,V'\left(\frac{hq}{2}\right)\frac{\partial Q}{\partial y} = 0.
\end{aligned}\right\} \tag{8}$$

Die Reihe der ersten Zeile ist ja die TAYLOR-Reihe für die Ableitung von $V$ an der Stelle $\dfrac{hq}{2}$. Die allgemeine Lösung dieser Gleichung
lautet
$$Q(y,q) = f(y^2 - W(q)), \tag{9}$$
wo $f$ eine beliebige Funktion ist und zur Abkürzung

$$W(q) = \frac{8\,m}{h^2}\left(V_0 - V\left(\frac{hq}{2}\right)\right) \tag{9a}$$

gesetzt ist.

Über das Überschreiten von Potentialschwellen bei chemischen Reaktionen.   209

Um die Grenzbedingungen zu befriedigen verfahren wir folgendermassen[1]): Zunächst bleibt (6) invariant, wenn man darin $x$ durch $-x$
und $p$ durch $-p$ ersetzt. Mit (7) ist daher auch $\iint Q(y,q)e^{i(xy-pq)}dy\,dq$
oder auch

$$P(x,p) = \int\limits_{0}^{\infty} \int\limits_{-\infty}^{\infty} Q(y,q)\sin(xy-pq)\,dy\,dq \tag{7a}$$

eine Lösung; (7a) hat gleichzeitig die erwünschte Eigenschaft an der
Stelle $-x$, $-p$ entgegengesetzt gleich wie an der Stelle $x$, $p$ zu sein.
Nunmehr setzen wir versuchsweise

$$Q(y,q) = C \cdot \delta(y^2 - W(q)) \tag{10}$$

mit der DIRACschen $\delta$-Funktion. Es ergibt sich nach einer einfachen
Umformung, indem man für $y^2 = z$ einsetzt und die Integration über $z$
ausführt

$$P(x,p) = \frac{\sqrt{W_2}\,e^{-\beta V_0}}{\pi\sqrt{2}} \int\limits_{0}^{\infty} \frac{\sin(pq - x\sqrt{W(q)})}{\sqrt{W(q)}}\,dq. \tag{11}$$

Für die Konstante $C$ wurde dabei ein bestimmter Wert genommen,
um die Grenzbedingung zu erfüllen. Es bedeutet wiederum

$$W_i = \frac{\delta^i W(0)}{\delta q^i}.$$

Man überzeugt sich leicht, dass (11) die Differentialgleichung (6)
befriedigt, das folgt auch aus seiner Herleitung. Da indessen (11) nicht
absolut konvergiert, ersetzen wir es durch

$$P(x,p) = \frac{\sqrt{W_2}\,e^{-\beta V_0}}{\pi\sqrt{2}} \int\limits_{0}^{\infty} \frac{\sin(pq - x\sqrt{W(q)})}{\sqrt{W(q)}}\,e^{-aq}\,dq \tag{11a}$$

und lassen nachträglich $a$ gegen Null gehen. Im folgenden wollen wir
mit (11a) rechnen. Um noch die Randbedingungen zu verifizieren,
nehmen wir an, dass $x$ sehr gross ist. Dann wird das Integrations-
gebiet mit endlich grossem $q$ nichts zum Integral beitragen, weil der
Integrand in diesem Gebiet sinusförmige sehr rasche Schwankungen
ausführt. Nur das Gebiet um $q = 0$ wird einen endlichen Beitrag
liefern, weil der Nenner für $q = 0$ verschwindet. In diesem Gebiet
können wir $W(q)$ durch $\frac{1}{2} \cdot W_2 q^2$ ersetzen und dann bis $\infty$ integrieren,
da das Integral doch gut konvergent ist. Nach bekannten Formeln
wird dann für $x = \infty$, $p = \infty$

---

[1]) Für seine freundliche Hilfe bei der Bestimmung der richtigen Lösung von (6)
sei Herrn J. v. NEUMANN auch an dieser Stelle bestens gedankt.

$$P(x, dx) = \frac{\sqrt{W_2}\, e^{-\beta V_0}}{\pi \sqrt{2}} \int_0^\infty \frac{\sin\left((p - x\sqrt{W_2/2})q\right)}{q\sqrt{W_2/2}}\, dq$$

$$= \begin{cases} \dfrac{1}{2}\, e^{-\beta V_0} & \text{für } \dfrac{p}{x} > \sqrt{W_2/2} = \sqrt{-m\, V_2} \\[2mm] -\dfrac{1}{2}\, e^{-\beta V_0} & \text{für } \dfrac{p}{x} < \sqrt{W_2/2} = \sqrt{-m\, V_2} \end{cases}$$

wie wir es gewünscht haben. Die gesamte Wahrscheinlichkeitsfunktion in der Nähe von $x = 0$, $p = 0$ erhalten wir, indem wir zu (11a) noch $\frac{1}{2}\, e^{-\beta V_0}$ addieren.

Nun müssen wir mit Hilfe der eben gewonnenen Wahrscheinlichkeitsfunktion die Zunahme des Stromes in der Nähe des Nullpunktes nach (3) berechnen und daraus die bisher fälschlicherweise angenommene Zunahme ($V_2 x e^{-\beta V_0}$ bis $x = 0$, von dort ab Null; wir sind ja immer in der Nähe des Nullpunktes, und Korrektionsglieder können wir in diesem Korrektionsglied vernachlässigen) abziehen. Zunächst haben wir für (3)

$$\sum_{n=0}^\infty \left(\frac{h}{2i}\right)^{2n} \frac{1}{(2n+1)!} \frac{\partial^{2n+1} V(x)}{\partial x^{2n+1}} \frac{\partial^{2n} P(x,0)}{\partial p^{2n}}$$

zu berechnen. Es ergibt sich dafür für kleine $x$, wenn wir für $P(x, p)$ (11a) benutzen

$$\left.\begin{aligned} &\frac{\sqrt{W_2}}{\pi \sqrt{2}}\, e^{-\beta V_0} \sum_n \left(\frac{h}{2i}\right)^{2n} \frac{V_{2n+2}\, x}{(2n+1)!} \int_0^\infty \frac{\sin(-x\sqrt{W(q)})}{\sqrt{W(q)}} (iq)^{2n} e^{-aq}\, dq \\ &= -\frac{\sqrt{2W_2}}{\pi h}\, e^{-\beta V_0} x \int_0^\infty V'\!\left(\frac{hq}{2}\right) \frac{\sin x\sqrt{W(q)}}{q\sqrt{W(q)}}\, e^{-aq}\, dq. \end{aligned}\right\} \tag{12}$$

Hierzu wäre noch wegen des zu (11a) hinzukommenden Gliedes $\frac{1}{2}\, e^{-\beta V_0}$ die Grösse $\frac{1}{2} V_2 x e^{-\beta V_0}$ zu addieren und dann, um die bisher fälschlicherweise angenommene Zunahme wieder abzuziehen, bis $x = 0$ der Betrag $V_2 x e^{-\beta V_0}$ zu subtrahieren. Im ganzen bleibt zu (12)

$$\frac{1}{2}\,\mathrm{sign}\, x \cdot x V_2 e^{-\beta V_0}$$

zu addieren, wo $\mathrm{sign}\, x = +1$ für $x > 0$ und $= -1$ für $x < 0$ ist. Wir können für dieses Glied auch

$$e^{-\beta V_0} \frac{x V_2}{\pi} \int_0^\infty \frac{\sin x q \sqrt{W_2/2}}{q}\, dq \tag{12a}$$

Über das Überschreiten von Potentialschwellen bei chemischen Reaktionen.    211

schreiben. Zum Schluss wäre die Summe von (12) und (12a) in der Umgebung von $x = 0$ über $x$ zu integrieren, um das restliche Korrektionsglied für (5a) zu erhalten. Die Integration über $x$ kann man aber auch von $-\infty$ bis $\infty$ erstrecken, da der Integrand nach beiden Seiten vom Nullpunkt sehr stark abfällt. Da der Integrand eine gerade Funktion von $x$ ist, kann man dies auch durch das doppelte von 0 bis $\infty$ genommene Integral ersetzen. Um weiterhin die Integrationsreihenfolge umkehren und zuerst über $x$ integrieren zu können, versehen wir den Integranden mit einem Faktor $e^{-bx}$, wobei wir später $b$ zu Null machen werden. Nach Ausführung der Integration über $x$ erhalten wir

$$\frac{\sqrt{2\,W_2}}{\pi\,m}\,e^{-\beta\,V_0}\int\limits_{0}^{\infty}\left(\frac{W'(q)\,b}{q\,(b^2 + W(q))^2} - \frac{W_2\,b}{(b^2 + q^2\,W_2/2)^2}\right)dq. \tag{13}$$

Dies muss noch nach $q$ von Null bis $\infty$ integriert werden. Da $b$ am Schlusse zu Null gehen soll, ist es klar, das nur das Gebiet um $q = 0$ etwas zum Integral beitragen wird — für endliche $q$ und kleine $b$ verschwindet der Integrand. Man überzeugt sich leicht, dass es genügt im ersten Glied $W(q)$ und $W'(q)$ bis zum zweiten Glied in eine Reihe zu entwickeln, das nächste Glied würde im Resultat schon mit einem Faktor $b$ behaftet sein. Dadurch geht (13) in das Integral einer rationalen gebrochenen Funktion über, das man entweder elementar oder auf komplexem Wege auswerten kann. Man erhält so

$$\frac{W_4\,e^{-\beta\,V_0}}{24\,m\,W_2} = \frac{h^2\,V_4\,e^{-\beta\,V_0}}{96\,m\,V_2} \tag{14}$$

als letztes Korrektionsglied für den Strom durch die Schwelle, so dass sich dieser nach (5a) zu

$$\int\limits_{0}^{\infty}\frac{p}{m}\,P(\infty, p)\,dp = e^{-\beta\,V_0}\left(\frac{1}{\beta} - \frac{h^2\,\beta}{24\,m}\,V_2 - \frac{h^2}{96\,m}\,\frac{V_4}{V_2}\right) \tag{14a}$$

ergibt.

Wir haben schon gesehen, dass (14) in den meisten Fällen, die praktisch von Interesse sind, sehr klein gegen das erste Korrektionsglied ist. Es ist aber sehr störend, zu sehen, dass es über alle Grenzen steigt, wenn $V_2$ zu Null geht. Es ist nicht leicht zu sehen, ob diesem Verhalten von (14) ein tatsächlicher Effekt zugrunde liegt, oder ob es nur im verwendeten Näherungsverfahren seine Begründung hat. Es ist ja klar, dass zwischen den verschiedenen Ableitungen von $V$ Beziehungen bestehen müssen, wenn es die Form einer einfachen Schwelle haben soll. Andererseits ist aber auch unsere Rechnung wegen

212                           E. Wigner

Konvergenzschwierigkeiten mathematisch keineswegs streng, so z. B. divergiert (11) für $p = 0$.

5. In der klassischen statistischen Mechanik kann man die Wahrscheinlichkeit der Überschreitung einer Schwelle sehr einfach folgendermassen berechnen: Man betrachtet den Fall vollständigen thermodynamischen Gleichgewichts. Dann ist die Anzahl der Atome mit dem Impuls $p$ auf der Schwelle $= e^{-\beta V_0} e^{-\beta p^2/m}$ und folglich die Gesamtzahl der Atome, die die Schwelle pro Zeiteinheit nach der einen Richtung hin passieren

$$\int_0^\infty \frac{p}{m}\, e^{-\beta V_0}\, e^{-\frac{\beta p^2}{2m}}\, dp = \frac{1}{\beta}\, e^{-\beta V_0}. \tag{15}$$

Nun kann man annehmen, dass dies auch die Zahl der Atome ist, die die Schwelle dann passieren, wenn diese von einem Strom von Atomen im Temperaturgleichgewicht nur von der einen Seite getroffen wird. Man sieht ja, dass im Falle einer einfachen Schwelle alle Atome, die die Schwelle nach rechts passieren, von der linken Seite stammen, und ihre Zahl wird gleich sein, unabhängig davon, ob von der rechten Seite Atome kommen oder nicht.

Diese Überlegung kann man in dieser Form in die Quantenmechanik nicht übertragen, weil man überhaupt nicht von der Wahrscheinlichkeit von Atomen reden darf, die an einer bestimmten Stelle sind und dabei eine vorgeschriebene Geschwindigkeit haben. Doch kann man immerhin versuchen, die der klassischen entsprechende Grösse mit Hilfe der Wahrscheinlichkeitsfunktion (4) zu berechnen. Man erhält so

$$\int_0^\infty \frac{p}{m}\, e^{-\beta V_0}\, e^{-\frac{\beta p^2}{2m}} \left(1 - \frac{h^2 \beta^2}{8m}\, V_2 + \frac{h^2 \beta^3}{24 m^2}\, V_2 p^2\right) dp = e^{-\beta V_0}\left(\frac{1}{\beta} - \frac{h^2 \beta}{24 m}\, V_2\right) \tag{15 a}$$

also tatsächlich das richtige erste und damit wichtigste Korrektionsglied von (14a).

Dies erlaubt es einem, zu erraten, was die Wahrscheinlichkeit der Schwellenüberschreitung im mehrdimensionalen Falle ist. An sich könnte man die Rechnung genau so wie im eindimensionalen Fall ausführen, nur erscheint sie noch etwas mühevoller. Es sei auch bemerkt, dass die in 3. verwendete vereinfachte Überlegung genau zu unserem Resultat (17) führt.

Im Falle mehrerer Dimensionen handelt es sich nicht so sehr um die Überschreitung einer Schwelle, als um die eines Grates, wie sie

Über das Überschreiten von Potentialschwellen bei chemischen Reaktionen.    213

aus der Theorie der einfachsten chemischen Reaktionen hinlänglich bekannt ist. Es spielen dabei zwei Effekte eine Rolle, die sich gegenseitig zu kompensieren versuchen: der bisher betrachtete Tunneleffekt, der die Wahrscheinlichkeit der Gratüberschreitung erhöht, und die Nullpunktsenergie, die eine grössere Energie für die Überschreitung eines engen Passes erheischt, als dem untersten Punkte des Passes entsprechen würde. Wir legen das Koordinatensystem so, dass das Potential in der Nähe des Passes die Gestalt

$$V(x_1 \cdots x_n) = V_0 + \frac{1}{2}(A_1 x_1^2 + A_2 x_2^2 + \cdots + A_n x_n^2) \tag{16}$$

habe. Dabei ist $A_1$ negativ, alle anderen Koeffizienten positiv. Die Wahrscheinlichkeitsfunktion lautet längs des Grates $(x_1 = 0)$

$$
\left.
\begin{aligned}
P(x_1 \cdots x_n; p_1 \cdots p_n) &= e^{-\frac{\beta}{2m}(p_1^2 + \cdots + p_n^2)} e^{-\beta V_0} e^{-\frac{\beta}{2}(A_2 x_2^2 + \cdots + A_n x_n^2)} \\
&\cdot \left\{ 1 - \frac{h^2 \beta^2}{8m}(A_1 + \cdots + A_n) + \frac{h^2 \beta^3}{24m}(A_2^2 x_2^2 + \cdots + A_n^2 x_n^2) \right. \\
&\left. + \frac{h^2 \beta^3}{24 m^2}(p_1^2 A_2 + \cdots + p_n^2 A_n) \right\}
\end{aligned}
\right\} \tag{16a}
$$

und mit $\frac{p_1}{m}$ multipliziert und über $x_2, x_3, \ldots, x_n$ und alle Momente integriert, ergibt dies für den Strom über den Grat

$$\frac{4\, m^{1/2}(n+1)\, \pi^{n-1}}{(2\,\beta)^{n+1}}\, \frac{e^{-\beta V_0}}{\sqrt{A_2 A_3 \cdots A_n}}\, \left\{ 1 - \frac{h^2 \beta^2}{24 m}(A_1 + \cdots + A_n) \right\}. \tag{17}$$

Es muss jedoch bemerkt werden, dass hierbei Zusatzglieder, wie (14) konsequent weggelassen worden sind. Diese treten hier nämlich nicht nur aus demselben Grunde wie im eindimensionalen Fall auf, sondern eigentlich schon bei der Integration von (16a), wenn man für $V$ auch höhere Glieder als in (16) angegeben mitbenutzt. Dies ist übrigens schon im klassischen Teil der Formel der Fall. Ausserdem muss bemerkt werden, dass die ganze Überlegung, die zu (15) führt, im mehrdimensionalen Fall auch in der klassischen Statistik nicht mehr ohne weiteres berechtigt ist, weil es nicht mehr immer wahr ist, dass die Atome, die den Grat von links überschreiten, von links stammen. Sie stammen, wenn das Potentialbild etwas kompliziert ist, zum Teil von rechts, gingen dann bereits nach links hinüber und passieren die Schwelle nun wieder nach rechts. Doch spielt dieser Effekt wohl in den meisten Fällen eine nur untergeordnete Rolle.

6. Es sei nunmehr (14a) auf die ECKARTsche Potentialschwelle angewandt und mit dem ECKARTschen Resultat verglichen. Der symmetrische Teil des ECKARTschen Potentials lautet

214                              E. Wigner

$$\frac{4 V_0}{(1 + e^{x/b})(1 + e^{-x/b})},\tag{18}$$

und seine Durchlässigkeit für eine Welle mit der Energie $E$ berechnet
ECKART zu

$$1 - \varrho = \frac{\mathrm{Cosh}\,\pi\sqrt{\frac{E}{C} - 1}}{\mathrm{Cosh}\,\pi\sqrt{\frac{V_0 - C}{C}} + \mathrm{Cosh}\,\pi\sqrt{\frac{E}{C}}},\tag{19}$$

wo $C = \frac{h^2}{32\,m\,b^2}$ ist. Die Durchlässigkeit einem Atomstrom in Tempe-
raturverteilung gegenüber ist daher

$$\int_0^\infty (1 - \varrho)\,e^{-\beta E}\,dE \approx e^{-\beta V_0}\left(\frac{1}{\beta} + \frac{h^2\beta V_0}{48\,m\,b^2} + \frac{h^2}{48\,m\,b^2}\right).\tag{20}$$

Die rechte Seite ergibt sich durch Integration, ist aber nur bis
auf Glieder mit $h^2$ richtig. Formel (14a) ergibt

$$e^{-\beta V_0}\left(\frac{1}{\beta} + \frac{h^2\beta V_0}{48\,m\,b^2} + \frac{h^2}{48\,m\,b^2}\right)\tag{20a}$$

also ein mit (20) identisches Resultat. Der Vergleich unserer Formel
in diesem Spezialfall mit der ECKARTschen mag wegen des bereits
betonten Umstandes nicht überflüssig erscheinen, dass die Ableitung
von (14a) mathematisch nicht streng ist.

7. Zum Schluss sei die Quantenkorrektion für die Umwandlungsgeschwindigkeit des Parawasserstoffs in normalen Wasserstoff berechnet. Da die Übereinstimmung zwischen den experimentellen
Werten[1]) und den ohne Quantenkorrektion berechneten[2]) bereits sehr
gut ist, kann es sich nur darum handeln, nachzusehen, ob die Übereinstimmung durch die Quantenkorrektion nicht zerstört wird, wie
dies zunächst vermutet werden könnte[2]).

In der Tat zeigt es sich, dass die Quantenkorrektion sehr beträcht-
liche Werte hat, und man kann gar nicht schematisch nach (17) rechnen,
weil die Näherung mit der zweiten Potenz von $h$ nicht ausreicht.
Glücklicherweise können die Glieder in zwei Gruppen eingeteilt werden:
solche, bei denen man mit (17) sehr wohl auskommt und solche, bei
denen man noch ganz im Quantengebiet ist, so dass man die Existenz
der höheren Schwingungen ganz vernachlässigen kann. Da die Rechnung etwas langwierig ist, sei sie hier nur skizzenhaft dargestellt.

---

[1]) A. FARKAS, Z. physikal. Ch. (B) **10**, 419. 1930. H. GEIB und P. HARTECK.
Z. physikal. Ch., BODENSTEIN-Festband, 849. 1931.    [2]) H. PELZER und E. WIGNER.
Z. physikal. Ch. (B) **15**, 445. 1932.

Über das Überschreiten von Potentialschwellen bei chemischen Reaktionen.    215

Die Rechnung wurde (loc. cit.) auch mit Hilfe der Wahrscheinlichkeitsfunktion ausgeführt, es wurde aber die klassische Wahrscheinlichkeitsfunktion verwandt. Dies erleichtert das Anbringen der Quantenkorrektionen, da die entsprechenden Ausdrücke immer nur mit einem Faktor $(1 + h^2 \ldots)$ multipliziert bzw. dividiert werden müssen.

Die erste Korrektion betrifft die Normierung der Wahrscheinlichkeitsfunktion. An Stelle·des Integrals über die klassische Wahrscheinlichkeitsfunktion des $H_2$ sollte das über die korrigierte Wahrscheinlichkeitsfunktion treten, die sich von ersterem durch den Faktor

$$1 - \frac{h^2 \beta^2}{24} \omega^2 \qquad (21)$$

unterscheidet, wo $\omega$ die $2\pi$fache Frequenz des $H_2$-Moleküls ist. Nun zeigt ein Vergleich der Grössenordnungen der beiden Glieder in (21), dass in diesem Fall die Näherung mit $h^2$ zweifellos noch nicht hinreichend ist. In diesem Fall ist es aber leicht, für (21) die genaue Zustandssumme

$$\beta h \omega \frac{e^{-\beta h \omega/2}}{1 - e^{-\beta h \omega}} \qquad (21a)$$

zu berechnen. Der Faktor $e^{-\beta \omega h/2}$ trägt nur zur Aktivierungswärme etwas bei, für den Rest kann man bei allen in Betracht kommenden Temperaturen $\frac{6500}{T}$ schreiben. In die Formel für die Reaktionsgeschwindigkeit geht das Reziproke hiervon ein.

Die Korrektion für die eigentliche Gratüberschreitung können wir durch den Faktor [vgl. (17)]

$$\left(1 - \frac{h^2 \beta^2}{24\,m} A_1\right)\left(1 - \frac{h^2 \beta^2}{24\,m} A_2\right)\left(1 - \frac{h^2 \beta^2}{24\,m} A_3\right)\left(1 - \frac{h^2 \beta^2}{24\,m} A_4\right) \qquad (22)$$

berücksichtigen. Dabei berechnen sich die $A$ zu $-0.44 \cdot 10^5$; $2.1 \cdot 10^5$; $0.89 \cdot 10^5$; $0.89 \cdot 10^5$. Bei·dem ersten Faktor können wir die Form (22) beibehalten, die anderen drei beziehen sich auf die Zustandswahrscheinlichkeit auf dem Grat (nicht auf die Überschreitung), bei ihnen ist die in (22) auftretende Näherung nicht zulässig, sie werden analog zu (21) durch

$$\frac{2700}{T} \frac{e^{-1350/T}}{1 - e^{-2700/T}} \frac{1100}{T} \frac{e^{-550/T}}{1 - e^{-1100/T}} \frac{1100}{T} \frac{e^{-550/T}}{1 - e^{-1100/T}} \qquad (22a)$$

ersetzt. Der erste Faktor kann wiederum für alle in Betracht kommenden Temperaturen $\frac{2700}{T}$ geschrieben werden.

216   E. Wigner, Überschreiten von Potentialschwellen bei chemischen Reaktionen.

Die numerische Berechnung ergibt folgende Tabelle:

| $T$ | $k_3$ | $C_{alt}$ | $C_{neu}$ |
|---|---|---|---|
| 283 | $8\text{\textperiodcentered}6\cdot10^4$ | $2\cdot10^6$ | $2\cdot10^6$ |
| 373 | $2\text{\textperiodcentered}7\cdot10^6$ | $2\text{\textperiodcentered}5\cdot10^6$ | $3\cdot10^6$ |
| 873 | $1\text{\textperiodcentered}2\cdot10^9$ | $2\cdot10^6$ | $3\cdot10^6$ |
| 1023 | $1\text{\textperiodcentered}9\cdot10^9$ | $1\text{\textperiodcentered}5\cdot10^6$ | $2\cdot10^6$ |

In der ersten Spalte stehen die Temperaturen, in der zweiten die von GEIB und HARTECK bzw. von A. FARKAS beobachteten Geschwindigkeitskonstanten in Liter/Mol in der dritten Spalte die von PELZER und WIGNER aus diesen berechnete Grösse $C = k_3 T^{-\frac{3}{2}} e^{W_0/RT}$ (mit $W_0 = 6600$ cal) in der letzten schliesslich diese Zahlen dividiert noch mit der Quantenkorrektion, und mit $W_0 = 5600$ anstatt mit $W_0 = 6600$ cal berechnet[1]). Der theoretische Wert[2]) für die letzte Spalte wäre $1\text{\textperiodcentered}0\cdot10^6$. Wir sehen, dass durch die Quantenkorrektion die Übereinstimmung zwischen den aus den Versuchen berechneten $C$ und seinem theoretischen Wert etwas verschlechtert wurde. Sie war aber, wie dies schon (loc. cit.) betont wurde, vor Anbringung der Quantenkorrektion sehr weitgehend zufällig, während das jetzt nicht mehr wesentlich der Fall sein dürfte.

Zum Schluss möchte ich es nicht versäumen, Herrn H. PELZER für seine liebenswürdige Hilfe zu danken, mit der er mich bei der Ausführung dieser Arbeit unterstützt hat.

---

[1]) Wegen der von (21a) und (22a) weggelassenen Faktoren berechnet sich hieraus die Höhe des Grates über dem tiefsten Punkt des Tales zu 9400 cal.    [2]) Der theoretische Wert war von PELZER und WIGNER wegen eines Fehlers in ihrer Formel (22) zweifach zu hoch angegeben worden.

# On the Quantum Correction
# for Thermodynamic Equilibrium

E. P. Wigner

Physical Review *40*, 749–759 (1932)

(Received March 14, 1932)

The probability of a configuration is given in classical theory by the Boltzmann formula exp $[-V/hT]$ where $V$ is the potential energy of this configuration. For high temperatures this of course also holds in quantum theory. For lower temperatures, however, a correction term has to be introduced, which can be developed into a power series of $h$. The formula is developed for this correction by means of a probability function and the result discussed.

## 1

IN classical statistical mechanics the relative probability for the range $p_1$ to $p_1+dp_1$; $p_2$ to $p_2+dp_2$; $\cdots$ ; $p_n$ to $p_n+dp_n$ for the momenta and $x_1$ to $x_1+dx_1$; $x_2$ to $x_2+dx_2$; $\cdots$ ; $x_n$ to $x_n+dx_n$ for the coordinates is given for statistical equilibrium by the Gibbs-Boltzmann formula

$$P(x_1, \cdots, x_n; p_1, \cdots, p_n)dx_1 \cdots dx_n dp_1 \cdots dp_n = e^{-\beta\epsilon}dx_1 \cdots dx_n dp_1 \cdots dp_n \quad (1)$$

where $\epsilon$ is the sum of the kinetic and potential energy $V$

$$\epsilon = \frac{p_1{}^2}{2m_1} + \frac{p_2{}^2}{2m_2} + \cdots + \frac{p_n{}^2}{2m_n} + V(x_1 \cdots x_n) \quad (2)$$

and $\beta$ is the reciprocal temperature $T$ divided by the Boltzmann constant

$$\beta = 1/kT. \quad (3)$$

In quantum theory there does not exist any similar simple expression for the probability, because one cannot ask for the simultaneous probability for the coordinates and momenta. Moreover, it is not possible to derive a simple expression even for the relative probabilities of the coordinates alone—as is given in classical theory by $e^{-\beta V(x_1 \cdots x_n)}$. One sees this by considering that this expression would give at once the square of the wave function of the lowest state $|\psi_0(x_1 \cdots x_n)|^2$ when $\beta = \infty$ is inserted and on the other hand we know that it is not possible, in general, to derive a closed formula for the latter.

The thermodynamics of quantum mechanical systems is in principle, however, given by a formula of Neumann,[1] who has shown that the mean value of any physical quantity is, (apart from a normalizing constant depending only on temperature), the sum of the diagonal elements of the matrix

$$Qe^{-\beta H} \quad (4)$$

where $Q$ is the matrix (operator) of the quantity under consideration and $H$ is the Hamiltonian of the system. As the diagonal sum is an invariant under

---

[1] J. von Neumann, Gött. Nachr. p. 273, 1927.

E. WIGNER

transformations, one can choose any matrix or operator-representation for the $Q$ and $H$. In building the exponential of $H$ one must, of course, take into account the non-commutability of the different parts of $H$.

## 2

It does not seem to be easy to make explicit calculations with the form (4) of the mean value. One may resort therefore to the following method.

If a wave function $\psi(x_1 \cdots x_n)$ is given one may build the following expression[2]

$$P(x_1, \cdots, x_n; p_1, \cdots, p_n)$$
$$= \left(\frac{1}{h\pi}\right)^n \int_{-\infty}^{\infty} \cdots \int dy_1 \cdots dy_n \psi(x_1 + y_1 \cdots x_n + y_n)^*$$
$$\psi(x_1 - y_1 \cdots x_n - y_n) e^{2i(p_1 y_1 + \cdots + p_n y_n)/h} \quad (5)$$

and call it the probability-function of the simultaneous values of $x_1 \cdots x_n$ for the coordinates and $p_1 \cdots p_n$ for the momenta. In (5), as throughout this paper, $h$ is the Planck constant divided by $2\pi$ and the integration with respect to the $y$ has to be carried out from $-\infty$ to $\infty$. Expression (5) is real, but not everywhere positive. It has the property, that it gives, when integrated with respect to the $p$, the correct probabilities $|\psi(x_1 \cdots x_n)|^2$ for the different values of the coordinates and also it gives, when integrated with respect to the $x$, the correct quantum mechanical probabilities

$$\left| \int_{-\infty}^{\infty} \cdots \int \psi(x_1 \cdots x_n) e^{-i(p_1 x_1 + \cdots + p_n x_n)/h} dx_1 \cdots dx_n \right|^2$$

for the momenta $p_1, \cdots, p_n$. The first fact follows simply from the theorem about the Fourier integral and one gets the second by introducing $x_k + y_k = u_k; x_k - y_k = v_k$ into (5).

Hence it follows, furthermore, that one may get the correct expectation values of any function of the coordinates or the momenta for the state $\psi$ by the normal probability calculation with (5). As expectation values are additive this even holds for a sum of a function of the coordinates and a function of the momenta as, e.g., the energy $H$. In formulas, it is

$$\int_{-\infty}^{\infty} \cdots \int \int_{-\infty}^{\infty} \cdots \int dx_1 \cdots dx_n dp_1 \cdots dp_n [f(p_1 \cdots p_n) + g(x_1 \cdots x_n)]$$
$$P(x_1 \cdots x_n; p_1 \cdots p_n)$$
$$= \int_{-\infty}^{\infty} \cdots \int \psi(x_1 \cdots x_n)^* \left[ f\left(\frac{h}{i} \frac{\partial}{\partial x_1}, \cdots, \frac{h}{i} \frac{\partial}{\partial x_n}\right) \right. \quad (6)$$
$$\left. + g(x_1 \cdots x_n) \right] \psi(x_1 \cdots x_n) dx_1 \cdots dx_n$$

for any $\psi, f, g$, if $P$ is given by (5).

[2] This expression was found by L. Szilard and the present author some years ago for another purpose.

Of course $P(x_1, \cdots, x_n; p_1, \cdots, p_n)$ cannot be really interpreted as the simultaneous probability for coordinates and momenta, as is clear from the fact, that it may take negative values. But of course this must not hinder the use of it in calculations as an auxiliary function which obeys many relations we would expect from such a probability. It should be noted, furthermore, that (5) is not the only bilinear expression in $\psi$, which satisfies (6). There must be a great freedom in the expression (5), as it makes from a function $\psi$ of $n$ variables one with $2n$ variables. It may be shown, however, that there does not exist any expression $P(x_1 \cdots x_n; p_1 \cdots p_n)$ which is bilinear in $\psi$, satisfies (6) and is everywhere (for all values of $x_1, \cdots, x_n, p_1, \cdots, p_n$) positive, so (5) was chosen from all possible expressions, because it seems to be the simplest.

If $\psi(x_1, \cdots, x_n)$ changes according to the second Schrödinger equation

$$ih\frac{\partial \psi}{\partial t} = - \sum_{k=1}^{n} \frac{h^2}{2m_k} \frac{\partial^2 \psi}{\partial x_k^2} + V(x_2, \cdots, x_n)\psi \tag{7}$$

the change of $P(x_1, \cdots, x_n; p_1, \cdots, p_n)$ is given by

$$\frac{\partial P}{\partial t} = - \sum_{k=1}^{n} \frac{p_k}{m_k}\frac{\partial P}{\partial x_k} + \sum \frac{\partial^{\lambda_1+\cdots+\lambda_n}V}{\partial x_1^{\lambda_1}\cdots\partial x_n^{\lambda_n}} \frac{(h/2i)^{\lambda_1+\cdots+\lambda_n-1}}{\lambda_1!\cdots\lambda_n!} \frac{\partial^{\lambda_1+\cdots+\lambda_n}P}{\partial p_1^{\lambda_1}\cdots\partial p_n^{\lambda_n}} \tag{8}$$

where the last summation has to be extended over all positive integer values of $\lambda_1, \cdots, \lambda_n$ for which the sum $\lambda_1+\lambda_2+ \cdots +\lambda_n$ is odd. In fact we get for $\partial P/\partial t$ by (5) and (7)

$$\frac{\partial P}{\partial t} = \frac{1}{(h\pi)^n}\int\cdots\int dy_1\cdots dy_n e^{2i(p_1y_1+\cdots+p_ny_n)/h}$$

$$\cdot\left\{ \sum_k \frac{ih}{2m_k}\left[ -\frac{\partial^2 \psi(x_1+y_1, \cdots, x_n+y_n)^*}{\partial x_k^2} \psi(x_1-y_1, \cdots, x_n-y_n) \right.\right.$$

$$+ \psi(x_1+y_1, \cdots, x_n+y_n)^* \frac{\partial^2 \psi(x_1-y_1, \cdots, x_n-y_n)}{\partial x_k^2} \tag{9}$$

$$+ \frac{i}{h}[V(x_1+y_1, \cdots, x_n+y_n)$$

$$- V(x_1-y_1, \cdots, x_n-y_n)]\psi(x_1+y_1, \cdots, x_n+y_n)^*\psi(x_1-y_1, \cdots x_n-y_n)\Big\}.$$

Here one can replace the differentiations with respect to $x_k$ by differentiations with respect to $y_k$ and perform in the first two terms one partial integration with respect to $y_k$. In the last term we can develop $V(x_1+y_1, \cdots, x_n+y_n)$ and $V(x_1-y_1, \cdots, x_n-y_n)$ in a Taylor series with respect to the $y$ and get

$$\frac{\partial P}{\partial t} = \frac{1}{(\pi h)^n}\int\cdots\int dy_1\cdots dy_n e^{2i(p_1y_1+\cdots+p_ny_n)/h}$$

$$\cdot\left\{ \sum_k \frac{p_k}{m_k}\left[ -\frac{\partial \psi(x_1+y_1, \cdots, x_n+y_n)^*}{\partial y_k} \psi(x_1-y_1, \cdots, x_n-y_n) \right.\right.$$

$$(10)$$

$$+ \psi(x_1 + y_1, \cdots, x_n + y_n)^* \frac{\partial \psi(x_1 - y_1, \cdots, x_n - y_n)}{\partial y_k} \Big]$$

$$+ \frac{i}{h} \sum_\lambda \frac{\partial^{\lambda_1 + \cdots + \lambda_n} V}{\partial x_1^{\lambda_1} \cdots \partial x_n^{\lambda_n}} \frac{y_1^{\lambda_1} \cdots y_n^{\lambda_n}}{\lambda_1! \cdots \lambda_n!} \psi(x_1 + y_1, \cdots, x_n + y_n)^*$$

$$\cdot \psi(x_1 - y_1, \cdots, x_n - y_n) \Big\},$$

which is identical with (8) if one replaces now the differentiations with respect to $y_k$ by differentiations with respect to $x_k$. Of course, (8) is legitimate only if it is possible to develop the potential energy $V$ in a Taylor series.

Eq. (8) shows the close analogy between the probability function of the classical mechanics and our $P$: indeed the equation of continuity

$$\frac{\partial P}{\partial t} = - \sum_k \frac{p_k}{m_k} \frac{\partial P}{\partial x_k} + \sum_k \frac{\partial V}{\partial x_k} \frac{\partial P}{\partial p_k}$$

differs from (8) only in terms of at least the second power of $h$ and at least the third derivative of $V$. Expression (8) is even identical with the classical when $V$ has no third and higher derivatives as, e.g., in a system of oscillators.

There is an alternative form for $\partial P/\partial t$, which however will not be used later on. It is

$$\frac{\partial}{\partial t} P(x_1, \cdots, x_n; p_1, \cdots, p_n) = - \sum_k \frac{p_k}{m_k} \frac{\partial}{\partial x_k} P(x_1, \cdots, x_n; p_1, \cdots, p_n)$$

$$(11)$$

$$+ \int_{-\infty}^{\infty} \cdots \int dj_1 \cdots dj_n P(x_1, \cdots, x_n; P_1 + j_1, \cdots, P_n + j_n) J(x_1, \cdots, x_n;$$

$$j_1, \cdots, j_n)$$

where $J(x_1, \cdots, x_n; j_1, \cdots, j_n)$ can be interpreted as the probability of a jump in the momenta with the amounts $j_1, \cdots, j_n$ for the configuration $x_1, \cdots, x_n$. The probability of this jump is given by

$$J(x_1, \cdots, x_n; j_1, \cdots, j_n)$$

$$= \frac{i}{\pi^n h^{n+1}} \int_{-\infty}^{\infty} \cdots \int dy_1 \cdots dy_n [V(x_1 + y_1, \cdots, x_n + y_n)$$

$$- V(x_1 - y_1, \cdots, x_n - y_n)] e^{-(2i/h)(y_1 j_1 + \cdots + y_n j_n)} \quad (11a)$$

that is, by the Fourier expansion coefficients of the potential $V(x_1, \cdots, x_n)$. This form clearly shows the quantum mechanical nature of our $P$: the momenta change discontinuously by amounts which would be half the momenta of light quanta if the potential were composed of light.[2a] To derive (11) one can insert both for $P$ and $J$ their respective values (5) and (11a) on the right hand side of (11). In the first term one can replace $p_k e^{2i(p_1 v_1 + \cdots + p_n v_n)/h}$ by

[2a] Cf. F. Bloch, Zeits. f. Physik **52**, 555 (1929).

$(h/2i)(\partial/\partial y_k)e^{2i(p_1y_1+\cdots+p_ny_n)/h}$ and then perform a partial integration with respect to $y_k$. Then one can replace the differentiation with respect to $y$ by differentiation with respect to $x$, upon which some terms cancel and the rest goes over to

$$\sum_k \frac{h}{2im} \int \cdots \int dy_1 \cdots dy_n \left[ \frac{\partial^2 \psi(x_1+y_1,\cdots,x_n+y_n)^*}{\partial x_k^2} \psi(x_1-y_1,\cdots,x_n-y_n) \right.$$

$$\left. - \psi(x_1+y_1,\cdots,x_n+y_n) \frac{\partial^2 \psi(x_1-y_1,\cdots,x_n-y_n)}{\partial x_k^2} \right] e^{2i(p_1y_1+\cdots+p_ny_n)/h} \quad (12)$$

which is just what we need for the left side of (11). By integrating the second term on the right side of (11)

$$\int \cdots \int dy_1 \cdots dy_n \psi(x_1+y_1 \cdots x_n+y_n)^* \psi(x_1-y_1 \cdots x_n-y_n)$$

$$\cdot \int \cdots \int dj_1 \cdots dj_n e^{(2i/h)[(p_1+j_1)y_1+\cdots+(p_n+j_n)y_n]}$$

$$\cdot \frac{i}{\pi^n h^{n+1}} \int \cdots \int dz_1 \cdots dz_n [V(x_1+z_1 \cdots x_n+z_n)$$

$$- V(x_1-z_1 \cdots x_n-z_n)] e^{-2i(z_1j_1+\cdots+z_nj_n)/h}$$

with respect to $z$ and $j$ one gets because of the Fourier theorem[3]

$$(i/h) \int \cdots \int dy_1 \cdots dy_n \psi(x_1+y_1 \cdots x_n+y_n)^* \psi(x_1-y_1 \cdots x_n-y_n)$$
$$e^{2i(p_1y_1+\cdots+p_ny_n)/h} \cdot [V(x_1+y_1 \cdots x_n+y_n) - V(x_1-y_1 \cdots x_n-y_n)] \quad (12a)$$

and this gives the second part of the left side of (11).

### 3

So far we have defined only a probability function for pure states, which gives us the correct expectation values for quantities $f(p_1\cdots p_n) + g(x_1\cdots x_n)$. If, however, we have a mixture,[4] e.g., the pure states $\psi_1$, $\psi_2$, $\psi_3$, $\cdots$ with the respective probabilities $w_1$, $w_2$, $w_3$, $\cdots$ (with $w_1+w_2+w_3+\cdots=1$) the normal probability calculation suggests a probability function

$$P(x_1,\cdots,x_n,p_1,\cdots,p_n) = \sum_\lambda w_\lambda P_\lambda(x_1,\cdots,x_n,\cdots,p_n) \quad (13)$$

where $P_\lambda$ is the probability function for $\psi_\lambda$. This probability function gives obviously the correct expectation values for all quantities, for which (5) gives correct expectation values and therefore will be adopted.

For a system in statistical equilibrium at the temperature $T=1/k\beta$ the relative probability of a stationary state $\psi_\lambda$ is $e^{-\beta E_\lambda}$ where $E_\lambda$ is the energy of $\psi_\lambda$. Therefore the probability function is a part from a constant

[3] Cf. e. g., R. Courant und D. Hilbert, Methoden der mathematischen Physik I. Berlin 1924. p. 62, Eq. (29).

[4] J. v. Neumann, Gött Nachr. 245, 1927. L. Landau, Zeits. f. Physik **45**, 430 (1927).

$P(x_1 \cdots x_n; p_1 \cdots p_n)$

$$= \sum_\lambda \int \cdots \int dy_1 \cdots dy_n \psi_\lambda(x_1 + y_1 \cdots x_n + y_n)^*$$

$$e^{-\beta E_\lambda}\psi(x_1 - y_1 \cdots x_n - y_n)e^{2i(p_1 y_1 + \cdots + p_n y_n)/h}. \quad (14)$$

Now

$$\sum_\lambda \psi_\lambda (u_1 \cdots u_n)^* f(E_\lambda) \psi_\lambda (v_1 \cdots v_n)$$

is that matrix element of the operator $f(H)$, ($H$ is the energy operator) which is in the $u_1 \cdots u_n$ row and $v_1 \cdots v_n$ column. Therefore (14) may be written as

$P(x_1 \cdots x_n; p_1 \cdots p_n)$

$$= \int_{-\infty}^{\infty} \cdots \int dy_1 \cdots dy_n e^{i[(x_1+y_1)p_1+\cdots+(x_n+y_n)p_n]/h} \left[e^{-\beta H}\right]_{x_1+y_1\cdots x_n+y_n; x_1-y_1\cdots x_n-y_n}$$

$$\cdot e^{-i[(x_1-y_1)p_1+\cdots+(x_n-y_n)p_n]/h}. \quad (15)$$

so that we have under the integral sign the $x_1+y_1 \cdots x_n+y_n$; $x_1-y_1 \cdots x_n-y_n$ element of the matrix $e^{-\beta H}$ transformed by the diagonal matrix $e^{i(p_1 x_1+\cdots+p_n x_n)/h}$. Instead of transforming $e^{-\beta H}$ we can transform $H$ first and then take the exponential with the transformed expression. By transforming $H$ we get the operator (the $p$ are numbers, not operators!)

$$H = e^{i(x_1 p_1+\cdots+x_n p_n)/h}\left(- \sum_k \frac{h^2}{2m_k} \frac{\partial^2}{\partial x_k^2} + V(x_1 \cdots x_n)\right)e^{-i(x_1 p_1+\cdots+x_n p_n)/h}$$

which is equal to

$$\tilde{H} = \epsilon + \sum_{k=1}^n \left(\frac{ihp_k}{m_k} \frac{\partial}{\partial x_k} - \frac{h^2}{2m_k} \frac{\partial^2}{\partial x_k^2}\right) \quad (16)$$

where

$$\epsilon = \sum_{k=1}^n \frac{p_k^2}{2m_k} + V(x_1, \cdots, x_n). \quad (17)$$

So we get for (15)

$P(x_1, \cdots, x_n; p_1, \cdots, p_n)$

$$= \int \cdots \int dy_1 \cdots dy_n \left[e^{-\beta \tilde{H}}\right]_{x_1+y_1\cdots x_n+y_n; x_1-y_1\cdots x_n-y_n}. \quad (18)$$

By calculating the mean value of a quantity $Q = f(p_1, \cdots, p_n) + g(x_1, \cdots, x_n)$ by (18) one has to obtain the same result as by using the original expression (4) of Neumann.

If we are dealing with a system, the behavior of which in statistical equilibrium is nearly correctly given by the classical theory, we can expand (18) into a power of $h$ and keep the first few terms only. The term with the zero power of $h$ is $\sum_r (-\beta)^r \epsilon^r / r!$ Now $\epsilon^r$ is the operator of multiplication with the $r$

power of (17). Its $x_1+y_1, \cdots, x_n+y_n; x_1-y_1, \cdots, x_n-y_n$ element is consequently

$$\epsilon(x_1 + y_1, \cdots, x_n + y_n)^r \delta(x_1 + y_1, x_1 - y_1) \cdots \delta(x_n + y_n, x_n - y_n).$$

As $\delta$ (also $\delta', \delta'', \cdots$) only depends on the difference of its two arguments, one can write $\delta(-2y_1) \cdots \delta(-2y_n)$ for the last factors and perform the integration by introducing $-2y_1, \cdots, -2y_n$ as new variables. The terms with the zero power of $h$, arising from the first part of (16) only, give thus

$$(1/2^n) \sum_r (-\beta)^r \epsilon(x_1, \cdots, x_n)^r/r! = e^{-\beta\epsilon}/2^n \tag{19}$$

which is just the classical expression.

The higher approximations of the probability function can be calculated in a very similar way. The terms of $e^{-\beta\tilde{H}}$, involving the first power of the second part of $\tilde{H}$ only, are

$$\sum_{r=0}^{\infty} \frac{(-\beta)^r}{r!} \sum_{\rho=1}^{r} \epsilon^{\rho-1} \sum_k \left( \frac{ihp_k}{m_k} \frac{\partial}{\partial x_k} - \frac{h^2}{2m_k} \frac{\partial^2}{\partial x_k^2} \right) \epsilon^{r-\rho} \tag{20}$$

By replacing all operators by symbolic integral-kernels one gets for the $x_1+y_1, \cdots, x_n+y_n; x_1-y_1, \cdots, x_n-y_n$ element of the operator (20)

$$\sum_r \frac{(-\beta)^r}{r!} \sum_{\rho=1}^{r} \epsilon(x_1 + y_1, \cdots, x_n + y_n)^{\rho-1}$$

$$\cdot \sum_k \left[ \frac{ihp_k}{m_k} \delta(-2y_1) \cdots \delta'(-2y_k) \cdots \delta(-2y_n) \right.$$

$$\left. - \frac{h^2}{2m_k} \delta(-2y_1) \cdots \delta''(-2y_k) \cdots \delta(-2y_n) \right] \epsilon(x_1 - y_1, \cdots, x_n - y_n)^{r-\rho}.$$

Now

$$\sum_{\rho=1}^{r} \epsilon_+^{\rho-1} \epsilon_-^{r-\rho} = \sum_{\rho=1}^{r} \epsilon_-^{r-1} \left( \frac{\epsilon_+}{\epsilon_-} \right)^{\rho-1} = \frac{\epsilon_+^r - \epsilon_-^r}{\epsilon_+ - \epsilon_-}$$

so that the summation over $\rho$ and $r$ can be performed in (21). By introducing again new variables $w_1, \cdots, w_n$ for $-2y_1, \cdots, -2y_n$ and performing the integration one has

$$\frac{1}{2^n} \sum_k \left[ \frac{ihp_k}{m_k} \frac{\partial}{\partial w_k} - \frac{h^2}{2m_k} \frac{\partial^2}{\partial w_k^2} \right]$$

$$\frac{e^{-\beta\epsilon(x_1, \cdots, x_k - w_k/2, \cdots, x_n)} - e^{-\beta\epsilon(x_k, \cdots, x_k + w_k/2, \cdots, x_n)}}{\epsilon(x_1, \cdots, x_k - \frac{1}{2}w_k, \cdots, x_n) - \epsilon(x_1, \cdots, x_k + \frac{1}{2}w_k, \cdots, x_n)}$$

where $w_k = 0$ must be inserted after differentiation. The first differential quotient vanishes at $w_k = 0$, as the expression to be differentiated is an even function of $w_k$. The second part gives

$$\frac{e^{-\beta\epsilon}}{2^n} \sum_k \frac{h^2}{m_k} \left( -\frac{\beta^2}{8} \frac{\partial^2\epsilon}{\partial x_k^2} + \frac{\dot\beta^3}{24} \left(\frac{\partial\epsilon}{\partial x_k}\right)^2 \right). \tag{21}$$

In principle it is possible to calculate in the same way the terms involving the higher powers of the second part of $\bar H$ also, the summation over $r$ and the quantities corresponding to our $\rho$ can always be performed in a very similar way. In practice, however, the computation becomes too laborious. Still it is clear, that if we develop our probability function for thermal equilibrium in a power series of $h$

$$P(x_1, \cdots, x_n; p_1, \cdots, p_n) = e^{-\beta\epsilon} + hf_1 + h^2 f_2 + \cdots \tag{22}$$

(we can omit the factor $1/2^n$ before $e^{-\beta\epsilon}$, as we are dealing with relative probabilities anyway) all terms will be quite definite functions of the $p$, $V$ and the different partial derivatives of the latter. Furthermore it is easy to see, that $f_k$ will not involve higher derivatives of $V$ than the $k$-th nor higher powers of $p$ than the $k$-th. These facts enable us to calculate the higher terms of (22) in a somewhat simpler way, than the direct expansion of (18) would be.

The state (22) is certainly stationary, so that it would give identically $\partial P/\partial t = 0$ when inserted into (8). By equating the coefficients of the different powers of $h$ in $\partial P/\partial t$ to zero one gets the following equations:

$$\sum_k -\frac{p_k}{m_k} \frac{\partial e^{-\beta\epsilon}}{\partial x_k} + \sum_k \frac{\partial V}{\partial x_k} \frac{\partial e^{-\beta\epsilon}}{\partial p_k} = 0 \tag{23, 0}$$

$$\sum_k -\frac{p_k}{m_k} \frac{\partial f_1}{\partial x_k} + \sum_k \frac{\partial V}{\partial x_k} \frac{\partial f_1}{\partial p_k} = 0 \tag{23, 1}$$

$$\sum_k -\frac{p_k}{m_k} \frac{\partial f_2}{\partial x_k} + \sum_k \frac{\partial V}{\partial x_k} \frac{\partial f_2}{\partial p_k} - \sum_k \frac{\partial^3 V}{\partial x_k^3} \frac{h^2}{24} \frac{\partial^3 e^{-\beta\epsilon}}{\partial p_k^3}$$
$$- \sum_{k \neq l} \frac{\partial^3 v}{\partial x_k^2 \partial x_l} \frac{h^2}{8} \frac{\partial^3 e^{-\beta\epsilon}}{\partial p_k^2 \partial p_l} = 0 \tag{23, 2}$$

and so on. The first of these equations is an identity because of (17), as it must be; (23, a), (23, 2), $\cdots$ will determine $f_1$, $f_2$, $\cdots$ respectively. All Eqs. (23, a) are linear inhomogeneous partial differential equations for the unknown $f$. From one solution $f_a$ of (23, a) one obtains the general solution by adding to it the general solution $F$ of the homogeneous part of (23, a), which is always

$$\sum_k -\frac{p_k}{m_k} \frac{\partial F}{\partial x_k} + \sum_k \frac{\partial V}{\partial x_k} \frac{\partial F}{\partial p_k} = 0.$$

This equation in turn is the classical equation for the stationary character of the probability distribution $F(x_1, \cdots, x_n; p_1, \cdots, p_n)$. It has in general only one solution which contains only a finite number of derivatives of $V$, namely

$$F(x_1, \cdots, x_n; p_1, \cdots, p_n) = F\left( \sum_k \frac{p_k^2}{2m_k} + v(x_1 \cdots x_n) \right) = F(\epsilon).$$

In fact, if it had other integrals, like

$$F(p_1, \cdots, p_n; V, \partial V/\partial x_1, \partial V/\partial x_2, \cdots) \tag{24}$$

then all mechanical problems would have in addition to the energy-integral further integrals of the form (24) which, of course, is not true.

One solution of (23, 1) is $f_1 = 0$ and the most general we have to consider is therefore $f_1 = F(\epsilon)$. We have to take however $F(\epsilon) = 0$ as $f_1$ has to vanish for a constant $V$. So we get $f_1 = 0$, as we know it already from the direct expansion of (18). The same holds consequently for $f_3, f_5, \cdots$, as the inhomogeneous part of the equation for $f_3$ only contains $f_1$, the inhomogeneous part of the equation for $f_5$ only $f_1$ and $f_3$, and so on.

For $f_2$ one easily gets

$$f_2 = e^{-\beta\epsilon}\left[\sum_k\left(-\frac{\beta^2}{8m_k}\frac{\partial^2 V}{\partial x_k{}^2} + \frac{\beta^3}{24m_k}\left(\frac{\partial V}{\partial x_k}\right)^2\right) + \sum_{k,l}\frac{\beta^3 p_k p_l}{24 m_k m_l}\frac{\partial^2 V}{\partial x_k \partial x_l}\right] \tag{25}$$

as a solution of (23, 2) and it is also clear, that this is the solution we need. The first two terms of $f_2$ we have already directly computed (21), the third arises from terms with the second power of the second part of $\tilde{H}$. Similarly $f_4$ is for one degree of freedom $(n = 1)$

$$\begin{aligned}
64 m^2\beta^{-2}\,e^{\beta\epsilon}f_4 = \;& H_4(q)\left[\beta^2 V''{}^2/72 - \beta V''''/120\right] \\
& + H_2(q)\left[\beta^3 V'{}^2 V''/18 - 2\beta^2 V''{}^2/15 - \beta^2 V'V'''/15 + \beta V''''/15\right] \\
& + H_0(q)\left[\beta^4 V'{}^4/18 - 22\beta^3 V'{}^2 V''/45 + 2\beta^2 V''{}^2/5 + 8\beta^2 V'V'''/15 \right. \\
& \left. - 4\beta V''''/15\right]
\end{aligned} \tag{26}$$

where $H_r$ is the $r$-th Hermitean polynomial and $q = \beta^{1/2}p/(2m)^{1/2}$.

It does not seem to be easy to get a simple closed expression for $f_k$, but it is quite possible to calculate all of them successively. A discussion of Eqs. (23) shows, that the $g$ in

$$P(x_1, \cdots, x_n; p_1, \cdots, p_n) = e^{-\beta\epsilon}(1 + h^2 g_2 + h^4 g_4 + \cdots) \tag{27}$$

are rational expressions in the derivatives of $V$ only (do not contain $V$ itself) and all terms of $g_k$ contain $k$ differentiations and as functions of the $p$ are polynomials of not higher than the $k$-th degree. The first term in (27) with the zero power of $h$ is the only one, which occurs in classical theory. There is no term with the first power, so that if one can develop a property in a power series with respect to $h$, the deviation from the classical theory goes at least with the second power of $h$ in thermal equilibrium. One familiar example for this is the inner energy of the oscillator, where the term with the first power of $h$ vanishes just in consequence of the zero point energy. The second term can be interpreted as meaning that a quick variation of the probability function with the coordinates is unlikely, as it would mean a quick variation, a short wave-length, in the wave functions. This however would have the consequence of a high kinetic energy. The quantum mechanical probability is therefore something like the integral of the classical expression $e^{-\beta\epsilon}$ over a finite range of coordinates of the magnitude $\sim h/\bar{p}$ where $\bar{p}$ is the mean momentum $\sim(kTm)^{1/2}$. The correction terms of (27) have, among other effects, the consequence that the probability for a particle being in a narrow hole is smaller than would be in classical statistics. From now on we will keep only the first two terms of (27).

### 4

From (25) one easily calculates the relative probabilities of the different configurations by integration with respect to the $p$:

$$\int \cdots \int dp_1 \cdots dp_n P(x_1 \cdots x_n; p_1 \cdots p_n)$$

$$= e^{-\beta V}\left[1 - \frac{h^2\beta^2}{12} \sum_k \frac{1}{m_k} \frac{\partial^2 V}{\partial x_k^2} + \frac{h^2\beta^3}{24} \sum_k \frac{1}{m_k}\left(\frac{\partial V}{\partial x_k}\right)^2\right]. \quad (28)$$

Hence the mean potential energy is

$$\bar{V} = \frac{\int V e^{-\beta V} dx}{\int e^{-\beta V} dx} + \frac{h^2\beta^2}{24} \frac{\int \sum_k \frac{1}{m_k} \frac{\partial^2 V}{\partial x_k^2} e^{-\beta V} dx \int V e^{-\beta V} dx}{\left(\int e^{-\beta V} dx\right)^2}$$

$$+ \frac{h^2\beta}{24} \frac{\int \sum_k \frac{1}{m_k} \frac{\partial^2 V}{\partial x_k^2}(1 - \beta V) e^{-\beta V} dx}{\int e^{-\beta V} dx} \quad (29)$$

where $dx$ is written for $dx_1 \cdots dx_n$ and the higher power terms of $h$ are omitted. Similarly the mean value of the kinetic energy is

$$\sum_k \frac{\overline{p_k^2}}{2m_k} = \frac{n}{2\beta} + \frac{h^2\beta}{24} \frac{\int \sum_k \frac{1}{m_k} \frac{\partial^2 V}{\partial x_k^2} e^{-\beta V} dx}{\int e^{-\beta V} dx}. \quad (30)$$

This formula also is correct only within the second power of $h$; in order to derive it one has to perform again some partial integrations with respect to the $x$. Eqs. (28), (29), (30) have a strict quantum mechanical meaning and it should be possible to derive them also from (4). One sees that the kinetic energy is in all cases larger than the classical expression $\frac{1}{2}nkT$.

### 5

One fact still needs to be mentioned. We assumed that the probability of a state with the energy $E$ is given by $e^{-\beta E}$. This is not true in general, since the Pauli principle forbids some states altogether. The corrections thus introduced by the Bose or Fermi statistics even give terms with the first power of $h$, so that it seems, that as long as one cannot take the Bose of Fermi statistics into account, Eq. (25) cannot be applied to an assembly of identical particles, as, e.g., a gas. There is reason to believe however, that because of the large radii of the atoms this is not true and the corrections due to Fermi and Bose statistics may be neglected for moderately low temperatures.

The second virial coefficient was first calculated in quantum mechanics by F. London on the basis of his theory of inneratomic forces.[5] He also pointed out that quantum effects should be taken into account at lower temperatures. Slater and Kirkwood[6] gave a more exact expression for the inneratomic potential of He and Kirkwood and Keyes[7] calculated on this basis the classical part of the second virial coefficient of He. H. Margenau[8] and Kirkwood[9] performed the calculations for the quantum-correction. The present author also tried to calculate it by the method just outlined. He got results, which differ from those of Margenau and Kirkwood in some cases by more than 100 percent.[10] It does not seem however to be easy to compare these results with experiment, as the classical part of the second virial coefficient is at low temperatures so sensitive to small variations of the parameters occurring in the expression of the interatomic potential, that it changes by more than 20 percent if the parameter in the exponential (2.43) is changed by $\frac{1}{2}$ percent and it does not seem to be possible to determine the latter within this accuracy.

[5] F. London, Zeits. f. Physik **63**, 245 (1930).

[6] J. C. Slater and J. G. Kirkwood, Phys. Rev. **37**, 682 (1931).

[7] J. G. Kirkwood and F. G. Keyes, Phys. Rev. **38**, 516 (1931).

[8] H. Margenau, Proc. Nat. Acad. **18**, 56, 230 (1932). Cf. also J. C. Slater, Phys. Rev. **38**, 237 (1931).

[9] J. G. Kirkwood, Phys. Zeits. **33**, 39 (1932).

[10] I am very much indebted to V. Rojansky for his kind assistance with these calculations. The reason for the disagreement between our results and those of Margenau and Kirkwood may be the fact that they did not apply any corrections for the continuous part of the spectrum.

In a paper which appeared recently in the Zeits. f. Physik (**74**, 295 (1932)) F. Bloch gets results which are somewhat similar to those of the present paper. (*Note added at proof.*)

# Contributions to the Theory of the Neutron*

## E. P. Wigner

(Original in Hungarian with German abstract.) Adalékok a Neutron Elmélltéhez. Magyar Tudományos Akadémia Matematikai és Természettudományi Értesitöye *49*, 142–146 (1932). Translated from Hungarian by Akos Sebestyén.

1. Pauli [1], striving to explain the continuous spectrum of $\beta$-particles, developed a theory of the neutron. It is well known that various atoms of a radioactive element emit electrons of various energies, despite the fact that both the energies of the disintegrating atom and of the products of the decay are always the same. The later condition can be concluded from the fact that the energies of the $\alpha$-particles preceding and following the $\beta$-particle are constant.

We can interpret this experimental fact in two ways, either the principle of conservation of energy has to be abandoned (Bohr) or we have to suppose that neutral particles called "neutrons" having a continuous energy-spectrum are emitted simultaneously with the $\beta$-particles so that the sum of the energy of the neutron and the electron emitted by an atom is constant.

L. M. Mott-Smith and G. L. Locher [2] have pointed out the possibility that cosmic radiation consistes of neutrons. Sometime later Chadwick's [3] has brought forward a number of arguments that the radiation emitted by Beryllium under the influence of $\alpha$-radiation and detected by Curie-Joliot and Joliot consisted of neutrons. Chadwick's[1] experiments represent the most definite proofs on the behalf of the neutron-theory at present.

The mass of the neutrons observed by Chadwick is undoubtedly identical to the mass of the protons as was already pointed out by Chadwick himself. Though directly from the experiments it follows only that the magnitude of the mass of these particles and that of the proton is the same. However from the atomic weights being always integer multiples of the mass of the proton it follows that only very light particles and those of the mass of the proton can exist in nuclei.

Chadwick's neutrons, which we shall call heavy neutrons in what follows, are not suitable to explain continuous $\beta$-radiation. In $\beta$-radiation the atomic weight remains constant whereas the loss of a heavy neutron would decrease the atomic weight by one. Moreover it would be possible to detect heavy neutrons emerging simultaneously with $\beta$-radiation, as experiments show that their ionising capability is not so small that this would be impossible. Hence if we want to explain the continuous $\beta$-radiation without violating the energy-principle we

---

* *Editor's Note:* At the time that Wigner wrote this article, the name neutrino had not been adopted for Pauli's light neutral particle. The reader is warned that the name neutron is often used here for what we now call neutrino.

have to postulate another kind of neutron – which we shall call light neutron in what follows – and whose mass is so small that it does not manifest itself in the atomic weight. For reasons of symmetry it is the simplest to suppose that its mass is equal to the mass of the electron.

Now the main question obviously is: Is there any reason to suppose that the ionising power of the light neutron is substantially smaller than that of the heavy neutron? Indeed it is possible to bring forward reasons for that. In general at big velocities the mean range of heavy particles is shorter than that of light particles; examples for this are the $\beta$ and $\alpha$-radiations. Born has shown generally that the collision cross-section of neutrons with other bodies is proportional to the square of the relative mass $(m_1 m_2/(m_1 + m_2))$ of the two colliding bodies. Bohr has pointed[1] out that this explains why the heavy neutrons collide mostly with nuclei; in this case the relative mass is about a thousand times larger than for collisions with electrons. The same rule shows also that the collision cross section of light neutrons is much smaller than the collision cross section of heavy neutrons. Namely it is more probable that the most effective collisions will take place with electrons. In this case the relative mass is not significantly smaller than for collisions with nuclei (the ratio is about 0,5) however the energy loss is much larger because in collisions with heavy bodies the light particle can lose only a small part of its energy due to the law of conservation of momentum.

Consequently the fact that demonstrations of rays of light neutrons emanating from $\beta$-radiating bodies have been so far unsuccessful is not an argument against light neutron radiation. Since every $\beta$-radiation is continuous we have to conclude that together with each electron a light neutron must leave he nucleus too, or in other words the number of electrons and light-neutrons must coincide in every nucleus.

Supposing this we get rid of a further grave difficulty which seemed to make the quantum mechanical treatment of nuclei hopeless. What I am thinking of is that neither the statistics nor the proper angular momenta of nuclei correspond to the rules [4] that can be deduced for the statistics and angular momenta of compound bodies from the basic principles of quantum mechanics using symmetry considerations only [5]. Experiments according to Heitler and Herzberg [6] show on the contrary that when particles in the nuclei are counted, electrons must be left out of consideration and only the atomic weight, or as it had to be understood then the number of protons, determine the integer or half integer properties of the statistics and angular momentum. If however we suppose that together with each electron a light neutron is also present in the nucleus then the total number of particles is even or odd according to even or odd number of protons and heavy neutrons as the number of light neutrons and electrons is always even. Hence the atomic weight, that is the sum of the numbers of protons and heavy neutrons determines the integer property of the statistics and the angular momentum.

---

[1]  Thanks are due to Mr. F. Sauter for acquainting me with this hitherto unpublished work of Bohr.

Perhaps it is needless to stress that these considerations are hypothetical. The only concrete evidence for the neutrons is provided by the fact that the properties of some radiations very abundant in energy are not consistent with the picture we have constructed on the ground of the properties of the electron, proton and rays of light known at present. However these properties have been detected in the case of radiations less abundant in energy and the extrapolation to radiations that are very energetic as the cosmic radiation or the Curie-Joliot radiation not known very well yet is by no means safe. Naturally we know hardly anything about neutrons yet consequently we may assume their characteristics to correspond to those of these radiations.

Note added in proof. Recently it became customery to suppose [7] that the neutron detected by Chadwick is not an elementary particle rather it consists of a proton and an electron tightly bound to it. This opinion is supported by Chadwick's observation who found the mass of the neutron to be lighter than the proton by about a mass of four electrons, and it is not likely that the masses of two elementary particles differ by 2/1000. According to Chadwick the mass of four electrons corresponds to the binding energy of the neutron.

If this conjecture holds its ground and the heavy neutron is not an elementary particle then the existence of Pauli's neutron an elementary particle with a mass of an electron becomes very unlikely.

However, besides that in this case we have no explanation for the anomalies of spin and statistics and for the continuous $\beta$-radiation, the stability of the hydrogen atom also contradicts this assumption. Namely it is possible to estimate the time needed for the transition of an H-atom from the state $2p$ to the lowest state, that is to a neutron, if such a state existed at all. This time is the reciprocal of the transition probability

$$\frac{64\pi^4 e^2 \nu^3}{3hc^3} r^2 , \tag{1}$$

where $\nu \sim 4mc^2/h$ is the frequency of the radiation, $r$ is the integral

$$r = \iiint z\psi_0(xyz)\psi_1(xyz)\, dx\, dy\, dz \tag{2}$$

$\psi_0$ being the wave function of the electron in the neutron state and $\psi_1$ is that in the $2p$ state. Since we only look for an estimate of (1) we may write

$$\psi_0 = \frac{1}{\sqrt{8\pi a_0^3}} e^{-r/2a_0} \tag{3}$$

where $a_0$ stands for the radius of the neutron and $\psi_1$ is known from the theory of the H-atom. For the transition probability we get

$$\frac{64\pi^4 2^{18} m^3 c^3}{3h^4}\left(\frac{a_0}{a_1}\right)^5 a_0^2 = \frac{2\pi mc^2}{3h}(2\alpha)^{13} \tag{4}$$

where on the right hand side we have substituted the radius of the neutron by $e^2/2mc^2$ and Sommerfeld's constant $\alpha$ is $\sim 1/137$. Thus for the transition probability we obtain $10^{-4}\,\text{sec}^{-1}$.

Naturally the hydrogen atom is not in the state $2p$ normally and the transition probability from the state $1s$ is zero. If however the H atom appears in a compound then the coefficient of the $2p$ state in its eigenfunction is not zero but is about 0,1 (for the probability of the $2p$ state this gives 0,01). This way the life of the H atom turns out to be about one month.

Hence in the afore-mentioned theory of the neutron not only the theory of the hydrogen atom must be abandoned but it must be accepted that the formula for the transition probability does not yield a correct value even in magnitude without comprehensive reason.

# References

[1] J. F. Carlson and J.R. Oppenheimer, Phys. Rev. *38*, 1787 (1931)
[2] Phys. Rev. *38*, 1399 (1931)
[3] Nature *129*, 469 (1932)
[4] R. de L. Kronig: Naturwissenschaften *16*, 355 (1928)
[5] J. Wigner: M. Tud. Értesitö (Reports of the Hung. Sci. Ac.) *46*, 576 (1929). See also P. Ehrenfest and J.R. Oppenheimer: Phys. Rev. *37*, 333 (1931)
[6] Naturwissenschaften, *17*, 673 (1929)
[7] Compare to W. Heisenberg: Zs. f. Physik *77*, 1 (1932)

(From the session of Division III of the Hungarian Scientific Academy, June 13, 1932.)

# BEITRÄGE ZUR THEORIE DES NEUTRONS.

## Von EUGEN WIGNER.

Wenn man annimmt, dass die kürzlich von CHADWICK beobachteten Neutronen elementare Partikel sind, wird es nahe-
liegend auch Neutronen von Elektronenmasse zu postulieren.
Wenn die Zahl dieser «leichten Neutronen» im Atomkern immer
gleich der Zahl der Kernelektronen ist, wird bei jedem $\beta$-Zerfall
ein leichtes Neutron frei. PAULI, von dem die Neutronenhypothese
herrührt, erklärt in dieser Weise den kontinuierlichen Charakter
der $\beta$-Strahlung, indem er annimmt, dass die Summe der Energien
des austretenden Neutrons und Elektrons immer dieselbe ist, und
so sowohl die Energie des Ausgangsstoffes, wie die des Zerfalls-
produkts scharf sein kann. Die Anwendung einer (bisher unpubli-
zierten) Überlegung von BOHR zeigt, dass das Ionisationsvermögen
der leichten Neutronen so klein sein muss, dass man sie mit den
heutigen Mitteln nicht beobachten kann.

Auch die statistischen und Spinanomalien der Kerne werden
durch die leichten Neutronen erklärt. Da die Summe der Anzahl
der Elektronen und der der leichten Neutronen in allen Kernen gerade ist, ändern sie weder die Statistik noch die Ganzzahligkeit
des Spins.

Wenn man dagegen voraussetzt, dass die CHADWICK-schen Neutronen lediglich Wasserstoffatome in einem bisher unbekannten,
besonders tiefen Zustand sind, muss man auch die gewöhnlichen
Formeln für die Übergangswahrscheinlichkeit aufgeben. Nach
diesen müssten nämlich die $H$-Atome einer Verbindung durch
Strahlung in etwa einem Monat in ein Neutron übergehen.

---

(Aus der Sitzung der III. Klasse der Ungarischen Akademie der Wissenschaften vom 13. Juni 1932.)

# Über die paramagnetische Umwandlung von Para-Orthowasserstoff. III

## E. P. Wigner

Zeitschrift für physikalische Chemie B *23*, 28–32 (1933)

(Eingegangen am 28. 6. 33.)

Es wird die Umwandlung von $p\,H_2$ in $o\,H_2$ (und umgekehrt) beim Stoss mit einem paramagnetischen Molekül berechnet. Die Stossausbeute der Umwandlung (sofern die Energie beim Zusammenstoss ausreicht) ist umgekehrt proportional zur absoluten Temperatur und quadratisch abhängig vom wirkenden Moment. Der Minimalabstand beim Zusammenstoss geht mit der — 6. Potenz in die Stossausbeute ein.

Im Anschluss an die vorangehenden Untersuchungen der Herren L. FARKAS und H. SACHSSE soll im folgenden die Geschwindigkeit der durch paramagnetische Gase bedingten Umwandlung von Parawasserstoff in Orthowasserstoff (und umgekehrt) abgeschätzt werden.

Die besondere Stabilität des Parawasserstoffes beruht auf einer Symmetrieeigenschaft des quantenmechanischen Energieoperators für das Wasserstoffmolekül: Er bleibt nicht nur dann ungeändert, wenn man sämtliche Koordinaten der beiden Protonen vertauscht, sondern näherungsweise auch schon dann, wenn man ihre CARTESISchen Koordinaten allein vertauscht, während die Spinkoordinaten ungeändert bleiben können. Diese näherungsweise Symmetrie[1] muss daher vom umwandelnden Faktor beeinträchtigt werden (bei den vorangehenden Versuchen vom inhomogenen magnetischen Feld der stossenden paramagnetischen Moleküle) und derjenige Teil des durch das Feld bedingten Zusatzoperators wird für die Umwandlung massgebend sein, der diese Symmetrieeigenschaft stört. Er entstammt offenbar aus der Wechselwirkung der Kernspin mit dem magnetischen Feld und lautet

$$\mu_P \sum_{\varkappa = xyz} \left( s_{1\varkappa}\,\mathfrak{h}_\varkappa(x_1 y_1 z_1) + s_{2\varkappa}\,\mathfrak{h}_\varkappa(x_2 y_2 z_2) \right). \tag{1}$$

[1] Für das unangeregte $H_2$-Molekül besteht diese Symmetrie viel genauer, als die übliche grobe Überschlagsrechnung zeigt, die nur einen Faktor $a^2$ im Koeffizienten ($a$ ist die Feinstrukturkonstante), also $a^4$ in der Wahrscheinlichkeit des antisymmetrischen Anteils im Parawasserstoff ergibt. Dies rührt daher, dass jene Matrixelemente der Wechselwirkung von Kernspin und Elektronenbewegung, die zwei Rotationsniveaus des unangeregten Moleküls verbinden, verschwinden. Vgl. auch K. F. BONHOEFFER und P. HARTECK, Z. physikal. Ch. (B) 4, 113, 127. 1929.

Hierbei ist $\mu_P$ das magnetische Moment des Protons, $\mathfrak{h}\,(x_1 y_1 z_1)$ bedeutet das magnetische Feld an der Stelle $x_1$, $y_1$, $z_1$ des ersten Protons, die $s_{1\varkappa}$ sind die PAULIschen Spinmatrizen desselben. Die Grössen mit dem Index 2 beziehen sich entsprechenderweise auf das zweite Proton. Hierbei wird das paramagnetische Gas als ein rasch veränderliches inhomogenes Magnetfeld beschrieben. Das ist eigentlich im Sinne der Quantenmechanik nicht berechtigt, weil es sich hierbei keineswegs um ein äusseres Feld handelt und insbesondere die Rückwirkung des $H_2$-Moleküls auf das paramagnetische Gas und ein eventuell vorhandenes virtuelles Moment unberücksichtigt bleibt. Besonders im Falle, dass das paramagnetische Gas selber Orthowasserstoff ist, wäre aus diesem Grunde eine eingehendere Behandlung notwendig. Im allgemeinen wird aber das Verwenden des Feldbegriffes keine allzu grossen Fehler mit sich bringen.

Wenn das Magnetfeld homogen wäre, so hätte (1) noch immer die Eigenschaft, dass es ungeändert bliebe, wenn man $x_1$, $y_1$, $z_1$ mit $x_2$, $y_2$, $z_2$ vertauscht, ein homogenes Feld hat keine umwandelnde Wirkung. Deshalb liegt es nahe, für (1)

$$\frac{1}{2}\,\mu_P \sum_{\varkappa = x, y, z} (s_{1\varkappa} + s_{2\varkappa})\,(\mathfrak{h}_\varkappa(x_1 y_1 z_1) + \mathfrak{h}_\varkappa(x_2 y_2 z_2)) \left.\vphantom{\begin{array}{c} a \\ b \end{array}}\right\} \tag{2}$$
$$+ (s_{1\varkappa} - s_{2\varkappa})\,(\mathfrak{h}_\varkappa(x_1 y_1 z_1) - \mathfrak{h}_\varkappa(x_2 y_2 z_2))$$

zu schreiben. Der umwandelnde Teil von (1) ist dann allein das zweite Glied in (2), das wir durch

$$Q = \frac{1}{2}\,\mu_P \sum_{\varkappa, \lambda} (s_{1\varkappa} - s_{2\varkappa})\,(x_{1\lambda} - x_{2\lambda})\,\frac{\delta \mathfrak{h}_\varkappa}{\delta x_\lambda} \tag{3}$$

ersetzen wollen. Um die Matrixelemente von $Q$ zu berechnen, können wir annehmen, dass das Magnetfeld in Entfernungen, in die die Protonen kommen können, durch das Feld eines Dipols ersetzt werden kann, das sich im Molekülschwerpunkt befindet. Dann kann man bei entsprechender Wahl des Koordinatensystems näherungsweise

$$\frac{\delta \mathfrak{h}_\varkappa}{\delta x_\lambda} = \frac{3\mu_a \delta_{\varkappa\lambda}}{r^4}$$

setzen, wo $\mu_a$ das magnetische Moment des paramagnetischen Moleküls bedeutet. Nunmehr kann man für $Q$ auch

$$Q = \frac{3\mu_P \mu_a}{2r^4}\left(s_z z + \frac{1}{2}\,(s_x + i s_y)\,(x - iy) + \frac{1}{2}\,(s_x - i s_y)\,(x + iy)\right) \tag{4}$$

schreiben, wo $x_\varkappa = x_{1\varkappa} - x_{2\varkappa}$ und $s_\varkappa = s_{1\varkappa} - s_{2\varkappa}$ ist. Die drei Glieder von (4) geben alle zu verschiedenen Übergängen Anlass, so dass sie

30                                 E. Wigner

getrennt behandelt werden können. Es sei hier das erste Glied herausgegriffen, das Endresultat muss aber dann noch mit 3 multipliziert werden. Das Matrixelement, das den Zustand mit der Rotationsquantenzahl $l$ und der magnetischen Quantenzahl $\eta$ verbindet mit dem Zustand mit den Quantenzahlen $l'$ und $\eta'$

$$M_{l\eta l'\eta'} = \left( \psi_{l\eta}, \frac{3\mu_P\mu_a}{2r^4} s_z z\, \psi_{l'\eta'} \right) \tag{5}$$

verschwindet für $\eta \neq \eta'$ oder wenn $l'$ nicht $l\pm1$ ist, so dass $\psi_{l\eta}$ und $\psi_{l'\eta'}$ verschiedenen Wasserstoffarten entsprechen müssen. Das Spinmoment in der $Z$-Richtung muss für den Orthozustand Null sein, dies gibt einen Faktor $\frac{1}{3}$ bei der Umwandlung von Ortho zu Para, wie er für das Gleichgewicht notwendig ist. Für die Umwandlung von Parawasserstoff in Orthowasserstoff fällt dieser Faktor weg. Man kann die Matrix (5) mit Hilfe der gewöhnlichen Intensitätsformeln berechnen, für $l' = l-1$ ergibt sich

$$M_{l\eta l-1\eta} = \frac{3a_{H_2}\mu_a\mu_P}{r^4}\sqrt{\frac{l^2-\eta^2}{(2l+1)(2l-1)}} \tag{6a}$$

und für $l' = l-1$

$$M_{l\eta l+1\eta} = \frac{3a_{H_2}\mu_a\mu_P}{r^4}\sqrt{\frac{(l+1)^2-\eta^2}{(2l+1)(2l+3)}}, \tag{6b}$$

wo $a_{H_2}$ der Abstand der $H$-Atome im $H_2$-Molekül ist.

Nunmehr kann man die Umwandlungswahrscheinlichkeit direkt berechnen. Ein Parawasserstoffmolekül mit der Rotationszahl $l$ stosse auf ein paramagnetisches Molekül, der Stossabstand sei $a_s$, die Relativgeschwindigkeit $v$. Wenn $v$ so gross ist, dass die kinetische Energie des Stosses die Energie, die zur Umwandlung nötig ist, übersteigt, so kann man nach KALLMANN und LONDON[1]) den Stoss ersetzen durch folgenden Prozess: Die Stosspartner werden einander unendlich rasch bis auf den Stossabstand genähert, die mittlere Zeit des Zusammenstosses beisammen gelassen, dann wieder unendlich rasch getrennt. Während des Zusammenseins lautet dann die SCHRÖDINGER-Gleichung für den Entwicklungskoeffizienten $c_{l+1}$ des entstehenden Orthozustandes mit der Rotationsquantenzahl $l+1$, wenn man als Nullpunkt der Energieskala diesen Zustand wählt

$$i\hbar\frac{\partial c_{l+1}}{\partial t} = M_{l\eta l+1\eta}\, e^{-i\nu_{l,l+1}t}. \tag{7}$$

Hierin ist $\hbar$ die PLANCKsche Konstante dividiert durch $2\pi$ und $\nu_{l,l+1}$ die Energiedifferenz zwischen den Zuständen $l$ und $l+1$ dividiert

---

[1]) H. KALLMANN und F. LONDON, Z. physikal. Ch. (B) 2, 207. 1929.

Über die paramagnetische Umwandlung von Para-Orthowasserstoff.    31

durch $\hbar$, also $\hbar\left(\dfrac{l+1}{J}\right)$, wobei $J$ das Trägheitsmoment des $H_2$ ist. Die Lösung von (7) ist

$$c_{l+1} = \frac{3\,a_{H_2}\mu_a\,\mu_P\,J}{\hbar^2\,a_s^4\,(l+1)} \cdot \sqrt{\frac{(l+1)^2 - \eta^2}{(2l+1)\,(2l+3)}} \left(e^{-\frac{i\,\hbar\,(l+1)\,t}{J}} - 1\right). \tag{8}$$

Hierin muss noch für $t$ die Zeit des Zusammenseins eingesetzt werden, wofür wir die Zeit $\dfrac{a_s}{3\,v}$ schreiben, in der das Matrixelement (5) von der Hälfte seines Maximalwertes zum Maximalwert ansteigt und dann wieder auf die Hälfte herabfällt. Durch Quadrieren und Mitteln über $\eta$ erhält man so für die Wahrscheinlichkeit des Orthozustandes $l+1$ nach dem Stoss

$$|c_{l+1}|^2 = \frac{12\,a_{H_2}^2\,\mu_a^2\,\mu_P^2\,J^2}{\hbar^4\,a_s^8\,(l+1)\,(2l+1)} \sin^2 \frac{\hbar\,(l+1)\,a_s}{6\,J\,v}, \tag{9a}$$

und entsprechend für die Wahrscheinlichkeit des Orthozustandes mit der Rotationsquantenzahl $l-1$

$$|c_{l-1}|^2 = \frac{12\,a_{H_2}^2\,\mu_a^2\,\mu_P^2\,J^2}{\hbar^4\,a_s^8\,l\,(2l+1)} \sin^2 \frac{\hbar\,l\,a_s}{6\,J\,v}. \tag{9b}$$

Die Gleichungen (9) zeigen, dass man im wesentlichen nur mit dem Übergang $0 \to 1$ (und vielleicht noch mit $2 \to 1$) zu rechnen braucht. Bei diesen kann man den Sinus durch den Winkel ersetzen und erhält so für die Stossausbeute des $0 \to 1$-Überganges

$$\frac{2\,\mu_a^2\,\mu_P^2\,J}{3\,\hbar^2\,a_s^6\,M\,v^2}, \tag{10}$$

sofern man nur diejenigen Stösse in Betracht zieht, deren Energie zur Umwandlung ausreicht. Setzt man für $Mv^2 = 3\,kT$ und auch für die übrigen Grössen die entsprechenden Werte ein, so erhält man mit $a_s = 1$ bis 2 Å Stossausbeuten von $10^{-11}$ bis $10^{-13}$, in Übereinstimmung mit den experimentellen Werten.

Was die Temperaturabhängigkeit anbetrifft, so ergibt der hier benutzte Mechanismus ebenfalls ein mit den Versuchen übereinstimmendes Resultat. Zunächst hängt (10) explizite nur dadurch von der Temperatur ab, dass die Stossausbeute umgekehrt proportional zum Quadrate der Stosszeit ist. Diese würde eine der absoluten Temperatur umgekehrt proportionale Reaktionsgeschwindigkeit bedingen, der aber ein Abfallen des Minimalabstandes $a_s$ (SUTHERLANDsche Temperaturabhängigkeit des Wirkungsquerschnittes) beim Zusammenstoss mit wachsender Molekülgeschwindigkeit entgegenwirkt.

32    E. Wigner, Paramagnetische Umwandlnng von Para-Orthowasserstoff.

Beim Vergleich mit der Erfahrung muss zudem noch in Betracht gezogen werden, dass die beobachtete Geschwindigkeit nur der Grössenordnung nach durch (10) wiedergegeben wird, in der Tat setzt sich diese aus der Summe der Einzelbeträge der verschiedenen Übergänge zusammen, die sich ihrerseits ebenfalls mit der Temperatur ändern. Die genauere Diskussion zeigt[1]), dass die schwache Temperaturabhängigkeit der beobachteten Stossausbeute auf ein Zusammenwirken dieser Faktoren beruht.

Der hier vorgeschlagene Mechanismus scheint schliesslich auch die Abhängigkeit vom magnetischen Moment des wirkenden paramagnetischen Moleküls richtig wiederzugeben.

---

[1]) Vgl. die vorangehenden Arbeiten von L. FARKAS und H. SACHSSE.

# On a Modification
## of the Rayleigh-Schrödinger Perturbation Theory

E. P. Wigner

Magyar Tudományos Akadémia Matematikai
és Természettudományi Értesitöje *53*, 477–482 (1935)

1. The RAYLEIGH—SCHRÖDINGER perturbation theory[1] gives an explicit power series in $\lambda$ for the characteristic values $F_n$ and the characteristic functions $\varphi_n$ of a Hermitean operator $H+\lambda V$

$$(H+\lambda V)\,\varphi_n = F_n\varphi_n \tag{1}$$

if the corresponding quantities $E_n$ and $\psi_n$ for the «unperturbed operator» $H$ are known

$$H\psi_n = E_n\psi_n. \tag{1a}$$

If the so-called matrix elements of $V$ are denoted, as usual, by

$$V_{nm} = (\psi_n,\ V\psi_m) = V_{mn}^\star \tag{2}$$

the first terms of the aforementioned series read

$$F_n^{(2)} = E_n + \lambda V_{nn} + \lambda^2 \sum_k{}' \frac{|\,V_{nk}\,|^2}{E_n-E_k} \tag{3a}$$

$$\varphi_n^{(1)} = \psi_n + \lambda \sum_k{}' \frac{V_{kn}}{E_n-E_k}\,\psi_k. \tag{3b}$$

Generally only these first terms of the series are used in actual calculations, the higher terms become increasingly complicated.

---

[1] J. W. S. RAYLEIGH, The theory of Sound. London and New York 1894, vol. 1, p. 113. E. SCHRÖDINGER, Collected papers on Wave Mechanics. London and Glasgow 1928, p. 64.

BY EUGENE WIGNER.

We shall fix our attention on the lowest energy value $F_1$. While it is evident that the first approximation for this $F_1^{(1)} = E_1 + \lambda V_{11}$ is always greater than its real amount — since it is the expectation value of a normalized wave function $\psi_1$; nothing like this holds for the second and higher approximations. It even happens quite often that the last series in $(3a)$ diverges in cases when the lowest energy value is finite itself. In these cases, of course, the RAYLEIGH—SCHRÖDINGER perturbation theory is inapplicable to the problem. The aim of the present paper is to give an approximation formula for $F_1$ which always yields values that are too high, and which can be proved to converge at least in certain simple cases. Such an expression is naturally provided for by the variational method which had been used frequently indeed in cases for which the general shape of the characteristic functions could be obtained by physical considerations.

The final result, the $\infty$-th approximation, will appear in the form of an infinite series. This infinite series was first found by L. BRILLOUIN[1] who obtained it by an intuitive consideration of the usual scheme. He has already pointed out in his important paper that his series converges much more rapidly than the power series of SCHRÖDINGER. He has not investigated, however, the successive approximations and their relations to the actual problem.

2. For the sake of convenience we shall denote further on $F_k^{(1)} = E_k + \lambda V_{kk}$ simply by $E_k$, assuming $V_{kk} = 0$ or that the diagonal part of $V$ is already put into $H$. In addition to this, we shall put $\lambda = 1$. The expectation value of $H + V - E_1$ for the wave function

$$\varphi_1 = \psi_1 + \sum_{k=2}^{\infty} a_k \psi_k \qquad (4a)$$

is ($R$ means that the real part of the ensuing expression must be taken)

$$F_1 - E_1 = \frac{\sum_k (E_k - E_1) |a_k|^2 + 2R \sum_k V_{1k} a_k + \sum_{kl} V_{kl} a_k^\star a_l}{1 + \sum_k |a_k|^2} \qquad (4b)$$

---

[1] L. BRILLOUIN, Journal de physique, 4, 1, 1933. The perturbed energy first appears as resonance denominator in J. E. LENNARD-JONES article, Proc. Roy. Soc. London, A. 129, 598, 1930.

where all summations must be extended, as always in this paper, from 2 to $\infty$. Expression (4b) must be made to a minimum by choosing the $a$'s properly.

To orient ourselves we proceed as follows. We neglect first the double sum in the numerator of (4b), i. e. assume $V_{kl} = 0$ for $k > 1$, $l > 1$. Differentiation of (4b) with respect to $a_k$ gives then

$$a_k^{(1)} = \frac{V_{k1}}{F_1^{(2)} - E_k}. \tag{5a}$$

This, inserted into (4b), gives

$$F_1^{(2)} - E_1 = \sum_k \frac{|V_{1k}|^2}{F_1^{(2)} - E_k}. \tag{5b}$$

This is an implicit equation for $F_1^{(2)}$, which can be solved e. g. by plotting both sides of (5b) against $F_1^{(2)}$. It is an exact solution of the problem if $V_{kl} = 0$ (i. e. it is $F_1^{(2)} = F_1$): the finding of the characteristic values of a JACOBI matrix.[1]

It appears to be natural, now, even if the $V_{kl}$ are different from zero, to try (5a) for $a_k^{(1)}$ but to use instead of (5b) an equation expressing again that $F_1^{(2)} - E_1$ is equal to the total fraction (4b), which, under the present conditions will no longer be equal to the right side of (5b). One obtains, instead of (5b)

$$F_1^{(2)} - E_1 = \sum_k \frac{V_{1k}V_{k1}}{F_1^{(2)} - E_k} + \sum_{kl} \frac{V_{1k}V_{kl}V_{l1}}{(F_1^{(2)} - E_k)(F_1^{(2)} - E_l)}. \tag{6}$$

The value of $F_1$ obtained from (6), since it is an expectation value of $H + V$, is always too high. For the next approximation one may try

$$a_k^{(2)} = a_k^{(1)} + \beta_k$$

neglect second power terms in the double sum of (4b), then minimize the resulting expression. One obtains, by denoting the total fraction this time with $F_1^{(3)} - E_1$

---

[1] (5b) goes over into the usual formula of the RAYLEIGH—SCHRÖDINGER theory if $E_1$ is doubly degenerate, and $E_3$, $E_4$ etc. are far away from $E_1 = E_2$. However, (5b) may be used as it stands, without first solving a «secular equation».

$$a_k^{(2)} = \frac{V_{k1}}{F_1^{(3)} - E_k} + \sum_l \frac{V_{kl}a_l^{(1)}}{F_1^{(3)} - E_k}$$

which is in this approximation equal to

$$a_k^{(2)} = \sum_k \frac{V_{k1}}{F_1^{(3)} - E_k} + \sum_{kl} \frac{V_{kl}V_{l1}}{(F_1^{(3)} - E_k)(F_1^{(3)} - E_l)}. \qquad (7a)$$

Inserting this into (4b) again, one obtains

$$\begin{aligned}
F_1^{(3)} - E_1 = &\sum_l \frac{V_{1l}V_{l1}}{F_1^{(3)} - E_l} + \sum_{lk} \frac{V_{1l}V_{lk}V_{k1}}{(F_1^{(3)} - E_l)(F_1^{(3)} - E_k)} + \\
&+ \sum_{lkj} \frac{V_{1l}V_{lk}V_{kj}V_{j1}}{(F_1^{(3)} - E_l)(F_1^{(3)} - E_k)(F_1^{(3)} - E_j)} + \\
&+ \sum_{lkji} \frac{V_{1l}V_{lk}V_{kj}V_{ji}V}{(F_1^{(3)} - E_l)(F_1^{(3)} - E_k)(F_1^{(3)} - E_j)(F_1^{(3)} - E_i)}.
\end{aligned} \qquad (7)$$

This, solved by a graphical or other method, **again gives certainly too high** values for $F_1$.

It is evident now, how the higher approximations look. If one inserts

$$a_k^{(n)} = a_k^{(n-1)} + \sum_{\mu_1 \ldots \mu_{n-1}} \frac{V_{k\mu_{n-1}} V_{\mu_{n-1}\mu_{n-2}} \ldots V_{\mu_2\mu_1} V_{\mu_1 1}}{(f - E_k)(f - E_{\mu_{n-1}}) \ldots (f - E_{\mu_1})} \qquad (8a)$$

into (4b), the right side, which is the expectation value of a normalized wave function minus $E_1$, becomes

$$\begin{aligned}
\frac{1}{N} [T_2 + T_3 + \cdots + T_{2n} + T_{2n+1} + (N-1)(f - E_1)] = \\
= \frac{(\varphi_1^{(n+1)}, (H+V)\varphi_1^{(n+1)})}{(\varphi_1^{(n+1)}, \varphi_1^{(n+1)})} - E_1
\end{aligned} \qquad (9)$$

with

$$N = 1 + \sum_k |a_k^{(n+1)}|^2 \qquad (9a)$$

and

$$T_{j+1} = \sum_{\mu_1 \ldots \mu_j} \frac{V_{1\mu_j} V_{\mu_j\mu_{j-1}} \ldots V_{\mu_2\mu_1} V_{\mu_1 1}}{(f - E_{\mu_j})(f - E_{\mu_{j-1}}) \ldots (f - E_{\mu_1})}. \qquad (9b)$$

Equ. (9) is an identity. If one chooses $f$ such that

$$T_2 + T_3 + \cdots + T_{2n+1} = f - E_1 \qquad (8)$$

one sees that the expectation value of the corresponding wave function is just

$$F_1^{(n+1)} = f$$

which can be taken as the $(n+1)$th approximation. One can satisfy oneself easily, that the infinite series, obtained by generalization of (8), together with a similar generalization of (8a) and with (4a), formally satisfie equation (1) with $F_1 = f$. They can diverge, of course, in spite of this. It is true, however, that all equations of the form (8) give too high values for $F_1$. One easily sees that the odd term of (8) are always negative,[1] (therefore improving on the energy value), while the even ones may be positive or negative.

Of course, (8) converges in many cases in which the usual RAYLEIGH—SCHRÖDINGER method diverges. It can converge even in the case of a continuous spectrum which has been made artificially discrete by setting finite boundaries to an originally infinite problem. There is one, as I think, rather general case, in which it can easily be shown to converge. This is when $H$ is an «even» operator, which remains unaltered under a transformation $Q$, satisfying the equation $Q^2 = 1$, and $V$ an odd operator which goes over into $-V$ under the same transformation. Then $V_{kl}$ is zero if $k$ and $l$ refer either both to even states, or both to odd states and the even terms of the series (7) drop out. The conditions of the convergence are in this case merely that $H + V$ should be bounded downwards (the lowest energy value finite) and applicable to all characteristic functions of $H$.

Of course, (8), (8a), (4a) still give a formal solution of the problem, if one inserts other numbers instead of 1. It is not so easy, however, to discuss the resulting equation in these cases.

---

[1] The $(2n-1)$st term in (8) is equal

$$T_{2n} = \sum_k (\alpha_k^{(n)} - \alpha_k^{(n-1)})^2 (f - E_k) \qquad (*)$$

which is, of course, negative. Equation (*) is material for the proof of convergence in cases to be mentioned later

3. As an example, I want to discuss the MATHIEU-equation

$$-\frac{d^2\varphi}{dx^2} + \sin x\,\varphi = F\varphi(x)$$

in which the total potential $\sin x$ will be regarded as perturbation. The unperturbed characteristic functions are

$$\psi_0 = \frac{1}{\sqrt{2\pi}}\,,\ \ \psi_1 = \frac{\sin x}{\sqrt{\pi}}\,,\ \ \psi_2 = \frac{\cos x}{\sqrt{\pi}}\,,\ \ \psi_4 = \frac{\sin 2x}{\sqrt{\pi}} \cdots$$

and the $V_{kl}$ matrix is

$$\begin{pmatrix}
0 & \frac{1}{\sqrt{2}} & 0 & 0 & 0 \cdots \\
\frac{1}{\sqrt{2}} & 0 & -\frac{1}{2} & 0 & 0 \cdots \\
0 & -\frac{1}{2} & 0 & \frac{1}{2} & 0 \cdots \\
0 & 0 & \frac{1}{2} & 0 & -\frac{1}{2} \cdots \\
0 & 0 & 0 & -\frac{1}{2} & 0 \cdots \\
\vdots & \vdots & \vdots & \vdots & \vdots
\end{pmatrix}$$

The equation for the lowest energy is

$$-F_1 = \frac{1}{2(1-F_1)} + \frac{1}{8(1-F_1)^2(4-F_1)} +$$

$$+ \frac{1}{32(1-F_1)^3(4-F_1)^2} + \frac{1}{32(1-F_1)^2(4-F_1)^2(9-F_1)} + \cdots$$

This equation can be solved by successive approximations very easily, one obtains $F_1 = -0{\cdot}37856$.

Another example of a similar simple case will be given by Mr. F. Seitz in the calculation of the Fermi energy of metallic electrons.

---

(From the meeting of the IIIrd class of the Hungarian Academy of Sciences on the 12th November 1934.)

---

Felelős kiadó: Wigner Jenő. — Franklin-Társulat nyomdája.

# Calculation of the Rate of Elementary Reactions of Light and Heavy Hydrogen

L. Farkas and E. P. Wigner

Transactions of the Faraday Society *32*, 708–723 (1936)

*Received* 15*th January,* 1936.

### 1.

The reactions which will be the subject of the present paper are as follows :

$$H + H_2 \xrightarrow{k_1} H_2 + H \qquad . \qquad . \qquad . \qquad . \quad (1)$$

$$D + D_2 \xrightarrow{k_2} D_2 + D \qquad . \qquad . \qquad . \qquad . \quad (2)$$

$$D + H_2 \xrightarrow{k_3} DH + H \qquad . \qquad . \qquad . \qquad . \quad (3)$$

$$H + D_2 \xrightarrow{k_4} HD + D \qquad . \qquad . \qquad . \qquad . \quad (4)$$

L. FARKAS AND E. WIGNER    709

The first two reactions determine the rate of homogeneous thermal conversion of light and heavy *para- ortho*-hydrogen.   The mechanism of the homogeneous thermal conversion of light *para*-hydrogen into ordinary hydrogen has been cleared up in an important paper by A. Farkas.[1] The rates of reaction (2) together with those of (3) and (4) have been measured recently by A. and L. Farkas [2] and compared with each other on the basis of the usual theory of collision numbers.   In the present paper we shall compute the reaction rate of *the elementary reactions* (1) *to* (4) on the basis of more recent theories [3], [4], [5] since, so far, these are the only examples providing both theoretical and experimental data for a comparison.

It will be interesting to calculate theoretically the number of interchanges which take place per unit time and volume (sec. litre) and unit concentration (mol./litre) of the reagents (H and $H_2$ e.g.).   We designate as interchange a reaction involving atoms and molecules of the general type

$$A + BC = AB + C$$

where A, B and C represent H or D atoms.   The rate of interchange can be measured directly, in principle, in cases (3) and (4).   In reaction (1) only the rate of conversion :

$$H + p\text{-}H_2 \rightarrow o\text{-}H_2 + H \qquad . \qquad . \qquad . \quad (1a)$$

can be measured directly.   In the number of interchanges, as defined before, however, the reaction :

$$H + p\text{-}H_2 \rightarrow p\text{-}H_2 + H \qquad . \qquad . \qquad . \quad (1b)$$

should also be counted.   The number of these is 1/3 of those which lead to the *ortho*-molecule, strictly according to their statistical weights.[6] The *total number of atomic interchanges is*, thus, 4/3 *times the number of para- ortho-conversions.*   Similarly *the number of interchanges in case* (2) *is three times greater than the number of ortho- para-conversions*, which can be measured directly in this case.[2]

The results for the rates of the first two interchange reactions are as follows :

TABLE I.

| T. | $k_1$. | $k_2$. |
|---|---|---|
| 283 [7] | $1 \cdot 3 \times 10^5$ | — |
| 373 [7] | $3 \cdot 1 \times 10^6$ | — |
| 873 | $1 \cdot 37 \times 10^9$ | $0 \cdot 70 \times 10^9$ |
| 930 | $1 \cdot 71 \times 10^9$ | $0 \cdot 88 \times 10^9$ |
| 973 | $2 \cdot 00 \times 10^9$ | $1 \cdot 02 \times 10^9$ |
| 1023 | $2 \cdot 38 \times 10^9$ | — |

[1] A. Farkas, *Z. Elektrochem.*, 1930, 36, 782 ; *Z. physik. Chem.* 1931, 10B, 419.
[2] A. Farkas and L. Farkas, *Proc. Royal Soc.*, A, 1935, 152, 124.
[3] Pelzer and Wigner, *Z. physik. Chem.*, 1932, 15B, 445.
[4] E. Wigner, *ibid.*, 19B, 203.
[5] Eyring, *J. Chem. Physics*, 1935, 3, 107 ; E. Evans and M. Polanyi, *Trans. Faraday Soc.*, 1935, 31, 875.
[6] A. Farkas, "*Ortho*-hydrogen, *para*-hydrogen and heavy hydrogen," Cambridge, 1935, p. 66.
[7] K. H. Geib and P. Harteck, *Z. physik. Chem.*, 1931, *Bodensteinband*, p. 849.

710    REACTIONS OF LIGHT AND HEAVY HYDROGEN

The ratio of the reaction rates of (1) and (2) according to Table I. is about 1·95. This differs somewhat from the ratio given previously [2] (which was 1·8) since the higher D atom concentration in 95 per cent. $o$-$D_2$ compared with that in ordinary $D_2$ is now taken into account. The increase of the H atom concentration in 50 per cent. $p$-$H_2$, which was used in the experiments on the conversion $p$-$H_2 \to o$-$H_2$, can be neglected. The increase of H and D atom concentration in $p$-$H_2$ and $o$-$D_2$ respectively is due to the comparatively low entropy of these gases.

*In the case of reactions* (3) *and* (4) *it is not possible* from the present data, *to calculate both* $k_3$ *and* $k_4$ *separately ; only one equation for* $k_3$ *and* $k_4$ *can be derived.* The reaction investigated was the formation of HD from $H_2$ and $D_2$ which had the mechanism :

$$H_2 \rightleftharpoons 2H \qquad . \qquad . \qquad . \qquad . \qquad . \quad (a)$$

$$D_2 \rightleftharpoons 2D \qquad . \qquad . \qquad . \qquad . \qquad . \quad (b)$$

$$HD \rightleftharpoons H + D \qquad . \qquad . \qquad . \qquad . \quad (c)$$

$$D + H_2 \underset{k_5}{\overset{k_3}{\rightleftharpoons}} DH + H \qquad . \qquad . \qquad . \qquad . \quad (3)$$

$$H + D_2 \underset{k_6}{\overset{k_4}{\rightleftharpoons}} HD + D \qquad . \qquad . \qquad . \qquad . \quad (4)$$

The content of HD in the reacting mixture was measured as function of time under different pressures and temperatures. The velocity constants of the three dissociation reactions will be denoted by $k_{da}$ $k_{db}$ $k_{dc}$ that of the reverse, recombination reaction (which occurs at triple collisions) by $k_{ra}$ $k_{rb}$ $k_{rc}$. The constants of the reactions (3) and (4) are $k_3$ and $k_4$ in the sense from left to right, and $k_5$ and $k_6$ in the reverse sense. Since the number of atoms is certainly low compared to the number of molecules, the number of the reactions (3) and (4) is greater than that of (a), (b) and (c) and the ratio [H]/[D] will be determined by them.

$$\frac{[H]}{[D]} = \frac{k_3\,[H_2] + k_6[HD]}{k_5[DH] + k_4[D_2]} = \frac{1}{\alpha} \cdot \qquad . \qquad . \qquad . \quad (5)$$

The total-number of atoms will be given of course by reactions (a), (b) and (c), these giving their stationary concentrations :

$$[H] = \left\{ \frac{k_{da}[H_2] + k_{db}\,[D_2] + k_{dc}[HD]}{k_{ra} + k_{rb}\alpha^2 + k_{rc}\alpha} \right\}^{\frac{1}{2}} . \qquad . \qquad . \quad (6)$$

Now we assume that the same fraction of collisions of H + H and D + D and H + D leads to a recombination. This is, of course, only approximately true. Thus $k_{ra} = k_{rb}$. Furthermore $k_{rc} = 2k_{ra}$, since the number of collisions between $n$ atoms H and $n$ atoms D is twice as great as between $n$ atoms H. From the dissociation equilibrium we obtain :

$$\frac{k_{da}}{k_{ra}} = K_a ; \quad \frac{k_{db}}{k_{rb}} = K_b ; \quad \frac{k_{dc}}{k_{rc}} = K_c \qquad . \qquad . \qquad . \quad (7)$$

which yields, inserted into (6)

$$[H] + [D] = \sqrt{K_a[H_2] + K_b[D_2] + 2K_c[HD]} \quad . \qquad . \quad (8)$$

With the figures for the dissociation constants at $T = 1000°$ we obtain

$$[H] + [D] = \sqrt{1 + \frac{1}{1·32^2}\frac{[D_2]}{[H_2]} + \frac{2[HD]}{1·32[H_2]\sqrt{K}}}$$

where $K = \dfrac{[HD]^2}{[H_2][D_2]}$. $K$ is approximately 4 for 1000° abs.

## L. FARKAS AND E. WIGNER                    711

The original concentration of $H_2$ will be denoted by $c$. Since the concentration of $D_2$ was in all experiments practically equal to that of $H_2$, the momentary concentration are $[HD] = 2cy$, that of

$$[H_2] = [D_2] = c(1 - y)$$

Thus we have finally the following equation :

$$[H] + [D] = \sqrt{K_a c} \cdot \sqrt{1 - y} \cdot \sqrt{1 + 0.57 + 1.57 \frac{y}{1 - y}} \approx 1.25 \sqrt{K_a c} \quad (9)$$

This means that the total concentration of atoms is hardly changing at all during the reaction. The numerical value for $[H] + [D]$ is the same as hitherto used.[2] We assume furthermore that the ratio $[H]/[D]$ is always $k_3/k_4$ as in the beginning of the reaction.

The kinetic equation for the rate of formation of HD is then

$$2c\frac{dy}{dt} = (k_3[D] + k_4[H])c(1 - y) - (k_5[H] + k_6[D])2cy \quad . \quad (10)$$

Integrated, this gives

$$y_\infty - y_t = \frac{a}{b} \cdot e^{-bt} \quad . \qquad . \qquad . \qquad . \quad (11)$$

where $a = \frac{1}{2}(k_3[D] + k_4[H])$ and $b = (\frac{1}{2}k_3 + k_5)[D] + (\frac{1}{2}k_4 + k_6)[H]$. Introducing the calculated values of $[H]$ and $[D]$ and expressing $k_5$ and $k_6$ by $k_3$ and $k_4$ and the equilibrium constant [2] of the reactions (3) and (4) we have

$$a = \frac{1.25 \cdot \sqrt{K_a c} k_3 k_4}{k_3 + k_4} \qquad . \qquad . \qquad . \qquad . \qquad . \quad (12)$$

$$b = \frac{1.25 \sqrt{K_a c}}{k_3 + k_4} \cdot (k_3 k_4 + 0.38 k_3{}^2 + 0.67 k_4{}^2) \qquad . \qquad . \quad (13)$$

The time of half-change of the reaction is determined by $b$. According to the experiments [2] this time of half-change is 1.5 times greater than that of reaction (1) at the same pressure and temperature (in the region 800°-1000°). Thus

$$b = \frac{k_1 \sqrt{K_a 2c}}{1.5} \qquad . \qquad . \qquad . \qquad . \quad (14)$$

The factor 2 enters under the square root because at the same pressure the concentration of $H_2$ is twice as great in pure $H_2$ (as used for reaction (1)) than it is in the mixture used for the reaction (3) and (4).

Hence we have

$$0.75 \, k_1 = \frac{k_3 k_4 + 0.38 k_3{}^2 + 0.67 k_4{}^2}{k_3 + k_4} \qquad . \qquad . \quad (15)$$

which is the only equation for $k_3$ and $k_4$ derivable under the above mentioned approximations from the present experiments. Into this equation we shall insert the theoretically calculated values of $k_3$ and $k_4$ and see whether it checks.

According to the usual collision number theory, the rate of reaction should vary with the temperature

$$k = 6.06 \cdot 10^{20} \left(\frac{d_a + d_m}{2}\right)^2 \sqrt{\frac{8RT(M_a + M_m)}{\pi M_a M_m}} \cdot S \cdot e^{-W/RT} \quad . \quad (16)$$

Here $d_a$ and $d_m$ are the diameters of the colliding particles (atom and molecule), $M_a$ and $M_m$ their masses. $S$ is a steric factor, $W$ the heat of activation. For reaction (1) in the temperature region 873°-1023° abs.

712    REACTIONS OF LIGHT AND HEAVY HYDROGEN

one obtains for $W = 5500$ cals., $S = 0.07$. In the low temperature region the figures are $W = 7000$ cals., $S = 0.13$. Although this is not very unsatisfactory it is obvious that the formula, using rather crude assumptions of the kinetic theory of gases and containing two arbitrary constants $S$ and $W$, cannot represent the true state of affairs, since a *strict calculation of the reaction velocity constant must be based on the general methods of statistical mechanics.*

## 2.

A real theory of reaction rates should contain no arbitrary constant at all. Attempts in this direction were made in the following way. London [8] first introduced the idea of the energy surface. The configuration of a set of atoms (the three H's in (1)) can be described by the distances between them. The energy of the atoms in a configuration is thus a function of their distances

$$E = F(r_{12}\, r_{13}\, r_{23})$$

from which their motion during the reaction can be determined to a good approximation by classical mechanics. Classical mechanics can be applied because of the comparatively large masses of the nuclei. The equation of motion, *e.g.*, in the $x$ co-ordinate is

$$M \cdot \frac{\partial x^2}{\partial t^2} = -\frac{\partial E}{\partial x}$$

It should be emphasised that $E$ can be calculated only by quantum mechanics and that it contains the energy which the electrons have if the configuration is characterised by $r_{12}, r_{13}, r_{23}$. Since this energy *does not depend on the masses of the nuclei, the energy surface is the same for reactions 1 to 4.* Of course there are, according to the different quantum states of the electrons, many such energy surfaces, but here, in case of chemical reactions, we shall be concerned with the lowest one only.

In Fig. 1 which is taken from the work of Eyring and Polanyi [9] the contour lines of the energy surface of three H atoms are plotted for a straight configuration $r_{13} = r_{12} + r_{23}$, atom 2 being in the middle. *The angle between the axes $r_{12}$ and $r_{23}$ is chosen in such a way that the free motion* (motion under the influence of the gravity) *of a small heavy ball on the surface of Fig. 1 corresponds to the motion of the atoms in a straight line.*[10] The valley on the right of the figure corresponds to a molecule formed by the atoms 2 and 3 ($r_{23}$ small) 1 being free. The valley towards the top of the figure corresponds to the molecule formed by the atoms 1 and 2, 3 being free. If the system performs the reaction

$$H^{(1)} + H_2^{(2,\ 3)} \rightarrow H_2^{(1,\ 2)} + H^{(3)}$$

the point representing the state of the system must go somehow from the top valley to the valley in the right. The dotted line represents one such path. The following is the picture given by London for exchange reactions.

[8] London, *Sommerfeld Festschrift*, S. Hirzel, 1928, p. 104.
[9] Eyring and Polanyi, *Z. physik. Chem.*, 1931, 12B, 279.
[10] Professor O. K. Rice has kindly pointed out that this angle is 60° for three equal masses, instead of 120°, since the sign in formula (25) *Z. physik. Chem.*, 1931, 12B, 279 is incorrect.

L. FARKAS AND E. WIGNER                    713

The energy of the system is the energy of the heavy ball and it is evident that for very low energies a passage from one valley into the other is impossible.   Only if the energy of the system exceeds a certain limit—that of point **x** in our figure—is such a passage possible.   *Points* **x** *are the highest points of that path leading from one valley into the other, for which the highest point is as low as possible.*   Generally this path has only one highest point—*its energy is the activation energy*—only in our symmetrical example there are two equally high points.   The *system in the activation point* is called the " *activated complex* " or simply " *complex.*" It is evident therefore that the activation energy is given, if the energy surface is known, and this was aimed at in the pioneer work of Eyring and Polanyi.[9]   It should be emphasised, however, that the present more accurate experimental data make more exact calculations desirable.   Of course Fig. I represents, as already mentioned, the energy surface only for the straight configurations.   For all configurations a three-dimensional surface would be necessary, but nothing would be changed fundamentally.

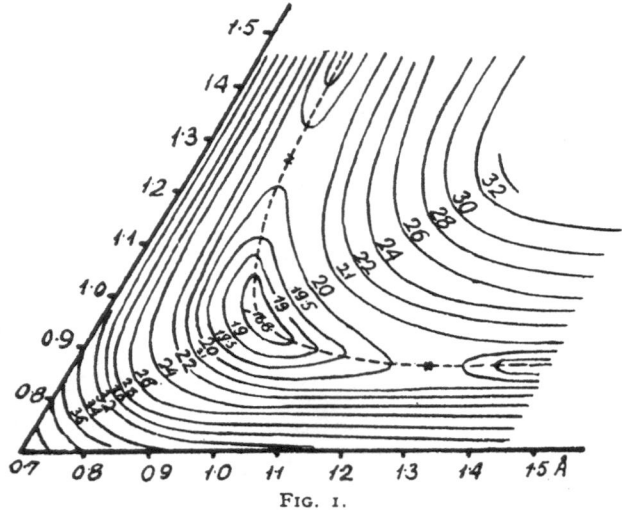

FIG. 1.

If we consider a system of three atoms in a finite box, the energy surface must be cut off at a certain point corresponding to the size of the box.   The ball on our energy surface would roll around for some time in one valley, would pass over into the other, etc.   The number of passages per unit time is equal to the number of atomic interchanges.

It is not easy to calculate this number exactly.   *It is* quite *possible,* however, as we shall see, *to calculate the number of passages over the lines through the activation points* **x**, *where the energy surface forms a saddle.*   Now, in our case *not every passage over a saddle will lead to an atomic interchange.   After every passage through one of the saddles the ball will roll around in the little hole of the middle of the figure and after a certain time may go on into the second valley or else may return to the initial one.   The relative probability of the first event will be the ratio of the number of atomic interchanges to the number of passages through the saddle.*

Although the number of passages through the saddles will be

714    REACTIONS OF LIGHT AND HEAVY HYDROGEN

calculated sufficiently accurately this ratio can be estimated only.  It is evidently rather near to $\frac{1}{2}$ in case the atoms 2 and 3 are equal, because the ball in the hole will go out in both directions with equal probability. If the atoms 2 and 3 are different (one D, the other H), though the potential surface is unaltered by this fact, the passage through both the barriers is for some quantum theoretical reasons not equally probable over the two saddles.  We shall return to this point later.  In this case the number of atomic interchanges will be only slightly smaller than the number of passages over the saddle which is more difficult to cross.

### 3.

The calculation of the number of passages through a given saddle can be performed by the so-called transition state method.[11]  *A small strip in phase space is considered perpendicular to the direction of the ball's motion during the reaction.  The probability of the ball being in this strip can be calculated by ordinary statistical mechanics.*  In the classical theory this probability is given by

$$\eta = \frac{\displaystyle\int_{\text{strip}} ds\, e^{-V/kT}}{\displaystyle\int_{\text{vol.}} d\tau\, e^{-V/kT}} \cdot$$

In the numerator the integration must be extended over the strip, in the denominator over the whole phase space, $V$ is the (*potential*) energy. The number of passages is obtained by multiplying this quantity by the mean component of the velocity of the ball which is perpendicular to the strip, and dividing by the width of the strip.

This *classical theory must be corrected*, of course, *for quantum effects. There is the zero point energy, the quantum-mechanical leakage through the barriers and finally the jump from one energy surface to the other, which must be taken into account.*

*The formula which takes the first two effects into account* * *and is valid for exchange reactions in which the activated complex is straight is* :

$$k = \gamma^{\frac{1}{2}} \left(\frac{M_a + M_m}{M_a M_m}\right)^{3/2} \frac{I_c}{I_m} \cdot \left(\frac{h}{2\pi}\right)^2 \cdot \sqrt{\frac{2\pi}{kT}} \cdot e^{-Q/RT} \cdot \frac{N_L}{1000}$$

$$\times \frac{\sinh \beta \nu_m}{(\sinh \beta \nu_d)^2 \sinh \beta \nu_s} \cdot (1 + \tfrac{1}{6}\beta^2 \nu_t^2) \quad (17)$$

Here, $M_a$ and $M_m$ are the masses of the colliding atom and molecule respectively.  $I_c$ is the moment of inertia of the activated complex, $I_m$ that of the molecule, $Q$ the activation energy, $N_L$ the Avogadro number, $\beta = \frac{h}{2kT}$, $\nu_m$ is the vibrational frequency of the molecule which corresponds to 12,500 cals./mole energy in $H_2$ and 8900 cals./mole in $D_2$,

[11] For the history of this method, *cf*. Evans and Polanyi, *Trans. Faraday Soc.*, 1935, **31**, 875.
 * This formula is identical with that derived by Pelzer and Wigner [3] and Wigner,[4] and is practically identical with that of Eyring.[5]  Eyring's formula, however, does not contain the zero point energy of the complex explicitly and also neglects the leakage through the barrier.

$\nu_d$ is the deformational frequency of the complex and $\nu_s$ the other stable frequency of the complex, whereas $\nu_i$ is the imaginary frequency of the same : $\nu_i$ can be calculated from the curvature of the potential surface in the same way as the real frequencies, except that the second derivative of the potential surface which is negative in this direction must be taken with the positive sign. $k$ is the reaction velocity constant in mole/litre sec.; $\gamma$ is the above discussed ratio of the number of atomic interchanges to the number of passages through one of the saddles.

The *first line of formula* (17) *takes into account the classical quantities only, the second one contains the effect of the zero point energy of both the molecule and the complex and the effect of quantized vibrational levels. In*

$$\frac{1}{2\sinh \beta \nu} = \frac{e^{-h\nu/2kT}}{1 - e^{-h\nu/kT}},$$

the numerator is due to the zero point energy $\frac{1}{2}h\nu$, the denominator is the sum of states for the vibrational states.* *The last factor* $(1 + \frac{1}{8}\beta^2\nu_i{}^2)$ *contains the "tunnel" effect. In consequence of the negative curvature of the energy surface perpendicular to the strip, some passages will occur by leakage even if the energy of the ball is not high enough to reach the activation point. This manifests itself in the last factor virtually as a negative zero point energy of the activated complex, as it facilitates the reaction.*

The deformation frequency is doubly degenerate and $\nu_d$ enters as a squared term into (17). A straight 3 atomic molecule has 4 vibrational ($\nu_d$, $\nu_d$, $\nu_s$, $\nu_i$) 2 rotational and 3 translational degrees of freedom, altogether 9 degrees of freedom.

If the energy surface of § 2 is known all quantities entering into (17) may be calculated, thus (17) differs considerably from the formula of kinetic theory. The activation energy Q is the difference between the potential energy of the activation point and the bottom of the valley from which the reaction starts (without zero point energies).

For the calculation of vibrational frequencies, the potential energy in the neighbourhood of the activation point must be known up to the second power terms of the distances. This will be assumed to have the form :

$$V = Q + p(r_{23} - 0.85)^2 - q(r_{12} - 1.25)^2 + s\delta^2$$
$$= Q + p\xi^2 - q\eta^2 - \nu\eta^2 + s\delta^2 \qquad . \qquad . \qquad (18)$$

the activation point having the co-ordinates $r_{23} = 0.85$ A., $r_{12} = 1.23$ A., $\delta$ is the deformation angle 123. The fact that no $(r_{12} - 1.25)(r_{23} - 0.85)$ term enters is not evident, *a priori;* but can be seen from the figure, $p$, $q$, $s$ are positive constants the values of which can be derived from the Fig. 1.

For the vibrations,[12] $\nu_i$ and $\nu_s$, $\delta = 0$ can be assumed and the molecule put into the $x$-axis. The kinetic energy can be written then if we denote by $M_1$, $M_2$ and $M_3$ the masses of the three atoms

$$K = \frac{1}{2}(M_1\dot{x}_1{}^2 + M_2\dot{x}_2{}^2 + M_3\dot{x}_3{}^2) = \frac{M_1\dot{x}_1 + M_2\dot{x}_2 + M_3\dot{x}_3)^2}{2(M_1 + M_2 + M_3)}$$
$$+ \frac{M_1(M_2 + M_3)(\dot{x}_2 - \dot{x}_1)^2}{2(M_1 + M_2 + M_3)} + \frac{M_3(M_1 + M_2)(\dot{x}_3 - \dot{x}_2)^2}{2(M_1 + M_2 + M_3)}$$
$$+ \frac{M_1M_3(\dot{x}_3 - \dot{x}_2)(\dot{x}_2 - \dot{x}_1)}{M_1 + M_2 + M_3} \qquad . \qquad . \qquad . \qquad . \qquad (19)$$

* The vibrational states are considered as "harmonic."
[12] Cross and Vleck, *J. Chemical Physics*, 1933, **1**, 350.

716    REACTIONS OF LIGHT AND HEAVY HYDROGEN

which gives, if one omits the first term, assuming that the centre of mass is at rest :

$$K = \frac{M_3(M_1 + M_2)}{2(M_1 + M_2 + M_3)}\dot{\eta}^2 + \frac{M_1(M_2 + M_3)}{2(M_1 + M_2 + M_3)}\dot{\xi}^2 + \frac{M_1 M_3 \dot{\xi}\dot{\eta}}{M_1 + M_2 + M_3}$$
$$= \mu_1\dot{\xi}^2 + \tfrac{1}{2}\mu_2\dot{\eta}^2 + \mu\dot{\xi}\dot{\eta} \qquad . \qquad . \qquad . \qquad . \qquad . \qquad . \quad (20)$$

If the last term were not present, the motion of the ball could be described in rectangular co-ordinates.

The equations of motion are hence $\dfrac{d}{dt} . \dfrac{\partial K}{\partial \dot{\xi}} = -\dfrac{\partial V}{\partial \xi}$, etc.

$$\mu_1\ddot{\xi} + \mu\ddot{\eta} = -2p\xi \qquad . \qquad . \qquad . \quad (21)$$
$$\mu\ddot{\xi} + \mu_2\ddot{\eta} = +2q\eta$$

Assuming
$$\xi = a^2 \cos 2\pi\nu t, \quad \eta = b^2 \cos 2\pi\nu t$$

we obtain :

$$4\pi^2\nu^2 = \frac{p\mu_2 - q\mu_1 \pm \sqrt{(q\mu_1 - p\mu_2)^2 + 4qp(\mu_1\mu_2 - \mu^2)}}{(\mu_1\mu_2 - \mu^2)} \qquad . \quad (22)$$

One of the two values of $4\pi^2\nu^2$ is positive, corresponding to the stable vibration $\nu_s$, the other negative corresponding to the unstable vibration $\nu_i$.

The values for $p$ and $q$ can be estimated from the Fig. 2 which shows the two cross-sections of the saddle, $p = 3 \cdot 10^5$ cals./mole Å.$^2$, $q = 4 \cdot 10^4$ cals./mole Å.$^2$. For the reaction (1) this value of $p$ and $q$ gives $\tfrac{1}{2}h\nu_s = 5300$ cals./mole, $\tfrac{1}{2}h\nu_i = 1840$ cals./mole. There is no objection on theoretical grounds against the first figure. A glance at Fig. 2b shows, however, that a negative zero point energy of 1840 cals. is not possible in this case, since the energy—distance curve of Fig. 2 does not continue to be even nearly parabolic thus far. We have arbitrarily decreased $\tfrac{1}{2}h\nu_i$ to 1000 cals./mole in the low temperature region. It should be mentioned that it is not possible to justify this procedure at present, but it hardly effects our results. Eckart[13] has shown how to calculate the non-classical penetration of potential barriers of rather different shapes and Bell[14] has used his results for the reaction $H + H_2$. It must be mentioned, however, that Eckart's results apply to a one-dimensional problem only and the fact that Bell omitted to take the zero point energy $\tfrac{1}{2}h\nu_s$ of the perpendicular vibration into account

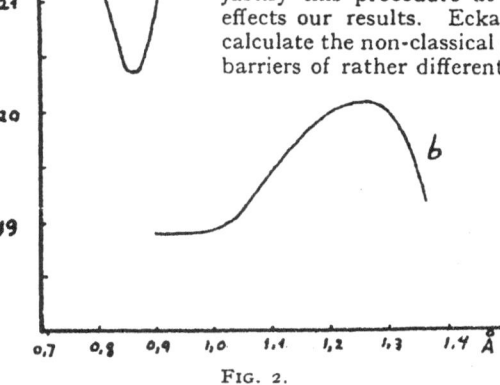

FIG. 2.

[13] Eckart, *Physic. Rev.*, 1930, **35**, 1303.
[14] Bell, *Proc. Roy. Soc.*, 1933, **139A**, 466.

L. FARKAS AND E. WIGNER                717

invalidates his results to some extent.    On the other hand the method for many dimensional barriers [3] is applicable only if the $\frac{h\nu}{kT}$ are not great, *i.e.*, in the high temperature region only.    Here (17) has been consequently, with $\frac{1}{2}h\nu_i = 1840$ cals./mole.    Since the formula, though not sensitive for $h\nu_i$, the tunnel effect being always comparatively small, is rather sensitive for $\frac{1}{2}h\nu_s$, the latter *has been determined from the measurements* (*cf.* section 4) and $p$ calculated backwards.    From this new $p$ the $\frac{1}{2}h\nu_s$ were obtained for the other reactions since the ratios of $h\nu_s$ and $h\nu_i$ are practically independent of the value of $p$.    These ratios, as obtained with aid of formula (22), are listed in Table II.

While the displacements are in the direction of the internuclear line for the vibrations $\nu_s$ and $\nu_i$, they are perpendicular to this in the deformation vibration $\nu_d$.    We shall denote these displacements by $y_1$, $y_2$, $y_3$ for the atoms 1, 2 and 3, the condition that the centre of gravity is at rest and $y = 0$, is then

$$M_1 y_1 + M_2 y_2 + M_3 y_3 = 0 \qquad . \qquad . \qquad . \quad (23)$$

We can put down the condition, furthermore, that the molecule has no angular momentum around the middle atom (in virtue of the previous equation this holds, then around any point) :

$$M_1 y_1 \, 1 \cdot 25 - M_3 y_3 \, 0 \cdot 85 = 0 \qquad . \qquad . \qquad . \quad (24)$$

The deformation angle $\delta$ is

$$\delta = \frac{y_1 - y_2}{1 \cdot 25} + \frac{y_3 - y_2}{0 \cdot 85}$$

and the potential energy $s\delta^2$.    The equations of motion are :

$$M_1 \ddot{y}_1 = - \frac{\partial V}{\partial y_1} = - \frac{2 s \delta}{1 \cdot 25}$$

and similar equations for $y_2$ and $y_3$, which are automatically fulfilled, however, if we assume (23) and (24) from the beginning.    These equations allow us to express $y_3$ and $y_2$ by $y_1$ and thus $\delta$ by $y_1$ alone.    The equation of motion one obtains thus, is :

$$\ddot{y}_1 = \alpha y_1 \quad . \qquad . \qquad . \qquad . \qquad . \qquad . \qquad . \qquad . \quad (25a)$$

$$\alpha = 2s \cdot \left[ \frac{1}{1 \cdot 25^2 \, M_1} + \frac{1}{0 \cdot 85^2 \, M_3} + \left( \frac{1}{1 \cdot 25} + \frac{1}{0 \cdot 85} \right)^2 \frac{1}{M_2} \right] \quad . \quad (25b)$$

the deformation frequency is consequently

$$\nu_d = \frac{1}{2\pi} \sqrt{\alpha} \quad . \qquad . \qquad . \qquad . \qquad . \quad (26)$$

The constant $s$ should have, according to the calculated energy surface, the value 9000 cals./mole.    This gives $h\nu_d = 2150$ cals./mole for reaction (1).    It should be mentioned, however, that the value of $s$ is rather sensitive to small changes of the constants from which the energy surface is derived, and not too much stress should be laid on this value of $h\nu_d$.

The ratios of $\nu_d$ for the different complexes are listed also in Table II. in which the symbol means (*e.g.*, D—DH) the activated complex of reaction (4) in which the H is nearer to the central D atom than the other D.

718    REACTIONS OF LIGHT AND HEAVY HYDROGEN

This completes the calculation of the frequencies. The ratio of the frequencies for reactions (1) and (2) is always $\sqrt{2}$, independent of the actual model assumed.

TABLE II.

| Symbol. | Relative. | | | $I_c$. |
|---|---|---|---|---|
| | $\nu_s$. | $\nu_i$. | $\nu_d$. | |
| H—HH . . . | 5·4 | 1·77 | 2·43 | 3·71 × 10⁻⁴⁰ |
| D—DD . . . | 3·8 | 1·25 | 1·72 | 7·42 × 10⁻⁴⁰ |
| D—HH . . . | 5·4 | 1·42 | 2·37 | 5·25 × 10⁻⁴⁰ |
| DH—H . . . | 4·6 | 1·38 | 2·39 | 4·90 × 10⁻⁴⁰ |
| H—DD . . . | 3·8 | 1·60 | 1·81 | 4·90 × 10⁻⁴⁰ |
| HD—D . . . | 4·7 | 1·30 | 1·91 | 5·52 × 10⁻⁴⁰ |

The moment of inertia of the complex $I_c$ which is necessary to calculate the velocity constant according to formula (17) can also be readily obtained from the energy surface. As already mentioned the distances of the end atoms from the central atom are 0·85 Å. and 1·25 Å. In the symmetric case there is only one momentum of inertia, in the asymmetric ones, the momentum of inertia of the complex is different in the two activation points.

If $x$ is the distance of the centre of gravity from the middle atom, the momentum of inertia is given by

$$I_c = M_1(1\cdot25 - x)^2 + M_2 x^2 + M_3(0\cdot85 + x)^2$$

and

$$x = \frac{M_1\, 1\cdot25 - M_3\, 0\cdot85}{M_1 + M_2 + M_3} \cdot \qquad . \qquad . \qquad . \quad (27)$$

$I_c$ does not depend very much on the dimension of the complex. The moment of inertia of the $H_2$ molecule is known from spectroscopical data. It is $0\cdot467 \times 10^{-40}$ g. cm.² and is twice as great for $D_2$.

4.

In fact the energy surface for reactions 1 to 4 is not known at present with sufficient accuracy, especially not for the calculation of the constants for which (7) is sensitive, i.e., $Q$, $\nu_s$ and $\nu_d$. *Therefore another procedure will be adopted in this section : a part of the experimental data will be used to calculate the constants of the energy surface relevant for these quantities.*

First of all we shall compare the reaction rates of (1) and (2) in the high temperature region. This ratio is, according to Table I., 1·65 at 950° abs. and we obtain the following equation :

$$\frac{k_1}{k_2} = 1\cdot95 = 2^{3/2}\frac{\sinh 3\cdot29}{\sinh 2\cdot34} \cdot \left(\frac{\sinh \beta\nu_d/\sqrt{2}}{\sinh \beta\nu_d}\right)^2 \frac{\sinh \beta\nu_s/\sqrt{2}}{\sinh \beta\nu_s} \frac{(1+\frac{1}{8}(3700\beta)^2)}{(1+\frac{1}{12}(3700\beta)^2)} \quad *$$

(28)

* This formula for the ratio of the velocity constants of the two isotopic reactions is essentially the same—if we neglect the tunnel effect—as the usual one, derived from the kinetic expression (16).

The difference in the " over all " activation energy of the two isotopic reactions is at low temperatures $0\cdot15(h\nu_m - h\nu_s - 2h\nu_d)$.

[Continuation of footnote on next page.

if $\nu_d$ and $\nu_s$ are the constants of the complex HHH and 3700 cals., according to the calculation on page 715, $\nu_t$ for HHH. Since $\beta\nu_d$ is certainly not greater than unity, we can put $\sinh x = x + \dfrac{x^3}{6}$ and thus we obtain finally :

$$\frac{\sinh \beta\nu_s/\sqrt{2}}{\sinh \beta\nu_s} = 0{\cdot}54$$

which gives $\beta\nu_s = 1{\cdot}9$ and $h\nu_s = 7200$ cals./mole.

In order to calculate the energy of the activation point $Q$ we shall use the experimental figures for the temperature dependence of $k_1$. From the figures of the Table I. yields

$$Tw_{exp.} = T\frac{d \ln k_1}{dT} = 3{\cdot}4 \text{ to } 3{\cdot}5 \quad . \qquad . \qquad . \quad (29)$$

whereas theoretically we obtain for the expression $T\dfrac{d \ln k_1}{dT}$ from equation (17)

$$T\frac{d \ln k_1}{dT} = -\frac{1}{2} + \frac{Q}{RT} - \beta\nu_m \coth \beta\nu_m + 2\beta\nu_d \coth \beta\nu_d + \beta\nu_s \coth \beta\nu_s$$
$$- \frac{\beta^2\nu_t{}^2}{3(1 + \frac{1}{6}\beta^2\nu_t{}^2)} \quad (30)$$

Inserting the values for $\nu_m$, $\nu_s$ and $\nu_t$ and putting $2\beta\nu_d \coth \beta\nu_d$ approximately * $2{\cdot}20$ we obtain by comparison with the experimental value for

$$Q/RT = 3{\cdot}4 \text{ and } Q = 6400 \text{ cals.}$$

According to this calculation the energy of the activation point is rather low and is not in good agreement with the calculated value [9] which is 20,000 cals. However the first approximation calculations of chemical bonds are generally uncertain by this amount. The remaining unknown frequency $\nu_d$ will be determined from the absolute value of the reaction rate of (1) at 950° abs.

$$1{\cdot}85 \times 10^9 = \frac{2{\cdot}4 \times 10^{11}}{30{\cdot}8} e^{-\frac{6400}{RT}} \cdot \frac{\sinh 3{\cdot}29}{\sinh 1{\cdot}90} \cdot \frac{1{\cdot}16}{(\sinh \beta\nu_d)^2} \quad . \quad (31)$$

This equation solved yields $\beta\nu_d = 0{\cdot}74$ and thus $h\nu_d = 2800$ cals./mole.

In the ratio of the two velocity constants the factor

$$- \frac{\sinh \beta\nu_m}{\sinh \beta\nu_m/\sqrt{2}} \cdot \frac{\sinh \beta\nu_s/\sqrt{2}}{\sinh \beta\nu_s} \cdot \left(\frac{\sinh \beta\nu_d/\sqrt{2}}{\sinh \beta\nu_d}\right)^2$$

$$(h\nu_m - h\nu_s - 2h\nu_d)\left(1 - \frac{1}{\sqrt{2}}\right)\frac{1}{2RT}.$$

tends at low temperatures to $e$

The factor $2^{3/2}$ in the ratio $\dfrac{k_1}{k_2}$ is partially compensated by the term

$$\left(\frac{\sinh \beta\nu_s/\sqrt{2}}{\sinh \beta\nu_s}\right)^2,$$

and becomes (at high temperatures) $\sqrt{2}$. In the kinetic formula this $\sqrt{2}$ arises from the lower number of collisions in the heavier isotope.

* For small $\beta\nu_d$, $2\beta\nu_d \coth \beta\nu_d \approx 2$.

720    REACTIONS OF LIGHT AND HEAVY HYDROGEN

## 5.

In this section we shall compare the theoretically calculated values for $k_3$ and $k_4$ with the experiments and calculate the reaction rates of (1) and (2) at low temperatures. In the case of $H + H_2 \rightleftharpoons H_2 + H$ the theoretical figures can be checked also by the experiments.

$k_3$ and $k_4$ can be estimated on basis of the values of $Q$, $\nu_s \nu_d$ and $\nu_i$ obtained in the preceding sections with aid of the Table II. In these cases, however, the molecular constants of the complexes in the two activation points are not equal, and consequently also the number of passages over the two saddles will be different. The number of passages over the two saddles of the reaction (3) is given by

$$(D - HH) : n_3' = 3 \cdot 6 \times 10^8 \left(\frac{4}{2 \cdot 2}\right)^{3/2} 11 \cdot 25 \, \frac{\sinh 12400\beta(1 + \frac{1}{3}2960^2\beta^2)}{(\sinh 2730\beta)^2 \sinh 7200\beta} \quad (32)$$

$$(DH - H) : n_3'' = 3 \cdot 6 \times 10^8 \left(\frac{4}{2 \cdot 2}\right)^{3/2} 10 \cdot 50 \, \frac{\sinh 12400\beta(1 + \frac{1}{3}2880^2\beta^2)}{(\sinh 2750\beta)^2 \sinh 6140\,\beta} \quad (33)$$

which yields for $n_3' = 2 \cdot 95 \times 10^9 \, \dfrac{\text{mole}}{\text{litre sec.}}$

and for

$$n_3'' = 3 \cdot 5 \times 10^9 \, \frac{\text{mole}}{\text{litre sec.}} \text{ at } 950° \text{ abs.}$$

Analogous formula for the reaction $H + D_2$ gives for the saddle $(H - DD)$

$$n_4' = 2 \cdot 0 \times 10^9 \, \frac{\text{mole}}{\text{litre sec.}},$$

and for the saddle $(HD - D)$

$$n_4'' = 1 \cdot 7 \times 10^9 \, \frac{\text{mole}}{\text{litre sec.}}$$

at the same temperature.

The number of atomic interchanges which in the symmetric cases was calculated with $\gamma = \frac{1}{2}$ from the number of passages over one of the saddles, in the case of the reactions (3) and (4) can only be estimated. Since for $n' = n''$ $\gamma = \frac{1}{2}$, and for the other limiting case if $n' \gg n''$ with $\gamma = 1$, $k = n''$, we shall put in both cases $\gamma = \frac{3}{4}$, and thus $k_3$ and $k_4$ become

$$k_3 = 2 \cdot 2 \times 10^9 \, \frac{\text{mole}}{\text{litre sec.}}$$

$$k_4 = 1 \cdot 27 \times 10^9 \, \frac{\text{mole}}{\text{litre sec.}}$$

In order to compare these theoretical values for $k_3$ and $k_4$ with the experiments we have to insert them into the formula (15) of section 1. The right side of the formula yields $1 \cdot 65 \times 10^9$ in agreement with the experimental value $1 \cdot 4 \times 10^9$. The agreement must be considered quite good, especially since the right side of the formula (15) is proportional to $\gamma$, which was chosen rather arbitrarily $\frac{3}{4}$. Unfortunately, lack of more experimental and theoretical data about the reactions (3) and (4) makes it impossible for us at present further to check formula (15) for a wide range of $H_2$ and $D_2$ concentrations, and also for different temperatures.

L. FARKAS AND E. WIGNER                    721

It is remarkable that the velocity of reaction (3) is the fastest among all elementary reactions of hydrogen. This is caused by the comparatively low zero point energy of the activated complexes D — HH and DH — H. This was predicted by Polanyi [15] on qualitative considerations.

In Fig. 3 the energy diagrams for all the reactions (1 to 4) are sketched.

In the *initial state the zero point energy difference of H and D is neglected* (the zero point energy difference of H and D is 84 cals., H having with this amount greater ionisation energy), *in the activated state only the positive part of the zero point energy is taken into account.* Qualitatively the energy levels of the complexes are in the same order like those which, in the paper of Farkas and Farkas were calculated on basis of rather crude assumptions about the force constants of the activated complexes. The main difference in the absolute values of the zero point energies arises from the inclusion of the deformation frequencies which were neglected in *cf.* 2.

Much less satisfactory is the agreement between the theoretically calculated rates for the reaction (1) and the experimental figures at low temperatures, as given in Table I. For the reaction (2) thus far no experimental data are known at low temperatures.*

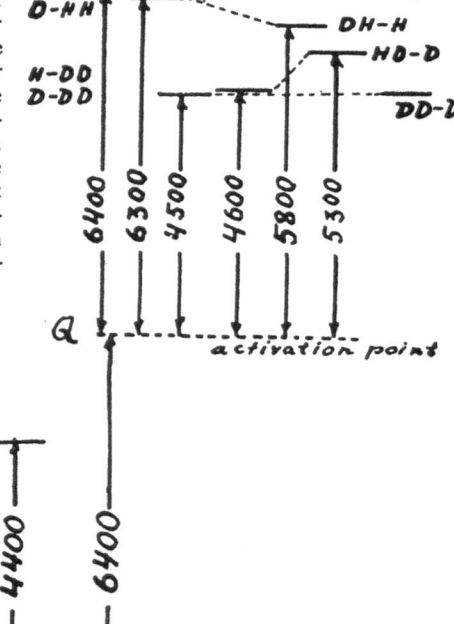

Fig. 3.

With the constants, $Q$, $\nu_s$, $\nu_d$ obtained in section 4 and $\nu_i$ which was derived from the rough energy surface, the calculation of the rates of the reaction (1) is as follows :

The tunnel effect in this region as mentioned on page 715 is sufficiently accurately given by $(1 + 1/6(2000\beta)^2)$ which factor has at T = 283° abs. the value 1·52 and, at $T = 373°$, 1·30. At this temperature region sinh $x$ can be replaced by $\frac{1}{2}e^x$ and we obtain, inserting the numerical values into (17),

---

[15] Polanyi, *Nature*, 1934, **133**, 26.

* According to the theoretical formula (17) the ratio of $k_1/k_2$ should at room temperature be 3·1.

$$5 \cdot 7 \times 10^{10} \times 1 \cdot 52 e^{-\frac{6600}{283\,RT}} = 6 \cdot 9 \times 10^5 \frac{\text{mole}}{\text{litre sec.}},$$

$$5 \cdot 0 \times 10^{10} \times 1 \cdot 30 e^{-\frac{6600}{373\,RT}} = 8 \cdot 8 \times 10^6 \frac{\text{mole}}{\text{litre sec.}},$$

instead of the experimental figures of $1 \cdot 3 \times 10^5 \frac{\text{mole}}{\text{litre sec.}}$ and

$3 \cdot 1 \times 10^6 \frac{\text{mole}}{\text{litre sec.}}$ at the corresponding temperatures.

The 6600 Cal. which appears in the exponential factor is the " over all " activation energy at low temperatures. This energy $W$ is in this region

$$W = Q - \tfrac{1}{2} h\nu_m + h\nu_d + \tfrac{1}{2} h\nu_s.$$

Theoretically, $k_1$ is thus given by :

$$k_1 = 6 \cdot 06 \times 10^{20} \times \left(\frac{M_a + M_m}{M_a M_m}\right)^{3/2} \frac{I_c}{I_m} \frac{h^2}{\sqrt{2\pi k}} T^{-1/2} e^{-\frac{W}{RT}} \left(1 - \frac{1}{6}\left(\frac{500}{T}\right)^2\right)^* \quad (34)$$

$T$ enters into this formula, besides the exponential factor essentially in the $-\tfrac{1}{2}$ power, in contrast to the kinetic formula, and also to the theoretical temperature dependence at high temperatures.

It is at present impossible to tell whether this discrepancy is caused by some errors in the experiments or by the imperfection of the theory. It must be emphasised, however, that it is impossible to fit both the low and the high temperature measurement on the reaction rate of (1) to the theoretical calculations by alteration of the constants $\nu_s$, $Q$, $\nu_d$ of section 4.

If one uses the low temperature measurements, instead of the temperature dependence of $k_1$ at high temperatures, to determine the vibrational frequencies, one obtains the following set of values: $h\nu_s = 10600$ cals./mole, $h\nu_d = 1500$ cals./mole and the activation energy becomes 7800 cals./mole. With these constants one can fit all measurements, except, however, the temperature dependence of $k_1$ and $k_2$ in the high temperature region.

Possibly the calculation of $Q$ from the temperature dependence of $k_1$ at high temperatures according to equation (30) is affected with the greatest inaccuracy. The temperature dependence of $k_1$ is namely derived from the data about the thermal conversion of $p\text{-H}_2 \to o\text{-H}_2$. The velocity of this reaction varies very much, since the thermal H atom concentration is enormously increased with rising temperature, and thus the directly measured figures on the temperature dependence of the reaction $p\text{-H}_2 \to o\text{-H}_2$ show a more than ten times higher temperature coefficient than that of $k_1$ itself. In $k_1$, for instance, the increase of the thermal H atom concentration has been taken already into account. In this connection it should be mentioned that the experimental figure for $\frac{T d \ln k_2}{dT}$ is practically the same as for reaction (1), namely $3 \cdot 5$, whereas

---

* By dividing this formula by the kinetic formula (16) (if we neglect the tunnel effect) we obtain for the steric factor the complicated expression

$$S = \frac{M_a + M_m}{M_a M_m} \times \frac{I_e}{I_m} \times \frac{h^2}{(d_a + d_m)^2\, 2\pi k T}.$$

Hence $S$ is, in the low temperature region, proportional to $T^{-1}$.

according to the formula (30) the theoretical value with the respective values for the molecular-constants of the complex DDD yields 4·12. This discrepancy may be due to the state of affairs just mentioned, which holds also for reaction (2). A small uncertainty in the larger temperature coefficients of the thermal *ortho-para*-conversions causes a comparatively large uncertainty in the temperature dependence of reactions (1) and (2).

If we assume that the measurements at high temperatures on the reactions (1) and (2) are fairly correct, the discrepancy at low temperature between the experimental data and the theoretical rate must be ascribed, as already mentioned, either to errors in the experiments or to some imperfection of the theoretical treatment. It is planned to revise the experimental data in a wide temperature range for both reactions, since it seems very important to decide whether the present theory of reaction velocities, as developed recently, is correct or not.

### Summary.

The velocity constants of the elementary reactions of hydrogen and deuterium

$$H + H_2 \rightarrow H_2 + H \qquad . \qquad . \qquad . \qquad . \quad (1)$$
$$D + D_2 \rightarrow D_2 + H \qquad . \qquad . \qquad . \qquad . \quad (2)$$
$$D + H_2 \rightarrow DH + H \qquad . \qquad . \qquad . \qquad . \quad (3)$$
$$H + D_2 \rightarrow HD + D \qquad . \qquad . \qquad . \qquad . \quad (4)$$

are calculated on basis of the formula

$$k = \gamma \tfrac{1}{2} \times \left(\frac{M_a + M_m}{M_a M_m}\right)^{3/2} \frac{I_c}{I_m} \times \left(\frac{h}{2\pi}\right)^3 \sqrt{\frac{2\pi}{kT}} \times e^{-Q/RT} \frac{N_L}{1000}$$
$$\times \frac{\sinh \beta \nu_m}{(\sinh \beta \nu_d)^2 \sinh \beta \nu_s} \cdot \left(1 + \tfrac{1}{8}\beta^2 \nu_i^2\right)$$

which was derived by Pelzer and Wigner and Wigner for this type of reaction. The constants $\nu_m$ and $I_m$ of the molecule are known from spectroscopic data. The molecular constants $I_c$, $\nu_a \nu_d \nu_i$ for the different complexes involved in reactions (1 to 4) are partly computed by the energy surface of these reactions and partly obtained by comparison of one part of the experimental data with the theory. The remaining experimental data on reactions (1 to 4) are then compared with the calculated velocity constants.

*Princeton University,*
*Princeton, N.J., U.S.A.* (E. W.)

*Daniel Sieff Research Institute,*
*Rehovoth (Palestine), and*
*The Hebrew University, Jerusalem,*
*Dep. of Physical Chemistry* (L. F.)

# On the Rate of Chemical Reactions

E. P. Wigner and H. Eyring

Scientific Monthly *44*, 564–567 (1937)

FROM less than 92 known elements, about a million compounds have been prepared by chemists. More than half of these compounds contain only four elements: carbon, hydrogen, oxygen and nitrogen. How is such a great number of combinations possible?

Certainly there is something strange about this. If we mix two elements, *e.g.*, sodium and chlorine, only three different ''compounds'' can arise; if we have more sodium than chlorine, practically all the latter will be bound to form rock salt and the remainder of the sodium will be left intact. If we have more chlorine than sodium, chlorine will be left over in addition to the rock salt. There are only three compounds: sodium, rock salt and chlorine. What an enormously different situation we encounter if we consider now the compounds containing only carbon and hydrogen! About ten thousand distinct compounds of these elements are known. The actual number of such compounds is probably only limited by the C and H atoms available to be built into the compounds. Certainly at any particular temperature not all these compounds are stable. At ordinary temperatures, methane is the only hydrocarbon which is stable, in addition to which we can have excess of carbon (graphite or diamond) or hydrogen. No two pure solids of the same composition can be stable over a range of temperature.

Why is it, then, that although a benzene molecule could break up into three acetylene molecules, pure benzene remains unchanged for an indefinite length of time? However, if one adds some suitable compound (a catalyst) the system proceeds very rapidly to equilibrium, the benzene decomposing into acetylene.

It obviously is the slowness of most reactions which allows this almost endless variety of organic substances to exist. This indeed is what makes life possible and prevents our immediate and complete combustion in the oxygen of the air. No wonder that the beginnings of the study of reaction rates are lost in antiquity.

The first measurement of a reaction rate has been made by Wilhelmy in 1850 and the problem of the mechanisms by which chemical reactions proceed has been a center of interest in chemistry ever since. Van't Hoff, in his famous ''Étude de dynamique chimique,'' recognized clearly the *dynamical* nature of chemical equilibria. That is to say, whenever a reaction like

$$2C_2H_6 = CH_4 + C_3H_8$$

comes to equilibrium, this means that the number of ethane molecules being formed in unit time is just balanced by the number of such molecules combining to give methane and propane. The well-known rapid increase of reaction rates with temperature results from the fact that only abnormally violent collisions can lead to reaction and the number of such collisions increases rapidly with temperature. The necessity of high energy concentrations for reactions was pointed out by Arrhenius toward the end of the last century.

2

At the present time, we believe that the fundamental features of the mechanism of reactions are well understood. This has been possible because of the pioneering work of many brilliant men, only a few of whom we can mention here. W. C. McLewis was the first to calculate the number of violent collisions. To the pioneering work of Marcelin, Polanyi, Bronsted and Herzfeld we owe the gradual development of the notion of the activated state. Rice, Ramsperger and Kassel completed the theory of unimolecular reactions along the lines suggested by Lindemann and Hinshelwood. Lewis and Smith, Daniels and Trautz applied kinetic theory to monomolecular processes and did much in the way of studying systematically many reactions. Later a considerable portion of reactions were found to go faster on the walls, and the understanding of these catalytic processes have been conspicuously advanced by the work of I. Langmuir and that of H. S. Taylor.

The rest of the development is an interesting example of the progress and obstacles and the way they are overcome in the advancement of science. As a matter of fact, we shall see that the calculation of absolute reaction rates is a simple example of the application of statistical mechanics. It is based on the concept that the atoms are moving under the influence of ordinary forces which are also responsible for the chemical valence. The difficulty of physicists in grasping this situation was that they believed on the basis of the older quantum theory that chemical reactions involved rather mysterious quantum jumps. Tolman and Herzfeld derived formulas very similar in principle to those which we shall discuss, on the basis of this theory. Chemists were not bothered by the notion of quantum jumps but were less familiar with the methods of statistical mechanics. As a matter of fact, Rodebush and Rice and Gershinowitz derived formulas which are similar to the ones to be discussed. They, like earlier workers, however, did not make the fullest use of our present knowledge of the nature of forces acting during a chemical reaction.

When the notion of ordinary forces acting between atoms was reestablished by the work of London, it soon became clear how the considerations just mentioned can be formulated on a general basis. This is all that has been done by Polanyi and Pelzer in collaboration with the present writers.

A nice way of representing the motion of the atoms is by a diagram in the "configuration space." Suppose we have three atoms moving with respect to each other in a straight line. The configuration space is two-dimensional then, the X coordinate being the distance of atoms 1 and 2, the Y coordinate the distance between 2 and 3. Every point in this configuration space corresponds to a configuration of the three atoms. The forces between the atoms can be derived from the potential energy for this configuration. If we make a landscape over the configuration space such that the height at any point is equal to the potential energy for this configuration, a ball rolling on this landscape will represent the motion of the three atoms under the influence of the forces between them. The relative position of the atoms will change in reality in such a way that the corresponding point in configuration space always coincides with the position of the ball. If we are interested in a system of more than three atoms or if their motion is not restricted to a line, we must use a configuration space of more dimensions.

Stable chemical compounds correspond to low regions in our landscape. If the ball is in such a low region and has little velocity, it will stay in this region forever. There may be several regions of comparatively low energy, corresponding to several apparently stable groups of molecules. *A reaction will then consist*

3

*in the passing over of our ball from one low region to another.*

For such a passage, it needs, first of all, enough energy. The average amount of energy which such systems have is proportional to the temperature. However, at any particular temperature, some systems will have less than the average energy and a very small number, very much more energy. Only the systems with exceptionally high energy will be able to pass from one low region to the other, and the fraction of the systems which have this unusual amount of energy increases very rapidly with the temperature. This fact accounts for the rough empirical rule that the reaction rate doubles with a 10° increase in temperature.

In order to make an actual calculation of the reaction rate of the reaction $H_2 + J_2 = 2HJ$, say, the fundamental idea of Gibbs may be utilized. Imagine a reaction vessel with a great number of $H_2$ and $J_2$ molecules. Let us subdivide the vessel into small compartments, each containing one single $H_2$ and $J_2$ molecule. The number of collisions between $H_2$ and $J_2$ molecules will be still the same as in the original vessel. However, every $H_2$ can react with one $J_2$ only so that we need to consider the configuration space of one pair of molecules only, instead of considering the coordinates of all the molecules simultaneously.

The instantaneous state in each compartment can be represented by a point in the configuration space. There will be one point in the configuration space for each compartment. As the molecules move in each compartment, the corresponding points in the configuration space will swarm like people in a mountainous region. To begin with, only one valley in configuration space is populated. The people scurry around apparently aimlessly. Most of them have too little energy to rise much above the floor of the valley. Even those who have enough energy to emerge from the crowd will go uphill at any arbitrary place and

only a few of the lucky ones will strike the path that leads into the neighboring valley. The number of these successful ones is all that concerns us. Their number gives us the number of reactions in our original vessel.

We can count these lucky people by multiplying their density at the top of the pass with the velocity with which they are traveling. Account must be taken, of course, of the fact that some of them which passed the crest of the hill, having encountered some obstacle, return without having descended into the new valley.

Both the density of the people in the pass and their velocity are given by standard formulas of statistical mechanics. Indeed, the velocity depends only on the temperature and the mass of the atoms involved. The density depends only on the temperature, the density of population in the valley and the height of the pass above the valley floor. Thus the whole reaction rate depends only on the nature of the landscape in the immediate neighborhood of the pass, the nature of the valley floor and the temperature. It has practically nothing to do, however, with the intervening country. The paths leading from one valley to the other may traverse many lower passes and intermediate valleys. All this will have little effect on the density of the highest pass and thus practically no effect on the reaction rate.

The intermediate valleys correspond to the intermediate compounds, frequently isolated by skillful chemists. These give us important information on the topography of the landscape. The mere fact that they can be isolated shows, however, that they inhabit low valleys and not the critical pass where density determines the rate of reaction.

Of all the quantities entering into the calculation of reaction rates the height of the pass above the valley is responsible for the greatest uncertainty. Except for a few cases, so far, it has always been necessary to derive this value from the

4

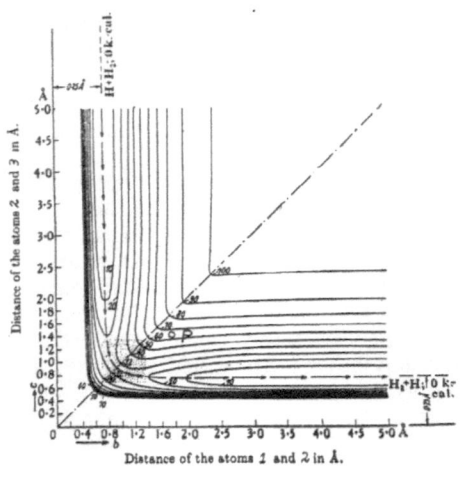

FIG. 2.   PHOTOGRAPHIC PICTURE OF THE LAND-
SCAPE OF FIG. 1, TAKEN FROM THE WORK OF
C. F. GOODEVE.

FIG. 1.   THE CIRCLE P IN THE FIGURE ABOVE
CORRESPONDS TO THE LINEAR CONFIGURATION OF
THE THREE ATOMS AS INDICATED BELOW.   THE
FIGURE ABOVE REPRESENTS THE ENERGY SURFACE
FOR ALL LINEAR CONFIGURATIONS.   THE HEIGHT
OF THE SURFACE ABOVE THE PLANE OF THE FIGURE
IS CHARACTERIZED BY THE CONTOUR LINES, AS IS
USUAL ON MAPS.

experiments on reaction rates themselves.
An improvement of our knowledge of the
landscape is highly desirable, therefore.
This will be achieved, no doubt, in two
different ways—partly by a further improvement in the theoretical calculation
of such surfaces, partly also by experimental investigations from which empirical rules can be derived.   The question of absolute rates thus reduces to the
construction of the appropriate energy
surfaces, after which it becomes a simple
problem of arithmetic.

# Calculation of the Rate
# of Elementary Association Reactions

E. P. Wigner

Journal of Chemical Physics 5, 720–725 (1937)

(Received June 18, 1937)

An upper limit for the rate of association reactions is found by determining the probability of a decrease of the relative energy of two atoms below zero energy, under the influence of a third body. The relation of this approximate calculation to the rigorous solution of the problem is discussed in Section 2. The results are applied to the recombination of $J$ atoms, measured by Rabinowitch and Wood. Numerically, the agreement is quite good; however, the calculated values are somewhat too low, which cannot be explained by an inaccuracy of the method. Reasons for the discrepancy other than the possible nonadiabatic character of the reaction are discussed.

## 1

THE transition state method[1] allows the calculation of the rate of reactions which involve an activation energy. The assumptions which one has to make in order to apply this method are (1) that the reaction shall not involve a jump in the quantum state of the electrons,[2] (2) that classical mechanics be applicable to the motion of the nuclei (or that the ordinary way of taking into account quantum effects be justifiable) and, (3) that the number of systems crossing the activated state be the number of systems reacting. This last condition is always satisfied at sufficiently low temperatures and if the energy surfaces are not too complicated, the temperature at which their validity ceases will be rather high.[3] The common feature of the reactions to

---

[1] Cf. for previous literature e.g. M. G. Evans and M. Polanyi, Trans. Faraday Soc. 31, 876 (1935).

[2] The possibility of chemical reactions without quantum jumps in the state of the electronic system has been first realized by F. London, *Sommerfeld Festschrift* (S. Hirzel, 1928), p. 104.

[3] Cf. for a closer discussion of these conditions, E. Wigner, Trans. Faraday Soc. (1937).

which the transition state method can be applied in its usual form[4] is that the transition over a state with high energy be the rate determining step.

However, some of the most simple reactions, like the association of atoms to a molecule in a three-body collision, cannot be calculated by the present form of the transition state method, although its original form[5] embraces these reactions also. The purpose of the present note is the derivation of a formula for the rate of reactions of the type $2J + A = J_2 + A$, or

$$A_1 + A_2 + A_3 = A_1 A_2 + A_3. \qquad (1)$$

The assumption of the validity of classical mechanics which will be necessary in the present derivation also, is better fulfilled than for the reactions with activation energy. Also the exact shape of the energy surface enters much less critically, because there is no activation energy $Q$, and a calculation of the absolute rate is possible at least approximately. On the other hand, there is no limiting case in which the formula (11) to be derived here is a consequence of well-established principles of statistical mechanics. Such a limiting case existed for reactions with activation energy as that of very low temperature and sufficiently heavy masses to assure the validity of classical mechanics.

Eyring, Gershinowitz and Sun[6] have given a very ingenious method by which the rate of (1) can be rigorously calculated if the energy surface can be divided into two regions, both of which are developable. This is very nearly true for that part of the energy surface for the $H + H + H = H_2 + H$ reaction which corresponds to the straight configuration of the atoms. However, it cannot be claimed that the majority of associations will proceed from a nearly straight configuration. For a bent configuration, the limits of the integration of Eq. (7) reference 6 should be changed and this will increase the probability of association for bent configurations. Eyring, Gershinowitz and Sun[6] tried to compen-

sate for neglecting to change the limits of their integral by considering the neighborhood of the isosceles triangle configuration also. Thus they counted some reactions twice, some not at all. This criticism of the very ingenious work of reference 6 should not be taken, of course, too seriously, since all that was claimed, and all that can be done at present, is to obtain the reaction rate within a not too large factor. That certainly has been accomplished by Eyring, Gershinowitz and Sun. The point was brought up partly to illustrate the point that an exact calculation of the rate of (1) is not feasible at present[7]—not even under simplifying assumptions for the energy surface—and partly to justify taking up the subject here again.

2

Theoretically,[8] of course, one could obtain the rate of (1) by considering the macrocanonic assembly of systems, containing an $A_1$, an $A_2$ and an $A_3$ atom in a great box of volume $v$. One can consider then those systems which are in the dissociated state at present but will go over into the right side of (1) within a time $t$. Whether a point in phase space corresponds to the left or right side of (1), may be ambiguous if all three atoms are close. However, if $t$ is sufficiently large, this ambiguity will extend to a very small proportion of all the systems reacting, most of the molecules formed will be sufficiently far away already from the third body to leave no doubt as to their associated character. On the other hand, $t$ shall not be taken long enough to allow for a second collision, i.e., $t$ shall be small compared with the time between collisions but very long compared with the collision time.

The systems which will react before $t$ will be contained in a region $R_t$ of phase space. The ends of this region may be a little uncertain, but the total probability of $R_t$ will be well defined. The

[4] H. Eyring, J. Chem. Phys. 3, 107 (1935); M. G. Evans and M. Polanyi (reference 1).

[5] M. Polanyi, Zeits. f. Physik 1, 90 (1920). The first complete outline of the theory has been given in R. Tolman's *Statistical Mechanics* (Chemical Catalog Co., 1927). Cf. also K. F. Herzfeld, *Kinetische Theorie der Wärme* (Müller-Pouillet's *Handbuch der Physik*, second edition, 1925).

[6] H. Eyring, H. Gershinowitz, C. E. Sun, J. Chem. Phys. 3, 786 (1935).

[7] This does not preclude the possibility of calculating the absolute rate of (1) more accurately than can be done in practice for reactions with activation energy. Although the transition state method should give nearly exact results for the latter, the parameters of the energy surface enter in such a critical way into the formulas that a calculation of the whole rate is hardly feasible at present. Cf. also reference 3.

[8] The second section serves only to establish the connection between the (impracticable) exact calculation of the rate and the approximate calculation of the ensuing sections.

rate constant $k$ is this probability, divided by $t$.

As we increase $t$ (keeping it still below the time between two three-body collisions), $R_t$ will assume the form of a longer and longer tube in phase space. We can put a surface $S$ across this tube and $k$ will be the number of systems crossing $S$. It does not matter where the surface $S$ lies, it only must have no wrinkles which could be crossed by the same system twice. All the systems will cross $S$ in the same direction.

The number of decomposing systems can be counted in a similar way. Furthermore, the continuation of $S$ may be taken for the surface for this counting. If we do this, and still allow no wrinkles in $S$, the number of systems crossing it in one direction will be the number of associations, the number crossing it in the other direction, the number of dissociations (which, naturally, must be equal to the number of associations). Apart from having no "wrinkles," $S$ can be quite arbitrary as long as it divides the whole phase space into two separate parts: those corresponding to the left and the right side of (1). If the distance of $A_3$ is very great from $A_1$ and $A_2$, $S$ must be the $H_0=0$ surface, where $H_0$ is the mutual energy (potential and kinetic) of $A_1$ and $A_2$.

If the distance of the third atom is small, and one still would take $H_0=0$ for $S$, this $S$ could possibly show wrinkles. These wrinkles would correspond to the relative energy of $A_1$ and $A_2$ falling below zero and rising again above this value during the same collision. No association would result then, although, counting the crossings through $S$ in one direction, one would count these crossings. However, one can define an $H_0$ *depending on all coordinates*, such that $H_0=0$ should have no wrinkles, i.e., no path in phase space shall cross it more than once during one collision. This $H_0$ will be equal to the relative energy of $A_1$ and $A_2$ for great distance of the $A_3$, will be different from this, however, if $A_3$ is near to $A_1$ and $A_2$. It will divide the whole phase space in two distinct parts and it will have the property that the total number of paths crossing it is as small as possible. This latter condition does not, of course, determine $S$ (and even less $H_0$), but it determines that minimum number, which is the reaction rate $k$.

### 3

We shall calculate now the number of systems crossing the $H_0=0$ surface in unit time. No special assumption concerning the position of this surface will be made in this section and the formulas (3) and (5) hold, therefore, quite generally. In the later sections, however, $H_0$ will be interpreted (for every position of $A_3$) as the energy of $A_1$ and $A_2$ in the coordinate system in which their center of mass is at rest. Formula (11) will give, therefore, only an approximate value (an upper limit) for the rate constant $k$.

If $H$ is the total Hamiltonian, we have

$$
\begin{aligned}
\frac{dH_0}{dt} &= \sum_i \left( \frac{\partial H_0}{\partial q_i}\frac{dq_i}{dt} + \frac{\partial H_0}{\partial p_i}\frac{dp_i}{dt} \right) \\
&= \sum_i \left( \frac{\partial H_0}{\partial q_i}\frac{\partial H}{\partial p_i} - \frac{\partial H_0}{\partial p_i}\frac{\partial H}{\partial q_i} \right) \qquad (2)\\
&= \sum_i \left( \frac{\partial H_0}{\partial q_i}\frac{\partial (H-H_0)}{\partial p_i} - \frac{\partial H_0}{\partial p_i}\frac{\partial (H-H_0)}{\partial q_i} \right).
\end{aligned}
$$

Those systems will cross the $H_0=0$ surface in the decreasing sense, which are nearer than $(dH_0/dt)/|\text{grad } H_0|$ from it, provided $dH_0/dt$ is negative. We have, therefore ($Z$ is the total integral of $e^{-H/kT}$ over phase space)

$$
k = \frac{v^2}{Z} \int \frac{dH_0/dt}{|\text{grad } H_0|} e^{-H/kT} d\sigma, \qquad (3)
$$

where $d\sigma$ is a surface element of the $H_0=0$ surface and the integration has to be extended over the region in which $dH_0/dt<0$.

This is simply the derivative of the integral

$$
I(E) = -\int \cdots \int_{\substack{H_0<E \\ dH_0/dt<0}} (dH_0/dt)e^{-H/kT}
$$
$$
\cdot dq_1 \cdots dq_n dq_1 \cdots dp_n \qquad (4)
$$

with respect to $E$ at the point $E=0$

$$
k = \frac{v^2}{Z}\left( \frac{dI(E)}{dE} \right)_{E=0}. \qquad (5)
$$

[The real $k$ is (rigorously) the minimum value of this expression for such $H_0(q_1, \cdots, q_n, p_1, \cdots, p_n)$ which go over into the proper energy of the two associating particles for a large distance of the third body but are not everywhere necessarily equal to it.] The integral $I(E)$ is to be extended over that region of phase space in which $H_0 < E$ and also (2) is negative.

### 4

As an example, the rate of a simple reaction (1) shall be calculated where all $A$ are atoms. The approximate $H_0$ which shall be used will be (for every distance of the $A_3$) the proper energy of the associating atoms. We shall use in what follows the following coordinates: the components of the vector $q_{12}$ pointing from $A_1$ to $A_2$, the components of $q_3$ connecting $A_3$ with the center of mass of $A_1$ and $A_2$, and finally the components of the vector $Q$ pointing to the center of mass of all three atoms. The corresponding momenta are $p_{12}$, $p_3$, $P$;

$$H_0 = \frac{1}{2m_r} p_{12}{}^2 + V_0, \qquad (6)$$

where $m_r$ is the relative mass of the associating atoms $1/m_r = 1/m_1 + 1/m_2$ and $V_0(r_{12})$ is the potential energy $V(r_{12}, r_{13}, r_{23})$ for $r_{13} = r_{23} = \infty$. Thus

$$H - H_0 = \frac{1}{2\mu} p_3{}^2 + \frac{1}{2M} P^2 + V - V_0, \qquad (7)$$

where $1/\mu = 1/(m_1 + m_2) + 1/m_3$ and $M$ is the mass of all three atoms together.

We have then by (2)

$$\frac{dH_0}{dt} = -\frac{1}{m_r}(p_{12}, \operatorname{grad}_{12}(V - V_0)) \qquad (8)$$

and,

$$I(E) = \int \cdots \int_{H_0 < E} (1/m_r)(p_{12}, \operatorname{grad}_{12}(V - V_0))$$

$$\cdot e^{-H/kT} dq_{12} dq_3 dQ dp_{12} dp_3 dP. \qquad (9)$$

The integration over $Q$ gives the total volume $v$ that over $p_3$ and $P$ gives $(2\pi\mu kT)^{\frac{3}{2}}(2\pi MkT)^{\frac{3}{2}}$. One can perform an orthogonal transformation for $p_{12}$ making $p_{12z}$ parallel to $\operatorname{grad}_{12}(V - V_0)$. The condition $dH_0/dt < 0$ means that the scalar product in (9) is positive, i.e., that $p_{12z}$ has the sign of the grad. The conditions $H_0 < E$ means $p_{12z}{}^2 + p_{12y}{}^2$

$+ p_{12z}{}^2 < 2m_r(E - V_0)$. Thus (9) becomes

$$I(E) = (2\pi kT)^3 v(\mu M)^{\frac{3}{2}} m_r{}^{-1} \int dq_{12} \int dq_3 \int_{H_0 < E,\ p_{12z} > 0} dp_{12}$$

$$\cdot p_{12z} |\operatorname{grad}_{12}(V - V_0)|$$

$$\cdot \exp(-p_{12}{}^2 / 2m_r kT - V/kT).$$

The integration over $p_{12z}$ can be performed immediately and that over $p_{12y}$ and $p_{12z}$ also if one introduces polar coordinates in the $p_{12y} p_{12z}$ plane. One obtains

$$I(E) = (2\pi kT)^4 kT v m_r (\mu M)^{\frac{3}{2}}$$

$$\cdot \int_{V_0 < E} dq_{12} \int dq_3 |\operatorname{grad}_{12}(V - V_0)|$$

$$\cdot \{\exp(-V/kT) - [1 + (E - V_0)/kT]$$

$$\cdot \exp[-(E + V - V_0)/kT]\}. \qquad (10)$$

For (5), we have to differentiate this with respect to $E$. Differentiating with respect to the limit of integration involves setting $V_0 = E$ in the integrand, which vanishes hereupon. Differentiating inside with respect to $E$, setting $E = 0$ and dividing by $Z/v^2 = v(2\pi kT)^{\frac{9}{2}}(m_1 m_2 m_3)^{\frac{3}{2}}$ gives $(m_1 m_2 m_3 = m_r \mu M)$

$$k = (2\pi m_r kT)^{-\frac{1}{2}} \int dq_{12} \int_{V_0 < 0} dq_3$$

$$\cdot |\operatorname{grad}_{12}(V - V_0)| (-V_0/kT) e^{-(V - V_0)/kT}. \qquad (11)$$

If the associating atoms are different and there are $n_1$ and $n_2$ of them in the volume $v$, and $n_3$ of the third body, the number of effective collisions will be $n_1 n_2 n_3$ times (11), i.e., the rate constant is given by (11). If the associating atoms are identical and their number $n$, the number of effective collisions will be $(n(n-1)/2)n_3$ times (11). Thus in case of the formation of a symmetrical molecule the well-known factor $\frac{1}{2}$ must be applied to (11). In addition to this, in most cases the reaction can occur only for definite relative orientations of the angular momenta of the atoms and this must be taken care of by another factor giving the probability of such an orientation. It may be pointed out again that (11) is only an upper limit for the rate of association and one could probably somewhat lower it by choosing $H_0$ more appropriately.

E. WIGNER

## 5

Equation (11) gives always an upper limit for the rate of association. However, if $V - V_0$ can assume negative values, this upper limit can be lowered somewhat quite easily. Indeed, in a system formed by three atoms, the total relative energy $H_r = H - P^2/2M$ never can be negative

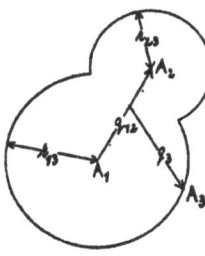

FIG. 1.

and it is not necessary, therefore, to include systems with negative energy into the assembly.

For large distance of the third body, naturally $H_r > H_0$. However, at $p_3^2/2\mu + V - V_0 = 0$ the $H_r = 0$ surface cuts the $H_0 = 0$ surface and goes above it. It is a better choice for $S$, then, to use the $H_r = 0$ surface where it lies above the $H_0 = 0$ surface, than this one, because no system can go through the $H_r = 0$ surface, $H_r$ being (apart from the influence of the walls) an integral of the motion.

The integration of (9) over $p_3$ should not be extended, therefore, over the region $p_3^2/2\mu + V - V_0 < 0$ in (9) and a factor

$$\int_{p_3^2 > 2\mu(V_0 - V)} \exp(-p_3^2/2\mu kT) dp_3$$

$$\Big/ \int_0^\infty \exp(-p_3^2/2\mu kT) dp_3 =$$

$$1 - F((V_0 - V)^{\frac{1}{2}}/(kT)^{\frac{1}{2}})$$

$$+ 2((V_0 - V)/\pi kT)^{\frac{1}{2}} e^{-(V_0 - V)/kT} \quad (12)$$

can be inserted to the integrand of (11) wherever $V_0 - V$ is positive. $F$ is the error integral

$$F(x) = \frac{2}{\pi^{\frac{1}{2}}} \int_0^x e^{-z^2} dx. \quad (13)$$

Another improvement can be made on (11) if the attractive region of $V_0$ is preceded by a

repulsive part. However, since it is a rather rare occurrence, it will not be discussed further.

## 6

If the third body is an inert gas, the total potential energy is the sum of the interactions of the three pairs of atoms. $V - V_0 = V_{13}(r_{13}) + V_{23}(r_{23})$ is positive in these cases and the correction (12) cannot be applied. The integrand of (11), as function of $q_3$, has a rather sharp maximum on two spherical portions, represented in Fig. 1. The radius $r_{13}$ of the sphere around $A_1$ is the sum $a_{13}$ of the collision radii of $A_1$ and $A_3$, similarly $r_{23}$ is the sum $a_{23}$ of the collision radii of $A_2$ and $A_3$. On the first sphere, one can introduce instead of $q_3$ polar coordinates around $A_1$. Integration over the polar angles gives the solid angle $\alpha_{13}$ of the first spherical portion. This is

$$\alpha_{13}(q_{12}) = \pi[(a_{13} + q_{12})^2 - a_{23}^2]/a_{13}q_{12} \quad (14a)$$

for $q_{12} < a_{13} + a_{23}$ and

$$\alpha_{13}(q_{12}) = 4\pi \quad (14b)$$

for $q_{12} > a_{13} + a_{23}$. On the sphere around $A_1$, one can set $V - V_0 = V_{13}(r_{13})$, and the derivative of this, with respect to $q_{13}$, at constant $q_3$, is $m_2/(m_1 + m_2)$ times the derivative with respect to $r_{13}$. Taking the factor $a_{13}^2 \alpha_{13}(q_{12})$ out of the integral, one must integrate $|\partial V_{13}/\partial r_{13}| \exp(-V_{13}/kT)$ with respect to $r_{13}$ between 0 and $\infty$, which gives just $kT$. One obtains, performing a similar transformation on the sphere around $A_2$ also,

$$k = -\frac{(2\pi m_r kT)^{-\frac{1}{2}}}{m_1 + m_2} \int_{V_0 < 0} dq_{12}$$

$$\cdot V_0[m_2 a_{13}^2 \alpha_{13}(q_{12}) + m_1 a_{23}^2 \alpha_{23}(q_{12})]$$

$$= -2\pi \left(\frac{2\pi m_r}{kT}\right)^{\frac{1}{2}} \int_{V_0 < 0} V_0 q_{12} \{(a_{13}^2 - a_{23}^2)$$

$$\cdot (a_{13}/m_1 - a_{23}/m_2) + 2(a_{13}^2/m_1 + a_{23}^2/m_2) q_{12}$$

$$+ (a_{13}/m_1 + a_{23}/m_2) q_{12}^2 \} dq_{12}. \quad (15)$$

In the last expression, (14a) has been used throughout because in most practical cases $V_0$ is very small for $q_{12} > a_{13} + a_{23}$.

This equation will be compared with the experimental results of Rabinowitch and W. C.

Wood[9] on the recombination of iodine atoms. For $V_0$ the Morse curve was used

$$V_0 = D(e^{-2\alpha(r-r_0)} - 2e^{-\alpha(r-r_0)}) \qquad (16)$$

with[10] $D = 2.46 \cdot 10^{-12}$ erg, $\alpha = 1.83 \cdot 10^8$ cm$^{-1}$, $r_0 = 2.66 \cdot 10^{-8}$ cm. This gives

$$-\int V_0 q_{12}^2 dq_{12} = 2.7 \cdot 10^{-35} \text{ erg cm}^3$$

and

$$-\int V_0 q_{12}^3 dq_{12} = 0.91 \cdot 10^{-42} \text{ erg cm}^4.$$

In this case, $a_{13} = a_{23} = r_J + r_3$ and the values shown in Table I were adopted for the radii.[11] The last line gives the values calculated by (14). These must be divided first by 2, because the associating molecules are identical, and next by 16, because only one of the ten states formed by two $^2P_{3/2}$ iodine atoms leads to the normal state of the molecule. The experimental results are compared with the calculated ones in Table II.

Although the disagreement is not very great, it is important to notice that the calculated values are lower throughout than the experimental ones. For the molecules $H_2$, $N_2$, $O_2$ this is, perhaps, not so serious, because (15) should hold for atoms only.[12] For He and A, however, (15) should definitely give an upper limit for the reaction rate. The calculated values should be indeed materially higher than the observed ones.

One possible source of error is contained in the radii of Table I. It would seem, however, that these hardly can be in error by a sufficient amount to explain the discrepancy. The second source of error lies in the potential curve (16). It cannot be expected, of course, that the Morse curve be accurate throughout. In fact, if one determines the $\alpha$ of (16) from the anharmonicity of the vibration instead of dissociation energy and vibrational frequency,[13] one obtains $1.47 \cdot 10^8$

cm$^{-1}$. This would compensate the discrepancy. However, the curve of Brown[14] for $V_0$ integrated directly, gives practically the same result as (16).

It is possible also that not only a $O_g^+$ state is strongly attractive among the 10 different states

TABLE I.

|          | He  | A    | H$_2$ | N$_2$ | O$_2$ |
|----------|-----|------|------|------|------|
| $r \cdot 10^8$ | 1   | 1.43 | 1.09 | 1.58 | 1.48 |
| $k \cdot 10^{31}$ | 4.5 | 6.5  | 5.4  | 7    | 6.7  |

TABLE II.

|              | He   | A    | H$_2$ | N$_2$ | O$_2$ |
|--------------|------|------|------|------|------|
| calc. $k \cdot 10^{31}$ | 0.14 | 0.20 | 0.17 | 0.22 | 0.21 |
| exp "  " | 0.18 | 0.36 | 0.40 | 0.66 | 1.05 |

arising from two $^2P_{3/2}$ iodine atoms. Although all states known so far are either repulsive or only very weakly attractive, this possibility must be left open. No such difficulty would come up at the calculation of the rate of recombination of H atoms, e.g., since all states are known in that case.

There is one more point which should be mentioned. For very large $q_{12}$, the Morse curve is not valid and the attraction due to the magnetic moment of the spins decreases only as the inverse third power of $q_{12}$. Since the square bracket of (15) becomes $8\pi m_2 a_{13}^2$ for very large $q_{12}$, the integral appears to diverge logarithmically. It does not appear sensible, however, to consider two atoms as forming a molecule if, in all probability, they will suffer a dissociating collision before completing a vibration. One is thus led to extending the integration over $q_{12}$ to the mean free path only. The contribution of the magnetic dipole forces then becomes negligible and the breaking down of (15) (caused by the incorrect situation of $S$) does not become serious.

[9] E. Rabinowitch and W. C. Wood. J. Chem. Phys. 4, 497 (1936).
[10] Cf. for the spectroscopic data, H. Sponer, *Molekül-spektren* (Berlin, 1936), Vol. I, p. 18.
[11] Cf. Landolt-Börnstein's Tables, 1st Ergänzungsband, p. 69. The radius of Xe, i.e., 1.75A, was used for $J$.
[12] In addition to this, the formation of an intermediate iodide of the third body may be considered. Cf. reference 9.
[13] For a discussion of this and related questions, cf. reference 10, Vol. 2, p. 103 f.

[14] W. G. Brown, Phys. Rev. 38, 1187 (1931).
[15] Cf. reference 14, also J. H. Van Vleck, Phys. Rev. 40, 544 (1932).

# The Transition State Method

## E. P. Wigner

Transactions of the Faraday Society *34*, 29–41 (1938)

*Received* 29*th June*, 1937.

## 1.

According to our present notions, the theory of reaction rates involves three steps. First, one should know the behaviour of all molecules present in the system during the reaction, how they will move, and which products they will yield when colliding with definite velocities, etc. Practically, this amounts in most cases to the construction of the energy surface for the reacting system. Professor Eyring told us about the results which can be obtained by the application of quantum mechanics to molecular systems for this part of the theory. The second step in the theory I would call the statistical part. It endeavours to solve the problem of the rate of elementary reactions. Assuming only the material on the left side of a chemical equation to be present in a vessel, and the molecules of these to have the Maxwell-Boltzman energy and velocity distribution, one wants to know how many molecules corresponding to the right side of the equation will be formed in unit time. The elementary properties of the molecules are supposed to be known in this second step and one wants to express the reaction rate of elementary reactions in terms of these. The present paper will be devoted entirely

to this second step.    The third step is the consideration of the co-opera-
tion of the various elementary reactions, which may occur beside and
must occur after each other in order to complete a real reaction.    In
especially favourable cases there is only one important chain of reactions
leading to the final products and this has one link which is so much slower
than all the others, that it is made responsible for the observed rate.
The others are then assumed to be so much faster that one has practically
equilibrium between the two sides of their chemical equations.

The fact that one has to subdivide the calculation of the reaction
rates into three different steps shows that it has not acquired the neatness
and finality of the theory of equilibria.    I should like, on this occasion,
to point again to one " elementary reaction," which, though often mentioned, never has been properly taken into account.    It is the production
of fast molecules, the establishing of the Maxwell-Boltzman velocity and
energy distribution.    This was tacitly assumed to be a fast reaction and
in complete equilibrium, before I began the discussion of the second step.
In many cases this will not be true.

Naturally, the treatment of the statistical step will be different for
the different kinds of elementary reactions.    It seems that one can
divide the elementary reactions into three groups.    I would class into
the first group those which do not change either the electronic quantum
state of the colliding particles nor their chemical formula and effect an
interchange of kinetic, vibrational and rotational energy only.    The
consideration of these is, as I just pointed out, a rather neglected dis-
cipline in chemistry.[1]    The physicist is strongly interested in them for
the so-called problems of mean free path (viscosity, heat conduction, etc.).
Into the second group, one could class all reactions which involve no
jump in the electronic quantum numbers but change the chemical constitution.    These are the most common types of chemical reactions [2] and
the transition state method applies to these alone.    The remaining third
class [3] deals with reactions which involve a jump in the electronic structure.    This type of reactions has been dealt with in several papers of
Professor Polanyi ; it is clearly the most general type and probably the
most difficult of all.

## 2.

The rest of this paper will be devoted to the consideration of the
second kind of reactions.    I shall endeavour rather to emphasise the
basic assumptions of the theory than to derive ready formulas.
Especially on account of some discrepancies with experiment, I think
that it may be useful to see that the transition state method is based, in
addition to well-established principles of statistical mechanics, on only
three assumptions, two of which are generally accepted.

The first assumption is that the comparatively slow motion of the
nuclei is followed by the rapid motion of the electrons to such an extent

[1] C. Zener, *Physic. Rev.*, 1931, **37**, 556, and O. K. Rice, *J. Am. Chem. Soc.*,
1932, **54**, 4558 have studied the excitation of vibrations and the changing of
rotational energy through collisions.    *Cf.* also J. M. Jackson and N. F. Mott,
*Proc. Roy. Soc., A,* 1932, **137**, 703.

[2] According to quantum mechanics, chemical reactions can occur without
a jump in the quantum state of the electrons.    This fact was first recognised
by F. London, *Sommerfeld Festschrift*, S. Hirzel, 1928, p. 104.

[3] Quantum mechanics was first applied to these by F. London, *Z. Physik*,
1932, **74**, 143, and by L. Landau, *Physik. Z. Sowjetunion*, 1932, **1**, 88 and 1932,
**2**, 146.    *Cf.* also O. K. Rice, *J. Amer. Chem. Soc.*, 1933, **1**, 375.

E. WIGNER                                                        31

that they are, for every position of the nuclei, in the lowest quantum state [2] (adiabatic assumption). The energy of this lowest state, together with the electrostatic energy of the nuclei, will be denoted by $E_1(X_1, X_2, \ldots, X_{3n})$ where $X_1, X_2, \ldots, X_{3n}$ are the co-ordinates of the nuclei. Under the adiabatic assumption the nuclei will move as if they were acted upon only by the potential $E_1(X_1, X_2, \ldots, X_{3n})$.

The second assumption is that the motion of the nuclei under this potential can be described sufficiently accurately by classical mechanics (or, if this is not the case, the usual way of taking the quantum effects into account is correct).

The last assumption can be best given at the end of the description of the transition state method, which will be the subject of the next section.

### 3.

For the calculation of the reaction rate it is sufficient to consider a system with as many atoms as occur in the chemical equation of the elementary reaction, since for a system with many atoms, the reaction rate will be simply proportional to the concentrations. Following Gibbs'

Fig. 1.—The regions (*l*) and (*r*), the activation point A and the surface F, in the section of the configuration space, corresponding to a linear configuration of the atoms H, H, Br. The ordinate is the distance of the two H atoms in A, the abscissa the distance of the middle H from the Br. The numbered lines are contour lines of the energy surface, constructed by Eyring and Polanyi for the reaction

$$H + HBr = H_2 + Br.$$

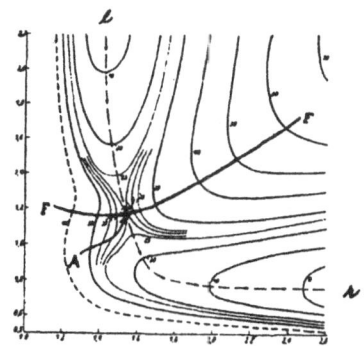

procedure, one will consider very many such systems, forming a macrocanonic ensemble. The configuration space of the system will have $3n$ co-ordinates and one can plot the function $E_1(X_1, X_2, \ldots, X_{3n})$ along a $(3n + 1)$th direction. Examples of such energy surfaces, for two-dimensional sections of the configuration space, are given in Figs. 1, 2, and 3. (The legends of Figs. 2 and 3 refer to questions to be brought up in section 5).

The configuration of every system of the macrocanonic ensemble will be represented by a point in the configuration space. At ordinary temperature the points of most systems will be in regions of low energy. One such low region will correspond to the left side of the chemical equation, another low region to the right side of the equation. In the first region (*l*), the co-ordinate differences are small for those atoms which are in the same molecule on the left side of the equation. In the second region (*r*), the co-ordinate differences are small for those atoms which form a molecule on the right side of the chemical equation. The transition between the two regions forms the reaction.

We assume now that the two low regions are separated by a higher region, the height of which is at least great compared with the ordinary thermal energy.   Then, all paths leading from the first valley into the second will have a highest point.   This highest point will be different for different paths.   The paths for which the highest point is as low as possible are said to pass through an activation point A, which indeed is the highest point of these paths.   The situation is better illustrated in the figures than can be done in words (Figs. 1, 2, and 3).

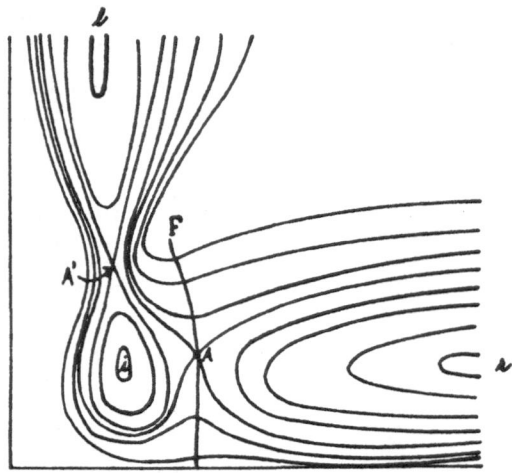

Fig. 2. — Reaction with an intermediate region ($i$).   At not too low temperatures, most points coming from ($r$) and entering ($i$) will return to ($r$), rather than to go through the narrow channel A' to ($l$).   In the reverse reaction, the points coming from ($l$) and passing A', will soon go over to ($r$), so that the concentration of points in ($i$) will not correspond to equilibrium.

Most reactions will have not only one activation point, but a whole super-surface $A$ of activation points (activation surface).   Indeed, if we displace or rotate a system in the activated state, the state thus obtained will still be an activated state.   All activation points have the same energy $Q_0$ and no system can react which has less energy than this amount.

Fig. 3. — The section of the activation surface with the plane of the Figure is the contour line 6, from A to its forking point. A path shows a point crossing the activation surface and returning to ($l$) again.

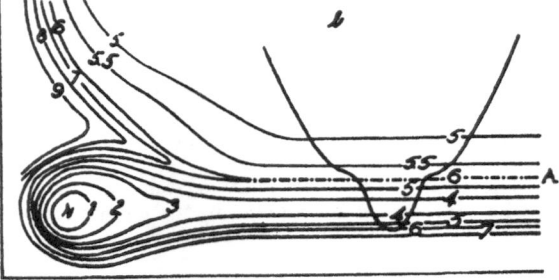

If we move on the activation surface $A$, the energy $E_1$ will remain $Q_0$. If we move in the direction from which the path defining the activated state comes, or if we move in the direction of this path, the energy will decrease, because the activation point is the highest one of this path. In all directions perpendicular to the direction of steepest descent, however, $E_1$ will *increase*.   Otherwise, the highest point of an adjoining path would be *lower* than $Q_0$. The activation surface $A$ forms a saddle-region; it is a maximum in one direction (in the direction of the path) but it is a minimum in all other directions.

We can now draw a $3n - 1$ dimensional surface $F$ through the activated points, perpendicular always to the direction of steepest descent. $A$ will have, in general, less than $3n - 1$ dimensions and lie entirely in $F$ (*cf.* Fig. 1). For a reaction in a system to occur, it is necessary that the representative point shall cross this $3n - 1$ dimensional surface. Most crossings will occur, of course, in the neighbourhood of the activation points, at any rate at such points the energy of which is not many $kT$ greater than the activation energy $Q_0$. It is easy to calculate the number of systems of our assembly which cross $F$ in one direction in unit time. It is simply equal to the density of points on the surface multiplied by the mean velocity perpendicular to the surface, and this integrated all over the surface :

$$n = \int_F \exp. (- E_1/kT)\bar{v}_p d\sigma \qquad . \qquad . \qquad . \quad (1)$$

For the practical evaluation of this integral one can expand $E_1$ into a power series of the co-ordinates which represent the distance from the activation surface A :

$$E_1 = Q_0 + \sum_{i=1}^{3n-s-1} Q_i(\xi_1, \ldots, \xi_s)\eta_i^2 \qquad . \qquad . \quad (2)$$

where $\xi_1, \ldots, \xi_s$ are $s$ co-ordinates within $A$ and $\eta_1, \ldots, \eta_{3n-s-1}$ are the other $3n - s - 1$ co-ordinates within $F$. All $a_i(\xi_1, \ldots, \xi_s)$ are necessarily positive and (1) can be evaluated, if (2) is sufficiently accurate, by elementary methods.

We can formulate now the third assumption for the validity of the formulas of the transition state method. If all paths crossing the surface $F$ in one direction originate in the low region $l$ and, crossing $F$ only once, end in the low region $r$, the reaction rate is simply given by the ratio of (1) to the number of systems in the region $l$, which is

$$\int_l \exp. (- E_1/kT)dX_1 \ldots dX_{3n} \qquad . \qquad . \quad (3)$$

Here again, a development of the form (2) is possible in most cases and this allows a simple evaluation of (3).

An especially simple way to evaluate the ratio of (1) and (3), *i.e.* the rate constant, has been developed by Eyring and by Evans and Polanyi.[4] Their considerations can be summarized as follows. Clearly, the ratio of (1) and (3) will be the same for all systems in which the potential in ($l$) and $F$ is the same, while it may be different in the intermediate region. One can replace, therefore, the real potential which is decreasing in the direction perpendicular to $F$, by a potential which increases in this direction. The region of $F$ will represent then a metastable molecule, the " activated molecule," the probability $C$ of which can be calculated by the well-known formulas for chemical equilibria. The activated molecule will have a vibrational frequency $\nu$ corresponding to the motion perpendicular to $F$, so that $C$ will contain a factor $kT/\nu$.

[4] M. C. Evans and M. Polanyi, *Trans. Faraday Soc.*, 1935, 31, 875 ; H. Eyring, *J. Chem. Physics*, 1935, 3, 107.

34          THE TRANSITION STATE METHOD

The activated molecule will pass through $F$ in one direction $\nu$ times per second, so that the rate constant becomes

$$k = C\nu \qquad . \qquad . \qquad . \qquad . \qquad . \qquad (4)$$

This, of course, in reality does not depend on $\nu$, *i.e.*, on the steepness of the increase of the artificial potential perpendicular to $F$. Equation (4) is very practical for actual calculations. It is, however, quite easy to evaluate (1) and (3) as they stand also.[5]

Most formulæ for $C$ take quantum effects into account also. Strictly speaking, it should be remembered that it is only in classical statistics permissible to make the above change in the potential.

Before discussing the theoretical foundations of the theory, I should like to quote a few characteristic instances in which it was possible to compare the theory with experiment.

The difficulty of this comparison lies mainly in the difficulty of obtaining accurate energy surfaces. I think, therefore, that at the present stage of the theory one must derive the most important quantity in (4), the activation heat $Q_0$, from the experiments themselves. The frequencies of the activated state can be derived from the calculated surface with less danger. If one determines these from the observed rate also, very few data may be left as an actual check of the theory. This procedure can be used only for reactions for which rich experimental material is available, especially if the isotopic effect is known.

The experimental difficulties in measuring reaction rates are only too well known. Since at least one quantity for (4), namely $Q_0$, is derived from the experiments, the values of the rate at at least two temperatures must be known. If these temperatures are too close, a relatively small error in the measurement of the rate may lead to an incorrect determination of $Q_0$ and a very great inaccuracy in the remaining factors.

The oldest example of the application of the transition state method is the reaction $H + H_2 = H_2 + H$ and the similar reactions with deuterium.[6] They have been measured partly directly, partly by the *ortho-para*-conversion.[7] The calculation of the corresponding energy surface[8] is the simplest example of its kind. Although in my opinion, the agreement between experiment and theory is not complete, the disagreement amounts to less than 50 per cent. with the most suitably chosen set of constants. A similar example is $Cl + H_2 = ClH + H$, where the agreement with the measurement of Rodebush[9] is within[10]

---

[5] In this way the transition state method was first applied to actual energy surfaces. *Cf.* H. Pelzer and E. Wigner, *Z. physik. Chem.*, B, 1932, 15, 445. The quantum corrections were included *ibid.*, 1932, 19, 203. Formulæ similar to those resulting from the transition state method[8] were obtained not much later by W. H. Rodebush, *J. Chem. Physics*, 1933, 1, 440 and by O. K. Rice and H. Gershinowitz, *ibid.*, 1934, 2, 853. However, these are very ingenious guesses rather than real derivations based on statistical mechanics. Consequently, the results are not quite identical with those of the transition state method.

[6] *Cf.* H. Pelzer and E. Wigner, and E. Wigner[4]; J. Hirschfelder, H. Eyring and B. Topley, *J. Chem. Physics*, 1936, 4, 170; L. Farkas and E. Wigner, *Trans. Faraday Soc.*, 1936, 32, 1.

[7] A. Farkas, *Z. Electrochem.*, 1930, 36, 782; *Z. physik. Chem.*, B, 1931, 10, 419; K. H. Geib and P. Harteck, *ibid.*, 1931; *Bodensteinband*, 849; A. Farkas and L. Farkas, *Proc. Roy. Soc.*, A, 1935, 152, 124.

[8] H. Eyring and M. Polanyi, *Z. physik. Chem.*, B, 1931, 12, 279; J. Hirschfelder, H. Eyring and N. Rosen, *J. Chem. Physics*, 1936, 4, 121.

[9] W. H. Rodebush and Klingelhoeffer, *J. Amer. Chem. Soc.*, 1933, 55, 130; M. Bodenstein, *Trans. Faraday Soc.*, 1931, 27, 8, 413.

[10] A. Wheeler, B. Topley and H. Eyring, *J. Chem. Physics*, 1936, 4, 178.

the experimental error.  The two temperatures at which the rate is measured are very close in this case (0° and 25°) and the limits of error very large.

Two four-body reactions have been investigated: $H_2 + I_2 = 2HI$ and $H_2 + ICl = HI + HCl$.  The discrepancy in these cases is a factor of 25 and 300 respectively.[11]  The construction of potential surfaces is quite difficult in these cases.

Perhaps the most interesting example is that of the reaction $2NO + O_2 = 2NO_2$ studied theoretically by H. Gershinowitz and H. Eyring.[12]  The energy surface is not known for this reaction.  It is clear, however, that in the activated state vibrational degrees of freedom take the place of several rotational degrees of freedom of the left side of the chemical equation.  Thus the phase space available is strongly diminished in the activated state and the reaction rate unusually low.  In spite of the absence of a detailed energy surface, Gershinowitz and Eyring could calculate the reaction rate with a surprising accuracy and also explain the negative temperature coefficient of this reaction.

I have given only the simplest examples of the comparison of theory and experiment, omitting entirely the highly interesting work of Polanyi and Evans on liquids,[13] of Bawn on " slow " reactions, and the extension of the transition state method to the ionogenic reactions by Ogg, Polanyi, and A. G. and M. G. Evans.[14]  This subject will be taken up probably by other speakers.

On the whole, I feel that the agreements are not as good as could be expected.  Doubtless the experiments are partly to blame for this, and the way of comparison I adopted here.  On the other hand, the disagreements call for a review of the theoretical foundations of the transition state method.

In the next two sections I will, therefore, discuss the validity of the three conditions heretofore stated.

### 4.

(a) Adiabatic Condition.—The electronic motion will be able to follow the motion of the nuclei, if that latter motion is not too rapid and if the electronic wave function does not change too radically for a small change in the nuclear positions.  The first of these is fulfilled practically always.  In all cases in which a non-adiabatic reaction is assumed, the rapid change of the wave function in the neighbourhood of an approximate crossing of energy levels is made responsible for it.  I think that for an energy surface which does not show singularities in abnormal curvatures, etc., the adiabatic condition cannot cause serious discrepancies.

[11] For $H_2 + I_2 = 2HI$ cf.[10], for $H_2 + ICl = HCl + HI$, Wm. Altar and H. Eyring, J. Chem. Physics, 1936, 4, 661.
[12] H. Gershinowitz and H. Eyring, J. Amer. Chem. Soc., 1935, 57, 985.  Cf. also O. K. Rice, J. Chem. Physics, 1936, 4, 53.  The measurements are due to M. Bodenstein, Z. Electrochem., 1918, 24, 183; M. Bodenstein and Lindner, Z. physik. Chem., 1922, 100, 87; M. Bodenstein and Ramstätter, ibid., 106; Briner, Pfeiffer and Malet, J. chim. physique, 1924, 21, 25.
[13] M. G. Evans and M. Polanyi, Trans. Faraday Soc., 1935, 31, 875; M. G. Evans and M. Polanyi, ibid., 1936, 32, 1333; W. F. K. Wynne-Jones and H. Eyring, J. Chem. Physics, 1935, 3, 492.  For the theory of slow reactions C. E. H. Bawn, Trans. Faraday Soc., 1935, 31, 1.
[14] R. A. Ogg and M. Polanyi, Trans. Faraday Soc., 1935, 31, 1375; A. G. Evans and M. G. Evans, ibid., 1400.  Cf. also C. F. Goodeve, ibid., 1934, 30, 60.

36          THE TRANSITION STATE METHOD

**(b) Quantum Effects.**—In the argument leading to (4) it was necessary to assume the validity of classical mechanics only for the activated state.   In most cases, however, some vibrations of the activated molecule will not be in the classical region.   Even this will not matter for those vibrations which either are too far from the bond to be changed, or, for some other reason, do not interfere with the motion of the important degree of freedom.   It appears indeed natural to use the usual quantum expressions for $C$ in (4), even for those degrees of freedom which interact strongly with the critical bond.   A reasonably rigorous argument shows [15] that this expression is (when modified by an additional factor to take into account tunnelling) valid up to the fourth power of $h$.   This may render (4) probable, but we must admit that it does not prove it beyond the third power terms in $h$.

Indeed, one can point out two cases in which (4) with the usual quantum expression for $C$ breaks down badly.   The first one is given in Fig. 4.   The zero-point energy corresponding to the vibration along OX will appear in (4), not, however, in reality, since the potential barrier is very thin.

In the second case, tunnelling under the high region of $E_1$ plays an important rôle.   A negative curvature of $E_1$ in the activated state can

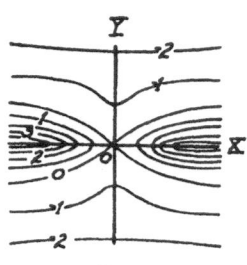

FIG. 4.

be described as imaginary $\nu$.   It may be interesting to note that (4) still holds up to the third power of $h$, if this imaginary frequency is used as one of the $\nu$ of the activated state.   Of course, the higher power terms of (4) are not correct in this case, partly, indeed, because the rate depends on the further continuation of the series of (2).

The reason that it is so much more difficult to apply the transition state method in quantum theory than it is in classical theory is twofold.   Firstly, there is no corresponding simple expression to Boltzmann's exp.$(- E_1/kT)$ for the probability of a configuration.   But an even greater difficulty is that one cannot speak about the mean velocity $\bar{v}$ at the activation point. (Heisenberg's indetermination principle.) There is thus, strictly speaking, no way to define the activated complex and its lifetime.   All that appears to be possible, in general, is to start out from the classical expression and develop the quantum corrections into a power series of $h$ (only even powers will occur).   This has been done to the second power, but the incentive to continue this procedure is not very great, partly because of the laborious mathematical work, partly because three terms will not be very much better than two are, and finally and mainly because our knowledge of energy surfaces hardly warrants an elaborate work.

One can form an idea of the magnitude of the corrections from the one dimensional case where, for several potentials, one has exact solutions of the Schrödinger equation which make a power series development unnecessary.   Bell [16] has calculated the penetration of Eckart's one dimensional barrier [17] (Fig. 5a) as function of the temperature.   His results for the reaction rate, the classical values, for the same conditions

[15] E. Wigner, *Z. physik. Chem.*, B, 1932, **19**, 203.
[16] R. P. Bell, *Proc. Roy. Soc.*, A, 1933, **139**, 466.
[17] C. Eckart, *Physic. Rev.*, 1930, **35**, 1303.

E. WIGNER    37

and the values of $A$ and $Q$ derived from Bell's results in the usual way by representing them as $Ae^{-Q/RT}$ are given in Table I. The values of $A$ and $Q$ become 1 and 14·5 Cal. for all temperatures in classical theory.

One sees that $Q$ and $A$ both decrease with decreasing temperature. This behaviour can be seen qualitatively without calculation: at very low temperature, the tunnelling rate becomes independent of temperature (which means $Q = 0$) and very slow (i.e., $A$ is very small). One

TABLE I.

| $T$. | $k_{\text{Bell}}$. | $k_{\text{Class}}$. | $A_{\text{Bell}}$. | $Q_{\text{Bell}}$. |
|---|---|---|---|---|
| 273 | $7 \cdot 10^{-10}$ | $2 \cdot 5 \cdot 10^{-12}$ | $10^{-7}$ | 3·5 |
| 323 | $4 \cdot 2 \cdot 10^{-9}$ | $1 \cdot 5 \cdot 10^{-10}$ | $1 \cdot 5 \cdot 10^{-3}$ | 8·5 |
| 373 | $3 \cdot 1 \cdot 10^{-8}$ | $3 \cdot 10^{-9}$ | $3 \cdot 10^{-2}$ | 10 |
| 473 | $7 \cdot 3 \cdot 10^{-7}$ | $2 \cdot 10^{-7}$ | 0·15 | 11·5 |
| 573 | $6 \cdot 4 \cdot 10^{-6}$ | $2 \cdot 9 \cdot 10^{-6}$ | 0·30 | 12·5 |
| 673 | $3 \cdot 4 \cdot 10^{-5}$ | $2 \cdot 10^{-5}$ | 0·55 | 13·2 |

is tempted to use this behaviour for the explanation of the very small $A$ values, occasionally observed. It must be remembered, however, that even a potential curve like that of Bell gives an appreciable tunnelling for very light atoms only (Bell's calculation is for hydrogen).

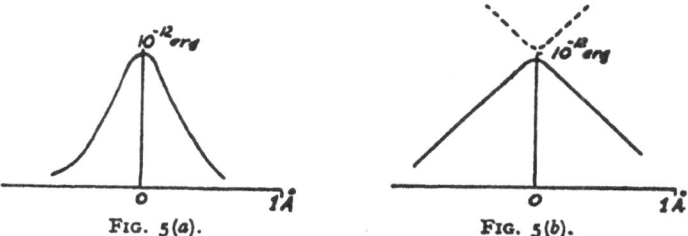

FIG. 5(a).                    FIG. 5(b).

A potential curve which may be realised more frequently. is given as a full line in Fig. 5b. Such a shape would occur always if two energy surfaces nearly cross (the other energy surface is dotted in Fig. 5b). A non-adiabatic reaction is possible in such cases, but we shall disregard this. For such a potential curve one may expect some tunnelling near the top even for somewhat heavier atoms.

The penetration probability can be exactly calculated if one neglects the small rounding off of the top of the curve and makes it quite sharp-edged. For energies $W$ which are not too near the classical activation energy $Q_0$, the penetration probability is [18]

$$R = 4W^{1/2}(Q_0 - W)^{1/2}Q_0^{-1} \exp. (- \alpha(Q_0 - W)^{3/2}), \quad \alpha = 4a\sqrt{2M}/3\hbar \quad (5)$$

where $M$ is the mass of the penetrating particle, $a$ is the sum of the reciprocal slopes of the two sides of the potential curve. At the temperature $T = 1/k\beta$ the rate of penetration, divided by the classical rate $e^{-\beta Q_0}/\beta$ is, assuming (5) correct throughout,

$$\frac{k}{k_{\text{class.}}} = \int_0^\infty R\beta e^{\beta(Q_0 - W)}dW . \quad . \quad . \quad . \quad (6)$$

[18] R. H. Fowler and L. Nordheim, *Proc. Rcy. Soc.*, A, 1928, 119, 173.

38                    THE TRANSITION STATE METHOD

For $1 \ll \beta^3/\alpha^2 \ll \beta Q_0$ this can be evaluated and gives

$$\frac{k}{k_{\text{class.}}} = \frac{32\sqrt{\pi}\beta^{3/2}}{9\alpha^2\sqrt{Q_0}}e^{4\beta^3/27\alpha^2} \quad . \qquad . \qquad . \qquad . \quad (7)$$

A plot of (7) against $T$ shows that for an atomic weight 10 and the potential curve of Fig. 5$b$, the quantum correction becomes insignificant above 0° C. It seems, therefore, that apart from reactions involving H, the tunnelling effect cannot be made responsible at ordinary temperatures for any large decrease of the temperature independent factor.

These considerations have a particular reference to the dissociation of $N_2O$ in which the temperature independent factor was found [19] to be about $10^{-3}$ times the value obtained by estimating (4). This was taken, for awhile, as an indication of the non-adiabatic nature of the reaction,[20] which was suggested by the fact that it violates the spin conservation law. However, it has been shown by Zener [21] on the basis of the interaction integrals obtained from the intensities of forbidden

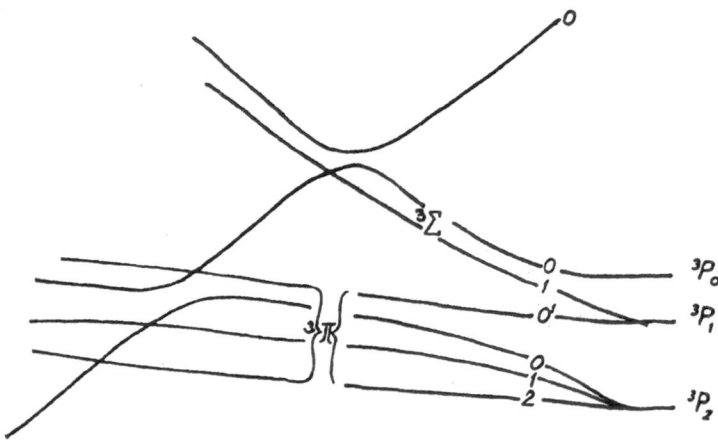

FIG. 6.

lines, that the non-adiabatic character of the reaction cannot be made responsible for a factor of the order of $10^{-3}$. We saw that the same holds of the tunnelling effect, so that it must be admitted that the experimental finding is unexplained at present. Perhaps some explanation can be obtained from a more detailed level scheme, like the one drawn schematically and very tentatively, in Fig. 6. If the lower level arising from the $^3P$ term of oxygen is a $^3\Pi_2$ level, this will strictly cross the $^1\Sigma_0$ level of the ground state for a straight configuration of the $N_2O$ molecule. Then, the reaction will certainly not be adiabatic for a linear

[19] M. Volmer and H. Kummerow, Z. physik. Chem., B, 1930, 9, 141 ; Nagasako and Volmer, ibid., 1930, 10, 414 ; Hunter, Proc. Roy. Soc., A, 1934, 144, 386.
[20] H. Eyring, Chem. Rev., 1932, 10, 103 ; H. Pelzer and E. Wigner,[5] G. Herzberg, Z. physik. Chem., B, 1932, 17, 68 ; A. E. Stearn and H. Eyring, J. Chem. Physics, 1935, 3, 778.
[21] C. Zener, private communication. The spectroscopic data were interpreted by A. F. Stevenson, Proc. Roy. Soc., A, 1932, 137, 298 ; E. U. Condon, Astrophys. J., 1934, 79, 217.

E. WIGNER    39

dissociation—not on account of the spin conservation law but because of the conservation law for angular momentum around the molecular axis. The reaction still can go adiabatically for a bent molecule, but it seems that the energy surface becomes so complicated that a larger effect of the tunnelling cannot be *a priori* excluded.

This example again emphasises how every calculation of reaction rates is bound to founder unless we have more detailed energy surfaces than are available at present for $N_2O$. One is also led to believe that the energy surfaces show more often a complicated structure, as exemplified in Fig. 6, than has been assumed hitherto. If this is true, the deviations from (4) become readily understandable, not only on account of the uncertainty in the quantum theoretical expression for $C$, but also because the last condition for the validity of (4) will be strongly violated.

## 5.

The last assumption of the transition state method was that all systems crossing the potential barrier are reacting systems. This appears to be quite natural for a simple energy surface of the kind illustrated in Fig. 1. Figs. 2 and 3 show, however, that especially for more complicated potential surfaces the system is even most likely to cross the activated state and to return then to the original configuration.

It appears thus that the formula (4) will lead, in general, to too high values of the reaction rate and should be corrected by a factor $\gamma$, smaller than one, which is the ratio of the number of the crossings through $F$ which end in $r$, coming from $l$ (or end in $l$, coming from $r$), to the number of all the crossings. The problem is then to calculate $\gamma$.

There is at least one case in which $\gamma$ is certainly 1. If the dimensionality $s$ of the activation surface $A$ is *less* than $3n - 1$ and if the temperature is very low, then most of the crossings through $F$ will go through a very narrow strip of $F$ around $A$. Even if the system does not go straight through into the final state, after having crossed $F$, but is reflected as in [21a] Fig. 2, the probability that it will find that very narrow strip through which it can return into the region from where it started, is very small. It will be the smaller, the thinner the strip is, *i.e.* the less the energy of the system surpasses $Q_0$. Since the mean energy excess of the molecules, which have an energy larger than $Q_0$, over this energy $Q_0$ is proportional to $T$, the probability of a return will go to zero with decreasing temperature (unless quantum effects should interfere), and $\gamma$ goes to 1. However, at higher temperatures $\gamma$ will decrease, and, apart from special cases (like that of an intermediate valley), it may be quite difficult [22] to find its accurate value. The complication indicated in Fig. 2 is not quite an adequate picture of the situation, since in the $3n$ dimensional configuration space, the picture may be even more involved.

[21a] On account of a mistake in drawing Fig. 2, the original substances correspond to $r$ in this case and the reaction products to $l$. (In all other examples it is the opposite way.)

[22] Mathematically, the problem is the same as that attacked by G. Lemaitre and M. S. Vallarta (*Physic. Rev.*, 1933, **43**, 87) and by G. Lemaitre, M. S. Vallarta and L. P. Bouckaert (*ibid.*, 1935, **47**, 434) for the calculation of the allowed cone of cosmic radiation. A complete solution involves in every case very much labour.

40                THE TRANSITION STATE METHOD

## 6.

Since one cannot calculate $Q_0$ at present sufficiently accurately, the comparison of (4) with experiment requires the experimental determination of the rate constant at at least two temperatures. This procedure tends to multiply the experimental errors and this may be the cause of a good part of the discrepancies mentioned in section 3. This difficulty would not arise in reactions without a heat of activation, like

$$A + B + C = AB + C \qquad . \qquad . \qquad . \quad (8)$$

However, the transition state method cannot be applied directly to this type.

Eyring, Gershinowitz and Sun [23] calculated the reaction rate of (8) if A, B, C are atoms by a very ingenious method for the case that the energy surface consists of two parts, both of which are developable. This is the case to a large extent for the $H + H + H = H_2 + H$ reaction. I should like to show how to calculate the rate constant without using this assumption, but using a symbolic application of the transition state method, as first considered by Polanyi.[24]

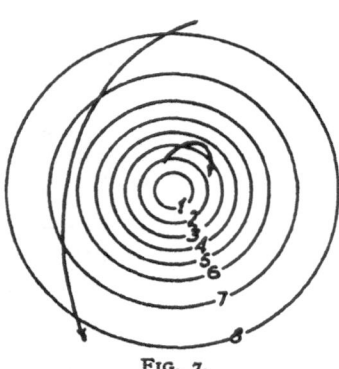

FIG. 7.

The energy surface of two attracting atoms shows clearly that one cannot define an activation surface in the usual way. If one puts it at a short distance, the atoms which cross it will not necessarily dissociate, if one puts it a large distance, most of the atoms crossing it are coming from the dissociated state to begin with (Fig. 7). One obtains the impression that it is not the passing through a surface in space which is important, but the passing of the relative energy of the two atoms (kinetic and potential) through the value 0. One is thus led to consider the energy (or something equivalent) as a co-ordinate instead of the usual space co-ordinates.

Following this line of thought I derived the formula for the velocity constant

$$k = \frac{2(2\pi\beta)^{3/2}}{\sqrt{m_r}} \iiint (-V_0) e^{-\beta(V - V_0)} g r_1 r_2^2 \sin \epsilon \, dr_1 \, dr_2 \, d\epsilon \qquad (9)$$
$$g = [(\partial V/\partial \epsilon)^2 + (r_2 \partial (V - V_0)/\partial r_1)^2]^{1/2}.$$

Here $\beta = 1/kT$; the relative mass of the two atoms A and B is $m_r$; the potential energy of these at the distance $r_1$ is $V_0$ if C is very distant. If C is near, the potential energy of all three is $V(r_1 r_2 \epsilon)$, where $r_1$ is the distance from A to B, the distance of their centre of mass C' from C is $r_2$ and $\epsilon$ is the angle between the lines C'C and AC'. The integration has to be extended over all the region in which $V_0$ is negative. The details of the derivation of (9) will be given elsewhere.[24a] $V - V_0$ is assumed to

[23] H. Eyring, H. Gershinowitz and C. E. Sun, *J. Chem. Physics*, 1935, 3, 786; H. Gershinowitz, *ibid.*, 1936, 4, 363.
[24] M. Polanyi and E. Wigner, *Z. physik. Chem.*, 1928; *Haber Band*, 439.
[24a] *J. Chem. Physics*, 1937, 5, September issue.

E. WIGNER    41

be positive throughout ; (9) will be slightly modified otherwise.   A factor
1/2 enters if A and B are identical atoms.

The formula (9) for the association reaction is much more complicated
than that for exchange reactions, *e.g.*, the reason for this is that the latter
could be simplified by an assumption like (2), since only the neighbourhood of the activation surface was important for these reactions.   This is
not true for the type (8).

Formula (9) tells us how often the relative energy of two atoms A and
B decreases in unit time below 0, if there is a third atom C together with
them in a box, assuming complete equilibrium.   Again, this will be
only an upper limit for the reaction rate, because the relative energy of
A and B may increase above zero immediately after it has fallen below,
thus necessitating the introduction of a factor $\gamma$ similar to that for exchange reactions.   It may be again expected, however, that $\gamma$ will be
of the order 1 if the interaction between C and the A-B attraction is not
very great.   There is one important difference between this and the
$\gamma$ of Fig. 2 : the $\gamma$ for (9) will not go to 1 with decreasing temperature, and
will be in general further from 1 than that for (4).

Equation (9) shows the same dependence on the mass of the reacting
particles as that of Eyring, Gershinowitz and Sun.   This is in agreement
with measurements on the association of hydrogen atoms.[25]   It also shows
that the " third body " C will be more efficient which exerts greater
forces on A and B.   For equal chemical properties, the efficiency will
be roughly proportional to the surface of C.   Even so far as numerical
values are concerned, the disagreement between (9) and experiment
is not very great, for the $2I + A = I_2 + A$ reaction, (9) yields
$k = 2 \cdot 0 \times 10^{-32}$ cm.$^6$/sec., the experimental value of E. Rabinowitch
and W. C. Wood [26] being $3 \cdot 6 \times 10^{-32}$.   The values with He as the third
body, instead of A, are $1 \cdot 4 \times 10^{-32}$ and $1 \cdot 8 \times 10^{-32}$, respectively.

*Department of Physics,*
   *University of Wisconsin,*
      *Madison, Wisconsin, U.S.A.*

[25] I. Amdur, *J. Amer. Chem. Soc.*, 1935, **57**, 856 ; W. Steiner, *Trans. Faraday
Soc.*, 1935, **31**, 623.
   [26] E. Rabinowitch and W. C. Wood, *J. Chem. Physics*, 1936, **4**, 497.

# Some Quantum-Mechanical Considerations in the Theory of Reactions Involving an Activation Energy

J. O. Hirschfelder and E. P. Wigner[*]

Journal of Chemical Physics 7, 616–628 (1939)

(Received February 28, 1939)

The activated complex or transition state method for calculating the absolute rate of a chemical reaction with an activation energy would be rigorously valid if classical mechanics applied to all degrees of freedom. In quantum mechanics, two kinds of limitations must be considered. First, because of Heisenberg's uncertainty principle, the transition state itself can be defined only if the potential surface is sufficiently flat around the highest point of the reaction path. Second, even if this condition is fulfilled, the transmission coefficient can differ from the value expected on the basis of classical mechanics, because a wave packet can be reflected both on its way up, and also on its way down the potential barrier separating the initial and final states. In fact, the transmission coefficient is, in many cases, a rapidly fluctuating function of the energy of the system. If the temperature distribution of the energy is sufficiently broad to cover several periods of this fluctuation, an average transmission coefficient can be defined which nearly agrees with the classical value. For the crossing of a one-dimensional potential barrier, the quantum corrections are surprisingly small. In problems with several degrees of freedom, the transmission coefficient is affected by the interchange of translational and vibrational energy. However, if the vibrational motion is fast as compared with the motion along the reaction path, these degrees of freedom can be treated on a par with the electronic coordinates. In this case, the formulas of Eyring, with a mechanically sensible transmission coefficient, are satisfactory. On the whole, we conclude that quantum-mechanical considerations invalidate the transition state method to a much smaller extent than could be presumed and it is only in the consideration of the relative rates of reactions between isotopes and reactions at very low temperatures that these effects may be important.

## I. THE TRANSITION STATE METHOD AND ITS VALIDITY

IT is well known that whenever classical mechanics is valid, the formula[1]

$$k = (P_t/P_i)(\delta/\bar{v})^{-1}\gamma = P_t\bar{v}\gamma/(P_i\delta) \qquad (1)$$

for the rate, $k$, of reactions with an activation energy is a direct consequence of statistical mechanics. In (1), $P_t$ is the probability of the transition state in thermal equilibrium. The transition state is a strip of width $\delta$ in configuration space. This strip lies across the deepest saddle on the energy mountain separating the two regions in configuration space which correspond, respectively, to the initial and final state of the reaction. $P_i$ is the probability of the system being in the initial state; $\bar{v}$ is the average velocity with which the configuration points cross the saddle; $\delta/\bar{v}$ their average time of sojourn in the transition state. Finally, $\gamma$ expresses the proba-

bility that a system which crosses the saddle at complete thermal equilibrium actually originated in the initial state and will proceed to the final state to complete the chemical reaction.[2] The transmission coefficient, $\gamma$, is the only quantity appearing in Eq. (1) which cannot be evaluated by well-known methods of statistical mechanics —it can be estimated only from the general shape of the potential mountain.

There is no reaction with an activation energy in which classical mechanics is valid for all parts of the reacting system. However, it appears logical[3] to replace the classical expressions for $P_t$ and $P_i$ in (1) by the corresponding quantum theoretical sums. The question as to what extent this is justifiable has been discussed repeatedly.[4] One difficulty is that the notion of an activated

---

[*] Now at Princeton University.

[1] This equation was first proposed by A. Marcelin, Ann. chim. phys. (9)3, 120, 185 (1915). Subsequent treatments of this nature have been given by A. March, Physik. Zeits. 18, 53 (1917); Pelzer and Wigner, Zeits. f. physik. Chemie B15, 445 (1932); E. Wigner, Zeits. f. physik. Chemie B19, 203 (1932); H. Eyring, J. Chem. Phys. 3, 107 (1935); Evans and Polanyi, Trans. Faraday Soc. 31, 875 (1935).

[2] A more detailed explanation of (1) will be found, e.g. Trans. Faraday Soc. 34, 29 (1938). The reaction is considered "completed" when the distance between the reaction products becomes large compared with molecular dimensions. This has a clear sense only in gas reactions where the products separate at once to very large distances (of the order of the mean free path) if they begin to separate at all.

[3] This was first done by E. Wigner (Zeits. f. physik. Chemie B19, 203 (1932). It was also the method adopted by H. Eyring in developing his formulae of great generality.

[4] See for example, E. Wigner, Trans. Faraday Soc. 34, 29 (1938).

or transition state is not strictly compatible with the laws of quantum mechanics. The transition state method can only be justified when the path of the reaction is sufficiently flat in the neighborhood of the transition state so that we can consider simultaneously the position and the velocity of the point representing the system in configuration space. In classical mechanics, $\delta$ is always considered to be so small that the potential is practically constant across the transition state. If $\delta$ is large, $P_t$ and the average length of time that the system spends in the transition state become rather complicated functions of $\delta$ and (1) no longer applies. In quantum mechanics, it is necessary to take the width of the transition state sufficiently great to allow reducing the uncertainty in the velocity, $\Delta v$, to a small fraction of the average thermal velocity, $\bar{v}$. Certainly, if we are to apply Maxwell's formula, $\exp(-mv^2/2kT)$, for the probability of a system with a mass $m$ to have the velocity $v$ at the temperature $T$, the spread in the velocity, $\Delta v$, due to the uncertainty principle must be so small that it makes little difference whether we use the upper bound, $v + \Delta v/2$, or the lower bound, $v - \Delta v/2$, in the exponential. This means that $mv\Delta v$ must be small compared with $kT$. Putting $(kT/m)^{\frac{1}{2}}$ for $v$, this gives $\Delta v \ll (kT/m)^{\frac{1}{2}}$. We must take the transition state sufficiently wide so that this last relation can be satisfied. On the other hand, this width must be sufficiently small so that it is still a good approximation to take $P_t$ proportional to $\delta$. Thus, if the energy surface in the vicinity of the saddle has a curvature in the direction of the reaction path corresponding to the parabola, $V = V_0 - aX^2$, we must confine the activated state to a strip across the path which is so thin that $\exp(-aX^2/kT)$ shall be nearly unity all over the strip. This means that $\delta \ll (kT/a)^{\frac{1}{2}}$. These limits for the maximum values of $\Delta v$ and of $\delta$ are only compatible with the quantum-mechanical uncertainty principle if:

$$\frac{h}{2\pi} < \delta m \Delta v \ll m(kT/a)^{\frac{1}{2}}(kT/m)^{\frac{1}{2}}.$$

Thus the transition state can only be defined when the curvature of the energy surface satisfies the condition:

$$(h/2\pi)(a/m)^{\frac{1}{2}} \ll kT. \tag{2}$$

Apart from an unimportant numerical factor, the left side of this equation is the zero point energy of a mass, $m$, in a potential, $+aX^2$. Thus, Eq. (1) and the transition states method can be justified only if this virtual zero point energy is smaller than $kT$. If Eq. (2) is satisfied and if the dependence of the potential on the coordinates along the strip does not change appreciably from one side of the strip to the other, the transition state can be defined sufficiently accurately for deriving Eq. (1) in spite of the limitations of quantum mechanics. Fortunately, this is fulfilled in many important cases.

The computation of $\gamma$ presents another difficulty in the quantum-mechanical applications. As long as the reacting system behaves classically, $\gamma$ can be estimated from simple geometrical and mechanical arguments. Usually the coordinates in configuration space are chosen so that the kinetic energy is the product of an effective mass and the sum of squares of the generalized velocities. In such coordinate systems, molecular collisions are kinematically equivalent to the rolling of a ball on a surface whose height is proportional to the potential energy of the corresponding molecular configuration. The question whether a given collision will result in a chemical reaction can be answered by examining the trajectory of the ball when it is shot in the corresponding manner. However, when quantum theory is used, the ball must be replaced by a wave packet. In some reactions $\gamma$ will retain its classical value. In others, due primarily to the diffraction of the wave packets, it will have a different magnitude. Thus $\gamma$ can be much smaller than unity in quantum-mechanical systems even for an energy surface for which a mechanical picture does not indicate an appreciable probability of reflection. Consequently, the use of Eq. (1) with seemingly reasonable values of $\gamma$ can lead to erroneous rates even when the condition of (2) is fulfilled.

If the condition (2) is not fulfilled, (1) must certainly be corrected to take into account the quantum-mechanical penetration of the potential barrier.[5] But it may not be possible to use a

[5] R. E. Langer; Born and Franck, C. Eckart, Phys. Rev. 35, 1303 (1930); R. P. Bell, Proc. Roy. Soc. A139, 466 (1933). That (2) is the condition for the absence of an appreciable amount of tunneling is evident also from the

penetration factor in any simple way to correct for the deviations from classical theory. If Eq. (2) is not fulfilled, a reflection after the crossing of the saddle (which is usually taken into account by choosing an appropriate value of $\gamma$) cannot be distinguished from a failure to cross the saddle. Thus it would no longer be possible to define a transmission coefficient.

## II. THE TRANSMISSION COEFFICIENT

A. We shall first carry out a rather formal consideration concerning transmission coefficients, which will be based on the assumption that there is a definite probability for a system, which has crossed the transition state, to be reflected back to the transition state. This probability will be denoted by $\rho_i$ and $\rho_f$ for crossings from left to right and right to left, respectively. It will be assumed that these probabilities are equal for systems originating from the two sides of the transition state and are independent of the number of times the system has already crossed the transition state.

The value of the transmission coefficient, $\gamma$, depends of course on the exact nature of the reaction surface. It is unity if all systems cross the transition state only once in their passage from the initial to the final state or from the final to the initial state. In the general case, some of the systems[6] which cross the transition state will cross it again in the other direction. This may happen repeatedly before the atoms of the system finally separate, either to form the same molecules which originally collided, or else to form the molecules corresponding to the completed reaction. Thus some trajectories which cross the transition state from left to right will not lead to a chemical reaction and others may require many crossings before the reaction is completed.

At complete thermal equilibrium, half of the systems in the transition state are moving from left to right and the other half from right to left.

formula for the first quantum correction to (1). (Cf. reference 3.)

[6] The term "system" is used here also for the point in configuration space which corresponds to the position of the atoms forming the system. The left side of the transition state is supposed to correspond to the atoms forming the molecules of the reacting substances, the right side corresponds to the reaction product.

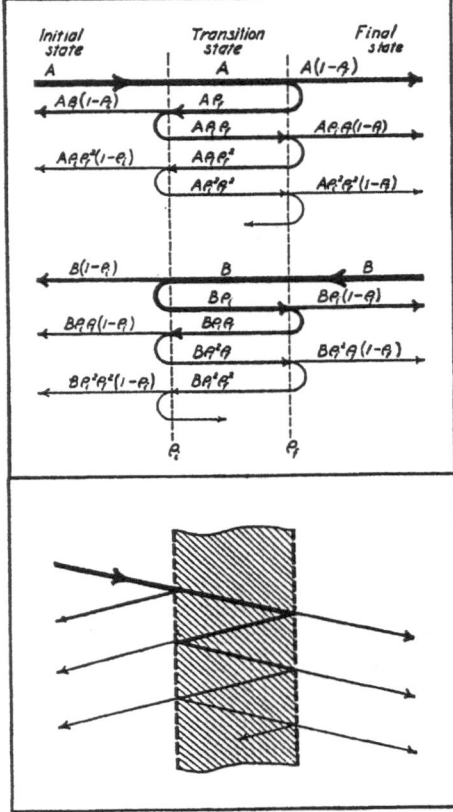

FIG. 1. The upper drawing shows the reflection of the systems on the two sides of the transition state. This is similar to the reflection of a beam of light at the air-glass and glass-air interfaces in passing through the plate glass shown in the lower figure.

The transmission coefficient, $\gamma$, is the fraction of the systems moving from left to right at complete thermal equilibrium which originally came from the initial state and which will proceed to the final state without first returning to the transition state. At complete thermal equilibrium, $A$ systems arrive, in unit time, at the transition state directly from the initial state and $B$ systems come from the final state. Fig. 1 shows schematically the flux of the different types of systems through the transition region. By simple addition, we see that the total number of systems crossing the transition state from left to right in unit time is

$$N_{l \to r} = A(1 + \rho_i\rho_f + \rho_i^2\rho_f^2 + \cdots)$$
$$+ B\rho_i(1 + \rho_i\rho_f + \rho_i^2\rho_f^2 + \cdots)$$
$$= (A + B\rho_i)(1 - \rho_i\rho_f)^{-1} \qquad (3)$$

and the number of those crossing it from right to left is

$$N_{r \to l} = A\rho_f(1 + \rho_i\rho_f + \rho_i^2\rho_f^2 + \cdots)$$
$$+ B(1 + \rho_i\rho_f + \rho_i^2\rho_f^2 + \cdots)$$
$$= (A\rho_f + B)(1 - \rho_i\rho_f)^{-1}. \qquad (4)$$

At equilibrium $N_{l \to r} = N_{r \to l}$, and hence

$$B = A(1 - \rho_f)/(1 - \rho_i). \qquad (5)$$

Substituting this expression for $B$ into Eq. (3),

$$N_{l \to r} = A/(1 - \rho_i). \qquad (6)$$

However, the number of systems which have originated in the initial state and which proceed to the final state is seen to be

$$N_{i \to f} = A(1 - \rho_f)(1 + \rho_i\rho_f + \rho_i^2\rho_f^2 + \cdots)$$
$$= A(1 - \rho_f)/(1 - \rho_i\rho_f). \qquad (7)$$

The transmission coefficient is the ratio of $N_{i \to f}$ to $N_{l \to r}$

$$\gamma = N_{i \to f}/N_{l \to r} = (1 - \rho_i)(1 - \rho_f)/(1 - \rho_i\rho_f). \qquad (8)$$

The assumption made at the beginning of this section that the systems crossing the transition state from left to right have the same probability for being reflected no matter whether they originated in the initial or in the final state (or how many times they have already crossed the transition state), is not always justified. We shall see in Section 3 that for a one-dimensional barrier, the transmission coefficient for any one value of the energy does not necessarily satisfy Eq. (8). However, the average transmission coefficient for many systems with slightly different energies will agree with the above expression.

We can compare the transmission of a wave packet through a potential barrier, as considered in this section, with the passage of a beam of light through a piece of plate glass partially silvered on both sides. (See lower part of Fig. 1.) Light of one particular frequency shows a very complicated diffraction pattern. If the direction of the beam is sharply defined, the amount of

light passing through the glass depends on the exact wave-length. For light of a range of frequencies, the transmission no longer depends critically on the wave-length and will be given by (8) (provided that the thickness of the plate is large compared with the wave-length). From this analogy we see that Eq. (8) can apply for the average transmission coefficient of an assembly of states in thermal distribution, not, however, for the transmission coefficient of a single quantum state. The example in Section 3 should make this more clear.

B. In classical theory, for low temperatures, $\rho_i$ and $\rho_f$ approach zero as it is very improbable that a system with barely enough energy to cross the activated state will find its way back.[4] For higher temperatures, the reflection coefficients may increase. This must be expected, in particular, if the vibrations in the transition state are less stiff than in the initial state, i.e., for the exceptionally fast reactions.

We shall consider an energy surface on which the energy and vibrational frequency change within a very short distance along the reaction path. The abruptness of the energy change tends to make $\gamma$ small; while the straightness of the reaction path tends to make it large. We shall suppose that the energy surface at the initial state has a smaller curvature perpendicular to the reaction path than at the final state, but that the minimum energy of the initial state is higher than that of the final state. Fig. 2 shows such an energy surface. All the systems of low energy pass from the initial to the final state, but part of the systems of high energy are reflected. If the potential energy is $V_i = A + a_i x^2$ for the initial and $V_f = a_f x^2$ for the final state, only those systems can be reflected which lie outside the point of intersection, $x' = A^{1/2}/(a_f - a_i)^{1/2}$. This is a negligibly small fraction of the systems at low temperatures and hence $\gamma = 1$. The reflection plays an important role only at temperatures so high that $kT > a_i x'^2$. This is equivalent to the condition:

$$A/kT < a_f/a_i - 1 = (\nu_f/\nu_i)^2 - 1, \qquad (9)$$

where $\nu_i$ and $\nu_f$ are the vibrational frequencies in initial and final states.

The existence of a limiting temperature, above which $\gamma$ becomes small, will always be true

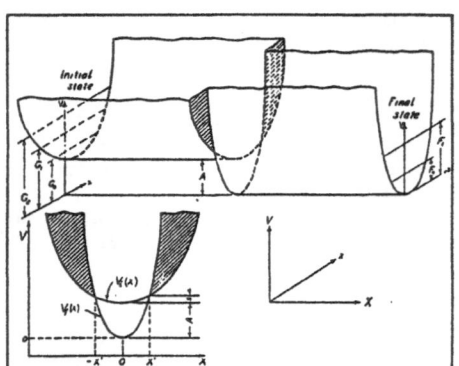

FIG. 2. Potential energy surface illustrating the variation of the transmission coefficient with temperature. From classical mechanical considerations, only those systems can be reflected which have sufficient energy to reach the baffle. The coordinate axes refer to the upper figure.

whenever the frequencies in the transition state are lower than in the initial state. The change in energy between the initial and the final state does not have to take place abruptly nor does the reaction path have to be straight. This becomes evident if we compare the rate of the reverse reaction, according to Eq. (1), with the total number of collisions in which the energy of the colliding particles is greater than $A$. Clearly, this latter quantity gives an upper limit for the rate.[7] The number of collisions involving energies greater than $A$ is

$$(kT/2\pi m)^{\frac{1}{2}}(A/kT+1)\exp\,(-A/kT). \quad (10)$$

This must be larger than the reaction rate according to Eq. (1):

$$k=\gamma(\nu_f/\nu_i)(kT/2\pi m)^{\frac{1}{2}}\exp\,(-A/kT). \quad (11)$$

Thus $\gamma$ must be less than unity if

$$A/kT<\nu_f/\nu_i-1. \quad (12)$$

The limiting temperature for a transmission coefficient of unity as given by Eq. (12) is about right if the reaction path is straight and the transverse vibration frequencies change slowly along the reaction path. The temperature given by (12) is much higher than that of Eq. (9). This corresponds to the fact that $\gamma$ is larger for

[7] K. F. Herzfeld, Zeits. f. Physik 8, 132 (1922); M. Polanyi, Zeits. f. Physik 1, 337 (1920).

energy surfaces having smooth rather than abrupt changes. In actual chemical reactions, $\gamma$ will depart from unity somewhere between these two temperatures. *Thus, in general, the transmission coefficient falls below 1 at high temperatures for classical mechanical reasons.* We shall see in the next section that at low temperatures, where $\gamma$ should be 1 according to classical mechanics, it is also smaller than 1 for quantum-mechanical reasons.

### III. ONE-DIMENSIONAL RATE PROBLEMS

The simplest examples of reaction rate involve motion in only one dimension. If the activated state method applies, we can divide the calculation of $\gamma$ into the two easier problems of determining the reflection coefficients $\rho_f$ and $\rho_i$ which refer to the passage from the activated state to the final and to the initial states, respectively. For one-dimensional problems these reflection coefficients are, of course, zero in classical mechanics.

In the special case of Eckart's potential functions,

$$V(X)=A(1+\exp\,(2\pi X/l))^{-1}, \quad (13)$$

decreases from $A$ to 0, as $X$ increases from large negative values (initial state) to large positive values (final state). The energy drop takes place around $X=0$ in a distance roughly equal to $l/2$ (Fig. 3). The reflection coefficient corresponding to the motion from negative to positive $X$ is:

$$\rho=\left[\frac{\exp\,(2\pi l(p-q)/h)-1}{\exp\,(2\pi l(p+q)/h)-1}\right]^2\exp\,(4\pi lq/h), \quad (14)$$

where $q=(2m(E-A))^{\frac{1}{2}}$ and $p=(2mE)^{\frac{1}{2}}$ are the momenta of the system before and after the potential drop. As this drop becomes abrupt ($l\to0$), the reflection coefficient becomes

$$\rho=(p-q)^2/(p+q)^2. \quad (15)$$

The explanation of this reflection is similar to that for the reflection of light at a glass to air interface. In quantum mechanics the dynamical system is represented by a wave packet with a wave-length which decreases from $h/q$ to $h/p$ as it passes across the potential drop. The energy change can therefore be expressed as a change in the index of refraction and the analogy with light is complete.

In Fig. 4 we have plotted the reflection coefficient for an H atom when passing across abrupt, steep and gradual Eckart potential drops of 10 kcal. The gradual drop, $l = 1.0A$, is still somewhat steeper than those which occur in the majority of chemical reactions. We see that the reflection coefficient approaches zero rapidly as the energy of the system is increased. One will expect, therefore, that the quantum corrections in one-dimensional problems are important only for reactions at very low temperatures involving light molecules or for isotope separations where slight differences in rates are significant.

Strictly speaking, this conclusion holds only if the barrier is so wide that (2) is satisfied, and consequently the method of Section 2 can·be used to express the transmission coefficient in terms of the reflection coefficients. Otherwise the transmission must be computed by quantum-mechanical methods. For the sake of concreteness, let us consider a potential barrier with abrupt energy changes (Fig. 5). We denote by $p_i$, $q$ and $p_f$ the momenta of the system in the initial, transition, and final states. Substituting the reflection coefficient (15) into (8), we would obtain for the transmission coefficient of the barrier:

$$\bar{\gamma} = 4qp_ip_f/(q^2 + p_ip_f)(p_i + p_f). \quad (16)$$

In this expression there is no dependence of the transmission coefficient on the width of the barrier, $d$.

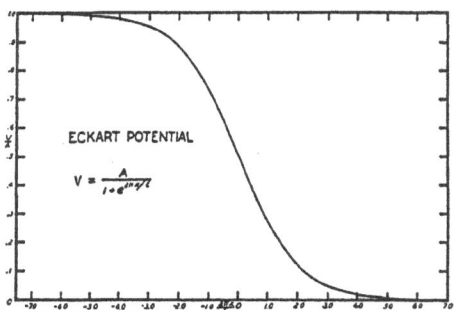

FIG. 3. A special case of the Eckart potential gives a function which changes smoothly from one constant value to another.

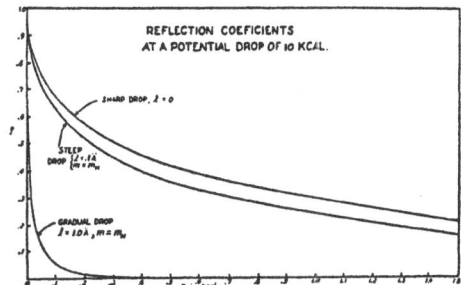

FIG. 4. Probability of reflection at abrupt, steep and gradual potential energy drops as a function of the energy of the system.

One can compute the exact value of the transmission coefficient by fitting together the wave function and its derivative at the discontinuities of the potential energy. One obtains in this way

$$\gamma = \frac{4q^2p_ip_f}{q^2(p_i + p_f)^2 \cos^2 \varphi + (q^2 + p_ip_f)^2 \sin^2 \varphi}, \quad (17)$$

where $\varphi = 2\pi qd/h$. This is evidently different from (16). If the width of the barrier is not large compared with the wave-length, quantum mechanics prevents us from even defining a transition state. If the barrier is wide enough to justify the activated state method, the exact transmission coefficient of Eq. (17) shows rapid fluctuations with the energy. Averaging $\gamma$ over a small energy range (which corresponds under these conditions to averaging over $\varphi$) one obtains an averaged transmission coefficient, which turns out to be equal to the transmission coefficient of Eq. (16):

$$\bar{\gamma} = \frac{1}{2\pi} \int_0^{2\pi} \gamma d\varphi. \quad (18)$$

It is only in this sense that Eq. (8) is valid. For any particular energy, the waves reflected at the two discontinuities of the potential interfere with each other to aid or to hinder the transmission. The effect of this interference averages out, however, for reasonably large energy ranges.

Figure 6 shows $\gamma$ and $\bar{\gamma}$ plotted as functions of the energy of the impinging particles. Here $A_1 = A_2 = A = 10$ kcal. and $d = 1.0A$. The exact

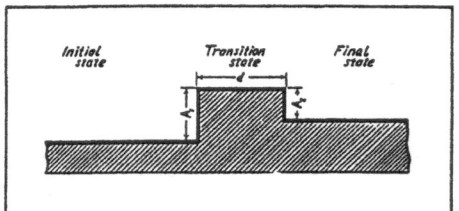

FIG. 5. One-dimensional energy barrier with abrupt changes in the potential.

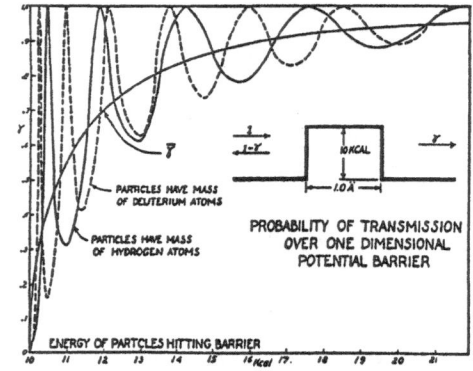

FIG. 6. Transmission coefficients for H and for D atoms passing over a one-dimensional potential energy barrier.

transmission coefficient, $\gamma$, oscillates about $\bar{\gamma}$. The oscillation of the transmission coefficient becomes quicker with increasing mass of the particles. It is interesting to note that the maximum in the transmission coefficient occurs for deuterium before it occurs for hydrogen so that at an energy of 10.2 kcal., the transmission coefficient for the former is about five times greater than for the latter. The situation is reversed, however, at 10.5 kcal.

It seems tempting, at first, to try to utilize the large ratios of transmission coefficients, as shown in Fig. 6, for the separation of isotopes. This is difficult, however, for several reasons. First, because the energy of the reacting systems is not restricted to one single value in actual experiments but covers—according to the Maxwell-Boltzmann distribution formula—a region of appreciable width so that the fluctuations shown in Fig. 6 average out to a large extent. Furthermore, the ratio of transmission coefficients fluctuates not only as a function of the energy but in many dimensional problems also as a function of the other parameters of the problem. As these parameters also cover a range of values

for the reacting systems, the fluctuations tend to average out to an even larger extent. Finally, the fluctuations are as pronounced as in Fig. 6 only if the top of the barrier has a flat portion, which is not very much shorter than the wavelength.

On the other hand, one can, in principle, always obtain large separations at very low temperatures, where "tunneling" becomes important. In quantum mechanics, systems with less energy than the activation energy may still react although only with a small probability.[8] The transmission coefficient is still given by (17) where now $q$ is imaginary.

In unit time the number of molecules with an energy between $E$ and $E+dE$ which hit the barrier is proportional to exp $(-E/kT)dE$. The ratio of the actual rate, $k$, to the classical rate, $k_{\text{class}}$, is then:

$$\frac{k}{k_{\text{class}}} = \frac{\int_0^A \gamma(E)\exp(-E/kT)dE + \int_A^\infty \gamma(E)\exp(-E/kT)dE}{\int_A^\infty \exp(-E/kT)dE}$$

$$= \frac{k_{\text{through}}}{k_{\text{class}}} + \frac{k_{\text{over}}}{k_{\text{class}}}.$$

(19)

[8] Similar calculations were made before particularly by R. P. Bell, J. Chem. Phys. 2, 164 (1934); Proc. Roy. Soc. A139, 466 (1933); H. Eyring and A. Sherman, J. Chem. Phys. 1, 345 (1933); C. E. H. Banri and G. Ogden, Trans. Faraday Soc. 30, 432 (1934).

Figure 7 shows the relative number of D atoms which react at 126°K as a function of their energy. There are two peaks in this curve. One corresponds to the reactions, $k_{through}$, due to penetration of the barrier and the other, $k_{over}$, corresponds to passage over the barrier. At a slightly higher temperature the penetration is unimportant; at a slightly lower temperature the passage over the barrier is unimportant. At high temperatures, where classical theory is valid, the rate of the crossing of the barrier is $2^{\frac{1}{2}}$ times higher for hydrogen than for deuterium, owing to the higher velocity of the former. Table I shows the rates of crossing, for H and for D, relative to the classical rate of H at 252°. Here $k_{trans}$ is the reaction rate which would be predicted on the basis of the transition state method using the quantum-mechanical value for the reflection coefficients, and Eq. (8)

$$\frac{k_{trans}}{k_{class}} = \frac{\int_A^\infty \bar{\gamma}(E) \exp\left(-E/kT\right) dE}{\int_A^\infty \exp\left(-E/kT\right) dE}.$$

At room temperatures, the penetration of this barrier is unimportant; the separation of the isotopes is small; and the transition state method is satisfactory. At low temperatures the transition state method is inapplicable since most reactions occur by penetration of the barrier. Here the reaction rates are thousands of times faster for the lighter isotope. The above table is in agreement with the calculations of Bell[8] who used an Eckart type of potential barrier. From this table we can make the following observations:[4] At very low temperatures an experimenter would find that a reaction of this nature has an anomalously small steric factor and shows practically no "activation energy." (But at temperatures so low that $k_{through}$ is greater than $k_{over}$, the rates are so small that they can hardly be measured.) At high temperatures he would obtain an activation energy agreeing with his classical expectations but the steric factor would still be small by a factor of two or three.

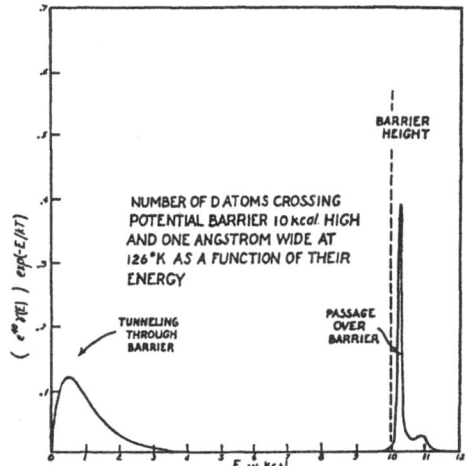

FIG. 7. Relative number of D atoms which succeed in crossing a potential energy barrier 10 kcal. high and 1A wide plotted as a function of their energy. The systems approaching the barrier have the energy distribution corresponding to 126°K.

## IV. THE SIMPLEST MANY-DIMENSIONAL RATE PROBLEM—EYRING'S FORMULATION

Reaction rate problems in many dimensions are more complicated than the one-dimensional examples principally because of the interchange of energy between the various degrees of freedom. In this section we consider the simplest of all many-dimensional problems—a straight reaction path along which the motion is slow and perpendicular to which the motion is fast.

If we designate by $X$ the coordinate along the reaction path and $x$ the coordinate perpendicular to the reaction path (considering only one such coordinate $x$ for the sake of simplicity), the potential energy of the system is $U(X, x)$. We

TABLE I.

|  | $T$ | $\frac{k_{THROUGH}}{k_{CLASS}}$ | $\frac{k_{OVER}}{k_{CLASS}}$ | $k$ | $k_{CLASS}$ | $k_{TRANS}$ |
|---|---|---|---|---|---|---|
| H | 84° | $1.7 \cdot 10^{13}$ | 0.15 | $7.1 \cdot 10^{-5}$ | $4.3 \cdot 10^{-18}$ | $3.2 \cdot 10^{-19}$ |
|  | 126° | $7.2 \cdot 10^{4}$ | 0.22 | $1.5 \cdot 10^{-4}$ | $2.1 \cdot 10^{-9}$ | $2.9 \cdot 10^{-10}$ |
|  | 252° | 0.014 | 0.35 | 0.364 | 1.0 | 0.39 |
| D | 84° | $1.5 \cdot 10^{8}$ | 0.22 | $4.6 \cdot 10^{-10}$ | $3.0 \cdot 10^{-18}$ | $2.3 \cdot 10^{-19}$ |
|  | 126° | 0.76 | 0.26 | $1.5 \cdot 10^{-9}$ | $1.5 \cdot 10^{-9}$ | $2.0 \cdot 10^{-10}$ |
|  | 252° | 0.0023 | 0.37 | 0.26 | 0.707 | 0.28 |

denote by $x_0$ the point $x$ at which $U(X, x)$ assumes its minimum value for a given $X$. Obviously $x_0$ is a function of $X$. $U(X, x)$, as a function of $x$ has approximately the shape of a parabola in the neighborhood of $x = x_0$. If the motion along the reaction path is so slow that the system can vibrate many times in the $x$ direction before it proceeds so far in the direction of $X$ that the curvature of the parabola is changed considerably, the motion perpendicular to the reaction path can be considered as simple harmonic with the frequency $\nu(X)$. The potential energy which is responsible for the motion of the slow coordinate $X$ is then

$$V_n(X) = U(X, x_0) + (n + \tfrac{1}{2}) h\nu(X). \qquad (20)$$

Here $n$ is the quantum number of the vibration along $x$. If $V_n(X)$ is a slowly varying function of $X$, the motion along the reaction path will obey the classical equations of motion.

It is essential in the derivation of Eyring's formulae[1] that the motion in the fast and in the electronic coordinates be adiabatic in the sense that neither the electronic quantum states nor $n$ undergo any changes. We can use the ordinary methods of perturbation theory to obtain the corresponding condition on the velocity, $v = dX/dt$, and on the change in the vibration frequency along the reaction path, $d\nu(X)/dX$,

$$d\nu/dt = v\,d\nu(X)/dX \ll \nu^2. \qquad (21)$$

Here $v$ can be much greater than the velocity corresponding to the average thermal energy since it must be fast enough to permit the crossing of the potential barrier. Thus (21) is more stringent than for the case of motion with ordinary thermal energies of the order of $kT$.

Under the above conditions, the motion of a system with the vibrational quantum number $n$ will obey classical mechanics and will be governed by the potential (20). Therefore, the number of systems with this vibrational quantum number which react in unit time will be given by (1) into which

$$P_{in} = \delta \exp(-V_n(X_0)/kT) \qquad (22)$$

must be substituted for $P_i$, where $V_n(X_0)$ is the highest value of $V_n(X)$, corresponding to the activated state.

The total reaction rate is the sum of the rates of all systems in all possible vibrational states. This is

$$\sum_n P_{in}\bar{v}\gamma/\delta P_i = P_i\bar{v}\gamma/\delta P_i, \qquad (23)$$

where

$$P_i = \sum_n P_{in} = \delta \exp(-V_0(X_0)/kT)$$

$$\times (1 - \exp(-h\nu(X_0)/kT))^{-1}. \qquad (24)$$

Thus under these assumptions we are led back to Eq. (1) except that the quantum theoretical probability appears in it, instead of the classical one.

It may be worth while to remark in this connection that if the reflection coefficients are zero and the molecules are distributed before the reaction according to thermal equilibrium, the thermal equilibrium will be maintained throughout the course of the reaction. Consider each vibrational state as forming a separate energy surface with respect to the motion along the reaction path. On the surface characterized by the vibrational quantum number $n$, the density at any point, $X$, is given by the barometric formula, $C_n \exp((V_n(\text{initial}) - V_n(X))/kT)$. The condition that the systems be at thermal equilibrium initially is that

$$C_n = C \exp((V_0(\text{initial}) - V_n(\text{initial}))/kT).$$

From this it follows that the ratio of the density on the $n$th to that on the zeroth vibrational state is $\exp((V_0(X) - V_n(X))/kT)$ for any value of $X$.

The consideration of this section shows that it is easy to justify Eyring's formulae with a mechanically sensible transmission coefficient if the coordinates can be divided into some which involve slow motion and others which involve fast motion. The slow motion can be described by classical mechanics. The fast motion can be treated on a par with the motion of the electrons: the corresponding quantum numbers undergo no changes during the reaction. This condition restricts the consideration of this section to potential surfaces where the frequencies along the reaction path change slowly.

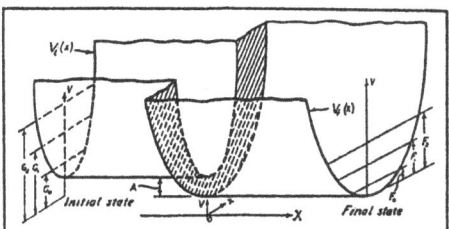

FIG. 8. Two-dimensional potential energy trough with straight reaction path and an abrupt change in the potential. Here the $G$'s are the vibrational energy levels in the initial state and the $F$'s in the final state.

## V. Abrupt Change of Potential Energy on Two-Dimensional Surfaces

The difficulty of solving partial differential equations restricts the number of many-dimensional rate problems which we can treat accurately. In this section we consider examples in which the reaction path is straight and the change in the potential occurs in a very short distance. It would also be desirable to consider the effect of curvature in the reaction path. The potential energy surfaces are constructed so that the Schrödinger equation is separable in each of a number of regions. In each region, the wave functions are obtained as solutions of ordinary (one-dimensional) differential equations. The wave functions are then pieced together so that the functions and their derivatives are continuous at the boundaries of every two regions. This is carried out by matrix methods and the complete mathematical treatment is given in the appendix.

Let us consider an energy surface similar to that shown in Fig. 8. For negative $X$, the potential energy is $V_i(X) = A + a_i x^2$; for positive $X$, the potential energy is $V_f(X) = a_f x^2$. This surface is the two-dimensional analog of the potential drop of Section 3. Eq. (8) enables us to obtain the average transmission coefficient from the reflection coefficients at two energy drops similar to that of Fig. 8. The case in which the energy of the system is large compared with the vibrational energy of any of the levels in the final state to which the waves are likely to be transmitted is particularly easy to treat.

The vibrational quantum number of the reflected systems can differ, to the approximation used in the appendix, only by 0 or $\pm 2$ from the vibrational quantum number of the incident systems. Eqs. (36) and (40) of the Appendix show that even the change by $\pm 2$ is quite improbable as $\epsilon$ is always a rather small quantity. The transmitted waves can have a vibrational quantum number differing from that in the initial state by an even number. In Fig. 9, we have plotted the probabilities of reflection and transmission into states with different vibrational quantum numbers. One sees by comparison with Fig. 4, that the reflection coefficient is rather similar to that of a system with equal translational energy in a one-dimensional problem. One sees, furthermore, that the vibrational quantum number has a marked persistence even in this case of an abrupt potential change. This means that the vibrational energy increases if the frequency is larger after the drop, and decreases in the opposite case. The changes in the vibrational frequency, inasmuch as they occur, have a tendency to counteract this. Thus the vibrational quantum number is more likely to decrease if the vibration is stiffer after the drop and is more likely to increase in the opposite case. A more detailed discussion is given at the end of the Appendix.

The persistence of the vibrational quantum number is doubtless in close connection with the slowness with which the equilibrium distribution of vibrations is established. At higher vibrational quantum numbers changes in the vibrational quantum number become more probable which is also in qualitative agreement with experiment.[9]

For the study of reaction rates, the reflection coefficients are most important. It is seen that in the case investigated here—large kinetic energy in the final state—these reflection coefficients are not increased over the value they assume in a similar one-dimensional problem. This leads to the conclusion that the transmission coefficient is even in this case not far from unity and Eyring's well-known formulae remain applicable. It would be desirable to obtain a more general solution of the problem considered in this section than that given in the Appendix and we intend to return to this question at a later time.

[9] Cf. F. Patat, Zeits. f. Elektrochem. 42, 265 (1936).

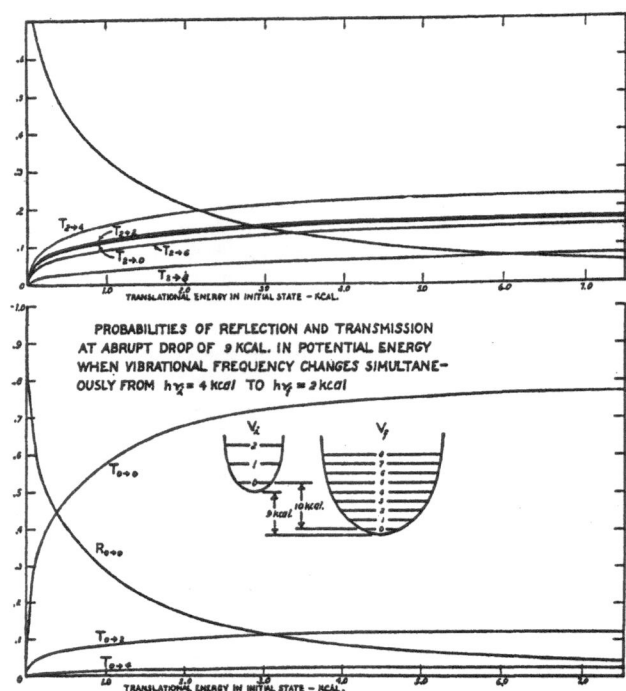

FIG. 9. Probabilities of reflection and of transmission at an abrupt potential drop of 9 kcal. when the vibrational frequency changes simultaneously from $h\nu_i = 4$ kcal. to $h\nu_f = 2$ kcal. The lower curves are for systems initially in the zeroth vibrational state; the upper curves are for systems initially in the 2nd vibrational state.

## APPENDIX

### Calculation of transmission and reflection coefficients for many dimensional problems

Let us denote by $X$ the coordinate along the reaction path and by $x$ the coordinate (or coordinates) perpendicular to it. For $X < 0$ the potential energy is $V_i(x)$; for $X > 0$ it is $V_f(x)$. The characteristic functions and characteristic values of the Schrödinger equation $H_i\psi = I\psi$ with the potential $V_i(x)$ shall be $\psi_\kappa(x)$ and $I_\kappa$, those of the Schrödinger equation $H_f\varphi = F\varphi$ with the potential $V_f(x)$ will be $\varphi_k(x)$ and $F_k$. Greek letters as subscripts refer to the initial state, roman letters to the final state. Expanding the $\varphi$ in terms of the $\psi$ we obtain

$$\varphi_k(x) = \sum_\kappa u_{k\kappa}\psi_\kappa(x) \tag{26}$$

and $u$ is a unitary matrix

$$(u^{-1})_{\kappa k} = u_{k\kappa}^*. \tag{27}$$

We shall consider an incident wave from the left having the form $\psi_\kappa(x) \exp(2i\pi q_\kappa X/h)$. It will give rise to a reflected beam in which the atoms have all possible quantum numbers $\lambda$ for their motion in the $x$ direction. Thus the wave function for $X < 0$ becomes

$$\psi_\kappa(x) \exp(2\pi i q_\kappa X/h)$$
$$+ \sum_\lambda R_{\kappa\lambda}\psi_\lambda(x)\exp(-2\pi i q_\lambda X/h) \tag{28}$$

with

$$q_\kappa^2/2m + I_\kappa = E, \tag{28a}$$

where $E$ is the total energy of the system. For

$X > 0$ we have only an outgoing wave

$$\sum_k T_{\kappa k}\varphi_k(x)\, \exp\,(2\pi i p_k X/h)$$

$$= \sum_{k\lambda} T_{\kappa k}u_{k\lambda}\psi_\lambda(x)\, \exp\,(2\pi i p_k X/h), \quad (29)$$

where again

$$p_k^2/2m + F_k = E. \quad (29a)$$

Naturally, the $p_\kappa$ and $q_k$ with high quantum numbers $\kappa$ and $k$ will be imaginary. The corresponding terms represent local disturbances in the neighborhood of the discontinuity. The imaginary $p$ and $q$, must be, if divided by $i$, positive. Terms with imaginary momenta of the opposite sign of $p$ or $q$ would correspond to waves which have progressively greater amplitudes as we depart from the discontinuity. The total number of systems with quantum number $\lambda$ reflected in unit time by the discontinuity is $R_{\kappa \to \lambda} = |R_{\kappa\lambda}|^2 q_\lambda/m$, if $q_\lambda$ is real (there is of course no reflected wave with imaginary $q_\lambda$). Similarly, the total number of systems transmitted in unit time with the quantum number $k$ into the final states is $T_{\kappa \to k} = |T_{\kappa k}|^2 p_k/m$ if $p_k$ is real, zero if $p_k$ is imaginary. There are $q_\kappa/m$ systems incident in unit time.[10]

Both the wave function and its derivative must be continuous at $X = 0$. Comparing the coefficients of $\psi_\lambda(x)$ of Eqs. (28) and (29) and of their derivatives this gives:

$$\delta_{\kappa\lambda} + R_{\kappa\lambda} = \sum_k T_{\kappa k}u_{k\lambda}, \quad (30)$$

$$\delta_{\kappa\lambda}q_\kappa - R_{\kappa\lambda}q_\lambda = \sum_k T_{\kappa k}u_{k\lambda}p_k. \quad (30a)$$

The formal solution of these equations can be obtained by matrix algebra. We can write for (30) and (30a)

$$1 + R = Tu, \quad (31)$$

$$q - Rq = Tpu. \quad (31a)$$

Here 1 is the unit matrix, $p$ and $q$ diagonal matrices while $R$, $T$ and $u$ are general matrices. We have, from (31) and (31a), $T = (1+R)u^{-1}$ $= (1-R)qu^{-1}p^{-1}$ and hence, with $\zeta = qu^{-1}p^{-1}u$,

$$R = (\zeta - 1)(\zeta + 1)^{-1}; \qquad T = 2\zeta(\zeta + 1)^{-1}u^{-1}. \quad (32)$$

We can use (32) to verify that the total number of incident systems is equal to the number of reflected and transmitted systems. The reflection and transmission matrices $\bar{R}$ and $\bar{T}$ for the problem in which the potential is $V_f$ in the initial and $V_i$ in the final state, can be calculated from $R$ and $T$

$$\bar{R} = -uRu^{-1}; \qquad \bar{T} = u(1-R). \quad (32a)$$

Naturally, (32) represents only a formal solution of our problem, and the matrix elements of $R$ and $T$ still remain to be calculated. We shall obtain them for the special case in which the total energy $E$ is very large compared to the energies $F_k$ of the levels to which there is an appreciable transition. This means that in the final state the translational energy $p_k^2/2m \gg F_k$ will be large compared with the vibrational energy, although this need not be true for the initial state.

In this case $F_k \ll E$ and[11] we can write, denoting by $f$ the diagonal matrix with the diagonal elements $F_k$,

$$\zeta = qu^{-1}p^{-1}u$$

$$= qu^{-1}(2mE)^{-\frac{1}{2}}(1+f/2E)u = \zeta_0 + \zeta_1, \quad (34)$$

where we have with $P = (2mE)^{\frac{1}{2}}$

$$\zeta_0 = P^{-1}q; \quad \zeta_1 = \tfrac{1}{2}P^{-1}E^{-1}qu^{-1}fu \quad (34a)$$

and $\zeta_1$ is small compared with $\zeta_0$. Now we can write, up to terms of second order in $\zeta_1$

$$R = (\zeta_0 - 1 + \zeta_1)(\zeta_0 + 1 + \zeta_1)^{-1}$$

$$= (\zeta_0 - 1 + \zeta_1)(\zeta_0 + 1)^{-1}[(\zeta_0 + 1 + \zeta_1)(\zeta_0 + 1)^{-1}]^{-1}$$

$$= (\zeta_0 - 1 + \zeta_1)(\zeta_0 + 1)^{-1}[1 - \zeta_1(\zeta_0 + 1)^{-1}]$$

$$= (\zeta_0 - 1)(\zeta_0 + 1)^{-1} + 2(\zeta_0 + 1)^{-1}\zeta_1(\zeta_0 + 1)^{-1}. \quad (35)$$

Hence

$$R_{\kappa\lambda} = \frac{q_\kappa - P}{q_\kappa + P}\delta_{\kappa\lambda} + \frac{PE^{-1}q_\kappa F_{\kappa\lambda}}{(q_\kappa + P)(q_\lambda + P)}, \quad (36)$$

where

$$F_{\kappa\lambda} = \sum_k u_{k\kappa}^* F_k u_{k\lambda} = \int \psi_\kappa^* H_f\psi_\lambda.$$

For the amplitude of the wave reflected without any change in the vibrational quantum number,

---

[10] The matrix method for problems of atomic scattering has been initiated by J. A. Wheeler, Phys. Rev. 52, 1107 (1937).

[11] In this calculation, the zero of the energy is at the bottom of the potential curve $V_f(x)$. $E$ and $P$ are numbers, the other symbols matrices.

the expression

$$R_{\kappa\kappa} = (q_\kappa - p_\kappa')(q_\kappa + p_\kappa')^{-1}; \quad p_\kappa'^2/2m + F_{\kappa\kappa} = E \quad (37)$$

is correct up to second-order terms in $F$. This shows a marked analogy to (15) with the difference, however, that $p_\kappa'$ enters instead of $p_\kappa$. The former is smaller than the latter, at least for $\kappa = 1$, as $F_{11} > F_1$. Hence, the reflection without change of the vibrational quantum number is smaller in the two-dimensional case than the whole reflection in the one-dimensional case. In order to calculate the $F_{\kappa\lambda}$ we can note that

$$F_{\kappa\lambda} = \int \psi_\kappa^* H_f \psi_\lambda = \int \psi_\kappa^* H_i \psi_\lambda - \int \psi_\kappa^* (V_i - V_f) \psi_\lambda$$

$$= I_\nu \delta_{\kappa\lambda} - (V_i - V_f)_{\kappa\lambda}. \quad (38)$$

If $V_i$ and $V_f$ have both the potential of a simple harmonic oscillator,

$$V_i = A + 2\pi^2 m \nu_i^2 x^2; \qquad V_f = 2\pi^2 m \nu_f^2 x^2, \quad (39)$$

(36) and (38) show that the vibrational quantum number can change only by 0 or $\pm 2$ in our approximation. Then

$$(V_i - V_f)_{\kappa\kappa} = A + \tfrac{1}{2}\epsilon h\nu_i(\kappa + \tfrac{1}{2}),$$

$$(V_i - V_f)_{\kappa,\kappa+2} = (V_i - V_f)_{\kappa+2,\kappa} \quad (40)$$

$$= \tfrac{1}{4}\epsilon h\nu_i(\kappa+1)^{\frac{1}{2}}(\kappa+2)^{\frac{1}{2}}$$

with $\epsilon = (\nu_i^2 - \nu_f^2)/\nu_i^2$. There is no change whatever in the vibrational quantum number if the frequencies in initial and final states are equal. In general, the probability of a change increases with increasing vibrational energy in the initial state, and also with increasing translational energy as long as the latter remains small compared with the total drop $A$ in the potential.[12]

[12] It should be remembered that the probability of reflection is $|R_{\kappa\lambda}|^2 q_\lambda/q_\kappa$, not $R_{\kappa\lambda}$.

For the transmission matrix we obtain from (31)

$$T_{\kappa k} = \sum_\lambda (\delta_{\kappa\lambda} + R_{\kappa\lambda}) u_{k\lambda}^*$$

$$= \frac{2q_\kappa}{q_\kappa + P} u_{k\kappa}^* + \sum_\lambda \frac{PE^{-1} q_\kappa F_{\kappa\lambda} u_{k\lambda}^*}{(q_\kappa + P)(q_\lambda + P)}. \quad (41)$$

The second term contains the factor $F/E$ and is much smaller than the first. We shall evaluate (41) also for the potentials (39). In this case $T_{\kappa k}$ will be zero unless the change $(k - \kappa)$ of the vibrational quantum number is even. This is because both $F_{\kappa\lambda}$ and $u_{k\kappa}$ vanish, unless the difference of their two indices is even. The first few $u$ are

$$u_{00} = (1-\epsilon)^{\frac{1}{4}}(1 - \epsilon + \tfrac{1}{4}\epsilon^2)^{-\frac{1}{2}},$$

$$u_{02} = -u_{20} = 2^{-\frac{1}{2}} u_{00} \epsilon (2-\epsilon)^{-1}, \quad (42)$$

$$u_{22} = u_{00}(1 - \epsilon - \tfrac{1}{4}\epsilon^2)(1 - \epsilon + \tfrac{1}{4}\epsilon^2)^{-1}.$$

If the change in vibrational frequency is not too large, $\epsilon$ will be small. Under these conditions, at least for low values of the indices, the diagonal elements of $u$ are nearly 1. The off-diagonal elements are much smaller, as $u_{k\kappa}$ contains the $\tfrac{1}{2}(k - \kappa)$'th power of $\epsilon$. Thus the probability for a change in $\kappa$ is small. For large values of the quantum numbers, the vibrational energy in the final state covers a wider range. In the classical region, it extends from the vibrational energy in the initial state $I_\nu$ to $I_\nu(1-\epsilon)$. Its mean value is still very nearly the same as if there was no change in vibrational quantum number: it is $I_\nu(1-\tfrac{1}{2}\epsilon)$ instead of $I_\nu(1-\epsilon)^{\frac{1}{2}}$.

The authors would like to express their appreciation to the Wisconsin Alumni Research Foundation for financial support throughout the course of this work.

# Some Remarks on the Theory of Reaction Rates*

E. P. Wigner

Journal of Chemical Physics 7, 646–652 (1939)

(Received April 21, 1939)

Effects connected with a more complicated nature of the energy surfaces of relatively simple reactions are discussed. The theory of elementary association reactions is examined and it is shown that better results are obtained in the rate of association of atoms when three additional attraction states are taken into consideration. The rates of association of iodine, bromine and chlorine atoms are calculated and compared with the experimental rates. The improvement to be expected from consideration of the angular momentum of the associating pair of atoms is discussed.

EVERY modern theory of reaction rates has to deal with two essentially different kinds of problems. The first problem concerns the forces acting between the atoms participating in the reaction, the second one is a statistical question and aims at a calculation of the rate constant, assuming that the answer to the first question is known. We shall discuss first the reactions with, then those without, activation energy from this point of view.

## I

The solution of the statistical problem is probably known at present,[1] for gas reactions with an activation energy, to such an extent that not very much further progress can be expected in this connection in the near future. To be sure, several points of the theory still remain to be cleared up,[2] particularly when quantum effects play a major role and also very few examples of nonadiabatic reactions have been considered so far in detail. However, the remaining questions are of a more special nature and appear to be so difficult that we must feel satisfied, for the time being, with a partial solution.

On the other hand, our knowledge of the forces between atoms is far from being complete enough to allow the accurate calculation of reaction rates with activation energies. Considering the difficulties involved in explaining the binding energies of well-known ordinary compounds, one will be hardly surprised to find that—apart from one single instance[3]—no accurate calculation of activation energies exists. It is more unsatisfactory that even the general character of the forces between the atoms of a reaction are known only in the very simplest cases. Although one could numerically predict rates of reactions with activation energy only if one were able to calculate the activation energy accurately, the knowledge of the general properties of the forces between atoms would be valuable in any case by enabling us to compare rates of different reactions on a more satisfactory basis.

The general character of the forces between atoms, i.e., the "energy surface," is well known only for a system of three atoms, at most, all of which are in $S$ states. There are several reasons to believe that the energy surfaces are much more complicated in other cases than these known energy surfaces are. Firstly, the number of ways atoms in $P$ and higher states can be orientated with respect to each other is much greater than for atoms in $S$ states. Second, it appears at present most probable that the reason for the discrepancies between experimental and theoretical reaction rates is due to the unexpectedly complicated shape of these potential surfaces. Strong indications in the same direction are given also by preliminary calculations of Magee[4] who considered a system formed by a $P$ and two $S$ atoms. Finally Knipp[5] has shown that complications can arise even in the case of two atoms inasmuch as states which are known not to be attractive at short distances become weakly

---

* Presented at the symposium on "Kinetics of Homogeneous Gas Reactions." See page 633.
[1] Cf. various papers in the Trans. Faraday Soc. (1937) in the general discussion on reaction kinetics.
[2] Cf. also J. O. Hirschfelder and E. Wigner, J. Chem. Phys. (in publication).

[3] Hirschfelder, Eyring and Topley, J. Chem. Phys. 4, 170 (1936).
[4] J. L. Magee, unpublished.
[5] J. Knipp, Phys. Rev. 53, 734 (1938).

attractive for larger separations, while states which are attractive at short distances are in many instances repulsive at large distances. Thus, there is no energy surface, as in the case of only $S$ atoms, which is lower than all other energy surfaces for all configurations.

Possibly, the explanation of a very puzzling phenomenon, observed by Polanyi[6] and collaborators, and first emphasized in this connection by Goodeve,[7] lies also in a similar direction. The phenomenon is the excitation of a Na atom, with a probability approaching 1, by an excited NaCl* molecule obtained by the reaction $Na_2 + Cl = Na + NaCl^*$. Since the whole system is, originally, on the lowest energy surface, it should remain there, according to the theory of the adiabatic nature of these reactions.[8] However, a system containing an excited Na atom is clearly not on the lowest energy surface. According to the customary ideas, the NaCl* molecule has only vibrational and rotational but no electronic excitation. The two ions in the molecule vibrate against each other very strongly—in the rather improbable case that the NaCl* contains all the reaction energy of 3.35 electron volts, the maximum distance of the Na and Cl ions becomes about 6.5A. This is, nevertheless, much less than the distance at which the molecular curve, corresponding to the ordinary heteropolar molecule, crosses[9] the molecular curve corresponding to the homopolar NaCl. This latter distance is more than 10A. Thus, for the energies available, the lowest curve is well separated from even the next curve. In the presence of an additional Na atom, i.e., during the collision of the NaCl* with Na, both molecular curves will be deformed and an overcrossing cannot be altogether excluded. However, even if such an overcrossing should occur, it hardly would lead with an appreciable probability to transitions to such highly excited energy surfaces which correspond to an excited Na atom.

It appears, on the whole, more probable that the original reaction $Na_2 + Cl = Na + NaCl^*$ leads

to the homopolar state of the NaCl molecule rather than the normal state. The distance at which the $Na_2 + Cl$ curve crosses the $Na_2^+ + Cl^-$ curve is greater than 6.7A so that there remains a large probability for a nonadiabatic behavior, i.e., for the transition to the higher, homopolar level. If this is true, the energy in NaCl is stored up to a great extent as electronic energy and it is well known that this can serve more readily than vibrational energy can, to the excitation of the electron in another atom. The weakest point of this explanation consists in the apparent absence of a reasonably deep minimum in the molecular curve of NaCl in the homopolar state. However, the evidence is not conclusive on this point.

This example should only emphasize the fact that complications can exist even in the energy surfaces of very simple systems and that these can lead to rather unexpected results. On the whole, the status of the theory of reaction rates with activation energy is that the statistical part of the theory is well developed, but our knowledge of the forces between atoms is adequate only in a few cases.

## II

The status of the theory of reactions without an activation energy is rather the opposite. The statistical part of the theory offers considerable difficulties, but the details of the behavior of the forces between atoms enter in a much less critical way than for reactions with activation energy. Nevertheless, the example to follow shows that they may form, even in this case, the most important link of missing information.

Perhaps the most simple among the reactions without activation energy are those leading to a recombination of atoms by collision with a third atom and I attempted, some time ago, to derive a formula for the rate of these reactions[10] and applied it to the rate of the reaction $I + I + X = I_2 + X$ which has been measured by Rabinowitch and Wood.[11]

The general idea of the calculation is illustrated in the schematic Fig. 1. In this, $r$ is the distance of the two I atoms, $p_r$ the corresponding momentum, $x$ a coordinate of the third body $X$,

[6] H. Beutler and M. Polanyi, Zeits. f. Physik 47, 379 (1928); Zeits. f. physik. Chemie 1B, 3 (1928); St. v. Bogdandy and M. Polanyi, *ibid.* 1B, 21 (1928); M. Polanyi and G. Schay, *ibid.* 1B, 30 (1938).

[7] C. F. Goodeve, Trans. Faraday Soc. 30, 60 (1934).

[8] F. London, *Sommerfeld Festschrift* (Leipzig, 1928), p. 104.

[9] F. London, Zeits. f. Physik 74, 143 (1932).

[10] E. Wigner, J. Chem. Phys. 5, 720 (1937).

[11] E. Rabinowitch and W. C. Wood, J. Chem. Phys. 4, 497 (1936).

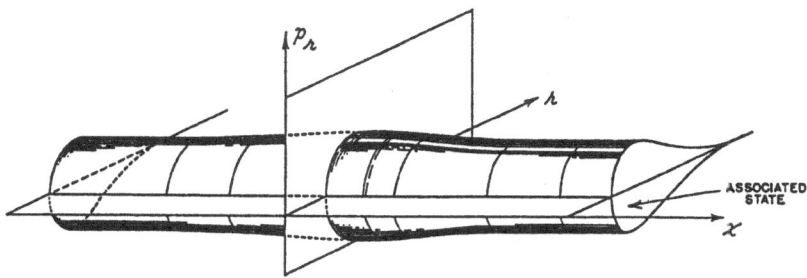

FIG. 1.

measured from the center of the two I atoms. The other 15 coordinates of the phase space are suppressed in the figure, for sake of simplicity.

The rate is defined as the number of pairs of I atoms passing in unit time from the dissociated into the associated state. For large values of $x$, positive or negative, there is no doubt which part of the figure corresponds to associated and which to dissociated state: the former is the region in which the relative energy $V(r) + \frac{1}{2}p_r^2/m_r$ of the two I atoms is negative, the latter in which it is positive. ($V(r)$ is the potential energy of the I atoms with respect to each other, $m_r$ is their relative mass.) The region corresponding to the associated state lies within the surface shown on the figure. The paths of the different systems are lines in the phase space and the reaction rate is the number of such lines crossing the surface and entering the associated region. At large values of $x$, the paths in the neighborhood of the surface are parallel to it and do not cross it since the relative energy of the two I atoms is a constant of the motion if the $X$ is far away. This is only another expression of the fact that the association can only occur in the neighborhood of the third body $X$, i.e., for small values of $x$.

There will be some paths which enter the associated region but leave it immediately afterwards—these, of course, should not be counted as associations. One will, therefore, draw the surface at small values of $x$ in such a way that the number of paths crossing it shall be as small as possible. This condition, while not changing the shape of the surface at large values of $x$ (where no crossings occur, anyway), will change

it at small values of $x$. It will be such that no paths will cross it in one direction just to cross it again soon afterwards in the opposite direction; all paths crossing it will stay inside while $x$ increases to large positive or negative values. The number of paths crossing this surface will give the real association rate, while the number of paths crossing any other surface, e.g., the surface shown on the figure, will be greater than the real association rate. Of course, the actual calculation must be done for the 18-dimensional phase space —not for the schematic space of the figure.

It is clear then, that the calculated rate should be greater than the observed one. Nevertheless, Table I taken from reference 10 and completed with the results of a similar calculation on Br atoms shows that this is not the case. The values are given in units of $10^{-32}$ (molecule/cc)$^{-2}$ sec.$^{-1}$, the different columns refer to the different "third bodies" $X$ used for the reaction. The computation has been carried out exactly as described in reference 10. For the radius of Br, that of Kr, 1.59A was assumed.[12] The rate of measurements for Br are also due to Rabinowitch and Wood.[13]

TABLE I.

| $X$ | He | A | $H_2$ | $N_2$ | $O_2$ |
|---|---|---|---|---|---|
| Br exp. | 0.76 | 1.3 | 2.2 | 2.5 | 3.2 |
| theor. | 1.6 | 2.1 | 1.7 | 2.3 | 2.2 |
| I exp. | 1.8 | 3.8 | 4.0 | 6.6 | 10.5 |
| theor. | 1.5 | 2.1 | 1.3 | 2.3 | 2.2 |
| Cl theor. | 1.9 | 2.5 | 2.2 | 2.8 | 2.7 |

[12] Cf. *Landolt-Börnstein's Tables*, Vol. I, p. 69.
[13] E. Rabinowitch and W. C. Wood, Trans. Faraday Soc. 32, 907 (1936).

It should be remarked, however, that the earlier work of Hilferding and Steiner[13a] gave substantially lower results.

Of course, the extremely large discrepancy for $H_2$ and $N_2$ is not alarming as it is not permissible to treat these as atoms. But a discrepancy obtains for the I association in the case of He and A also, and every improvement of the calculation makes this necessarily only worse. In fact, an improvement of the calculation, to be discussed later, decreases the calculated values approximately by a factor 2.

Several possible reasons for this discrepancy have been discussed, among others the possible existence of other attractive states of the two I atoms, in addition to the one leading to the normal $I_2$ molecule. It now seems that this is the cause of the discrepancy.

One should remember that the system formed by two I atoms and a He atom can be in 16 different quantum states, corresponding to the four possible orientations of the angular momenta of each I atom with respect to each other and to the internuclear axis. Only one-sixteenth of all pairs of I atoms will be in any specified one of these quantum states and if only one of these states were attractive, the association could occur only for one-sixteenth of all pairs of I atoms. In this case—and this was assumed at the calculation of Table I—the association rate would be 16 times smaller than it would be for 16 equally attractive states.

E. Rabinowitch[14] has considered the interpretation of his and Wood's experiments in a very interesting fashion from the point of view of the simple collision theory. He pointed out that there are indications[15] that among the molecular curves obtained by approaching two I atoms in their normal states, there are at least four which are attractive. One of these is simple and corresponds to the normal state of $I_2$, the other three are doubly degenerate each but, of

[13a] Zeits. f. physik. Chemie B30, 399 (1935).
[14] E. Rabinowitch, Trans. Faraday Soc. 33, 283 (1937).
[15] Cf. R. S. Mulliken, Phys. Rev. 46, 549 (1934); W. G. Brown, ibid. 42, 355 (1932); H. Cordes, Zeits. f. Physik 97, 603 (1935). Note added in proof.—Professor Mulliken has kindly informed me that he does not consider the evidence leading to the second lowest curve ($^1\Pi_u$) of Cordes to be satisfactory. In fact, theoretical considerations lead him to expect this state to be repulsive. In this case, the theoretical values of Table II should be reduced by about 30 percent.

TABLE II.

|         | He   | A   | $H_2$ | $N_2$ |
|---------|------|-----|-------|-------|
| I  exp. | 1.8  | 3.8 | 4.0   | 6.6   |
| theor.  | 6.1  | 7.7 | 6.4   | 8.1   |
| Br exp. | 0.76 | 1.3 | 2.2   | 2.5   |
| theor.  | 4.25 | 5.4 | 4.5   | 5.9   |

course, less strongly attractive than the first one, Rabinowitch concludes that no "statistical factor," corresponding to our $\frac{1}{16}$ should be taken into account for the consideration of the association rate. This is, perhaps, somewhat exaggerated, as there are certainly several repulsive states of two I atoms and these hardly can be expected to lead to a recombination. On the other hand, the rate of recombination into the three less strongly attractive states should also be calculated and added to the rate of recombination into the normal state of $I_2$. The values of the integrals over the three additional molecular curves of $I_2$ have been obtained by a graphical integration of Cordes' estimated curves. It gave $35 \times 10^{-36}$ erg cm$^3$ for the integral $-\int V(r)r^2dr$ and $129 \times 10^{-44}$ erg cm$^4$ for the integral of $-\int V(r)r^3dr$ for the three less strongly attractive states together. These numbers should be multiplied by two on account of the statistical weight of these states. The corresponding integrals for the normal state of $I_2$ are $33.5 \times 10^{-36}$ erg cm$^3$ and $114 \times 10^{-44}$ erg cm$^4$ if taken from Cordes' curve and $27 \times 10^{-36}$ and $91 \times 10^{-44}$, respectively, if computed by Morse's formula, as outlined in reference 10. A similar calculation has been done for Br also, although on an even less secure basis. The rates calculated for the number of transitions into all four states are shown in Table II in the same units as in Table I.

Considering the crudeness of the calculation and the fact that the rate could be calculated in terms of constants obtained from entirely different sources, the agreement is as good as can be expected. The relation between calculated and observed values is as it should be; the latter ones are several times smaller than the former ones. Even a larger ratio between calculated and observed values would not be disturbing as the former ones must be, by the very nature of their derivation, too large and every improvement of the calculation—e.g., the one to be outlined below—will bring the two values nearer to each other.

The apparently better agreement for $H_2$ and $N_2$ has no significance as the formulas apply only for atoms. It is somewhat surprising that the ratio of the calculated rates for Br and I is so much smaller than the experimental ratio.

An improvement of the calculation can be obtained on the basis of the following consideration: A molecule becomes stabilized even if its energy is greater than zero if it contains mainly rotational and only to a smaller extent vibrational energy.[16] Denoting the angular momentum by $\mu$, a diatomic molecule is (apart from quantum-mechanical effects) stable if its total energy is smaller than $\mu^2/2m_r r_\mu^2$ and the separation of its atoms smaller than $r_\mu$ where $m_r$ is the relative mass of the atoms in the molecule and $r_\mu$ the distance at which $V(r)+\mu^2/2m_r r^2$ assumes its maximum value. In this way, a molecule can accumulate substantially greater energy amounts than sufficient for its dissociation. This fact gives us the possibility to place the boundary dividing the free state from the associated state not at the place where the relative energy of the pair is zero, but at the place where, at a separation $r < r_\mu$, the relative energy $\frac{1}{2}m_r v_r^2 + V(r)$ becomes smaller

[16] Cf. G. Herzberg, Erg. exact. Naturw. 10, 207 (1931).

than $\mu^2/2mr_\mu^2$. *A priori*, both procedures are justified. The more useful procedure will be, however, that one which gives the smaller upper limit for the rate. Naturally, even the rate obtained with the new calculation will be only an upper limit for the actual rate, as the relative energy can rise above $\mu^2/2mr_\mu^2$ immediately after it falls below this value. The reason that the old calculation gave higher rates than the new one is that it considered the number of pairs of atoms with a positive total energy to have the equilibrium value. It has, however, a lower value in reality, since the above-described metastable states with positive energy are more quickly impoverished by collisions of the second kind than they are replenished by the processes which we are considering now. The old calculation counts as associations these collisions of the second kind also (which is correct) and assumes that the metastable molecules are present in the amount which is in equilibrium with the free atoms. It is the avoiding of this last assumption which constitutes the improvement of the calculation as outlined above.

It is clear, however, that the computation becomes much more involved and its details will not be given here. The calculated rates are reduced by almost a factor 2 and the agreement of Table II much improved.

## Discussion

**E. P. Wigner,** *Princeton University:* I wish to make some remarks concerning the adiabatic hypothesis. The adiabatic hypothesis states that the atoms can be considered to move under the influence of a potential. This means, physically, that the electronic motion adjusts itself quickly enough to the changing positions of the nuclei so that no change in the electronic excitation occurs. Clearly, if the electrons become excited, the corresponding excitation energy is unavailable for the motion of the nuclei, and these do not seem to move any more as under the influence of a steady potential.

The conditions for the validity of the adiabatic hypothesis have been investigated repeatedly from the theoretical standpoint. I wish to point to some information, obtained on the basis of experimental evidence, concerning the validity of this hypothesis.

The infraction of the adiabatic hypothesis can be visualized in the following way. Given two atoms in their normal states, there will be a potential between them. If one or both atoms are excited, the forces will be different and we have, therefore, several potential curves (or "energy surfaces" in case of more than two atoms). If we start out with a certain energy and the two atoms in their normal states, the excess energy is present as the kinetic energy of the nuclei. These will approach each other with increasing velocity at first, as long as they are in the attractive region of the potential; then they will approach each other with decreasing velocity, come to a momentary standstill and separate again, finally attaining the same velocity with which they collided. They will both remain unexcited. This sequence of events will occur if the process is adiabatic. Assuming nonadiabatic behavior, there will be a certain probability that one of the atoms is excited and the velocity of separation is smaller by an amount equal to the excitation energy. The nonadiabatic behavior is, therefore, a jump from one energy level to another.

Knowing these energy surfaces, it is not difficult to give the approximate condition for adiabatic behavior. During the period $h/E$ of the electronic motion the change in the potential $V$ must be small as compared with the energy necessary to excite an electron: $(h/E)vd\,V/dx < E$. Landau and Zener developed an expression for the transition probability from one level to the other in the case of certain simple shapes of the potentials and this expression will play an important part in the paper of Noyes and Henriques. It shows that the transition probability is a very sharp function of the quantities involved. It can increase by an extremely large factor for a comparatively small increase in the parameters.

The first set of relevant experiments which may be brought up in this connection concern the excitation of atoms by impact of other atoms. In this case $E$ is relatively large; and we must expect adiabatic behavior. Although the experimental material is unfortunately rather scarce, it is evident that, in agreement with the expectation, the excitation by atomic impact has a very low yield. Beeck, Dopel, Gailer, and Holzer have performed such excitation experiments and the excitation seems to be quite unobservable up to velocities of about 150 volts where it sets in, according to Holzer, with a very small cross section.

We know much more about the reverse process, i.e., the so-called collisions of the second kind, than we do about excitation by atomic impact. This process may be readily observed as the phenomenon of the quenching of fluorescence by foreign gases. In this case the system is first on the upper curve and goes over to the lower curve (when energy is plotted against internuclear distances).

In the most simple cases the cross section is very low for this process, about 1/30 of the gas kinetic cross section. Both experiments on Hg and on Na illustrate this, if the luminescence is quenched with noble gases. This means, of course, that the cross section for the reverse process is extremely small according to well-known considerations using the principle of de-

tailed balance. It is smaller in the ratio of the kinetic energies before and after the collision, which is about 100 to 1. Thus the calculated cross section for the excitation process is only 1/3000 of the gas kinetic cross section and the adiabatic hypothesis is well obeyed. However, the cross section for excitation increases strongly with increasing excess energy of the impinging atom.

The quenching cross section for molecules is already very much greater than for atoms of noble gases. The smallest one is that of $N_2$ against Hg and is $\frac{1}{5}$ of the gas kinetic cross section. The other known cross sections are all gas kinetic or larger. Two circumstances must be remembered, however; first, that the cross section for the reverse process, in which we are mainly interested, is about 100 times smaller—increasing, though slowly, with energy. Second, that the observed cross section is really the sum of very many cross sections since the electronic excitation energy originally present can be distributed in very many more ways among vibrational, rotational, etc. degrees of freedom in the final state than it could be distributed in the initial state. Thus the adiabatic hypothesis is reasonably well obeyed in these cases also.

The situation becomes radically different if we consider not transitions from the lowest to a higher surface, but transitions between a pair of closely spaced surfaces. The cross sections are in this case often greater than the gas-kinetic, cross section by a very large factor and Beutler, among others, has shown that they increase very strongly with decreasing distance of the levels. This is, of course, in qualitative agreement with our formula, but it says at the same time that under these conditions we are very far from a fulfillment of the adiabatic hypothesis. It is customary in this connection to point out that CO and $O_2$ which strongly quench Hg radiation have probably electronic excitation levels at 4.75 and 4.68 volts, while the Hg excitation energy is 4.86 volts.

I think that these considerations show that the adiabatic hypothesis can be expected to hold in many cases—not, however, generally. In general, the more complicated a system is, the more electronic excitation levels will be present and the chances for a resonance will be correspondingly increased. This can either hinder the reaction—a case which has been much discussed—or it can make reactions possible which would not be possible otherwise. All cases of chemiluminescence belong to this latter class since, in these cases, there is an electronic excitation in the reaction products while there is none in the initial substances. One group of such reactions already considered in more detail—the chemiluminescence of Na induced by the reaction between Na and a halogen—has been carefully investigated by Polanyi and co-workers, especially Schay, and discussed recently very thoroughly from another point of view by Evans and Polanyi.[17]

**James Franck,** *University of Chicago:* I do not understand why one should stress the difference between an adiabatic and nonadiabatic process. In chemistry the important thing is the relative positions of the nuclei with respect to each other.

In the impacts between two particles, a transition can occur when the potential curves (energy plotted against internuclear distance) come near to each other. A mixture of helium and mercury at high temperatures gives an emission of light, but at low temperatures helium does not quench excited mercury atoms. If the nuclei are close enough together, the curves will cross and electrons can move to a new state.

**John L. Magee,** *University of Wisconsin:* In connection with the reaction

$$Na+Cl_2 \rightarrow NaCl+Na^*$$

which Professor Wigner has discussed I should like to point out that the luminescence is to be expected even for an adiabatic process. This is seen when the splitting of the energy surfaces due to effect of the $p$ state of the chlorine atom is considered. The reaction occurs in a plane since it involves only three atoms, and so the electronic states of the reaction system are classified as even or odd with respect to this plane. This symmetry is preserved throughout the reaction. The

resulting NaCl molecule is in a $\Sigma$ state which is even. Thus for reactions which occur in an odd state the sodium atom which is formed must be excited. The even state is doubly degenerate, and thus for adiabatic behavior one of these states will also lead to excitation of the sodium atom. Professor Wigner points out that the surface for the system $Na_2^+ - Cl^-$ crosses the surface for $Na_2 - Cl$ at a large distance and the other even state which would adiabatically lead to a normal sodium atom is probably nonadiabatic due to this crossing.

The relative importance of these three surfaces for the reaction will depend upon the relative activation energies. At any rate the luminescence is not very surprising, since two of the possible electronic states must lead to an excited sodium atom adiabatically and the third most likely yields the same products.

**Gregory Breit,** *University of Wisconsin:* The motion of the nuclei is just as important as the shape of the energy surface in determining whether or not the reactions are adiabatic.

**E. P. Wigner,** *Princeton University:* I think that the point of view which I presented is very similar to Professor Franck's. Indeed, the purpose of the approximate formula which I mentioned is to provide a more quantitative background to the phenomena which were just brought up. There are only two minor additions which I would like to make to his remark: First, I believe that it is very infrequent that the lowest energy level is crossed or even very strongly approached by another one and that such a crossing or very strong approach, whenever it occurs, is due to special conditions such as different symmetry properties of the crossing levels or radically different electronic arrangements. Second, I do not think that such crossings are absolutely necessary for nonadiabatic behavior. This will be likely to occur whenever the above-mentioned inequality is infringed. In practice, of course, this will happen most easily if $E$ is small, i.e., if the levels come very close to each other.

[17] M. G. Evans and M. Polanyi, Trans. Faraday Soc. 35, 178 (1939).

# On the Calculation of the Distribution Function

C. W. Ufford and E. P. Wigner

Physical Review *61*, 524–527 (1942)

(Received January 22, 1942)

The distribution function for particles confined to a large circle is found for a repulsive potential of long range. The distribution function is calculated exactly from the Boltzmann-Gibbs equation, and compared with the solution of an implicit equation of the type used by Debye and Hückel in the theory of electrolytes. The solution of the Debye-Hückel equation agrees, for our particular potential, only fairly well with the exact distribution function.

## I.

ALTHOUGH the problem of classical statistical mechanics is, in principle, solved by the Boltzman-Gibbs equation

$$P(x_1, y_1, z_1, \cdots x_n, y_n, z_n)dx_1 \cdots dz_n$$
$$= \exp(-\beta V)dx_1 \cdots dz_n \quad (1)$$

for the probability of the configuration characterized by the rectangular coordinates $x_1, y_1, z_1, \cdots, x_n, y_n, z_n$, the answering of questions of immediate physical interest meets, in most cases, serious mathematical difficulties. In (1), $\beta = 1/kT$ and $V$ is the potential energy

$$V = \sum_{i<k} v(|r_i - r_k|), \quad (1a)$$

which we shall assume to be the sum of interactions between pairs.

Many of the quantities of immediate physical importance depend on the distribution function $g(r)$, i.e., on the probability of a distance $r$ between, say, the particles 1 and 2. This is given by

$$g(|r_1 - r_2|) = \text{const.} \int \cdots \int \exp(-\beta V)$$
$$\times dx_3 dy_3 dz_3 \cdots dx_n dy_n dz_n. \quad (2)$$

However, one can evaluate (2) easily only in the case of dilute matter, i.e., gases. For condensed material, the integrations in (2) are so difficult that indirect methods had to be devised for the evaluation of the distribution function. These methods have two features in common: (a) they do not give $g$ directly but only as the solution of an implicit equation the solving of which, though in most cases not very easy, is still less difficult than the integrations of (2); and (b) the implicit equation for $g$ is not exact, i.e., its solution is only an approximation to the rigorous expression (2). The first implicit equation was given by Debye and Hückel[1] in their theory of electrolytes. More rigorous equations are due to Kirkwood[2] and Mayer.[3]

The present authors became interested in these methods because integrals of the type (2) occur in the theory of the so-called correlation energy. The present note deals with a special problem in which all particles are confined to a large circle of length $L$; and in which the potential between a pair of particles is given by

$$\beta v(x) = -\ln \sin^2 \pi x/L, \qquad (3)$$

where $x$ is the distance between the particles, measured along the circle. We found that for this interaction the distribution function can be evaluated exactly by (2), and that the implicit equation corresponding to Debye and Hückel's theory also can be solved. The comparison of the two results gives an indication of the accuracy of the Debye-Hückel equation (5) for the case of a repulsive potential of long range. In the case of long range forces the Debye-Hückel equation can be expected to give good results.

In the Debye-Hückel theory the distribution function is given, apart from a constant, by an average potential $U$

$$\ln g(x) = -\beta U(x) + C. \qquad (4)$$

The average potential contains the potential $v(x)$ of the original particle around which we investigate the distribution of the others, and the average field of the other particles. There are $g(x')dx'$ particles at a distance $x'$ from the original particle; and their potential, at $x$, is

[1] P. Debye and E. Hückel, Physik. Zeits. 24, 185, 305 (1923); P. Debye, Physik. Zeits. 25, 97 (1924).
[2] J. G. Kirkwood and Elizabeth Monroe, J. Chem. Phys. 9, 514 (1941); J. G. Kirkwood, J. Chem. Phys. 3, 300 (1935).
[3] J. E. Mayer and E. Montroll, J. Chem. Phys. 9, 2 (1941); J. E. Mayer and S. F. Harrison, J. Chem. Phys. 6, 87 (1938).

given by $g(x')dx'v(x-x')$. Hence

$$U(x) = v(x) + \int g(x')v(x-x')dx' \qquad (4a)$$

and the implicit equation becomes

$$\ln g(x) = -\beta v(x) - \beta \int g(x')v(x-x')dx' + C. \qquad (5)$$

The constant $C$ is determined by the condition that the integral of $g$ must be $n-1$,

$$\int g(x)dx = n-1. \qquad (5a)$$

In most cases $g$ is a constant if $x$ is large compared to the average distance between neighboring particles. One can, therefore, substitute

$$g(x) = g_0(1+h(x)), \qquad (6)$$

in which $h(x)$ vanishes for large $x$. Substitution of (6) into (5) gives a somewhat different form to the implicit equation

$$\ln(1+h(x)) = -\beta v(x)$$
$$- g_0\beta \int h(x')v(x-x')dx' + C', \qquad (6a)$$

since the integral of $g_0 v(x-x')$ is independent of $x$.

## II.

The calculations remain to be done. For evaluating (2) we can write

$$\exp(-\beta V) = \prod_{i<k} \sin^2 \pi(x_i-x_k)/L$$
$$= 2^{-n(n-1)} |\prod_{i<k} (\exp 2\pi i x_i/L - \exp 2\pi i x_k/L)$$
$$\times \exp(-\pi i(x_i+x_k)/L)|^2. \qquad (7)$$

The factor $\exp(-\pi i(x_i+x_k)/L)$ may be omitted since its absolute value is 1. If we write

$$\omega_i = \exp 2\pi i x_i/L, \qquad (7a)$$

we get

$$\exp(-\beta V) = 2^{-n(n-1)} |\prod_{i<k} (\omega_i-\omega_k)|^2, \qquad (7b)$$

the product on the right can be written as a determinant $\Delta = |\Delta_{il}|$ where

$$\Delta_{il} = \omega_i^l = \exp 2\pi i l x_i/L \quad (l=0, 1, 2, \cdots n-1). \qquad (8)$$

This gives

$$\exp(-\beta V) = 2^{-n(n-1)}\Delta\Delta^*. \qquad (9)$$

The determinant $|\Delta_{il}|$ is, in case of odd $n$, apart from a factor $\exp -\pi i(n-1)(x_1 + \cdots + x_n)/L$, the wave function of a one-dimensional degenerate Fermi gas.

One can integrate (9) immediately by developing both $\Delta$ and $\Delta^*$ with respect to the two rowed minors of the first two rows. The minor of the $l_1 l_2$ columns in $\Delta$ is

$$\exp 2\pi i(l_1 x_1 + l_2 x_2)/L$$
$$-\exp 2\pi i(l_2 x_1 + l_1 x_2)/L. \quad (10)$$

Its cofactor is orthogonal to the cofactor of every two rowed minor of $\Delta^*$, excepting the cofactor of the minor of the $l_1 l_2$ columns. The product of the cofactors of the $l_1 l_2$ minors of $\Delta$ and $\Delta^*$ gives $(n-2)!$ if integrated over $x_3, \cdots, x_n$. Thus

$$\int \cdots \int \exp(-\beta V) dx_3 \cdots dx_n$$
$$= \sum_{l_1 < l_2} 2^{-n(n-1)}(n-2)! \, |\exp 2\pi i(l_1 x_1 + l_2 x_2)/L$$
$$-\exp 2\pi i(l_2 x_1 + l_1 x_2)/L|^2. \quad (10a)$$

This is, according to (2), proportional to $g(x_1 - x_2)$. The sum in (10a) can be readily evaluated and gives, with (5a),

$$g(x) = \frac{n}{L}\left(1 - \frac{\sin^2 \pi n x/L}{n^2 \sin^2 \pi x/L}\right)$$
$$\approx \frac{n}{L}\left(1 - \frac{\sin^2 \pi n x/L}{\pi^2 n^2 x^2/L^2}\right). \quad (11)$$

The last expression is valid if $x$ remains of the order of magnitude of the average distance of two particles, i.e., is not very much larger than $L/n$. Equation (11) gives the rigorous expression for $g$; it is the lowest curve in Fig. 1. The abscissae in this figure are $n\pi x/L$ so that the average distance of two particles corresponds to the abscissa $\pi$ in the graph. The ordinates are multiplied by $L/n$.

That the solution of the Debye-Hückel Eq. (5) is different from (11), can be seen by inserting (11) and (3) into the right side of (5). If $g'(x)$ is the value of $g(x)$ obtained by these substitutions,

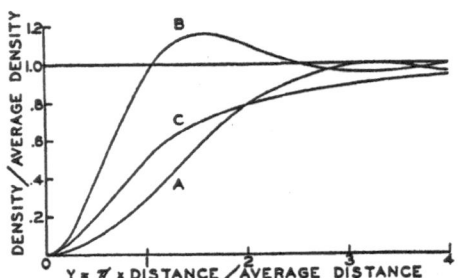

FIG. 1. Distribution functions. The ordinates are the distribution function $g(x)/(n/L)$ representing the density of particles divided by the average density. The abscissae are $y = \pi n x/L$, $\pi$ times the distance divided by the average distance of nearest neighbors. The distance is equal to the average distance for the abscissa $y = \pi$. Curve $A$ represents the rigorous value of the distribution function $g(x)/(n/L)$ given by (11) from the Boltzmann-Gibbs equation. Curve $B$ represents the distribution function, $g'(x)$ of (12a), obtained by inserting the rigorous value (11) in the right-hand side of the Debye-Hückel Eq. (5). Curve $C$ represents the distribution function (13) obtained by solving the Debye-Hückel equation.

this gives

$$\ln g'(x) = -\frac{\sin 2\pi n x/L}{\pi n x/L} + 2Ci(2\pi n x/L) \quad (12)$$

or

$$g'(x) = \exp\left[-\frac{\sin 2\pi n x/L}{\pi n x/L} + 2Ci(2\pi n x/L)\right]. \quad (12a)$$

$g'(x)$ is the uppermost curve in Fig. 1. If (11) were a solution of (5), the value of $g'(x)$ in (12a) would be equal to the rigorous value of $g(x)$ given by (11); and the uppermost curve of Fig. 1 would coincide with the lowest curve. One sees that this is far from being true. The evaluation of the integral obtained by inserting (11) and (3) into (5), i.e., the derivation of (12), requires a somewhat lengthy calculation and will not be given in detail. $Ci$ in (12) is the cosine integral as defined in Jahnke-Emde.[4]

One would expect, on the basis of the lowest and highest curves in Fig. 1, that the solution of (5) would be represented by a curve similar to the middle one of the figure. This corresponds to

$$g(x) = \frac{n}{L}\left(1 - \frac{1}{1 + n^2 \sin^2 \pi x/L}\right) \quad (13)$$

[4] E. Jahnke and F. Emde, *Tables of Functions* (B. G. Teubner, 1933), second edition.

and the following calculation shows that (13) is indeed the solution of (5) for the potential (3). In order to show this, one can first verify (5a), using the assumption $n \gg 1$, and then calculate

$$I = -\beta \int g(x')v(x-x')dx'$$

$$= \frac{n}{L} \int_0^L \left(1 - \frac{1}{1+n^2 \sin^2 \pi x'/L}\right)$$

$$\times \ln \sin^2 \pi(x-x')/Ldx'. \quad (13a)$$

The first term is a well known definite integral; its value is $-2n \ln 2$. In order to integrate the second term, one can change the limits of integration and integrate from $-\frac{1}{2}L$ to $\frac{1}{2}L$. Then the first factor will be very small, except for very small $x'$ for which the sine can be replaced by its argument. Thus (13a) becomes

$$I = -2n \ln 2 - \frac{n}{L} \int_{-\frac{1}{2}L}^{\frac{1}{2}L} \frac{\ln \sin^2 \pi(x-x')/L}{1+\pi^2 n^2 x'^2/L^2} dx'. \quad (13b)$$

If $x$ is of the order of $L$, the remaining integral will have two maxima: at $x'=0$ and at $x'=x$. However, the integral over the second maximum is inversely proportional to $n$ and can be neglected. The integral over the first maximum becomes, if one replaces $x-x'$ by $x$ in the numerator

$$I = -2n \ln 2 - \ln \sin^2 \pi x/L \quad \text{for } x \gg L/n. \quad (13c)$$

If $x$ is not large as compared with $L/n$, the sine can be replaced by its argument in (13b) since the denominator becomes very large for $x' \gg L/n$. This gives, with $y' = \pi n x'/L$ and $y = \pi n x/L$,

$$I = -2n \ln 2 - \frac{1}{\pi} \int_{-\infty}^{\infty} \frac{\ln (y-y')^2/n^2}{1+y'^2} dy'$$

$$= -2n \ln 2 + \ln n^2 - \frac{1}{\pi} \int_{-\infty}^{\infty} \frac{\ln (y-y')^2}{1+y'^2} dy' \quad (13d)$$

and

$$\frac{\partial I}{\partial y} = -\frac{2}{\pi} \int_{-\infty}^{\infty} \frac{dy'}{(1+y'^2)(y-y')} = -\frac{2y}{1+y^2}, \quad (13e)$$

since the main value of the integral has to be

taken in (13e). We now have

$$I = -2n \ln 2 + \ln n^2 - \ln (1+y^2)$$

$$-\frac{1}{\pi} \int_{-\infty}^{\infty} \frac{\ln y'^2}{1+y'^2} dy'. \quad (13f)$$

In the last integral the lower limit can be replaced by 0 if a factor 2 is inserted. Substitution then of $z$ for $\ln y$ shows that the integral in (13f) vanishes and we have

$$I = -2n \ln 2 + \ln n^2 - \ln (1+\pi^2 n^2 x^2/L^2)$$

$$\text{for } x \ll L. \quad (13g)$$

Since $n \gg 1$,

$$I = -\beta \int g(x')v(x-x')dx' = -2n \ln 2$$

$$+ \ln n^2 - \ln (1+n^2 \sin^2 \pi x/L) \quad (14)$$

holds for both regions (13c) and (13g). Insertion of (13), (14), and (3) into (5) shows that (5) is indeed satisfied by (13). $C$ in (5) has the value $\ln 2^{2n} n/L$. The solution (13) of (5) is shown in Fig. 1 lying for small abscissae between the other two curves.

### III.

One sees that the solution of the Debye-Hückel equation agrees fairly well with the correct distribution function if the potential is given by (3). In particular, the total volume of the "hole" in the distribution function is 1 in both cases. Nevertheless, (5) gives too high a probability of a close approach of two particles since (13) is too high for small abscissae. The failure of (5) in this respect could have been foreseen and depends only on the repulsive character of our potential (3). What is essentially neglected in (5) is that the particle under consideration, particle 2 in our notation, has an effect on the rest of the particles. In our case, it pushes them away from itself so that the distribution of the particles 3, 4, $\cdots$, $n$ will, in reality, show a hole around the point where particle 2 is. Thus particle 2 will be under a lower potential than the average potential $U$ given by (4a). However, the difference between the actual and the average potential is smallest for small $x$ because, where the second particle is close to the first, there are very few particles to be shoved away from its neighborhood.

# Electromagnetic Field Expansions
# in Loss-Free Cavities Excited Through Holes*

T. Teichmann and E. P. Wigner[†]

Journal of Applied Physics 24, 262–267 (1953)

(Received April 30, 1952)

It is shown that the electromagnetic field in a loss-free cavity excited through holes cannot be completely expressed in terms of the short-circuit modes of the cavity satisfying the condition that the tangential component of the electric field is zero on the boundary of the cavity including the openings. For a complete expansion it is necessary to add an irrotational magnetic field, which contributes a term inversely proportional to the frequency, to the usual admittance matrix. If the cavity is presumed to include a reasonable portion of the guides feeding the openings, this irrotational component becomes almost diagonal.

## I. INTRODUCTION

IN the course of comparing some similar aspects of the theory of nuclear reactions and of electromagnetic transmission through loss-free microwave junctions, the writers were led to a re-examination of the problem of expanding arbitrary electromagnetic fields in these junctions in terms of normal modes.

The case of a completely closed cavity with perfectly conducting walls was disposed of by Weyl[1,2] many years ago. Using the theory of linear integral equations, he showed that an arbitrary electromagnetic field satisfying homogeneous *short-circuit* boundary conditions on the perfectly conducting surface $S$ of a hollow simply connected cavity $V$ (i.e., a field $E$, $H$ satisfying $n \times E = 0$ on $S$, $n$ being the unit normal to $S$) can be completely expanded in terms of the normal short-circuit modes of the cavity. These modes satisfy the equations

$$\text{curl} E_\alpha = -j\omega_\alpha H_\alpha \quad \text{in } V \qquad (1.1)$$
$$\text{curl} H_\alpha = j\omega_\alpha E_\alpha$$

and the boundary condition

$$n \times E_\alpha = 0 \quad \text{on } S. \qquad (1.2)$$

Throughout this discussion one is, of course, dealing with the temporal Fourier transforms of the actual fields $(j = \sqrt{-1})$.

Weyl attacked this problem mainly to study the asymptotic distribution of proper frequencies. In recent years the expansion aspect of the problem has acquired new interest, because of the large part played by cavities (with holes, however) in modern microwave system,[3,4] and also because of attempts to describe stellar electromagnetic fields in terms of normal modes.[5]

In the first of these problems, which is the one of interest here, it usually proves convenient to separate the effect of the guides and of the junction proper (see Fig. 1). The field in the guides is characterized by the various guide modes, while in the junction proper it must be characterized by some *complete* set of vectors. The junction couples the various modes and guides, and the generalized "currents" and "voltages" in the guides are then related by the so-called *admittance* matrix,

* This work was assisted in part by the U. S. Atomic Energy Commission and the Higgins Scientific Trust Fund.
† Now at the Research and Development Laboratories, Hughes Aircraft Company, Culver City, California.
[1] H. Weyl, J. Math. 143, 177 (1913).
[2] H. Weyl, rend. circ. mat. Palermo 39, 1 (1916).

[3] E. U. Condon, Revs. Modern Phys. 14, 341 (1942); J. C. Slater, Revs. Modern Phys. 18, 441 (1946).
[4] K. Franz, Elek. Nachr.-Tech. 21, 8 (1944).
[5] W. Elsasser, "Boundary-Value Problems of Hydromagnetics" Tech. Rept. ONR Contract N-7-onr-388 (Sept. 1950).

which is a function of frequency in which the proper frequencies of the junction play a basic role.[3,6] It seems to have been generally assumed that the short-circuit modes of the junction, regarded as a closed cavity, form a complete set with respect to an arbitrary admissible electromagnetic field (excited by the wave-guide fields through the openings). While these modes can indeed be made basic to any such expansion, they are *not* complete. Instead, it is necessary to add an irrotational magnetic field; similarly, in the case of open-circuit modes (i.e., modes satisfying the boundary condition n×H = 0 on S) it is necessary to add an irrotational electric field in order to permit expansion of the relevant fields. Since these facts do not appear to be widely known, this paper is intended to elucidate them and to estimate the effect of the additional field on the admittance matrix.

## II. CHARACTERISTIC FIELDS OF THE CAVITY

The problem one has to consider is how to characterize the electromagnetic field adequately in a hollow simply connected cavity with perfectly conducting walls, which is excited via holes out in these walls.

The simplest way to attack this is to introduce a set of orthogonal vectors (in Hilbert space) and add further such vectors until the set is complete, i.e., permits an expansion of all the fields of interest. The natural starting point is, thus, the set of normal vectors, satisfying short-circuit boundary conditions on the entire "surface" $S$ of the cavity, consisting of the conducting portion $S^1$, and the holes (wave-guide openings, henceforth called terminal planes) $T_1, T_2, \cdots$, assumed to be plane. The total open area is denoted by $T = \sum_M T_M = T_1 + T_2 + \cdots$. These normal modes thus satisfy

$$\text{curl}\mathbf{E}_\alpha = -j\omega_\alpha \mathbf{H}_\alpha$$
$$\text{curl}\mathbf{H}_\alpha = j\omega_\alpha \mathbf{E}_\alpha \qquad (2.1)$$

in the cavity $V$, and the boundary condition

$$\mathbf{n}\times\mathbf{E}_\alpha = 0 \qquad (2.2)$$

on $S$. The $\omega_\alpha$ are real, and there is no loss of generality in taking $\mathbf{E}_\alpha$ real, and $\mathbf{H}_\alpha$ imaginary. $\mathbf{E}_\alpha, \mathbf{H}_\alpha$ may be normalized to satisfy

$$\int_V (\mathbf{E}_\alpha \cdot \mathbf{E}_\lambda - \mathbf{H}_\alpha \cdot \mathbf{H}_\lambda)dV = \delta_{\alpha\lambda}. \qquad (2.3)$$

The solutions occur in pairs $\omega_\alpha, \mathbf{E}_\alpha, \mathbf{H}_\alpha$, and $\omega_{-\alpha} = -\omega_\alpha$, $\mathbf{E}_{-\alpha} = \mathbf{E}_\alpha$, $\mathbf{H}_{-\alpha} = -\mathbf{H}_\alpha$. It will generally prove convenient to combine these.

It is clear at this stage where the difficulty may arise. The above equations assert that the $\mathbf{E}_\alpha$ are the proper

A comprehensive treatment of the background material is given by J. C. Slater, reference 3. A treatment giving the notation and some of the results stated below in more detail may also be found in T. Teichmann, J. Appl. Phys. 23, 701 (1952).

vectors of the boundary value problem

$$\Delta\mathbf{E}_\alpha + \omega_\alpha^2 \mathbf{E}_\alpha = 0 \qquad (2.4)$$

with the boundary condition (2.2). If $\omega_\alpha \neq 0$ this implies that the $\mathbf{H}_\alpha$ are the proper vectors of the problem

$$\Delta\mathbf{H}_\alpha + \omega_\alpha^2 \mathbf{H}_\alpha = 0, \qquad (2.5)$$

with the boundary condition

$$(\mathbf{n}\cdot\mathbf{H}_\alpha) = 0 \quad \text{on } S. \qquad (2.6)$$

This last condition follows from Eqs. (2.1) and (2.2) if $\omega_\alpha \neq 0$, but not if $\omega_\alpha = 0$. In this latter case, neither (2.6) nor the vanishing of the divergences of $\mathbf{H}$ and $\mathbf{E}$ follows from (2.1) and (2.2). However, since their curls vanish, they can be represented as gradients of scalars,

$$\mathbf{E}_\beta = \text{grad}w \quad \mathbf{H}_\beta = \text{grad}u. \qquad (2.7)$$

The zero-frequency fields (2.7) appear to lack physical reality, and it is customary to assume that they are not needed for the expansion of the actual field $\mathbf{E}, \mathbf{H}$ in the cavity. That this is not so will be demonstrated in two steps. We shall show, first, that the $\mathbf{E}_\beta, \mathbf{H}_\beta$ of (2.7) is orthogonal in the sense of (2.3) to every $\mathbf{E}_\alpha, \mathbf{H}_\alpha$ field with $\omega_\alpha \neq 0$, as long as w assumes a constant value on the surface $S$ of the cavity ($u$ can be arbitrary). The second step will show that a

$$\mathbf{E}_\beta = 0 \quad \mathbf{H}_\beta = \text{grad}u \qquad (2.8)$$

field is, in general, not orthogonal to the actual field, although a

$$\mathbf{E}_\beta = \text{grad}w \quad \mathbf{H}_\beta = 0 \qquad (2.9)$$

field is. Hence, the expansion of the actual field will in general contain a term of the character (2.8).

The first step will be carried out first for the field (2.8). We have

$$\int \mathbf{H}_\beta \cdot \mathbf{H}_\alpha dV = \frac{j}{\omega_\alpha} \int \mathbf{H}_\beta \cdot \text{curl}\mathbf{E}_\alpha dV$$

$$= \frac{j}{\omega_\alpha}\left\{ \int \text{div}(\mathbf{E}_\alpha \times \mathbf{H}_\beta)dV + \int \mathbf{E}_\alpha \cdot \text{curl}\mathbf{H}_\beta dV \right\}. \qquad (2.10)$$

However, the second term of this vanishes because $\mathbf{H}_\beta$ is a gradient, and the first term is equal to the surface integral of

$$(\mathbf{E}_\alpha \times \mathbf{H}_\beta) \cdot \mathbf{n} = (\mathbf{n}\times\mathbf{E}_\alpha) \cdot \mathbf{H}_\beta, \qquad (2.11)$$

which vanishes by (2.2). A similar calculation for $\int \mathbf{E}_\beta \cdot \mathbf{E}_\alpha dV$ leads to the surface integral of $(\mathbf{H}_\alpha \times \mathbf{E}_\beta) \cdot \mathbf{n}$ which will vanish only if $(\mathbf{H}_\alpha \times \mathbf{E}_\beta)$ is parallel to the surface $S$. Since $\mathbf{H}_\alpha$ can be any vector parallel to $S$, $(\mathbf{H}_\alpha \times \mathbf{E}_\beta)$ is parallel to the surface $S$. Since $\mathbf{H}_\alpha$ can be any vector parallel to $S$, $(\mathbf{H}_\alpha \times \mathbf{E}_\beta)$ will be parallel to $S$ only if $\mathbf{E}_\beta$ is perpendicular thereto, i.e., if $w$ is constant on $S$. The condition on $\mathbf{E}_\beta$ is, therefore, the same (2.2) as on $\mathbf{E}_\alpha$.

The second step of our demonstration is very similar to the first one, with the difference, however, that the actual field $E$, $H$ replaces $E_\alpha$, $H_\alpha$. We can assume, without loss of generality, that the actual field $E$, $H$ is periodic with a finite period $\omega$ (this excludes static fields), this then replaces $\omega_\alpha$ in (2.10). The transformation (2.11) finally gives

$$\int_V H_\beta \cdot H dV = \frac{j}{\omega} \int_S (n \times E) \cdot H_\beta dS. \quad (2.12)$$

This is not zero in general because $n \times E$ vanishes only on the conducting portion $S^1$ of $S$, not on $T_1$, $T_2$, $\cdots$. On the other hand, a similar calculation with the field (2.9) leads to the surface integral of

$$-(n \times H) \cdot E_\beta = (n \times E_\beta) \cdot H, \quad (2.13)$$

and this vanishes because $E_\beta$ is perpendicular to $S$.

The component (2.8) is most significant in the description of the actual field if the linear dimensions of the cavity are small as compared with the wavelength of the radiation which one wishes to describe, i.e., for which one wishes to obtain the admittance. In general, the significance of the static component (2.9) can be best gauged by means of the final formula and this will be done following (3.35). Naturally, no (0, grad$u$) term would be necessary if one wanted to expand only fields which can exist in the cavity $S$ the walls of which are actually all conducting. However, this is not the object of the admittance matrix.

One could avoid using the static portion (0, grad$u$) by replacing the boundary condition (2.2) by an *impedance* boundary condition relating the tangential components of $E_\alpha$ and $H_\alpha$, viz.,

$$(n \times E_\alpha) = \mathfrak{A}(n \times H_\alpha), \quad (2.14)$$

where $\mathfrak{A}$ is a two-dimensional dyadic[7] with zero trace (to insure orthogonality). For $\mathfrak{A} \neq 0$, the boundary value problem thus defined has a double infinity of proper solutions (as compared to the case $\mathfrak{A} = 0$), because the boundary conditions have, as it were, an additional degree of freedom. In the limit $\mathfrak{A} = 0$, these normal modes coalesce in pairs to give the normal modes of the original problem with short-circuited boundary conditions. The field grad$u$ is thus required to make up for this deflection of half the modes. However, this method appears rather artificial.

## III. EFFECT OF THE IRROTATIONAL COMPONENT

In light of the considerations of the previous section, the electromagnetic field $E$, $H$, excited in the cavity

may be represented by the series

$$E = \sum_\alpha f_\alpha E_\alpha,$$
$$H = \sum_\alpha f_\alpha H_\alpha + \text{grad}u. \quad (3.1)$$

If this field is of frequency $\omega$, $E$, $H$ satisfy the equations

$$\text{curl}E = -j\omega H \quad \text{div}E = 0,$$
$$\text{curl}H = j\omega E \quad \text{div}H = 0. \quad (3.2)$$

Thus,

$$\text{div}(E \times H_\alpha - E_\alpha \times H) = j(\omega - \omega_\alpha)(E \cdot E_\alpha - H \cdot H_\alpha) \quad (3.3)$$

and application of Gauss' theorem, and the orthonormality relations (2.3) leads to the result

$$j(\omega_\alpha - \omega)f_\alpha = \int_T H_\alpha \cdot (E \times n)dT \quad (3.4)$$

[since $(E \times n) = 0$ on $S^1$]. The $f_\alpha$ are therefore completely determined by the frequency and the transverse component of $E$ on the terminal planes. To determine $u$, one notes that $\Delta u = 0$, so that $u$ is completely determined by the value of its normal derivative over $S$. Since $n \cdot \text{grad}u = n \cdot H$ on $S$, one has

$$u(p) = \int_T N(p, q)(n \cdot H)(q)dT(q), \quad (3.5)$$

where $p$, $q$ are points of $V$ or $S$, and $N(p, q)$ is the Neumann function[8] of the cavity, determined as follows:

(i) $$N(p, q) = (4\pi r_{pq})^{-1} + \nu(p, q)$$

with $\nu(p, q)$ regular and harmonic (i.e. $\Delta\nu = 0$) in $V$;

(ii) $$\int_S N(p, q)dS(q) = 0;$$

(iii) $$\frac{\partial N(p, q)}{\partial n_q} = -\frac{1}{S} \quad \text{for } q \text{ on } S. \quad (3.6)$$

In the last equation $S$ denotes the area of $S$.

In order to estimate the relative contribution of the irrotational component, one proceeds most conveniently by expressing $E$ and $H$ on the terminal planes $T_N$ in the "canonical" form, i.e., in terms of the electric and magnetic modes of the wave guide, which has the terminal plane in question as cross section. Thus one writes

$$E = \sum_N \sum_s v_{Ns}F_{t, Ns} + i_{Ns}Z_{Ns}F_{z, Ns},$$
$$H = \sum_N \sum_s i_{Ns}G_{t, Ns} + v_{Ns}Z_{Ns}^{-1}G_{z, Ns}. \quad (3.7)$$

---

[7] We have been informed that boundary conditions of this nature have been considered already by J. Schwinger in an as yet unpublished article.

[8] S. Bergman and M. Schiffer, Duke Math. J. 14, 609 (1947). This article also contains an account of various properties of these Neumann functions which are used below. See also O. D. Kellog, *Foundations of Potential Theory* (J. Springer, Berlin, 1929), Chap. IX. Neumann's function is also called Green's function of the second kind.

265   ELECTROMAGNETIC FIELD EXPANSIONS IN LOSS-FREE CAVITIES

$\mathbf{F}_{Ns}$, $\mathbf{G}_{Ns}$ are the electric and magnetic fields of the $s$th modes in the $N$th guide (i.e., on the $N$th terminal plane or "hole" $T_N$), and the suffixes $t$, $z$ denote the transverse and longitudinal components of these fields (with respect to the normal $T_N$). $Z_{Ns}$ is the impedance of this mode. A full description of these modes may be found in Condon's work (reference 3); the properties relevant to the present discussion are listed below.

When we introduce the scalar solutions $U_{Ns}$ of the equation

$$(grad_{t,N})^2 U_{Ns} + K_{Ns}^2 U_{Ns} = 0 \qquad (3.8)$$

on the terminal plane $T_N$, subject to the boundary conditions

$$(grad_{t,N} U_{Ns})_n = 0 \quad s \text{ odd}, \quad s = 1, 3, \cdots \quad (3.8m)$$

$$U_{Ns} = 0 \quad s \text{ even}, s = 2, 4, \cdots \quad (3.8e)$$

on the boundary $c_N$ of $T_N$, and placing

$$k_{Ns}^2 = \omega^2 - K_{Ns}^2, \qquad (3.9)$$

the wave-guide modes are given as follows.

Magnetic modes, $s$ odd

$$\mathbf{G}_{t,Ns} = grad_t U_{Ns}$$

$$\mathbf{G}_{z,Ns} = -(jK_{Ns}^2 k_{Ns}^{-1}) U_{Ns} \mathbf{e}_N \qquad (3.10m)$$

$$\mathbf{F}_{t,Ns} = (\mathbf{e}_N \times \mathbf{G}_{t,Ns})$$

$$\mathbf{F}_{z,Ns} = 0$$

$$Z_{Ns} = \omega k_{Ns}^{-1}. \qquad (3.11m)$$

Electric modes, $s$ even

$$\mathbf{F}_{t,Ns} = grad_{t,N} U_{Ns}$$

$$\mathbf{F}_{z,Ns} = -(jK_{Ns}^2 k_{Ns}^{-1}) U_{Ns} \mathbf{e}_N \qquad (3.10e)$$

$$\mathbf{G}_{t,Ns} = (\mathbf{F}_{t,Ns} \times \mathbf{e}_N)$$

$$\mathbf{G}_{z,Ns} = 0$$

$$Z_{Ns} = \omega^{-1} k_{Ns}. \qquad (3.11e)$$

$\mathbf{e}_N$ denotes the unit normal to the terminal plane $T_N$. It is directed away from the cavity and is equal to n on $T_N$. $\mathbf{F}_{Ns}$ and $\mathbf{G}_{Ns}$ are defined to be zero on all parts of $T$ except $T_N$, where they are given by the previous equations; hence they satisfy the orthonormality relations

$$\int_{T_L} \mathbf{G}_{t,Ns} \cdot \mathbf{G}_{t,Mr} dT_L = \int_{T_L} \mathbf{F}_{t,Ns} \cdot \mathbf{F}_{t,Mr} dT_L$$

$$= \int_{T_L} (\mathbf{F}_{t,Ns} \times \mathbf{G}_{t,Mr}) \cdot \mathbf{e}_L dT_L \quad (3.12)$$

$$= \delta_{LM} \delta_{NM} \delta_{rs},$$

the $U_{Ns}$ being normalized by

$$\int_{T_N} (grad_{t,N} U_{Ns})^2 dT_N = K_{Ns}^2 \int_{T_N} U_{Ns}^2 dT_N = 1. \quad (3.13)$$

The $U_{Ns}$ may be taken to be real without introducing any essential restriction.

Putting the expression (3.7) for $E$ into Eq. (3.4), and using the above results one finds that

$$f_\alpha = \frac{1}{\omega_\alpha - \omega} \sum_N \sum_s \gamma_{\alpha Ns} v_{Ns}, \qquad (3.14)$$

where

$$\gamma_{\alpha Ns} = -j \int_{T_N} \mathbf{H}_\alpha \cdot \mathbf{G}_{t,Ns} dT_N \qquad (3.15)$$

is real and independent of frequency. Similarly, combining Eqs. (3.7) and (3.5) yields

$$-u(p) = \frac{j}{\omega} \sum_N \sum_s{}' v_{Ns} \left[ K_{Ns}^2 \int_{T_N} N(p,q) U_{Ns}(q) dT_N(q) \right], \qquad (3.16)$$

$\sum_s{}'$ means that the sum is to be taken only over *odd* values of $s$, since only the magnetic modes contribute to the normal component of $H$ at the terminal planes. The factors

$$\left[ K_{Ns}^2 \int_{T_N} N(p,q) U_{Ns}(q) dT_N(q) \right]$$

are independent of the frequency $\omega$, so that for fixed values of the voltages $v_{Ns}$, $u(p)$ varies as $\omega^{-1}$.

Using these values for the $f_\alpha$ and $u(p)$ in the expansion (3.1) of $H$ and comparing this with the expression (3.7) which holds for the same quantity on the terminal planes $T_N$, it follows simply that

$$\int_{T_M} \mathbf{H} \cdot \mathbf{G}_{t,Mr} dT_M = i_{Mr}$$

$$= \sum_N \sum_s (A_{Mr,Ns} + C_{Mr,Ns}) v_{Ns} \quad (3.17)$$

$$= \sum_N \sum_s Y_{Mr,Ns} v_{Ns},$$

where

$$A_{Mr,Ns} = j \sum_\alpha \frac{\gamma_{\alpha Mr} \gamma_{\alpha Ns}}{\omega_\alpha - \omega} = 2j\omega \sum_{\alpha > 0} \frac{\gamma_{\alpha Mr} \gamma_{\alpha Ns}}{\omega_\alpha^2 - \omega^2}. \quad (3.18)$$

The last part follows from the remark following Eq.

(2.3). Similarly,

$$G_{Mr, Ns} = -j\omega^{-1}K_{Ns}^2$$

$$\times \int_{T_M} dT_M(p) \, \text{grad}_{t, M} U_{Mr}(p) \cdot \text{grad}_{t, M}$$

$$\times \int_{T_N} dT_N(q)N(p, q)U_{Ns}(q) \quad (3.19)$$

for $r$, $s$ both odd, and zero otherwise. A direct application of Green's theorem enables $C_{Mr, Ns}$ to be written in the more symmetrical form

$$C_{Mr, Ns} = -j\omega^{-1}K_{Mr}^2 K_{Ns}^2$$

$$\times \int_{T_M} \int_{T_N} dT_N(p) dT_N(q)$$

$$\times U_{Mr}(p)N(p, q)U_{Ns}(q). \quad (3.20)$$

In matrix notation one may therefore write the equation

$$i = Yv, \quad (3.21)$$

in which the admittance matrix‡

$$Y = A + C \quad (3.22)$$

is composed of two parts, the contribution $A$ of the cavity short-circuit modes and the purely inductive (irrotational) part $C$.

Before proceeding to estimate the matrix elements of $C$ and to compare them with those of $A$, it is instructive to show how these two terms contribute to the total energy

$$W_H = \int_V |\mathbf{H}|^2 dV. \quad (3.23)$$

From Eq. (3.1) and the results of Sec. II,

$$W = \sum_\alpha |f_\alpha|^2 + \int_V |\text{grad} u|^2 dV. \quad (3.24)$$

When Eqs. (3.14) and (3.18) are used the sum on the right is given simply by

$$-j\sum_N \sum_s \sum_M \sum_r v_{Ns}A_{Ns, Mr}v_{Mr}^*,$$

or in matrix notation

$$\sum_\alpha |f_\alpha|^2 = -jv\dot{A}v^*. \quad (3.25)$$

The dot denotes differentiation with respect to $\omega$, $\dot{f} = df/d\omega$, etc. The contribution of the irrotational

‡ Because the transverse components of the wave-guide modes have been defined to be independent of frequency, the admittance matrix here derived does not contain the inconvenient factors $Z^{\frac{1}{2}}$ which occur in reference 6 and introduce an unnecessary additional frequency dependence which is quite independent of the properties of the cavity.

part is

$$\int_V |\text{grad} u|^2 dV$$

$$= \omega^{-2} \sum_N \sum_s \sum_N \sum_r v_{Ns}v_{Mr}^* K_{Ns}^2 K_{Mr}$$

$$\times \int_V dV(p) \, \text{grad} \int_{T_N} dT_N(q)N(p, q)U_{Ns}(q) \cdot \text{grad}$$

$$\times \int_{T_M} dT_M(y)N(p, y)U_{Mr}(y).$$

It is a quite simple consequence of potential theory that

$$\int_V dV(p) \, \text{grad}N(p, q) \cdot \text{grad}N(p, y) = N(q, y)$$

(see reference 8, for example), and hence the expression on the right reduces to

$$-j\sum_N \sum_s \sum_M \sum_r v_{Ns}C_{Ns, Mr}v_{Mr}^*,$$

or, in matrix notation

$$\int_V |\text{grad} u|^2 dV = -jv\dot{C}v^*. \quad (3.26)$$

Consequently,

$$W = -jv(\dot{A} + \dot{C})v^*. \quad (3.27)$$

If the excitation of the cavity is such that there is no normal magnetic field over the terminal planes $\dot{C}$ will not contribute anything to $W$, because the $v$'s for which $C$ is finite are then zero.

In order to calculate $C_{Mr, Ns}$ exactly it is necessary to know the Neumann function $N(p, q)$ of the cavity, and to carry out the double integration on the right of Eq. (3.20). Both the determination of $N(p, q)$ and the subsequent integration may generally be expected to be rather laborious. However, if suitable assumptions are made about the position of the terminal planes $T_N$, $C_{Mr, Ns}$ may be estimated without too much difficulty, and the approximate form turns out to be rather simple.

One therefore supposes that the terminal planes $T_N$ are displaced outward by an amount $d_N$ from the "natural" boundary of the cavity, along the wave guides which they generate. The cavity now consists of the "natural" cavity plus sections of guide of length $d_N$, (see Fig. 1). The calculation now rests on the fact that

$$\phi_{Ns}(p) = \int_{T_N} N(p, q)U_{Ns}(q)dT_N(q) \quad (3.28)$$

is the potential at the point $p$ due to a normal field $U_{Ns}(q)$ over $T_N$, and zero elsewhere on $S$. In the region

267 ELECTROMAGNETIC FIELD EXPANSIONS IN LOSS-FREE CAVITIES

between the natural boundary and $T_N$, it is possible to determine $\phi_{N_s}(p)$ directly, by using the fact that the $U_{N_s}$ form a complete set of solutions of the boundary value problem [given by Eqs. (3.8) and (3.8m)], and the well-known potential theoretic results that $\phi_{N_s}(p)$ cannot have any maxima or minima in $V$, but only at the boundary. A simple calculation then yields the result that

$$\phi_{N_s}(p) = K_{N_s}^{-1}(1 - \theta_{N_s}e^{-2K_{N_s}d_N})^{-1}$$
$$\times (e^{-K_{N_s}z_N} + \theta_{N_s}e^{-2K_{N_s}d_N}e^{K_{N_s}z_N})U_{N_s}. \quad (3.29)$$

Here $z_N$ denotes the distance of $p$ from the terminal plane $T_N$. $\theta_{N_s}$ is a constant of absolute magnitude less than unity,

$$|\theta_{N_s}| < 1. \quad (3.30)$$

It then follows from the formulas (3.20) and the orthonormality relations (3.12) and (3.13) that

$$C_{Nr, N_s} = -\frac{j}{\omega}K_{N_s}\delta_{rs} \cdot (1 - \theta_{N_s}e^{-2K_{N_s}d_N})^{-1}$$
$$\times (1 + \theta_{N_s}e^{-2K_{N_s}d_N}). \quad (3.31)$$

It is clear that if $d_N$ is made sufficiently large

$$C_{Nr, N_s} \simeq -\frac{j}{\omega}K_{N_s}\delta_{rs}. \quad (3.32)$$

The extension to matrix elements $C_{Mr, N_s}$ with $M \neq N$ is now also plain. If $C_{Mr, N_s}{}^0$ represent the matrix elements of $C$ when the terminal planes $T_N$ coincide with the natural boundary, then (3.29) shows that

$$C_{Mr, N_s} \simeq C_{Mr, N_s}{}^0 e^{-(K_{Mr}d_M + K_{N_s}d_N)}. \quad (3.33)$$

For large $d_M$ and $d_N$ these matrix elements also become negligible. Hence, if the cavity is defined to include a reasonable length of each entering guide, one has

$$C_{Mr, N_s} \simeq -\frac{j}{\omega}K_{N_s}\delta_{MN}\delta_{rs}. \quad (3.34)$$

Since $C$ has a frequency dependence of the form $1/\omega$, while $A$ has an infinity of poles extending to infinity, it follows that $C$ can contribute appreciably to $Y$ only for low frequencies (small $\omega$). This is, of course, consistent with the interpretation of $C$ as the manifestation of the quasi-static part of $H$.

The above considerations show that the usual admittance matrix

$$A_{Mr, N_s} = 2j\omega \sum_{\alpha > 0} \frac{\gamma_{\alpha Mr}\gamma_{\alpha N_s}}{\omega_\alpha^2 - \omega^2}$$

must be augmented by the additional "irrotational" (i.e., purely inductive) matrix $C_{Mr, N_s}$, given by formula (3.20). If the cavity is chosen so as to include a portion of each wave guide larger than the diameter, then

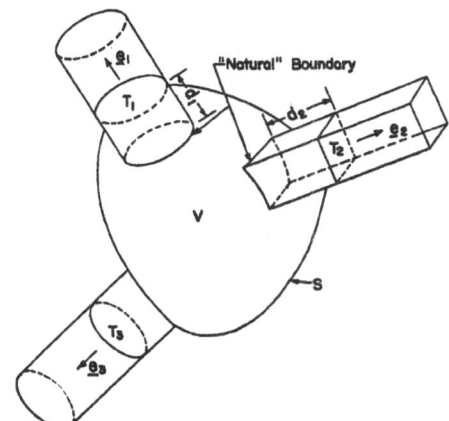

FIG. 1. Sketch of cavity showing terminal planes and natural boundary.

$C_{Mr, N_s}$ becomes approximately diagonal and has the form (3.34).

One may obtain a rough estimate of the relative contributions of the two terms by using the asymptotic relation between the $\gamma_{\alpha N_s}^2$ and the $\omega_\alpha$ derived in reference 6, viz.,

$$\pi < \gamma_{\alpha N_s}^2 > \simeq D_\alpha, \quad (3.35)$$

where $<\gamma_{\alpha N_s}^2>$ denotes an average of $\gamma_{\alpha N_s}^2$ over several values of $\alpha$, and $D_\alpha$ the average spacing between the $\omega_\alpha$ in this region. Applying this result to all the individual terms $\alpha$, except the lowest $\alpha = 1$, and supposing that $\omega \simeq \omega_1$, one finds (see reference 6 for details)

$$A_{N_s, N_s} \simeq \frac{2j\omega\gamma_{1N_s}^2}{\omega_1^2 - \omega^2} + \frac{j}{\pi}\log\left(1 + 2\frac{\omega_1}{D_1}\right). \quad (3.36)$$

Thus, the nonresonant contribution of $A_{N_s, N_s}$ for $\omega \simeq \omega_1$ is approximately equal to $j/\pi$, while $C_{N_s, N_s}$ contributes $-jK_{N_s}/\omega_1$ which is of the order $-jV^{\frac{1}{3}}T_N^{-\frac{1}{2}}$ for the lowest wave-guide mode $s$, $T_N$ being the area of the $N$th terminal plane. This latter term will be larger than the nonresonant part of $A_{N_s, N_s}$ except if the cavity is very shallow in the direction of the guide. Since the nonresonant part of $Y$ at any resonance (pole) $\omega_\alpha$ serves only to change the width and position of the transmission resonance slightly, one may say that at the lowest resonance $\omega_1$, these changes are determined largely by the inductive component $C$; at the higher modes, $C$ becomes small because of its $1/\omega$ frequency dependence, and the changes in width and resonance position are then due mainly to the cumulative effect of the other (nonresonant) terms of $A$ (see also reference 6, Sec. III).

We are very much indebted to Professor R. H. Dicke for many valuable discussions.

# The Problem of Multiple Scattering

E. P. Wigner

Physical Review *94*, 17–25 (1954)

(Received November 16, 1953)

Multiple scattering can be regarded as a succession of elementary events. The distribution function for the particles which have gone through $n+1$ elementary events is the convolution of two functions. The first of these expresses the scattering law; the second one is the distribution function for particles which have gone through $n$ events. It is well known that such convolutions can be calculated very easily by means of Fourier transforms if the elementary event is the traversal of a free path in an arbitrary direction. In this case, the Fourier transform of the convolution is the product of the Fourier transforms of the convolvents. In the case of more general scattering laws, integrals over products of the distribution function, and of representations of the group which leaves the scattering law invariant, play the same role which the Fourier transforms play in the aforementioned case. From the present point of view, the exponential in the Fourier transform is a representation of the displacement group. It is shown that one can solve several problems of multiple scattering on the basis of the above observation. These problems include the scattering of a point particle without change of energy but an arbitrary angular distribution, and several more involved problems.

## INTRODUCTION

GROSJEAN[1] has given, by direct calculation, a solution of the problem of multiple scattering in an infinite medium. The reason for taking up the same problem again is that the method of calculation to be presented is somewhat more transparent and also applicable to a wider range of problems. It is based on the symmetry of the problem considered and uses the theory of group representations.

By multiple scattering we mean a succession of "elementary events" which change the state of the system and which are statistically independent of each other. In the theory of multiple scattering, the elementary event is, in general, a collision and the subsequent traversal of a free path. It changes the state of the system from the one in which it is before a collision to the state in which it is before the next one. The statistical independence of elementary events will be guaranteed if the description of the state of the system is complete, i.e., if it extends to all the parameters. In many cases, this is not necessary. Thus, if one is interested only in the velocity distribution after a certain number of collisions, and if the medium in which the particle moves is homogeneous and extends over all space, one can suppress the position coordinates in the description of the state and consider, as elementary events, the changes in the velocity vector caused by the subsequent collisions. It is clear that the

statistical independence of subsequent elementary events is preserved under the conditions specified even though the description of the state is incomplete in this case. The same holds in the case of spherically symmetric scattering with respect to the direction of the velocity if one is interested only in the density distribution of the particles, irrespective of the direction of their velocities. In the most important case of multiple scattering, in which a complete specification of the state of the system is necessary, the elementary act can contain, just as well, a free path with a *subsequent* collision. In this case, the elementary event changes the state from the one *after* a collision to the state *after* the next collision.

The condition under which we can give an explicit solution of the problem is as follows. (1) The probability that an elementary event changes the state in a certain way is invariant under a group $G$. (2) Every state of the system is obtainable from a single fixed state by the operations of the group $G$. In the case of multiple scattering in an infinite homogeneous and isotropic medium, the group $G$ contains all rotations and displacements in space, i.e., is the Euclidean group. The conditions of homogeneity and isotropy of the scattering medium express the fact that the scattering law, i.e., the law of the elementary event, is invariant under $G$. If the moving particle has no structure and does not change its energy as a result of the collisions, its state is completely characterized by its position and the direction of its velocity. Hence, every state of the particle can be obtained from a standard one by a displacement and a rotation, i.e., by an element of $G$. As standard state one can choose, for instance, the one in which the particle is at the origin of the coordinate system and its velocity is parallel to the $Z$ axis. In the second case mentioned in the preceding paragraph, in which one is interested only in the velocity distribution, the scattering law will have the symmetry of the rotation group and $G$ will be this group. If the collision does not

---

[1] C. C. Grosjean, dissertation, Columbia University, 1951 (unpublished). This thesis also has rather extensive references to earlier literature. Among earlier papers, those of W. Bothe [Z. Physik **54**, 161 (1929)] and of S. Goudsmit and J. L. Saunderson [Phys. Rev. **54**, 773 (1939) and **58**, 36 (1940)] anticipate most nearly Grosjean's results. The last article solves, in particular, the same problem which is treated by Eqs. (16) of the present note. Grosjean's article goes further than these by being able to give a rigorous expression for the probability of a given displacement, not only for the probability of a given deflection. This corresponds to Eqs. (17) of the present note. For later developments of Grosjean's method, see also C. C. Grosjean, Koninkl. Vlaam. Acad. Wetenschap. Letter en Shone Kunsten België Jaarboek **13**, No. 36 (1951) and Physica **19**, 29 (1953).

change the energy of the moving particle, its state is given by the direction of the velocity vector and any such direction can be obtained from a standard one by rotation, i.e., by an element of $G$. Hence, our method· will be applicable to both these cases and a similar discussion shows that it is applicable also in the last case mentioned above,—that of spherically symmetric scattering if the particle again does not change its energy in the course of the collisions. Further and more general examples will be given later.

## EQUATIONS OF MULTIPLE SCATTERING

### A.

Even though the method of solution to be adopted will be essentially the same in all cases to be considered, it seems worth while to distinguish two cases at this point. In the first case, there is only one group element which carries the standard state $e$ of the particle into a given state. In this case, the group element in question can be denoted by $s$ itself, the unit element $e$ corresponding to the standard state. This is the situation if the particle is restricted to move in a plane, no matter whether its "state" has to be characterized only by the direction of its velocity, or by this direction and the position of the particle. Let us denote in this case the probability distribution (per unit volume element of the invariant group space) after $n$ elementary events by $f_n(s)$. The volume element is simply $d\varphi$, with $\varphi$ the angle between direction of the velocity and the standard direction, if one excludes the consideration of the position of the particle. It is $d\varphi dx dy$, with $x$ and $y$ the rectangular coordinates of the particle, if one is interested also in the probability of the position of the particle. If the energy also changes in the course of collisions, it becomes a fourth parameter of $s$ and the volume element will depend on the energy dependence of the law of scattering.

The probability that an elementary event change the state $s$ into a unit volume element at $t$ will be denoted by $P(s,t)$. The invariance of this probability under the operations of the group means that for every group element $u$

$$P(s,t) = P(us,ut). \qquad (1)$$

This equation already uses our assumption that there is a correspondance between states and the elements of the group which leaves the law of the elementary event invariant. Writing, in particular, $u = s^{-1}$ in (1), one obtains

$$P(s,t) = P(e,s^{-1}t) = P(s^{-1}t). \qquad (1a)$$

This equation expresses the fact that the scattering law remains the same no matter whether one uses $e$ or $s$ as the basic state. The probability distribution $f_{n+1}(t)$ is given therefore by

$$f_{n+1}(t) = \int f_n(s)P(s,t)ds = \int f_n(s)P(s^{-1}t)ds, \qquad (2)$$

where $ds$ signifies the invariant group integration. (Right and left invariant group integrals are identical in all the groups to be considered.) It is very natural that the probability distribution after $n+1$ collisions be given by the convolution of the probability distribution after $n$ collisions and the scattering law.

Because of the somewhat abstract nature of the derivation of (2), it may be worth while to illustrate it on the aforementioned examples. If, in two-dimensional scattering, the position of the particle is suppressed, the group $G$ is simply the group of rotations in two dimensions and can be characterized by an angle $\varphi$. The state $(\varphi_0)$ of the particle is obtained from the standard state, with the velocity directed parallel to the $X$ axis, by a rotation with $\varphi_0$; i.e., it is the state in which the velocity includes an angle $\varphi_0$ with the $X$ axis. If $t$ and $s$ in (2) are characterized by the angles $\varphi$ and $\varphi_0$ this equation becomes

$$f_{n+1}(\varphi) = \int f_n(\varphi_0)P(\varphi - \varphi_0)d\varphi_0, \qquad (3)$$

since the angle which characterizes the group element $s^{-1}t$ becomes $\varphi - \varphi_0$. According to its definition,

$$P(\varphi') = \sigma(\varphi')/\sigma, \qquad (3a)$$

where $\sigma = \int \sigma(\varphi')d\varphi'$ is the total cross section. Hence, (3) gives the usual way to calculate the angular distributions successively.

If one wishes to obtain the spatial distribution of the particle as well as its velocity, the group $G$ becomes the group of motions in two dimensions. It can be characterized by a displacement by $x$, $y$ and a rotation by $\varphi$ (the latter preceding the former). The group element,

$$s = \begin{Vmatrix} \cos\varphi_0 & -\sin\varphi_0 & x_0 \\ \sin\varphi_0 & \cos\varphi_0 & y_0 \\ 0 & 0 & 1 \end{Vmatrix} = \begin{Vmatrix} 1 & 0 & x_0 \\ 0 & 1 & y_0 \\ 0 & 0 & 1 \end{Vmatrix}$$
$$\cdot \begin{Vmatrix} \cos\varphi_0 & -\sin\varphi_0 & 0 \\ \sin\varphi_0 & \cos\varphi_0 & 0 \\ 0 & 0 & 1 \end{Vmatrix},$$

characterizes the state in which the particle is at $x_0$, $y_0$ with the direction of the velocity includes an angle $\varphi_0$ with the $X$ axis. A direct method to obtain $P$ in terms of the cross section will be given in the section Calculation of $P(s,t)$. In the present case we wish to verify only a posteriori that (2) gives the well-known equation (3e) if

$$P(x',y',\varphi') = \sigma(\varphi')r^{-1}e^{-\sigma r}\delta(\psi - \varphi'), \qquad (3b)$$

where $r$ and $\psi$ are abbreviations for the polar coordinates of $x' = r\cos\psi$ and $y' = r\sin\psi$. The $\delta$ function in (3b) expresses the fact that the displacement is always in the direction of the velocity. For the verification of (2), let us denote the parameters of the group element $s$ by $x_0$, $y_0$, $\varphi_0$, those of $t$ by $x$, $y$, $\varphi$; the parameters of $s^{-1}t$

then become

$$x' = (x-x_0)\cos\varphi_0 + (y-y_0)\sin\varphi_0,$$
$$y' = -(x-x_0)\sin\varphi_0 + (y-y_0)\cos\varphi_0, \quad (3c)$$
$$\varphi' = \varphi - \varphi_0.$$

The polar coordinates of $x'$, $y'$ are $r$ and $\psi = \alpha - \varphi_0$, where $r$ and $\alpha$ are polar coordinates for $x - x_0 = r\cos\alpha$, $y - y_0 = r\sin\alpha$. Hence

$$P(x',y',\varphi') = \sigma(\varphi - \varphi_0)r^{-1}e^{-\sigma r}\delta(\alpha - \varphi),$$

and with the invariant volume element for $ds$ given above, (2) becomes

$$f_{n+1}(x,y,\varphi) = \int\int\int dx_0 dy_0 d\varphi_0 f_n(x_0,y_0,\varphi_0)$$
$$\times \sigma(\varphi - \varphi_0)r^{-1}e^{-\sigma r}\delta(\alpha - \varphi). \quad (3d)$$

This becomes the well-known equation,

$$f_{n+1}(x,y,\varphi) = \int\int\int dr d\alpha d\varphi_0 f_n(x_0,y_0,\varphi_0)$$
$$\times \sigma(\varphi - \varphi_0)e^{-\sigma r}\delta(\alpha - \varphi)$$
$$= \int\int dr r \varphi_0 f_n(x - r\cos\varphi, y - r\sin\varphi, \varphi_0)$$
$$\times \sigma(\varphi - \varphi_0)e^{-\sigma r}, \quad (3e)$$

if one replaces the integration variables $x_0$, $y_0$ by $r$ and $\alpha$ and carries out the integration over $\alpha$. It may be arguable that the well-known (3e) is simpler than (2) but we shall see that (2) suggests more directly a method of solution than does (3e).

### B.

The preceding calculation applies in the case in which $G$ has only one element which transforms the standard state $e$ into a given state $s$. This is not true if, for instance, a particle without structure moves in three-rather than two-dimensional space. The reason therefore is that there are rotations—those about the direction of the velocity in the standard state—which leave the standard state unchanged. The rest of the present section contains a discussion of the group theoretic description of the state and of the scattering law. While the former discussion is essential, the discussion following (5a) is not necessary if one is interested only in using the present method. The result of this discussion is (9) which becomes evident at any rate if one calculates $P$ explicitly.

Let us denote the number of parameters which characterize the states by $n_1$, the number of parameters of the group $G$ by $n$. There will be then an $n-n_1$ parametric subgroup $E$, the elements of which carry the standard state $\delta$ into itself:

$$e_1\delta = \delta \text{ if } e_1 \text{ contained in } E. \quad (4)$$

Two group elements, $s$ and $s'$, will carry $\delta$ into the same state $\delta$ if

$$\delta = s\delta = s'\delta; \quad \delta = s^{-1}s'\delta. \quad (4a)$$

$s^{-1}s'$ is an element of the subgroup $E$, i.e., if $s$ and $s'$ are in the same left coset of $E$. One can say that in the more general case now considered the left cosets of $E$ correspond to the different states of the particle rather than the group elements themselves.

We shall see that expressions of the form (2) are very easily evaluated and we wish to define, therefore, also in the present case a probability function $f(s)$, depending on the group elements rather than on left cosets. This $f(s)$, if integrated over the elements of the cosets which correspond to a set of states, will give the probability of the states of the set. In order to express this analytically, it is useful to introduce two types of parameters, $\zeta$ and $\epsilon$, for the group $G$. There are $n_1$ parameters $\zeta$ and they have the same value for all elements of a left coset and serve to distinguish these cosets, i.e., to characterize the states of the particle. There are $n-n_1$ parameters $\epsilon$ and they distinguish the various elements of the left cosets. One can choose, for instance, the parameters $\zeta$ and $\epsilon$ in such a way that the element with parameters $\zeta$ and $\epsilon$ become

$$s(\zeta,\epsilon) = s(\zeta,0)e(\epsilon), \quad (4b)$$

where $e(\epsilon)$ are the elements of the subgroup $E$ and $s(\zeta,0)$ is a continuous function of $\delta(\zeta)$. If we denote the probability of the states $\delta(\zeta)$ per unit range of $\zeta$ by $F(\zeta)$, we shall demand of $f(s)$

$$F(\zeta) = \int f(s)g(\zeta,\epsilon)d\epsilon, \quad (5)$$

where the integration over $\epsilon$ is to be extended over the subgroup $E$, $g(\zeta,\epsilon)$ is the weight factor which makes $g(\zeta,\epsilon)d\zeta d\epsilon$ the invariant integral, and $s$ is the group element with the parameters $\zeta$, $\epsilon$. This equation expresses the postulate that the integral of $f$ over a coset give the probability of the state to which the coset corresponds.

Clearly, (5) does not determine $f(s)$ completely and we must further postulate that it have the same value for each element of a coset, i.e., be a constant along the path of integration of (5)

$$f(s) = f(se_1) \text{ if } e_1 \text{ contained in } E. \quad (5a)$$

All our probability functions shall have the property (5a).

In a similar way, we shall try to replace the transition probability $P(\delta_1,\delta_2) = P(\zeta_1,\zeta_2)$ by a function $P(s,t)$ of the group elements $s,t$. This has to be done in such a way that if

$$F_{n+1}(\zeta_2) = \int F_n(\zeta_1)P(\zeta_1,\zeta_2)d\zeta_1, \quad (6)$$

and if $F_n$ and $f_n$ correspond to each other in the sense of (5), (5a), then

$$f_{n+1}(s_2) = \int f_n(s_1)P(s_1,s_2)ds_1 \qquad (6a)$$

also correspond to $F_{n+1}$ in the same sense, no matter what the functional form of $F_n$ is. Since $f_{n+1}$, as a probability function, will have to satisfy (5a), we can conclude at once that

$$P(s_1,s_2) = P(s_1,s_2 e_2) \text{ if } e_2 \text{ contained in } E. \qquad (7a)$$

Expressing now both $F$ in (6) by the corresponding $f$, we find

$$\int f_{n+1}(s_2)g(\zeta_2,\epsilon_2)d\epsilon_2 = \int\int f_n(s_1)g(\zeta_1,\epsilon_1)P(\zeta_1,\zeta_2)d\epsilon_1 d\zeta_1$$

$$= \int f_n(s_1)P(\zeta_1,\zeta_2)ds_1,$$

where $s_2$ and $s_1$ are the group elements with the parameters $\zeta_2,\epsilon_2$ and $\zeta_1,\epsilon_1$, respectively, and the second line follows from the definition of $g(\zeta,\epsilon)d\zeta d\epsilon$ as the invariant group integral. The condition that the $f_{n+1}$ obtained from (6) satisfy (6a) is, therefore,

$$\int ds_1 \int f_n(s_1)P(s_1,s_2)g(\zeta_2,\epsilon_2)d\epsilon_2 = \int ds_1 f_n(s_1)P(\zeta_1,\zeta_2).$$

Since $f_n$ is not an arbitrary function of $s_1$ but satisfies (5a), the last equation does not fully determine $P(s_1,s_2)$ and we are free to let it depend on $\epsilon_1$ in an arbitrary fashion. The simplest choice is

$$\int P(s_1,s_2)g(\zeta_2,\epsilon_2)d\epsilon_2 = P(\zeta_1,\zeta_2), \qquad (7)$$

according to which we have, in addition to (7a),

$$P(se_1,s_2) = P(s_1,s_2) \text{ if } e_1 \text{ contained in } E. \qquad (7b)$$

According to (7a) and (7b), $P(s_1,s_2) = P(s_1',s_2')$ if $s_1$ and $s_1'$ are in the same left coset of $E$ and the same holds of $s_2$, $s_2'$. This is indeed the most natural convention; it renders all quantities $f(s)$, $P(s,t)$ functions only of the $\zeta$ parameters of their (group element) variables. It follows that $P(s_1,s_2)$ can be taken out of the integral sign of (7) and this equation, together with (7a), (7b), completely determines $P(s_1,s_2)$ once $P(\zeta_1,\zeta_2)$ is given.

The invariance of the law governing the elementary event can be expressed in the following way. Let us assume that the distribution will be given by $f_{n+1}(s_2)$ if it was $f_n(s_1)$ one event before. Then if the distribution is $h_n(s_1) = f_n(us_1)$, it will go over into $h_{n+1}(s_2) = f_{n+1}(us_2)$ after another event. This will hold for every group element $u$. One easily verifies that the $h$ satisfy (5a)

if the $f$ do. Expressing the above condition analytically, one finds that

$$f_{n+1}(us_2) = h_{n+1}(s_2) = \int h_n(s_1)P(s_1,s_2)ds_1$$

$$= \int f_n(us_1)P(s_1,s_2)ds_1 \qquad (8)$$

must be a consequence of

$$f_{n+1}(t) = \int f_n(s)P(s,t)ds. \qquad (8a)$$

Since $ds_1$ indicates the invariant group integration, one can replace in (8) $us_1$ by $s$. Writing, furthermore, $t$ for $us_2$, (8) goes over into

$$f_{n+1}(t) = \int f_n(s)P(u^{-1}s,u^{-1}t)ds, \qquad (8b)$$

and this will be a consequence of (8a) if

$$\int f_n(s)P(s,t)ds = \int f_n(s)P(u^{-1}s,u^{-1}t)ds. \qquad (8c)$$

Writing out the integration in terms of $\zeta,\epsilon$,

$$\int f_n(\zeta)P(s,t)g(\zeta,\epsilon)d\zeta d\epsilon$$

$$= \int f_n(\zeta)P(u^{-1}s,u^{-1}t)g(\zeta,\epsilon)d\zeta d\epsilon,$$

where $s$ is the group element with the parameters $\zeta,\epsilon$. Since the last equation must hold for all $f_n$,

$$P(s,t)\int g(\zeta,\epsilon)d\epsilon = P(u^{-1}s,u^{-1}t)\int g(\zeta,\epsilon)d\epsilon. \qquad (8d)$$

Both $P(s,t)$ and $P(u^{-1}s,u^{-1}t)$ could be placed before the integral sign because they are independent of $\epsilon$. Setting then $u = s$, one again obtains (1a) for $P$ and also obtains, from (6a), the same Eq. (2) which holds in the case in which there is only one group element which transforms the standard state into a given state. The only difference between the two cases is, therefore, the additional condition (5a) on the distribution functions, and the conditions (7a), (7b) on $P(s,t)$, which must be satisfied in order to make (2) valid also if there are group elements which leave the standard state unchanged and hence all the elements of a left coset transform the standard state into the same state. It is worth noting that (7a), (7b) give

$$P(e_1 s e_2) = P(s) \text{ for } e_1, e_2 \text{ contained in } E. \qquad (9)$$

Hence, $P$ depends in general on even fewer independent variables than $f$. The example of the following section will illustrate this point.

## CALCULATION OF $P(s,t)$

In the case of the motion of a particle in two-dimensional space without energy change, we obtained the expressions (3a), (3b) for $P$ by inspection; insertion of these expressions into (2) gave the relatively simple and well-known equations for multiple scattering for this case and thus verified (3a) and (3b) *a posteriori*. We shall not follow this procedure in the case of the scattering of a particle in three-dimensional space without energy change, principally because the explicit form of the equation of multiple scattering in three dimensions [i.e., the analog of (3e)] is very complicated. In fact, one of the simplifications which the method here presented introduces is the avoiding of these explicit equations and the possibility of evaluating (2) without writing down this equation in any other form. In order to use and interpret the solution, it is necessary, however, to express $P(s,t)$ in terms of the quantities commonly used (i.e., the differential cross section) and to establish the connection between the $f_n(s)$ and the commonly used angular distribution $F(\Omega)$. These connections are given, in principle, by (7), (7a), (7b), and by (5), (5a) but the corresponding relation will be explicitly evaluated now for the aforementioned case, i.e., scattering of a particle without change of energy under disregard of its position. It will be given also for the case in which the position of the particle is considered also.

The group $G$ in the first case is the three-dimensional rotation group. As standard state, we choose the one in which the velocity is parallel to the $Z$ axis. The subgroup $E$ which leaves the standard state invariant consists of the rotations about $Z$. It is reasonable to choose the polar angles $\varphi$, $\vartheta$ of the velocity direction as the variables $\zeta$ describing the state. Equation (4b) then becomes

$$s(\varphi,\vartheta,\epsilon) = s(\varphi,\vartheta,0)Z(\epsilon), \tag{10}$$

where $Z(\epsilon)$ is a rotation by $\epsilon$ about the $Z$ axis and we can choose for $s(\varphi,\vartheta,0)$ any rotation which turns the $Z$ axis into the $\varphi$, $\vartheta$ direction. If we choose

$$s(\varphi,\vartheta,0) = Z(\varphi)X(\vartheta), \tag{10a}$$

the angles $\varphi$, $\vartheta$, $\epsilon$ become the Eulerian parameters of the group elements. The distribution function $f$ will depend, by (5a), only on $\varphi$ and $\vartheta$ and we have, by (5) for the probability of a unit range in $\varphi$ and $\vartheta$:

$$F(\varphi,\vartheta) = \int_0^{2\pi} f(\varphi,\vartheta)\sin\vartheta d\epsilon = 2\pi f(\varphi,\vartheta)\sin\vartheta, \tag{10b}$$

$\sin\vartheta d\varphi d\vartheta d\epsilon$ being the invariant volume element expressed in terms of Eulerian parameters.

According to (9), the function $P(s)$ depends only on the parameter $\vartheta$—the simplicity of this condition is a consequence of (10a), i.e., of the use of Eulerian parameters to describe the group elements. A similar simpli-

fication can be accomplished, however, also in most other cases by choosing three sets of parameters $\zeta_1$, $\zeta_2$, $\epsilon$ defined by the equation

$$s(\zeta_1,\zeta_2,\epsilon) = e(\zeta_1)\delta(\zeta_2)e(\epsilon), \tag{11}$$

where $e(\zeta_1)$ and $e(\epsilon)$ cover the subgroup $E$ and the $\delta(\zeta_2)$ are so chosen that all elements of the group are obtained by letting $\zeta_2$ vary over a suitable domain.

Since $P(s,t)$ depends only on $s^{-1}t$, it is sufficient to determine $P(e,t) = P(t)$ where $e$ is the unit element. We then have, from (7),

$$\int_0^{2\pi} P(\vartheta)\sin\vartheta d\epsilon = \sigma(\vartheta)\sin\vartheta/\sigma. \tag{10c}$$

Since we have set $s_1 = e$ the right side represents the probability that an elementary event change the velocity, originally parallel to $Z$, into a velocity at $\vartheta$, $\varphi$ within unit range of $d\vartheta$ and $d\varphi$. Since the differential cross section $\sigma(\vartheta)$ is usually defined per unit solid angle, the probability of the transition to unit $d\vartheta d\varphi$ range becomes $\sigma(\vartheta)\sin\vartheta$. It is an obvious consequence of the spherical symmetry of the problem that this probability is independent of $\varphi$. This does not give an additional condition, however, because the symmetry of $P(s)$ is fully described already in (9).

Summarizing, we have

$$f(s) = f(\varphi,\vartheta,\epsilon) = (2\pi\sin\vartheta)^{-1}F(\varphi,\vartheta), \tag{12}$$

$$P(s) = P(\varphi,\vartheta,\epsilon) = (2\pi)^{-1}\sigma(\vartheta)/\sigma. \tag{12a}$$

It may be useful to repeat that $F(\varphi,\vartheta)$ is not the probability per unit solid angle but per unit $d\varphi d\vartheta$ interval. No $\sin^{-1}\vartheta$ appears in the expression for $f(s)$ if $F$ is given in terms of the distribution function per unit solid angle.

The expressions for the case in which one wishes to consider not only the direction of the velocity but also the position of the particle can be derived with equal ease. The group $G$ for this case is the group of motions in three dimensions (Euclidean group) defined in the same fashion as the group of motions in two dimensions was. The distribution function $F(x,y,z,\varphi,\vartheta)$ depends on the position of the particle as well as the direction of its velocity. The group elements depend on six parameters: the Eulerian angles $\varphi,\vartheta,\epsilon$ of the rotation, and the three components $x,y,z$ of the subsequent displacement. The subgroup $E$ contains only the rotations about $Z$; it is the same group as in the preceding example. The connection between the group theoretic $f$ and the usual $F$ is, in complete analogy to (12)

$$f(s) = f(x,y,z,\varphi,\vartheta,\epsilon) = (2\pi\sin\vartheta)^{-1}F(x,y,z,\varphi,\vartheta). \tag{13}$$

The form of $P(s)$ is simpler if we consider as fundamental process the traversal of a free path and a subsequent collision (i.e., if we adopt the second point of view of the Introduction). The probability of the transition from the standard state into unit interval at the state $s = (x, y, z, \varphi, \vartheta)$ then becomes $\delta(x)\delta(y)e^{-\epsilon s}\sigma(\vartheta)\sin\vartheta$. Hence, application of (7) to the case $s_1 = e$, $s_2 = s$ gives

$$P(s) \int \sin\vartheta d\epsilon = \delta(x)\delta(y)e^{-\epsilon s}\sigma(\vartheta)\sin\vartheta$$

or

$$P(s) = P(x, y, z, \varphi, \vartheta, \epsilon) = (2\pi)^{-1}\sigma(\vartheta)e^{-\epsilon s}\delta(x)\delta(y). \quad (13a)$$

One convinces oneself easily that this expression satisfies (9).

The above expressions could have been foreseen without the detailed derivation given above. In fact, the numerical factors in (12) to (13a) need not be known for actual calculations since the successively calculated distribution functions can easily be normalized a posteriori.

### EVALUATION OF THE EXPRESSION (2)

The reason that (2) can be evaluated particularly simply if $s$ and $t$ are elements of a group is that the group integral of the product of the convolute and the matrix of a representation is the product of two matrices which are obtained from the convolvents in a similar fashion. In fact, we obtain from $D(s)D(u) = D(su)$:

$$\int f_{n+1}(t)D(t)dt = \int \int f_n(s)P(s^{-1}t)D(t)dsdt$$

$$= \int \int f_n(s)P(u)D(su)dsdu$$

$$= \int f_n(s)D(s)ds \cdot \int P(u)D(u)du. \quad (14)$$

The second line is obtained by substituting $t = su$ and noting the invariance of the group integral with respect to such a substitution.[2] Since $D(s)$ is, in general, a matrix, all expressions in (14) are matrices, with the number of dimensions of $D$. Written out in more detail, (13) reads

$$\int f_{n+1}(s)D(s)_{e\lambda}dt$$

$$= \sum_\mu \int f_n(s)D(s)_{e\mu}ds \int P(s)D(s)_{\mu\lambda}ds. \quad (14a)$$

This equation holds no matter whether the representation $D$ is reducible or irreducible. If $D$ is the regular representation, one is led back to (2); the simplest results are obtained if $D$ is irreducible. If (14) holds for a function $f_{n+1}$ and all irreducible representations, this $f_{n+1}$ satisfies (2).

[1] Equation (14) must have been known already to G. Frobenius.

It follows from the above that

$$\int f_n(s)D(s)ds = \Phi^{(n)} = \Phi^{(0)}\Pi^n;$$

$$\Pi = \int P(s)D(s)ds. \quad (14b)$$

Again, the $\Phi$ and $\Pi$ are matrices with as many dimensions as $D$. The evaluation of (14b) is often made very much easier by the fact that $f_0$ and $P$ satisfy (5a) and (9), i.e., actually do not depend on all the group variables. In particular, if one assumes the representations in the form in which the matrices corresponding to elements of $E$ are in the reduced form, only those $\int f_0(s)D(s)_{\lambda\nu}ds$ will be different from zero in which $\nu$ corresponds to a unit representation of $E$ and both $\mu$ and $\nu$ must correspond to such representations if $\int P(s)D(s)_{\mu\nu}ds$ is to be finite.

In the first case discussed in the last section, it follows either from (9) or more directly from (12a), that assuming the customary form of the irreducible representations[3] only the 0,0 element of $\Pi(l)$ is different from zero. One has

$$\Pi(l)_{\nu\mu} = \int P(s)D^{(l)}(s)_{\nu\mu}ds$$

$$= \int_0^{2\pi}\int_0^\pi\int_0^{2\pi} \frac{\sigma(\vartheta)}{2\pi\sigma}e^{i\nu\varphi}d^{(l)}_{\nu\mu}(\vartheta)e^{i\mu\epsilon}d\varphi \sin\vartheta d\vartheta d\epsilon$$

$$= (2\pi/\sigma)\delta_{\nu 0}\delta_{\mu 0}\int_0^\pi \sigma(\vartheta)d_{00}^{(l)}(\vartheta)\sin\vartheta d\vartheta. \quad (15)$$

Hence, in all powers of the "matrix" $\Pi(l)$ only the 0–0 element is different from zero and this is the corresponding power of (15). Similarly, it follows from (5a), or more directly from (12), that only the 0 column of

$$\Phi^{(n)}(l) = \int f_n(s)D^{(l)}(s)ds \quad (15a)$$

contains elements which are different from 0. This is, of course, consistent with (14b) and (15).

Let us consider, for instance, the angular distribution obtained by means of Born's first approximation, for Rutherford scattering on a shielded nucleus. For a potential proportional to $r^{-1}e^{-\alpha r}$ the angular distribution becomes, with $\beta = \alpha^2/2k^2$, where $k$ is the wave number of the particle,

$$\frac{\sigma(\vartheta)}{\sigma} = \frac{\beta(2+\beta)}{4\pi(1+\beta-\cos\vartheta)^2}. \quad (16)$$

[3] See, e.g., E. P. Wigner, *Gruppentheorie und ihre Anwendung* (Friedr. Vieweg, Braunschweig, 1931), Chap. XV.

In order to evaluate the integral (15), it is most convenient to write $\cos\vartheta = x$ and to use $d_{00}$ in the form

$$d_{00}{}^{(l)}(\vartheta) = P^l(\vartheta) = \frac{1}{2^l l!} \frac{d^l(x^2-1)^l}{dx^l}. \tag{16a}$$

This gives for $\Pi(l)_{00}$ (we shall omit the indices 0 for convenience)

$$\Pi(l) = 2\pi \int_{-1}^{1} \frac{\beta(2+\beta)}{4\pi(1+\beta-x)^2} \frac{1}{2^l l!} \frac{d^l(x^2-1)^l}{dx^l} dx$$

$$= \frac{\beta(2+\beta)}{2^{l+1} l!} \int_{-1}^{1} \frac{(l+1)!(1-x^2)^l}{(1+\beta-x)^{l+2}} dx. \tag{16b}$$

The second line is obtained by $l$-fold partial integration. For $l=0$, this gives (as always) $\Pi(0)=1$ for $l=1$,

$$\Pi(1) = 1 + \beta - \tfrac{1}{2}\beta(2+\beta)\ln(2+\beta)/\beta. \tag{16c}$$

This is the forward bias after one collision; after $n$ collisions the forward bias in the $n$th power of this. (16c) has been obtained already in the earlier articles of reference 1.

Let us consider, as a second example, the motion of a particle without energy change in two-dimensional space. The unitary irreducible representations of the corresponding group—the Euclidean group of the plane—are infinite-dimensional and can be characterized by a continuous variable $k$:

$$D^{(k)}(r,\psi,\varphi)_{m'm} = e^{im'\psi} J_{m'-m}(kr) e^{im(\varphi-\psi)}. \tag{17}$$

The notation is the same as in (3b): the group element $(r,\psi,\varphi)$ is given by the matrix of the equation preceding (3b) if one sets therein $x=r\cos\psi$, $y=r\sin\psi$, $\varphi_0=\varphi$. The very simple derivation of (16) is given in the Appendix.[4] $J_n$ is the Bessel function of order $n$. We define again

$$\Pi(k)_{m'm} = \int P(s) D^{(k)}(s)_{m'm} ds$$

$$= \int \sigma(\varphi) r^{-1} e^{-\sigma r} \delta(\psi-\varphi) e^{im'\psi} J_{m'-m}(kr)$$
$$\times e^{im(\varphi-\psi)} r\, dr\, d\psi\, d\varphi; \tag{17a}$$

the $P(s)$ was taken from (3b). The integration over $\psi$ can be carried out at once, and one obtains

$$\Pi(k)_{m'm} = \int_0^{2\pi} \sigma(\varphi) e^{im'\varphi} d\varphi \int_0^{\infty} e^{-\sigma r} J_{m'-m}(kr) dr$$

$$= \frac{\sigma_{m'}}{(k^2+\sigma^2)^{\frac{1}{2}}} \xi^{m'-m}, \tag{17b}$$

where

$$\sigma_m = \int_0^{2\pi} \sigma(\varphi) e^{im\varphi} d\varphi; \quad \xi = \frac{\sigma}{k}-\left(1+\frac{\sigma^2}{k^2}\right)^{\frac{1}{2}}. \tag{17c}$$

One sees that $\Pi(k)$ is again of rank 1. One easily proves by induction that

$$[\Pi(k)]^{n+1} = \frac{[2\pi\sigma(0)]^n}{[\sigma^2+k^2]^{\frac{1}{2}n}} \Pi(k), \tag{17d}$$

where $\sigma(0)$ is the differential cross section in the forward direction:

$$2\pi\sigma(0) = \sum \sigma_m. \tag{17e}$$

It would have been possible, of course, to transform (17) in such a way that, similar to (15), only one element of the $\Pi(k)$ matrix would have been different from zero. However, the rather arbitrarily chosen form of (17) has caused no difficulty.

In order to obtain the expansion matrices $\Phi^{(n)}(k)$ of $f_n$ from those of $[\Pi(k)]^n$, we have to multiply the matrix expanding $f_0$ with $[\Pi(k)]^n$. If the particle is, originally, in the standard state, its expansion matrices are all unit matrices and $\Phi^{(n)}(k) = [\Pi(k)]^n$. Hence, these matrices play a role similar to that of the source solution of diffusion equations. In order to return, from the $\Phi$, to the distribution function $f$, one has to invert equations of the form (15a). This is a trivial matter if the group is closed and can be done easily also for the representations (17) by means of well-known expansion formulas for Bessel and trigonometric functions. The problem of inverting Eqs. (15a) for not closed groups has been attacked recently in a rather general form by Harish-Chandra and by Segal.[5] In many cases, such as dealt with in Eqs. (16), the expansion matrices have more immediate physical significance than the distribution function $f$ itself.

### CALCULATION OF TOTAL DENSITY FUNCTIONS

In many cases one is interested in the total distribution of all particles, irrespective of the number of elementary events they have passed through. Usually, this will be finite only if the elementary process can lead to absorption.

If the probability of absorption $1-\gamma$ is independent of the state of the particle before the elementary event, the total distribution function will be

$$f = \sum \gamma^n f_n. \tag{18}$$

The corresponding momentum matrix becomes

$$\Phi(l) = \int f(s) D^{(l)}(s) ds = \sum \gamma^n \Phi^{(n)}(l)$$

$$= \Phi^{(0)}(l) \sum \gamma^n \Pi(l)^n = \Phi_0(l)[1-\gamma\Pi(l)]^{-1}. \tag{18a}$$

The right side of (18a) can be written down at once in all cases considered in the preceding sections.

[4] For representations of similar groups, see also E. Inonu and E. P. Wigner, Nuovo cimento 9, 705 (1952).

[5] I. E. Segal, Ann. Math. 52, 272 (1950) and Harish-Chandra, Proc. Natl. Acad. Sci. U.S. 37, 813 (1951).

## APPLICABILITY OF THE METHOD

It has been realized already by Grosjean that one can successively evaluate the distribution functions for multiple scattering, if one wishes to take the energy changes into account, most easily if the probability of a certain fractional change in energy is only a function of the scattering angle. This is the case for elastic scattering on atoms at rest. It seems natural to ask, therefore, whether the calculations presented above, if applied to a larger than the Euclidean group, could be given a physical interpretation. We shall not try to solve this problem in general but point only to the simplest generalization of the Euclidean group, that formed by the matrices

$$s = \begin{Vmatrix} cr & a \\ 0 & 1 \end{Vmatrix}. \tag{19}$$

In (19), $r$ is a real orthogonal matrix, $a$ is a real vector, $c > 0$. For $c = 1$, the $s$ form the Euclidean group; dropping this restriction increases the number of group parameters by 1. Clearly, $c$ will be connected in some way with the velocity of the particle.

The condition of invariance means that the transition probability from $s_1$ to $s_2$ be equal, for all $u$, to the transition probability from $u s_1$ to $u s_2$. It is enough to demand this for a set of $u$ from which all group elements can be obtained by multiplication. We choose the following $u$:

$$u_1 = \begin{Vmatrix} 1 & a \\ 0 & 1 \end{Vmatrix}, \quad u_2 = \begin{Vmatrix} r & 0 \\ 0 & 1 \end{Vmatrix}, \quad u_3 = \begin{Vmatrix} c1 & 0 \\ 0 & 1 \end{Vmatrix}. \tag{19a}$$

The above condition, applied to $u = u_1$ shows that the transition probability from the state $(c_1, r_1, a_1)$ to $(c_2, r_2, a_2)$ is equal to the transition probability from $(c_1, r_1, a_1 + a)$ to $(c_2, r_2, a_2 + a)$. This shows that a plays the role of the position vector of the particle, or is proportional to the position vector. The invariance condition is also satisfied with respect to $u_2$ if $r$ continues to describe the direction of the velocity of the particle. Application to $u_3$ gives, finally, that the transition probabilities

$$(c_1, r_1, a_1) \to (c_2, r_2, a_2) \text{ and } (c c_1, r_1, c a_1) \to (c c_2, r_2, c a_2) \tag{19b}$$

are equal for all $c$. In other words, the probability of a path of length $|c(a_1 - a_2)|$ at the value $c c_1$ of the first parameter is as great as the probability of the path $|a_1 - a_2|$ at the value $c_1$ of the first parameter. Expressed still differently, the mean free path is proportional to the first parameter. For the first parameter itself, the change from $c_1$ to $c_2$ is as probable as from $c c_1$ to $c c_2$. The simplest interpretation of this is that

$$c = C v^n; \tag{19c}$$

that is, $c$, and hence also the mean free path, is proportional to the $n$'th power of the velocity where $n \neq 0$ but can be arbitrary otherwise. Together with the preceding conditions, condition (19b) then stipulates that the angular distribution of the scattering be independent of the velocity and that the probability of a certain fractional change in energy depend only on the change in the direction of the velocity.

It seems likely that many other multiple scattering distributions can be calculated accurately, many others approximately, following the procedure outlined above. So far, no case has been encountered in which the calculation of the $n$th power of $\Pi$, or of the reciprocal in (18a), would have been at all difficult. On the other hand, it does not appear to be possible to use the rotation (or Euclidean) group to calculate the multiple scattering of a particle with spin, by using the variable $\epsilon$ of (12) to describe the spin's motion. It seems that a more powerful method is needed to overcome the difficulties of this problem.

## APPENDIX

The irreducible representations of the group of motions in two dimensions can most easily be given in the Hilbert space of functions $f(\alpha)$ the variable $\alpha$ of which is restricted to the interval $0 \leq \alpha < 2\pi$. The operators $P_s = P_{r, \psi, \varphi}$ of the representation in this space are defined by

$$P_s f(\alpha) = P_{r, \psi, \varphi} f(\alpha) = e^{-ik(r \cos\psi \cos\alpha + r \sin\psi \sin\alpha)} f(\alpha - \varphi)$$
$$= e^{-ikr \cos(\psi - \alpha)} f(\alpha - \varphi). \tag{A1}$$

The group element $s$ is a rotation by $\varphi$ about the origin, followed by a displacement with a vector the polar coordinates of which are $r$ and $\psi$. Borrowing concepts from quantum mechanics, one can say that $f$ describes states in which the absolute value of the momentum is $k$ but the direction of the momentum is variable and given by $\alpha$. Rotation of this state by $\varphi$ replaces $\alpha$ by $\alpha - \varphi$; displacement by the vector $r \cos\psi$, $r \sin\psi$ multiplies it with the exponential in (A1). The magnitude of $k$ characterizes the representation; it can assume any positive value.

One can write (A1) also in the form

$$P_s f(\alpha) = \int \Delta^{(k)}(r, \psi, \varphi)_{\alpha\alpha'} f(\alpha') d\alpha', \tag{A2}$$

with a singular representation matrix:

$$\Delta^{(k)}(r, \psi, \varphi)_{\alpha\alpha'} = e^{-ikr \cos(\psi - \alpha)} \delta(\alpha - \varphi, \alpha'), \tag{A2a}$$

and the calculation of the text can be carried out also using this form of $\Delta^{(k)}$. The form of $D^{(k)}$ given in the text is obtained by using, instead of $f(\alpha)$, its Fourier expansion:

$$f(\alpha) = \sum_m f_m e^{-im\alpha}. \tag{A3}$$

Substitution of this into (A1) gives

$$P_{r,\psi,\varphi}\sum_m f_m e^{-im\alpha} = \sum_m e^{-ikr\cos(\psi-\alpha)}e^{im(\varphi-\alpha)}f_m$$

$$= \sum_{mm'} D^{(k)}(r,\psi,\varphi)_{m'm}f_m e^{-im'\alpha}, \quad (A4)$$

where

$$D^{(k)}(r,\psi,\varphi)_{m'm} = (2\pi)^{-1}\int_0^{2\pi} e^{-ikr\cos(\psi-\alpha)}e^{im(\varphi-\alpha)}e^{im'\alpha}d\alpha$$

$$= (2\pi)^{-1}\int_0^{2\pi} e^{-ikr\cos\beta-i(m'-m)\beta}d\beta$$
$$\times e^{i(m'-m)\psi+im\varphi}. \quad (A5)$$

Because of

$$J_n(z) = (2\pi)^{-1}i^{-n}\int_0^{2\pi} e^{-iz\cos\beta-in\beta}d\beta, \quad (A5a)$$

this differs from the expression given in the text only by the factor $i^{m'-m}$ which can be eliminated by a similarity transformation. It is worth noting that the transition from (17b) to (17c), i.e., the evaluation of the second integral of (17b), can be best carried out using the form (A5a) for $J$.

A similar calculation is possible also in the three-dimensional case but it is more laborious and will not be given here. It is, essentially, contained in the Appendix to Grosjean's last article.

# Derivations of Onsager's Reciprocal Relations

E. P. Wigner

Journal of Chemical Physics 22, 1912–1915 (1954)

(Received May 24, 1954)

Two derivations of Onsager's reciprocal relations are presented in simplified form, and the range of their validity is discussed.

## 1. GENERAL STATEMENTS

THERE appear to be two groups of derivations of Onsager's reciprocal relations.[1] In the first of these, it is assumed that the macroscopic laws of motion hold for the averages of the macroscopic coordinates (such as temperature gradient, concentration gradient, etc.) even if their values are microscopic, i.e., in the usual range of fluctuations.[2] The second group assumes a definite statistical law for the path representing the system in phase space.[3] Both types of derivations will be reviewed briefly below.

Common to both derivations is the concept of macroscopic coordinates $Q_1, \cdots, Q_n$ which are functions of the microscopic coordinates $q_1, \cdots, q_N, p_1, \cdots, p_N$. The latter are the coordinates of the phase space $\Gamma$; the space of the former will be called $\gamma$. A typical set of macroscopic coordinates are the components of the temperature gradient. This can be defined as

$$(2/3k)\sum q_i(p_i^2/2m_i - 3kT/2),$$

and similar definitions apply to other macroscopic variables.

[1] L. Onsager, Phys. Rev. 37, 405 (1931); 38, 2265 (1931). These articles contain also extensive references to earlier literature. See also I. Prigogine, *Etude Thermodynamique des Phenomenes Irreversibles* (Editions Desoer, Liege, 1947) and S. de Groot, *Thermodynamics of Irreversible Processes* (North Holland Publishing Company, Amsterdam, 1951). For a review of the recent work on the subject see E. M. Montroll and M. S. Green, "Statistical Mechanics of Transport and Nonequilibrium Processes" (to be published in Annual Reviews of Physical Chemistry).

[2] For the first type of derivation, see in addition to Onsager's articles, particularly R. T. Cox, Revs. Modern Phys. 22, 238 (1950); 24, 312 (1952). This article uses quantum concepts, in particular the notion of the transition probability. It is assumed that the system changes its state by means of spontaneous quantum jumps. This assumption eliminates the induction period discussed in the text [after Eq. (14a)] and entails the validity of differential equations of the form (4a) for the macroscopic variables. It is, therefore, not entirely clear that the reasoning contained in these articles would remain valid in all details if quantum mechanical concepts were used. In quantum mechanics, the probability of a state decreases, initially, as $1-ct^2$ rather than $1-ct$ and the latter, linear, law takes over only after an induction period. Ideas similar to those of Cox have been put forward also by H. B. Callen and R. Greene, [Phys. Rev. 86, 702 (1952); 99, 1387 (1952)], based on work by H. B. Callen and T. A. Welton [Phys. Rev. 83, 34 (1951)] and by J. Jackson [Phys. Rev. 87, 471 (1952)]. These papers also assume the "law of transition probabilities."

[3] The basis of the second type of consideration seems to have been provided by J. G. Kirkwood [J. Chem. Phys. 14, 180 (1946)]. For the actual derivation see M. S. Green [Princeton dissertation (1951) and J. Chem. Phys. 20, 1281 (1952)] and N. Hashitzume [Progr. Theoret. Phys. 8, 461 (1952)]. It was developed with particular elegance by L. Onsager and S. Machlup [Phys. Rev. 91, 1505, 1512 (1953)].

There corresponds to every domain in $\gamma$ a domain in phase space but the ratio of the volume of the latter domain to that of the former is not constant. The integral of $e^{-H/kT}$ over the domain in $\Gamma$ space gives the unnormalized probability of the domain in $\gamma$ space to which it corresponds. This probability has a sharp maximum at the "equilibrium values" of the $Q$ which will be assumed to be 0. Second, it will be assumed that, for small but macroscopic values of the $Q$, the entropy of the system can be expanded into a quadratic function of the $Q$:

$$S(\mathbf{Q}) - S(0) = -k\sum s^{ik}Q_iQ_k = -k(\mathbf{Q}, s\mathbf{Q}), \quad (1)$$

in which $s = (s^{ik})$ is a symmetric positive definite matrix and $\mathbf{Q}$ stands for $Q_1, \cdots, Q_n$. The normalized probability of a unit volume element at $\mathbf{Q}$ is then given by

$$P(\mathbf{Q}) = Z^{-1}\exp(\mathbf{Q}, -s\mathbf{Q}). \quad (1a)$$

This is, by the connection between entropy and extension in $\Gamma$ space adopted here, the integral of $e^{-H/kT}$ over that part of $\Gamma$ space which corresponds to the unit volume element at $\mathbf{Q}$ in $\gamma$ space, divided by the integral $Z = \int e^{-H/kT}dpdq$ over the whole $\Gamma$ space. $Z$ can be defined also by the condition that

$$\int P(\mathbf{Q})d\mathbf{Q} = 1; \quad Z = \pi^{-n/2}|\det s|^{\frac{1}{2}}. \quad (1b)$$

One can easily verify, furthermore, that

$$2\int Q_iQ_kP(\mathbf{Q})d\mathbf{Q} = \sigma_{ik} \quad (1c)$$

forms the reciprocal matrix to $s$

$$\sigma = s^{-1}. \quad (1d)$$

Only $\sigma$ will occur in subsequent work so that (1c) can be taken as the definition of $\sigma$. The relation to the entropy is given, however, by (1) and (1d).

It will be assumed, finally, that the $Q$ contain the time variable only in even order, i.e., that they are invariant with respect to time inversion and that the same holds also for the whole system, i.e., its Hamiltonian. Let us denote then by $\Gamma_Q$ the domain in $\Gamma$ space which corresponds to a unit volume element at $\mathbf{Q}$ in $\gamma$. Similarly, $\Gamma_{Q'}$ corresponds to a unit volume element at $\mathbf{Q}'$. In proper units, the volumes of $\Gamma_Q$ and $\Gamma_{Q'}$ are equal to $P(\mathbf{Q})$ and $P(\mathbf{Q}')$. As time flows, the systems represented by the points in $\Gamma_Q$ will change. Let us denote by $\Gamma_{Q \to Q'}$ the part of the domain of $\Gamma_Q$ the points of which move, in the time interval $t$, into the domain $\Gamma_{Q'}$. Then

$$T_t(Q \to Q') = \text{volume of } \Gamma_{Q \to Q'}/\text{volume of } \Gamma_Q$$
$$= \text{volume of } \Gamma_{Q \to Q'}/P(Q) \quad (2)$$

is the transition probability from $Q$ to $Q'$. Since every point in $\Gamma_Q$ will arrive, after time $t$, in some $\Gamma_{Q'}$,

$$\int T_t(Q \to Q')dQ' = 1. \quad (2a)$$

Let us consider now the domain $\Gamma_{Q \to Q'}{}^*$ which can be obtained from $\Gamma_{Q \to Q'}$ by replacing all points $q_1, \cdots, q_N$, $p_1, \cdots, p_N$ of $\Gamma_{Q \to Q'}$ by $q_1, \cdots, q_N, -p_1, \cdots, -p_N$, or, more generally, by applying the operation of time inversion to $\Gamma_{Q \to Q'}$. Clearly, the volume of $\Gamma_{Q \to Q'}$ is equal to the volume of $\Gamma_{Q \to Q'}{}^*$.

$$\text{volume of } \Gamma_{Q \to Q'} = \text{volume of } \Gamma_{Q \to Q'}{}^*. \quad (3)$$

Furthermore, because of the invariance of the $Q$ with respect to time inversion, $\Gamma_{Q \to Q'}{}^*$ is also contained in $\Gamma_Q$. We note, finally, that the points which are at time 0 in $\Gamma_{Q \to Q'}{}^*$ were, at time $-t$, in $\Gamma_{Q'}$ just as the points which are at time 0 in $\Gamma_{Q \to Q'}$ will be, at the time $t$, in $\Gamma_{Q'}$. In fact, the points which are, at time 0, in $\Gamma_{Q \to Q'}{}^*$ formed, at time $-t$, the domain $\Gamma_{Q' \to Q}$. Hence, because of Liouville's theorem,

$$\text{volume of } \Gamma_{Q' \to Q} = \text{volume of } \Gamma_{Q \to Q'}{}^*. \quad (3a)$$

It follows (see Fig. 1) that the volumes of $\Gamma_{Q \to Q'}$ and of $\Gamma_{Q' \to Q}$ are equal, or because of (2)

$$P(Q)T_t(Q \to Q') = P(Q')T_t(Q' \to Q), \quad (3b)$$

i.e., that *the principle of detailed balance holds in $\gamma$ space.* This conclusion is based solely on the postulate of the invariance of the $Q$ and of $H$ with respect to time inversion and, of course, on Liouville's theorem and on the independence of $H$ from time.

## 2. ASSUMPTION OF THE MICROSCOPIC VALIDITY, ON THE AVERAGE, OF THE MACROSCOPIC EQUATIONS

It is in conformity with the idea of a relaxation phenomenon to demand, at least for macroscopic values of the $Q$, that the average values of the $Q$, at time $t$, be linear functions of the initial values

$$\int T_t(Q \to Q')Q_i' dQ' = \sum \epsilon(t)_i{}^k Q_k. \quad (4)$$

In fact, from a purely macroscopic point of view, one would expect (4) to hold even for infinitely short times, i.e., one would expect equations of the form

$$(d/dt)\bar{Q}_i(t) = -\sum l_{ik}\bar{Q}_k \quad (4a)$$

to be valid. This would not be consistent, however,

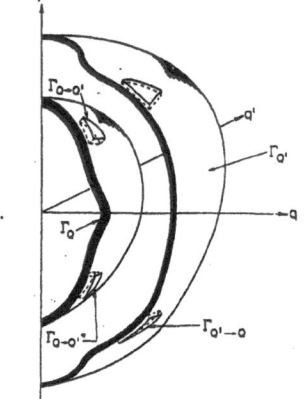

Fig. 1.

with the microscopic point of view; in fact, because of the invariance of the $Q$ with respect to time inversion, the average of each $dQ/dt$ over $\Gamma_Q$ vanishes. However, the integrated form (4) of (4a) can be expected to be valid after a certain induction period. This induction period $t$ is so long that $dQ/dt$ already undergoes substantial changes during this period, i.e., $|d^2Q/dt^2|t$ is, for most points of $\Gamma_Q$, much larger than $|dQ/dt|$. We shall use, below, only the integrated form (4).

However, the integrated form will be assumed to be valid for all, even microscopically small $Q$. The reciprocal relations then follow at once: multiplication of (4) with $P(Q)Q_l$ and integration yields, because of (1c),

$$\iint P(Q)T_t(Q \to Q')Q_l Q_i' dQ dQ' = \tfrac{1}{2}\sum \epsilon(t)_i{}^k \sigma_{kl}. \quad (5)$$

One sees by means of (3b) that the left side is symmetric in $l$ and $i$,

$$\iint P(Q)T_t(Q \to Q')Q_l Q_i' dQ dQ'$$

$$= \iint P(Q')T_t(Q' \to Q)Q_l Q_i' dQ dQ'$$

$$= \iint P(Q)T_t(Q \to Q')Q_i Q_l' dQ' dQ.$$

The last part follows by interchanging the variables $Q$ and $Q'$. It follows that the matrix $\epsilon\sigma$ is symmetric

$$\epsilon\sigma = (\epsilon\sigma)' = \sigma\epsilon' \quad \text{or} \quad \sigma^{-1}\epsilon\sigma = \epsilon'. \quad (5a)$$

The last form shows that (5a) holds also if any function of $\epsilon$ is substituted for $\epsilon$; (5a) embodies the reciprocal relations. Because of $\sigma = s^{-1}$, one can write it also in the form

$$s\epsilon = \epsilon's = (s\epsilon)'. \quad (5b)$$

It was first noted by Casimir[4] that the validity of the assumption, that (4) holds even for very small $Q$, is very difficult to assess. In fact, the significance of (4) for such values of $Q$ as are assumed in the course of ordinary fluctuations, is quite different from the macroscopic case. In the latter case, the $T_t(Q \rightarrow Q')$, as function of $Q'$, has a rather sharp maximum, the width of which is much smaller than its distance from the point $Q'=0$. If the $Q$ are microscopic to begin with, most of the systems represented by $\Gamma_Q$ will have gone through all the planes $Q_i'=0$ several times during the time interval $t$; after the time $t$ the system will have only a rather faint memory of its initial state $Q$. It is not at all clear that this memory should be subject to the same macroscopic laws which govern the values of the $Q$ in the macroscopic case. In the macroscopic case the value of the macroscopic variables at time $t$ is given by the right side of (4) for practically all systems represented by the points of $\Gamma_Q$.

The above point can be illustrated on the example of the stopping of a macroscopic particle by the viscosity of the medium in which it is moving. If the velocity of the particle is many times in excess of the equilibrium average value, a Stokes' flow will establish itself around it. The energy loss will be determined by the properties of this flow. On the other hand, if the energy is of the order $kT$, it is very likely that the atoms around it will not assume any ordered motion. One must recall that, in thermal equilibrium, there are no correlations between the velocities of any particles: $e^{-H/kT}$ is, as far as the velocities are concerned, a product of functions, each depending on one velocity only. Hence, on the average, there is no Stokes' flow; an ordered motion of the gas is contingent on the macroscopic body having moved, some time before the observation, with an even higher than its velocity at the time of observation.

The reader interested in mathematical implications may note that, according to (3b),

$$K(Q,Q') = P(Q)^{\frac{1}{2}} T_t(Q \rightarrow Q') P(Q')^{-\frac{1}{2}} \quad (5c)$$

is the kernel of a symmetric operator in the real Hilbert space of square integrable functions of the $Q_1, \cdots, Q_n$. The normalization property (2a) (which was not used in the above derivation) then expresses the fact that $P(Q)^{\frac{1}{2}}$ is a characteristic function of $K$ with characteristic value 1. The macroscopic equations (4) show that the subspace of the functions $Q_i P(Q)^{\frac{1}{2}}$ is invariant under $K$.

### 3. MOTION OF THE SYSTEM CAN BE REPRESENTED AS A BROWNIAN MOTION IN THE SPACE OF THE MACROSCOPIC VARIABLES

A rather different and very suggestive derivation of the reciprocal relations[2] is based on the explicit assumption that the point representing the state of the system in $\gamma$ space performs a Brownian motion. More specifi-

[4] H. B. G. Casimir, Revs. Modern Phys. 17, 343 (1945).

cally, it is assumed that

$$T_t(Q \rightarrow Q')$$
$$= C \exp - \{(Q,\alpha Q) + 2(Q',\beta Q) + (Q',\gamma Q')\}. \quad (6)$$

The matrices $\alpha$ and $\gamma$ are symmetric. They, as well as $\beta$, depend on $t$. The normalization condition (2a) can be satisfied only if $\alpha = \beta'\gamma^{-1}\beta$. Hence, writing $\epsilon = -\gamma^{-1}\beta$, one obtains

$$T_t(Q \rightarrow Q') = C \exp - (Q' - \epsilon Q, \gamma(Q' - \epsilon Q)). \quad (6a)$$

This shows that $\gamma$ must be positive definite. One sees that the most probable value of the $Q'$ after the time $t$, as well as their average value, is given by $Q' = \epsilon Q$. Hence, $\epsilon$ plays the same role in the present derivation as it did in the preceding section.

Clearly, (6a) constitutes a particularly simple assumption for $T_t$ which entails, furthermore, the validity of (4). Hence, the reciprocal relations will surely follow from (6a) and (3b), if (6a) is assumed to be valid for all $Q$. However, and this is the point of the present consideration (in addition to the suggestive nature of the picture which it presents), it will be necessary to assume (6a) only for macroscopic $Q$. Instead, the form (1a) will be explicitly assumed for $P(Q)$. One may recall, however, that (1a), as well as (3b), followed from the general principles of statistical mechanics.

Inserting therefore (1a) and (6a) into (3b), one obtains

$$\exp - \{(Q,sQ) + (Q' - \epsilon Q, \gamma(Q' - \epsilon Q))\}$$
$$= \exp - \{(Q',sQ') + (Q - \epsilon Q', \gamma(Q - \epsilon Q'))\}. \quad (6b)$$

One has to assume the validity of this equation only for a few different values $Q$, $Q'$, to conclude that

$$s + \epsilon'\gamma\epsilon = \gamma \quad \epsilon'\gamma = \gamma\epsilon. \quad (7)$$

Inserting the second equation into the first one gives $s + \gamma\epsilon^2 = \gamma$ or

$$\gamma = s(1 - \epsilon^2)^{-1}, \quad (7a)$$

and substitution of this into the second one

$$\epsilon's(1 - \epsilon^2)^{-1} = s(1 - \epsilon^2)^{-1}\epsilon. \quad (7b)$$

Since $1 - \epsilon^2$ and $\epsilon$ commute, this is equivalent to (5b), i.e., the reciprocal relations.

It might appear from the above that the assumption (6) is used only in a very mild form. This is indeed true as far as the necessity is concerned to use (6) for $Q$ which are close to equilibrium: It is not necessary to use (6) for such $Q$. However, at least one of the pairs $Q \rightarrow Q'$ or $Q' \rightarrow Q$ in (7) is such that the corresponding transition probability is extremely small. The assumption that (6) can be used in such a case is a rather stringent one.

Again, the reader may be interested to note the consequences of the Markoffian nature of the process, i.e., the consequences of the condition that $T_{t+t'}(Q \rightarrow Q'')$ is the integral, over $dQ'$, of $T_t(Q \rightarrow Q')T_{t'}(Q' \rightarrow Q'')$. This condition gives two equations for the matrices $\epsilon(t)$ and $\gamma(t)$ which occur in (6a). The first of these is obtained most easily by means of (4) which is valid also in this case

$$\epsilon(t+t') = \epsilon(t')\epsilon(t). \tag{8}$$

It follows from this that

$$\epsilon(t) = e^{-lt}, \tag{8a}$$

where $l$ is a positive definite matrix. The condition for $\gamma$ reads

$$\gamma(t+t')^{-1} = \gamma(t')^{-1} + \epsilon(t')\gamma(t)^{-1}\epsilon'(t'). \tag{9}$$

This is satisfied by the $\gamma$ given by (7a)

$$\gamma(t)^{-1} = (1-e^{-2lt})\sigma, \tag{9a}$$

if the reciprocal relations (5a) are assumed to hold. Both $l$ and $\sigma$ are independent of $t$. However, (9a) is not a consequence of (9) but follows from it only if (3b), (1a) and (1d) are also assumed to be valid. Actually, as was explained after (4a), none of the equations (8) or (9) holds for very small $t$.

It might appear from the above discussion that both derivations of Onsager's relations assume additional conditions which are hard to justify rigorously. Actually, as was pointed out by Casimir[4] (and in a particular case already by Onsager)[1], there are cases in which the deviation from the equilibrium decays exponentially even after it has become so small that it is in the normal range of fluctuations. According to Casimir, this will always be the case for processes which can be described by means of Boltzmann's distribution function $f(x,y,z,p_x,p_y,p_z; t)$ in the space of the variables of single atoms and if the deviation of $f$ from the equilibrium $f$ is sufficiently small at every point of this space to justify the usual linear approximation. In such a case, the first derivation[1,2] is fully justified. No similar argument can be made on the basis of considerations in $\Gamma$ space, because the deviation from the equilibrium distribution in $\Gamma$ space is large for all nonequilibrium systems.

# Lower Limit for the Energy Derivative of the Scattering Phase Shift

E. P. Wigner

Physical Review *98*, 145–147 (1955)

(Received December 10, 1954)

It is shown that the derivative of the scattering phase shift with respect to energy, $d\eta/dE$, must exceed a certain limit if the interaction of scattered particle and scatterer vanishes beyond a certain distance. This limitation of $d\eta/dE$ is, fundamentally, a consequence of the principle of causality; it is derived, however, from a property of the derivative matrix $R$.

## I

THE cross section and its angular dependence, as functions of energy, do not seem to determine in general the phase shifts uniquely.[1] It may be useful, therefore, to derive certain general rules about the energy dependence of phase shifts which may facilitate the choice between the apparently equivalent sets of phase shifts. The relations to be derived here are based, fundamentally, on what has come to be called "the principle of causality." It states that the scattered wave cannot leave the scatterer before the incident wave has reached it. However, the calculation to be carried out will make use of a single property of the derivative matrix $R$ which was given already by Eisenbud and the present writer.[2]

Before carrying out the very simple calculation, the general nature of the result will be illustrated by means of Eisenbud's interpretation of the energy derivative of the phase shift as time delay.[3] Let us consider, for sake of simplicity, a scattering center of radius $a$, i.e., assume that the incident particle behaves like a free particle outside of a sphere of this radius. Let us consider then an incident beam which is the superposition of two monoenergetic beams of energy $h(\nu+\nu')$ and $h(\nu-\nu')$, respectively. The corresponding wave numbers will be denoted by $k+k'$ and $k-k'$. Hence,

$$\psi_{\text{inc}} = r^{-1}\left(e^{-i(k+k')r - i(\nu+\nu')t} + e^{-i(k-k')r - (\nu-\nu')t}\right). \quad (1)$$

Both $k'$ and $\nu'$ are infinitesimally small so that (1) is a substitute for a wave packet,[4] the center of which is at the point where the two spherical waves of (1) are in phase, i.e., where

$$2k'r + 2\nu't = 0. \quad (1a)$$

Since $\nu'/k' = d\nu/dk$ is the velocity of the particle, the incident particle indeed moves with a velocity $v$ toward

the scattering center. If $\eta+\eta'$ and $\eta-\eta'$ are the phase shifts which correspond to the energy values $h(\nu+\nu')$ and $h(\nu-\nu')$, the outgoing wave will be

$$\psi_{\text{out}} = r^{-1}\big(e^{i(k+k')r - i(\nu+\nu')t + 2i(\eta+\eta')}$$
$$+ e^{i(k-k')r - i(\nu-\nu')t + 2i(\eta-\eta')}\big). \quad (1b)$$

The two waves of (1b) are in phase where

$$2k'r - 2\nu't + 4\eta' = 0,$$

i.e., where

$$r = -2\eta'/k' + (\nu'/k')t = -2d\eta/dk + (d\nu/dk)t. \quad (1c)$$

One sees that the outgoing wave is retarded by a stretch $2d\eta/dk$; it arrives at a point $r - 2d\eta/dk$ at the time it would have arrived at $r$ without the action of the scattering center. The causality principle as formulated in the first paragraph gives no upper value for the retardation $2d\eta/dk$: if the particle is temporarily captured by the scattering center, there is no reason for it not being retained an arbitrarily long time. However, the "retardation" cannot assume arbitrarily large negative values, in classical theory it could not be less than $-2a$. It will be seen that the wave nature of the particles does permit some infringement of the relation

$$d\eta/dk > -a. \quad (2)$$

It will be shown that, nevertheless, (2) is essentially preserved also in quantum theory. It does hold, in particular, for large $k$.

The relation (2) gives a simple physical interpretation to the qualitative behavior of the energy dependence of $\eta$. Close to resonances, where the incident particle is in fact captured and retained for some time by the scattering center, $d\eta/dk$ will assume large positive values. On the other hand, $d\eta/dk$ will be close to $-a$ at energy values at which the incident particle hardly enters the scatterer. One would expect (on the basis of Liouville's theorem or the completeness relations) that the two effects, on the whole, balance each other, i.e., that the integral of $d\eta/dE$ over the whole energy range is close to zero, at least if the scattering can be described by a nonsingular potential. This is indeed the case: $\eta=0$ at $E=\infty$ and $\eta=b\pi$ for $E=0$, where $b$ is the number of bound states.[5] Hence, if the cross section

---

[1] See, e.g., the articles of Fermi, Metropolis, and Alei, Phys. Rev. **95**, 1581 (1954); de Hoffmann, Metropolis, Alei, and Bethe, Phys. Rev. **95**, 1586 (1954); and of R. L. Martin, Phys. Rev. **95**, 1606 (1954) for the ambiguities in the case of pion-nucleon scattering.

[2] E. P. Wigner and L. Eisenbud, Phys. Rev. **72**, 29 (1947), see (i) on page 35.

[3] L. Eisenbud, dissertation, Princeton, June 1948 (unpublished).

[4] Instead of the superposition of only two monoenergetic waves, one can use a regular wave packet in this consideration; $r$ in (1a) and (1c) then becomes the coordinate of the center of mass of that wave packet.

[5] I am much indebted to Dr. V. Bargmann for this observation.

shows a resonance behavior, one will expect $\eta$ to decrease slowly between resonances and increase fast at resonances, increases and decreases almost exactly balancing if considered over the whole energy spectrum.

## II

In order to derive a rigorous minimum value for $d\eta/dk$, it will be necessary to assume that there is a radius $a$ outside of which the wave function is that of a free particle. This is a somewhat artificial assumption which will be hardly ever satisfied rigorously. One can, therefore, expect that the condition (4b) to be derived under this assumption will be violated occasionally. However, if $a$ is chosen reasonably large the violations will be uncommon and will extend over narrow energy intervals.

We define as internal region the inside of a sphere of radius $a$ and call the rest of space the external region. The wave function $\psi$ in the external region is then given by

$$r\psi = I(r) - e^{2i\eta}I^*(r), \qquad (3)$$

where $I(r)$ is the radial part of an incoming spherical wave (of arbitrary angular momentum), its conjugate complex $I^*(r)$ is the outgoing wave; $\eta$ is called the phase shift. For $\eta = 0$, $\psi$ must be regular at $r = 0$, which fixes the phase of $I$ except for $I$'s sign.[6] We shall normalize $I$ so that

$$I(r)I'(r)^* - I(r)^*I'(r) = 2i. \qquad (3a)$$

$I(r)$ then represents a wave with flux $m/h$. The prime denotes the derivative with respect to $r$. Both $I$ and $\eta$ also depend on the energy; we shall use, instead, the wave vector $k$ as variable and denote differentiation with respect to $k$ by a dot.

The value of the reciprocal logarithmic derivative $R$ of $\psi$, with respect to $r$, at $r = a$, becomes

$$R = \frac{I - e^{2i\eta}I^*}{I' - e^{2i\eta}I'^*}. \qquad (3b)$$

The $I$, $I'$, etc., without argument, denote the value of the corresponding expression at $r = a$. Calculation of $e^{2i\eta}$ from (3a) yields

$$e^{2i\eta} = (I - I'R)(I^* - I'^*R)^{-1}. \qquad (3c)$$

It may be remarked, parenthetically, that this expression remains valid if, in addition to scattering, transmutations are also possible. The left side has to be replaced then by the collision matrix, $R$ by the derivative matrix, and $I$, $I'$, etc., by diagonal matrices, the diagonal elements of which are the $I(a)$, $I'(a)$, etc., for the channel to which the row of the collision matrix refers.

---

[6] It is necessary, in collision theory, to fix also the sign of $I(r)$ in terms of the decomposition of the plane wave into spherical waves. This is not necessary, however, for the derivation of (4b).

## III

In the present case, all quantities in (3c) are numbers and logarithmic differentiation of (3c) with respect to $k$ gives

$$\dot\eta = \frac{1}{2i}\left[\frac{\dot I - \dot I'R - I'\dot R}{I - I'R} - \frac{\dot I^* - \dot I'^*R - I'^*\dot R}{I^* - I'^*R}\right]. \qquad (4)$$

The term proportional to $\dot R$ in (4) is

$$\frac{-I'(I^* - I'^*R) + I'^*(I - I'R)}{2i|I - I'R|^2}\dot R = \frac{\dot R}{|I - I'R|^2}. \qquad (4a)$$

The last part follows from (3a). The partial fraction expansion[2] of $R$ shows directly that $\dot R = (h^2k/m)dR/dE > 0$ for $k > 0$ so that omission of the term containing $\dot R$ will make the left side of (4) larger than the right side. Elimination of $R$ from the resulting inequality by means of (3b) yields

$$d\eta/dk = \dot\eta > \tfrac{1}{2}\operatorname{Re}[\dot I'I^* - \dot II'^* + (\dot II' - \dot I'I)e^{-2i\eta}]. \qquad (4b)$$

$\operatorname{Re}[\cdots]$ denotes the real part of the expression contained in the bracket. The computation leading from (4) to (4b) involves the use of (3a) and the equation obtained from it by differentiation with respect to $k$.[6]

It remains to point out that the partial fraction expansion of $R$, though obtained in reference 2 by direct calculation, has been shown[7] to follow also from the causality condition described at the beginning of this note. This establishes the relation between (4b) and the qualitative consideration based on Eisenbud's early work. The connection becomes even clearer if one inserts, into (4b), the expressions for $I$ for $l = 0$ or $l = 1$:

$$I_0(r) = k^{-\frac12}e^{-ikr}; \quad I_1(r) = k^{-\frac12}[(kr)^{-1} + i]e^{-ikr}. \qquad (5)$$

These give, in the former case,

$$\dot\eta_0 > -a + (2k)^{-1}\sin 2(\eta_0 + ka); \qquad (5a)$$

while one has, in the latter case,

$$\dot\eta_1 > -a + (k^2a)^{-1}[1 - \cos 2(\eta_1 + ka)] \\ - (2k)^{-1}\sin 2(\eta_1 + ka). \qquad (5b)$$

It will be noted that, as long as $ka \ll 1$, these equations actually entail $\dot\eta > 0$ for a reasonably large interval of $\eta$. However, at very low $k$, (5b) merely expresses the fact that $\eta_0$ and $\eta_1$ are proportional to $k$ and $k^3$, respectively, without limiting the proportionality constant.

Naturally, the proper limitation of $a$ constitutes the principal difficulty in using (4b) to select the actual

---

[7] In addition to Eisenbud's doctoral dissertation, see W. Schutzer and J. Tiomno, Phys. Rev. 83, 249 (1951) and, in particular N. G. Van Kampen, Phys. Rev. 89, 1072 (1953); 91, 1267 (1953). Also J. S. Toll, doctoral dissertation, Princeton University, 1952 (unpublished), and the more recent article of Gell-Mann, Goldberger, and Thirring, Phys. Rev. 95, 1612 (1954). This article also contains references to the early literature.

phase shifts among the sets which reproduce the cross section. An estimate of $a$ can be obtained only from an at least qualitative theory of the nature of the interaction. It is commonly believed, for instance, that the pion-nucleon scattering proceeds via absorption and re-emission of the pion by the nucleon. If this is the case, $a$ will be of the order of the pion Compton wavelength, i.e., $\hbar/\mu c$. However, it would be quite difficult to tell to what extent $a=2\hbar/\mu c$ is a permissible choice or whether it is necessary to assume $a=3\hbar/\mu c$ or an even larger $a$, giving less and less stringent forms to (4b). An alternative form of applying the relations of this paper, which might be somewhat more free of this ambiguity, would be to plot $R$, as calculated from (3b), against the energy and to judge whether any possible deviation of the $R$ obtained this way from a regular $R$ function can be blamed on having assumed a too low value for $a$.

# Review of Collision Theory

## E. P. Wigner

In: Transfert d'energie dans les gaz. Douzième conseil de chimie 1962 (R. Stoops., ed.). Interscience Publishers, New York 1963, pp. 211–239, 303–306, 308, 453–457, 515–516. Translated into Hungarian: Áttekintés az ütközések elméletérol. Fizikai Szemle *14*, 35–44 (1964)

Our appreciation of the significance of collision theory has increased immensely in the post-war years. This appreciation has not yet had an adequate effect on the textbooks for courses in quantum mechanics and the evident need for learning the subject has created a veritable flood of treatises and reviews. As a result, the same basic ideas and applications have been described many times and there is a certain repetitiousness in the literature. This will apply also to the present review, the purpose of which is, as I understand my assignment, to serve as a background for the deliberations of this conference. Collision theory is one of the pillars which support the structure of transport theory. However, one may as well admit at the outset that, from the point of view of first principles, this pillar is not fully integrated into the structure. Transport theory should be a many body theory both in classical and in quantum physics, i.e., one should at least start from the many body Hamiltonian and many body wave functions. The reduction of the many body problem to the problems we commonly consider is plausible, in fact practically surely correct in the case of a dilute gas. However, it has not been carried out rigorously to date [1] and it is surely impossible to carry it out rigorously except in the limiting case of a dilute gas.

The problem which I just touched upon will not be part of this review, which will deal, rather, solely with collision theory. It will consist of three parts : a shorter first part dealing with the fundamental aspects of collision theory, and a somewhat longer third part dealing with its methods and applications. Sandwiched between these two parts will be some remarks on R-matrix theory.

## I. FUNDAMENTAL CONCEPTS

The conceptual significance of the collision matrix, originally introduced by Wheeler [2], was recognized by Heisenberg [3]. In order to define the collision matrix, it is not necessary to give the wave function for those parts of the configuration space in which the colliding particles are in intimate contact. Rather, it suffices to specify the state of the system long before, and long after, the collision. Equivalently, one can specify the wave function asymptotically, i.e., for those parts of the configuration space in which the colliding particles, and the collision products, are well separated. In this region, the wave function for a stationary state will have two parts : that describing the incident beam and that the separating collision products. Let us denote the incident beam which contains particles of the type $c$ by $\psi_{ic}$. The letter $c$ describes the nature of the colliding particles (whether H and $Cl_2$, or H and an electron) and their state of excitation. The wave function in the aforementioned part of the configuration space then has the form

$$\psi_{ic} - \sum_{c'} S_{cc'}\psi_{oc'} \tag{1}$$

where $\psi_{oc'}$ is an outgoing wave, describing separating particles, the nature and state of excitation of which is characterized by the index $c'$. The quantities $S_{cc'}$ form the collision matrix. It has as many rows and columns as there are states $c'$ which can be produced by the collision of the particles characterized by $c$. It is a function of the energy E of the system and also the directions into which the particles which constitute the pairs $c$ and $c'$ travel. As far as the end result is concerned, the collision matrix describes the collision completely, and it describes it not only for the collision in which the particles $c$ collide but also for collisions in which any of the possible collision products of this collision collide. The row index $c$ can assume the same values which the column index $c'$ can assume.

Heisenberg's early publications do not provide a method for calculating the collision matrix without considering the whole process of the collision, i.e., the state vector at times at which the colliding systems are in intimate contact. As a matter of fact, the development immediately following Heisenberg's suggestion led first to Dyson's formula [4] for the S matrix

$$S = \sum_n (i\hbar)^{-n}\theta \int \ldots \int V(t_1) \ldots V(t_n)dt_1 \ldots dt_n/n!. \tag{1a}$$

212

Here $V(t)$ is the operator of the interaction, in the interaction (or Dirac) representation, between the colliding bodies; $\theta$ signifies that the time ordered product of the ensuing expression has to be taken. The integrations have to be extended over all time. In field theory, $V(t)$ is an integral of the interaction operator over all space so that the term on the right side of (1a) contains, effectively, a space-time integral over $n$ quartets of variables. Hence, (1a) can be written in such a way that its relativistic invariance is at once evident. Clearly, (1) is but another form of the Born series [5], i.e., the expansion of the scattering and collision amplitudes (which are essentially the matrix elements of S) into a power series with respect to the interaction operator. There is only one such power series and, even if they have different forms, two power series for the same amplitude must be equal term by term.

The principal significance of (1a) derives unquestionably from the fact that it lends itself to calculations, transformations and modifications in which the relativistic invariance is never lost sight of. However, (1a) became a milestone in another way also : it marks the renewal of the attempts to use the Born series both for the actual calculation of scattering and collision amplitudes, and also for the derivation of general theorems concerning these amplitudes. Prior to the establishment and detailed investigation of (1a), it was commonly believed that the first term of the Born series (the so-called Born approximation) may be useful for the calculation of cross sections in particularly simple cases, or that it may serve as a general orientation, that the use of the whole series would be, however, impracticable and may also yield incorrect results.

There is one particular transformation of (1a) which, though perhaps not of conceptual significance, is sufficiently important to be mentioned at this point. It represents a return from (1a) to a form more closely resembling the Born series. Since

$$V(t) = e^{iH_0t/\hbar} \, V \, e^{-iH_0t/\hbar} \tag{2}$$

the matrix element of $V(t)$ between two states $\psi_i = \int a(E) \, \psi(E, \alpha) dE$ and $\psi_f = \int a'(E') \, \psi(E', \alpha') dE'$ (where $\alpha$ refers to the quantum numbers which are needed, in addition to the energy, to characterize the state completely), becomes

$$<\psi_i|V(t)|\psi_f>$$
$$= \int e^{iE't/\hbar} \, a'(E')^* \, e^{-iEt/\hbar} \, a(E) dE dE' \, < E', \alpha'|V|E, \alpha>. \tag{3}$$

Hence, the integration over $t$ can be carried out and gives for the $n = 1$ term of (1a)

$$B_1 = (2\pi/i) \int \delta(E - E') a'(E')^* a(E) dE' dE < E', \alpha' |V| E, \alpha>. \quad (3a)$$

This is the first significant term of the Born series, the so-called Born approximation. Similarly, the integrations over all the variables $t_1, \ldots, t_n$ can be carried out in the $n$th term and one obtains for this term the limit $\varepsilon \to 0$ of

$$B_n = 2\pi i \sum_{\alpha_i} (-)^n \int dE_1 \ldots dE_{n-1} \, dE dE' \, a'(E')^* a(E) \, \delta(E - E')$$
$$<E', \alpha' |V| E_1 \alpha_1>$$

$$\frac{<E_1\alpha_1|V|E_2\alpha_2>}{E_1 - E - i\varepsilon} \quad \frac{<E_2\alpha_2|V|E_3\alpha_3>}{E_2 - E - i\varepsilon} \quad \cdots \quad \frac{<E_{n-1}\alpha_{n-1}|V|E\alpha>}{E_{n-1} - E - i\varepsilon}. \quad (4)$$

The expression $<\psi_i|S-1|\psi_f> = \Sigma B_n$ is called the Lippman-Schwinger series.[6] The individual terms $B_n$ for which (4) gives a convergent result are relativistically invariant if $V$ is the interaction of a relativistic field theory, but this is not evident any more in the form (4).

Heisenberg's suggestion to calculate the S matrix without considering the process of collision in detail was realized more nearly by the dispersion theoretic approach.[7] This was considered, in some detail, at last year's conference, Goldberger and Mandelstam having been the main reporters. Since that time, we have learned a great deal more about the Regge poles, which are also an outgrowth of dispersion theory. The subject has been summarized so often that I feel a bit embarrassed to give yet another review.

The most symmetric coordinates for the description of a collision are the four-momenta of the colliding particles and the *negative* four-momenta of the particles *resulting* from the collision. In collisions of two particles which result again in two particles, there are four such momenta which may be denoted by $p_1$, $p_2$, $p_3$, $p_4$. The time components of two of these are positive, those of the other two negative. The sum of the four-momenta

$$\Sigma p_i = 0 \quad (5)$$

because of the conservation laws for linear momentum and energy. Because of its relativistic invariance, S depends only on the invariants which can be formed from the $p_1$. Because of (5) and

$$p_i^2 = m_i^2, \quad (6)$$

where $m_i$ is the mass of the particle $i$, there are two independent invariants. (The square of a four-vector will always denote its Lorentz product with itself.) The most symmetric set of invariants are the Lorentz-lengths of the four vectors which contain two of the $p_i$ with positive, two with negative, sign. It is customary to write

$$[(p_1 + p_2 - p_3 - p_4)/2]^2 = (p_1 + p_2)^2 = (p_3 + p_4)^2 = s \quad (6a)$$

$$[(p_1 - p_2 + p_3 - p_4)/2]^2 = (p_1 + p_3)^2 = (p_2 + p_4)^2 = t \quad (6b)$$

$$[(p_1 - p_2 - p_3 + p_4)/2]^2 = (p_2 + p_3)^2 = (p_1 + p_4)^2 = u. \quad (6c)$$

The three invariants $s$, $t$, $u$ are not independent but obey the identity

$$s + t + u = \Sigma m_i^2 \quad (7)$$

The $s$, $t$, $u$ are the variables used most conveniently to formulate the dispersion relations. If the $p_i$ are the four four-vectors of an actual collision, one of the variables $s$, $t$, $u$ is positive, the two others negative. If $p_1$ and $p_2$ are the four-momenta of the colliding particles, $s$ is the positive one and gives the square of the energy in the center of mass coordinate system $(p_1{}^0 + p_2{}^0)^2$; $t$ and $u$ are negative squares of differences between final and initial momenta so that one of these is the square of the " momentum transfer ".

In order to formulate the dispersion relations, the definition domain of S($s$, $t$, $u$), in which $s$, $t$, $u$ are real and obey the preceding equations, is extended to all complex values of its variables. The dispersion relations are then postulates concerning the region of analyticity of S and its behavior as $s$, $t$ or $u$ go to infinity. In addition, S must be unitary for physically possible $s$, $t$, $u$, i.e., for real $s$, $t$, $u$ which satisfy the aforementioned requirements. It is surmised that these postulates, coupled with some information concerning the low or high energy ($s$, $t$, or $u$) behavior of S suffice to determine S completely.

Since we are interested principally in the low energy (that is, non-relativistic) behavior of the scattering cross sections, it is not necessary to use the full apparatus of dispersion theory and Mandelstam representation to obtain significant results. The variables commonly used for describing low energy scattering are the wave number $k$, giving the kinetic energy $\hbar^2 k^2/2\mu$ in the center of mass system ($\mu$ is the reduced mass of the colliding particles) and the

momentum transfer $p$, giving the scattering angle $\vartheta$ by $p = 2k\hbar \sin \vartheta/2$.
Hence, one can write

$$S(\vec{k}, \vec{k}') = \delta^3(\vec{k} - \vec{k}') + \delta(E - E')S(k, p). \qquad (8)$$

The postulate of the dispersion theory is, then, that for fixed, real $p$, the scattering function $S(k, p)$ is analytic in the upper half plane of the complex variable $k$, except that it can have poles on the imaginary axis corresponding to possible bound states. Further, $S(k, p)$ tends to the same value $S_1(p)$ no matter how $k$ tends to infinity in the upper half plane. The limit, $S_1(p)$, is given by the first term of the Born series. One has, further,

$$S(k, p) = S(k, -p) = S(-k^*, p^*)^*. \qquad (8a)$$

If there are no bound states, it follows from the behavior of S in the upper half plane that

$$S(k, p) = S_1(p) + \frac{1}{\pi i} P \int \frac{S(k', p)}{k' - k} \, dk' \qquad (9)$$

$$= S_1(p) + \frac{1}{\pi i} \int_0^\infty S(k + x, p) - S(k - x, p) dx/x$$

where $P\int$ denotes a principal value integral which is to be extended from $-\infty$ to $\infty$. Its meaning is evident from the second line of (9). Together with the requirement that S be unitary, (9) suffices to determine $S(k, p)$ if $S_1(p)$ is known. Gell-Mann has given a method to construct $S(k, p)$ from $S_1(p)$ on the basis of (9) and the unitary condition.[8]

Actually, (9) has been proved only for two particular types of potentials : (a) those which decrease with increasing distance faster than any exponential [9], and (b) for a sum of Yukawa potentials with arbitrary ranges and strengths.[10] It is known that (9) is not valid for certain other potentials.[11] It should be possible, however, to approximate any reasonable potential by a potential of the type (a) or (b) so that (9) should be of practical usefulness. To my knowledge, however, no practical use has been made of (9) in actual calculations of atomic collisions [12] so that we lack practical experience in the use of this relation.

The third important development of a conceptual nature is Regge's theory of complex angular momentum. The relation of this theory to the Mandelstam representation is not yet quite clear and, for reasons which will become evident, I wish to postpone its discussion to the next section. This section will deal with a semi-phenomenological theory in which I was very greatly interested some years ago and on which I wish to relate a few new results. Some of these are elaborations of Regge's theory.

## II. R-MATRIX THEORY

R-matrix theory [13] describes collisions in configuration space. It is necessary to subdivise this space into two regions: the so-called inner and external regions. In the latter, the colliding particles or the separating reaction products are either not interacting with each other any more, or their interaction can be described by a potential. The wave function and its normal derivative can be developed on the surface dividing internal and external regions into a set of orthogonal functions $\psi_c$; the expansion coefficients will be denoted by $v_c$ and $d_c$. The R matrix, which is a function of the energy, gives the linear relation between these sets of quantities

$$v_c = \sum_{c'} R_{cc'} d_{c'} . \tag{10}$$

The orthogonal set $\psi_c$ is so chosen that it describes a definite pair $c$ of colliding or separating particles. As a result, the wave function in the external region becomes

$$\Psi = \sum_c r_c^{-1} [a_c I_c(r_c) - b_c O_c(r_c)] \psi_c \tag{11}$$

where $I_c$ and $O_c = I_c{}^*$ are incoming and outgoing waves in the " channel " $c$; they depend on the energy and on the separation $r_c$ of the pair $c$. The factor $r_c^{-1}$ is introduced so that the condition that the radial functions $I_c$ and $O_c$ carry unit flux have the simple form

$$I_c \frac{d}{dr_c} O_c - O_c \frac{d}{dr_c} I_c = \frac{2im_c}{\hbar} \tag{11a}$$

where $m_c$ is the reduced mass of the pair $c$. The first essential equation of the theory is then the expression for the collision matrix

$$S = (I - I'R)(O - O'R)^{-1} \tag{12}$$

valid in this simple form if the surface between internal and external regions is so located that the wave function already vanishes at this surface in closed channels.[14]   The I, I′, O, O′ are diagonal matrices; the diagonal elements of I are $I_c(a_c)$ where $a_c$ is the separation of the pair $c$ at the surface dividing internal and external regions and I′, O = I*, O′ = I′* are similar matrices with diagonal elements $dI/dr_c$, O, $dO/dr_c$, all at $r_c = a_c$.   The second essential equation of the theory is the expression for the elements of the R matrix

$$R_{cc}' = \sum_\lambda \frac{\gamma_{\lambda c}\ \gamma_{\lambda c}'}{E_\lambda - E} \tag{12a}$$

where the $\gamma_{\lambda c}$ and $E_\lambda$ are energy independent constants, the so-called reduced widths and resonance energies.

R matrix theory has two principal weaknesses.   The first of these is the necessity to introduce a largely arbitrary dividing surface between internal and external regions.   Since the constants $\gamma_{\lambda c}$ and $E_\lambda$ depend to some extent on the location of this surface, they are not truly characteristic of the collision system.   This dependence of the $\gamma_{\lambda c}$ and $E_\lambda$ on the surface is, naturally, compensated by the dependence of the I, I′, O, O′ in (12) on the $a_c$ since S does not depend on the surface.   The second weakness became evident when the dispersion relations were developed.   These are sharper conditions on the analyticity of S than can be obtained from R matrix theory.[15]   The reason is that the collisions are described, in dispersion theory, by energy (or wave number $k$) and momentum transfer $p$, whereas R matrix theory is formulated in terms of energy (or $k$) and angular momentum.   The validity of dispersion theoretic relations made it evident that there must be interrelations between the $E_\lambda$ and $\gamma_{\lambda c}$ for various angular momenta.   However, prior to Regge's work, there appeared no means to discover such interrelations.

As to the first problem, the results of the arbitrariness of the surface dividing internal and external regions, there is no complete answer.   Several attempts [16] have been made to eliminate this arbitrariness, but they have achieved only partial success.   However, the investigation of this problem has led to a precise formulation of the so-called sum rules [17] which appeared only as an approximate relation in earlier calculations

$$\sum_\lambda \left( \sum_{\mu \neq \lambda} \frac{(\gamma_{\lambda c'}\gamma_{\mu c''} + \gamma_{\mu c'}\gamma_{\lambda c''})\gamma_{\lambda c}\,\gamma_{\mu c}}{E_\mu - E_\lambda} \right) = \frac{\hbar^2}{2m_c}\, \delta_{cc'}\, \delta_{cc''}. \tag{13}$$

218

The detailed derivation of (13) will be published elsewhere. It shows a certain similarity to the Born-Heisenberg-Jordan formulation of the Reiche-Thomas-Kuhn sum rules inasmuch as the sum on the left is conditionally convergent and contains oppositely equal terms. Hence, if the order of summation is reversed, i.e., if one sums first over $\lambda$, then over $\mu$, the result is oppositely equal to the right side of (13). Similarly, the commutation relation $pq - qp = \hbar/i$, which was derived from the optical sum rule, gives a non-zero (in fact infinite) value to the trace of a commutator which also contains pairs of oppositely equal terms. The point which is remarkable about (13) is that it is valid for all locations of the aforementioned surface in configuration space. It may be worthwhile to repeat here the equations which give the dependence of the constants $E_\lambda$, $\gamma_{\lambda c}$ on the location of the dividing surface, as characterized by the $a_c$. These are

$$\frac{\partial E_\lambda}{\partial a_c} = \frac{2m_c}{\hbar^2} (V_c - E_\lambda)\gamma_{\lambda c}{}^2 \tag{13a}$$

where $V_c$ is the potential energy acting between the pair $c$ at $r = a_c$ and

$$\frac{\partial \gamma_{\lambda c'}}{\partial a_c} = -\frac{m_c}{\hbar^2} \gamma_{\lambda c'}\gamma_{\lambda c}{}^2 + \frac{2m_c}{\hbar^2}(E_\lambda - V_c) \sum_{\mu \neq \lambda} \frac{\gamma_{\lambda c}\gamma_{\mu c}\gamma_{\mu c'}}{E_\mu - E_\lambda}. \tag{13b}$$

The second point which I wish to mention is a generalization of the R matrix formalism which appears to be relevant also for Regge's theory.[18] This will be sketched only for the one-dimensional case, i.e., for pure scattering. Let us consider the *double* characteristic value problem

$$\left(-\frac{\hbar^2}{2m}\frac{d^2}{dr^2} + V\right)\psi = Ev^2\psi + Lw^2\psi. \tag{14}$$

In the application of the method which one has in mind, $v = 1$, $w = 1/r$, $L = -(\hbar^2/2m)\, l\,(l+1)$. The only relevant limitation of the development which follows from the mathematical point of view is that E and L are the coefficients of positive definite self-adjoint operators; $\psi$ is a function of $r$ but depends also on E and L. The essential point is that, if we again define R by (10), that is, in the present case as the reciprocal logarithmic derivative of $\psi$ at $r = a$, the expansion (12a) remains valid for E (even though its coefficient is not 1 but $v^2$) and a similar expansion is valid in terms of L. In order to obtain this result, one has to modify the derivation of (12a) but slightly. One defines $X_\lambda$ by the equation

$$\left(-\frac{\hbar^2}{2m}\frac{d^2}{dr^2} - V - Lw^2\right)X_\lambda = E_\lambda v^2 X_\lambda \tag{14a}$$

and the boundary condition $(d/dr)X_\lambda = 0$ which is supposed to leave $v$ and $w$ self-adjoint. The $vX_\lambda$ are then the characteristic functions of a self-adjoint operator $v^{-1}Qv^{-1}$ where Q is the operator on the left of (14a) which is, however, a function of L. If this has only a discrete spectrum, any function, and in particular $v\psi$, can be expanded in terms of the $vX_\lambda$. If $v^{-1}Qv^{-1}$ has also a continuous spectrum — this is the case if $L > \hbar^2/8m$ in the example mentioned after (14) — an integral has to be added to the usual linear sum of characteristic functions.  However, we shall disregard here this possibility. The expansion coefficients $b_\lambda$ of the $vX_\lambda$ in the expansion of $v\psi$ then can be obtained by the usual manipulation of (14) and (14a) and become

$$b_\lambda = \left(\frac{\hbar^2}{2m}\right)^{1/2}\frac{\gamma_\lambda}{E_\lambda - E}\left[\int |vX_\lambda(r, L)|^2 dr\right]^{-1/2}\frac{d}{dr}\psi(a, L). \tag{15}$$

The factor in the brackets is necessary because we have not normalized the $X_\lambda$. The $\gamma_\lambda^2$ are the reduced widths of the usual theory; they are given by

$$\gamma_\lambda(L) = \left(\frac{\hbar^2}{2m}\right)^{1/2}\frac{X_\lambda(a, L)}{[\int |vX_\lambda(r, L)|^2 dr]^{1/2}}. \tag{15a}$$

Hence,

$$R(E, L) = \sum_\lambda \frac{[\gamma_\lambda(L)]^2}{E_\lambda(L) - E} \tag{16}$$

so that, for real L, the R(E, L) is an R function of E.  Similarly,

$$R(E, L) = \sum_\lambda \frac{\beta_\lambda(E)^2}{L_\lambda(E) - L} \tag{17}$$

and R(E, L) is also an R function of L for real E.  Furthermore, since, when deriving (17) one uses the same $X_\lambda$ as when (16) was derived,

$$E_\lambda[L_\lambda(E)] = E \quad \text{or} \quad L_\lambda[E_\lambda(L)] = L \tag{18}$$

for all E and L respectively.   Similarly

$$\gamma_\lambda(L)[\int|v X_\lambda(r, L)|^2 dr]^{1/2} = \beta_\lambda(E_\lambda(L))[\int|w X_\lambda(r, L)|^2 dr]^{1/2} \qquad (18a)$$

and there is a similar relation in which E and L are interchanged

Equation (18) simply means that the poles of R occur for the same E, L combinations in both expansions (16) and (17). Similarly, (18) gives the relation between the residues of R, depending on whether they are considered as functions of E or L. There are some further interesting properties of $E_\lambda(L)$ and of $L_\lambda(E)$ which will not be discussed, however. What one would like to obtain is, of course, a joint expansion of R in terms of E and L of which (16) and (17) are specializations. This does not seem to be quite easy, however. Such an expansion might well give the relation between the R(E) for different L, the absence of which was mentioned as the second failing of R matrix theory. At any rate, the preceding discussion can be amplified considerably and shows that at least some of the properties of the complex angular momenta of Regge are shared by a great variety of other parameters.

The question naturally arises concerning the minimum information, in two-particle scattering, which determines all the scattering cross sections. This question was solved, from one point of view, already by Levinson, Jost and Bargmann [19] who showed that if there are no bound states, the scattering cross section for any single angular momentum already determines the interaction potential. The situation is only slightly more complicated if there are bound states. Once the potential is determined, the scattering for any angular momentum can be obtained. The opposite question, that of obtaining the potential if the phase shifts for all angular momenta but only a single energy are given, was first raised and partially answered by Wheeler. [20]   The result recently obtained by Regge [21] in this connection is truly surprising : the scattering cross sections, for every value of the energy and angular momentum, are determined once the phase shifts for a single energy and any infinite arithmetical series of angular momenta are given. It is most unlikely that this follows from R matrix theory, even if it is extended in the way indicated. The reason is, probably, that R matrix theory remains valid if the interaction cannot be described by a potential but is a more general integral operator. Regge's result is, most likely, conditioned on a potential type interaction, i.e., an interaction operator which commutes with the position operators.

## III. METHODS AND APPLICATIONS

By methods of collision theory one means approximation methods which then can be applied to specific problems. It is almost tautological to say that no approximation method is perfect and it is usually not difficult to find a case in which it fails grossly. What is surprising, rather, is the wide applicability of approximation methods and the often almost uncanny instinct which some physicists have in the choice of these methods.[22] Some of the methods to be discussed were suggested by the mathematical structure of the basic theory, some were designed to meet particular conditions. There are three principal areas in which collision theory plays a decisive role : atomic collisions, nuclear collisions, and collisions between " elementary particles ". Methods have been developed to meet the conditions in each of these fields. These methods have found, however, applications also outside the area of their origin.

The choice of approximation methods to be discussed in a short review may be as difficult as the choice of a method suited to a particular problem. I am well aware that my discussion will appear dilettantish to those who have worked with the method under discussion. However, before proceeding with the discussion, it may be well to make another remark.

As a result of the development and increasing use of computing techniques, the straight two-particle scattering problem, for a definite angular momentum, can be solved for any definite angular momentum with sufficient accuracy. This assumes, of course, that the interaction between the particles is known. Hence, the only significant two-particle problem which remains concerns the high angular momenta and there will be some remarks on this question. The other significant problems are, then: scattering of composite, particles and transmutations. It would seem logical, then, to devote the rest of this report to these questions. However, a detailed discussion of these necessarily more specialized questions would render the remainder of this report too diffuse. Since, further, most of the points to be made can be well illustrated on simple two-body scattering, all examples have been taken from this most elementary field.

### (a) The Born approximation [5]

This is the first term of the Born series, discussed before, and it is surprising how close it often comes to reality. The most noted case in point is, of course, the unrelativistic Coulomb scattering.

222

It should be noted, nevertheless, that the Born approximation often does give very inaccurate results. This will be the case, in general, if the actual wave function differs considerably from a plane wave. This will occur if the interaction is strong enough to distort the incident plane wave appreciably. In regions of configuration space in which the potential is strongly repulsive, the actual wave function will have a much smaller value than the plane wave. Hence, the scattering and reaction due to the interaction in these regions will be much smaller than calculated for an undistorted plane wave, that is, using the Born approximation. A strongly attractive potential affects the amplitude of the wave function less drastically and is less likely to lead to gross errors. Nevertheless, section (d) contains implicitly an example for this opposite behavior : it calculates the Born approximation for the scattering by an $r^{-p}$ potential in order to obtain the scattering by a potential which behaves as $r^{-p}$ at infinity but has no singularity at $r = 0$. This latter singularity does not affect the Born approximation for the higher angular momenta although it would affect the actual scattering. However, in practice, strongly repulsive potentials are much more likely to cause a failure of the Born approximation than attractive ones: they give a too-large cross section. In fact, the most common sign for the inaccuracy of the Born approximation is that it gives a very large cross section.

Whether or not the cross section is very large can be judged more easily if one uses the Born approximation for the spherical waves with definite angular momenta. In this case there is a natural measure, the maximum possible cross section $4\pi(2l + 1)/k^2$, with which to compare the approximate cross section. This is given by

$$\sigma = 4\pi(2l + 1)m^2I_l^2/\hbar^4k^4 \qquad (19)$$

$$I_l = \int_0^\infty 2V(r)[j_l(kr)]^2dr \qquad (20)$$

the $j_l$ being $kr$ times the spherical Bessel functions: $j_0(kr) = \sin kr$, etc. One will expect the Born approximation to be valid or to fail, depending on whether or not $2mI/\hbar^2k$ is small as compared with one. However, this is not a safe criterion as it seems altogether not easy to give general criteria for the accuracy of the Born approximation. Thus, the scattering in a Coulomb field in a given direction is correctly given by the Born approximation even though the integral for I diverges in this case. In this case, the total cross section is infinite and it is this circumstance that is reflected in (20) becoming infinite. On the other hand, in the case of a rapidly·decreasing potential (19) is surely grossly inaccurate if $2mI/\hbar^2 k$ is of the order of 1 or larger. Conversely, if $2mI/\hbar^2 k$ is small, one can reasonably assume that (19) is accurate. There is one well-known exception to this : a resonance is not reproduced by (19) so that (19) surely can give far too small cross sections over narrow energy intervals.

A further point worth remembering is that, for a potential which is attractive in some, repulsive in other, regions, I, and hence the Born cross section, may well vanish for a definite $k$. This is a consequence of the real nature of I and is a rather general occurrence in Born approximation. The actual cross section, however, vanishes only under exceptional circumstances [23] and both theoreticians and experimental physicists have learned to view with suspicion the zeros in the cross sections, or in the angular distributions, predicted by the Born approximation and some of its generalizations.

It hardly needs to be repeated that the preceding discussion, restricted as it was to the simplest possible case, does not do justice to the immense area of applicability of the Born approximation. In fact, the principal value of Born's method may well lie in the ease with which it gives, for almost every collision problem, at least some guidance on what to expect. This is only indicated by the simplicity of the formula (19) which, however, could well serve to illustrate its strengths as well as its weaknesses.

### (b) Distorted wave Born method [24]

The Born approximation becomes inaccurate if the potential modifies the wave function considerably. Thus, in the hard sphere case ($V = \infty$ for $r < a$) the Born approximation gives an infinite cross section even if $a$ is small, so that the actual cross section is quite moderate. The reason is evident : the calculation assumes that the incident wave penetrates to $r < a$ and is very strongly scattered there. Evidently, the accuracy of the method could be improved considerably if at least the bulk of the effect of the potential could be taken into account to begin with. This is the idea of the distorted wave Born approximation.

The distorted wave method can be used to calculate scattering; more commonly it is used to calculate reaction cross sections. The scattering amplitude consists of two parts : the one inherent in the distortion, and the other due to the part of the interaction which is not taken into account by the distortion.  Thus, in the afore-mentioned case of a hard sphere, the $l = 0$ part of the distorted wave would be sin $k(r - a)$ for $r > a$, and this would completely describe the effect of the hard sphere.  However, if the potential extended beyond the $r = a$ point, this could be taken into account approximately by the first term of the power series development of the solution of the wave equation, in terms of the residual potential.

The most important areas of applications of the distorted wave method are, however, problems of reaction and transmutation. When calculating these, it is important not to permit the incident wave to have a considerable amplitude in regions of configuration space where the interaction causing the reaction is strong but where the incident wave can hardly penetrate.  If one wants to preserve symmetry between initial and final state (and hence preserve the principle of detailed balance), not only the wave function of the colliding particles but also that of the reaction products must be distorted, that is at least crudely adjusted to the prevailing potential.

The method as described in the preceding paragraph assumes that the interaction can be separated into a part which modifies the wave function of the colliding particles (and may have a similar effect on the wave function of the reaction products) but does not affect their identity, and another part which is responsible for the reaction or transmutation processes.  The simplest way to obtain an expression for the first part of the interaction is to calculate its expectation value for the wave function of the colliding particles (or of the reaction products) as function of their distance and use this expectation value as distorting potential.  The starting point for the Born series is, then, the scattering wave function in this potential.  From what was said about the hard sphere case it should be evident that this will give a considerable improvement over the undistorted wave method and will be, in fact, a good approximation if the second part of the interaction, leading to reactions, does not modify the wave function seriously.

If the wave function is considerably modified by the part of the interaction which leads to reaction or transmutation, it may be possible to bar the incident particles from the region which they cannot reach, because they are transmuted before, by introducing an absorptive (complex) potential.  This is a rather customary

225

procedure in nuclear physics and has proved very fruitful. [25] However, no motivation from first principles seems to be available. [26] It is clear, in particular, that there must be a consistency condition between the magnitude of the absorptive part of the distorting potential and the result of the subsequent calculation of reaction probability. Similarly, in order to preserve the principle of detailed balance, the reaction products should also move in a complex potential. Consistency conditions of this nature have been applied in the past only sparingly and it is even customary to treat the distorting potential as an adjustable quantity. This is justifiable if one wishes to obtain a first orientation, but not in a definitive theory.

The distorted wave Born approximation has played an important role in the theory of atomic collisions much before its recent use for nuclear phenomena. In particular, Zener's calculation [27] of the excitation of vibrations in molecules by atomic impact is based on this method, as are also various modifications and refinements of his work. In this case, the method seems to give satisfactory results which is, unquestionably, at least in part due to the fact that the cross sections are small, i.e., the transfer of translational energy to vibrations is often a notoriously slow process. However, the distorted wave Born method has been used also to calculate the excitation of rotational degrees of freedom and applied also to some other processes to which it is, probably, less applicable. It is almost as easily handled as the straight Born approximation and is unquestionably more accurate. The report of O.K. Rice reviews these applications.

### (c) Classical path, quantum interaction

This is, in fact, a variant of the previous method, based on the fact that in many cases the motion of the colliding bodies can be better described by classical theory than by a crude approximation to the solution of the distorted wave equation. Recently, this method, originally devised for describing chemical processes [28] has been applied with great success to transmutations occurring in the course of heavy ion collisions. [29]

If the paths of two atoms cross at a reasonable distance and with sufficient relative velocity so that the energy transfer cannot be expected to affect the paths substantially, the paths can be calculated by means of classical theory with reasonable accuracy. One then

226

obtains the interaction operator affecting the internal coordinates
of the two atoms as function of their position or, on the basis of
the preceding calculation, as function of time.   The transitions
caused by this time-dependent interaction can then be calculated by
means of first order perturbation theory.

The method is particularly adapted to calculate electronic excitations or vibrational or rotational transitions.   The use of classical
theory for the calculation of orbits is unquestionable if the impact
parameter is so large that the orbits could be traced out by the
motion of wave packets.   In order to have appreciable transition
probabilities in spite of large impact parameters, the interaction
energy must decrease slowly with increasing distance of the particles.
This condition is satisfied most accurately if one of the particles is
charged and its field, varying at the site of the other particle, induces
transitions therein.   The calculation of this effect is probably the
most important application of the method in the theory of heavy
ion collisions.   However, it can be expected also to give a good
description of the collision excitation of molecules with dipole
moments.

## (d) The WKB method

This method is intermediate between the distorted wave Born
approximation and the preceding method, inasmuch as it takes the
quantum character of the motion of the colliding particles at least
approximately into account.   The principles of the method, as
usually applied [30], are probably too well known to need further
comment.   There is a number of subtle points in connection with
the replacement of $l(l + 1)$ by $(l + 1/2)^2$ and the behavior of the
wave function near the classical turning point.   I would like to call
attention, however, to a modification of this method, using the semiclassical distribution function.[31]   It is possible, apparently, to avoid
some of the ambiguities in this way.   However, as far as known to
me, the method has been applied so far only to two-body scattering.

There is a remarkable difference between the classical and the
quantum theory of scattering which has not been much emphasized,
perhaps because it has no obvious visualization.   In classical theory,
the cross section is always infinite if the forces have an infinite range,
no matter how fast they go to zero with increasing distance.   The
reason is that two particles, no matter at what distance they pass by

each other, suffer some deflection. Hence, the cross section is always infinite for very small scattering angles, and it is often infinite also [32] for a scattering angle $\pi$. In quantum mechanics this is not the case; if the potential drops fast enough, the total cross section remains finite; for an even faster drop, even the differential scattering cross section at very small angles is finite. It may be of some interest to ascertain how fast the potential must drop so that these conditions are satisfied.

The total cross section is

$$\sigma_t = \sum_l \frac{4\pi}{k^2}(2l+1)\sin^2\eta_{il} \tag{21}$$

where $\eta_l$ is the phase shift of the partial wave with angular momentum $l\hbar$. The forward scattering cross section is

$$\sigma(O) = \frac{1}{2k^2}\sum_l \left|(2l+1)(e^{2i\eta_l}-1)\right|^2. \tag{22}$$

Hence, since at any rate $\eta_l \to 0$ for large $l$, the total cross section will be finite if the series $\Sigma (2l+1)\eta_l^2$ converges, the zero angle differential cross section will be finite if $\Sigma (2l+1)\eta_l$ converges.

It is not difficult to show that, for large $l$, the Born approximation gives a correct phase shift. This follows essentially from the magnitude of the Born scattering cross section.

$$\sigma_l = 4\pi(m^2/\hbar^4k^4)(2l+1)I_l^2 \tag{23}$$

where $I_l$ is given by (20). Hence, one will write

$$\sin \eta_l = -(m/\hbar^2k)I_l. \tag{24}$$

If this is very small, the condition for the validity of the Born approximation is satisfied and one can prove its actual validity also more rigorously. We shall use a potential $r^{-p}$ because the calculation is easiest for this potential. Actually, this potential gives infinite scattering, because of its singularity at $r=0$, but it can be shown that I would be asymptotically the same for all potentials which become $r^{-p}$ at large distances. Hence, we calculate [33]

$$\int_0^\infty \frac{[j_l(r)]^2}{r^p}\, dr = \frac{\pi(p-2)!\,(l-p/2+1/2)!}{2^p[(p/2-1)!\,]^2(l+p/2-1/2)!}$$

$$\approx \frac{\pi}{2}\left(\frac{p-2}{p/2-1}\right)(2l+1)^{-p+1}. \tag{25}$$

Hence, the total cross section will be finite if the potential decreases faster than $r^{-2}$; the differential cross section at zero angle will be finite if it decreases faster than $r^{-3}$. It is interesting that, in the present case, the Born cross section is essentially correct where it is largest, namely at zero angle. Actually, for a potential which goes to zero as $r^{-p}$, the differential cross section at a very small scattering angle $\vartheta$ is proportional to $\vartheta^{2p-6}$ as long as $1 \leq p < 3$.

## (e) Variational principles

The great accuracy of the variational calculation of discrete energy levels, particularly that of the normal state, suggests the use of this principle also for collision processes. The early investigations of Hulthén [34] and of Tamm [35] were concerned only with scattering but Lippmann and Schwinger [36] extended the method to arbitrary collision processes.

There is at least one great difference between the application of the variational principle to the calculation of the energy of the normal state and its application to obtain scattering cross sections: the sign of the error which is positive in the former case remains unknown in the latter. As a result, in the case of collisions it is, as a rule, not possible to tell which of the trial functions gave the more accurate result. This impairs the usefulness of the method to some extent : its effectiveness is so striking in many cases that it is surprising that it is not used more generally.

The problem is to find an expression for the phase shift, or the collision matrix element, or something equivalent, in terms of a trial wave function suggested by simplicity or physical intuition. This expression should be, furthermore, insensitive to deviations of the trial wave function from the accurate one. More precisely, it is demanded that the error in the wave function enter only quadratically into the error in the phase shift or collision matrix element. This can be always accomplished by adding the first correction to the original value of the physical quantity in question. The remaining error will then be of second order. Let us consider a simple scattering problem

$$(H - E)\psi = (-\hbar^2/2m)\psi'' + V\psi - E\psi = 0 \qquad (26)$$

and assume that $\psi$ assumes, for large $r$, the form $\sin kr + c \cos kr$. The constant $c = \tan \eta$ can then be regarded as the tangent of an angle $\eta$ which is called the phase shift because, for large $r$, the wave function becomes proportional to $\sin(kr + \eta)$. Similarly, let us consider a trial wave function $\psi_t$ which we hope is not too different from $\psi$. At large $r$, it should approach the form $\sin kr + c_t \cos kr$, corresponding to a trial phase shift $\eta_t$ for which $c_t = \tan \eta_t$. If $\psi_t$ is inserted into (26), the right side will not vanish; we denote it by $e_t$

$$(H - E)\psi_t = -(\hbar^2/2m)\psi_t'' + V\psi_t - E\psi_t = e_t. \tag{26a}$$

One then easily derives the equation

$$(\hbar^2/2m)k(c - c_t) + \int_0^\infty \psi e_t dr = 0. \tag{27}$$

The integral converges because $\psi_t$ is a solution of (26) for large $r$ so that $e_t$ goes to zero. One can give (27) the form

$$(\hbar^2/2m)(k \tan \eta - k \tan \eta_t) + \int_0^\infty \psi_t e_t dr = \int_0^\infty (\psi_t - \psi)e_t dr. \tag{27a}$$

Both integrals converge. The expression on the right side is indeed second order in the deviation of $\psi_t$ from $\psi$ so that it can be assumed to be small.

If one believes that the trial wave function is already reasonably close to the true wave function, (27a) gives directly

$$\tan \eta = \tan \eta_t - (2m/\hbar^2 k) \int_0^\infty \psi_t e_t dr \tag{27b}$$

this being correct to the second order in the deviation of $\psi_t$ from $\psi$. This is the simplest and most common way to use the variational expression (27a).

If one is not ready to assume that a trial wave function of sufficient accuracy can be guessed, one can use an analysis which shows a certain similarity to the application of the variational principle to the determination of the energy of stationary states. Nevertheless, there are significant differences which render the variational principle for scattering less effective than for the determination of the energy of bound states. The principal reason for this is that the latter principle represents, at least for the norm state, a minimum principle whereas the sign of the right side of (27a) can be positive as well as negative. Similar to the trial wave function for bound states, one assumes for the scattering problem also a trial function of considerable flexibility

$$\psi_t(r) = \sin kr - c_t f(r) \cos kr + \sum_1^N t_n u_n(r) \tag{28}$$

where $f(r)$ is some function, such as $1 - e^{-\varkappa r}$, which vanishes at $r = 0$ and approaches 1 at large $r$. The $u_n(r)$ also vanish at $r = 0$ so that $\psi_t$ satisfies the boundary condition $\psi_t(0) = 0$. At large $r$, the $u_n(r)$ approach zero since these functions are square integrable and the asymptotic behavior of $\psi_t$ becomes $\sin kr + c_t \cos kr$. The functional form of $f(r)$ and the $u_n$, as well as their number N, are dictated by physical intuition; the $u_n$ are introduced to permit the trial function to approach the real $\Psi$ also for small $r$.

If $\psi_t(r)$ happens to be the true $\psi$, both sides of (27a) will vanish. If the $t_n$ and $\eta_t$ differ from the values which give the true $\psi$ by small amounts, the two sides of (27a) will differ from zero by an expression which is second order in $\eta - \eta_t$ and the deviations of the $t_n$ from their real values. It is reasonable, therefore, to determine the $t_n$ and $\eta_t$ by the condition that

$$L = (\hbar^2/2m)k\, c_t - \int_0^\infty \psi_t e_t dr \tag{29}$$

be an extremum. This will, indeed, lead to correct equations whenever $\psi(r)$ can be expressed in the form (28) with suitably chosen $t_n$ and $\eta_t$.

If one calculates the integral in (29), one obtains terms which are second order in the $c_t$ and $t_n$, something like $a c_t^2 + c_t \Sigma a_n t_n + \Sigma a_{nm} t_n t_m$. In addition, however, one also obtains linear terms in $c_t$ and the $t_n$, coming from the $\sin kr$ term in (28), and even a term which is independent of the $c_t$ and $t_n$. Hence, if one sets the derivative of (29) with respect to the $c_t$ and $t_n$ equal to zero, one obtains N + 1 linear inhomogeneous equations for the $c_t$ and $t_n$ which have, in general, a unique solution. This is different from the case of the application of the variational principle to the discrete spectrum, that is to bound stationary states. This leads to homogeneous linear equations and a secular equation for E to assure the compatibility of these homogeneous equations. In the present case, of course, the energy need not be adjusted. Once the $t_n$ and $c_t$ are obtained, the corresponding value of $\tan \eta = c$ can be obtained by evaluating L and using (27)

$$\tan \eta = c = (2m/\hbar^2 k)\mathrm{L}. \tag{30}$$

Again, this is different from the stationary case for which the energy is obtained as $\int \psi_t \, \mathrm{H} \, \psi_t$. The calculation of the parameters ($c_t$ and $t_n$ in the scattering case) is in principle simpler for the scattering than for the stationary case. On the other hand, whereas one can show by the minimum principle that the addition of further terms to the trial wave function of the stationary case always improves the result (lowers the energy), the addition of further terms $t_{N+1} u_{N+1}(r), \ldots$ may render the value (30) for $\eta$ less accurate. One can, of course, say that the earlier, better value was more accurate fortuitously, but the fact is that it is difficult to judge the accuracy of the $\eta$, obtained by the variational method, unless $e_t = 0$ in which case, of course, $\eta = \eta_t$ and the trial wave function was accurate.

There are other variational methods [37] which permit an estimate of the error in $\eta$. These are based on the calculation of the expression

$$\int [(\mathrm{H} - \mathrm{E})\psi_t]^2 dr \tag{31}$$

which, of course, does obey a minimum principle. However, the calculation of this expression seems to be very cumbersome, just as in the stationary case. Under certain conditions (e.g., at very low energy), the L of (29) also obeys a minimum principle. [38] In such a case, it should be on a par with the variational method for the energy of stationary states. However, no systematic theory could be developed so far to obtain general variational expressions of this nature.

Kato, and also Spruch and Kelly [37] observed that the phase shift increases if the potential is decreased (or made more negative). This suggests the use of a method similar to Bazley's [39] to obtain upper and lower limits for the phase shift, by approximating V, both from above and below, by potentials for which the scattering problem can be solved accurately. Unfortunately, there are few such potentials.

232

The variational method, as all other methods, was discussed in its application to simple scattering. However, most of the preceding considerations can be extended also to problems of reaction and energy transfer. Unfortunately, one often has the impression that the variational principle is used more often to demonstrate its power than to solve new problems. It is surprising how often (27b) gives a close value for $\tau_i$, even for a very simple $\psi_t$.

## THE IMPULSE APPROXIMATION [40]

Dr. Karplus drew attention during the meeting to the impulse approximation. This approximation is particularly simple to apply to the collision of a single particle with a system of other particles, the latter forming a bound state. It was originally devised to describe the collision of a single nucleon with a complex nucleus. It could be used also to describe the collision of an electron, or of a proton, or perhaps even of an atom, with an atom or a molecule.

The impulse approximation is somewhat similar to the Born approximation inasmuch as it is particularly useful at high energies. There are $n$ terms in the potential energy between an incident particle and a complex containing $n$ particles. These correspond to the interactions of the incident particle with each of the particles of the complex. Each term in the potential energy gives rise to a term in the scattering amplitude of the Born approximation. This is true also of the impulse approximation, the only difference being that the terms of the Born approximation are approximate expressions for the scattering by a single particle whereas the impulse approximation uses the accurate expressions for these quantities. However in both approximations the total scattering amplitude is simply a sum of the scattering amplitudes due to the particles of the complex. One reason for this not giving the accurate value is that the incident particle can be scattered by two or more particles in succession and the second and higher members of the Born series indeed contain terms which correspond to such events. Another reason is that the particle which scattered the incident particle will continue to interact with the other particles of the complex also after having collided with the incident particle, and the impulse approximation does not take this interaction into account. In fact the properties of the bound state of the complex enter the calculation only by determining the momentum distribution of the constituents.  The total scattering amplitude is the superposition of the scattering amplitudes which correspond to the momentum spectra of the constituents.

It follows from the preceding description of the impulse approximation that it can be used only if the scattering of the incident particle by the individual constituents of the complex is already known.  However, if this is the case, it can be used just about as easily as the Born approximation.  Unfortunately, no very detailed and conclusive discussion of the impulse approximation is available but the consensus is that it gives useful results only if the energy of the incident particle is much higher than the binding energies of the constituents of the complex.  Only if this is the case, can the scattering by bound particles be treated as if they were free; as was mentioned before, the bound nature of the scattering particles manifests itself only by their having a spectrum of momenta, rather than a single momentum.

The impulse approximation has not played a significant role so far in the theories of atomic or molecular collisions.

## ACKNOWLEDGMENT

I wish to express my appreciation of the friendly criticisms of Drs. Blankenbecler, Engelsberg, Goldberger and Treiman.  They were most helpful.

234

# REFERENCES

(¹) Professor I. Prigogine informed me, during the Conference, that he and his collaborators are engaged in the study of this question.

(²) J.A. Wheeler, *Phys. Rev.*, **52**, 1107 (1937).

(³) W. Heisenberg, *Zeits. f. Physik*, **120**, 513, 673 (1943).

(⁴) F. Dyson, *Phys. Rev.*, **75**, 486, 1736 (1949).

(⁵) M. Born, *Zeits. f. Physik*, **37**, 803 (1926).

(⁶) B.A. Lippmann and J. Schwinger, *Phys. Rev.*, **79**, 469 (1950).

(⁷) The literature of dispersion relations is so large that one hardly can hope to mention even the most important papers. See, however, the reports of M.L. Goldberger and of S. Mandelstam to the Solvay Congress of 1961 (12ᵉ Conseil de Physique) or the forthcoming book, " Collision Theory ", by M.L. Goldberger and K.M. Watson.

(⁸) M. Gell-Mann, Proceedings of the Sixth Annual Rochester Conference (Interscience Publishers, New York, 1956), pages III - 30 ff.

(⁹) N.N. Khuri, *Phys. Rev.*, **107**, 1148 (1957).

(¹⁰) See the articles of R. Blankenbecler, M.L. Goldberger, N.N. Khuri and S.B. Treiman, *Annals of Phys.*, **10**, 60 (1960) and of A. Klein, *Jour. Math. Phys.*, **1**, 41 (1960) which actually prove even further-going relations.

(¹¹) The only potential for which it is easy to prove this seems to be $c/r^2$.

(¹²) See, however, E. Gerjuoy, *Rev. Mod. Phys.*, **33**, 544 (1961), particularly page 547; also E. Gerjuoy and N.A. Krall, *Phys. Rev.*, **127**, 2105 (1962).

(¹³) For reviews of this theory see A.M. Lane and R.G. Thomas, *Rev. Mod. Phys.*, **30**, 257 (1958) and E. Vogt's article, " Resonance Reactions, Theoretical " in " Nuclear Reactions " (North Holland Publishing Co., Amsterdam 1959).

(¹⁴) For the treatment of the case in which this condition is not met, see T. Teichmann and E.P. Wigner, *Phys. Rev.*, **87**, 123 (1952) or the first article of Reference (¹³).

(¹⁵) The meromorphic nature of S as function of the complex $k$ and the position of its poles can easily be established by the methods of R matrix theory, This has been done already by W. Schutzer and J. Tiomno, *Phys. Rev.*. **83**, 249 (1951); see also E.P. Wigner, *Rev. Mexicana de Fis.*, **1**, 91 (1952). However, whereas it is known from dispersion theory that, for constant momentum transfer, S goes to a constant value as $k \rightarrow \infty$ in the upper half plane, no similarly far-reaching statement could be derived so far from R matrix theory.

(¹⁶) See e.g., H. Feshbach, *Annals of Phys.*, **5**, 357 (1959), R.E. Peierls, *Proc. Roy. Soc.*, **A253**, 16 (1959), J. Humblet and L. Rosenfeld, *Nuclear Physics*, **26**, 529 (1961).

(¹⁷) T. Teichmann, *Phys. Rev.*, **77**, 506 (1950); see also Reference (¹⁴) or the second article of Reference (¹³).

(¹⁸) T. Regge, *Nuovo Cim.*, **14**, 951 (1959), **18**, 947 (1960).

(¹⁹) N. Levinson, *Kgl. Danske Vid. Selskab Mat. Fys. Medd.*, **25**, No. 9 (1949); V. Bargmann, *Phys. Rev.*, **75**, 301 (1949); R. Jost and W. Kuhn, *Kgl. Danske Vid. Selskab Mat. Fys. Medd.*, **27**, No. 9 (1953). For Yukawa potentials and their superpositions, see A. Martin and G. Targonski, *Nuovo Cim.*, **20**, 1182 (1961).

[20] J.A. Wheeler, *Phys. Rev.*, **99**, 630 (1955).

[21] T. Regge and G.A. Viano, *Nuovo Cim.*, **25**, 709 (1962).

[22] See, for instance, N.F. Mott and H.S.W. Massey, " The Theory of Atomic Collisions " (1st Ed. Oxford University Press, 1933, or 2nd Ed. 1952); H.S.W. Massey and E.H.S. Burhop, " Electronic and Ionic Impact Phenomena " (Oxford University Press, 1952).

[23] E.P. Wigner, *Proc. Nat. Acad.*, **32**, 302 (1946). Actually, for pure scattering in the absence of any reaction, the cross section due to a single angular momentum, as function of energy, can and does go through zero. However, the Born approximation shows a similar behavior also in the presence of reactions and not only for a single angular momentum but also for the total scattering cross section in a definite direction.

[24] See P.M. Morse and E.C.G. Stückelberg, *Ann. d. Phys.*, **9**, 579 (1931); H.S.W. Massey and C.B.O. Mohr, *Proc. Roy. Soc.*, **132A**, 605 (1931); **140A**, 613 (1933). See also H.S.W. Massey, *Rev. Mod. Phys.*, **28**, 199 (1956), and the books quoted in Reference [22].

[25] See, for instance, C.A. Levinson and M.K. Banerjee. *Ann. of Phys.*, **2**, 471, 499 (1957); **3**, 67 (1958). J.R. Lamarsh and H. Feshbach, *Phys. Rev.*, **104**, 1633 (1956). The optical model also has a long history which will not be related here. The modern work dates, however, from the paper of H. Feshbach, C.E. Porter and V.F. Weisskopf, *Phys. Rev.*, **96**, 448 (1954).

[26] The article of A.M. Lane, R.G. Thomas, and E.P. Wigner, *Phys. Rev.*, **98**, 693 (1955) attempts to derive the results, not the equations, of the optical model.

[27] C. Zener, *Phys. Rev.*, **37**, 556 (1931); **38**, 277 (1931); J.M. Jackson and N.F. Mott, *Proc. Roy. Soc.*, **137A**, 703 (1932). Also R.N. Schwartz, Z.I. Slawsky and K.F. Herzfeld, *Jour. Chem. Phys.*, **20**, 1591 (1952); R.N. Schwartz and K.F. Herzfeld, *ibid.*, **22**, 767 (1954).

[28] F. London, *Zeits. f. Phys.*, **74**, 143 (1932).

[29] See, for instance, Proceedings of the Second Conference on Reactions between Complex Nuclei (John Wiley and Sons, New York, 1960), particularly Sessions A, B, D.

[30] See, for instance, J. de Boer and R.B. Bird, *Physica*, **20**, 185 (1954) and J.A. Wheeler and K.W. Ford, *Annals of Phys.*, **7**, 259 (1959).

[31] Sang-Il Choi and J. Ross, *Journ. Chem. Phys.*, **33**, 1324 (1960). See, however, also *Proc. Nat. Acad.*, **48**, 803 (1962), by the same authors.

[32] See Reference [29], page 309.

[33] See G.N. Watson, " Theory of Bessel Functions " (Cambridge University Press, 1958), page 243 ff.

[34] L. Hulthen, *Ark. Mat. Astr. och Fysik*, **35A**, No. 25 (1948); W. Kohn, *Phys. Rev.*, **74**, 1763 (1948).

[35] I. Tamm, *J. Exp. Theor. Phys. U.S.S.R.*, **18**, 337 (1948), **19**, 74 (1949).

[36] B.A. Lippmann and J. Schwinger, *Phys. Rev.*, **79**, 469 (1950); M.L. Goldberger, *ibid.*, **82**, 757 (1951); **84**, 929 (1951); M. Kolsrud, *ibid.*, **112**, 1436 (1958). This article also contains an example in which the second term of the Born series is very much smaller than the first one (less than 4 %) but this is nevertheless very far from the exact value.

[37] T. Kato, *Progr. Theoret. Phys.*, **6**, 295, 394 (1950); L. Spruch and M. Kelly. *Phys. Rev.*, **109**, 2144 (1957).

[38] L. Rosenberg, L. Spruch and T.F. O'Malley, *Phys. Rev.*, **118**, 184 (1960).

[39] N.W. Bazley, *Proc. Nat. Acad.*, **45**, 850 (1959). The method used is due to A. Weinstein, *Mémorial des Sci. Math.*, Paris No. 88, 1937.

[40] G.F. Chew, *Phys. Rev.*, **80**, 196 (1950); Y. Fujimoto and Y. Yamaguchi, *Progr. Theor. Phys.*, **6**, 166 (1951); G.F. Chew and G.C. Wick, *Phys. Rev.*, **85**, 636 (1952); G.F. Chew and M.L. Goldberger, *Phys. Rev.*, **87**, 778 (1952).

# Discussion du rapport de M. Wigner

**M. Karplus.** — I should like to emphasize Prof. Wigner's reference to the way properties of the bound state are treated in the impulse approximation. Because of the binding, the single particle in the bound system which acts as target is not in a state of known linear momentum (at rest, for example). Instead it is in a superposition of momentum states, a superposition which is described by the fourier transform of the target particle's spatial wave function. The total scattering amplitude from this one target particle, then, is obtained by superposing the amplitudes which correspond to the various momenta in the momentum spectrum of the target particle, each weighted by the amplitude for occurrence of this momentum. The total amplitude for scattering by the bound system is finally obtained by adding the impulse approximation amplitudes for each of the particles in the system considered as target. As Prof. Wigner has mentioned, many other important effects of binding are neglected.

The impulse approximation in nuclear physics has been particularly useful when the momentum transferred to one target particle in the collision is large compared to the momenta in the initial distribution. In atomic and molecular problems this relationship usually does not hold.

**M. Shuler.** — Prof. Wigner has just given us a beautiful and lucid talk on scattering theory. As he has pointed out, all the successful applications are in the fields of quantum electrodynamics, nuclear physics and electron-atom (or molecule) scattering. As far as I know from my own work and from the literature, there have been essentially no successful applications of modern quantum scattering theory to inelastic molecular collisions such as, for instance, the dissociation reaction

$$AB + M \rightarrow A + B + M.$$

I wonder if one of the practical problems in the calculation of inelastic atomic and molecular scattering is not the absence of a convenient and reasonable expansion parameter. In most of the successful applications of scattering theory, such as scattering of electrons or photons from atoms or molecules, there is such an expansion parameter — either the mass ratio of the scattering centers ($m_e/m_{atom}$) or an energy ratio. I would appreciate to hear Prof. Wigner's views on that.

**M. Wigner.** — Perhaps in my desire to report on the general aspects on collision theory, I have underemphasized the theoretical treatment of phenomena of particular interest to our conference. As for these problems, interesting results have been obtained not only on the theory of the excitation of vibrational motion but also on the electronic excitation by atomic collisions and a number of other phenomena which I am sure will be discussed in the course of our conference.

As to the main point made by Dr. Shuler, I agree fully. Except for the very simplest systems, all successful approximation methods are based on the circumstance that some parameter or quantity is small. In addition to those enumerated by Dr. Shuler, I might call attention again to the case in which the only process of high probability is that of scattering. Furthermore, it may be possible to treat this accurately. All the other processes, such as vibrational or electronic transitions, are then unlikely and their probability is effectively Dr. Shuler's small parameter. I realize that I have now restated the conditions for the validity of two of the approximation schemes which I discussed : the distorted wave Born approximation and the classical path, quantum interaction method. However, the validity of the other approximation methods, with the possible exception of that of the variational method, can also be shown to depend on the small value of some parameter.

As to Dr. Shuler's first question, it is true that the most recent theoretical treatment of the association reaction with which I am familiar, that of Keck, is not based on collision theory, but uses the transition state method. As a matter of fact, I consider it the main result of the transition state method that it shows that rate processes can be calculated without a detailed description of the collisions. This was so far possible only in classical, that is non-quantum, theory and quantum effects could be taken into account only approximately. Theoretically, this is a serious limitation in our treatment of association reactions. In practice, the limitation is less significant because quantum effects are, as a rule, not very important in association reactions.

**M. Widom.** — The Born series goes in inverse powers of Planck's constant. One knows the practical usefulness, however, of expressions for the cross-section which go in increasing powers, the leading term of which is the classical theory. The question then arises, whether the analytical properties of such series might not lead to results which sumplement in a valuable way the conclusions based on the analytical properties of the Born series.

**M. Wigner.** — Dr. Widom's suggestion seems to me to be a very interesting one. As a matter of fact, some hint in the direction which he mentions is contained in the written report which I presented, even though I omitted this part in my verbal presentation. $\hbar^2$ is the coefficient of a positive definite operator in the Hamiltonian so that some of the remarks contained in the section on R matrix theory apply to it. However, this point has not been worked out to date. Further, lest too much significance be attributed to the considerations in my report to which I am referring, I wish to re-emphasize that the behavior of S in the complex plane at infinity, which plays such an important role from the point of view of the dispersion relations, does not follow from R matrix theory.

# Quantum-Mechanical Distribution
# Functions Revisited

E. P. Wigner

In: Perspectives in Quantum Theory; Essays in Honor of Alfred Landé
(W. Yourgrau, A. van der Merwe, eds.).
M.I.T. Press, Cambridge, MA 1971, pp. 25–36

## I.

There seems to be a revival of interest in the formulation of quantum mechanics in terms of distribution functions, i.e. functions $P(q, p)$ of positional and momentum coordinates $q$ and $p$. The function $P$ represents a quantity which is, in the classical limit, the phase-space distribution function; it gives the probability that the coordinates and momenta have the values $q$ and $p$. The existence and properties of such distribution functions are closely related to the question of the possibility of the reformulation of quantum mechanics in terms of classical concepts and are, therefore, close to Landé's interests. The exact definition of $P$ when the quantum conditions, in particular the uncertainty relations, play a role is not clear. One wishes to define, for every normalized state vector $\psi$, a corresponding distribution function $P$. This is postulated to satisfy certain conditions which the classical phase-space distribution also satisfies, but these conditions do not determine $P$ completely. The choice between the $P$ which satisfy a set of specified conditions is made aiming at greatest mathematical simplicity and in such a way that the $P$ chosen becomes a useful tool for carrying out calculations of quantities which are not easily obtainable otherwise.

Two reasonable conditions which can be imposed on $P$ are: (a) that it be a Hermitian form of $\psi(x)$, i.e. that the $P$ which describes the same state as the wave function $\psi$ be given by

$$P(q, p) = (\psi, M(q, p)\psi), \qquad [4.1]$$

where $M$ is a self-adjoint operator depending on $p$ and $q$ and (b) that $P$, if integrated over $p$, give the proper probabilities for the different values of $q$ as

$$\int P(q, p) \, dp = |\psi(q)|^2, \qquad [4.2a]$$

and, if integrated over $q$, give the proper probabilities for the momentum as

$$\int P(q, p) \, dq = (2\pi\hbar)^{-1} \left| \int \psi(x) e^{-ipx/\hbar} \, dx \right|^2. \qquad [4.2b]$$

All integrations, unless otherwise noted, are to be extended over the whole range of the variables, from $-\infty$ to $+\infty$. It may be observed also that, even though the conditions which $P$ should satisfy are formulated above for a one-dimensional problem, they can be easily extended to an arbitrary, many-dimensional one by making $q$ and $p$ vectors with as many components as has the configuration space of the $\psi$ for which $P$ should give an alternate description.

A somewhat milder form of condition (b) is that $P$ should give the proper expectation values for all operators which are sums of a function of $p$ and a function of $q$ as

$$\iint P(p, q)[f(p) + g(q)] \, dp \, dq = \left( \psi, \left[ f\left(\frac{\hbar}{i}\frac{\partial}{\partial x}\right) + g(x) \right] \psi \right). \qquad [4.2]$$

It will suffice to use condition (b) in the form [4.2].

A third, very natural, condition on $P$ would be that $P$ is nonnegative for all values of $p$ and $q$:

$$P(p, q) \geq 0. \qquad [4.3]$$

It has been stated in an article by the present writer where a $P$ satisfying conditions (a) and (b) was given[1] (actually for the many-dimensional case) that these conditions are incompatible with [4.3]. It has been variously suggested to the writer that this proof be published, and this will be the first subject of the present article. It is sufficient, for this purpose, to consider the one-dimensional case.

In order to carry out the proof it will be shown that the assumption that a $P$ satisfying all three conditions [4.1], [4.2], and [4.3] can be defined for every $\psi$ leads to a contradiction. Actually, in order to obtain the contradiction, it will suffice to consider such $\psi$ which

27  QUANTUM-MECHANICAL DISTRIBUTION FUNCTIONS

are linear combinations $a\psi_1 + b\psi_2$ of any two fixed functions such that $\psi_1$ vanishes for all $x$ for which $\psi_2$ is finite. We start with the following lemma:

LEMMA 1

If $\psi(x)$ vanishes in an interval $I$, and if $g(q)$ is zero outside this interval, and nowhere negative therein, one has for the $P$ corresponding to the $\psi$ above

$$\int P(q,p)g(q)\,dq = 0 \tag{4.4}$$

for all $p$ (except for a set of measure zero). This follows from [4.2] with $f = 0$: the integral of [4.4] with respect to $p$,

$$\iint P(q,p)g(q)\,dp\,dq = [\psi, g(x)\psi] = 0, \tag{4.4a}$$

vanishes because the right side of [4.2] vanishes. However, the integrand with respect to $p$, that is the left side of [4.4], is nonnegative for the $g$ postulated as long as [4.3] holds for $P$. It follows that the integrand with respect to $p$ must vanish except for a set of $p$ of measure zero. Since, furthermore, [4.4] is valid for every function $g(q)$ which satisfies the conditions of Lemma 1, we can conclude in a similar way that

LEMMA 2

If $\psi(x)$ vanishes in an interval $I$, the corresponding $P(q, p)$ vanishes (except for a set of measure zero) for all values of $q$ in that interval.

Let us consider now two functions $\psi_1(x)$ and $\psi_2(x)$ which vanish outside of two nonoverlapping intervals $I_1$ and $I_2$, respectively. Because of [4.1], the distribution function $P_{ab}(q, p)$ which corresponds to $a\psi_1 + b\psi_2$ will have the form

$$P_{ab}(q,p) = |a|^2 P_1 + a^*b P_{12} + ab^* P_{21} + |b|^2 P_2. \tag{4.5}$$

Setting $b = 0$, we note that $P_1$ is the distribution function for $\psi_1$ and similarly $P_2$ is the distribution function for $\psi_2$. Let us consider

[4.5] for the $q$ outside the interval $I_1$. Since $P_1$ vanishes almost everywhere for such $q$, the distribution function [4.5] cannot be positive for all $a$ and $b$ unless both $P_{12}$ and $P_{21}$ vanish (except for a set of measure zero in $q$ and $p$) if $q$ is outside $I_1$. The same can be concluded as long as $q$ is outside $I_2$. Hence, we have, instead of [4.5], almost everywhere

$$P_{ab}(q,p) = |a|^2 P_1(q,p) + |b|^2 P_2(q,p). \qquad [4.6]$$

This means that the distribution function $P_{ab}$ is (almost everywhere) independent of the complex phase of $a/b$. This seems, at least intuitively, absurd if $P_{ab}$ is to give the proper momentum distribution, i.e. is to satisfy [4.2b].

In order to demonstrate this, let us denote the Fourier transforms of $\psi_1$ and $\psi_2$ by $\varphi_1(p)$ and $\varphi_2(p)$. Equation [4.2b] then reads

$$|a|^2 \int P_1(q,p)\,dq + |b|^2 \int P_2(q,p)\,dq$$
$$= |a|^2|\varphi_1(p)|^2 + |b|^2|\varphi_2(p)|^2 + 2Re\ ab^*\varphi_1(p)\varphi_2(p)^*. \quad [4.7]$$

Since this must be valid for all $a$, $b$, we must have identically in $p$

$$\varphi_1(p)\varphi_2(p)^* = 0. \qquad [4.7a]$$

This is, however, impossible since $\varphi_1$ and $\varphi_2$, being Fourier transforms of functions restricted to finite intervals, are analytic functions (in fact, entire functions) of their arguments and cannot vanish over *any* finite interval. This then completes the proof that no nonnegative distribution function can fulfill both postulates (a) and (b).

## II.

Since a bilinear distribution function which gives the proper probabilities for position and momentum (i.e. satisfies the two conditions of the preceding section) cannot be everywhere nonnegative, this last requirement will be abandoned for the rest of this article.[2] The question then arises of what other natural requirements on the distribution function can be made and to what

29  QUANTUM-MECHANICAL DISTRIBUTION FUNCTIONS

degree these determine the correspondence between wave functions and distributions. When investigating this question, it will be convenient to relax the mathematical rigor of the preceding section. In particular, the operator in (I) will be replaced by a kernel so that we will have

$$P(q, p) = \iint K(q, p; x, x')\psi(x)^*\psi(x')\, dx\, dx'. \qquad [4.8]$$

The kernel $K$ will be permitted to be a singular function, though.

The first and most natural requirement on the correspondence [4.8] is that it be Galilei invariant. This means, in one dimension, first, that it be displacement invariant

$$P(q + a, p) = \iint K(q, p; x, x')\psi(x + a)^*\psi(x' + a)\, dx\, dx'.$$

Substituting $x$ and $x'$ for $x + a$ and $x' + a$ on the right side, and considering that [4.8] must be valid for all $a$, one easily infers that $K$ can depend, in addition to $p$, only on the differences $q - x$ and $q - x'$, or on $2q - x - x'$ and $x - x'$, not on all three quantities $q, x, x'$. This was to be expected.

The second requirement of Galilei invariance is that if $\psi(x)$ is replaced by $\psi(x)e^{ip'x/\hbar}$, the new distribution function becomes $P'(q, p) = P(q, p - p')$

$$P'(q, p) = \iint K(q, p; x, x')\psi(x)^*\psi(x')e^{ip(x'-x)/\hbar}\, dx\, dx'$$

$$= P(q, p - p') = \iint K(q, p - p'; x, x')\psi(x)^*\psi(x')\, dx\, dx'. \qquad [4.9]$$

Since this must be valid for all $\psi$, we have

$$K(q, p; x, x')e^{ip'(x'-x)/\hbar} = K(q, p - p'; x, x'). \qquad [4.10]$$

Substituting $p' = p$, one finds

$$K(q, p; x, x') = e^{ip(x-x')/\hbar}K(q, 0; x, x')$$
$$= e^{ip(x-x')/\hbar}K^0(2q - x - x', x - x'). \qquad [4.11]$$

Since $K$ depends only on the differences $2q - x - x'$ and $x - x'$, a function $K^0$ of these can be substituted for $K(q, 0; x, x')$. We now have

$$P(q, p) = \iint K^0(2q - x - x', x - x')e^{ip(x-x')/\hbar}\psi(x)^*\psi(x')\, dx\, dx'. \qquad [4.12]$$

We shall find, next, the restrictions on $K^0$ implied by [4.2] and [4.2a]. We have, according to [4.2a],

$$\int P(q, p)\, dp$$
$$= \iiint K^0(2q - x - x', x - x')e^{ip(x-x')/\hbar}\psi(x)^*\psi(x')\, dx\, dx'\, dp$$
$$= 2\pi\hbar \int K^0(2q - 2x, 0)|\psi(x)|^2\, dx = |\psi(q)|^2, \qquad [4.13]$$

from which it follows that

$$2\pi\hbar K^0(2q - 2x, 0) = \delta(q - x), \qquad [4.14]$$

or

$$K^0(q, 0) = \frac{1}{\pi\hbar}\delta(q). \qquad [4.14a]$$

Similarly, [4.2b] gives

$$\int P(q, p)\, dq$$
$$= \iiint K^0(2q - x - x', x - x')e^{ip(x-x')/\hbar}\psi(x)^*\psi(x')\, dx\, dx'\, dq$$
$$= (2\pi\hbar)^{-1} \iint \psi(x)^*e^{ipx/\hbar}\psi(x')e^{-ipx'/\hbar}\, dx\, dx', \qquad [4.15]$$

from which it follows that

$$\int K^0(2q - x - x', x - x')\, dq = (2\pi\hbar)^{-1}, \qquad [4.16]$$

or

$$\int K^0(q, x)\, dq = (\pi\hbar)^{-1}. \qquad [4.16a]$$

31   QUANTUM-MECHANICAL DISTRIBUTION FUNCTIONS

We can next introduce the requirement that the correspondence [4.8] be invariant with respect to space and time reflections. The former requirement demands that a simultaneous change of sign of all the variables of $K$ leave it invariant. This means that $K^0$ is left unchanged if the signs of both of its variables are changed. Time-inversion invariance demands that replacement of $\psi$ by $\psi^*$ change $p$ into $-p$. It follows that

$$K(q, -p; x, x') = K(q, p; x', x) = K(q, p; x, x')^*. \qquad [4.16b]$$

The last part of [4.16b] follows from the Hermitian nature of $K$, required to give a real $P$. The first part of [4.16b] renders $K^0$ even in its second variable—it is therefore even in both of its variables. The second part of [4.16b] renders $K^0$ real. Therefore, $K^0$ must be a real and even function of both of its variables. Evidently, these conditions do not fully determine $K^0$ so that a great deal of arbitrariness remains in the definition of $P$.

A further condition which can be imposed is that the correspondence between $P$ and the wave function be symmetric in $p$ and $q$ (apart from the sign of $i$), so that one has, in addition to [4.12], also

$$P(q, p) = \iint K(p, q; k, k')^* \varphi(k)^* \varphi(k') \, dk \, dk', \qquad [4.17]$$

where

$$\varphi(k) = (2\pi\hbar)^{-1/2} \int \psi(x) e^{-ikx/\hbar} \, dx. \qquad [4.17a]$$

This condition will be satisfied if

$$\iint K(q, p; x, x') \psi(x)^* \psi(x') \, dx \, dx'$$

$$= \iiiint K(p, q; k, k')^* (2\pi\hbar)^{-1} \psi(x)^* \psi(x') e^{i(kx - k'x')/\hbar} \, dx \, dx' \, dk \, dk'. $$

$$[4.18]$$

It follows that

$$2\pi\hbar K(q, p; x, x') = \iint K(p, q; k, k')^* e^{i(kx - k'x')/\hbar} \, dk \, dk', \qquad [4.18a]$$

or, with [4.11], and since $K^0$ is real

$$2\pi\hbar K^0(2q - x - x', x - x')e^{ip(x - x')/\hbar}$$

$$= \iint K^0(2p - k - k', k - k')e^{i[q(k' - k) + kx - k'x']/\hbar} \, dk \, dk'. \qquad [4.19]$$

Introducing $\rho = 2p - k - k'$ and $\sigma = k - k'$ as integration variables and writing $\xi$ for $2q - x - x'$ and $\eta$ for $x - x'$, the equation simplifies greatly, and one has

$$K^0(\xi, \eta) = (4\pi\hbar)^{-1} \iint K^0(\rho, \sigma)e^{-i(\rho\eta + \sigma\xi)/2\hbar} \, d\rho \, d\sigma. \qquad [4.20]$$

This means that $K^0$ belongs to the characteristic value 1 of the operator $Q$ with the kernel

$$Q(\xi, \eta; \rho, \sigma) = (4\pi\hbar)^{-1}e^{-i(\xi\sigma + \eta\rho)/2\hbar}. \qquad [4.21]$$

The operator $Q$ shows great similarity to that of the Fourier transformation. In particular, its square transforms a function $f(\rho, \sigma)$ into $f(-\rho, -\sigma)$, so that, if $f$ is even in $\rho, \sigma$,

$$K^0 = (1 + Q)f \qquad [4.22]$$

will satisfy [4.20]. One can verify, furthermore, that if $f(\rho, 0) = \delta(\rho)/\pi\hbar$, then $Qf$ satisfies [4.16a], and conversely, if $f$ satisfies

[4.16a] so that $\int f(\rho, \sigma) \, d\rho = (\pi\hbar)^{-1}$, then $Qf(\xi, 0) = \delta(\xi)/\pi\hbar$.

Hence, if $f$ satisfies both conditions, the same will be true for $Qf$.

It follows that the condition of the symmetry between $q$ and $p$, as expressed by [4.18], still leaves a great deal of freedom in the choice of $K$. This is essentially restricted to one-quarter of a Hilbert space.

**33**  QUANTUM-MECHANICAL DISTRIBUTION FUNCTIONS

The original choice of the correspondence between $P$ and the wave function was[3]

$$P(q, p) = (\hbar\pi)^{-1} \int \psi(q + x)^* \psi(q - x) e^{2pix/\hbar} \, dx. \qquad [4.23]$$

This corresponds to

$$K^0(q, x) = (\hbar\pi)^{-1} \delta(q), \qquad [4.23a]$$

and this is, evidently, the simplest way to satisfy all the preceding requirements. However, this is not the only possible choice: another, relatively simple selection is

$$K^0(\xi, \eta) = \frac{1}{\pi^2 \hbar} \frac{|\eta|}{\xi^2 + \eta^2}, \qquad [4.24]$$

leading to

$$
\begin{aligned}
P(q, p) &= \frac{1}{\pi^2 \hbar} \iint \frac{|x - x'| e^{ip(x - x')/\hbar}}{(2q - x - x')^2 + (x - x')^2} \psi(x)^* \psi(x') \, dx \, dx' \\
&= \frac{1}{2\pi^2 \hbar} \iint \frac{|x - x'| e^{ip(x - x')/\hbar}}{x^2 + x'^2} \psi(q + x)^* \psi(q + x') \, dx \, dx'.
\end{aligned}
\qquad [4.25]
$$

**III.**

The question now arises whether there is any simple criterion which permits a unique coordination of distribution functions to wave functions. The only one known to this writer is to demand that, in the force-free case, the equation of motion of $P$ be the classical one

$$\frac{\partial P}{\partial t} = -\frac{p}{m} \frac{\partial P}{\partial q}. \qquad [4.26]$$

This, together with the postulate of invariance [4.11], and the conditions (a) and (b) of the first section specify $P$ uniquely. The corresponding $K$ will turn out to be given by [4.23a] so that the $q$-$p$ symmetry condition [4.18] is automatically satisfied.

One finds from

$$\frac{\partial \psi(x)}{\partial t} = \frac{\hbar i}{2m} \frac{\partial^2 \psi}{\partial x^2} = \frac{\hbar i}{2m} \psi''$$    [4.27]

and [4.12] that

$$\frac{\partial P(q, p)}{\partial t} = \frac{\hbar i}{2m} \iint K^0 (2q - x - x', x - x') e^{ip(x-x')/\hbar}$$

$$[\psi(x)^* \psi''(x') - \psi''(x)^* \psi(x')] \, dx \, dx'.$$    [4.28]

Partial integrations with respect to $x$ and $x'$ in the first and second terms, respectively, give

$$\frac{\partial P(q, p)}{\partial t} = \frac{\hbar i}{m} \iint \left( 2K^0_{12} + \frac{2ip}{\hbar} K^0_1 \right) e^{ip(x-x')/\hbar} \psi(x)^* \psi(x') \, dx \, dx',$$    [4.29]

where the lower indices of $K^0$ denote differentiations with respect to the first and second variables, respectively,

$$K^0_{12}(\xi, \eta) = \frac{\partial^2 K^0(\xi, \eta)}{\partial \xi \, \partial \eta} \qquad K^0_1(\xi, \eta) = \frac{\partial K^0(\xi, \eta)}{\partial \xi}.$$    [4.29a]

The variables of $K^0$ in [4.29], however, remain $2q - x - x'$ and $x - x'$. On the other hand,

$$-\frac{p}{m} \frac{\partial P(q, p)}{\partial p} = -\iint \frac{p}{m} 2K^0_1 e^{ip(x-x')/\hbar} \psi(x)^* \psi(x') \, dx \, dx'$$    [4.30]

gives just the second term of [4.29]. It follows that

$$K^0_{12} = \frac{\partial^2 K^0(\xi, \eta)}{\partial \xi \, \partial \eta} = 0,$$    [4.31]

or $K^0(\xi, \eta) = f(\xi) + g(\eta)$. Condition [4.16a] now shows that $g(\eta)$ must be independent of $\eta$, and it can be assumed to be zero since a constant can be absorbed into $f$. Then, [4.14a] shows that $f(\xi) = \delta(\xi)/\pi\hbar$ so that we recovered [4.23a], the old distribution function.

## 35    QUANTUM-MECHANICAL DISTRIBUTION FUNCTIONS

Naturally, in the presence of forces, the distribution function does not satisfy the classical equation of motion

$$\frac{\partial P}{\partial t} = -\frac{p}{m}\frac{\partial P}{\partial q} + \frac{\partial V}{\partial q}\frac{\partial P}{\partial p}, \qquad [4.32]$$

$V$ being the potential. There are many typical quantum phenomena which are in conflict with [4.32]. However, the departure from [4.32] contains at least the second power of $\hbar$. The full equation of motion was given before.[4]

In addition to the correction terms to [4.32], there is a departure from classical theory inasmuch as not every $P(q, p)$ is realizable. Thus, one has, in addition to the normalization integral

$$\iint P(q, p)\, dq\, dp = 1, \qquad [4.33]$$

the added condition

$$\iint [P(q, p)]^2\, dq\, dp = 2\pi\hbar, \qquad [4.33a]$$

which is an expression for the uncertainty relation. In the case where $P$ represents a mixture rather than a pure state, the left side of [4.33a] is actually smaller than $2\pi\hbar$. Equation [4.33a] is a special case of a more general one[5] which gives the transition probability between two states, $\psi_1$ and $\psi_2$, in terms of the corresponding distribution functions

$$\left|\int \psi_1(x)^*\psi_2(x)\, dx\right|^2 = (2\pi\hbar)^{-1} \iint P_1(q, p)P_2(q, p)\, dq\, dp, \qquad [4.34]$$

reemphasizing the fact that the distribution functions have to be able to assume negative values. Equation [4.34] is a consequence of the unitary nature of the kernel $(2\pi\hbar)^{-1/2}K$ given in [4.23a] for the old distribution function.

36  EUGENE P. WIGNER

## References

1.
Some of the papers dealing with this subject are: E. Wigner, Phys. Rev. **40**, 749–759 (1932); J. E. Moyal, Proc. Cambridge Phil. Soc. **45**, 99–124 (1949); H. Mori, I. Oppenheim, and J. Ross, *Studies in Statistical Mechanics*, ed. by J. De Boer and G. E. Uhlenbeck (North-Holland, 1962), Vol. I, Part C, pp. 213–298; W. E. Brittin and W. R. Chappell, Rev. Mod. Phys. **34**, 620 (1962). This article also contains more complete references to the earlier literature; J. C. T. Pool, J Math. Phys. **1**, 66–76 (1966); L. Cohen, J. Math. Phys. **1**, 781–786 (1966); K. Imre, E. Özizmir, M. Rosenbaum, P. F. Zweifel, J. Math. Phys. **8**, 1097–1108 (1967). Optical phenomena were treated by means of similar concepts by R. J. Glauber in *Fundamental Problems in Statistical Mechanics*, ed. by E. C. D. Cohen (North-Holland, 1968), Vol. II, pp. 140–187; and J. R. Klauder and E. C. G. Sudarshan, *Fundamentals of Quantum Optics* (W. A. Benjamin, 1968). This book also contains extensive references to earlier literature; G. S. Agarwal and E. Wolf, Phys. Rev. D **2**, 2161–2225 (1970); G. Nienhuis, J. Math. Phys. **11**, 239 (1970).
2.
Positive (more precisely: nonnegative) distribution functions were defined, however, by K. Husimi, Proc. Phys. Math. Soc. Japan **22**, 264 (1940); Y. Kano, J. Math. Phys. **6**, 1913 (1965). See also a forthcoming paper by K. E. Cahill and R. J. Glauber in the Phys. Rev., and W. A. Smith and E. P. Wigner, Bull. Am. Phys. Soc. **14**, 59 (1969). See also F. Bopp, Ann. Inst. Henri Poincaré **15**, 81 (1956).
3.
Ref. 1.
4.
Ibid.
5.
This formula must have been known rather generally, but no explicit reference to it has so far been found.

# Quantum-Mechanical Distribution Functions: Conditions for Uniqueness

R. F. O'Connell and E. P. Wigner[1]

Physics Letters *83A*, 145–148 (1981)

Received 18 March 1981

We add to the postulate, that the distribution function give the proper probabilities for the position and momentum variables (actually only the former is needed) and that its connection with the wave function which it represents have the natural invariances, another one. This is that the integral of the product of two distribution functions be equal, except for a universal constant (which turns out to be $2\pi\hbar$), to the transition probability between the two states they represent. We then show that it follows from these conditions that the distribution function is the one defined earlier by one of us (E.W.).

Quantum-mechanical distribution functions provide a framework for an exact reformulation of non-relativistic quantum mechanics in terms of classical concepts [1,2]. This has led to various discussions of their properties (refs. [3–6], and references therein, provide a representative sample of recent papers on this subject) and to their application to a variety of problems in physics [7][*1].

The distribution functions $P(q, p)$ are functions of positional and momentum coordinates $q$ and $p$, and various definitions for $P$ appear in the literature. Thus the question arises as to what conditions are necessary for providing a unique definition for $P$. One of us has previously examined this question and provided one such set of conditions [2]. In addition to those listed below — (a) to (e) — the relevant new condition was that, for free particles, the time dependence of $P$ becomes the non-relativistic classical one. It is our purpose here to replace this by a perhaps more general criterion.

As in ref. [2], we list as our basic constraints on $P$:

(a) that it be a hermitian, that is bilinear, form of the wave function $\psi$, and

(b) that, if integrated over $p$, it give the proper probabilities for the different values of $q$, and similarly with $p \leftrightarrow q$.

Next, on grounds of simplicity and naturalness, it was required

(c) that the correspondence between $P$ and the wave function $\psi$ be Galilei invariant, i.e. invariant with respect to displacements and non-relativistic transitions, to moving coordinate systems;

(d) that $P$ be real for all $\psi$.

These conditions led to the result that [2, eq. (4.12)]

$$P(q, p) = \iint K^0(2q - x - x', x - x') \, e^{ip(x - x')/\hbar} \, \psi(x)^* \psi(x') \, dx \, dx' , \qquad (1)$$

the explicit form of $K^0$ to be determined by further conditions. Our considerations here will be restricted to one dimension but it will be clear that they can be generalized to higher dimensions.

For our present deliberations, it turns out to be more convenient to replace $K^0$ by a more symmetric-looking function $f$, as follows:

---

[1] Permanent address: Department of Physics, Joseph Henry Laboratory, Princeton University, Princeton, NJ 08540, USA.
[*1] See, for example, the various papers cited in ref. [1] of ref. [2] above. Some more recent applications are discussed in refs. [8–10].

Volume 83A, number 4                    PHYSICS LETTERS                    25 May 1981

$$P(q,p) = \iint f(q - x, q - x') \, e^{ip(x-x')/\hbar} \, \psi(x)^* \psi(x') \, dx \, dx' \; . \tag{2}$$

As in ref. [2], we next introduce the requirement:
    (e) that the correspondence given by eq. (2) be invariant with respect to space and time reflections.
    This leads to the result that

$$f(\alpha, \beta) = f(\beta, \alpha)^* = f(\beta, \alpha) = f(-\alpha, -\beta) = f(-\beta, -\alpha) \; . \tag{3}$$

    A new requirement — which will turn out to be the only other condition necessary to determine $P$ uniquely — is:
    (f) that the transition probability between two states $\psi$ and $\phi$ is given, in terms of the corresponding distribution functions, $P_\psi$ and $P_\phi$ say, as follows:

$$\left| \int \psi(x)^* \phi(x) \, dx \right|^2 = k \iint P_\psi(q,p) P_\phi(q,p) \, dq \, dp \; , \tag{4}$$

where $k$ is an arbitrary constant, the same for all $\psi$ and $\phi$, which is determined by the condition that the total integral of $P$ be unity. As we will see below, $k$ turns out to be $2\pi\hbar$.
    We will now show that we have enough information to determine the function $f$ uniquely.
    Using eqs. (2) and (4) we obtain

$$\int \psi(u)^* \phi(u) \, \psi(v) \phi(v)^* \, du \, dv = k \int dp \, dq \, e^{ip(x-x')/\hbar} f(q - x, q - x') \psi(x)^* \psi(x')$$

$$\times e^{ip(y-y')/\hbar} f(q - y, q - y') \phi(y)^* \phi(y') \, dx \, dx' \, dy \, dy' \; . \tag{5}$$

Carrying out the $p$ integration introduces a $2\pi\hbar\delta\left[(x - x') + (y - y')\right]$ factor, which prompts us to introduce a new variable $\xi$ such that $x' = x - \xi, y' = y + \xi$, so that the right side of eq. (5) becomes

$$2\pi\hbar k \int dq \, f(q - x, q - x + \xi) f(q - y, q - y - \xi) \psi(x)^* \psi(x - \xi) \phi(y)^* \phi(y + \xi) \, dx \, dy \, d\xi \; .$$

In this, we replace the variable $x$ by $u$ and $\xi$ by $u - v$. Hence, we obtain, instead of eq. (5),

$$\int \psi(u)^* \psi(v) \phi(u) \phi(v)^* \, du \, dv$$

$$= 2\pi\hbar k \int dq \, f(q - u, q - v) f(q - y, q - y - u + v) \psi(u)^* \psi(v) \phi(y)^* \phi(y + u - v) \, du \, dv \, dy \; . \tag{6}$$

Since this is valid for any function $\psi$, one can conclude rather easily that the factors of $\psi(u)^* \psi(v)$ in the two integrals must be equal, i.e.

$$\phi(u)\phi(v)^* = 2\pi\hbar k \iint dq \, dy \, f(q - u, q - v) f(q - y, q - y - u + v) \phi(y)^* \phi(y + u - v) \; . \tag{7}$$

To proceed further, we replace the integration variable $y$ by $y + v$ to obtain

$$\phi(u)\phi(v)^* = 2\pi\hbar k \int dq \, dy \, f(q - u, q - v) f(q - v - y, q - u - y) \, \phi(u + y) \phi(v + y)^* \; , \tag{8}$$

from which we conclude that

$$2\pi\hbar k \int dq \, f(q - u, q - v) f(q - u - y, q - v - y) = \delta(y) \; . \tag{9}$$

Volume 83A, number 4                     PHYSICS LETTERS                          25 May 1981

Using eq. (3), we have interchanged the variables in the second $f$ of eq. (8).

We can, finally, regard $f$ as a function of the sum and the difference of its variables,

$$f(\alpha, \beta) = \kappa(\tfrac{1}{2}(\alpha + \beta), \alpha - \beta). \tag{10}$$

In terms of this $\kappa$, if we substitute $w$ for $q - \tfrac{1}{2}(u + v)$ and $z$ for $v - u$, eq. (9) reads

$$2\pi\hbar k \int \kappa(w, z) \kappa(w - y, z)\, dw = \delta(y), \tag{11}$$

from which it follows that $\kappa$ is a $\delta$ function of its first variable. Hence

$$f(\alpha, \beta) = g(\alpha - \beta)\delta(\alpha + \beta - c). \tag{12}$$

It follows from eq. (3), $f(\alpha, \beta) = f(-\beta, -\alpha)$, that

$$g(\alpha - \beta)\delta(\alpha + \beta - c) = g(-\beta + \alpha)\delta(-\alpha - \beta - c), \tag{12a}$$

i.e. that $c = 0$. But the function $g$ remains to be determined.

From requirement (b), we have

$$\int P(q, p)\, dp = |\psi(q)|^2. \tag{13}$$

Hence, from eqs. (2), (12) and (13) we obtain $g(q) = (\pi\hbar)^{-1}$. Thus

$$f(\alpha, \beta) = (\pi\hbar)^{-1}\delta(\alpha + \beta), \tag{14}$$

which, when substituted in eq. (2), leads to the result

$$P(q, p) = (\pi\hbar)^{-1} \iint \psi(x)^* \psi(x') \delta(2q - x - x') e^{ip(x-x')/\hbar}\, dx\, dx'$$

$$= (\pi\hbar)^{-1} \int \psi(x)^* \psi(2q - x) e^{ip(2x-2q)/\hbar}\, dx. \tag{15}$$

Thus, if we set $x = q + y$, we finally obtain

$$P(q, p) = (\pi\hbar)^{-1} \int \psi(q + y)^* \psi(q - y) e^{2ipy/\hbar}\, dy. \tag{16}$$

Using this result in eq. (4) leads to the result that $k = 2\pi\hbar$ [*2].

In summary, we have shown that, if in addition to the postulates (a) to (e), already discussed in ref. [2], we add the postulate that the integral of the product of two distribution functions gives, up to a universal constant (which turned out to be $2\pi\hbar$), the transition probability between the corresponding two states, we are led uniquely to the distribution function given by eq. (16) — which is the original distribution function given in ref. [1].

One of us (R.F. O'C) was partially supported by the Department of Energy under Contract No. DE-AS05-79ER10459.

[*2] Hence eq. (4) is equivalent to eq. (4.34) of ref. [2], except that the latter equation — and also eq. (4.33a) of ref. [2] — has the $2\pi\hbar$ factor in the wrong side of the equation, as is evident from dimensionality considerations.

*References*

[1] E.P. Wigner, Phys. Rev. 40 (1932) 749.
[2] E.P. Wigner, in: Perspectives in quantum theory, eds. W. Yourgrau and A. van der Merwe (Dover Publ., New York, 1979) p. 25.

Volume 83A, number 4                    PHYSICS LETTERS                    25 May 1981

[3] G. George and I. Prigogine, Physica 99A (1979) 369.
[4] N.L. Balazs, Physica 102A (1980) 236.
[5] D. Bohm and B.J. Hiley, On a quantum algebraic approach to a generalized phase space, Found. Phys., to be published.
[6] J.P. Amiet and P. Huguenin, Mécaniques classique et quantique dans l'espace de phase, to be published.
[7] E.P. Wigner, Z. Phys. Chem. (Leipzig) 19 (1932) 203.
[8] J.H. Shirley and S. Stenholm, J. Phys. A10 (1977) 613.
[9] V.V. Dodonov, V.I. Man'ko and V.V. Rudenko, Sov. Phys. JETP 51 (1980) 443.
[10] G.J. Iafrate, H.L. Grubin and D.K. Ferry, Bull. Am. Phys. Soc. 26 (1981) 458.

# Some Properties of a Non-negative
# Quantum-Mechanical Distribution Function

R. F. O'Connell and E. P. Wigner[1]

Physics Letters *85A*, 121–126 (1981)

Received 11 July 1981

We consider the distribution function obtained by smoothing the original distribution function, defined in an earlier publication, with a ground-state harmonic oscillator wave function. We derive its time dependence and show that, in particular, the field-free result does not correspond to the classical result. We point out that the non-negative property of the smoothed function follows immediately from the fact that the integral of the product of two of the original distribution functions is equal, except for a factor $2\pi\hbar$, to the transition probability between the two states they represent.

*1. Definition of a non-negative ("smoothed") quantum mechanical distribution function.* In a recent publication [1], we examined the conditions which are necessary for providing a unique definition for a quantum-mechanical distribution function $P(q, p)$, where $q$ and $p$ are positional and momentum coordinates. The conditions we listed lead uniquely to the result that, for every normalized state vector $\psi$,

$$P(q, p) = (\pi\hbar)^{-1} \int \psi(q + y)^* \psi(q - y) e^{2ipy/\hbar} \, dy , \tag{1}$$

which is the original distribution function given in ref. [2]. Some of the properties of $P$ are [1–3]:

(a) It is a hermitian, that is bilinear, form of the wave-function $\psi$. Hence it is real for all $q$ and $p$. The hermitian operator is, of course a function of $q$ and $p$.

(b) If integrated over $p$, it gives the proper probabilities for the different values of $q$, and similarly with $p \leftrightarrow q$.

(c) The transition probability between two states $\psi$ and $\phi$ is given, in terms of the corresponding distribution functions, $P_\psi$ and $P_\phi$ say, as follows:

$$\left| \int \psi(x)^* \phi(x) \, dx \right|^2 = 2\pi\hbar \iint P_\psi(q, p) P_\phi(q, p) \, dq \, dp . \tag{2}$$

As pointed out in ref. [2] and proven in ref. [3], conditions (a) and (b) are incompatible with the requirement that the distribution function be everywhere nonnegative. In fact it is clear from condition (c) that the distribution function given by eq. (1) has to be able to assume negative values.

Starting with Husimi [4], many authors obtained non-negative distribution functions by dropping condition (a). In most cases this is essentially achieved, for all points $(q, p)$, by smoothing $P(q', p')$ with a density function $D(q', p')$ and integrating over all $p'$ and $q'$.

A natural and popular choice of the density function $D$ is a gaussian distribution [4–8], recently considered anew by Cartwright [9].

Our considerations here will be restricted to one dimension but it will be clear that they can be generalized to higher dimensions. Consider a linear harmonic oscillator, centered at the point $q$ and moving with an average momentum $p$. Then, as is well known, the ground-state wave-function, at the position $q$ and average momentum $p$

[1] Permanent address: Department of Physics, Joseph Henry Laboratory, Princeton University, Princeton, NJ 08540, USA.

121

Volume 85A, number 3                          PHYSICS LETTERS                          21 September 1981

becomes, as function of $q'$,

$$\psi_{q,p}(q',\alpha) = (\pi\alpha)^{-1/4} e^{-(q'-q)^2/2\alpha} e^{ipq'/\hbar} , \tag{3}$$

where $(\Delta q')^2 = \alpha/2$. Using this expression for $\psi_{q,p}$ in eq. (1), it may easily be verified that the corresponding distribution function, $P_{q,p}(q',p',\alpha)$ say, is

$$P_{q,p}(q',p',\alpha) = (\pi\hbar)^{-1} e^{-(q'-q)^2/\alpha} e^{-\alpha(p'-p)^2/\hbar^2}. \tag{4}$$

This function has the property that

$$(\Delta q')^2 = \alpha/2 \quad \text{and} \quad (\Delta q')(\Delta p') = \hbar/2 . \tag{4a}$$

This is the density function which, following Husimi [4], we will use to smooth $P(q',p')$. Thus, the resultant smoothed distribution function, $P_s(q,p,\alpha)$ say, is simply

$$P_s(q,p,\alpha) = \iint P(q',p')P_{q,p}(q',p',\alpha)\, dp'\, dq' . \tag{5}$$

Carrying out an explicit calculation, Cartwright [9] showed that $P_s(q,p,\alpha)$ is everywhere non-negative. However, it is clear that since the rhs of eq. (5) is a particular case of the rhs of eq. (2), and since the lhs of eq. (2) is non-negative it follows immediately that

$$P_s(q,p,\alpha) \geqslant 0 , \tag{6}$$

for all $q$ and $p$.

Before turning to a consideration of other properties of $P_s$, not heretofore considered in the literature, it is convenient to use eqs. (4) and (5) to write explicitly

$$P_s(q,p,\alpha) = (\pi\hbar)^{-1} \iint P(q',p')e^{-(q'-q)^2/\alpha} e^{-\alpha(p'-p)^2/\hbar^2}\, dq'\, dp' , \tag{7}$$

where the $P$ without the index is the old distribution function (1). It may be of some interest to remark here already that $P_s$ is an "entire function" of $q$ and $p$, i.e. that the range of these variables may be extended to the whole complex plane without encountering any irregularities. First of all, by simple differentiation, it is easy to verify that

$$\alpha \partial P_s/\partial\alpha = (\alpha/4)\partial^2 P_s/\partial q^2 - (\hbar^2/4\alpha)\partial^2 P_s/\partial p^2 . \tag{8}$$

*2. Time dependence of the smoothed distribution function.* We next consider the equation of motion of $P_s$. The time dependence of $P_s$ may be decomposed into two parts,

$$\partial P_s/\partial t = \partial_k P_s/\partial t + \partial_v P_s/\partial t , \tag{9}$$

the first part resulting from the $(i\hbar/2m)\partial^2/\partial q^2$, the second from the potential energy $V/i\hbar$ part of the expression for $\partial\psi/\partial t$.

It has already been shown [2] that the time dependence of $P(q',p')$ corresponds to the classical result in the field-free case, i.e.

$$\partial_k P(q',p')/\partial t = -(p'/m)\partial P(q',p')/\partial q' , \tag{10}$$

and in the presence of a potential we have the extra contribution

$$\partial_v P(q',p')/\partial t = \int dj\, P(q';p'+j)J(q',j) , \tag{11}$$

where

Volume 85A, number 3                          PHYSICS LETTERS                          21 September 1981

$$J(q',j) = (i/\pi\hbar^2) \int [V(q'+y) - V(q'-y)] \, e^{-2iyj/\hbar} \, dy \tag{12}$$

is the probability of a jump of the momentum by an amount $j$ if the positional coordinate is $q'$.

Applying the results to eq. (7), we have, first of all, that

$$\frac{\partial_k P_s(q,p,\alpha)}{\partial t} = -(\pi\hbar)^{-1} \iint \frac{p'}{m} \frac{\partial P(q',p')}{\partial q'} \{e^{-(q'-q)^2/\alpha} \, e^{-\alpha(p'-p)^2/\hbar^2}\} \, dq' \, dp' \tag{13}$$

$$= (\pi\hbar)^{-1} \iint \frac{p'}{m} P(q',p') \frac{\partial}{\partial q'} \{e^{-(q'-q)^2/\alpha} \, e^{-\alpha(p'-p)^2/\hbar^2}\} \, dq' \, dp' \, ,$$

the right side having been obtained by partial integration. But since the $\partial/\partial q'$, as applied to the expression in the curly bracket can be replaced by $-\partial/\partial q$, we also have

$$\frac{\partial_k P_s(q,p,\alpha)}{\partial t} = -(\pi\hbar)^{-1} \frac{\partial}{\partial q} \iint \frac{p'}{m} P(q',p') \{e^{-(q'-q)^2/\alpha} \, e^{-\alpha(p'-p)^2/\hbar^2}\} \, dq' \, dp' \, . \tag{14}$$

The classical expression for $\partial_k P_s(q,p,\alpha)/\partial t$ would be

$$-\frac{p}{m} \frac{\partial P_s(q,p,\alpha)}{\partial q} = -(\pi\hbar)^{-1} \frac{\partial}{\partial q} \iint \frac{p}{m} P(q',p') \{e^{-(q'-q)^2/\alpha} \, e^{-\alpha(p'-p)^2/\hbar^2}\} \, dq' \, dp' \, . \tag{14a}$$

The difference between the two expressions (14) and (14a) is

$$(\pi\hbar)^{-1} \frac{\partial}{\partial q} \iint \frac{p-p'}{m} P(q',p',\alpha) \{e^{-(q'-q)^2/\alpha} \, e^{-\alpha(p'-p)^2/\hbar^2}\} \, dq' \, dp' \tag{15}$$

$$= -(\pi\hbar)^{-1} \frac{\hbar^2}{2\alpha m} \frac{\partial^2}{\partial q \partial p} \iint P(q',p',\alpha) \{e^{-(q'-q)^2/\alpha} \, e^{-\alpha(p'-p)^2/\hbar^2}\} \, dq' \, dp' = -\frac{\hbar^2}{2\alpha m} \frac{\partial^2}{\partial q \, \partial p} P(q,p,\alpha) \, .$$

Hence we obtain

$$\frac{\partial_k P_s(q,p,\alpha)}{\partial t} = -\frac{1}{m} \left(p + \frac{\hbar^2}{2\alpha} \frac{\partial}{\partial p}\right) \frac{\partial}{\partial q} P_s(q,p,\alpha) \, . \tag{16}$$

It is thus clear that, in the field-free case, the time dependence of $P_s(q,p,\alpha)$, in contrast to that of $P(q,p)$, is not given by the classical expression but contains a correction term of order $\hbar^2$. But this is not a quantum effect: the same expression would appear in the time derivative of the classical distribution function if this were "smoothed" as in (7).

We turn now to a consideration of

$$\frac{\partial_v P_s(q,p,\alpha)}{\partial t} = (\pi\hbar)^{-1} \iint \frac{\partial_v P(q',p')}{\partial t} \{e^{-(q'-q)^2/\alpha} \, e^{-\alpha(p'-p)^2/\hbar^2}\} \, dq' \, dp' \, . \tag{17}$$

Thus, utilizing eqs. (11) and (12) in eq. (17), we obtain

$$\frac{\partial_v P_s(q,p,\alpha)}{\partial t} = \frac{i}{(\pi\hbar)(\pi\hbar^2)} \iiiint dp' \, dq' \, dj \, dy \, P(q',p'+j)[V(q'+y) - V(q'-y)] \tag{18}$$

$$\times \{e^{-(q'-q)^2/\alpha} \, e^{-\alpha(p'-p)^2/\hbar^2} \, e^{-2ijy/\hbar}\} \equiv -\frac{2}{\pi^2\hbar^3} \, \mathrm{Im} \, I \, ,$$

where

Volume 85A, number 3                    PHYSICS LETTERS                    21 September 1981

$$I = \iiiint dp'\, dq'\, dj\, dy\, P(q';p'+j)\, V(q'+y)\, \{e^{-(q'-q)^2/\alpha}\, e^{-\alpha(p'-p)^2/\hbar^2}\, e^{-2ijy/\hbar}\} \;. \tag{19}$$

Replacing $y$ by a new variable $z = y + q'$ and $p'$ by $p'' = p' + j$ this becomes

$$I = \iiiint P(q',p'')\, e^{-(q+i\alpha j/\hbar - q')^2/\alpha}\, e^{-\alpha(p+j-p'')^2/\hbar^2}\, V(z)\, e^{-\alpha j^2/\hbar^2 + 2ij(q-z)/\hbar}\, dp''\, dq'\, dj\, dz \;. \tag{20}$$

Because of eq. (7) and the possibility of extending $P_s$ also to complex values of $q$ and $p$, this can be written also as

$$I = \pi\hbar \iint dj\, dz\, P_s(q + i\alpha j/\hbar\,,\, p+j,\, \alpha)\, V(z)\, e^{-\alpha j^2/\hbar^2 + 2ij(q-z)/\hbar} \;. \tag{21}$$

As was observed already before, $P_s$ is an entire function so that it remains uniquely defined in spite of the complex nature of one of its arguments. It now follows from eqs. (9), (16), (18) and (21) that

$$\frac{\partial P_s(q,p,\alpha)}{\partial t} = \frac{\partial_k P_s(q,p,\alpha)}{\partial t} + \frac{\partial_v P_s(q,p,\alpha)}{\partial t} = -\frac{1}{m}\left(p + \frac{\hbar^2}{2\alpha}\frac{\partial}{\partial p}\right)\frac{\partial P_s(q,p,\alpha)}{\partial q} \tag{22}$$

$$- i(\pi\hbar)^{-1} \iint [P_s(q - i\alpha j, p + \hbar j, \alpha)e^{-2ij(q-z)} - P_s(q + i\alpha j, p + \hbar j, \alpha)e^{2ij(q-z)}]\, V(z)\, e^{-\alpha j^2}\, dj\, dz \;,$$

the integration variable $j$ of (21) having been replaced by $\hbar j$.

Eq. (22) is, probably, the shortest expression for the time derivative of $P_s$. It may be observed that the time derivative of the smoothed distribution is not very simple even in classical theory − not even if we restrict ourselves, as was done in all the preceding discussion, to the non-relativistic limit.

The extension of $P_s$ into the complex plane in (22) can be made unnecessary by expanding the $P_s$ under the integral sign into a power series of $i\alpha j$. The exponentials of $ij(q-z)$ are then replaced, for the successive $q$ derivatives of $P_s$, by $\sin 2j(q-z)$, $\cos 2j(q-z)$, $-\sin 2j(q-z)$, $-\cos 2j(q-z)$, again $\sin 2j(q-z)$, and so on. For the factor accompanying the $n$th derivative of $P_s$ this can be written as $\sin[2j(q-z) + \tfrac{1}{2}n\pi]$. Hence one obtains

$$\frac{\partial P_s(q,p,\alpha)}{\partial t} = -\frac{1}{m}\left(p + \frac{\hbar^2}{2\alpha}\frac{\partial}{\partial p}\right)\frac{\partial P_s(q,p,\alpha)}{\partial q}$$

$$- \frac{1}{\pi\hbar}\sum_n \frac{1}{n!} \iint \frac{\partial^n P_s(q, p + \hbar j, \alpha)}{\partial q^n}\, \sin[2j(q-z) + \tfrac{1}{2}n\pi]\, V(z)\, e^{-\alpha j^2}\, dj\, dz \;. \tag{22a}$$

There are two, apparently different but mathematically identical, expressions for the time derivative of the old distribution function $P$ of (1): the one given by eq. (11) of ref. [2] (eqs. (10) and (11) of the present article), the other by eq. (8) of the same article. The latter expression is a power series in $\hbar^2$ and shows the analogy to the classical expression for $\partial P/\partial t$ more clearly. The next section will derive the analogue of this for $\partial P_s/\partial t$, using the old expression (8) of ref. [2] for $\partial P/\partial t$.

*3. Other expression for the time derivative of $P_s$.* The alternate expression for $\partial P_s/\partial t$ is also a sum of the two parts, as given by (9). Since the first part is again given by (16), only the second part, i.e. $\partial_v P_s/\partial t$ will be recalculated. According to eq. (8) of ref. [2]

$$\frac{\partial_v P(q,p)}{\partial t} = \sum_\lambda \frac{1}{\lambda!}\left(\frac{\hbar}{2i}\right)^{\lambda-1} \frac{\partial^\lambda V(q)}{\partial q^\lambda} \frac{\partial^\lambda P(q,p)}{\partial p^\lambda} \;. \tag{23}$$

The summation over $\lambda$ is to be extended over all odd positive integers. We have, therefore,

Volume 85A, number 3                          PHYSICS LETTERS                          21 September 1981

$$\frac{\partial_\nu P_s(q,p,\alpha)}{\partial t} = \frac{1}{\pi\hbar}\sum_\lambda \frac{1}{\lambda!}\left(\frac{\hbar}{2i}\right)^{\lambda-1}$$

$$\times \iint \frac{\partial^\lambda V(q')}{\partial q'^\lambda}\frac{\partial^\lambda P(q',p')}{\partial p'^\lambda}\{e^{-(q'-q)^2/\alpha}\,e^{-\alpha(p'-p)^2/\hbar^2}\}\,dq'\,dp'. \tag{24}$$

By partial integrations with respect to $p'$ and replacement of the $p'$ derivatives of the last factor by its negative $p$ derivatives one obtains

$$\frac{\partial_\nu P_s(q,p,\alpha)}{\partial t} = \frac{1}{\pi\hbar}\sum_\lambda \frac{1}{\lambda!}\left(\frac{\hbar}{2i}\right)^{\lambda-1}\frac{\partial^\lambda}{\partial p^\lambda}\iint \frac{\partial^\lambda V(q')}{\partial q'^\lambda}P(q',p')\{e^{-(q'-q)^2/\alpha}\,e^{-\alpha(p'-p)^2/\hbar^2}\}\,dq'\,dp'. \tag{25}$$

It is natural now to expand the derivatives of $V$ into a power series of $q' - q$,

$$\frac{\partial^\lambda V(q')}{\partial q'^\lambda} = \sum_\mu \frac{1}{\mu!}\frac{\partial^{\lambda+\mu}V(q)}{\partial q^{\lambda+\mu}}(q'-q)^\mu. \tag{26}$$

We then use the identity [*1]

$$x^\mu e^{-x^2/\alpha} = \frac{(-)^\mu \mu!}{2^\mu}\sum_\kappa \frac{\alpha^{\mu-\kappa}}{\kappa!(\mu-2\kappa)!}\frac{d^{\mu-2\kappa}e^{-x^2/\alpha}}{dx^{\mu-2\kappa}}, \tag{27}$$

in which the summation over $\kappa$ is to be extended from 0 until $\mu - 2\kappa$ becomes negative. It may be observed that if $\mu$ is even, the last term of (27) contains the function $e^{-x^2/\alpha}$ itself, if $\mu$ is odd the last term contains the first derivative of this. Introducing now (27) into (25) with $x = q' - q$ and replacing the $q'$ derivative of the last factor by the negative $q$ derivative [which introduces a factor $(-)^\mu$], one obtains finally

$$\frac{\partial_\nu P_s(q,p,\alpha)}{\partial t} = \frac{1}{\pi\hbar}\sum_{\lambda\mu\kappa}\frac{(\hbar/2i)^{\lambda-1}\alpha^{\mu-\kappa}}{\lambda!2^\mu\kappa!(\mu-2\kappa)!}\frac{\partial^{\lambda+\mu}V(q)}{\partial q^{\lambda+\mu}}\frac{\partial^\lambda}{\partial p^\lambda}\frac{\partial^{\mu-2\kappa}}{\partial q^{\mu-2\kappa}}\iint P(q',p')e^{-(q'-q)^2/\alpha}\,e^{-\alpha(p'-p)^2/\hbar^2}\,dp'\,dq'. \tag{28}$$

Hence

$$\frac{\partial P_s(q,p,\alpha)}{\partial t} = -\frac{1}{m}\left(p+\frac{\hbar^2}{2\alpha}\frac{\partial}{\partial p}\right)\frac{\partial}{\partial q}P_s + \sum_{\lambda\mu\kappa}\frac{(i\hbar)^{\lambda-1}\alpha^{\mu-\kappa}}{2^{\lambda+\mu-1}\lambda!\kappa!(\mu-2\kappa)!}\frac{\partial^{\lambda+\mu}V(q)}{\partial q^{\lambda+\mu}}\frac{\partial^\lambda}{\partial p^\lambda}\frac{\partial^{\mu-2\kappa}}{\partial q^{\mu-2\kappa}}P_s(q,p,\alpha). \tag{29}$$

It may be good to recall that $\lambda$ assumes all odd — but only odd — values from 1 to infinity, $\kappa$ all integer values from 0 as long as $\mu - 2\kappa$ remains non-negative, $\mu$ all integer values from 0 to infinity.

The second part of (29) is surely not simple, neither is the corresponding expression in the classical limit ($\hbar = 0$). Eq. (29) was derived because it is easier to derive from the approximate expressions for $\partial P_s/\partial t$ than it is from (22). If the "smoothing" over the coordinate is very narrow, i.e. if $\alpha$ is assumed to be very small, the second part of $\partial P_s/\partial t$ naturally goes over into the expression for $\partial P/\partial t$ as given by eq. (23). If $\hbar$ is also assumed to be small, it goes over into the classical expression. But many other approximate expressions for $\partial P_s/\partial t$ can be derived from (29), as were also from (23).

The preceding calculations and the resulting equations are apparently restricted to the case of a single dimension. However, it is not difficult to generalise them for the general case of several dimensions. As to the more simple expression (22) for $\partial P_s/\partial t$, the variables $q, p, j$ and $z$ have to be treated as vectors of as many dimensions as there are dimensions in coordinate (or momentum) space. The exponents $j(q - z)$ are to be replaced by the scalar products of the vectors $j$ and $q - z$, the $j^2$ by $j \cdot j$. In the case of (29) the situation is a bit more complex: the indices $\kappa, \lambda, \mu$ must be replaced by as many sets of indices $\kappa_n, \lambda_n, \mu_n$ as there are space (or momentum) dimensions and the summations extended over all these indices. The restrictions are then that the *sum* of all the $\lambda$ must be odd and that all the $\mu_n - 2\kappa_n$ must be non-negative, as are also all $\kappa_n, \mu_n, \lambda_n$.

[*1] We could not find this equation in the literature, even though it must have been known. It is not difficult to verify it — for $\alpha = 1$ the factors of $\exp(-x^2)$ on the right side are Hermite polynomials.

Volume 85A, number 3                    PHYSICS LETTERS                    21 September 1981

One of us (R.F. O'C.) was partially supported by the Department of Energy under Contract No. DE-AS05-79ER10459. The other (E.P.W.) was Visiting Professor at Louisiana State University.

*References*

[1]  R.F. O'Connell and E.P. Wigner, Phys. Lett. 83A (1981) 145.
[2]  E.P. Wigner, Phys. Rev. 40 (1932) 749.
[3]  E.P. Wigner, in: Perspectives in quantum theory, eds. W. Yourgrau and A. van der Merwe (Dover Publ., 1979) p. 25.
[4]  K. Husimi, Proc. Phys. Math. Soc. Japan 22 (1940) 264.
[5]  F. Bopp, Ann. Inst. H. Poincaré 15 (1956) 81.
[6]  Y. Kano, J. Math. Phys. 6 (1965) 1913.
[7]  J. McKenna and J.R. Klauder, J. Math. Phys. 5 (1964) 878.
[8]  J. McKenna and H.L. Frisch, Phys. Rev. 145 (1966) 93.
[9]  N.D. Cartwright, Physica 83A (1976) 210.

# Distribution Functions in Physics: Fundamentals

M. Hillery, R. F. O'Connell, M. O. Scully, and E. P. Wigner

Physics Reports *106*, 121–167 (1984)

Received December 1983

Abstract:

This is the first part of what will be a two-part review of distribution functions in physics. Here we deal with fundamentals and the second part will deal with applications. We discuss in detail the properties of the distribution function defined earlier by one of us (EPW) and we derive some new results. Next, we treat various other distribution functions. Among the latter we emphasize the so-called P distribution, as well as the generalized P distribution, because of their importance in quantum optics.

## 1. Introduction

It is well known that the uncertainty principle makes the concept of phase space in quantum mechanics problematic. Because a particle cannot simultaneously have a well defined position and momentum, one cannot define a probability that a particle has a position $q$ and a momentum $p$, i.e. one cannot define a true phase space probability distribution for a quantum mechanical particle. Nonetheless, functions which bear some resemblance to phase space distribution functions, "quasiprobability distribution functions", have proven to be of great use in the study of quantum mechanical systems. They are useful not only as calculational tools but can also provide insights into the connections between classical and quantum mechanics.

The reason for this latter point is that quasiprobability distributions allow one to express quantum mechanical averages in a form which is very similar to that for classical averages. As a specific example let us consider a particle in one dimension with its position denoted by $q$ and its momentum by $p$. Classically, the particle is described by a phase space distribution $P_{Cl}(q, p)$. The average of a function of the position and momentum $A(q, p)$ can then be expressed as

$$\langle A \rangle_{Cl} = \int dq \int dp \, A(q, p) \, P_{Cl}(q, p).  \tag{1.1}$$

The integrations in this equation are from $-\infty$ to $+\infty$. This will be the case with all integrations in this paper unless otherwise indicated. A quantum mechanical particle is described by a density matrix $\hat{\rho}$ (we will designate all operators by a ^) and the average of a function of the position and momentum operators, $\hat{A}(\hat{q}, \hat{p})$ is

$$\langle \hat{A} \rangle_{quant} = \text{Tr}(\hat{A}\hat{\rho})  \tag{1.2}$$

(Tr $\hat{Q}$ means the trace of the operator $\hat{Q}$). It must be admitted that, given a classical expression $A(q, p)$, the corresponding self-adjoint operator $\hat{A}$ is not uniquely defined – and it is not quite clear what the purpose of such a definition is. The use of a quasiprobability distribution, $P_Q(q, p)$, however, does give such a definition by expressing the quantum mechanical average as

$$\langle \hat{A} \rangle_{quant} = \int dq \int dp \, A(q, p) \, P_Q(q, p)  \tag{1.3}$$

where the function $A(q, p)$ can be derived from the operator $\hat{A}(\hat{q}, \hat{p})$ by a well defined correspondence rule. This allows one to cast quantum mechanical results into a form in which they resemble classical ones.

The first of these quasiprobability distributions was introduced by Wigner [1932a] to study quantum corrections to classical statistical mechanics. This particular distribution has come to be known as the

Wigner distribution,[†] and we will designate it as $P_w$. This is, and was meant to be, a reformulation of Schrödinger's quantum mechanics which describes states by functions in configuration space. It is non-relativistic in nature because it is not invariant under the Lorentz group; also, configuration space quantum mechanics for more than one particle would be difficult to formulate relativistically. However, it has found many applications primarily in statistical mechanics but also in areas such as quantum chemistry and quantum optics. In the case where $P_Q$ in eq. (1.3) is chosen to be $P_w$, then the correspondence between $A(q, p)$ and $\hat{A}$ is that proposed by Weyl [1927], as was first demonstrated by Moyal [1949]. Quantum optics has given rise to a number of quasiprobability distributions, the most well-known being the $P$ representation of Glauber [1963a] and Sudarshan [1963], which have also found extensive use. As far as the description of the electromagnetic field is concerned, these do exhibit (special) relativistic invariance. Other distribution functions have also been proposed (Husimi [1940]; Margenau and Hill [1961]; Cohen [1966]) but have found more limited use, although, more recently, extensive use has been made of the generalized $P$ representations by Drummond, Gardiner and Walls [1980, 1981]. In this paper we will discuss the basic formalism of these quasiprobability distributions and illustrate them with a few simple examples. We will defer any detailed consideration of applications to a later paper.

We now proceed to the basic problem: how do we go about constructing a quantum mechanical analogue of a phase space density? Let us again consider, for simplicity, a one particle system in one dimension which is described by a density matrix $\hat{\rho}$. In this paper we will work, for simplicity, in one dimension; the generalization to higher dimensions will be given in a few cases but is in most circumstances obvious. It is possible to express the position and momentum distributions of the particle as

$$P_{pos}(q) = \text{Tr}(\hat{\rho}\delta(q - \hat{q})) \tag{1.4a}$$

$$P_{mom}(p) = \text{Tr}(\hat{\rho}\,\delta(p - \hat{p})) \tag{1.4b}$$

where $\delta(q - \hat{q})$ is the operator which transforms $|q'\rangle$ as follows:

$$\delta(q - \hat{q})\,|q'\rangle \equiv |q\rangle\langle q|q'\rangle = \delta(q - q')\,|q'\rangle \tag{1.5}$$

and similarly for $\delta(p - \hat{p})$. We introduce the function $\rho(q', q'')$ defined by

$$\rho(q', q'') = \langle q'|\hat{\rho}|q''\rangle = \sum_\lambda w_\lambda\,\psi_\lambda(q')\,\psi_\lambda(q'')^* \tag{1.6}$$

where $w_\lambda$ is the probability of the system being in the state $\psi_\lambda$, and the $\{\psi_\lambda\}$ form a complete set. Then

$$P_{pos}(q) = \rho(q, q) \tag{1.7a}$$

and

$$P_{mom}(p) = (2\pi\hbar)^{-1} \int dx \int dx'\, \rho(x, x') \exp\{ip(x' - x)/\hbar\}. \tag{1.7b}$$

[†] We use this designation here and throughout the paper despite the strenuous objections of one of us since the majority of us feel we should adhere to what is now common nomenclature.

*M. Hillery et al., Distribution functions in physics: Fundamentals*    125

To show that this corresponds to the usual definition we will examine $P_{pos}(q)$. We have that, in the Dirac bracket notation,

$$P_{pos}(q) = \text{Tr}(\hat{\rho}\,\delta(q - \hat{q})) = \int dq' \,\langle q'|\hat{\rho}\,\delta(q - \hat{q})|q'\rangle$$

$$= \int dq' \,\delta(q - q')\,\langle q'|\hat{\rho}|q'\rangle = \langle q|\hat{\rho}|q\rangle \tag{1.8}$$

which is a more conventional expression for the position density. A first guess for some kind of a phase space density might then be

$$P_1(q, p) = \text{Tr}(\hat{\rho}\,\delta(q - \hat{q})\,\delta(p - \hat{p})). \tag{1.9}$$

On the other hand, we might choose instead

$$P_2(q, p) = \text{Tr}(\hat{\rho}\,\delta(p - \hat{p})\,\delta(q - \hat{q})). \tag{1.10}$$

But these expressions are not equal and although either of them, or a combination of both, could be used to evaluate expectation values of functions of $\hat{q}$ and $\hat{p}$ (provided the operators are ordered properly, the ordering for $P_1$ being different than that for $P_2$), they do not possess what we regard as desirable properties (see section 2). In fact, they are, in general, not real.

The association of distribution functions with operator ordering rules (or, equivalently, the association of operators with classical expressions) is one which will recur throughout this paper. Each of the distribution functions which we will discuss can be used to evaluate expectation values of products of operators ordered according to a certain rule. We will consider distribution functions which can be used to compute expectation values of products of the position and momentum operators $\hat{q}$ and $\hat{p}$, and also distribution functions which can be used to compute expectation values of products of the creation and annihilation operators, $\hat{a}^+$ and $\hat{a}$. The latter are useful in problems involving electromagnetic fields. Because the creation and annihilation operators are simply related to $\hat{q}$ and $\hat{p}$ there is a relation between these two types of distribution functions. The Wigner distribution, for example, has proved useful in both the $\hat{a}$, $\hat{a}^+$ and $\hat{p}$, $\hat{q}$ contexts. The basic criterion for the choice of a distribution function for a particular problem is convenience.

In the next two sections we will continue to examine distribution functions expressed in terms of both the position and momentum variables. The Wigner function, $P_w$, will be discussed first in section 2 for not only was it the first quantum mechanical phase space distribution to be considered, but also it satisfies a number of properties which make it quite useful in applications. First of all, we will discuss its properties and then show that Wigner's distribution function gives the same expectation value for every function of $p$ and $q$ as does the corresponding operator, as proposed by Weyl [1927], for the density matrix which describes the same state to which the distribution function corresponds. As was mentioned before, this was first observed by Moyal [1949]. Next we derive an equation, in many different forms, for the time dependence of $P_w$. Finally, we apply the formalism we have developed to the calculation of $P_w$ for the eigenstates of the harmonic oscillator and also for the case of a canonical ensemble of harmonic oscillators at temperature $T$.

In section 3 we discuss distribution functions other than $P_w$ which correspond to operator ordering schemes different from that of Weyl–Wigner. Then in section 4 we treat distribution functions in terms of creation and annihilation operators, with emphasis on normal, symmetric and anti-normal ordering. In particular, we emphasize the normal ordering from which arises the well-known $P$ distribution of quantum optics. We also discuss the generalized $P$ representations. Finally, in section 5 we present our conclusions.

Applications will be treated in a future paper but we would be remiss not to mention the recent extensive review of quantum collision theory using phase space distributions (Carruthers and Zachariasan [1983]) and the work on relativistic kinetic theory – in addition to extensive discussions on the Wigner–Weyl correspondence – by the Amsterdam group (Suttorp and de Groot [1970]; Suttorp [1972]; de Groot [1974]; de Groot, van Leeuwen and van Weert [1980]). Also, a *brief* overview of some applications is presented in O'Connell [1983a,b].

## 2. Wigner distribution

### 2.1. Properties

In a 1932 paper (Wigner [1932a]) the distribution

$$P_w(q, p) = \frac{1}{\pi\hbar} \int_{-\infty}^{\infty} dy \, \langle q - y|\hat{\rho}|q + y\rangle \, e^{2ipy/\hbar} \tag{2.1}$$

was proposed to represent a system in a mixed state represented by a density matrix $\hat{\rho}$. In the case of a pure state, $\psi$, it follows from eq. (1.6) that $\rho(q', q'') = \psi(q')\psi^*(q'')$ and hence

$$P_w(q, p) = \frac{1}{\pi\hbar} \int_{-\infty}^{\infty} dy \, \psi^*(q + y) \, \psi(q - y) \, e^{2ipy/\hbar} \,. \tag{2.2a}$$

The latter result refers to one dimension. In the case of more than one dimension, the $\pi\hbar$ must be replaced by $(\pi\hbar)^{-n}$, where $n$ is the number of the variables of $\psi$ (or the number of variables of the rows or columns of $\hat{\rho}$) and $q$, $y$ and $p$ are $n$-dimensional vectors, with $py$ the scalar product of the two. The integration is then over all components of $y$. Explicitly, eq. (2.2a) generalizes to

$$P_w(q_1, \ldots q_n; p_1, \ldots p_n) = (\pi\hbar)^{-n} \int_{-\infty}^{\infty} \cdots \int_{-\infty}^{\infty} dy_1 \cdots dy_n \, \psi^*(q_1 + y_1, \ldots q_n + y_n)$$
$$\times \psi(q_1 - y_1, \ldots q_n - y_n) \exp[2i(p_1 y_1 + \cdots + p_n y_n)/\hbar] \,. \tag{2.2b}$$

It was mentioned that this choice for a distribution function was by no means unique and that this particular choice was made because it seemed to be the simplest of those for which each Galilei transformation corresponds to the same Galilei transformation of the quantum mechanical wave functions. In later work Wigner [1979] returned to this issue by considering properties which one would want such a distribution to satisfy. He then showed that the distribution given by eq. (2.1) was the only

M. Hillery et al., Distribution functions in physics: Fundamentals    127

one which satisfied these properties. A subsequent paper by O'Connell and Wigner [1981a] considered a somewhat different list of properties and showed that these, too, led to the expression in eq. (2.1).

The properties for a distribution function, $P(q, p)$, which were considered of special interest, for the case of a pure state (generalization to the case of a mixed state is straightforward), are as follows (O'Connell and Wigner [1981a]):

(i) $P(q, p)$ should be a Hermitean form of the state vector $\psi(q)$, i.e. $P$ is given by

$$P(q, p) = \langle \psi | \hat{M}(q, p) | \psi \rangle \tag{2.3}$$

where $\hat{M}(q, p)$ is a self-adjoint operator depending on $p$ and $q$. Therefore, $P(q, p)$ is real.

(ii)

$$\int \mathrm{d}p\, P(q, p) = |\psi(q)|^2 = \langle q | \hat{\rho} | q \rangle \tag{2.4}$$

$$\int \mathrm{d}q\, P(q, p) = \langle p | \hat{\rho} | p \rangle \tag{2.5}$$

$$\int \mathrm{d}q \int \mathrm{d}p\, P(q, p) = \mathrm{Tr}(\hat{\rho}) = 1 . \tag{2.6}$$

(iii) $P(q, p)$ should be Galilei invariant, i.e. if $\psi(q) \rightarrow \psi(q + a)$ then $P(q, p) \rightarrow P(q + a, p)$ and if $\psi(q) \rightarrow \exp(ip'q/\hbar)\, \psi(q)$ then $P(q, p) \rightarrow P(q, p - p')$.

(iv) $P(q, p)$ should be invariant with respect to space and time reflections, i.e. if $\psi(q) \rightarrow \psi(-q)$ then $P(q, p) \rightarrow P(-q, -p)$ and if $\psi(q) \rightarrow \psi^*(q)$ then $P(q, p) \rightarrow P(q, -p)$.

It should be admitted, however, that neither of these transformations is relativistic and also that they do not yet involve the spin variable.

(v) In the force-free case the equation of motion is the classical one

$$\frac{\partial P}{\partial t} = -\frac{p}{m} \frac{\partial P}{\partial q} . \tag{2.7}$$

(vi) If $P_\psi(q, p)$ and $P_\phi(q, p)$ are the distributions corresponding to the states $\psi(q)$ and $\phi(q)$ respectively then

$$\left| \int \mathrm{d}q\, \psi^*(q)\, \phi(q) \right|^2 = (2\pi\hbar) \int \mathrm{d}q \int \mathrm{d}p\, P_\psi(q, p)\, P_\phi(q, p) . \tag{2.8}$$

Property (vi) has two interesting consequences. If we set $\phi(q) = \psi(q)$ we get

$$\int \mathrm{d}q \int \mathrm{d}p\, [P_\psi(q, p)]^2 = \frac{1}{2\pi\hbar} \tag{2.9}$$

and, in the case of a mixed state, the right-hand side of eq. (2.9) is multiplied by $\sum_\beta w_k^2$ where the $w_k$ are the probabilities for the different states (the characteristic values of $\hat{\rho}$). This implies that $P_\psi(q, p)$ is not too highly peaked and rules out such distributions as $P_\psi(q, p) = \delta(q - \hat{q})\, \delta(p - \hat{p})$ which would be possible classically. We can also choose $\phi$ and $\psi$ so that they are orthogonal. We then have that

$$\int dq \int dp \, P_\psi(q, p) \, P_\phi(q, p) = 0 \tag{2.10}$$

which implies that $P(q, p)$ cannot be everywhere positive. This conclusion is actually rather general. Wigner [1979] has shown that any distribution function as long as it satisfies properties (i) and (ii) assumes also negative values for some $p$ and $q$.

(vii)

$$\int dq \int dp \, A(q, p) \, B(q, p) = (2\pi\hbar) \, \text{Tr}(\hat{A}\hat{B}) \tag{2.11}$$

where $A(q, p)$ is the classical function corresponding to the quantum operator $\hat{A}$, and is given, according to Wigner's prescription, by

$$A(q, p) = \int dz \, e^{ipz/\hbar} \langle q - \tfrac{1}{2}z|\hat{A}|q + \tfrac{1}{2}z \rangle \tag{2.12}$$

so that $\int \int dq \, dp \, A(q, p) = 2\pi\hbar \, \text{Tr}(\hat{A})$. A similar relation exists between $B(q, p)$ and $\hat{B}$.

The proof of eq. (2.11) will be shown below to follow as a particular case of a more general relation (eq. (2.23)) for $F(q, p)$, in terms of $A(q, p)$ and $B(q, p)$, where $\hat{F} = \hat{A}\hat{B}$. From eq. (2.12), it is at once evident that the phase space description $A(q, p)$ of the operator $\hat{A}$ is real if $\hat{A}$ is self-adjoint (Hermitean) and is imaginary if $\hat{A}$ is skew Hermitean. Since in neither case does $A(q, p)$ vanish, it is evident that if it is real, its operator $\hat{A}$ is self-adjoint, if it is imaginary $\hat{A}$ is skew symmetric. It is also evident that the phase space description of the Hermitean adjoint $\hat{A}^+$ of $\hat{A}$ is the complex conjugate of the similar description of $\hat{A}$. Similarly, if the phase space descriptions of two operators are complex conjugates of each other, then the operators are Hermitean adjoints of each other.

By comparison of eqs. (2.1) and (2.12), it is clear that $P(q, p)$, derived from the density matrix, is $(2\pi\hbar)^{-1}$ times the phase space operator which corresponds to the same matrix. Also, for $\hat{A} = \hat{\rho}$ and $\hat{B}$ equal to the unit matrix, eq. (2.6) immediately follows from eq. (2.11). Furthermore, for $\hat{B} = \hat{\rho}$, eq. (2.11) reduces to

$$\int dq \int dp \, A(q, p) \, P_w(q, p) = \text{Tr}(\hat{\rho} \, \hat{A}(\hat{q}, \hat{p})), \tag{2.13}$$

which is equivalent to eqs. (1.2) and (1.3). This result was originally obtained (Wigner [1932]) for the special case of $\hat{A}$ being the sum of a function of $\hat{p}$ only and a function of $\hat{q}$ only but Moyal [1949] showed it was actually true in the case where $\hat{A}$ is any function of $\hat{q}$ and $\hat{p}$, if $\hat{A}(\hat{q}, \hat{p})$ is the Weyl operator (discussed below in section 2.2) for $A(q, p)$. In addition, if we take $\hat{A} = \hat{B} = \hat{\rho}$ in eq. (2.11) and use the fact that, if $\hat{\rho}$ represents a pure state, $\text{Tr}(\hat{\rho})^2 = \text{Tr} \, \hat{\rho} = 1$, we obtain eq. (2.9) again.

(viii) If we define the Fourier transform of the wave function

$$\phi(p) = (2\pi\hbar)^{-1} \int dq \, \psi(q) \, e^{-iqp/\hbar}, \tag{2.14}$$

M. Hillery et al., Distribution functions in physics: Fundamentals                    129

then eq. (2.2a) can be re-written in the form

$$P(q, p) = (\pi\hbar)^{-1} \int dp' \, \phi^*(p + p') \, \phi(p - p') \, e^{-2iqp'/\hbar} \,, \tag{2.15}$$

exhibiting the basic symmetry under the interchange $q \leftrightarrow p$.

It may be worth observing also that the contraction of the distribution function from $n$ to $n-1$ variables

$$\int \int P(q_1, \ldots q_{n-1}, q_n; p_1, \ldots p_{n-1}, p_n) \, dq_n \, dp_n$$

$$= (\pi\hbar)^n \int \cdots \int \rho(q_1 - y_1, \ldots q_{n-1} - y_{n-1}, q_n - y_n; q_1 + y_1, \ldots q_{n-1} + y_{n-1}, q_n + y_n)$$

$$\times \exp[2i(p_1 y_1 + \cdots + p_{n-1} y_{n-1} + p_n y_n)/\hbar] \, dy_1 \cdots dy_{n-1} \, dy_n \, dq_n \, dp_n$$

$$= (\pi\hbar)^{n-1} \int \cdots \int \rho(q_1 - y_1, \ldots q_{n-1} - y_{n-1}, q_n - y_n; q_1 + y_1, \ldots q_{n-1} + y_{n-1}, q_n + y_n)$$

$$\times \exp[2i(p_1 y_1 + \cdots + p_{n-1} y_{n-1})/\hbar] \, \delta(y_n) \, dy_1 \cdots dy_{n-1} \, dy_n \, dq_n$$

$$= (\pi\hbar)^{n-1} \int \cdots \int [\rho(q_1 - y_1, \ldots q_{n-1} - y_{n-1}, q_n; q_1 + y_1, \ldots q_{n-1} + y_{n-1}, q_n) \, dq_n]$$

$$\times \exp[2i(p_1 y_1 + \cdots + p_{n-1} y_{n-1})/\hbar] \, dy_1 \cdots dy_{n-1} \tag{2.16}$$

gives the distribution function which corresponds to the properly contracted $\rho$ (in square brackets). Actually, this is true also for the other distribution functions which will be considered in section 3.

Wigner in his 1971 paper also showed that properties (i)–(v) determined the distribution function uniquely. O'Connell and Wigner [1981a] showed that properties (i)–(iv) and (vi) also accomplish this. In both cases the distribution function was that given by eq. (2.1).

Finally, we draw attention to two restrictions on the distribution function discussed above. First of all, as already mentioned, it is non-relativistic. Secondly, not all functions $P(q, p)$ are allowed, as we will now demonstrate by turning to the question of the admissability of $P$ and asking what condition is necessary so that $P$ implies the existence of the density function $\hat{\rho}$, the expectation values of which are, naturally, positive or zero. Our starting-point is eq. (2.2a) from which it follows that

$$\int dp \, e^{-2ipy/\hbar} \, P(q, p) = \rho(q - y, q + y) \,. \tag{2.17}$$

Hence, changing variables to $u = q + y$ and $v = q - y$, we obtain

$$\rho(v, u) = \int dp \, e^{-ip(u-v)/\hbar} \, P(\tfrac{1}{2}(u + v), p) \,. \tag{2.18}$$

We remark that since $p$ on the right-side of eq. (2.18) is a dummy variable it is clear that it could be replaced by $q$.

Now the condition for $P(q, p)$ to be a permissible distribution function is that the corresponding

130                              *M. Hillery et al., Distribution functions in physics: Fundamentals*

density matrix be positive definite, i.e.

$$\int dx \int dx' \, \psi^*(x) \, \rho(x, x') \, \psi(x') \geq 0 \tag{2.19a}$$

for all $\psi$. Using eq. (2.18) and eq. (2.19a), it follows that the condition that $P(q, p)$ be permissible is that

$$\int dq \int dp \, P(q, p) \, P'(q, p) \geq 0 \tag{2.19b}$$

for *any* $P'(q, p)$ which corresponds to a pure state. This is evident already from eq. (2.8). It also follows from eq. (2.11) and the fact that $\mathrm{Tr}(\hat{\rho}\hat{\rho}') \geq 0$. Eq. (2.19b) holds, of course, for any $P'$ which is itself permissible but the permissibility of $P$ follows already if it is valid for *all* $P'$ which correspond to a pure state.

Eight properties of the distribution function were discussed above, eqs. (2.3) to (2.16), with the emphasis on the use of this function to form another description of a quantum mechanical state, i.e. be a substitute for the density matrix. Just as eq. (2.1) permits one to give a phase space formulation to the density matrix $\hat{\rho}$, we emphasize that eq. (2.12) permits one also to give a phase space formulation to any matrix – or operator – and it may be useful to consider the properties of eq. (2.12).

In particular, we wish to derive an expression for the function $F(q, p)$ which corresponds to the product $\hat{F} = \hat{A}\hat{B}$ of two operators $\hat{A}$ and $\hat{B}$ to which the $q, p$ functions $A(q, p)$ and $B(q, p)$ correspond. We assume that the operators $\hat{A}$ and $\hat{B}$ are matrices, the rows and columns of which can be characterized by a single variable, but the generalization to a many-dimensional configuration space is obvious. We can write, therefore

$$\hat{F}(x, x'') = \int \hat{A}(x, x') \, \hat{B}(x', x'') \, dx' . \tag{2.20}$$

Analogous to eqs. (2.17) and (2.18), eq. (2.20) can be written as (taking $\hbar = 1$ for this derivation)

$$\int dp_1 \, F(\tfrac{1}{2}(x + x''), p_1) \, e^{-ip_1(x''-x)} = (2\pi)^{-1} \int \int \int dx' \, dp' \, dp'' \, A(\tfrac{1}{2}(x + x'), p') \, e^{-ip'(x'-x)}$$

$$\times B(\tfrac{1}{2}(x' + x''), p'') \, e^{-ip''(x''-x')} . \tag{2.21}$$

Substituting $x = q + q'$, $x'' = q - q'$, multiplying with $e^{-2iq'p}$ and integrating over $q'$ one obtains

$$F(q, p) = 2 \, (2\pi)^{-2} \int \int \int \int dq' \, dx' \, dp' \, dp'' \, A(\tfrac{1}{2}(q + q' + x'), p') \, B(\tfrac{1}{2}(q - q' + x'), p'')$$

$$\times \exp\{-iq'(2p - p' - p'') - i(p'' - p')(q - x')\} . \tag{2.22}$$

Introducing finally new variables $y = \tfrac{1}{2}(q + x')$, $y' = \tfrac{1}{2}q'$, $p' = \rho - \rho'$, $p'' = \rho + \rho'$, one obtains

$$F(q, p) = 16 \, (2\pi)^{-2} \int \int \int \int dy \, dy' \, d\rho \, d\rho' \, A(y + y', \rho - \rho') \, B(y - y', \rho + \rho')$$

$$\times \exp\{-4i \, y'(p - \rho) - 4i \, \rho'(q - y)\}$$

$$= 16 \, (2\pi)^{-2} \int \int \int \int dy \, dy' \, d\rho \, d\rho' \, A(q + y + y', p + \rho - \rho') \, B(q + y - y', p + \rho + \rho') \, e^{4i(\rho y' + y \rho')} . \tag{2.23}$$

M. Hillery et al., *Distribution functions in physics: Fundamentals*     131

This expression for $F(q, p)$, which is a new result, also shows the similarity of the roles of $p$ and $q$ in Hamiltonian mechanics. In the next subsection, another expression (eq. (2.59)) for $F(q, p)$ will be presented.

If we integrate $F(q, p)$ in eq. (2.23) over $q$ and $p$, we obtain

$$\int \int F(q, p) \, dq \, dp = 16 \, (2\pi)^{-2} \int \int \int \int dy \, dy' \, d\rho \, d\rho' \, A(y + y', \rho - \rho') B(y - y', \rho + \rho')$$

$$\times \exp\{4iy'\rho + 4i\rho'y\} \, (4\pi^2/16) \, \delta(y') \, \delta(\rho') \, . \tag{2.24}$$

Hence

$$\int \int F(q, p) \, dq \, dp = \int \int A(q, p) \, B(q, p) \, dq \, dp \, . \tag{2.25}$$

Since the left-hand side of this equation is the same as $(2\pi\hbar) \, \text{Tr}(\hat{F})$, it is clear that eq. (2.25) is the same as eq. (2.11). In the case of $n$ dimensions, the $n$th power of $(4/\pi^2)$ appears in the expression corresponding to eq. (2.23).

Eq. (2.23) provides also a means to ascertain, in terms of the phase space descriptions of $\hat{A}$ and $\hat{B}$, whether these two operators commute. Naturally, the condition for the commutative nature is

$$\int \int \int \int dy \, dy' \, d\rho \, d\rho' \, A(y + y', \rho - \rho') B(y - y', \rho + \rho') \exp\{-4i \, y'(p - \rho) - 4i \, \rho'(q - y)\}$$

$$= \int \int \int \int dy \, dy' \, d\rho \, d\rho' \, B(y + y', \rho - \rho') A(y - y', \rho + \rho') \exp\{-4i \, y'(p - \rho) - 4i \, \rho'(q - y)\} \, . \tag{2.26}$$

Since this is valid for all $p$ and $q$, the integration over the variables which are their factors in the exponent (i.e. $y'$ and $\rho'$) can be omitted. This gives as condition for the commutability of $\hat{A}$ and $\hat{B}$ (we replace $y$, $y'$ by $q$, $q'$ and $\rho$, $\rho'$ by $p$, $p'$):

$$\int \int dp \, dq \, [A(q + q', p - p') B(q - q', p + p') - A(q - q', p + p') B(q + q', p - p')]$$

$$\times \exp\{4i(q'p + p'q)\} = 0 \, , \tag{2.27}$$

a somewhat unexpected expression.

The last quantum mechanical relation that will be translated into phase space language is the equation $\hat{A}\hat{\rho} = \lambda\hat{\rho}$ specifying that the wave functions of which $\hat{\rho}$ consists are characteristic functions (eigenfunctions) of $\hat{A}$ with the characteristic value $\lambda$. Whether $\hat{\rho}$ contains only one or more such characteristic functions depends whether or not its phase space representation, $P_w$, satisfies eq. (2.9), i.e. whether its square integral is equal to or smaller than $(2\pi\hbar)^{-1}$.

The $\hat{A}\hat{\rho} = \lambda\hat{\rho}$ relation, with $\hat{\rho}$ represented by $P_w$, reads, according to eq. (2.23), in phase space language:

$$(4/\pi^2) \int \int \int \int dy\, dy'\, d\rho\, d\rho'\, A(y+y', \rho-\rho')\, P_w(y-y', \rho+\rho')\, \exp\{4i\, y'(\rho-p) + 4i\, \rho'(y-q)\}$$

$$= \lambda\, P_w(q, p).\tag{2.28}$$

In order to simplify this, one can multiply with $\exp\{4i(q'p + p'q)\}$ and integrate over $p$ and $q$ to obtain, substituting also $q$ and $p$ for the integration variables $y$ and $\rho$,

$$\int \int dq\, dp\, A(q+q', p-p')\, P_w(q-q', p+p')\, \exp\{4i(q'p + p'q)\}$$

$$= \lambda \int \int dq\, dp\, P_w(q, p)\, \exp\{4i(q'p + p'q)\}.\tag{2.29}$$

Both eqs. (2.27) and (2.29) are a good deal more complicated than the quantum mechanical equations for which they substitute. It is questionable whether they are really useful. We thought that they should be derived in spite of this because the final form is considerably simpler than the original one and because they clearly demonstrate the essential phase space equivalence of $q$ and $p$. It may be worth remarking finally that in the case of several dimensions all variables should be considered as vectors, and products like $q'p$ or $p'q$ should be replaced by scalar products of these vectors.

## 2.2. Associated operator ordering

We will now discuss the connection between a classical function of $q$ and $p$ and a quantum mechanical operator which is supposed to correspond to it. The result of the measurement of a quantum mechanical operator is well defined: it is supposed to transfer the state of the system on which the measurement is carried out into one of the characteristic vectors of the operator in question, and the probabilities with which the different characteristic vectors would result from the measurement are also well defined. They are the squares of the scalar products of the normalized initial state of the system on which the measurement is carried out and of the operator's normalized characteristic vector into which the state of the system is transformed. It must be admitted, even in this case, that, given an arbitrary operator, it is in many cases difficult, in others impossible, to construct an apparatus which can carry out the measurement, i.e. the desired change of the state of the system on which the measurement is to be carried out.

But as far as the measurement of a classical function of $p$ and $q$ is concerned, no similar postulate exists which can be formulated in classical terms. But Weyl did propose the association of a quantum mechanical operator to every function of $q$ and $p$ and defined the measurement of the classical quantity as being identical with the above described quantum mechanical measurement of the operator which he associated to the classical function of $q$ and $p$. This association will be described below. What is remarkable, however, and what has been first pointed out by Moyal [1949], is the close connection between Weyl's proposal and the distribution function as defined above. In particular, the expectation value of the result of the measurement of the operator $\hat{A}$, which Weyl associates with the classical function $A(q, p)$ if carried out on a system in the state $\psi$,

$$\langle \psi | \hat{A} | \psi \rangle = \int dq \int dp\, P_w(q, p)\, A(q, p)\tag{2.30}$$

M. Hillery et al., Distribution functions in physics: Fundamentals    133

is equal to the expectation value of the classical function $A(q, p)$ to which $\hat{A}$ corresponds assuming that the system is described by the distribution function $P_w(q, p)$ which corresponds to $\hat{\rho}$. This is the content of eq. (2.30) and it is valid, as will be demonstrated below, for every state vector $\psi$ and also for any density matrix $\hat{\rho}$

$$\text{Tr}(\hat{\rho}\hat{A}) = \int dq \int dp \, P_w(q, p) \, A(q, p) . \tag{2.31}$$

Actually eq. (2.31) is an easy consequence of eq. (2.30) and only the latter will be proved below.

In order to prove eq. (2.30), we start with Weyl's expansion of $A(q, p)$ into a Fourier integral (taking $\hbar = 1$ for the purpose of this proof):

$$A(q, p) = \int d\sigma \int d\tau \, \alpha(\sigma, \tau) \, e^{i(\sigma q + \tau p)} . \tag{2.32}$$

Weyl then defines the operator which corresponds to the exponential in the integrand on the right-hand side of eq. (2.32) as $\exp\{i(\sigma\hat{q} + \tau\hat{p})\}$. The operator which corresponds to $A(q, p)$ is then given by

$$\hat{A}(\hat{q}, \hat{p}) = \int d\sigma \int d\tau \, \alpha(\sigma, \tau) \exp\{i(\sigma\hat{q} + \tau\hat{p})\} . \tag{2.33}$$

If we substitute this result for $\hat{A}$ into the left-hand side of eq. (2.30) and replace $A(q, p)$ on the right-hand side by the right-hand side of eq. (2.32), it becomes evident that all we have to prove is that

$$\langle\psi|\exp\{i(\sigma\hat{q} + \tau\hat{p})\}|\psi\rangle = \int dq \int dp \, P_w(q, p) \exp\{i(\sigma q + \tau p)\}$$

$$= \frac{1}{2\pi} \int dy \int dq \int dp \, \psi^*(q + \tfrac{1}{2}y) \, \psi(q - \tfrac{1}{2}y) \exp\{ipy + i(\sigma q + \tau p)\} . \tag{2.34}$$

The integration over $p$ gives $2\pi \, \delta(y + \tau)$ and hence the right-hand side of eq. (2.34) becomes

$$\int dq \, \psi^*(q - \tfrac{1}{2}\tau) \, \psi(q + \tfrac{1}{2}\tau) \, e^{i\sigma q} .$$

In order to evaluate the left-hand side of (2.34) we note that according to the Baker–Hausdorff theorem (Messiah [1961]), if the commutator $\hat{D} = [\hat{A}, \hat{B}]$ commutes with $\hat{A}$ and $\hat{B}$ then

$$e^{\hat{A}+\hat{B}} = e^{\hat{A}} \, e^{\hat{B}} \, e^{-\hat{D}/2} . \tag{2.35a}$$

It then follows that

$$e^{i(\sigma\hat{q}+\tau\hat{p})} = e^{i\sigma\hat{q}} \, e^{i\tau\hat{p}} \, e^{i\sigma\tau/2} . \tag{2.35b}$$

Hence, the left-hand side of eq. (2.34) becomes

$$e^{i\sigma\tau/2}\langle\psi|e^{i\sigma\hat{q}}\,e^{i\tau\hat{p}}|\psi\rangle\,.$$

Next, using the fact that

$$e^{i\tau\hat{p}}\,|\psi(x)\rangle = |\psi(x+\tau)\rangle \tag{2.36}$$

and transferring the $e^{i\sigma\hat{q}}$ to the left-hand side, this becomes

$$e^{i\sigma\tau/2}\langle e^{-i\sigma x}\,\psi(x)|\psi(x+\tau)\rangle = \int dx\; e^{i(\sigma x+\sigma\tau/2)}\,\psi^*(x)\,\psi(x+\tau)\,, \tag{2.37}$$

which is equal to the expression obtained above for the right-hand side of eq. (2.34). Thus, we have proved eq. (2.34) and hence also eq. (2.30).

In summary, if a classical function

$$A(q,p) = \int d\sigma \int d\tau\, e^{(i/\hbar)(\sigma q+\tau p)}\,\alpha(\sigma,\tau) \tag{2.38}$$

goes over to the quantum operator

$$\hat{A}(\hat{q},\hat{p}) = \int d\sigma \int d\tau\, e^{(i/\hbar)(\sigma\hat{q}+\tau\hat{p})}\,\alpha(\sigma,\tau) \tag{2.39}$$

then the relation between $A(q,p)$ and $\hat{A}$ is that given by Wigner in eq. (2.12). Furthermore, it is clear that if, for all $A(p,q)$

$$\int dq \int dp\, P(q,p)\, A(q,p) = \int dq \int dp\, P'(q,p)\, A(q,p) \tag{2.40}$$

then $P'$ is identical with $P$.

In addition, we mention that under the Weyl correspondence the classical quantity $q^n p^m$ becomes

$$q^n p^m \rightarrow \frac{1}{2^n}\sum_{r=0}^{n}\binom{n}{r}\hat{q}^{n-r}\,\hat{p}^m\,\hat{q}^r \tag{2.41}$$

as can be seen by considering the $\sigma^n\tau^m$ coefficient in $(\sigma\hat{q}+\tau\hat{p})^{n+m}$.

Finally, we would like to mention the role played by the characteristic function. This is a description of the state $\hat{\rho}$ by means of a function of two new variables, $\sigma$ and $\tau$,

$$C(\sigma,\tau) = \mathrm{Tr}(\hat{\rho}\,\hat{C}(\sigma,\tau))\,, \tag{2.42}$$

where

$$\hat{C}(\sigma,\tau) = e^{(i/\hbar)(\sigma\hat{q}+\tau\hat{p})}\,. \tag{2.43}$$

*M. Hillery et al., Distribution functions in physics: Fundamentals*     135

Here we are following the nomenclature of Moyal which has now become standard in describing this quantity as a "characteristic function". This description stems from statistical terminology, and, in particular, should not be confused with the sometime usage of "characteristic function" in quantum mechanics to denote an eigenfunction.

$C(\sigma, \tau)$ is just the Fourier transform of $P(q, p)$. To see this we note that the function corresponding to $\hat{C}(\sigma, \tau)$ is just $\exp\{(i/\hbar)(\sigma q + \tau p)\}$. Making use of eq. (2.11) gives

$$C(\sigma, \tau) = \text{Tr}(\hat{\rho}\,\hat{C}(\sigma, \tau)) = \int \mathrm{d}q \int \mathrm{d}p \; e^{(i/\hbar)(\sigma q + \tau p)} \, P_w(q, p) \tag{2.44}$$

so that

$$P_w(q, p) = \left(\frac{1}{2\pi\hbar}\right)^2 \int \mathrm{d}\sigma \int \mathrm{d}\tau \; e^{(-i/\hbar)(\sigma q + \tau p)} \, C(\sigma, \tau). \tag{2.45}$$

We can use the characteristic function to compute expectation values of Weyl-ordered products of $p$ and $q$. We have that

$$\left(\frac{\hbar}{i}\right)^{m+n} \frac{\partial^m}{\partial \sigma^m} \frac{\partial^n}{\partial \tau^n} \, C(\sigma, \tau)\bigg|_{\sigma=\tau=0} = \int \mathrm{d}q \int \mathrm{d}p \, q^m p^n \, P_w(q, p), \tag{2.46}$$

the right-hand side of which is just the average of the Weyl-ordered product $q^m p^n$.

## 2.3. Dynamics

We would now like to derive equations for the time-dependence of $P_w$. As before, our detailed considerations will be confined to one dimension but some results will also be quoted for the multi-dimensional case. The time-dependence of $P_w$ may be decomposed into two parts (Wigner [1932a])

$$\frac{\partial P_w}{\partial t} = \frac{\partial_k P_w}{\partial t} + \frac{\partial_v P_w}{\partial t} \tag{2.47}$$

the first part resulting from the $(i\hbar/2m)\,\partial^2/\partial q^2$ part, the second from the potential energy $V/i\hbar$ part of the expression for $\partial \psi/\partial t$.

From the definition of $P_w$, given by eq. (2.2a), it follows that

$$\frac{\partial_k P_w}{\partial t} = -\frac{i}{2\pi m} \int \mathrm{d}y \left[ \frac{\partial^2 \psi^*(q+y)}{\partial y^2} \psi(q-y) - \psi^*(q+y) \frac{\partial^2 \psi(q-y)}{\partial y^2} \right] e^{2ipy/\hbar}, \tag{2.48}$$

where we have taken advantage of the functional dependence of $\psi$ to replace $\partial^2/\partial q^2$ by $\partial^2/\partial y^2$. Next we perform one partial integration with respect to $y$ to obtain

$$\frac{\partial_k P_w}{\partial t} = -\frac{p}{\pi \hbar m} \int \mathrm{d}y \left[ \frac{\partial \psi^*(q+y)}{\partial y} \psi(q-y) - \psi^*(q+y) \frac{\partial \psi(q-y)}{\partial y} \right] e^{2ipy/\hbar}, \tag{2.49}$$

since the boundary term does not contribute. Switching back from $\partial/\partial y$ to $\partial/\partial q$, we finally obtain

$$\frac{\partial_k P_w}{\partial t} = -\frac{p}{m}\frac{\partial P_w(q,p)}{\partial q}. \tag{2.50}$$

This is identical with the classical (Liouville) equation for the corresponding part of $\partial P/\partial t$, as was mentioned at eq. (2.7). We next calculate

$$\frac{\partial_v P_w}{\partial t} = \frac{i}{\pi\hbar^2}\int dy\,\{[V\psi^*(q+y)]\,\psi(q-y) - \psi^*(q+y)\,[V\psi(q-y)]\}\,e^{2ipy/\hbar}$$

$$= \frac{i}{\pi\hbar^2}\int dy\,[V(q+y) - V(q-y)]\,\psi^*(q+y)\,\psi(q-y)\,e^{2ipy/\hbar}. \tag{2.51}$$

Assuming that $V$ can be expanded in a Taylor series, we write

$$V(q+y) = \sum_{\lambda=0}^{\infty}\frac{y^\lambda}{\lambda!}\,V^{(\lambda)}(q) \tag{2.52}$$

where $V^{(\lambda)}(q) = \partial^\lambda V/\partial q^\lambda$. It follows that

$$\frac{\partial_v P_w}{\partial t} = \frac{2i}{\pi\hbar^2}\int dy\,\sum_\lambda\frac{y^\lambda}{\lambda!}\,V^{(\lambda)}(q)\,\psi^*(q+y)\,\psi(q-y)\,e^{2ipy/\hbar}, \tag{2.53}$$

where now the summation over $\lambda$ is restricted to all odd positive integers. It is clear that in the powers $y^\lambda$ in the integrand we can replace $y$ by $(\hbar/2i)(\partial/\partial p)$. It then follows that

$$\frac{\partial_v P_w}{\partial t} = \sum_\lambda\frac{1}{\lambda!}\left(\frac{\hbar}{2i}\right)^{\lambda-1}\frac{\partial^\lambda V(q)}{\partial q^\lambda}\frac{\partial^\lambda P_w(q,p)}{\partial p^\lambda}, \tag{2.54}$$

$\lambda$ again being restricted to odd integers. An alternative form for $\partial_v P_w/\partial t$ is given by

$$\frac{\partial_v P_w}{\partial t} = \int dj\,P_w(q,p+j)\,J(q,j), \tag{2.55}$$

where

$$J(q,j) = \frac{i}{\pi\hbar^2}\int dy\,[V(q+y) - V(q-y)]\,e^{-2ipy/\hbar}$$

$$= \frac{i}{\pi\hbar^2}\int dy\,[V(q+y) - V(q-y)]\,\sin(2jy/\hbar) \tag{2.55a}$$

is the probability of a jump in the momentum by an amount $j$ if the positional coordinate is $q$. The first part of eq. (2.55a) may be verified by inserting the Fourier expansion, with respect to $y$, of $V(q+y) -$

M. Hillery et al., Distribution functions in physics: Fundamentals    137

$V(q-y)$ into eq. (2.51). The second part is obtained by replacing the exponential by $\cos + i \sin$ and noting that the expansion in the square bracket is odd so that the integral of the cos part vanishes.

In the multi-dimensional case where $P_w = P_w(q_1, \ldots q_n; p_1, \ldots p_n)$, the corresponding results are

$$\frac{\partial P_w}{\partial t} = -\sum_{k=1}^{n} \frac{p_k}{m_k} \frac{\partial P_w}{\partial q_k} + \sum \frac{\partial^{\lambda_1 + \cdots + \lambda_n} V}{\partial q_1^{\lambda_1} \cdots \partial q_n^{\lambda_n}} \frac{(\hbar/2i)^{\lambda_1 + \cdots + \lambda_n - 1}}{\lambda_1! \cdots \lambda_n!} \frac{\partial^{\lambda_1 + \cdots + \lambda_n} P_w}{\partial p_1^{\lambda_1} \cdots \partial p_n^{\lambda_n}}, \tag{2.56}$$

where the last summation has to be extended over all positive integer values of $\lambda_1, \ldots \lambda_n$ for which the sum $\lambda_1 + \lambda_2 + \cdots + \lambda_n$ is odd.

The lowest term of eq. (2.56) in which only one $\lambda$ is 1 and the others vanish, and which has no $\hbar$ factors, is identical with the corresponding term of Liouville's equation. Hence eq. (2.56) reproduces the classical (but non-relativistic) equation if $\hbar$ is set equal to zero. The $\hbar^2$ terms give the quantum correction if this is very small. We will obtain a somewhat similar equation for the $1/T$ dependence of the distribution function of the canonical ensemble, which also is useful if the temperature $T$ is not too low so that the quantum correction is small.

Eq. (2.56) is the generalization of eq. (2.50) and eq. (2.54) for an $n$-dimensional configuration space. The same generalization of eq. (2.50) with eq. (2.55) is

$$\frac{\partial P_w}{\partial t} = -\sum_k \frac{p_k}{m_k} \frac{\partial P_w}{\partial q_k} + \int dj_1 \cdots \int dj_n \, P_w(q_1, \ldots q_n; p_1 + j_1, \ldots p_n + j_n) \, J(q_1, \ldots q_n; p_1, \ldots p_n), \tag{2.57}$$

where $J(q_1, \ldots q_n; j_1, \ldots j_n)$ can be interpreted as the probability of a jump in the momenta with the amounts $j_1, \ldots j_n$ for the configuration $q_1, \ldots q_n$. The probability of this jump is given by

$$J(q_1, \ldots q_n; j_1, \ldots j_n) = \frac{i}{\pi^n \hbar^{n+1}} \int dy_1 \cdots \int dy_n \, [V(q_1 + y_1, \ldots q_n + y_n) - V(q_1 - y_1, \ldots q_n - y_n)]$$

$$\times \exp\{-(2i/\hbar)(y_1 j_1 + \cdots + y_n j_n)\} \tag{2.58}$$

that is, by the Fourier expansion coefficients of the potential $V(q_1, \ldots q_n)$.

From eq. (2.56) it is clear that the equation of motion is the same as the classical equation of motion when $V$ has no third and higher derivatives as, for example, in the case of a uniform electric field or for a system of oscillators. However, there is still a subtle difference in that the possible initial conditions are restricted. This comes about because not all $P(q, p)$ are permissible (see eq. (2.19b)).

While we consider that the above form for the equations of motion (Wigner [1932]) are the simplest to use in practice, we will now discuss some other forms which occur frequently in the literature.

Before doing so it is useful to take note of another relation, in addition to that given by eq. (2.23), which expresses the Weyl function corresponding to an operator $\hat{F} = \hat{A}\hat{B}$ in terms of the Weyl functions corresponding to $\hat{A}$ and $\hat{B}$. This relation was first derived by Groenewold [1946] and was also discussed by Imre, Ozizmir, Rosenbaum and Zweifel [1967]. They find that the function corresponding to $\hat{F}$ is

$$\hat{A}\hat{B} = \hat{F} \rightarrow F(q, p) = A(q, p) \, e^{(\hbar \Lambda / 2i)} B(q, p)$$

$$= B(q, p) \, e^{-(\hbar \Lambda / 2i)} A(q, p), \tag{2.59}$$

where

$$\Lambda = \frac{\overleftarrow{\partial}}{\partial p} \frac{\overrightarrow{\partial}}{\partial q} - \frac{\overleftarrow{\partial}}{\partial q} \frac{\overrightarrow{\partial}}{\partial p} \tag{2.60}$$

and the arrows indicate in which direction the derivatives act. Also $(\partial/\partial p)(\partial/\partial q)$ is considered as the multi-dimensional scalar product of $\partial/\partial p$ and $\partial/\partial q$, or, in other words, it is equal to $(\partial/\partial p_i)(\partial/\partial q_i)$, where $i = (1, \ldots n)$ and $n$ denotes the number of dimensions and, as usual, repeated indices denotes summation.

To derive this result we first note that

$$\langle q''|\hat{A}|q'\rangle = \int d\sigma \, \exp\{(i/\hbar)\,\sigma(q'+q'')/2\}\,\alpha(\sigma, q'-q''), \tag{2.61}$$

where $\alpha$ is defined by eq. (2.32). This result follows from eq. (2.33) by taking the matrix element of both sides. A similar result follows for $\langle q''|\hat{B}|q'\rangle$ except that $\alpha$ is replaced by $\beta$, the Fourier transform of $B(q, p)$:

$$\hat{B}(\hat{q}, \hat{p}) = \int d\sigma \int d\tau \, \exp\{(i/\hbar)\,(\sigma\hat{q} + \tau\hat{p})\}\,\beta(\sigma, \tau). \tag{2.62}$$

We can now calculate $F(q, p)$. We have from eq. (2.12) that

$$
\begin{aligned}
F(q, p) &= \int dz \, e^{(i/\hbar)pz} \left\langle q - \frac{z}{2} \middle| \hat{A}\hat{B} \middle| q + \frac{z}{2} \right\rangle \\
&= \int dz \int dq' \, e^{(i/\hbar)pz} \left\langle q - \frac{z}{2} \middle| \hat{A} \middle| q' \right\rangle \left\langle q' \middle| \hat{B} \middle| q + \frac{z}{2} \right\rangle \\
&= \int dz \int dq' \int d\sigma \int d\sigma' \, e^{(i/\hbar)\,\sigma(q'+q-z/2)/2} \, e^{(i/\hbar)\,\sigma'(q'+q+z/2)/2} \\
&\quad \times \alpha\left(\sigma, q' - q + \frac{z}{2}\right) \beta\left(\sigma', q - q' + \frac{z}{2}\right) e^{(i/\hbar)pz}.
\end{aligned}
\tag{2.63}
$$

We now define two new variables of integration $\tau = q' - q + (z/2)$ and $\tau' = q - q' + (z/2)$ so that

$$F(q, p) = \int d\tau \int d\tau' \int d\sigma \int d\sigma' \, e^{(i/\hbar)\,(\sigma q + \tau p)} \, \alpha(\sigma, \tau) \, e^{(i/\hbar)\,(\sigma'\tau - \sigma\tau')/2} \, e^{(i/\hbar)\,(\sigma'q + \tau'p)} \, \beta(\sigma', \tau'). \tag{2.64}$$

It is possible to replace the exponential factor $\exp\{(i/\hbar)\,(\sigma'\tau - \sigma\tau')/2\}$ by $\exp(\hbar\Lambda/2i)$ so that eq. (2.64) becomes

$$F(q, p) = A(q, p) \, e^{\hbar\Lambda/2i} \, B(q, p) \tag{2.65}$$

i.e. just the first expression appearing on the right-hand side of eq. (2.59). The second expression also follows readily from eq. (2.64).

*M. Hillery et al., Distribution functions in physics: Fundamentals*    139

We can also make use of eq. (2.64) to find an alternative expression for $F(q, p)$ involving the Bopp operators (Bopp [1961] and Kubo [1964])

$$Q = q - \frac{\hbar}{2i} \frac{\partial}{\partial p}, \qquad P = p + \frac{\hbar}{2i} \frac{\partial}{\partial q}. \tag{2.66}$$

We first note that

$$\exp\left\{\frac{i}{\hbar}\left[\sigma\left(q - \frac{\hbar}{2i}\frac{\partial}{\partial p}\right) + \tau\left(p + \frac{\hbar}{2i}\frac{\partial}{\partial q}\right)\right]\right\} = \exp\left\{\frac{i}{\hbar}(\sigma q + \tau p)\right\} \exp\left\{\frac{1}{2}\left(\tau\frac{\partial}{\partial q} - \sigma\frac{\partial}{\partial p}\right)\right\} \tag{2.67}$$

so that

$$\exp\left\{\frac{i}{\hbar}\left[\sigma\left(q - \frac{\hbar}{2i}\frac{\partial}{\partial p}\right) + \tau\left(p + \frac{\hbar}{2i}\frac{\partial}{\partial q}\right)\right]\right\} e^{(i/\hbar)(\sigma'q + \tau'p)} = e^{(i/\hbar)(\sigma q + \tau p)} e^{(i/\hbar)(\sigma'\tau - \sigma\tau')/2}. \tag{2.68}$$

Using this result in eq. (2.64) we then have that

$$F(q, p) = \int d\tau \int d\tau' \int d\sigma \int d\sigma' \, e^{(i/\hbar)(\sigma Q + \tau P)} \, \alpha(\sigma, \tau) \, e^{(i/\hbar)(\sigma'q + \tau'p)} \, \beta(\sigma', \tau'). \tag{2.69}$$

From eq. (2.33) we see that the expression

$$\bar{A}(Q, P) \equiv \int d\tau \int d\sigma \, e^{(i/\hbar)(\sigma Q + \tau P)} \, \alpha(\sigma, \tau) \tag{2.70}$$

is just the Weyl-ordered operator $\hat{A}(\hat{q}, \hat{p})$ with $\hat{q} \to Q$ and $\hat{p} \to P$. $\bar{A}(Q, P)$ is also an operator but not on the Hilbert space on which $\hat{A}(\hat{q}, \hat{p})$ is an operator; it operates on functions in phase space. We can, therefore, express $F(q, p)$ as

$$F(q, p) = \bar{A}(Q, P) \, B(q, p). \tag{2.71}$$

In a similar manner one can show that

$$F(q, p) = \bar{B}(Q^*, P^*) \, A(q, p), \tag{2.72}$$

where

$$Q^* = q + \frac{\hbar}{2i} \frac{\partial}{\partial p}, \qquad P^* = p - \frac{\hbar}{2i} \frac{\partial}{\partial q}. \tag{2.73}$$

It is now possible to make use of the fact that the Wigner distribution is the function which is associated with $(1/2\pi\hbar)\hat{\rho}$. The equation of motion for $\hat{\rho}$ is just

$$i\hbar \, \partial\hat{\rho}/\partial\tau = [\hat{H}, \hat{\rho}]. \tag{2.74}$$

This implies that we have for the Wigner function

$$i\hbar \, \partial P_w/\partial t = H(q, p) \, e^{\hbar\Lambda/2i} \, P_w(q, p) - P_w(q, p) \, e^{\hbar\Lambda/2i} \, H(q, p)$$

or

$$\hbar \, \partial P_w/\partial t = -2 H(q, p) \sin(\hbar\Lambda/2) P_w(q, p) \, , \qquad (2.75)$$

where $H(q, p)$ is the function corresponding to the Hamiltonian operator for the system, $\hat{H}$. Actually, this is an abbreviated form of eq. (2.56) as can be verified by expanding the sin into a power series. Note that if we take the $\hbar \to 0$ limit of this equation we obtain the classical Liouville equation

$$\partial P_w^c/\partial t + \{P_w^c, H\} = 0 \, , \qquad (2.76)$$

where $\{ \}$ denote Poisson brackets and the superscript c on $P_w$ indicates the classical limit. For an $H(q, p)$ which is at most quadratic in $q$ and $p$, e.g. a free particle or an harmonic oscillator, eqs. (2.75) and (2.76) coincide. In these systems, then, the difference between a classical and a quantum ensemble is the restriction on the initial conditions in the case of latter (cf. eq. (2.19)).

We also want to quote two alternate forms of eq. (2.75). The first follows immediately from our discussion of the Bopp operators. We have, using eqs. (2.65), (2.71), (2.72) and (2.75), that

$$i\hbar \, \partial P_w/\partial t = [\bar{H}(Q, P) - \bar{H}(Q^*, P^*)] P_w(q, p) \, , \qquad (2.77)$$

a result first obtained by Bopp [1961]. Analogous to the definition of $A(\hat{Q}, P)$, given by eq. (2.70), $\bar{H}(Q, P)$ is the Weyl-ordered operator with $\hat{q} \to Q$ and $\hat{p} \to P$, where $Q$ and $P$ are defined in eq. (2.66). These equations do not exhaust the possible formulations of the dynamics of the Wigner function. One can also make use of propagation kernels. This approach is discussed by Moyal [1949] and Mori, Oppenheim and Ross [1962].

We turn now to a consideration of a canonical ensemble. If $\beta = 1/kT$ where $k$ is Boltzmann's constant and $T$ is the temperature, then the density matrix of the canonical ensemble is

$$\hat{\rho} = \frac{1}{Z(\beta)} \, e^{-\beta\hat{H}} \equiv \frac{1}{Z(\beta)} \, \hat{\Omega} \qquad (2.78)$$

and $Z(\beta) = \mathrm{Tr}(e^{-\beta\hat{H}})$. The unnormalized density matrix, $\hat{\Omega}$, then satisfies the equation

$$\partial\hat{\Omega}/\partial\beta = -\hat{H}\hat{\Omega} = -\hat{\Omega}\hat{H} \, , \qquad (2.79)$$

subject to the initial condition $\hat{\Omega}(\beta = 0) = \hat{I}$ where $\hat{I}$ is the identity operator. Eq. (2.79) is referred to as the Bloch [1932] equation for the density matrix of a canonical ensemble. Using the product rule given by eq. (2.59) we have that

$$\partial\Omega(q, p)/\partial\beta = -H(q, p) \, e^{\hbar\Lambda/2i} \, \Omega(q, p) = -H(q, p) \, e^{-\hbar\Lambda/2i} \, \Omega(q, p) \, , \qquad (2.80)$$

$\Lambda$ being given by eq. (2.60) so that

M. Hillery et al., Distribution functions in physics: Fundamentals    141

$$\partial\Omega(q, p)/\partial\beta = -H(q, p)\cos(\hbar\Lambda/2)\,\Omega(q, p)\,.\tag{2.81}$$

This is the Wigner translation of the Bloch equation, which was entensively studied by many authors and was first derived in this form by Oppenheim and Ross [1957]. It is useful in the calculation of quantum mechanical corrections to classical statistical mechanics. The initial condition for this equation is just the Wigner function corresponding to $\hat{\Omega}(\beta = 0) = \hat{I}$. Inserting $I$ in eq. (2.12) we find that the initial condition is just $\Omega(q, p)|_{\beta=0} = 1$.

It is also worth noting that $P_w(q, p)$ does not satisfy the Wigner translation of the Bloch equation simply because of the fact that it must be multiplied by the $\beta$-dependent factor $(2\pi\hbar)\,Z(\beta)$ in order to obtain $\Omega(q, p)$.

Finally, we emphasize that all equations from eq. (2.59) onwards hold in the multi-dimensional case, where we simply interpret $(q, p)$ to be $(q_1, \ldots q_n; p_1, \ldots p_n)$ and the simple products in the exponents as scalar products. The solution of eq. (2.81) in the multi-dimensional case, is to order $\hbar^2$ (Wigner [1932a]),

$$\Omega_w(q, p) = e^{-\beta H(q, p)}\left\{1 + (2\pi\hbar)^2\left[\sum_k\left(-\frac{\beta^2}{8m_k}\frac{\partial^2 V}{\partial q_k^2} + \frac{\beta^3}{24m_k}\left(\frac{\partial V}{\partial q_k}\right)^2\right) + \sum_{k,l}\frac{\beta^3 p_k p_l}{24 m_k m_l}\frac{\partial^2 V}{\partial q_k\partial q_l}\right]\right\}\,.\tag{2.82}$$

Actually, the Wigner translation of the Bloch equation, eq. (2.18) above, can be simplified further into a form, analogous to that of eq. (2.56), which is more convenient for applications. This is achieved by writing the cos term as the real part of the operator

$$\hat{0} \equiv \exp\left[\frac{i\hbar}{2}\left(\frac{\overleftarrow{\partial}}{\partial p}\frac{\overrightarrow{\partial}}{\partial q} - \frac{\overleftarrow{\partial}}{\partial q}\frac{\overrightarrow{\partial}}{\partial p}\right)\right]\,,\tag{2.83}$$

where we have used the explicit form for $\Lambda$ given in eq. (2.60), again noting that the arrows indicate in which direction the derivatives act and that the gradient operators are $3N$-dimensional. Next we decompose $\hat{0}$ by means of the Baker–Hausdorff theorem (eq. (2.35a)), and using the fact that

$$\frac{\partial}{\partial p}\frac{\partial}{\partial q}H(q, p) = 0\tag{2.84}$$

it follows that we may write

$$\hat{0} = \exp\left[\frac{i\hbar}{2}\frac{\overleftarrow{\partial}}{\partial p}\frac{\overrightarrow{\partial}}{\partial q}\right]\exp\left[-\frac{i\hbar}{2}\frac{\overleftarrow{\partial}}{\partial q}\frac{\overrightarrow{\partial}}{\partial p}\right]\,,\tag{2.85}$$

where we have neglected terms which do not contribute in the present context. Again because of eq. (2.84), and also using the fact that we are only interested in the real part, it follows that the only terms in $\hat{0}$ which contribute are

$$\exp\left[-\frac{i\hbar}{2}\frac{\overleftarrow{\partial}}{\partial q}\frac{\overrightarrow{\partial}}{\partial p}\right] - \frac{\hbar^2}{8}\frac{\overleftarrow{\partial}}{\partial p_i}\frac{\overrightarrow{\partial}}{\partial p_j}\frac{\overleftarrow{\partial}}{\partial q_i}\frac{\overrightarrow{\partial}}{\partial q_j}\tag{2.86}$$

where $i, j = 1, \ldots n$ (and as usual, it is understood that $(\partial/\partial q)(\partial/\partial p)$ stands for $(\partial/\partial q_i)(\partial/\partial p_i)$). From henceforth, we will assume that we are dealing with a system of $(n/3)$ identical particles of mass $m$.

Hence, since $H = (p^2/2m) + V$, it follows that

$$H \hat{0} \, \Omega = \left\{ \exp\left[ -\frac{i\hbar}{2} \frac{\overleftarrow{\partial}}{\partial q} \frac{\overrightarrow{\partial}}{\partial p} \right] V - \frac{\hbar^2}{2m} \frac{\overrightarrow{\partial}^2}{\partial q^2} + \frac{p^2}{2m} \right\} \Omega, \tag{2.87}$$

where it is to be understood that the $(\overleftarrow{\partial}/\partial q)$ term in the exponential operates only on $V$ and not on $\Omega$ (whereas the $\overrightarrow{\partial}^2/\partial q^2$ term operates on $\Omega$). Also, the $\overrightarrow{\partial}/\partial p$ term has no effect on $V$ and thus operates only on $\Omega$. Since all arrows now operate to the right, they will be omitted from henceforth so that we finally obtain

$$\frac{\partial \Omega(q, p)}{\partial \beta} = -\left\{ \frac{p^2}{2m} + \cos\left( \frac{\hbar}{2} \frac{\partial}{\partial q} \frac{\partial}{\partial p} \right) V - \frac{\hbar^2}{8m} \frac{\partial^2}{\partial q^2} \right\} \Omega \tag{2.88a}$$

$$= \left\{ -H + 2\sin^2\left( \frac{\hbar}{4} \frac{\partial}{\partial q} \frac{\partial}{\partial p} \right) V + \frac{\hbar^2}{8m} \frac{\partial^2}{\partial q^2} \right\} \Omega, \tag{2.88b}$$

where the $\partial/\partial q$ term in the cos and sin terms is to be understood as operating only on $V$. Such a form was given for the first time by Alastuey and Jancovici [1980] and, in fact, their result also takes account of the presence of a magnetic field. We recall that $(\partial/\partial p)(\partial/\partial q)$ is considered as the multi-dimensional scalar product of $\partial/\partial p$ and $\partial/\partial q$, or, in other words it is equal to $(\partial/\partial p_i)(\partial/\partial q_i)$ where $i$ goes from 1 to $n$ and $n$ denotes the number of dimensions. Hence, the explicit form of eq. (2.88a) is

$$\frac{\partial \Omega(q, p)}{\partial \beta} = -\left\{ \frac{p_i p_i}{2m} - \frac{\hbar^2}{8m} \frac{\partial^2}{\partial q_i \, \partial q_i} + \sum \frac{(i\hbar/2)^{\lambda_1 + \lambda_2 + \cdots + \lambda_n}}{\lambda_1! \lambda_2! \cdots \lambda_n!} \frac{\partial^{\lambda_1 + \cdots + \lambda_n} V}{\partial q_1^{\lambda_1} \cdots \partial q_n^{\lambda_n}} \frac{\partial^{\lambda_1 + \cdots + \lambda_n}}{\partial p_1^{\lambda_1} \cdots p_n^{\lambda_n}} \right\} \Omega \tag{2.89}$$

where the last summation is to be extended over all positive integer values, as well as zero values, of $\lambda_1, \lambda_2, \ldots \lambda_n$, for which the sum $\lambda_1 + \lambda_2 + \cdots + \lambda_n$ is even. This form for the Wigner translation of the Bloch equation is the most convenient from the point of view of applications.

One of the earliest applications of these results was to the quantum corrections of the classical equations of state and to similar corrections to chemical reaction rates (Wigner [1932b, 1938]) and they have been extensively used in statistical mechanics (Oppenheim and Ross [1957]; Mori, Oppenheim and Ross [1962]; Nienhuis [1970], for example). However, we will defer a detailed discussion of applications to Part II of our review, to be published at a later date.

### 2.4. An example

We would now like to use some of the formalism which we have developed to actually calculate some distribution functions. The system which we will consider is the harmonic oscillator and we will consider both pure and mixed states. We will find the Wigner functions corresponding to the eigenstates of the harmonic oscillator and also the function corresponding to a canonical ensemble of harmonic oscillators at temperature $T$.

The eigenstates of the harmonic oscillator are (Landau and Lifshitz [1965])

$$U_n(q) = \left( \frac{\alpha^2}{4} \right)^{1/4} \left( \frac{1}{2^n n!} \right)^{1/2} e^{-\alpha^2 q^2/2} H_n(\alpha q), \tag{2.90}$$

M. Hillery et al., Distribution functions in physics: Fundamentals    143

where $H_n$ is the $n$th Hermite polynomial and $\alpha = (m\omega/\hbar)^{1/2}$. Substituting this expression into the definition of the distribution function, eq. (2.2a), we find that

$$U_n^*(q+y) \, U_n(q-y) = \left(\frac{\alpha^2}{\pi}\right)^{1/2} \frac{1}{2^n n!} \exp\{-\alpha^2[(q+y)^2 + (q-y)^2]/2\} \, H_n(\alpha(q+y)) \cdot H_n(\alpha(q-y))$$

(2.91)

so that

$$P_w(q,p) = \frac{1}{\pi\hbar} \frac{\alpha}{\sqrt{\pi}} \frac{1}{2^n n!} e^{-\alpha^2 q^2} \int dy \; e^{2ipy/\hbar} \; e^{-\alpha^2 y^2} H_n(\alpha(q+y)) \, H_n(\alpha(q-y)) \,.$$

(2.92)

We now note that

$$\alpha^2 y^2 - 2ipy/\hbar = \alpha^2(y - ip/\alpha^2\hbar)^2 + p^2/\alpha^2\hbar^2$$

(2.93)

and define a new variable

$$z = \alpha(y - ip/\alpha^2\hbar) \,.$$

(2.94)

We then have that

$$P_w(q,p) = \frac{1}{\sqrt{\pi}} \frac{1}{\pi\hbar} \frac{1}{2^n n!} e^{-\alpha q^2} e^{\beta^2} \int dz \; e^{-z^2} H_n(\alpha q + z + \beta) \, H_n(\alpha q - z - \beta) \,,$$

(2.95)

where $\beta = ip/\alpha\hbar$. Noting the $H_n(-x) = (-1)^n H_n(x)$ we find

$$P_w(q,p) = \frac{1}{\sqrt{\pi}} \frac{1}{\pi\hbar} \frac{(-1)^n}{2^n n!} e^{-\alpha q^2} e^{\beta^2} \int dz \; e^{-z^2} H_n(\alpha q + z + \beta) \, H_n(z + \beta - \alpha q) \,.$$

(2.96)

The above integral can be done (Gradshteyn and Ryzhik [1980]) and is

$$\int dz \; e^{-z^2} H_n(z + \beta + \alpha q) \, H_n(z + \beta - \alpha q) = 2^n \sqrt{\pi} \, n! \, L_n(2(\alpha^2 q^2 - \beta^2)) \,,$$

(2.97)

where $L_n$ is the $n$th Laguerre polynomial. Re-expressing $\alpha$ and $\beta$ in terms of $q$ and $p$ we have

$$\alpha^2 q^2 - \beta^2 = \frac{2}{\hbar\omega} \left(\frac{p^2}{2m} + \frac{1}{2} m\omega^2 q^2\right) = \frac{2}{\hbar\omega} H(q,p)$$

(2.98)

so that (Groenewold [1946]; Takabayaski [1954]; Dahl [1982])

$$P_w(q,p) = (1/\pi\hbar)(-1)^n \, e^{-2H/\hbar\omega} L_n(4H/\hbar\omega) \,.$$

(2.99)

Before discussing this result we will first calculate the distribution for an ensemble of oscillators at temperature $T$ (Imre, Ozizmir, Rosenbaum and Zweifel [1967]). Here we proceed by way of the Wigner translation of the Bloch equation (eq. (2.88b)) which for this system results in

$$\frac{\partial \Omega(q, p)}{\partial \beta} = \left\{ -\left(\frac{p^2}{2m} + \frac{1}{2} m\omega^2 q^2\right) + 2\sin^2\left(\frac{\hbar}{4} \frac{\partial}{\partial q} \frac{\partial}{\partial p}\right) V + \frac{\hbar^2}{8m} \frac{\partial^2}{\partial q^2} \right\} \Omega. \tag{2.100}$$

Because $V$ is quadratic in $q^2$ it is clear that only the leading order term in the $\sin^2$ expansion will contribute, and since $\partial^2 V/\partial q^2 = m\omega^2$, it follows that the Wigner translation of the Bloch equation for the oscillator reduces to

$$\frac{\partial \Omega}{\partial \beta} = -\left(\frac{p^2}{2m} + \frac{1}{2} m\omega^2 q^2\right)\Omega + \frac{\hbar^2}{8}\left(\frac{1}{m} \frac{\partial^2 \Omega}{\partial q^2} + m\omega^2 \frac{\partial^2 \Omega}{\partial p^2}\right). \tag{2.101}$$

To solve this equation we make the Ansatz

$$\Omega(q, p) = \exp\{-A(\beta) H + B(\beta)\} \tag{2.102}$$

where $A(0) = B(0) = 0$, and $H = (p^2/2m) + \frac{1}{2}m\omega^2 q^2$. Substituting this into eq. (2.101) gives us

$$\left(-\frac{dA}{d\beta} H + \frac{dB}{d\beta}\right)\Omega = -H\Omega + \frac{\hbar^2}{8}\left[\frac{1}{m} \omega^2(-mA + m\omega^2 q^2 A^2) + m\omega^2\left(-\frac{A}{m} + \frac{p^2}{m}\right)A^2\right]\Omega$$

$$= -H\Omega + \frac{(\hbar\omega)^2}{4}\left(-A + HA^2\right)\Omega. \tag{2.103}$$

This equation can be re-expressed in the form

$$H(q, p)\left[-\frac{dA}{d\beta} + 1 - \frac{(\hbar\omega)^2}{4} A^2\right] + \left[\frac{dB}{d\beta} + \frac{(\hbar\omega)^2}{4} A\right] = 0. \tag{2.104}$$

Because this equation must hold for all $q$ and $p$, and the terms in the brackets are independent of $q$ and $p$, they must vanish independently, i.e.

$$\frac{dA}{d\beta} + \frac{(\hbar\omega)^2}{4} A^2 - 1 = 0 \tag{2.105}$$

$$\frac{dB}{d\beta} + \frac{(\hbar\omega)^2}{4} A = 0. \tag{2.106}$$

Eq. (2.105) can be integrated directly. One has that

$$\int \frac{dA}{1 - (\hbar\omega/2)^2 A^2} = \int d\beta \tag{2.107}$$

or

$$\beta = \frac{1}{\hbar\omega} \ln\left[\left(1 + \frac{\hbar\omega}{2} A\right) \bigg/ \left(1 - \frac{\hbar\omega}{2} A\right)\right]. \tag{2.108}$$

Inverting this equation gives us that

$$A(\beta) = (2/\hbar\omega)\tanh(\hbar\omega\beta/2) .$$ (2.109)

This can now be substituted into eq. (2.106) to give

$$B(\beta) = -\frac{\hbar\omega}{2}\int_0^\beta d\beta' \tanh\left(\frac{\hbar\omega\beta'}{2}\right) = -\ln\cosh\left(\frac{\hbar\omega\beta}{2}\right) .$$ (2.110)

Therefore, we have

$$\Omega(q, p) = \mathrm{sech}(\hbar\omega\beta/2)\exp[-(2/\hbar\omega)\tanh(\hbar\omega\beta/2)\,H(q, p)] .$$ (2.111)

To complete our derivation we need to normalize the above expression. As was noted before the Wigner function is the function which corresponds to the operator $(\hat{\rho}/2\pi\hbar)$. From eq. (2.78) we then have

$$P_w(q, p) = \frac{1}{2\pi\hbar}\frac{1}{Z(\beta)}\Omega(q, p)$$ (2.112)

as $\Omega(q, p)$ is just the function corresponding to $e^{-\beta\hat{H}}$. We also have from eq. (2.11) (setting $\hat{A} = e^{-\beta\hat{H}}$ and $\hat{B} = \hat{I}$)

$$Z(\beta) = \mathrm{Tr}(e^{-\beta\hat{H}}) = \frac{1}{2\pi\hbar}\int dq\int dp\,\Omega(q, p) .$$ (2.113)

Substituting eq. (2.111) into eq. (2.113) we find

$$Z(\beta) = \tfrac{1}{2}[\sinh(\hbar\omega\beta/2)]^{-1} .$$ (2.114)

Finally we obtain for $P_w(q, p)$, from eqs. (2.111), (2.112) and (2.114),

$$P_w(q, p) = (1/\pi\hbar)\tanh(\hbar\omega\beta/2)\exp[-(2/\hbar\omega)\tanh(\hbar\omega\beta/2)\,H(q, p)] .$$ (2.115)

We now want to compare the two expressions (eq. (2.99) and eq. (2.115)) for $P_w$ for the pure and mixed states, respectively. Examining the first few Laguerre polynomials

$$L_0(x) = 1$$
$$L_1(x) = 1 - x$$ (2.116)
$$L_2(x) = 1 - 2x + x^2$$

we see that for the ground state of the oscillator $P_w(q, p) > 0$ while for excited states $P_w(q, p)$ can assume negative values. The result for the canonical ensemble, however, is always positive. It does not have the oscillatory structure which is present in the expressions given by eq. (2.99). The incoherence induced by a finite temperature leads to a much smoother distribution function.

*2.5. Statistics and second-quantized notation* (Klimintovich [1958]; Brittin and Chappell [1962]; Imre, Ozizmir, Rosenbaum and Zweifel [1967])

When one is dealing with more than one particle one has to include the effects of quantum statistics. To illustrate how these effects come in to the Wigner function we will first consider an example. We will then show how the Wigner function can be expressed in second-quantized notation. In this form it is easier to take the effects of statistics into account, but two of us have an article in preparation (O'Connell and Wigner [1983]) which not only will take the effect of statistics into account, but will also include spin effects.

Let us consider two identical particles in one dimension in a harmonic potential well. We will further assume that the particles are bosons. The Hamiltonian for the system is

$$\hat{H} = \frac{1}{2m}(\hat{p}_1^2 + \hat{p}_2^2) + \tfrac{1}{2}m\omega^2(\hat{q}_1^2 + \hat{q}_2^2). \tag{2.117}$$

Suppose that we want to find the Wigner distribution for a canonical ensemble of these systems at a temperature $T$. We would again like to use the Wigner translation of the Bloch equation but now we must be more careful; the initial condition is no longer so simple.

To see this we first find the density matrix for the system. The eigenstates of the Hamiltonian given by eq. (2.117) are

$$\phi_{n_1 n_2}(q_1, q_2) = \begin{cases} \dfrac{1}{\sqrt{2}}(U_{n_1}(q_1)\,U_{n_2}(q_2) + U_{n_2}(q_1)\,U_{n_1}(q_2)) & \text{if } n_1 > n_2 \\[2mm] U_{n_1}(q_1)\,U_{n_1}(q_2) & \text{if } n_1 = n_2, \end{cases} \tag{2.118}$$

where $U_n(q)$ is given by eq. (2.90). This state has an energy $E_{n_1 n_2}$ given by

$$E_{n_1 n_2} = \hbar\omega\,(n_1 + n_2 + 1). \tag{2.119}$$

The unnormalized density matrix for this system is just

$$\hat{\Omega} = \sum_{n_1 \geq n_2} \exp(-\beta E_{n_1 n_2})\,|\phi_{n_1 n_2}\rangle\langle\phi_{n_1 n_2}|. \tag{2.120}$$

In the $\beta \to 0$ limit this becomes

$$\hat{\Omega}(\beta = 0) = \sum_{n_1 \geq n_2} |\phi_{n_1 n_2}\rangle\langle\phi_{n_1 n_2}|. \tag{2.121}$$

Taking matrix elements we find

$$\langle q_1', q_2'|\hat{\Omega}(\beta = 0)|q_1, q_2\rangle = \sum_{n_1 > n_2} \tfrac{1}{2}(U_{n_1}(q_1')\,U_{n_2}(q_2') + U_{n_2}(q_1')\,U_{n_1}(q_2'))$$

$$\times (U_{n_1}^*(q_1)\,U_{n_2}^*(q_2) + U_{n_2}^*(q_1)\,U_{n_1}^*(q_2)) + \sum_n U_n(q_1')\,U_n(q_2')\,U_n^*(q_1)\,U_n^*(q_2)$$

$$= \tfrac{1}{2} \sum_{n_1, n_2} (U_{n_1}(q_1')\,U_{n_2}(q_2')\,U_{n_1}^*(q_1)\,U_{n_2}^*(q_2) + U_{n_1}(q_1')\,U_{n_2}(q_2')\,U_{n_1}^*(q_2)\,U_{n_2}^*(q_1)). \tag{2.122}$$

We can now make use of the identity

M. Hillery et al., Distribution functions in physics: Fundamentals    147

$$\sum_n U_n(q_1') \, U_n^*(q_1) = \delta(q_1 - q_1') \tag{2.123}$$

to give

$$\langle q_1', q_2' | \hat{\Omega}(\beta = 0) | q_1, q_2 \rangle = \tfrac{1}{2} [\delta(q_1' - q_1) \, \delta(q_2' - \dot{q}_2) + \delta(q_1' - q_2) \, \delta(q_2' - q_1)] , \tag{2.124}$$

as was to be expected. If we operate on an arbitrary two particle state, $|\psi\rangle$, with $\hat{\Omega}(\beta = 0)$ we have that

$$\langle q_1', q_2' | \hat{\Omega}(\beta = 0) | \psi \rangle = \int dq_1 \int dq_2 \, \langle q_1', q_2' | \hat{\Omega}(\beta = 0) | q_1, q_2 \rangle \, \langle q_1, q_2 | \psi \rangle$$

$$= \tfrac{1}{2} [\psi(q_1', q_2') + \psi(q_2', q_1')] . \tag{2.125}$$

If $\psi$ is symmetric the result on the right-hand side of eq. (2.125) is $\psi$, if $\psi$ is anti-symmetric the result is 0.

Therefore, $\hat{\Omega}(\beta = 0)$ is just the projection operator, $\hat{P}_s$ say, onto the state of symmetric two-particle wave functions. This result is also true for an arbitrary number of particles, $N$. Our result that $\hat{\Omega}(\beta = 0)$ is $\hat{P}_s$ was derived for bosons. Similarly, if the particles are fermions $\hat{\Omega}(\beta = 0)$ is $\hat{P}_A$, the projection onto the space of anti-symmetric $N$-particle wave functions, but in this case, the spin variable should also be included.

Returning now to our example we want to find the initial condition for the Wigner translation of Bloch equation, i.e. we must find the function corresponding to $\hat{P}_s$. Making use of the two-particle extension of eq. (2.12) we find

$$\Omega(q_1, q_2, p_1, p_2) = \int dy_1 \int dy_2 \exp\{(i/\hbar)(p_1 y_1 + p_2 y_2)\} \langle q_1 - \tfrac{1}{2} y_1, q_2 - \tfrac{1}{2} y_2 | \hat{P}_s | q_1 + \tfrac{1}{2} y_1, q_2 + \tfrac{1}{2} y_2 \rangle$$

$$= \int dy_1 \int dy_2 \exp\{(i/\hbar)(p_1 y_1 + p_2 y_2)\} \tfrac{1}{2} [\delta(y_1) \, \delta(y_2)$$

$$+ \delta(q_2 - q_1 + \tfrac{1}{2}(y_2 + y_1)) \, \delta(q_1 - q_2 + \tfrac{1}{2}(y_1 + y_2))]$$

$$= \tfrac{1}{2} + \pi\hbar \, \delta(q_1 - q_2) \, \delta(p_1 - p_2) . \tag{2.126}$$

The corresponding result for fermions has a minus sign in front of the second term. This initial condition is considerably more complicated than the initial condition, $\Omega(q, p) = 1$, which was obtained in the one-dimensional case. The situation rapidly becomes worse with larger numbers of particles.

Second-quantized notation provides, in principle, a convenient way to deal with the problems imposed by quantum statistics. We will consider a Fock space and designate the vacuum state of this space by $|0\rangle$, and the quantized field operators at the point $r$ by $\hat{\psi}^+(r)$ and $\hat{\psi}(r)$. The interpretation of the field operators is that $\hat{\psi}^+(r)$ adds a particle at point $r$ to the system whereas $\hat{\psi}(r)$ removes a particle at point $r$. They are defined as

$$\hat{\psi}(r) = \sum_p \frac{1}{\sqrt{V}} e^{ip \cdot r} \, \hat{a}_p \tag{2.127a}$$

$$\hat{\psi}^+(r) = \sum_p \frac{1}{\sqrt{V}} e^{-ip \cdot r} \, \hat{a}_p^+ \tag{2.127b}$$

where the so-called annihilation and creation operators, $\hat{a}_p$ and $\hat{a}_p^+$, respectively (discussed in detail in section 4), act to remove or create a particle of momentum $p$ in a box of volume $V$. For bosons these operators obey the commutation relation

$$[\hat{\psi}(r), \hat{\psi}^{\dagger}(r')] = \delta^{(3)}(r - r') \tag{2.128a}$$

$$[\hat{\psi}(r), \hat{\psi}(r')] = 0 \tag{2.128b}$$

and for fermions the anti-commutation relation

$$\{\hat{\psi}(r), \hat{\psi}^{\dagger}(r')\} = \delta^{(3)}(r - r') \tag{2.129a}$$

$$\{\hat{\psi}(r), \hat{\psi}(r')\} = 0 . \tag{2.129b}$$

To every $N$-particle state $|\Psi_N\rangle$ in the Fock space corresponds an $N$-particle wave function given by (Schweber [1961])

$$\Psi_N(r_1, \ldots r_N) = \frac{1}{\sqrt{N!}} \langle 0|\hat{\psi}(r_N) \cdots \hat{\psi}(r_1)|\Psi_N\rangle . \tag{2.130}$$

The distribution function for the state $|\Psi_N\rangle$ then, is given by

$$
\begin{aligned}
P(r_1, \ldots r_N; p_1, \ldots p_N) &= \left(\frac{1}{2\pi\hbar}\right)^{3N} \int d^3y_1 \cdots \int d^3y_N \, \exp\{(i/\hbar)(p_1 \cdot y_1 + \cdots + p_N \cdot y_N)\} \\
&\quad \times \Psi_N^*(r_1 + \tfrac{1}{2}y_1, \ldots r_N + \tfrac{1}{2}y_N)\, \Psi_N(r_1 - \tfrac{1}{2}y_1, \ldots r_N - \tfrac{1}{2}y_N) \\
&= \left(\frac{1}{2\pi\hbar}\right)^{3N} \frac{1}{N!} \int d^3y_1 \cdots \int d^3y_N \, \exp\{(i/\hbar)(p_1 \cdot y_1 + \cdots + p_N \cdot y_N)\} \\
&\quad \times \langle 0|\hat{\psi}(r_N - \tfrac{1}{2}y_N) \cdots \hat{\psi}(r_1 - \tfrac{1}{2}y_1)|\Psi_N\rangle \langle \Psi_N|\hat{\psi}^{\dagger}(r_1 + \tfrac{1}{2}y_1) \cdots \hat{\psi}^{\dagger}(r_N + \tfrac{1}{2}y_N)|0\rangle .
\end{aligned}
\tag{2.131}
$$

This expression readily extends to $N$-particle density matrixes, $\hat{\rho}_N$, so that

$$
\begin{aligned}
P(r_1, \ldots r_N; p_1, \ldots p_N) &= \left(\frac{1}{2\pi\hbar}\right)^{3N} \frac{1}{N!} \int d^3y_1 \cdots \int d^3y_N \, \exp\{(i/\hbar)(p_1 \cdot y_1 + \cdots + p_N \cdot y_N)\} \\
&\quad \times \langle 0|\hat{\psi}(r_N - \tfrac{1}{2}y_N) \cdots \hat{\psi}(r_1 - \tfrac{1}{2}y_1)\, \hat{\rho}_N \, \hat{\psi}^{\dagger}(r_1 + \tfrac{1}{2}y_1) \cdots \hat{\psi}^{\dagger}(r_N + \tfrac{1}{2}y_N)|0\rangle ,
\end{aligned}
\tag{2.132}
$$

where, in the case of a pure state,

$$\hat{\rho}_N = |\Phi_N\rangle \langle \Phi_N| \tag{2.133}$$

with $|\Phi_N\rangle$ denoting the $N$-particle ket basis vector. An $N$-particle density matrix has the property that if $\Phi_{N'}$ and $\Phi_{N''}$ are $N'$-particle and $N''$-particle states respectively, then $\langle \Phi_{N''}|\hat{\rho}_N|\Phi_{N'}\rangle = 0$ unless $N' = N'' = N$. Therefore, eq. (2.132) can be expressed as

M. Hillery et al., Distribution functions in physics: Fundamentals    149

$$P(r_1, \ldots r_N; p_1, \ldots p_N) = \left(\frac{1}{2\pi\hbar}\right)^{3N} \frac{1}{N!} \int d^3y_1 \cdots \int d^3y_N \, \exp\{(i/\hbar)(p_1 \cdot y_1 + \cdots + p_N \cdot y_N)\}$$

$$\times \mathrm{Tr}(\hat{\psi}(r_N - \tfrac{1}{2}y_N) \cdots \hat{\psi}(r_1 - \tfrac{1}{2}y_1)\,\hat{\rho}_N\,\hat{\psi}^\dagger(r_1 + \tfrac{1}{2}y_1) \cdots \hat{\psi}^\dagger(r_N + \tfrac{1}{2}y_N))$$

$$= \left(\frac{1}{2\pi\hbar}\right)^{3N} \frac{1}{N!} \int d^3y_1 \cdots \int d^3y_N \, \exp\{(i/\hbar)(p_1 \cdot y_1 + \cdots + p_N \cdot y_N)\}$$

$$\times \mathrm{Tr}(\hat{\rho}_N\,\hat{\psi}^\dagger(r_1 + \tfrac{1}{2}y_1) \cdots \hat{\psi}^\dagger(r_N + \tfrac{1}{2}y_N)\,\hat{\psi}(r_N - \tfrac{1}{2}y_N) \cdots \hat{\psi}(r_1 - \tfrac{1}{2}y_1)) . \quad (2.134)$$

This is the desired expression for the Wigner function in second-quantized form (Brittin and Chappell [1962]; Imre, Ozizmir, Rosenbaum and Zweifel [1967]).

It is also possible to derive expressions for the reduced distribution functions in terms of the quantized field operators (Brittin and Chappell [1962]; Imre, Ozizmir, Rosenbaum and Zweifel [1967]). The distribution function of order $N$, reduced to the $j$th order, is defined as

$$P_j(r_1, \ldots r_j; p_1, \ldots p_j) = \int d^3r_{j+1} \cdots \int d^3r_N \int d^3p_{j+1} \cdots \int d^3p_N P(r_1, \ldots r_N; p_1, \ldots p_N) \quad (2.135)$$

and this definition will be used for the rest of this section. This can also be expressed, by making use of eq. (2.134), as

$$P_j(r_1, \ldots r_j; p_1, \ldots p_j) = \left(\frac{1}{2\pi\hbar}\right)^{3N} \frac{1}{N!} \int d^3r_{j+1} \cdots \int d^3r_N \int d^3p_{j+1} \cdots \int d^3p_N \int d^3y_1 \cdots \int d^3y_N$$

$$\times \exp\{i(p_1 \cdot y_1 + \cdots + p_N \cdot y_N)/\hbar\}\,\mathrm{Tr}(\hat{\rho}_N \hat{\psi}^\dagger(r_1 + \tfrac{1}{2}y_1) \cdots \hat{\psi}^\dagger(r_N + \tfrac{1}{2}y_N)\,\hat{\psi}(r_N - \tfrac{1}{2}y_N) \cdots \hat{\psi}(r_1 - \tfrac{1}{2}y_1))$$

$$= \left(\frac{1}{2\pi\hbar}\right)^{3j} \frac{1}{N!} \int d^3r_{j+1} \cdots \int d^3r_N \int d^3y_1 \cdots \int d^3y_j \, \exp\{i(p_1 \cdot y_1 + \cdots + p_j \cdot y_j)/\hbar\}\,\mathrm{Tr}(\hat{\rho}_N\,\hat{\psi}^\dagger(r_1 + \tfrac{1}{2}y_1)$$

$$\cdots \hat{\psi}^\dagger(r_j + \tfrac{1}{2}y_j)\,\hat{\psi}^\dagger(r_{j+1}) \cdots \hat{\psi}^\dagger(r_N)\,\hat{\psi}(r_N) \cdots \hat{\psi}(r_{j+1})\,\hat{\psi}(r_j - \tfrac{1}{2}y_j) \cdots \hat{\psi}(r_1 - \tfrac{1}{2}y_1)) .$$

$$(2.136)$$

In order to analyze this expression further we first note that

$$\int d^3r \, \hat{\psi}^\dagger(r)\,\hat{\psi}(r) = \hat{N}, \quad (2.137)$$

where $\hat{N}$ is just the number operator. We then have that, for both bosons and fermions

$$[\hat{\psi}(r), \hat{N}] = \hat{\psi}(r) . \quad (2.138)$$

Therefore,

$$\int d^3r_{N-1} \, \hat{\psi}^\dagger(r_{N-1})\,\hat{N}\,\hat{\psi}(r_{N-1}) = \hat{N}\,(\hat{N} - 1) \quad (2.139)$$

and

$$\int d^3r_{j+1} \cdots \int d^3r_{N-1}\, \hat{\psi}^\dagger(r_{j+1}) \cdots \hat{\psi}^\dagger(r_{N-1})\, \hat{N}\, \hat{\psi}(r_{N-1}) \cdots \hat{\psi}(r_{j+1}) = \hat{N}\,(\hat{N}-1) \cdots (\hat{N}-N+j+1)\,.$$

(2.140)

Eq. (2.136) becomes

$$P_j(r_1, \ldots r_j; p_1, \ldots p_j) = \left(\frac{1}{2\pi\hbar}\right)^{3j} \frac{1}{N!} \int d^3y_1 \cdots \int d^3y_j\, \exp\{i(p_1 \cdot y_1 + \cdots + p_j \cdot y_j)/\hbar\}$$
$$\times \mathrm{Tr}(\hat{\rho}_N \hat{\psi}^\dagger(r_1 + \tfrac{1}{2}y_1) \cdots \hat{\psi}^\dagger(r_j + \tfrac{1}{2}y_j)\, \hat{N}\,(\hat{N}-1) \cdots (\hat{N}-N+j+1)\, \hat{\psi}(r_j - \tfrac{1}{2}y_j) \cdots \hat{\psi}(r_1 - \tfrac{1}{2}y_1))\,.$$

(2.141)

Because $\hat{\rho}_N$ is an $N$-particle density matrix we have that

$$\hat{N}(\hat{N}-1) \cdots (\hat{N}-N+j+1)\, \hat{\psi}(r_j - \tfrac{1}{2}y_j) \cdots \hat{\psi}(r_1 - \tfrac{1}{2}y_1)\, \hat{\rho}_N = (N-j)!\, \hat{\psi}(r_j - \tfrac{1}{2}y_j) \cdots \hat{\psi}(r_1 - \tfrac{1}{2}y_1)\, \hat{\rho}_N\,,$$

(2.142)

so that our final expression for the reduced Wigner function is

$$P_j(r_1, \ldots r_j; p_1, \ldots p_j) = \left(\frac{1}{2\pi\hbar}\right)^{3j} \frac{(N-j)!}{N!} \int d^3y_1 \cdots \int d^3y_j\, \exp\{i(p_1 \cdot y_1 + \cdots + p_j \cdot y_j)/\hbar\}$$
$$\times \mathrm{Tr}(\hat{\rho}_N \hat{\psi}^\dagger(r_1 + \tfrac{1}{2}y_1) \cdots \hat{\psi}^\dagger(r_j + \tfrac{1}{2}y_j)\, \hat{\psi}(r_j - \tfrac{1}{2}y_j) \cdots \hat{\psi}(r_1 - \tfrac{1}{2}y_1))\,.$$

(2.143)

It is now possible to formulate the dynamics of this theory in a way which is independent of the number of particles. We first go to the Heisenberg picture in which the field operators become time dependent. We then consider the operators

$$\hat{F}_j(r_1, \ldots r_j; p_1, \ldots p_j) = \left(\frac{1}{2\pi\hbar}\right)^{3j} \int d^3y_1 \cdots \int d^3y_j\, \exp\{i(p_1 \cdot y_1 + \cdots + p_j \cdot y_j)/\hbar\}$$
$$\times \hat{\psi}^\dagger(r_1 + \tfrac{1}{2}y_1; t) \cdots \hat{\psi}^\dagger(r_j + \tfrac{1}{2}y_j; t) \cdots \hat{\psi}(r_j - \tfrac{1}{2}y_j; t) \cdots \hat{\psi}(r_1 - \tfrac{1}{2}y_1; t)\,.$$

(2.144)

The distribution functions for an $N$-particle theory are then just

$$\hat{P}_j(r_1, \ldots r_j; p_1, \ldots p_j) = \frac{(N-j)!}{N!}\, \mathrm{Tr}(\hat{\rho}_N \hat{F}_j(r_1, \ldots r_j; p_1, \ldots p_j))$$

(2.145)

We see that in this formulation all of the dynamical information is contained in the operators $\hat{F}_j$ which contain no reference to a specific particle number and also contain the information about the statistics of the particles. Thus, in principle, the second-quantized formalism should be a useful starting-point for the incorporation of statistics into problems involving a system of identical particles. However, it must be admitted that – to our knowledge – no application has been made along these lines.

## 3. Other distribution functions

We now want to examine certain other distributions besides the one considered so far. These may arise out of a desire to make use of an operator ordering scheme other than that proposed by Weyl or a desire to have a distribution function with certain properties. For example, we may want to make use of

M. Hillery et al., Distribution functions in physics: Fundamentals    151

symmetric ordering

$$q^m p^n \to \tfrac{1}{2}(\hat{q}^m \hat{p}^n + \hat{p}^n \hat{q}^m),\tag{3.1}$$

in which case we would use the distribution function (Margenau and Hill [1961]; Mehta [1964])

$$P_s(q, p) = \frac{1}{4\pi\hbar} \operatorname{Re}\left\{\psi(q)\int\limits_{\infty}^{\infty} \mathrm{d}y\, e^{-(i/\hbar)py}\, \psi^*(q - y)\right\}.\tag{3.2}$$

On the other hand, we may want to consider a distribution which is always greater than or equal to zero. We will discuss a distribution which has this property shortly.

A scheme for generating distribution functions was proposed by Cohen [1966] and further examined by Summerfield and Zweifel [1969]. They give the rather general expression

$$P_g(q, p) = \left(\frac{1}{2\pi\hbar}\right)^2 \int \mathrm{d}\sigma \int \mathrm{d}\tau \int \mathrm{d}u\, \exp\{-(i/\hbar)\,[\sigma(q - u) + \tau p]\}\, g(\sigma, \tau)\, \psi^*\!\left(u - \frac{\tau}{2}\right) \psi\!\left(u + \frac{\tau}{2}\right)\tag{3.3a}$$

$$= \int \mathrm{d}q' \int \mathrm{d}p'\, \bar{g}(q - q', p - p')\, P_w(q', p')\tag{3.3b}$$

for the distribution function of the pure state $\psi(q)$, where

$$\bar{g}(q, p) = \int \mathrm{d}\sigma \int \mathrm{d}\tau\, \exp\{-(i/\hbar)(\sigma q + \tau p)\}\, g(\sigma, \tau).\tag{3.4}$$

Thus the function $P$ is simply the original function $P_w$ smeared with another function $g$. The basic requirement which leads to eq. (3.3) is that $P$ transform correctly with respect to space displacement, $\psi(q) \to \psi(q - a)$, and transition to a uniformly moving coordinate system, $\psi(q) \to \exp(-im v q)\,\psi(q)$. These requirements were formulated in giving the form eq. (3.3a) to $P_g$ – and the satisfaction of the requirements can easily be verified; eq. (3.3b) then follows.

Cohen also pointed out that it is possible to obtain distributions whose dependence upon the wave function of the system is other than bilinear simply by choosing $g(\sigma, \tau)$ to depend upon $\psi(q)$. For example, one can choose

$$g(\sigma, \tau) = \int \mathrm{d}q\, \psi\!\left(q - q_0\frac{\sigma\tau}{\hbar}\right) \psi^*\!\left(q + q_0\frac{\sigma\tau}{\hbar}\right),\tag{3.5}$$

where $q_0$ is an arbitrary value of $q$. This choice for $g(\sigma, \tau)$ satisfies $g(0, \tau) = g(\sigma, 0) = 1$ so that the correct marginal distributions are obtained. On the other hand, we now have the rather awkward situation that the function-operator correspondence depends upon the wave function. An even simpler choice is, of course

$$P_g(q, p) = (\pi\hbar)^{-1}\,|\psi(q)|^2\,|\phi(p)|^2,\tag{3.6}$$

where $\phi(p)$, the Fourier transform of $\psi(q)$ is defined by eq. (2.14). The conditions on $\bar{g}(q, p)$ which must be satisfied so that the correct marginal distributions are obtained are

$$\int dq\, \tilde{g}(q, p) = (2\pi\hbar)^2\, \delta(p) \tag{3.7a}$$

$$\int dp\, \tilde{g}(q, p) = (2\pi\hbar)^2\, \delta(q). \tag{3.7b}$$

One choice of $\tilde{g}(q, p)$ which does not satisfy eqs. (3.7) but which is interesting nonetheless is given by

$$\tilde{g}(q, p; \alpha) = \left(\frac{1}{\pi\hbar}\right) e^{-q^2/\alpha}\, e^{-\alpha p^2/\hbar^2}. \tag{3.8}$$

The use of this smearing function was first proposed by Husimi [1940] and has been investigated by a number of authors since (Bopp [1956]; Kano [1965]; McKenna and Frisch [1966]; Cartwright [1976]; Prugovecki [1978]; O'Connell and Wigner [1981b]). It leads to a distribution function, $P_H(q, p)$, where the subscript H denotes Husimi, which is non-negative for all $p$ and $q$. One can see this by noting that $\tilde{g}(q - q', p - p'; \alpha)$ is just the Wigner distribution function which one obtains from the displaced (in both position and momentum) harmonic oscillator ground state wave function

$$\psi_{q,p}(q'; \alpha) = (\pi\alpha)^{-1/4}\, e^{-(q'-q)^2/2\alpha}\, e^{ipq'/\hbar}, \tag{3.9}$$

which we will call $P_{q,p}$ (O'Connell and Wigner [1981b]). If the Wigner distribution in question, $P_\phi$, corresponds to a wave function $\phi(q)$ we have

$$P_H(q, p) = \int dq' \int dp'\, P_{q,p}(q', p')\, P_\phi(q', p') = \left(\frac{1}{2\pi\hbar}\right) \left| \int dq'\, \psi_{q,p}^*(q')\, \phi(q') \right|^2 \geq 0, \tag{3.10}$$

where we have used eq. (2.8). Note that in order to get a positive distribution function we had to violate condition (ii) on our list of properties of the Wigner function. Property (vi) is also violated as was shown by Prugovecki [1978] and by O'Connell and Wigner [1981a].

We will encounter $P_H(q, p)$ again in the next section in a somewhat different form. It is the "$Q$" or "anti-normally-ordered" distribution function of quantum optics. It is one of a number of distributions which are useful in the description of harmonic oscillators, and, hence, modes of the electromagnetic field. We now proceed to examine these distribution functions.

## 4. Distribution functions in terms of creation and annihilation operators

The harmonic oscillator is a system that is ubiquitous in physics, so that it is not surprising that quantum distribution functions have been developed which are tailored to its description. It is in the description of the modes of the electromagnetic field that these distribution functions have found their widest application.

It should be emphasized that many problems in quantum optics require a fully quantized treatment not only of the atoms but also of the field. For example, an analysis of experiments dealing with photon counting or a derivation of the fluctuations in intensity of a laser near threshold both require the quantum theory of·radiation (Scully and Lamb [1967]; De Giorgio and Scully [1970]; Graham and Haken

M. Hillery et al., Distribution functions in physics: Fundamentals      153

[1970]). The latter is developed within the framework of annihilation and creation operators for bosons (see below) but it is then possible to go to a description in terms of c-numbers (while fully retaining the quantum aspects of the situation) by means of distribution functions. In most cases, this greatly facilitates the calculation while, at the same time, it contributes to a better understanding of the connection between the quantum and classical descriptions of the electromagnetic field.

A number of studies of these distribution functions have been done (Mehta and Sudarshan [1965]; Lax and Louisell [1967]; Lax [1968]; Cahill and Glauber [1969]; Agarwal and Wolf [1970]; Louisell [1973]). We will rely most heavily upon the papers by Cahill and Glauber [1969] in our treatment. Their discussion considers a continuum of possible operator ordering schemes, and hence distributions (an even larger class is considered in Agarwal and Wolf [1970]) but we will consider only three of these. A final section will discuss distributions defined on a 4-dimensional, rather than a 2-dimensional, phase space.

We will describe the system in terms of its annihilation and creation operators

$$\hat{a} = \left(\frac{1}{2\hbar}\right)^{1/2} \left(\lambda \hat{q} + \frac{i}{\lambda} \hat{p}\right) \tag{4.1a}$$

$$a^+ = \left(\frac{1}{2\hbar}\right)^{1/2} \left(\lambda \hat{q} - \frac{i}{\lambda} \hat{p}\right), \tag{4.1b}$$

satisfying

$$[\hat{a}, \hat{a}^+] = 1 . \tag{4.2}$$

As mentioned before, it is assumed that the field operators we consider obey Bose statistics. Each pair of $\hat{a}$, $\hat{a}^+$ refers to a certain function of position. These functions form an orthonormal set which is countable if the basic domain is assumed to be finite, and continuous if infinite. We deal with a very large, but finite, system so that the system is only approximately relativistically invariant (exact invariance is achieved for an infinitely large system, but this would make the calculation in other ways difficult).

The various functions of $\hat{a}$ and $\hat{a}^+$ are investigated individually because the corresponding $\hat{a}$ and $\hat{a}^+$ do not interact with the $\hat{a}$ and $\hat{a}^+$ of another member of the set. They interact with the matter which is in the basic domain. Thus, for example, when we apply this formalism to the case of the electromagnetic field, we investigate each mode (corresponding to a definite momentum and definite direction of polarization) separately, and the operators associated with different modes commute (no interaction between modes). In addition, there will be a distribution function corresponding to each mode.

The $\hat{a}$ and $\hat{a}^+$ operators act on the basis vectors $|n\rangle$, the so-called "particle number states", and have the properties:

$$\hat{a} |n\rangle = \sqrt{n} |n - 1\rangle \tag{4.2a}$$

$$\hat{a}^+ |n\rangle = \sqrt{n + 1} |n + 1\rangle \tag{4.2b}$$

$$\hat{a}^+ a |n\rangle = n |n\rangle \tag{4.2c}$$

$$\hat{a} |0\rangle = 0 . \tag{4.2d}$$

In addition, one can prove that

154                        *M. Hillery et al., Distribution functions in physics: Fundamentals*

$$[\hat{a}, (\hat{a}^+)^n] = n(\hat{a}^+)^{n-1}. \tag{4.2e}$$

If we are considering an oscillator of mass $m$ and angular frequency $\omega$ we take $\lambda = (m\omega)^{1/2}$ and if we are considering a mode of the electromagnetic field of angular frequency $\omega$ we set $\lambda = (\hbar^{1/2}\omega/c)$.

We also want to consider a special class of states known as coherent states (Schrödinger [1926]; Glauber [1963a]; Glauber [1963b]; Sudarshan [1963]; Glauber [1965]). To define these we first define for each complex number $\alpha$ the unitary displacement operator:

$$\hat{D}(\alpha) = e^{(\alpha\hat{a}^+ - \alpha^*\hat{a})} = e^{-|\alpha|^2/2} e^{\alpha\hat{a}^+} e^{-\alpha^*\hat{a}}, \tag{4.3}$$

where the last expression is obtained by use of the Baker–Hausdorff theorem (eq. (2.35a)) and the commutation relation given by eq. (4.2). The operator $\hat{D}(\alpha)$ has the property that

$$\hat{D}^{-1}(\alpha)\, \hat{a}\, \hat{D}(\alpha) = \hat{a} + \alpha \tag{4.4a}$$

$$\hat{D}^{-1}(\alpha)\, \hat{a}^+\, \hat{D}(\alpha) = \hat{a}^+ + \alpha^*. \tag{4.4b}$$

The proof of eq. (4.4) readily follows from eqs. (4.2e) and (4.3). We now define the coherent state (Glauber [1963a]; Glauber [1963b]; Sudarshan [1963]), which we denote by $|\alpha\rangle$, as

$$|\alpha\rangle \equiv \hat{D}(\alpha)\,|0\rangle = e^{-|\alpha|^2/2} \sum_{n=0}^{\infty} \frac{\alpha^n}{n!} (\hat{a}^+)^n |0\rangle = e^{-|\alpha|^2/2} \sum_{n=0}^{\infty} \frac{\alpha^n}{(n!)^{1/2}} |n\rangle, \tag{4.5}$$

where $|0\rangle$ is the ground state of the oscillator. This state has the property that it is an eigenstate of the annihilation operators with eigenvalue $\alpha$. Again, this can be verified by using eq. (4.2e). Perhaps it should be emphasized that the symbol $\alpha$ always refers to a complex eigenvalue whereas $|\alpha\rangle$ always denotes a state, just as $n$ denotes a real eigenvalue and $|n\rangle$ a state, the so-called "number state". Also, just as $|n\rangle$ refers to a definite state of excitation of a system of *one* mode, $|\alpha\rangle$ also refers to a state of *one* mode.

The $|\alpha\rangle$ states are not orthogonal but they are complete (in fact overcomplete). Explicitly,

$$\langle\beta|\alpha\rangle = \exp[-\tfrac{1}{2}(|\alpha|^2 + |\beta|^2) + \beta^*\alpha], \tag{4.6}$$

which follows immediately from eq. (4.5) and the fact that the number states are orthonormal. Furthermore, it is possible to express the identity operator as

$$I = \frac{1}{\pi} \int d^2\alpha\, |\alpha\rangle\langle\alpha|, \tag{4.7a}$$

where $d^2\alpha = d(\text{Re}\,\alpha)\,d(\text{Im}\,\alpha) = \tfrac{1}{2}d\alpha\,d\alpha^*$. The proof of eq. (4.7a) follows by setting $\alpha = r\,e^{i\theta}$, so that $d^2\alpha = r\,dr\,d\theta$, and then using eq. (4.5) to get

$$\frac{1}{\pi} \int d^2\alpha\, |\alpha\rangle\langle\alpha| = \frac{1}{\pi} \sum_{m=0}^{\infty} \sum_{n=0}^{\infty} \frac{|m\rangle\langle n|}{(m!)^{1/2}(n!)^{1/2}} \int_0^{\infty} dr\, e^{-r^2} r^{m+n+1} \int_0^{2\pi} e^{i(m-n)\theta}\, d\theta = 2 \sum_n \frac{|n\rangle\langle n|}{n!} \int_0^{\infty} dr\, e^{-r^2} r^{2n+1}, \tag{4.7b}$$

*M. Hillery et al., Distribution functions in physics: Fundamentals*                                    155

where we have used the fact that the angular integral simply equals $2\pi\delta_{mn}$. The latter radial integral equals $n!/2$, so that using the fact that $\Sigma_n |n\rangle\langle n| = 1$, eq. (4.7a) readily follows. A direct consequence of eq. (4.7a) is that the trace of any operator $\hat{A}$ is just

$$\text{Tr}(\hat{A}) = \frac{1}{\pi}\int d^2\alpha\,\langle\alpha|\hat{A}|\alpha\rangle. \tag{4.8}$$

It is also of use to compare the expression for the displacement operator $\hat{D}(\alpha)$ to our previous results and use this comparison to derive an expansion for a general operator $\hat{A}$ in terms of $\hat{D}^{-1}(\alpha)$. This will be of use later. First we note that if we set $\alpha = (2\hbar)^{-1/2}(\lambda\tau + i\lambda^{-1}\sigma)$ then (see eq. (4.1))

$$\hat{D}(\alpha) = \exp\{(i/\hbar)(\sigma\hat{q} + \tau\hat{p})\} = \hat{C}(\sigma,\tau), \tag{4.9}$$

where $\hat{C}$ was defined earlier by eq. (2.43). Thus from eqs. (2.42) and (4.9) it is clear that the characteristic function is the expectation value of the displacement operator. This in conjunction with eqs. (4.5), (4.6) and (4.8) gives

$$\text{Tr}(\hat{D}(\alpha)\,\hat{D}^{-1}(\beta)) = \pi\,\delta^{(2)}(\alpha - \beta), \tag{4.10}$$

where the $\delta$ function here is $\delta^2(\xi) = \delta(\text{Re }\xi)\,\delta(\text{Im }\xi)$. Suppose that we can expand the operator $\hat{A}(\hat{a}, \hat{a}^+)$ as

$$\hat{A} = \frac{1}{\pi}\int d^2\xi\, g(\xi)\,\hat{D}^{-1}(\xi). \tag{4.11}$$

Using eq. (4.10) we find that

$$g(\xi) = \text{Tr}(\hat{A}\,\hat{D}(\xi)). \tag{4.12}$$

It can be shown (Cahill and Glauber [1969]) that if $\hat{A}$ is Hilbert–Schmidt (i.e. $\text{Tr}(\hat{A}^+\hat{A}) < \infty$) then the function $g(\xi)$ is square integrable.

The three types of ordering of the operators $\hat{a}$ and $\hat{a}^+$ which we wish to consider are defined as follows:

(i) Normal ordering – A product of $m$ annihilation operators and $n$ creation operators is normally ordered if all of the annihilation operators are on the right, i.e. if it is in the form $(\hat{a}^+)^n \hat{a}^m$.

(ii) Symmetric ordering – A product of $m$ annihilation operators and $n$ creation operators can be ordered in $(n + m)!/n!\,m!$ ways. The symmetrically ordered product of these operators, denoted by $\{(\hat{a}^+)^n \hat{a}^m\}$, is just the average of all of these differently ordered products. For example

$$\{\hat{a}^+\hat{a}\} = \tfrac{1}{2}(\hat{a}^+\hat{a} + \hat{a}\hat{a}^+) \tag{4.13a}$$

$$\{\hat{a}^+\hat{a}^2\} = \tfrac{1}{3}(\hat{a}^+\hat{a}^2 + \hat{a}\hat{a}^+\hat{a} + \hat{a}^2\hat{a}^+) \tag{4.13b}$$

$$\{\hat{a}^{+2}\hat{a}^2\} = \tfrac{1}{6}(\hat{a}^{+2}\hat{a}^2 + \hat{a}^+\hat{a}\hat{a}^+\hat{a} + \hat{a}^+\hat{a}^2\hat{a}^+ + \hat{a}\hat{a}^{+2}\hat{a} + \hat{a}\hat{a}^+\hat{a}\hat{a}^+ + \hat{a}^2\hat{a}^{+2}). \tag{4.13c}$$

(iii) Anti-normal ordering – A product of $m$ annihilation operators and $n$ creation operators is anti-normally ordered if all of the annihilation operators are on the left, i.e. if it is of the form $\hat{a}^m(\hat{a}^+)^n$.

For each operator ordering we have a rule which associates a function of $\alpha$ and $\alpha^*$ with a given operator. The rule is as follows: for any operator ordering scheme the product of $m$ annihilation and $n$ creation operators, ordered according to that scheme, is associated with the function $(\alpha^*)^n\alpha^m$. For example, if we are considering normal ordering the product $(\hat{a}^n)^n\hat{a}^m$ is associated with $(\alpha^*)^n\alpha^m$; if anti-normal ordering is being considered then $\hat{a}^m(\hat{a}^+)^n$ is associated with $\alpha^m(\alpha^*)^n$. We will now make the meaning of our rule more explicit by considering each of these orderings and its associated distribution function.

### 4.1. Normal ordering

Let us suppose that we can expand a given operator $\hat{A}(\hat{a}, \hat{a}^+)$ in a normally ordered power series

$$\hat{A} = \sum_{n,m=0}^{\infty} c_{nm} (\hat{a}^+)^n \hat{a}^m \,. \tag{4.14}$$

Let us further suppose that we can express the density matrix as

$$\hat{\rho} = \int d^2\alpha \, P(\alpha) |\alpha\rangle \langle\alpha| \tag{4.15}$$

where $P(\alpha)$ is a c-number and the state $|\alpha\rangle$ is given by eq. (4.5). $P(\alpha)$ is called the $P$-representation of the density matrix (or the distribution function representing the density matrix) of the *particular* mode under study. It should be emphasized that both the real and imaginary parts of $\alpha$ are used as the variables of the distribution function. Also, it is probably worthwhile mentioning again that our discussion is restricted to a system of bosons and thus the distribution functions under study are not applicable to, for instance, a gas of neutrinos. Also, we are dealing with a very large but countable set since we assumed that the basic domain is finite.

From eqs. (4.10) and (4.7a) and because $\langle\alpha|\hat{a}^{+n}\hat{a}^m|\alpha\rangle = \alpha^{*n}\alpha^m$, it follows that

$$\text{Tr}(\hat{A}\hat{\rho}) = \frac{1}{\pi} \int d^2\beta \int d^2\alpha \, P(\alpha) \langle\beta|\hat{A}|\alpha\rangle \langle\alpha|\beta\rangle = \int d^2\alpha \, P(\alpha) \langle\alpha|\hat{A}|\alpha\rangle$$

$$= \int d^2\alpha \, P(\alpha) \left[ \sum_{n,m=0}^{\infty} c_{nm} (\alpha^*)^n \alpha^m \right] \equiv \int d^2\alpha \, P(\alpha) A_N(\alpha, \alpha^*), \tag{4.16}$$

with

$$A_N(\alpha, \alpha^*) = \langle\alpha|\hat{A}|\alpha\rangle = \sum_{n,m=0}^{\infty} c_{nm} (\alpha^*)^n \alpha^m \,. \tag{4.17}$$

Therefore, we associate the operator $\hat{A}(\hat{a}, \hat{a}^+)$ with the function $A_N(\alpha, \alpha^*)$ in the evaluation of expectation values with the $P$-representation.

We now want to derive two expressions for $P(\alpha)$ in terms of the density matrix. It is not always possible to find a useful representation of $\hat{\rho}$ of the form given by eq. (4.15). For some density matrices $P(\alpha)$ would have to be so singular that it would not even be a tempered distribution (Cahill [1965]; Klauder and Sudarshan [1968]). This difficulty will be apparent in our formal expression for $P(\alpha)$.

Let us now choose for the operator $A$

$$\hat{A} = e^{\xi \hat{a}^+} e^{-\xi^* \hat{a}}.$$    (4.18)

The corresponding function is then $A_N(\alpha, \alpha^*) = \exp(\xi \alpha^* - \xi^* \alpha)$. Inserting these expressions into eq. (4.16) we find that

$$\chi_N(\xi) \equiv \text{Tr}(\hat{\rho} \, e^{\xi \hat{a}^+} e^{-\xi^* \hat{a}}) = \int d^2\alpha \, P(\alpha) \, e^{\xi \alpha^* - \xi^* \alpha}.$$    (4.19)

The function $\chi_N(\xi)$ is known as the normally ordered characteristic function. The right-hand side of eq. (4.19) is just a Fourier transform in a somewhat disguised form. In fact one has that if

$$f(\alpha) = \frac{1}{\pi} \int d^2\xi \, e^{\alpha \xi^* - \alpha^* \xi} \, \tilde{f}(\xi)$$    (4.20a)

then

$$\tilde{f}(\xi) = \frac{1}{\pi} \int d^2\alpha \, e^{\xi \alpha^* - \xi^* \alpha} f(\alpha),$$    (4.20b)

and vice versa. Therefore, we have for $P(\alpha)$

$$P(\alpha) = \frac{1}{\pi^2} \int d^2\xi \, e^{\alpha \xi^* - \alpha^* \xi} \chi_N(\xi).$$    (4.21)

The problem with this expression is that $\chi_N(\xi)$ can grow rather rapidly. In fact we have that because $\exp(\xi \hat{a}^+ - \xi^* \hat{a})$ is unitary

$$|\chi_N(\xi)| = e^{|\xi|^2/2}|\text{Tr}(\hat{\rho} \, e^{\xi \hat{a}^+ - \xi^* \hat{a}})| \le e^{|\xi|^2/2},$$    (4.22)

which suggests the type of behavior which is possible. For example, if $\hat{\rho} = |n\rangle\langle n|$, where $|n\rangle$ is the eigenstate of the number operator with eigenvalue $n$, then for large $|\xi|$ we have $|\chi_N(\xi)| \sim |\xi|^{2n}$. This representation, then, is not appropriate for all density matrices, but, nonetheless, is useful in many of the cases of interest.

Finally, we will derive an expression for $P(\alpha)$ in terms of a series expansion for the density matrix. Let us suppose that we can express the density matrix as an *anti-normally* ordered series

$$\hat{\rho} = \sum_{n,m=0}^{\infty} \rho_{nm} \hat{a}^m (\hat{a}^+)^n.$$    (4.23)

If we again consider the expression for $\hat{A}(\hat{a}, \hat{a}^+)$ given by eq. (4.14) we find that

$$\text{Tr}(\hat{\rho}\hat{A}) = \sum_{n,m=0}^{\infty} \sum_{r,s=0}^{\infty} \rho_{nm} c_{rs} \, \text{Tr}(\hat{a}^m (\hat{a}^+)^n (\hat{a}^+)^r \hat{a}^s).$$    (4.24)

The trace in eq. (4.24) can be expressed as

$$\text{Tr}(\hat{a}^m (\hat{a}^+)^{n+r} \hat{a}^s) = \text{Tr}((\hat{a}^+)^{n+r} \hat{a}^{s+m})$$

so that

$$\text{Tr}(\hat{\rho}\hat{A}) = \frac{1}{\pi} \int d^2\alpha \sum_{n,m=0}^{\infty} \sum_{r,s=0}^{\infty} \rho_{nm} c_{rs} \alpha^{*n} \alpha^m \alpha^{*r} \alpha^s . \tag{4.25}$$

Comparing this with eq. (4.16) we see that

$$P(\alpha) = \frac{1}{\pi} \sum_{n,m=0}^{\infty} \rho_{nm} \alpha^{*n} \alpha^m . \tag{4.26}$$

The difficulties which we had when considering eq. (4.21) suggest that we will have similar problems with eq. (4.26). In fact the problem goes back to eq. (4.23). The class of operators for which a meaningful anti-normally ordered expansion exists is highly restricted. One can see this by considering the representation for an operator given by eq. (4.11). Expand $\hat{D}^{-1}(\xi) = \exp(\xi^*\hat{a}) \exp(-\xi\hat{a}^+) \exp(\frac{1}{2}|\xi|^2)$ in an anti-normally ordered power series and insert it back into eq. (4.11). This gives us an anti-normally-ordered power series for $\hat{A}$:

$$\hat{A} = \sum_{n,m=0}^{\infty} d_{nm} \hat{a}^m (\hat{a}^+)^n , \tag{4.27}$$

with the coefficients given by

$$d_{nm} = \frac{1}{n! \, m!} \frac{1}{\pi} \int d^2\xi \, \text{Tr}(\hat{A}\hat{D}(\xi)) \, e^{|\xi|^2/2} (-\xi)^n (\xi^*)^m . \tag{4.28}$$

For these coefficients to exist $\text{Tr}(\hat{A}\hat{D}(\xi))$ must be a very rapidly decreasing function of $|\xi|$. Our previous remarks indicate that this will not be true in general for Hilbert–Schmidt operators and, in fact, will not be true in general for operators of trace class (operators, $\hat{A}$, for which $\text{Tr}([\hat{A}^\dagger\hat{A}]^{1/2}) < \infty$) such as density matrices.

It should be mentioned that normally-ordered power series expansions are far better behaved. A derivation similar to the one above gives for the coefficients $c_{nm}$ in eq. (4.14)

$$c_{nm} = \frac{1}{n! \, m!} \frac{1}{\pi} \int d^2\xi \, \text{Tr}(\hat{A}\hat{D}(\xi)) \, e^{-|\xi|^2/2} (-\xi)^n (\xi^*)^m . \tag{4.29}$$

This clearly exists for a much wider class of operators than does $d_{nm}$. The $c_{nm}$'s exist, in fact, for all Hilbert–Schmidt operators and the series converges in the sense that if one takes its matrix element between two coherent states, $\langle\alpha|$ on the left and $|\beta\rangle$ on the right, the resulting series converges to $\langle\alpha|A|\beta\rangle$.

### 4.2. Symmetric ordering

Before proceeding with a discussion of the distribution function for this case we would like to consider a few properties of the ordering scheme itself. We first note that

M. Hillery et al., Distribution functions in physics: Fundamentals     159

$$(\xi_1 \hat{a}^+ + \xi_2 \hat{a})^n = \sum_{l=0}^{n} \xi^{(n-l)} \xi_2^l \binom{n}{l} \{(\hat{a}^+)^{n-l} \hat{a}^l\},$$  (4.30)

which implies that

$$e^{(\xi_1 \hat{a}^+ + \xi_2 \hat{a})} = \sum_{l,m=0}^{\infty} \frac{1}{l!\,m!} \xi_1^m \xi_2^l \{(\hat{a}^+)^m \hat{a}^l\}.$$  (4.31)

Our operator-function correspondence is now done in a way analogous to that of the preceding section. Expand an operator $\hat{A}(\hat{a}, \hat{a}^+)$ in a symmetrically ordered power series

$$\hat{A} = \sum_{n,m=0}^{\infty} b_{nm} \{(\hat{a}^+)^n \hat{a}^m\}.$$  (4.32)

The function corresponding to the operator is then

$$A_s(\alpha, \alpha^*) = \sum_{n,m=0}^{\infty} b_{nm} (\alpha^*)^n \alpha^m.$$  (4.33)

Under this correspondence we see from eq. (4.30) that the function $\hat{D}(\xi)$ goes to

$$\hat{D}(\xi) = e^{\xi \hat{a}^+ - \xi^* \hat{a}} \rightarrow e^{\xi \alpha^* - \xi^* \alpha}.$$  (4.34)

Comparison with eq. (4.9) shows us that this is nothing other than Weyl ordering expressed in a different form. The distribution function, therefore, should be the Wigner function. As before we define this as the Fourier transform of the characteristic function $\chi(\xi)$ (see eqs. (2.42)–(2.45)) and we use the real and imaginary parts of $\alpha = \alpha_r + i\alpha_i$ as the variables of the distribution function, so that, analogous to eq. (4.1), $\alpha = (2\hbar)^{-1/2} (\lambda q + (i/\lambda)p)$, where $\lambda = (m\omega)^{1/2}$. Thus

$$W(\alpha) = \frac{1}{\pi} \int d^2\xi \, e^{(\alpha \xi^* - \alpha^* \xi)} \chi(\xi) = \frac{1}{\pi} \int \int d\xi_r \, d\xi_i \, e^{2i(\alpha_i \xi_r - \alpha_r \xi_i)} \text{Tr}[\hat{\rho} \, e^{\xi \hat{a}^+ - \xi^* \hat{a}}]$$

$$= \frac{1}{\pi} \int \int d\xi_r \, d\xi_i \, \text{Tr}\left[\hat{\rho} \, \exp\left\{2i\xi_r\left(\alpha_i - \frac{\hat{p}}{(2\hbar)^{1/2}\lambda}\right) - 2i\xi_i\left(\alpha_r - \frac{\lambda \hat{q}}{(2\hbar)^{1/2}}\right)\right\}\right],$$  (4.35)

where

$$\chi(\xi) = \text{Tr}(\hat{\rho} \, \hat{D}(\xi)).$$  (4.36)

It may be verified, using eqs. (2.42), (2.45) and (4.35) that

$$W(\alpha) = (2\pi\hbar) \, P_w(q, p) = 2 \int_{-\infty}^{\infty} dy \sum_n \sum_m \langle q - y | n \rangle \langle n | \hat{\rho} | m \rangle \langle m | q + y \rangle \, e^{2ipy/\hbar}$$

$$= \beta \int_{-\infty}^{\infty} dy \sum_n \sum_m \left(\frac{1}{2^n n!}\right)^{1/2} \left(\frac{1}{2^m m!}\right)^{1/2} e^{-\beta^2(q^2+y^2)} e^{2ipy/\hbar} H_n(\beta(q+y)) H_m(\beta(q-y)) \psi_n^* \psi_m,$$

(4.37)

160                     M. Hillery et al., Distribution functions in physics: Fundamentals

where, in the derivation of the last line from the previous line, we have used eq. (2.90) and where $\beta = (m\omega/\hbar)^{1/2}$ and $\langle n|\hat{\rho}|m\rangle = \psi_n^* \psi_n$.

Examination of eqs. (4.11) and (4.34) shows us that the function $A_s$ which corresponds, by eqs. (4.32) and (4.33), to the operator $\hat{A}(\hat{a}, \hat{a}^+)$ can also be represented as

$$A_s(\alpha, \alpha^*) = \frac{1}{\pi} \int d^2\xi \, \mathrm{Tr}(\hat{A}\hat{D}(\xi)) \, e^{\xi^*\alpha - \xi\alpha^*} . \tag{4.38}$$

We would now like to use this to show that

$$\mathrm{Tr}(\hat{\rho}\hat{A}) = \frac{1}{\pi} \int d^2\alpha \, A_s(\alpha, \alpha^*) \, W(\alpha) . \tag{4.39}$$

Evaluating the right-hand side we see that

$$\frac{1}{\pi} \int d^2\alpha \, A_s(\alpha, \alpha^*) \, W(\alpha) = \frac{1}{\pi^2} \int d^2\alpha \int d^2\xi \, \mathrm{Tr}(\hat{A}\hat{D}(\xi)) \, e^{\xi^*\alpha - \xi\alpha^*} \, W(\alpha) . \tag{4.40}$$

Making use of the relation

$$\delta^{(2)}(\alpha) = \frac{1}{\pi^2} \int d^2\xi \, e^{\alpha\xi^* - \alpha^*\xi} , \tag{4.41}$$

we find that

$$\frac{1}{\pi} \int d^2\alpha \, e^{\xi^*\alpha - \xi\alpha^*} \, W(\alpha) = \chi(-\xi) . \tag{4.42}$$

We also have from eqs. (4.11) and (4.36)

$$\frac{1}{\pi} \int d^2\xi \, \chi(-\xi) \, \hat{D}(\xi) = \frac{1}{\pi} \int d^2\xi \, \chi(\xi) \, \hat{D}^{-1}(\xi) = \hat{\rho} , \tag{4.43}$$

so that

$$\frac{1}{\pi} \int d^2\alpha \, A_s(\alpha, \alpha^*) \, W(\alpha) = \frac{1}{\pi} \int d^2\xi \, \mathrm{Tr}(\hat{A}\hat{D}(\xi)) \, \chi(-\xi) = \mathrm{Tr}(\hat{A}\hat{\rho}) , \tag{4.44}$$

which proves eq. (4.39) and shows that $A_s(\alpha, \alpha^*)$ and $W(\alpha)$ can be used to calculate the expectation values of symmetrically ordered operators.

We would also like to say a word about symmetrically ordered power series. Comparison of eqs. (4.11) and (4.31) allows us to calculate the coefficients appearing in eq. (4.32)

$$b_{nm} = \frac{1}{n! \, m!} \frac{1}{\pi} \int d^2\xi \, \mathrm{Tr}(\hat{A}\hat{D}(\xi)) \, (-\xi)^n \, (\xi^*)^m . \tag{4.45}$$

*M. Hillery et al., Distribution functions in physics: Fundamentals*    161

These coefficients, then, will exist for all operators which have the property that all moments of $\mathrm{Tr}(\hat{A}\hat{D}(\xi))$ are finite. While this behavior is not as good as that for a normally ordered power series it is certainly better than that of anti-normally ordered series.

It is also of interest to examine the behavior of $W(\alpha)$. First we note that

$$\frac{1}{\pi}\int \mathrm{d}^2\xi\, |\chi(\xi)|^2 = \frac{1}{\pi}\int \mathrm{d}^2\xi\, \mathrm{Tr}[\mathrm{Tr}(\hat{\rho}\hat{D}(\xi))\,\hat{D}^{-1}(\xi)\hat{\rho}] = \mathrm{Tr}\,\hat{\rho}^2 \le 1 \tag{4.46}$$

so that $\chi(\xi)$ is a square integrable function. As $W(\alpha)$ is just the Fourier transform of $\chi(\xi)$ it too is square integrable. Therefore, $W(\alpha)$ is far better behaved than $P(\alpha)$ and will exist for all density matrices.

It is also possible to express the Wigner distribution in terms of the $P$ representation. If we can represent the density matrix as in eq. (4.15) we then have that

$$\chi(\xi) = \int \mathrm{d}^2\beta\, P(\beta)\,\langle\beta|\,e^{\xi\hat{a}^+ - \xi^*\hat{a}}\,|\beta\rangle = \int \mathrm{d}^2\beta\, P(\beta)\,e^{\xi\beta^* - \xi^*\beta - |\xi|^2/2}\,. \tag{4.47}$$

Taking the Fourier transform of $\chi(\xi)$ gives us, with the use of eqs. (4.35) and (4.37),

$$W(\alpha) = \frac{1}{\pi}\int \mathrm{d}^2\xi \int \mathrm{d}^2\beta\, P(\beta)\,e^{\alpha\xi^* - \alpha^*\xi}\,e^{\xi\beta^* - \xi^*\beta}\,e^{-|\xi|^2/2}$$

$$= \frac{1}{\pi}\int \mathrm{d}^2\beta \int \mathrm{d}^2\xi\, P(\beta)\,e^{(\alpha - \beta)\xi^* - (\alpha^* - \beta^*)\xi - |\xi|^2/2}$$

$$= 2\int \mathrm{d}^2\beta\, P(\beta)\,e^{-2|\alpha - \beta|^2}\,. \tag{4.48}$$

### 4.3. Anti-normal ordering

Let us suppose that we have an operator given by an anti-normally ordered power series as in eq. (4.27). The function corresponding to the $\hat{A}$ of eq. (4.27) is then

$$A_{\mathrm{a}}(\alpha, \alpha^*) = \sum_{n,m=0} d_{nm}\, \alpha^m\, (\alpha^*)^m\,. \tag{4.49}$$

By analogy with our discussion of the $P$ representation (eq. (4.26)) we can then express $\hat{A}(\hat{a}, \hat{a}^+)$ as

$$\hat{A}(\hat{a}, \hat{a}^+) = \frac{1}{\pi}\int \mathrm{d}^2\alpha\, A_{\mathrm{a}}(\alpha, \alpha^*)\,|\alpha\rangle\,\langle\alpha|\,. \tag{4.50}$$

We then have that

$$\mathrm{Tr}(\hat{\rho}\hat{A}) = \frac{1}{\pi}\int \mathrm{d}^2\alpha\, A_{\mathrm{a}}(\alpha, \alpha^*)\,\mathrm{Tr}(\hat{\rho}|\alpha\rangle\,\langle\alpha|) = \int \mathrm{d}^2\alpha\, A_{\mathrm{a}}(\alpha, \alpha^*)\,Q(\alpha)\,, \tag{4.51}$$

where we have set (Kano [1965])

$$Q(\alpha) = \frac{1}{\pi} \langle \alpha | \hat{\rho} | \alpha \rangle .$$  (4.52)

This distribution can also be expressed in terms of a characteristic function

$$\chi_A(\xi) = \text{Tr}(\hat{\rho}\, e^{-\xi^* \hat{a}}\, e^{\xi \hat{a}^+}) .$$  (4.53)

We have that

$$\chi_A(\xi) = \frac{1}{\pi} \int d^2\alpha \, \langle \alpha | \, e^{\xi \hat{a}^+} \, \hat{\rho} \, e^{-\xi^* \hat{a}} | \alpha \rangle = \frac{1}{\pi} \int d^2\alpha \, e^{\xi \alpha^* - \xi^* \alpha} \, \langle \alpha | \hat{\rho} | \alpha \rangle$$  (4.54)

so that

$$Q(\alpha) = \frac{1}{\pi^2} \int d^2\xi \, e^{\alpha \xi^* - \alpha^* \xi} \, \chi_A(\xi)$$

$$= \frac{1}{\pi^3} \int \int d^2\xi \, d^2\beta \, e^{\alpha \xi^* - \alpha^* \xi} \, e^{\beta^* \xi - \beta \xi^*} \, \langle \beta | \hat{\rho} | \beta \rangle = \frac{1}{\pi} \langle \alpha | \hat{\rho} | \alpha \rangle .$$  (4.55)

Again by considering our derivation of the $P$ representation we can derive an alternate expression for $A_a(\alpha, \alpha^*)$. Examining eq. (4.21) we see that

$$A_a(\alpha, \alpha^*) = \frac{1}{\pi} \int d^2\xi \, e^{\alpha \xi^* - \alpha^* \xi} \, \text{Tr}(\hat{A}\, e^{\xi \hat{a}^+}\, e^{-\xi^* \hat{a}}) .$$  (4.56)

The "function" $A_a(\alpha, \alpha^*)$ has, of course, all of the singularity problems of the $P$ representation.

The distribution function, $Q(\alpha)$, has, on the other hand, no singularity problems at all. It exists for all density matrixes, is bounded, and is even greater than or equal to zero for all $\alpha$. The problems in this ordering scheme arise in the representation of the operators.

As a final remark, we note that all of the distribution functions can be written in terms of the Wigner distribution function (McKenna and Frisch [1966]; Agarwal and Wolf [1970]; Haken [1975]; O'Connell [1983b]), by use of integrals or derivatives.

### 4.4. Examples

We would now like to calculate $Q(\alpha)$ and $P(\alpha)$ for a single mode of the radiation field of angular frequency $\omega$. The system which we will consider will be a canonical ensemble at temperature $T \equiv (k\beta)^{-1}$. Our discussion will follow that given in Nussenzveig [1973].

We first consider the anti-normal distribution function $Q(\alpha)$. The density matrix for this system is

$$\hat{\rho} = (1 - e^{-\beta\hbar\omega}) \sum_{n=0}^{\infty} e^{-n\beta\hbar\omega} |n\rangle \langle n| .$$  (4.57)

M. Hillery et al., Distribution functions in physics: Fundamentals    163

For $Q(\alpha)$ we than have from eq. (4.52)

$$
\begin{aligned}
Q(\alpha) &= \frac{1}{\pi} (1 - e^{-\beta\hbar\omega}) \sum_{n=0}^{\infty} e^{-n\beta\hbar\omega} \langle\alpha|n\rangle\langle n|\alpha\rangle \\
&= \frac{1}{\pi} (1 - e^{-\beta\hbar\omega}) \sum_{n=0}^{\infty} e^{-n\beta\hbar\omega} e^{-|\alpha|^2} \frac{|\alpha|^{2n}}{n!} \\
&= \frac{1}{\pi} (1 - e^{-\beta\hbar\omega}) \exp[-|\alpha|^2 (1 - e^{-\beta\hbar\omega})] \, .
\end{aligned}
\tag{4.58}
$$

To obtain $P(\alpha)$ we make use of our result for $Q(\alpha)$. We first find $\chi_A(\xi)$ from eq. (4.54). If we set

$$
s = (1 - e^{-\beta\hbar\omega}), \qquad \xi = x + iy, \qquad \alpha = r + ik
\tag{4.59}
$$

then

$$
\begin{aligned}
\chi_A(\xi) &= \frac{s}{\pi} \int d^2\alpha \; e^{\xi\alpha^* - \xi^*\alpha} e^{-s|\alpha|^2} = \frac{s}{\pi} \int dr \int dk \; \exp\{-2i(kx - ry) - s(r^2 + k^2)\} \\
&= \frac{s}{\pi} \int dr \int dk \; \exp\{-s(r - iy/s)^2 - s(k + ix/s)^2\} \exp\{-(x^2 + y^2)/s\} = e^{-|\xi|^2/s} \, .
\end{aligned}
\tag{4.60}
$$

To calculate $\chi_N(\xi)$, given by eq. (4.19), we now use the general relation

$$
\chi_N(\xi) = \mathrm{Tr}(\hat{\rho} \; e^{\xi a^+} e^{-\xi^* a}) = \mathrm{Tr}(\hat{\rho} \; e^{-\xi^* a} e^{\xi a^+}) e^{|\xi|^2} = e^{|\xi|^2} \chi_A(\xi) \, .
\tag{4.61}
$$

Therefore, we see that

$$
\chi_N(\xi) = \exp\{-|\xi|^2(1 - s)/s\} \, .
\tag{4.62}
$$

If we set $\lambda = (1 - s)/s = (e^{\beta\hbar\omega} - 1)^{-1}$ then from eq. (4.21) we have

$$
\begin{aligned}
P(\alpha) &= \frac{1}{\pi^2} \int d^2\xi \; e^{\alpha\xi^* - \alpha^*\xi} e^{-\lambda|\xi|^2} = \frac{1}{\pi^2} \int dx \int dy \; \exp\{2i(kx - ry) - \lambda(x^2 + y^2)\} \\
&= \frac{1}{\pi\lambda} e^{-|\alpha|^2/\lambda} = \frac{1}{\pi} (e^{\beta\hbar\omega} - 1) \exp[-|\alpha|^2 (e^{\beta\hbar\omega} - 1)] \, .
\end{aligned}
\tag{4.63}
$$

For this system $P(\alpha)$ is a well-behaved function, a Gaussian in fact, and has no singularities. It is even positive definite. $Q(\alpha)$ is also well behaved, but this comes as no surprise. Our general discussion had ensured that this would be the case.

### 4.5. Distribution functions on four-dimensional phase space

We would now like to briefly discuss some distribution functions which are functions on a four- rather than a two-dimensional phase space. The first of these, the $R$ representation, was discussed by

Glauber in his 1963 paper. It is very well behaved but has found little use in applications. More recently a new class of these distributions, the generalized $P$ representations, has been used to study the photon statistics of various non-linear optical devices [Walls, Drummond and McNeil [1981]; Drummond and Gardiner [1980]; Drummond, Gardiner and Walls [1981]).

The $R$ representation of the density matrix is obtained by using the coherent state resolution of the identity twice. One has

$$\hat{\rho} = \frac{1}{\pi^2} \int d^2\alpha \int d^2\beta \, \exp\{-\tfrac{1}{2}(|\alpha|^2 + |\beta|^2)\} \, R(\alpha^*, \beta) |\alpha\rangle \langle\beta| , \tag{4.64}$$

where $|\alpha\rangle$ is defined in eq. (4.5) and $|\beta\rangle$ has the corresponding meaning, and

$$R(\alpha^*, \beta) = \exp\{\tfrac{1}{2}(|\alpha|^2 + |\beta|^2)\} \langle\alpha|\hat{\rho}|\beta\rangle . \tag{4.65}$$

This representation has no singularity problems. Also it exists and is unique for all density matrices provided that $R(\alpha^*, \beta)$ is an analytic function of $\alpha^*$ and $\beta$ (Glauber [1963b]). It can be used to evaluate normally ordered products. One has

$$\langle (\hat{a}^+)^n \, \hat{a}^m \rangle = \mathrm{Tr}[\hat{\rho} \, (\hat{a}^+)^n \, \hat{a}^m] = \frac{1}{\pi^2} \int d^2\alpha \int d^2\beta \, \exp\{-(|\alpha|^2 + |\beta|^2) + \beta^*\alpha\} \, R(\alpha^*, \beta) \, \alpha^m \, (\beta^*)^n . \tag{4.66}$$

The generalized $P$ representations (Drummond and Gardiner [1980]; Drummond, Gardiner and Walls [1981]) are again functions of two complex variables but are not necessarily defined for all values of these variables. To define these representations we define the operator

$$\hat{\Lambda}(\alpha, \beta) = |\alpha\rangle \langle\beta^*| / \langle\beta^*|\alpha\rangle \tag{4.67}$$

and an integration measure $d\mu(\alpha, \beta)$. It is the choice of this measure which determines the distribution function. We will consider two different choices. The density matrix is then

$$\hat{\rho} = \int_D d\mu(\alpha, \beta) \, P(\alpha, \beta) \, \hat{\Lambda}(\alpha, \beta) , \tag{4.68}$$

where D is the domain of integration. Normally ordered products are then given by

$$\langle (\hat{a}^+)^n \, (\hat{a})^m \rangle = \int d\mu(\alpha, \beta) \, P(\alpha, \beta) \, \beta^n \, \alpha^m . \tag{4.69}$$

Our first integration measure is $d\mu(\alpha, \beta) = d\alpha \, d\beta$ where $\alpha$ and $\beta$ are to be integrated on some contours C and C' respectively. This gives rise to what is called the complex $P$ representation. Let us consider the case in which C and C' are contours which enclose the origin. One can then show (Drummond and Gardiner [1980]) that if the density matrix is of the form

$$\hat{\rho} = \sum_n \sum_m c_{nm} |n\rangle \langle m| , \tag{4.70}$$

*M. Hillery et al., Distribution functions in physics: Fundamentals*     165

where both sums are finite then $P(\alpha, \beta)$ exists and is analytic when neither $\alpha$ nor $\beta$ is 0. Whether $P(\alpha, \beta)$ exists for a general density matrix is not known. The complex $P$ representation is also not unique; if one complex $P$ representation exists for a given density matrix, then an infinite number of representation exist.

The second measure which we wish to consider is $d\mu(\alpha, \beta) = d\alpha^2 \, d\beta^2$. Because the coherent states are linearly dependent such a representation is not unique. In fact we have encountered one representation of this type already, the $R$ representation. It is possible to choose $P(\alpha, \beta)$ so that it is real and non-negative (Drummond and Gardiner [1980]), i.e.

$$P(\alpha, \beta) = (1/4\pi^2) \exp\{-\tfrac{1}{4}|\alpha - \beta^*|^2\} \langle \tfrac{1}{2}(\alpha + \beta^*)|\hat{\rho}|\tfrac{1}{2}(\alpha + \beta^*)\rangle . \tag{4.71}$$

This representation, the positive $P$ representation, is defined for all density matrices.

These two distributions have been used in problems in which non-classical photon states (states which are more like number states than coherent states) are produced. Under these conditions the above defined generalized $P$ representations are better behaved than the original $P$ representation. For example, the $P$ representation corresponding to a density matrix $\hat{\rho} = |n\rangle\langle n|$ contains derivatives of delta functions up to order $2n$. On the other hand, the complex $P$ representation for this state (again defined on two contours C and C' encircling the origin) is just (Drummond and Gardiner [1980])

$$P(\alpha, \beta) = -(1/4\pi^2) \, n! \, e^{\alpha\beta} \, (1/\alpha\beta)^{n+1} \tag{4.72}$$

while the positive $P$ representation is, from eq. (4.70)

$$P(\alpha, \beta) = (1/4\pi^2)(1/n!) \exp\{-\tfrac{1}{4}|\alpha - \beta^*|^2\} \exp\{-\tfrac{1}{2}|\alpha + \beta^*|^2\} |\tfrac{1}{2}(\alpha + \beta^*)|^{2n} . \tag{4.73}$$

Both of these functions are far less singular than the original $P$ representation.

The original motivation for the introduction of these generalized $P$ distributions was connected with their practical applicability to the solution of quantum mechanical master equations (Drummond and Gardiner [1980]; Drummond, Gardiner and Walls [1981]). In general, using a coherent state basis, it is possible to develop phase-space Fokker–Planck equations that correspond to quantum master equations for the density operator (Haken [1970]; Louisell [1973]). From this equation observables are obtained in terms of moments of the $P$ function. However, for various problems, as for example the analysis of recent experiments on atomic fluorescence (Kimble, Dagenais and Mandel [1978]) where we are dealing with non-classical photon statistics (Carmichael and Walls [1976]), the Glauber–Sudarshan $P$ function is singular whereas the generalized $P$ function discussed above is not. Also, use of the latter leads to Fokker–Planck equations with positive semi-definite diffusion coefficients whereas the former gives rise to non-positive-definite diffusion coefficients. In particular, the generalized $P$ representations were applied successfully to non-linear problems in quantum optics (two-photon absorption; dispersive bistability; degenerate parametric amplifier) and chemical reaction theory (Drummond and Gardiner [1980]; Drummond, Gardiner and Walls [1981]; Walls and Milburn [1982]). On the other hand, the usefulness of the Wigner distribution in quantum optics has been demonstrated in a paper by Lugiato, Casagrande and Pizzuto [1982] who consider a system of $N$ two-level atoms interacting with a resonant mode radiation field and coupled to suitable reservoirs. The presence of an external CW coherent field injected into the cavity is also included, which allows for the possibility of treating optical bistability (which occurs when a non-linear optical medium, interacting with a coherent driving field, has more than one stable steady state) as well as a laser with injected signal.

## 5. Conclusion

We have given what we hope is a useful summary of some of the formalism surrounding the use of quantum mechanical quasiprobability distribution functions. To be of use, however, the formalism should either provide insight or convenient methods of calculation. In our next paper dealing with applications we hope to show that this particular formalism does both in that it has proven to be a tool of great effectiveness in many areas of physics.

## Acknowledgments

R.F.O'C. acknowledges support from the Dept. of Energy, Division of Material Sciences, under Contract No. DE-AS05-79ER10459. He would also like to thank the Max-Planck Institute for Quantum Optics for hospitality, during the summers of 1981 and 1982, at which time part of this work was carried out. M.O.S. acknowledges support from the Office of Scientific Research, under Contract No. AFOSR-81-0128-A.

## References

Agarwal, G.S. and E. Wolf, 1970, Phys. Rev. D10, 2161, 2187, 2206.
Alastuey, A. and B. Jancovici, 1980, Physica 102A, 327.
Bloch, F., 1932, Zeits. f. Physik 74, 295.
Bopp, F., 1956, Ann. Inst. H. Poincaré 15, 81.
Bopp, F., 1961, Werner Heisenberg und die Physik unserer Zeit (Vieweg, Braunschweig) p. 128.
Brittin, W.E. and W.R. Chappell, 1962, Rev. Mod. Phys. 34, 620.
Cahill, K.E., 1965, Phys. Rev. 138, B1566.
Cahill, K.E. and R.J. Glauber, 1969, Phys. Rev. 177, 1857, 1883.
Carmichael, H.J. and D.F. Walls, 1976, J. Phys. B9, 1199.
Carruthers, P. and F. Zachariasan, 1983, Rev. Mod. Phys. 55, 245.
Cartwright, N.D., 1976, Physica 83A, 210.
Cohen, L., 1966, J. Math. Phys. 7, 781.
Dahl, J.P., 1982, Physica Scripta 25, 499.
De Giorgio, V. and M.O. Scully, 1970, Phys. Rev. A2, 1170.
De Groot, S.R. and L.G. Suttorp, 1972, Foundations of Electrodynamics (North-Holland, Amsterdam).
De Groot, S.R., 1974, La transformation de Weyl et la fonction de Wigner: une forme alternative de la mécanique quantique (Les Presses Universitaires de Montréal, Montréal).
De Groot, S.R., W.A. van Leeuwen and C.G. van Weert, 1980, Relativistic Kinetic Theory (North-Holland, Amsterdam).
Drummond, P.D. and C.W. Gardiner, 1980, J. Phys. A13, 2353.
Drummond, P.D., C.W. Gardiner and D.F. Walls, 1981, Phys. Rev. A24, 914.
Glauber, R.J., 1963a, Phys. Rev. Lett. 10, 84.
Glauber, R.J., 1963b, Phys. Rev. 131, 2766.
Glauber, R.J., 1965, in: Quantum Optics and Electronics, eds. C. DeWitt, A. Blandin and C. Cohen-Tannoudji (Gordon and Breach, New York).
Gradshteyn, I.S. and M. Ryzhik, 1980, Table of Integrals, Series and Products (Academic Press, New York) p. 838.
Graham, R. and H. Haken, 1970, Z. Physik 237, 31.
Groenewold, H.J., 1946, Physica 12, 405.
Haken, H., 1970, Handbuch der Physik, Vol. XXV/2c (Springer, Berlin).
Haken, H., 1975, Rev. Mod. Phys. 47, 67.
Husimi, K., 1940, Proc. Phys. Math. Soc. Japan 22, 264.
Imre, K., E. Ozizmir, M. Rosenbaum and P. Zweifel, 1967, J. Math. Phys. 8, 1097.
Kano, Y., 1965, J. Math. Phys. 6, 1913.
Kimble, H.J., M. Dagenais and L. Mandel, 1978, Phys. Rev. A18, 201.
Klauder J.R. and E.C.G. Sudarshan, 1968, Quantum Optics (Benjamin, New York) p. 178.
Klimintovich, I.L., 1958, Sov. Phys. JETP 6, 753.

*M. Hillery et al., Distribution functions in physics: Fundamentals*     167

Kubo, R., 1964, J. Phys. Soc. Japan **19**, 2127.
Landau, L.D. and E.M. Lifshitz, 1965, Quantum Mechanics (Pergamon, Oxford).
Lax, M., 1968, Phys. Rev. **172**, 350.
Lax, M. and W.H. Louisell, 1967, J. Quant. Electron. QE3, 47.
Louisell, W.H., 1973, Quantum Statistical Properties of Radiation (Wiley, New York).
Lugiato, L.A., F. Cassagrande and L. Pizzuto, 1982, Phys. Rev. **A26**, 3438.
Margenau, H. and R.N. Hill, 1961, Prog. Theoret. Phys. (Kyoto) **26**, 722.
McKenna, J. and H.L. Frisch, 1966, Phys. Rev. **145**, 93.
Mehta, C.L., 1964, J. Math. Phys. **5**, 677.
Mehta, C.L. and E.C.G. Sudarshan, 1965, Phys. Rev. **138**, B274.
Messiah, A., 1961, Quantum Mechanics (North-Holland, Amsterdam) Vol. 1, p. 442.
Mori, H., I. Oppenheim and J. Ross, 1965, in: Studies in Statistical Mechanics, eds. J. de Boer and G.E. Uhlenbeck (North-Holland, Amsterdam)
     Vol. 1, p. 213.
Moyal, J.E., 1949, Proc. Cambridge Phil. Soc. **45**, 99.
Nienhuis, G., 1970, J. Math. Phys. **11**, 239.
Nussenzveig, H.M., 1973, Introduction to Quantum Optics (Gordon and Breach, New York) p. 71.
O'Connell, R.F., 1983a, Found. Phys. **13**, 83.
O'Connell, R.F., 1983b, in: Proc. Third New Zealand Symp. on Laser Physics, eds. J.D. Harvey and D.F. Walls (Springer-Verlag, Berlin and New
     York), Lecture Notes in Physics No. 182.
O'Connell, R.F. and E.P. Wigner, 1981a, Phys. Lett. **83A**, 145.
O'Connell, R.F. and E.P. Wigner, 1981b, Phys. Lett. **85A**, 121.
O'Connell, R.F. and E.P. Wigner, 1983, in preparation.
Oppenheim, I. and J. Ross, 1957, Phys. Rev. **107**, 28.
Prugovecki, E., 1978, Ann. Phys. (NY) **110**, 102.
Schrödinger, E., 1926, Naturwissenschaften **14**, 664.
Schweber, S.S., 1961, An Introduction to Relativistic Quantum Field Theory (Harper and Row, New York) pp. 133, 140.
Scully, M.O. and W.E. Lamb Jr., 1967, Phys. Rev. **159**, 208.
Sudarshan, E.C.G., 1963, Phys. Rev. Lett. **10**, 277.
Summerfield, G.C. and P.F. Zweifel, 1969, J. Math. Phys. **10**, 233.
Suttorp, L.G. and S.R. de Groot, 1970, Nuovo Cimento **65A**, 245.
Takabayashi, T., 1954, Prog. Theor. Phys. **11**, 341.
Walls, D.F., P.D. Drummond and K.J. McNeil, 1981, in: Optical Bistability, eds. C.M. Bowden, M. Ciftan and H.R. Robl (Plenum, New York).
Walls, D.F. and G.J. Milburn, 1982, in: Proc. NATO ASI in Bad Windsheim, West Germany, ed. P. Meystre (Plenum, New York).
Weyl, H., 1927, Z. Phys. **46**, 1.
Wigner, E., 1932a, Phys. Rev. **40**, 749.
Wigner, E., 1932b, Z. Phys. Chem. **B19**, 203.
Wigner, E., 1938, Trans. Faraday Soc. **34**, 29.
Wigner, E.P., 1979, in: Perspectives in Quantum Theory, eds. W. Yourgrau and A. van der Merwe (Dover, New York) p. 25.

# Manifestations of Bose and Fermi Statistics on the Quantum Distribution Function for Systems of Spin-0 and Spin-$\frac{1}{2}$ Particles

R. F. O'Connell and E. P. Wigner*

Physical Review A *30*, 2613–2618 (1984)

(Received 17 April 1984)

We consider the manifestations of Bose and Fermi statistics on the original quantum distribution function. Initially, we consider some general symmetry properties of the density matrix. Next, we convert these results into distribution-function language for the case of a system of spin-0 particles. Finally, we consider a system of spin-$\frac{1}{2}$ particles, for which we treat the combined effects of both spin and statistics.

## I. INTRODUCTION

Our purpose here is to consider some manifestations of Bose and Fermi statistics on the original quantum distribution function.[1] In our earlier work we also used this distribution function to calculate quantum corrections to the various properties of a Boltzmann gas. The first consideration to the effects of Bose or Fermi statistics was given by Uhlenbeck and Gropper,[2] who calculated the equation of state of both a Bose and a Fermi nonideal gas. Explicit spin effects were ignored in these considerations, as they were also in the later work of Kirkwood.[3]

After a lapse of nearly 20 years, the subject was considered anew by Green,[4] who wrote a relation between the density matrix determined on the basis of classical statistics and the corresponding density matrices for particles satisfying Bose and Fermi statistics. This relation made use of symmetrization operators, which in turn was expressed as the product of a number of cyclic operators. However, we wish to establish more explicit expressions than are given by Green's equation.

In their consideration of the quantum theory of transport in gases, Ross and Kirkwood[5] wrote down the symmetrized pair distribution function in terms of singlet distribution functions by making use of symmetry operators. The latter were discussed in detail in Appendix B of their paper and they concluded that the symmetry operation on the original quantum distribution function is represented by an integral operator. A different approach to the subject was introduced by Schram and Nijboer,[6] who introduced the symmetry requirements by means of a restricted summation of states in Hilbert space. Finally, these authors were able to express the symmetrized distribution functions in terms of an integral involving the original unsymmetrized distribution function, permutation operators, and various δ functions. They then went on to express the partition function as a summation involving permutation operators—but did not obtain a very explicit result.

A particularly important contribution to the subject was the article of Stratonovich.[7] He recorded the distribution function for particles of arbitrary spin and introduced the concept of second quantization in the phase space by essentially replacing the wave functions appearing in the distribution function by operators. The second-quantization approach has been considered further by other authors and used as a method for the incorporation of Fermi and Bose statistics.[8-10]

In a recent series of papers dealing with a one-component plasma (also called "jellium"—a system of identical particles embedded in a uniform neutralizing background of opposite charge), Jancovici[10] and Alastuey and Jancovici[11,12] calculated exchange quantum corrections in the *near-classical* limit. They found, both for three-dimensional[10,11] and so-called two-dimensional systems,[11] that the exchange free energy is negligibly small—a conclusion not changed by the presence of a strong magnetic field.[12]

In this paper we wish to consider anew the general subject of the manifestations of Bose and Fermi statistics on the quantum distribution function. In Sec. II we consider some general properties of the density matrix describing a mixture of several states. Then, in Sec. III we apply these results to a calculation of the effect of statistics on a system of spin-0 particles. Finally, in Sec. IV we extend the description to a system of spin-$\frac{1}{2}$ particles, for which we treat the combined effects of both spin and statistics. Apart from the obvious potential application of our results to a calculation of the equation of state of Bose and Fermi gases, we also consider that they should prove useful in the analysis of the spin-spin correlation experiments[13] which are of interest in connection with tests of the Bell inequalities.[14]

## II. DENSITY MATRIX FOR A MIXTURE OF STATES OF PARTICLES SUBJECT TO BOSE OR FERMI STATISTICS

Consider a system of $n$ identical particles $(1, 2, \ldots, i, \ldots, n)$ and let $x_i$ and $x_i'$ denote all the coordinates of the $i$th particle, including its spin. Our starting point is the density matrix $\rho$ for a system of identical particles. This represents, by definition, a system in which a variety of wave functions $\psi_\mu$ is present with probabilities $w_\mu$. We

2614                    R. F. O'CONNELL AND E. P. WIGNER                    30

have, therefore, by the definition of $\rho$,

$$\rho = \sum_\mu w_\mu \psi_\mu^*(x_1, x_2, \ldots, x_i, \ldots) \psi_\mu(x'_1, x'_2, \ldots, x'_i, \ldots)$$

$$\equiv \rho(x'_1, x'_2, \ldots, x'_i, \ldots; x_1, x_2, \ldots, x_i, \ldots) . \quad (1)$$

No matter to which permutation we subject *both* the $x'$ and the $x$, the value of $\rho$ will remain unchanged. This is evident if the particles are of Bose type—in this case both $\psi$ factors remain unchanged. It is also true if the particles are of Fermi type since in this case either both $\psi$ factors remain unchanged (if the permutation is "even") or both $\psi$ factors are multiplied by $-1$ (if the permutation is "odd"), so that $\rho$ remains the same in this case also. It will be assumed, therefore, that all $\rho$, representing only identical particles, are unchanged if both sets of variables—the $x'$ and the $x$ of Eq. (1)—are subject to the same permutation. Naturally, if the $\rho$ refers to a system of several types of particles, this remark is valid only for permutations which interchange only variables referring to identical particles.

If only the so-called "column variables"—those after the semicolon (;) in the argument of $\rho$—are interchanged,

$\rho$ will remain unchanged if the particles are of Bose type or if the permutation is even. In these cases all the $\psi_\mu$ of Eq. (1) remain unchanged. If the particles are of Fermi type and if the permutation is odd, the sign of $\rho$ will be changed. The same applies, of course, for the so-called "row variables" of $\rho$—those before the semicolon (;) in the argument of $\rho$.

We will demonstrate now that if $\rho$ is symmetric in the sense previously specified, i.e., invariant under the simultaneous and identical interchange of both row and column variables, and if it has the right symmetry property with respect to the interchange of one single pair of two column variables (or any pair of two row variables), it then has the right symmetry property with respect to any interchange of variables.

What will be proved actually is that if $\rho$ is invariant with respect to any simultaneous and identical permutation of both row and column variables (which was demonstrated for Bose- and Fermi-particle density matrices) and if, in addition, it is symmetric or antisymmetric with respect to the interchange of the first two column variables,

$$\rho(x'_1, x'_2, \ldots, x'_i, \ldots, x'_j, \ldots; x_1, x_2, \ldots, x_i, \ldots, x_j, \ldots)$$
$$= \pm \rho(x'_1, x'_2, \ldots, x'_i, \ldots, x'_j, \ldots; x_2, x_1, \ldots, x_i, \ldots, x_j, \ldots), \quad (2a)$$

then it has the same symmetry or antisymmetry with respect to the interchange of any pair of column variables, in particular the interchange of the $i$th and $j$th column variables, so that

$$\rho(x'_1, x'_2, \ldots, x'_i, \ldots, x'_j, \ldots; x_1, x_2, \ldots, x_i, \ldots, x_j, \ldots)$$
$$= \pm \rho(x'_1, x'_2, \ldots, x'_i, \ldots, x'_j, \ldots; x_1, x_2, \ldots, x_j, \ldots, x_i, \ldots). \quad (2b)$$

It follows then from the possibility of arranging any permutation of the $x$'s by a succession of the interchange of two of them that any even permutation of the column indices will leave the density matrix unchanged and any odd permutation will do so also for the case of Bose particles and change the sign for fermions. It further follows from the self-adjoint nature of $\rho$, that is, $\rho(x'; x) = [\rho(x; x')]^*$, that the same is true for the row variables. Hence the symmetry of $\rho$ with respect to any identical interchange of row and column variables, plus the validity of Eq. (2a), establishes the fact that $\rho$ represents a set of identical bosons or fermions, depending on whether Eq. (2a) is valid with the $+$ or $-$ sign. This theorem facilitates the establishment of Bose or Fermi statistics in density matrices and hence also in distribution functions derived from them.

In order to derive Eq. (2b) from Eq. (2a) we first notice that the interchange of both the $x$ and the $x'$ sets of variables at positions 1 and $i$ and at positions 2 and $j$ leaves $\rho$ unchanged—as a result of the symmetry requirement. Hence

$$\rho(x'_1, x'_2, \ldots, x'_i, \ldots, x'_j, \ldots; x_1, x_2, \ldots, x_i, \ldots, x_j, \ldots)$$
$$= \rho(x'_i, x'_j, \ldots, x'_1, \ldots; x'_2, \ldots; x_i, x_j, \ldots, x_1, \ldots, x_2, \ldots). \quad (3)$$

We now interchange the first two column variables, which gives, according to the assumptions made,

$$\rho(x'_1, x'_2, \ldots, x'_i, \ldots, x'_j, \ldots; x_1, x_2, \ldots, x_i, \ldots, x_j, \ldots)$$
$$= \pm \rho(x'_i, x'_j, \ldots, x'_1, \ldots, x'_2, \ldots; x_j, x_i, \ldots, x_1, \ldots, x_2, \ldots). \quad (4)$$

If we now interchange again, in both row and column, the first and $i$th variables and also the second and $j$th variables, the desired equation (2b) is obtained. This shows that if Eq. (2a) is valid—or a similar equation for another pair—and if $\rho$ is invariant under the identical permutation of both row and column indices, then the density matrix takes care of the Bose or Fermi requirements of the state represented by it—Bose or Fermi depending on whether Eq. (2a) with a $+$ or $-$ is valid. This will facilitate the derivation of the proper conditions for the distribution functions. Finally, we repeat that if there are several systems of identical particles present, then Eq. (2b) is valid for each set of coordinates referring to the same type of particles.

### III. SPIN-0 PARTICLES

It should be observed that in the preceding argument *the x' and the x can stand for all coordinates of a particle, including its spin.* However, in the translation of Eq. (2a) into the language of the distribution function, which follows, we distinguish between position and spin coordinates. The effect of the spin can be treated separately.

We now consider the corresponding distribution functions $P(q_1,p_1, \ldots, q_i,p_i, \ldots, q_n,p_n)$, which are functions of position and momentum coordinates $q_1, \ldots, q_i, \ldots, q_n$ and $p_1, \ldots, p_i, \ldots, p_n$, respectively. In the classical limit, $P(q,p)$ is the phase-space distribution function which gives the probability that the coordinates and momenta have the values $q$ and $p$. Specifically,[1]

$$P(q_1,p_1, \ldots, q_i,p_i, \ldots, q_n,p_n)$$
$$= (\pi\hbar)^{-3n} \int \cdots \int dy_1 \cdots dy_i \cdots dy_n$$
$$\times \rho(q_1-y_1, \ldots, q_i-y_i, \ldots, q_n-y_n; q_1+y_1, \ldots, q_i+y_i, \ldots, q_n+y_n)$$
$$\times \exp[2i(p_1y_1+ \cdots +p_iy_i+ \cdots +p_ny_n)/\hbar] . \tag{5}$$

In Eq. (5), $q_i$, $p_i$, and $y_i$ are considered to be three-dimensional vectors, $py$ is the scalar product of $p$ and $y$, and $dy_i$ means integration over all three vector components. All integrations in this paper are from $-\infty$ to $\infty$.

From Eq. (5) it follows that

$$\int \cdots \int dp_1 \cdots dp_i \cdots dp_n \, P(q_1,p_1, \ldots, q_i,p_i, \ldots, q_n,p_n) \exp[-2i(p_1y_1+ \cdots +p_iy_i+ \cdots +p_ny_n)/\hbar]$$
$$= \rho(q_1-y_1, \ldots, q_i-y_i, \ldots, q_n-y_n; q_1+y_1, \ldots, q_i+y_i, \ldots, q_n+y_n) . \tag{6}$$

Hence, changing variables to $u_i$ ($=q_i-y_i$) and $v_i$ ($=q_i+y_i$) $i=1,2, \ldots, n$, we obtain

$$\rho(u_1, \ldots, u_i, \ldots, u_n; v_1, \ldots, v_j, \ldots, v_n)$$
$$= \int \cdots \int dp_1 \cdots dp_i \cdots dp_n \, P(\tfrac{1}{2}(u_1+v_1),p_1, \ldots, \tfrac{1}{2}(u_i+v_i),p_i, \ldots, \tfrac{1}{2}(u_n+v_n),p_n)$$
$$\times \exp\{+i[p_1(u_1-v_1)+ \cdots +p_i(u_i-v_i)+ \cdots +p_n(u_n-v_n)]/\hbar\} . \tag{7}$$

At this stage we note that since Eq. (3) refers only to two of the particles, we number these 1 and 2 and, in fact, omit the coordinates of the other particles from the equations in order to make them more simple. We will therefore express the equation [see Eqs. (2a) and (2b); we omit the $\pm$ sign since we are dealing here with spin-0 particles]

$$\rho(q_1',q_2';q_1,q_2)=\rho(q_1',q_2';q_2,q_1) , \tag{8}$$

in which the $q$ stands for all three space coordinates, in terms of the corresponding distribution functions. We then have [see Eq. (5)]

$$P(q_1,p_1,q_2,p_2)=(\pi\hbar)^{-6} \int \int \rho(q_1-y_1,q_2-y_2;q_1+y_1,q_2+y_2) \exp[2i(p_1y_1+p_2y_2)/\hbar] dy_1 \, dy_2 \tag{9}$$

and also [see Eq. (7)]

$$\rho(x_1',x_2';x_1,x_2)=\int \int dp_1 \, dp_2 \, P(\tfrac{1}{2}(x_1+x_1'),p_1,\tfrac{1}{2}(x_2+x_2'),p_2) \exp[ip_1(x_1'-x_1)/\hbar+ip_2(x_2'-x_2)/\hbar] . \tag{10}$$

The condition (8) can therefore be written by equating the right side of (10) with the same expression in which, however, $x_1$ and $x_2$ are interchanged.

In order to simplify the resulting equation, we introduce new variables instead of the $x$ and the $p$:

$$\tfrac{1}{4}(x_1+x_1'+x_2+x_2')=q, \quad \tfrac{1}{4}(x_1-x_1'+x_2-x_2')=Y ,$$
$$\tfrac{1}{4}(x_1+x_1'-x_2-x_2')=q_1, \quad \tfrac{1}{4}(-x_1+x_1'+x_2-x_2')=q_2 \tag{11}$$

and

$$\tfrac{1}{2}(p_1+p_2)=p, \quad \tfrac{1}{2}(p_1-p_2)=p' . \tag{12}$$

Since the interchange of $x_1$ and $x_2$ is represented by the interchange of $q_1$ and $q_2$, this leads to the equation, instead of (2a)($i=1, j=2$),

$$\int \int dp \, dp' \, P(q+q_1,p+p',q-q_1,p-p')e^{4i(p'q_2-pY)/\hbar} = \int \int dp \, dp' \, P(q+q_2,p+p',q-q_2,p-p')e^{4i(p'q_1-pY)/\hbar} \tag{13}$$

Since $Y$ appears only in the exponent, the integration with respect to $p$ can be eliminated, leading to (after replacing the dummy variable $p'$ by $p_1$)

$$\int dp_1 P(q+q_1,p+p_1,q-q_1,p-p_1)e^{4ip_1q_2/\hbar} = \int dp_1 P(q+q_2,p+p_1,q-q_2,p-p_1)e^{4ip_1q_1/\hbar} .$$ (14a)

It must be admitted that this equation, for the distribution function, postulating the Bose statistics for a system of spin-0 particles, is much more complicated than the corresponding equation (2a) for the density matrix.

The essential equivalence of the position and momentum variables in this equation can be demonstrated by multiplying it with $\exp[-4i(p'q_1+p''q_2)/\hbar]$ and integrating over $q_1$ and $q_2$. The resulting equation is

$$\int dq_1 P(q+q_1,p+p'',q-q_1,p-p'')e^{-4ip'q_1/\hbar} = \int dq_2 P(q+q_2,p+p',q-q_2,p-p')e^{-4ip''q_2/\hbar} ,$$ (14b)

in the right-hand side of which, naturally, $q_2$ can be replaced by $q_1$. The equation then becomes the analog of Eq. (14a) with the roles of $p$ and $q$ interchanged, but the signs in the exponentials reversed. Equally easy an equation can be obtained in which $q$ and $p$ play essentially the same roles, but the two sides of the equation are quite different.

This is achieved by multiplying Eq. (14a) with $e^{-4ip_2q_1/\hbar}$ and integrating over $q_1$. On the right side this gives a factor $(\pi\hbar/2)^3\delta(p_1-p_2)$ (we must not forget that the $q_1p_1$ in the exponent is a three-dimensional scalar product), hence one obtains

$$P(q+q_2,p+p_2,q-q_2,p-p_2) = \left[\frac{2}{\hbar\pi}\right]^3 \int\int P(q+q_1,p+p_1,q-q_1,p-p_1)e^{4i(p_1q_2-q_1p_2)/\hbar}dp_1\,dq_1 ,$$ (15)

which is essentially symmetric in position and momentum coordinates. Naturally, by another Fourier transformation, this can be transformed into the analog of Eq. (14a), with the position- and momentum-coordinate roles interchanged.

As a check, we use Eq. (15) to replace the function $P$ on the right side of Eq. (15), obtaining

$$P(q+q_2,p+p_2,q-q_2,p-p_2)$$
$$=(2/\pi\hbar)^6 \int\int\int\int P(q+q_2',p+p_2';q-q_2',p-p_2')$$
$$\times\exp[4i(q_2p_1-q_1p_2)/\hbar+4i(q_1p_2'-q_2'p_1)/\hbar]dq_1\,dp_1\,dq_2\,dp_2 .$$ (16)

Carrying out the integrations over $q_1$ and $p_1$ we obtain factors $(\pi\hbar/2)^3\delta(p_2-p_2')$ and $(\pi\hbar/2)^3\delta(q_2-q_2')$, as a result of which we verify that the right side of Eq. (16) reduces to the left side, proving the self-consistency of Eq. (15).

### IV. SYSTEM OF SPIN-$\frac{1}{2}$ PARTICLES

In the case of a pure state $\psi(q)$ we extend our previous definition of the single-particle quantum distribution function as follows:

$$P(q,p,\kappa)=(\pi\hbar)^{-3}\sum_{m,m'=1,-1}\sigma_{mm'}^\kappa \int[\psi(q+y,m)]^*\psi(q-y,m')e^{2ipy/\hbar}dy$$ (17a)

in the case of a pure state and

$$P(q,p,\kappa)=(\pi\hbar)^{-3}\sum_{m,m'=1,-1}\sigma_{mm'}^\kappa \int\rho(q-y,m';q+y,m)e^{2ipy/\hbar}dy$$ (17b)

in the case of a mixture of states, where $\kappa$ takes on the values 0, $x$, $y$, and $z$; the $\sigma^0$ is the unit matrix and the others are the Pauli matrices for the spin, the rows and columns being labeled 1 and $-1$. Specifically,

$$\sigma^0=\begin{bmatrix}1&0\\0&1\end{bmatrix}, \quad \sigma^y=\begin{bmatrix}0&-i\\i&0\end{bmatrix},$$
$$\sigma^x=\begin{bmatrix}0&1\\1&0\end{bmatrix}, \quad \sigma^z=\begin{bmatrix}1&0\\0&-1\end{bmatrix}.$$ (18)

If Eq. (17a) is integrated over momentum $p$, one obtains

$$\int P(q,p,0)dp = |\psi(q,1)|^2+|\psi(q,-1)|^2=\rho(q,1;q,1)+\rho(q,-1;q,-1) ,$$ (19a)

$$\int P(q,p,x)dp = \{[\psi(q,1)]^*\psi(q,-1)+[\psi(q,-1)]^*\psi(q,1)\} ,$$ (19b)

$$\int P(q,p,y)dp = -i\{[\psi(q,1)]^*\psi(q,-1)-[\psi(q,-1)]^*\psi(q,1)\} ,$$ (19c)

$$\int P(q,p,z)dp = |\psi(q,1)|^2-|\psi(q,-1)|^2 .$$ (19d)

Thus Eq. (19a) gives the total probability of finding the particle at position $q$, irrespective of its spin state, whereas Eqs. (19b)–(19d) give the difference in the probabilities of finding the particle with spin up and spin down, referred to the $x$, $y$, and $z$ directions, respectively—always at $q$. Hence, the normalization of $P$, though it involves integration over $q$ and $p$, restricts the value of $\kappa$, actually to zero.

In the case of a two-particle system, Eq. (17) generalizes to

$$P(q_1,p_1,\kappa_1;q_2,p_2,\kappa_2)=(\pi\hbar)^{-6}\sum_{m_1,m_1'}\sum_{m_2,m_2'}\sigma^{\kappa_1}_{m_1,m_1'}\sigma^{\kappa_2}_{m_2,m_2'}\int\int[\psi(q_1+y_1,m_1;q_2+y_2,m_2)]^*$$

$$\times\psi(q_1-y_1,m_1';q_2-y_2,m_2')\exp[2i(p_1y_1+p_2y_2)/\hbar]dy_1\,dy_2 \tag{20}$$

in the case of a pure state and to

$$P(q_1,p_1,\kappa_1;q_2,p_2,\kappa_2)$$

$$=(\pi\hbar)^{-6}\sum_{m_1,m_1'}\sum_{m_2,m_2'}\sigma^{\kappa_1}_{m_1'm_1'}\sigma^{\kappa_2}_{m_2'm_2'}\int\int dy_1\,dy_2\exp[2i(p_1y_1+p_2y_2)/\hbar]$$

$$\times\rho(q_1-y_1,m_1',q_2-y_2,m_2';q_1+y_1,m_1,q_2+y_2,m_2) , \tag{21}$$

in the case of a mixture of states. They follow from the fact that the $4\times4$ matrix $M_{\kappa;mm'}\equiv2^{-1/2}\sigma^\kappa_{mm'}$ is unitary. One has, therefore,

$$\sum_{m,m'}\sigma^\kappa_{mm'}(\sigma^\lambda_{mm'})^*=\sum_{m,m'}\sigma^\kappa_{mm'}\sigma^\lambda_{m'm}=2\delta_{\kappa\lambda} . \tag{22}$$

The second expression is equal to the first because all $\sigma^\lambda$ are self-adjoint. Similarly, we have

$$\sum_\kappa\sigma^\kappa_{mm'}(\sigma^\kappa_{nn'})^*=\sum_\kappa\sigma^\kappa_{mm'}\sigma^\kappa_{n'n}=2\delta_{mn}\delta_{m'n'} . \tag{23}$$

It then follows from Eq. (17b) that

$$\rho(x',n';x,n)=\tfrac{1}{2}\sum_\kappa\sigma^\kappa_{n'n}\int dp\,P(\tfrac{1}{2}(x+x'),p,\kappa)e^{ip(x'-x)/\hbar} \tag{24a}$$

or, with Eq. (10), for the two-particle situation,

$$\rho(x_1',m_1',x_2',m_2';x_1,m_1,x_2,m_2)=\tfrac{1}{4}\sum_{\kappa_1,\kappa_2}\sigma^\kappa_{m_1'm_1}\sigma^\kappa_{m_2'm_2}\int\int dp_1\,dp_2\,P(\tfrac{1}{2}(x_1+x_1'),p_1,\kappa_1;\tfrac{1}{2}(x_2+x_2'),p_2,\kappa_2)$$

$$\times\exp[ip_1(x_1'-x_1)/\hbar+ip_2(x_2'-x_2)/\hbar] , \tag{24b}$$

from whence it follows that the generalization of Eq. (14a), to the case of spin-$\tfrac{1}{2}$ particles, is

$$\sum_{\kappa_1}\sum_{\kappa_2}\sigma^{\kappa_1}_{m_1'm_1}\sigma^{\kappa_2}_{m_2'm_2}\int P(q+q_1,p+p_1,\kappa_1;q-q_1,p-p_1,\kappa_2)e^{4ip_1q_2/\hbar}dp_1$$

$$=-\sum_{\kappa_1}\sum_{\kappa_2}\sigma^{\kappa_1}_{m_1'm_2}\sigma^{\kappa_2}_{m_2'm_1}\int P(q+q_2,p+p_1,\kappa_1;q-q_2,p-p_1,\kappa_2)e^{4ip_1q_1/\hbar}dp_1 , \tag{25}$$

with the minus sign arising from the fact that we are dealing with fermions. This equation will be simplified by expressing the product of the two $\sigma$ matrices on the left side,

$$[\sigma(\kappa_1,\kappa_2)]_{m_1'm_2';m_1m_2}=\sigma^{\kappa_1}_{m_1'm_1}\sigma^{\kappa_2}_{m_2'm_2} , \tag{26a}$$

by the product of the $\sigma$ matrices on the right side,

$$[\tau(\kappa_1,\kappa_2)]_{m_1'm_2';m_1m_2}=\sigma^{\kappa_1}_{m_1'm_2}\sigma^{\kappa_2}_{m_2'm_1} . \tag{26b}$$

We can then write

$$\tau(\kappa_1,\kappa_2)=\sum_{\lambda_1,\lambda_2}B(\kappa_1\kappa_2;\lambda_1\lambda_2)\sigma(\lambda_1,\lambda_2) , \tag{27a}$$

and since the $\sigma(\kappa_1,\kappa_2)$ as functions of the $m$ are linearly independent of each other (in fact orthogonal), we can write, instead of Eq. (25),

$$\int P(q+q_1,p+p_1,\kappa_1;q-q_1,p-p_1,\kappa_2)e^{4ip_1q_2/\hbar}dp_1$$

$$=-\sum_{\lambda_1,\lambda_2}B(\lambda_1,\lambda_2;\kappa_1,\kappa_2)\int P(q+q_2,p+p_1,\lambda_1;q-q_2,p-p_1,\lambda_2)e^{4ip_1q_1/\hbar}dp_1 . \tag{27b}$$

Equation (27b) is the basic result sought, being a consequence of the antisymmetry of the wave function for spin-$\frac{1}{2}$ particles. The matrix $B$ is much simplified by the fact that the rotational transformation properties of the $\sigma$ and $\tau$ matrices are the same—they represent either scalars or vectors or tensors. Hence, for example, the matrix element of $B$ which connects the scalar component of $\sigma(\lambda_1,\lambda_2)$ with the vector components of $\tau(\kappa_1,\kappa_2)$ vanishes. It may be stated additionally, that $B$ is unitary and its square is the unit matrix so that its characteristic values are all 1 or $-1$. It is not difficult to calculate $B$—its elements are, of course, the same for the different component parts of the vector (tensor) components of $\sigma(\lambda_1,\lambda_2)$ which are connected to the vector (tensor) components of $\tau(\kappa_1,\kappa_2)$.

As can be easily verified, the explicit form of the relation (27a) is

$$\tau(0,0)=\tfrac{1}{2}\sigma(0,0)+\tfrac{1}{2}[\sigma(x,x)+\sigma(y,y)+\sigma(z,z)]$$

$$\equiv \tfrac{1}{2}[\sigma(0,0)+\sigma(\vec{r},\vec{r})] \; ; \tag{28a}$$

$$\tau(\vec{r},\vec{r})\equiv\tau(x,x)+\tau(y,y)+\tau(z,z)$$

$$=\tfrac{3}{2}\sigma(0,0)-\tfrac{1}{2}[\sigma(x,x)+\sigma(y,y)+\sigma(z,z)] \; ; \tag{28b}$$

$$\tau(x,y)+\tau(y,x)=\sigma(x,y)+\sigma(y,x); \tau(y,z)+\tau(z,y)$$

$$=\sigma(y,z)+\sigma(z,y),\ldots ; \tag{28c}$$

$$\tau(0,x)+\tau(x,0)=\sigma(0,x)+\sigma(x,0),\ldots,\tau(0,z)+\tau(z,0)$$

$$=\sigma(0,z)+\sigma(z,0) \; ; \tag{28d}$$

$$\tau(0,x)-\tau(x,0)=-i[\sigma(y,z)-\sigma(z,y)],\ldots ; \tag{28e}$$

$$\tau(y,z)-\tau(z,y)=i[\sigma(0,x)-\sigma(x,0)],\ldots ; \tag{28f}$$

$$2\tau(x,x)-\tau(y,y)-\tau(z,z)$$

$$=2\sigma(x,x)-\sigma(y,y)-\sigma(z,z),\ldots ; \tag{28g}$$

$$\tau(x,x)=\tfrac{1}{2}[\sigma(x,x)-\sigma(y,y)-\sigma(z,z)]+\tfrac{1}{2}\sigma(0,0)$$

$$\equiv\sigma(x,x)-\tfrac{1}{2}\sigma(\vec{r},\vec{r})+\tfrac{1}{2}\sigma(0,0) \; . \tag{28h}$$

Clearly, these equations are not all independent of each other—Eqs. (28b) and (28c) follow from Eq. (28h) and show only the effects of the invariances more explicitly.

Evidently, the consequences of the wave functions antisymmetry for spin-$\frac{1}{2}$ particles are much more complicated than for the density matrix even though the preceding equations could be given a more simple form. Nevertheless, it should not be truly difficult to give Eq. (27b)—the basic equation—a more explicit form and also a more simple one by introducing, instead of the indices $\kappa$ and $\lambda$ the indices referring to the left sides of the equations (28). Another possibility is to accept the matrix $B$ as a basic quantity. One way this could be done would be to decompose $B$ into three parts: those referring to scalar, vector, and irreducible-tensor expressions in terms of the $\tau$ appearing in Eq. (28). The first part of $B$ would be two dimensional, the second part would contain three separate three-dimensional matrices [referring to $\tau(0,x)$, $\tau(x,0)$, $\tau(y,z)-\tau(z,y)$, and the other two similar triplets], and the last one would be a five-dimensional unit matrix referring to the five components of the five-dimensional representation of the rotation group [cf. Eqs. (28c) and (28g)]. But even if this is done, it must be admitted that the equations will be more complicated than the equations postulating Fermi statistics for the density matrix.

### ACKNOWLEDGMENT

This research was partially supported by the Division of Materials Science, U. S. Department of Energy, under Contract No. DE-AS05-79ER10459 and Grant No. DE-FG05-84ER45135.

*Permanent address: Joseph Henry Laboratory, Department of Physics, Princeton University, Princeton, NJ 08540.

[1] E. Wigner, Phys. Rev. 40, 749 (1932).

[2] G. E. Uhlenbeck and L. Gropper, Phys. Rev. 41, 79 (1932).

[3] J. G. Kirkwood, Phys. Rev. 44, 31 (1933).

[4] H. S. Green, J. Chem. Phys. 19, 955 (1951).

[5] J. Ross and J. G. Kirkwood, J. Chem. Phys. 22, 1094 (1954).

[6] K. Schram and B. R. A. Nijboer, Physica (The Hague) 25, 733 (1959).

[7] R. L. Stratonovich, Zh. Eksp. Teor. Fiz. 31, 1012 (1956) [Sov. Phys.—JETP 4, 891 (1957)].

[8] Yu. L. Klimontovich, Zh. Eksp. Teor. Fiz. 33, 982 (1957) [Sov. Phys.—JETP 6, 753 (1958).

[9] W. E. Brittin and W. R. Chappell, Rev. Mod. Phys. 34, 620 (1962).

[10] B. Jancovici, Physica (Utrecht) 91A, 152 (1978).

[11] A. Alastuey and B. Jancovici, Physica (Utrecht) 97A, 349 (1979).

[12] A. Alastuey and B. Jancovici, Physica (Utrecht) 102A, 327 (1980).

[13] M. O. Scully, Phys. Rev. D 28, 2477 (1983).

[14] J. Bell, Rev. Mod. Phys. 38, 447 (1966).

# Covariant Phase-Space Representation
# for Harmonic Oscillators

Y. S. Kim and E. P. Wigner

Physical Review A *38*, 1159–1167 (1988)

(Received 25 February 1988)

It is shown that, in the phase-space representation of quantum mechanics, the uncertainty relation can be stated in terms of the integral invariant of Poincaré. The uncertainty relation for spreading free wave packets is discussed as an illustrative example. This phase-space approach can be extended to the relativistic regime. It is shown that Lorentz boosts are area-preserving canonical transformations in the phase space of the light-cone variables. The harmonic oscillator is discussed in detail as an illustrative example for the covariant realization of the uncertainty principle.

## I. INTRODUCTION

The phase-space representation of quantum mechanics[1] is of current interest. We have shown in our previous paper[2] that the light-cone coordinate system is the natural language for the Lorentz-covariant phase-space representation of quantum mechanics. The localized light wave was discussed as an illustrative example. However, the light wave depends on only one of the two light-cone variables.

The purpose of this paper is to discuss a physical example depending on both of the light-cone variables. The covariant harmonic oscillator will serve this purpose. The harmonic oscillator occupies a very prominent place in the physics of phase space. Unless we know how to deal with the covariance of the harmonic oscillator, we are not likely to understand the covariance of phase space.

In the phase-space representation, the uncertainty relation is stated in terms of the integral invariant of Poincaré,[3] which is called the "error box" in the current literature.[4] The area of the error box cannot be smaller than Planck's constant. In order to illustrate the advantage of using the phase-space representation, we start with the problem of wave-packet spreads. In the Schrödinger picture, the uncertainty product increases as time progresses or regresses. On the other hand, the volume of the error box remains constant for the spreading wave packet, even though its shape is deformed.[5] This means that the wave-packet spread is a canonical transformation in phase space.

The phase-space representation of nonrelativistic quantum mechanics has a built-in symmetry which is mathematically equivalent to that of the $(2+1)$-dimensional Lorentz group. Since the position and momentum variables are $c$ numbers in the phase-space representation, it is possible to formulate canonical transformations in a manner identical to the case in classical mechanics. Indeed, the group of linear canonical transformations is Sp(2) which is isomorphic to the $(2+1)$-dimensional Lorentz group.[3,6]

In this paper, we shall show that Lorentz boosts are canonical transformations in the phase space of the light-cone variables. Thus, from the mathematical point of view, the Lorentz covariance does not add new complications in phase space. However, from the physical point of view, we are dealing with a realization of the uncertainty principle which is different from that in the Schrödinger picture of quantum mechanics. The phase-space representation enables us to state the uncertainty relation in a Lorentz-covariant manner.

In Sec. II, we study the phase-space representation for wave-packet spreads and compare it with the case of harmonic oscillators. It is pointed out that the uncertainty relation can be stated in terms of the area of the error box in phase space. The area of the error box remains unchanged as time progresses or regresses. It is shown that the error boxes for the spreading wave packet and harmonic oscillator coincide with each other in the large-time and weak-spring-constant limits, respectively. In Sec. III, the wave-packet spread is formulated in terms of homogeneous linear canonical transformations in phase space.

Sections IV and V consist mostly of reviews which are needed for the covariant formulation given in Sec. VI. Section IV deals with the phase-space representation for nonrelativistic harmonic oscillators. Section V is based on the covariant harmonic-oscillator formalism which constitutes a representation of the Poincaré group, and which effectively describes the basic phenomena of relativistic hadrons in the quark model.[7]

In Sec. VI, we combine Secs. IV and V to formulate the covariant phase-space representation of harmonic oscillators. The conclusion of the present paper is that the concept of the covariant error box in phase space gives the physical basis for the phenomenology based on the mathematics of the covariant harmonic-oscillator formalism.

## II. WAVE-PACKET SPREADS

Let us start with a one-dimensional harmonic oscillator in its ground state. The uncertainty product remains invariant when the spring constant is gradually reduced. On the other hand, when the oscillator force is suddenly removed, the uncertainty product becomes dependent on time. Although these two different cases are two different manifestations of one physical principle, the wave-packet spread has been one of the agonizing features of the present form of nonrelativistic quantum mechanics.

In the phase-space representation of quantum mechanics, the uncertainty relation can be stated in a time-independent manner for both the harmonic oscillator and the spreading wave packet. If $\psi(x,t)$ is the physical solution of the Schrödinger equation, the phase-space distribution function is defined as[1]

$$P(x,p,t) = \frac{1}{\pi} \int \psi^*(x+y,t)\psi(x-y,t)e^{2ipy}dx \; , \quad (2.1)$$

which we shall hereafter call the PSD function. This is a function of $t$, $x$, and $p$, which are $c$ numbers. This function is real but is not necessarily positive everywhere in the phase space of $x$ and $p$. We can, however, recover the positive distribution functions in the position and momentum coordinates

$$\rho(x,t) = |\psi(x,t)|^2 = \int P(x,p,t)dp \; ,$$

$$\sigma(p,t) = |\phi(p,t)|^2 = \int P(x,p,t)dx \; , \qquad (2.2)$$

where $\phi(p)$ is the momentum wave function.

The time-dependent Schrödinger equation leads to the differential equation[1]

$$\frac{\partial}{\partial t}P(x,p,t) = -\left[\frac{p}{m}\right]\frac{\partial}{\partial x}P(x,p,t) + \sum_{n=0}^{\infty}\left[\frac{1}{2}\right]^{2n}\frac{1}{(2n+1)!}\left[\left[\frac{\partial}{\partial x}\right]^{2n+1}V(x)\right]\left[\frac{\partial}{\partial x}\right]^{2n+1}P(x,p,t) \; , \qquad (2.3)$$

where $m$ is the mass of the particle, and $V(x)$ is the potential. In the case of the harmonic oscillator with $V(x) = Kx^2/2$, the above differential equation becomes

$$\frac{\partial}{\partial t}P(x,p,t) = -\left[\frac{p}{m}\right]\frac{\partial}{\partial x}P(x,p,t) + Kx\frac{\partial}{\partial p}P(x,p,t) \; . \tag{2.4}$$

If the particle is free, this differential equation becomes

$$\frac{\partial}{\partial t}P(x,p,t) = -\left[\frac{p}{m}\right]\frac{\partial}{\partial x}P(x,p,t) \; . \tag{2.5}$$

The solution of the above differential equation is[5]

$$P(x,p,t) = P(x - pt/m, p, 0) \; . \tag{2.6}$$

The time evolution of this solution is illustrated in Fig. 1. Indeed, the error box undergoes a shear. The volume of the error box is invariant under time evolution. This is a more precise statement of the uncertainty relation than is given in the Schrödinger picture.

If we start with a free-particle wave packet with a Gaussian momentum wave function

$$g(k) = \left[\frac{b}{\pi}\right]^{1/4} e^{-bk^2/2} \tag{2.7}$$

at $t = 0$, the time-dependent Schrödinger wave function becomes

$$\psi(x) = \left[\frac{b}{\pi}\right]^{1/4}\left[\frac{1}{b+it/m}\right]^{1/2} e^{-x^2/2(b+it/m)} \; . \tag{2.8}$$

If we construct the PSD function for the above wave function, its form is

$$P(x,p,t) = \frac{1}{\pi}\exp\{-[(x-pt/m)^2/b + bp^2]\} \; . \tag{2.9}$$

This distribution is concentrated within the region where the exponent is less than 1 in magnitude. This region is described by the tilted ellipse described in Fig. 2. This is the error box for the spreading wave packet.

Since $x$ and $p$ are $c$ numbers in the phase-space representation, the PSD function $P(x,p,t)$ can be canonically transformed in phase space, as is done in classical mechanics. The concept of the error box is already in the Poisson-bracket formalism of classical mechanics. Its volume is invariant under canonical transformations. This is called the integral invariant of Poincaré. However, classical mechanics does not give the lower limit on the size of the error box.

Let us go back to the wave-packet spread. How is the spreading Gaussian wave packet different from the ground-state harmonic oscillator whose mass and spring constant become adiabatically weak? This is a situation

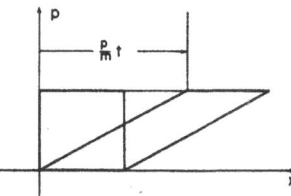

FIG. 1. Shear in phase space. Every point in phase space moves horizontally in the $x$ direction with velocity proportional to $p$. This is an area-preserving transformation.

very familiar to us, and is described in Fig. 2, in which the $x$ axis expands and the $p$ axis contracts. This deformation is also an area-preserving canonical transformation, and is commonly called "squeeze" in the literature.[8]

The spread of the Gaussian wave packet is also illustrated in Fig. 2. This is consistent with the shear effect given in Fig. 1. It is possible to take the "projection" of this elliptic distribution to the $x$ and $p$ spaces using the formulas for $\rho(x)$ and $\sigma(p)$ given in Eq. (2.2). These probability densities lead to the uncertainty product in the Schrödinger picture, which expands as time progresses or regresses.

The error box for the spreading wave packet in the infinite-time limit coincides with the error box in the limit of zero spring constant. In the phase-space representation, the magnitude of uncertainty is the same for both cases. In the Schrödinger representation, one is finite while the other is infinite.

## III. WAVE-PACKET SPREADS IN TERMS OF CANONICAL TRANSFORMATIONS

In order to study canonical transformation properties of the shear described in Fig. 1, let us write the solution of Eq. (2.5) given in Eq. (2.6) as

$$\begin{bmatrix} x' \\ p' \end{bmatrix} = \begin{bmatrix} 1 & t/m \\ 0 & 1 \end{bmatrix} \begin{bmatrix} x \\ p \end{bmatrix} . \tag{3.1}$$

On the other hand, the group of homogeneous linear canonical transformations is Sp(2),[3,6] which is locally isomorphic to the $(2+1)$-dimensional Lorentz group.[9] Indeed, the group of homogeneous linear canonical transformations consists of rotations around the origin generated by

$$L = \begin{bmatrix} 0 & -i/2 \\ i/2 & 0 \end{bmatrix} , \tag{3.2}$$

and squeezes[8] along the $x$ axis and along the $x = p$ line generated by

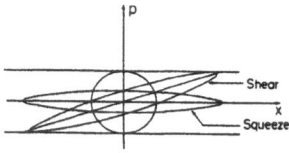

FIG. 2. Ground-state harmonic oscillator and the spread of the Gaussian wave packet in phase space. In terms of the integral invariant of Poincaré, the uncertainty relation can be stated in the same manner for both cases. Their projections to the $x$ and $p$ axes are different. In the oscillator case, the $p$ distribution contracts while the $x$ distribution expands. On the other hand, in the case of spreading wave packets, the $p$ distribution does not change. This is why the uncertainty product for spreading wave packets increases as time progresses or regresses.

$$B_1 = \begin{bmatrix} i/2 & 0 \\ 0 & -i/2 \end{bmatrix} , \quad B_2 = \begin{bmatrix} 0 & i/2 \\ i/2 & 0 \end{bmatrix} , \tag{3.3}$$

respectively. These generators satisfy the commutation relations

$$[B_1, B_2] = -iL, \quad [B_1, L] = -iB_2, \quad [B_2, L] = iB_1 . \tag{3.4}$$

This set of commutation relations is identical to the Lie algebra for the $(2+1)$-dimensional Lorentz group.[9] On the other hand, the shear transformation of Eq. (2.9) is generated by the matrix

$$N = \begin{bmatrix} 0 & i \\ 0 & 0 \end{bmatrix} , \tag{3.5}$$

with

$$e^{-i(t/m)N} = \begin{bmatrix} 1 & t/m \\ 0 & 1 \end{bmatrix} .$$

Then, where does the matrix $N$ stand among the generators of canonical transformations given in Eqs. (3.2) and (3.3)?

In order to answer this question, let us construct first the squeeze operator

$$S(\eta) = e^{-i\eta B_1} = \begin{bmatrix} \exp(\eta/2) & 0 \\ 0 & \exp(-\eta/2) \end{bmatrix} . \tag{3.6}$$

This operator expands the $x$ axis while contracting $p$. This is of course a canonical transformation. Thus the following operators also generate homogeneous linear canonical transformations:

$$B_1' = B_1 = S(\eta)B_1 S(-\eta), \quad B_2'(\eta) = e^{-\eta}S(\eta)B_2 S(-\eta) , \tag{3.7}$$

$$L'(\eta) = e^{-\eta}S(\eta)LS(-\eta) . \tag{3.8}$$

In the limit of large $\eta$, the above operators become $B_1$, $N$, and $N$, respectively. These operators do not form a closed Lie algebra of any group, but give a partial view of a more complete group-theoretical picture that both the Lorentz-boosted O(3) and O(2,1) become a group locally isomorphic to the two-dimensional Euclidean group in the large-$\eta$ limit.[7,10]

In the present case, the O(3)-like group is generated by $L$, $L_1$, and $L_2$, where

$$L_1 = \begin{bmatrix} \frac{1}{2} & 0 \\ 0 & -\frac{1}{2} \end{bmatrix}, \quad L_2 = \begin{bmatrix} 0 & \frac{1}{2} \\ \frac{1}{2} & 0 \end{bmatrix} . \tag{3.9}$$

The O(2,1)-like group is generated by $L_1$, $B_2$, and $B_3$, where

$$B_3 = \begin{bmatrix} 0 & \frac{1}{2} \\ -\frac{1}{2} & 0 \end{bmatrix} . \tag{3.10}$$

The mathematics of this process is called the contraction of O(3) and O(2,1) to E(2).[7,10] The physical content of this process is that a unified description can be given for the O(3)-like, E(2)-like, and O(2,1)-like internal space-time symmetries of massive, massless, and imaginary-mass particles, respectively, as Einstein's $E = (m^2 + p^2)^{1/2}$ unifies the energy-momentum relation for all relativistic particles. This point has been discussed in the literature,[10,11] and is illustrated in the first and second rows of Fig. 3.

## IV. HARMONIC OSCILLATORS

The Hamiltonian for the one-dimensional nonrelativistic harmonic oscillator with unit frequency can be written as

$$H = \tfrac{1}{2}(P^2 + x^2) . \tag{4.1}$$

The normalized solutions of the Schrödinger equation are

$$\psi_n(x) = [1/(\sqrt{\pi} 2^n n!)]^{1/2} H_n(x) \exp(-x^2/2) , \tag{4.2}$$

where $H_n(x)$ is the Hermite polynomial of $n$th order. These wave functions are in the energy eigenstates. It is possible to evaluate the quantum PSD function

$$P_n(x,p) = \frac{1}{\pi} \int \psi_n^*(x+y) \psi_n(x-y) e^{2ipy} dy . \tag{4.3}$$

The result of the calculation is[12]

$$P_n(x,p) = \left[\frac{n!}{\pi}\right] [\exp(-r^2)] \sum_{k=0}^{\infty} (-1)^k 2^{n-k} r^{2(n-k)} / \{[(n-k)!]^2 k!\} , \tag{4.4}$$

where $r^2 = (x^2 + p^2)$.

The above form is defined in the two-dimensional phase space spanned by $x$ and $p$ axes. Since it depends on $x$ and $p$ only through the variable $r$, the function is invariant under rotations around the origin. We can thus write $P_n(x,p)$ as

$$P_n(x,p) = P_n(r) . \tag{4.5}$$

As is expected, this function is positive in some regions and is negative in other regions in phase space. It vanishes on the circles on which the polynomial contained in $P_n(r)$ of Eq. (4.5) is zero.

In order to study this more systematically for the harmonic oscillator, let us see whether $P_n(x,p)$ of Eq. (3.5)

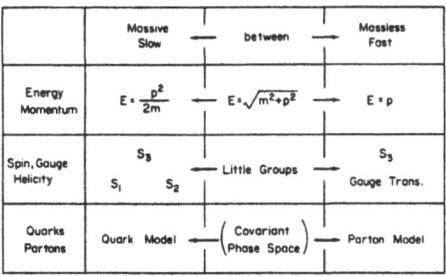

FIG. 3. Slow and fast particles. Einstein's $E = (P^2 + m^2)^{1/2}$ unifies the energy-momentum relations for massive (nonrelativistic) particles and for massless particles. The second row indicates that the little group of the Poincaré group unifies the internal space-time symmetries of massive and massless particles, as is discussed in Ref. 11. The third row states that the covariant phase-space representation forms the physical basis for the covariant harmonic-oscillator formalism which has been shown to give a unified picture of quark model and the parton picture at the phenomenological level. This point is discussed in Sec. VI of the present paper.

can be interpreted in terms of another equation. The PSD function given in Eq. (3.5) indeed satisfies the differential equation[13]

$$-\frac{1}{2\rho}\left[\frac{d}{d\rho}\rho\left[\frac{d}{d\rho}P_n(r)\right]\right] + \tfrac{1}{2}\rho^2 P_n(r) = (2n+1)P_n(r) , \tag{4.6}$$

where $\rho = \sqrt{2}r$. This is the radial part of the rotation-invariant Schrödinger equation for the harmonic oscillator in two-dimensional space spanned by the variables $\sqrt{2}x$ and $\sqrt{2}p$. If we use $R_k(\rho)$ for the normalized radial equation for the $k$th excited state with the eigenvalue $(k+1)$ with the orthogonality relation

$$\int_0^{\infty} \rho R_n(\rho) R_m(\rho) d\rho = \delta_{nm} , \tag{4.7}$$

then the PSD function is

$$P_n(r) = (1/\sqrt{4\pi}) R_{2n}(\rho) . \tag{4.8}$$

Therefore $P_n$ satisfies the orthogonality relation

$$\int P_m^*(x,p) P_n(x,p) dx\, dp = 2\pi \int P_n(r) P_m(r) r\, dr$$
$$= \delta_{nm} . \tag{4.9}$$

For the ground state, the PSD function is

$$P_0(x,p) = \frac{1}{\pi} \exp[-(x^2 + p^2)] . \tag{4.10}$$

If $n = 1$,

$$P_1(x,p) = \frac{1}{\pi}(x^2 + p^2 - \tfrac{1}{2}) \exp[-(x^2 + p^2)] . \tag{4.11}$$

We can then use the Schmidt orthogonalization procedure to construct the PSD function for higher values of $n$.

In this paper, we are interested in homogeneous linear canonical transformations in phase space consisting of rotations and squeezes along a given direction. The above

PSD functions are rotationally invariant. The squeeze along the $x$ direction means that the coordinate variable $x$ is multiplied by a real positive number $b$, while the $p$ variable is divided by $b$. The integral measure $(dx\,dp)$ remains invariant during this process.

In this case, the group of homogeneous linear canonical transformations is generated by

$$L = -\frac{i}{2}\left[x\frac{\partial}{\partial p} - p\frac{\partial}{\partial x}\right],$$

$$B_1 = \frac{i}{2}\left[x\frac{\partial}{\partial x} - p\frac{\partial}{\partial p}\right], \quad B_2 = \frac{i}{2}\left[x\frac{\partial}{\partial p} + p\frac{\partial}{\partial x}\right], \quad (4.12)$$

which satisfy the commutation relations for the generators of the $(2 + 1)$-dimensional Lorentz group given in Eq. (3.4). While this group is one of the fundamental symmetry groups in the physics of phase space,[6] we shall see in Sec. VI that the Lorentz boost is a squeeze in the phase space of the light-cone coordinate position and momentum variables.

## V. COVARIANT HARMONIC OSCILLATORS

The covariant harmonic oscillator has a long history. It was Dirac who suggested in 1945 the use of harmonic oscillators to construct representations of the Lorentz group.[14] Yukawa in 1953 studied the possibility of using the oscillator for studying relativistic composite particles.[15] However, its physical relevance was not revealed until the successful calculation of the proton form factor by Fujimura, Kobayashi, and Namiki in 1970.[16] In the 1971 paper of Feynman, Kislinger, and Ravndal, the authors point out the need for relativistic bound-state models, such as the harmonic oscillator model, in order to supplement the traditional Feynman-diagram approach which is not always effective in dealing with covariant bound-state problems.[17]

The covariant oscillator formalism has been extensively discussed in the literature.[7] It serves as one of the physical representations of the Poincaré group. At the same time, the formalism allows us to explain the peculiarities of Feynman's parton picture in terms of the bound-state quark model.[7,18]

Let us start with the differential equation of Feynman et al.[17] for a hadron of two quarks bound together by a harmonic-oscillator potential of unit strength

$$\left\{-2\left[\left(\frac{\partial}{\partial x_a^\mu}\right)^2 + \left(\frac{\partial}{\partial x_b^\mu}\right)^2\right] + (\tfrac{1}{16})(x_a^\mu - x_b^\mu)^2 + m_0^2\right\}\varphi(x_a, x_b) = 0 , \qquad (5.1)$$

where $x_a$ and $x_b$ are space-time coordinates for the first and second quarks, respectively. This partial differential equation has many different solutions depending on the choice of variables and boundary conditions.

In order to simplify the above differential equation, let us introduce new coordinate variables[17]

$$X = (x_a + x_b)/2, \quad x = (x_a - x_b)/2\sqrt{2} . \qquad (5.2)$$

The four-vector $X$ specifies where the hadron is located in space-time, while the variable $x$ measures the space-time separation between the quarks. In terms of these variables, Eq. (5.1) can be written as

$$\left[\frac{\partial^2}{\partial X_\mu^2} - m_0^2 + \frac{1}{2}\left(\frac{\partial^2}{\partial x_\mu^2} - x_\mu^2\right)\right]\varphi(X,x) = 0 . \qquad (5.3)$$

This equation is separable in the $X$ and $x$ variables. Thus

$$\varphi(X,x) = f(X)\psi(x) , \qquad (5.4)$$

and $f(X)$ and $\psi(x)$ satisfy the following differential equations, respectively:

$$\left[\frac{\partial^2}{\partial X_\mu^2} - m_0^2 - (\lambda + 1)\right]f(X) = 0 , \qquad (5.5)$$

$$\frac{1}{2}\left[\frac{\partial^2}{\partial x_\mu^2} - x_\mu^2 + (\lambda + 1)\right]\psi(x) = 0 . \qquad (5.6)$$

Equation (5.5) is a Klein-Gordon equation, and its solution takes the form

$$f(X) = \exp(\pm iP_\mu X^\mu) , \qquad (5.7)$$

with

$$-P^2 = -P_\mu P^\mu = M^2 = m_0^2 + (\lambda + 1) ,$$

where $M$ and $P$ are the mass and four-momentum of the hadron, respectively. The eigenvalue $\lambda$ is determined from the solution of Eq. (5.6). We are using the same notation for the operator and the eigenvalue for the hadronic four-momentum. This should not cause any confusion since we are dealing only with free hadronic states with a definite four-momentum.

As for the four-momenta of the quarks $p_a$ and $p_b$, we can combine them into the total four-momentum and momentum-energy separation between the quarks[17]

$$P = p_a + p_b, \quad q = \sqrt{2}(p_a - p_b) , \qquad (5.8)$$

where $P$ is the hadronic four-momentum conjugate to $X$. The internal momentum-energy separation $q$ is conjugate to $x$ provided that there exist wave functions which can be Fourier transformed. If the momentum-energy wave functions can be obtained from the Fourier transformation of the space-time wave function, the differential equation in the $q$ space is the same as the harmonic oscillator equation for the $x$ space given in Eq. (5.6).

Since the three-dimensional harmonic oscillator is quite familiar to us, we are naturally led to consider the separation of the space and time variables in Eq. (5.6). However, the ($x\,t$) system is not the only coordinate system in which the differential equation is separable. If the hadron moves along the $z$ direction with velocity parameter $\beta$, the hadronic rest frame is important. In this frame, the coordinate variables are

$$x'=x,\quad y'=y ,$$
$$z'=(z-\beta t)/(1-\beta^2)^{1/2} , \qquad (5.9)$$
$$t'=(t-\beta z)/(1-\beta^2)^{1/2} .$$

The differential equation of Eq. (5.6) is separable also in these variables:

$$\frac{1}{2}\left[-\nabla'^2+\frac{\partial^2}{\partial t'^2}-[(\mathbf{x}')^2-t'^2]\right]\psi(x)=(\lambda+1)\psi(x) .$$
$$(5.10)$$

The solution of this equation consists of a product of four one-dimensional oscillator wave functions. The $x'$ and $y'$ components are not affected by the boost along the $z$ direction. Thus we can drop them from our consideration. As for the $t'$ component, the excitation contributes a negative number to $\lambda$. However, this excitation can be suppressed on the grounds that the time and energy variables are $c$ numbers. Indeed, the time-energy uncertainty relation is a $c$-number relation.[19] This suppression of timelike excitations can be achieved by the subsidiary condition

$$P_\mu\left[x^\mu-\frac{\partial}{\partial x_\mu}\right]\psi_\beta(x)=0 . \qquad (5.11)$$

Then the wave function takes the form

$$\psi_\beta^n(z,t)=[1/(\pi 2^n n!)]^{1/2}H_n(z')\exp[-\tfrac{1}{2}(z'^2+t'^2)] .$$
$$(5.12)$$

This normalizable wave function describes the internal space-time structure of the hadron moving along the $z$ direction. If $\beta=0$, then the wave function becomes

$$\psi_0^n(z,t)=[1/(\pi 2^n n!)]^{1/2}H_n(z)\exp[-\tfrac{1}{2}(z^2+t^2)] .$$
$$(5.13)$$

Thus

$$\psi_\beta^n(z,t)=\psi_0^n(z',t') . \qquad (5.14)$$

## VI. COVARIANT PHASE-SPACE REPRESENTATION FOR HARMONIC OSCILLATORS

One of the most outstanding problems in modern physics is how to formulate covariantly interactions between two elementary particles. For instance, we still do not know exactly what force is responsible for keeping the quarks inside a hadron. It may be possible to make the interaction invariant by postulating that it depends on the distance in the coordinate system at rest with the temporary center of mass of the particles, or that it depends on the two positions at the time when their relativistic distance is zero—when one is on the light cone of the other.[20]

In this paper, we take the light-cone approach. While it is not possible to solve all the problems at this time, we can discuss the uncertainty principle applicable to the space-time separation between the quarks in a harmonic system, using the light-cone variables. The covariant harmonic oscillator discussed in Sec. V serves as a theoretical tool for this purpose. The harmonic-oscillator wave function consists of a Gaussian factor and Hermite polynomials. Since the Gaussian factor determines the localization property of the wave function, let us study first the ground-state wave function, whose form is

$$\psi_0^0(z,t)=\left[\frac{1}{\pi}\right]^{1/2}\exp[-(z^2+t^2)/2] . \qquad (6.1)$$

We have dropped the $x$ and $y$ variables which are not affected by the Lorentz boost along the $z$ direction.

This wave function can be written in the light-cone coordinate system.[21] If the hadron moves along the $z$ direction, the light-cone variables are defined to be

$$u=(t+z)/\sqrt{2},\quad v=(t-z)/\sqrt{2} . \qquad (6.2)$$

Their Fourier conjugate variables are[18]

$$q_u=(q_z-q_0)/\sqrt{2},\quad q_v=(q_z+q_0)/\sqrt{2} . \qquad (6.3)$$

The major advantage of using these variables is that the Lorentz boost of Eq. (5.9) takes a very simple form:[18]

$$u'=\left[\frac{1-\beta}{1+\beta}\right]^{1/2}u,\quad v'=\left[\frac{1+\beta}{1-\beta}\right]^{1/2}v ,$$

$$q_u'=\left[\frac{1+\beta}{1-\beta}\right]^{1/2}q_u,\quad q_v'=\left[\frac{1-\beta}{1+\beta}\right]^{1/2}q_v . \qquad (6.4)$$

Under this transformation, the products $uq_u$ and $uq_v$ remain invariant.

In terms of the light-cone variables, the wave function of Eq. (6.1) can be written as

$$\psi_0^0(z,t)=\psi_0^0(u,v)$$

$$=\left[\frac{1}{\pi}\right]^{1/2}\exp[-(u^2+v^2)/2] . \qquad (6.5)$$

If the system is boosted, the wave function becomes

$$\psi_\beta^0(z,t)=\left[\frac{1}{\pi}\right]^{1/2}\exp[-(u'^2+v'^2)/2]$$

$$=\left[\frac{1}{\pi}\right]^{1/2}\exp\left[-\left[\frac{1}{2}\right]\left[\frac{1-\beta}{1+\beta}u^2+\frac{1+\beta}{1-\beta}v^2\right]\right] .$$
$$(6.6)$$

This wave function undergoes a Lorentz deformation as $\beta$ increases.[18] The momentum-energy wave function is

$$\varphi_\beta^0(q_u,q_v)=\left[\frac{1}{2\pi}\right]\int \psi_\beta^0(x,t)e^{-i(q_z z - q_0 t)}dz\, dt \ . \qquad (6.7)$$

The evaluation of this integral leads to

$$\varphi_\beta^0(q_u,q_v)=\left[\frac{1}{\pi}\right]^{1/2}\exp\left[-\left[\frac{1}{2}\right]\left[\frac{1+\beta}{1-\beta}q_u^2\right.\right.$$
$$\left.\left.+\frac{1-\beta}{1+\beta}q_v^2\right]\right] \ . \qquad (6.8)$$

For the ground state, the PSD function can now be defined as

$$P_\beta^0(u,q_u;v,q_v)=\left[\frac{1}{\pi}\right]^2\int [\psi_\beta^0(u+x,v+y)]^* \psi_\beta^0(u-x,v-y)\exp[2i(q_u x+q_v y)]dx\, dy \ . \qquad (6.9)$$

After the evaluation of this integral, the PSD function becomes

$$P_\beta^0(u,q_u;v,q_v)=\left[\frac{1}{\pi}\right]^2\exp\left[-\left[\frac{1}{2}\right]\left[\frac{1-\beta}{1+\beta}u^2+\frac{1+\beta}{1-\beta}q_u^2\right]\right]\exp\left[-\left[\frac{1}{2}\right]\left[\frac{1+\beta}{1-\beta}v^2+\frac{1-\beta}{1+\beta}q_v^2\right]\right] \ . \qquad (6.10)$$

The above PSD function is defined in two independent phase spaces consisting of $(u,q_u)$ and $(v,q_v)$, respectively. When the hadron is at rest with $\beta=0$, the above PSD function is localized in the regions

$$(u^2+q_u^2)<1, \quad (v^2+q_v^2)<1 \ . \qquad (6.11)$$

These localization regions are described in Fig. 4. When the hadron moves, these regions undergo elliptic deformations.

This PSD function reproduces the distributions $|\psi_\beta^0(u,v)|^2$ and $|\varphi_\beta^0(q_u,q_v)|^2$ after the appropriate integrals:

$$|\psi_\beta^0(u,v)|^2=\int P_\beta^0(u,v;q_u,q_v)dq_u\, dq_v \ ,$$
$$|\varphi_\beta^0(q_u,q_v)|^2=\int P_\beta^0(u,v;q_u,q_v)du\, dv \ . \qquad (6.12)$$

As for the excited states, there are no timelike oscillations in the hadronic rest frame, and the oscillations in the transverse direction are not affected. Therefore, the only factor we have to consider is the Hermite polynomial $H_n(z')$ to be multiplied to the ground-state wave function. The $n$th excited-state wave function is given in Eq. (5.13). In terms of the $u'$ and $v'$ variables, $H_n(z')$ can be written as[22]

$$H_n(z')=H_n((u'+v')/\sqrt{2})$$
$$=\left[\frac{1}{2}\right]^{n/2}\sum_{m=0}^{n}\begin{bmatrix}n\\m\end{bmatrix}H_{n-m}(u')H_m(v') \ . \qquad (6.13)$$

Thus the explicit form of the physical wave function becomes

$$\psi_\beta^n(u,v)=\left[\frac{1}{2}\right]^{n/2}\left[\frac{1}{\pi n!}\right]^{1/2}\left[\sum_{m=0}^{n}\begin{bmatrix}n\\m\end{bmatrix}H_{n-m}(u')H_m(v')\right]\exp[-(u'^2+v'^2)] \ . \qquad (6.14)$$

This means that we need off-diagonal PSD functions for the one-dimensional harmonic oscillator, such as

$$P_{nm}(x,p)=\frac{1}{\pi}\int \psi_n^*(x+y)\psi_m(x-y)e^{2ipy}dy \ , \qquad (6.15)$$

to evaluate the PSD function for covariant harmonic oscillator. It is possible to evaluate this integral using the generating function of Hermite polynomials. The result is[23]

$$P_{nm}(x,p)=\frac{(n!m!)^{1/2}}{\pi}\left[\sum_{k=0}^{s}\frac{(-1)^k[\sqrt{2}(x+ip)]^{n-k}[\sqrt{2}(x-ip)]^{m-k}}{k!(n-k)!(m-k)!}\right]\exp[-(x^2+p^2)] \ , \qquad (6.16)$$

where $s$ is $n$ or $m$, whichever is smaller. We can then go back to Eqs. (6.9) and (6.14) to complete the evaluation of the PSD function. The localization and deformation properties of the PSD function for excited states are essentially the same as those of the ground-state oscillator.

Let us go back to the localization problem. Unlike the case of light waves,[2] we have to deal with two phase spaces. If the hadron is at rest with $\beta=0$, the localization region can be specified by a circle in both the phase spaces of $(u,q_u)$ and $(v,q_v)$. If the hadron moves, the $u$ and $q_v$ distributions expand, while those of $v$ and $q_u$ be-

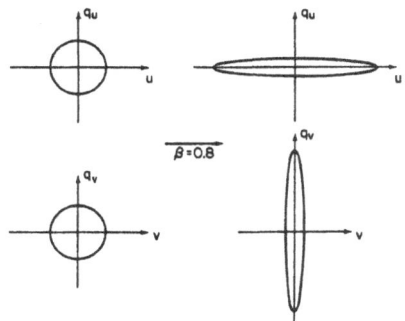

FIG. 4. Lorentz deformations in the light-cone phase space consisting of two pairs of conjugate variables. The major (minor) axis in the $uv$ coordinate system is conjugate to the minor (major) axis in the $q_u q_v$ coordinate system. The Lorentz boost is an area-preserving canonical transformation in both phase spaces. For the case of localized light waves, which was discussed in Ref. 2, there is only one phase space. The covariant phase space given in Ref. 2 is the lower half of this figure consisting of $v$ and $q_v$.

come contracted.

These deformations are canonical transformations, and are illustrated in Fig. 4. If the hadron's speed becomes close to the speed of light with $\beta \to 1$,

$$t = z, \quad u = \sqrt{2}z ,$$
$$q_t = q_z, \quad q_v = \sqrt{2}q_z , \qquad (6.17)$$

according to Eqs. (6.2)–(6.4). The simultaneous expansions of $z$ and $q_z$ are observed universally in high-energy laboratories. This is called Feynman's parton picture,[24] and the calculation based on the ground-state oscillator gives a good agreement with the observed proton structure function.[25] Indeed, the parton phenomenon is a manifestation of the Lorentz covariance of the uncertainty relation which can best be stated in terms of the phase-space representation.

## VII. CONCLUDING REMARKS

As we pointed out in Ref. 2, the phase-space representation of quantum mechanics serves useful purposes in many branches of modern physics. In this paper, we emphasized the fact that it can give a more precise interpretation of the uncertainty principle, as is manifested in the case of wave-packet spread. The phase-space representation allows us to formulate the uncertainty relation in a covariant manner. For this purpose, we have discussed in this paper the covariant phase-space representation for harmonic oscillators.

The major advantage of using the covariant oscillator formalism is that there is an experimental observation of the effect of covariance, as is explained in Sec. VI. Indeed, the covariant phase space is the physical basis for the covariant harmonic oscillator. This is illustrated in the third row of Fig. 3.

## ACKNOWLEDGMENT

We would like to thank Mr. Salman Habib for helpful discussions on the PSD functions for the spreading wave packet.

[1] E. P. Wigner, Phys. Rev. 40, 749 (1932); in Perspectives in Quantum Theory, edited by W. Yourgrau and A. van der Merwe (MIT Press, Cambridge, MA, 1971); M. Hillery, R. F. O'Connell, M. O. Scully, and E. P. Wigner, Phys. Rep. 106, 121 (1984).

[2] Y. S. Kim and E. P. Wigner, Phys. Rev. A 36, 1293 (1987).

[3] H. Goldstein, Classical Mechanics, 2nd ed. (Addison-Wesley, Reading, MA, 1980).

[4] C. M. Caves, K. S. Thorne, R. W. P. Drever, V. D. Sandberg, and M. Zimmerman, Rev. Mod. Phys. 52, 341 (1980).

[5] J. Kijowski, Rep. Math. Phys. 6, 361 (1974); H. W. Lee, Am. J. Phys. 50, 438 (1982); A. Royer, in Proceedings of the First International Conference on the Physics of Phase Space, College Park, Maryland, 1986, edited by Y. S. Kim and W. W. Zachary (Springer-Verlag, Heidelberg, 1987).

[6] V. Guillemin and S. Sternberg, Symplectic Techniques in Physics (Cambridge University Press, London, 1984); R. G. Littlejohn, Phys. Rep. 138, 193 (1986); D. Han, Y. S. Kim, and M. E. Noz, Phys. Rev. A 37, 807 (1988).

[7] Y. S. Kim and M. E. Noz, Theory and Applications of the Poincaré Group (Reidel, Dordrecht, 1986).

[8] D. F. Walls, Nature (London) 306, 141 (1983); W. Schleich and

J. A. Wheeler, ibid. 326, 574 (1987).

[9] E. P. Wigner, Ann. Math. 40, 149 (1939); V. Bargmann and E. P. Wigner, Proc. Natl. Acad. Sci. U.S.A. 34, 211 (1948).

[10] E. Inonu and E. P. Wigner, Proc. Natl. Acad. Sci. U.S.A. 39, 510 (1953); D. Han, Y. S. Kim, and D. Son, Phys. Lett. 131B, 327 (1983).

[11] D. Han, Y. S. Kim, and D. Son, J. Math. Phys. 27, 2228 (1986); Y. S. Kim and E. P. Wigner, ibid. 28, 1175 (1987).

[12] S. Shlomo, J. Phys. A 16, 3463 (1983).

[13] H. J. Groeneworld, Physica 12, 405 (1946); J. E. Moyal, Proc. Cambridge Philos. Soc. 45, 99 (1949); T. Takabayasi, Prog. Theor. Phys. 11, 341 (1954); J. P. Dahl, Phys. Scr. 25, 499 (1982).

[14] P. A. M. Dirac, Proc. Roy. Soc. London, Ser. A 183, 284 (1945).

[15] H. Yukawa, Phys. Rev. 91, 415 (1953); 91, 416 (1953).

[16] K. Fujimura, T. Kobayashi, and M. Namiki, Prog. Theor. Phys. 43, 73 (1970).

[17] R. P. Feynman, M. Kislinger, and F. Ravndal, Phys. Rev. D 3, 2706 (1971).

[18] D. Han, Y. S. Kim, and M. E. Noz, Phys. Rev. A 35, 1682 (1987).

[19]P. A. M. Dirac, Proc. Roy. Soc. London, Ser. A **114**, 234 (1927); **114**, 710 (1927); E. P. Wigner, in *Aspects of Quantum Theory, In Honour of P. A. M. Dirac's 70th Birthday*, edited by A. Salam and E. P. Wigner (Cambridge University Press, London, 1972). See also P. E. Hussar, Y. S. Kim, and M. E. Noz, Am. J. Phys. **53**, 142 (1985).

[20]H. van Dam and E. P. Wigner, Phys. Rev. **138**, B1576 (1965); E. P. Wigner, in *Proceedings of the First International Conference on the Physics of Phase Space, College Park, Maryland, 1986*, edited by Y. S. Kim and W. W. Zachary (Springer-Verlag, Heidelberg, 1987).

[21]P. A. M. Dirac, Rev. Mod. Phys. **21**, 392 (1949).

[22]W. Magnus and F. Oberhettinger, *Formulas and Theorems for the Functions of Mathematical Physics* (Chelsea, New York, 1949).

[23]P. Carruthers and F. Zachariasen, Rev. Mod. Phys. **55**, 245 (1983).

[24]R. P. Feynman, in *High Energy Collisions*, Proceedings of the Third International Conference, Stony Brook, New York, edited by C. N. Yang *et al.* (Gordon and Breach, New York, 1969).

[25]P. E. Hussar, Phys. Rev. D **23**, 2781 (1981); P. E. Hussar and M. L. Haberman, Z. Phys. C (to be published).

# The General Properties of the Distribution Function and Remarks on Its Weakness

Y. S. Kim and E. P. Wigner

In: The Physics of Phase Space (Y. S. Kim, W. W. Zachary, eds.).
Lecture Notes of Physics, vol. 278. Springer, New York 1987, pp. 162–170

(Reset by Springer-Verlag for this volume)

## Introduction

It was an unexpected pleasure to hear about the conference on the quantum mechanics of phase space, and I very much appreciate the pleasure to be invited thereto. I will be able to contribute very little to it that is not contained in the Physics Report article [Phys. Rep. *106*, 121 (1984)] by M. Hillery, R. F. O'Connell, M. Scully, and myself – an article to which I have actually contributed, in contrast to Dr. Scully, very little. But I will admit that the underlying reformulation of the Schrödinger equation was started by me, in 1932 (Phys. Rev. 40, p. 749). I was interested in the thermodynamic behavior of macroscopic objects which is given with high accuracy, at ordinary or high temperatures, by classical statistical mechanics. At low temperatures quantum effects can become important and this manifested itself also in the "equation of state" (temperature and density dependence of the pressure) of the He gas. It was natural, therefore, to develop a substitute for the classical expression for the density in phase space (to be described below) which forms the basis for the calculation of the thermodynamic behavior in the temperature region in which classical mechanics can be assumed to be valid, and which easily provides a good approximation in the temperature region not too far away from the validity of classical physics. This means a probability function of the position and momentum variables $q$ and $p$, defined in terms of the wave function $\psi$ or the density matrix $M$, a probability function which is a hermitean expression of the wave function, hence linear in the density matrix. It is not too difficult to calculate and does give accurate results for the equation of state, and I hope also for other quantities.

It must be admitted, of course, that the interpretation of the phase space density function is much less direct in the situation in which quantum effects play an important role than it is in the area of classical physics. The variables of the phase space are the $3N$ position and $3N$ momentum coordinates of the $N$-particle system to which the density function of the phase space refers. We'll write $n$ for $3N$. The classical phase space function's value at a point of phase space is the probability that the position and momentum coordinates of the $N$ particles have the values given by the coordinates of corresponding points in phase space. If the phase space function has to be so closely defined that quantum effects play a role, this interpretation is not possible because there is no state in which both position and momentum coordinates have definite values. In fact the states of the system are not specified in terms of these coordinates. It follows that the interpretation of the density function of phase space is much less straightforward in the region in which quantum effects play a significant role than it is in classical theory – that is if the probabilities do not change significantly within distances in which the products of the $p$ and $q$ are not far from $h$. All this shows that the definition – and hence also the meaning – of the phase space functions is not as unique in the quantum region as it is in the region of the classical theory. The next section will therefore discuss the meaning and the properties of the quantum distribution function as defined in 1932. This definition does not take care of the existence of the spin and the extension of the theory to the description of the spin state will be discussed afterwards.

### Properties of the Proposed Quantum Distribution Function

As is apparent from the preceding discussion, the quantum distribution function to be discussed does not have such a simple meaning as the classical phase space function. It may be useful, therefore, to describe its basic properties before discussing its applications.

Let us first define the distribution function $P$ to be considered. It will be defined as a function of $n$ position and $n$ momentum variables $q$ and $p$, the $n$ being three times the number of particles ($n = 3N$). If the state of the system is given by a position-dependent wave function $\psi(x_1, x_2, \cdots, x_n)$, the distribution function is

$$P(q_1, \cdots, q_n; p_1, \cdots, p_n) = \frac{1}{(\pi\hbar)^n} \int \cdots \int dy_1 \cdots dy_n \psi^*(q_1 + y_1, \cdots, q_n + y_n)$$
$$\times \psi(q_1 - y_1, \cdots, q_n - y_n) e^{2i(p_1 y_1 + \cdots + p_n y_n)/\hbar} . \qquad (1)$$

This is, clearly, a nonrelativistic definition – as is fundamentally also that of the classical distribution function – but has, similar to that, some useful properties. These will remain also after the introduction of the spin variables. Before enumerating its useful properties, it may be good to give the $P$ for a density matrix $M(q_1, \cdots; q_n; q'_1, \cdots, q'_n)$. This can be decomposed into orthogonal and normalized wave functions $\psi_1, \psi_2, \cdots$ which appear with probabilities $w_1, w_2, \cdots$. If the distribution function is assumed to be an additive function of these:

$$M(q_1, \cdots, q_n; q'_1, \cdots, q'_n) = \sum w_k \psi_k(q_1, \cdots, q_n) \psi_k^*(q'_1, \cdots, q'_n), \qquad (2)$$

it is natural to define the corresponding $P$ as

$$P(q, p) = \frac{1}{(\pi\hbar)^n} \int dy \, M(q - y, q + y) e^{2i(p \cdot y)/\hbar} . \qquad (2a)$$

In this equation, as in many later ones, the symbols $q$, $y$ and $p$ represent $n$ variables each, $\int dy$ means integration over the $n$ variables $y$, and $(p \cdot y)$ is the scalar product $\sum p_k y_k$. The notation used in these equations renders several future equations much simpler. Similar to the meaning of $dy$, the $dq$ will mean integration over the $n$ variables $q$ and $dp$ means integration over the $n$ variables $p$. These notations simplify several of the following equations.

Let me now come to the properties of the distribution functions (1) and (2a) which I consider to be of significance. The proofs will be given for (1) but, because of the definition (2) of $M$, it will be evident that they apply also for the more general form (2a).

1. $P(q, p)$, if integrated over $p$ (that is over $p_1$, $p_2, \cdots, p_n$), gives the probability of the configuration $q$ that is the probability that the position coordinates are $q_1, q_2, \cdots, q_n$. This is easily verified.

2. Similarly, if $P(q, p)$ is integrated over the $q$, it gives the probability that the momentum coordinates have the values $p_1, p_2, \cdots, p_n$.

These two properties can be easily verified, and it is clear also that they are less significant than the basic property of the classical $P$ which represents the probability for both the positions to be given by $q_1, \cdots, q_n$, that is by $q$, and the momenta by $p$. But they do show that the average value of the classical energy, being the sum of two functions, one of the momentum the other of positions, can be easily obtained.

3. These two observations suggest that $q$ and $p$ play similar roles in the definition of the distribution function $P$. Indeed, if the $\psi(q)$ is expressed in terms of its Fourier transform, the wave function $\chi(p)$ of the momentum coordinates:

$$\psi(q) = \int \chi(p) e^{ip \cdot q/\hbar} dp \,, \tag{3}$$

where we neglect constant factors temporarily, we obtain for $P(q, p)$

$$P(q,p) = \iiint \chi^*(p') e^{-ip' \cdot (q+y)/\hbar} \chi(p'') e^{ip'' \cdot (q-y)/\hbar} e^{2ip \cdot y/\hbar} dp' dp'' dy \,. \tag{3a}$$

The factors involving $y$ give a delta function $\delta(2p - p' - p'')$ so that, again disregarding a constant factor, we can set $p' = p + z$, $p'' = p - z$ and the integration over $z$ will replace the integration over $p'$ and $p''$. Hence (3a) becomes

$$P(q,p) = (\pi\hbar)^{-n} \int dz \chi^*(p + z) \chi(p - z) e^{-2iz \cdot q/\hbar} \,. \tag{3b}$$

The numerical constant before the integral sign follows from the fact that the integral of $P$ remains 1 and that the $\chi$ are also normalized. Eq. (3b) is a close analogue of (1), except for the fact that $i$ is replaced by $-i$ – which is natural – and shows that position and momentum coordinates play essentially the same role in the definition of our distribution function – just as they do in classical theory.

4. The transformation properties of $P$ are the classical ones with respect to any of the classical transformations. The substitution of $q + a$ for $q$ clearly gives $P(q + a, p)$ from $P(q,p)$ – and this remains true even if $a$ is not the same vector for all particles. If $\psi$ is replaced – we use (1) in this discussion

– by $e^{i\kappa \cdot q/\hbar}\psi$ the distribution function so obtained assumes the values of the original distribution function if $p + \kappa$ is substituted for $p$ – actually $\kappa$ can be an arbitrary $n$ dimensional vector, but naturally independent of the $q$.

The past three points are natural demands and are easily verified.

5. The so-called transition probability between two states, $\phi$ and $\psi$ for instance, is , as a rule, not really observable. If the system is in the state $\psi$ and an observation is made as a result of which the system's state vector becomes $\phi$, the probability of this result of the observation is (if both $\phi$ and $\psi$ are normalized) $|(\phi, \psi)|^2$, the absolute square of the scalar product of the two state vectors. The observation in question is not possible for every state vector $\phi$, but the existence of the scalar product, or at least the measurability of its absolute value, is often assumed for all $\phi$. If the original state of the system is best given by a density matrix $M$, the probability that the measurement transfers it into the state $\phi$ is given by the scalar product of $\phi$ and $M\phi$, that is $(\phi, M\phi)$. It is worth noting therefore that if $P_\psi$ and $P_\phi$ correspond, by (1), to $\psi$ and $\phi$, the so-called transition probability between them becomes

$$|(\psi, \phi)|^2 = (2\pi\hbar)^n \iint dpdq \, P_\psi(q,p) P_\phi(q,p) \,. \tag{4}$$

It follows from (4) also that if $P_M$ and $P_N$ are the distribution functions which correspond to the density matrices M and N, then

$$\text{Trace}\,(MN) = (2\pi\hbar)^n \iint dpdq \, P_M(q,p) P_N(q,p) \,. \tag{4a}$$

All the preceding observations are easily verified and are also contained in the aforementioned article of Hillery, O'Connell, Scully and myself – most are in fact also contained in the aforementioned 1932 article. Apparently, there is a great deal of arbitrariness in the definition (1) of the distribution function but R. F. O'Connell has shown that some of the preceding properties already fully determine it. This was not known when (1) was originally proposed but is well worth remembering.

The last observation, Eq. (4), also shows that most distribution functions, though real, are not everywhere positive. For two orthogonal wave functions, $(\psi, \phi) = 0$, the integral over $P_\psi P_\phi$ must vanish. They can not be both positive everywhere – for most $\psi$ and $\phi$ neither is. But, as Heisenberg pointed out, there is no state for which both $p$ and $q$ have definite values. Transition probabilities are observable, at least many of them, and it is satisfactory that, according to (4), the expressions for these cannot be negative. The fact that most functions of $p$ and $q$ do not represent possible states renders the quantum distribution function to be a less simple quantity than is the classical distribution function, since, in classical theory, all everywhere non-negative distribution functions are conceivable. This point will be mentioned again later, together with the fact that the condition which an arbitrary function of $p$ and $q$ must obey in order to be a possible distribution function is not simple. Clearly, it must be possible to write it in the form (2a) with a positive definite (or non-negative) self-adjoint matrix $M$, but this is not a simple condition.

6. The preceding observations on the properties of our quantum mechanical distribution functions gave properties which the classical distribution functions also had - in fact the properties of the latter were more general. We now come to an equation which shows the quantum mechanical nature of our distribution function – the equation of its time dependence. Essentially the same equation will be used afterwards to determine the distribution function for the thermodynamic equilibrium.

The equation for $\partial P/\partial t$ has two types of terms. The first type originates from the kinetic energy terms $-(\hbar^2/2m)\partial^2/\partial q^2$ of the expresssion for $i\hbar\partial\psi/\partial t$, the second one from the potential energy terms. Both are easily determined and were long ago. Here only the first one will be reproduced in full detail. It gives for $(\pi\hbar)^n(\partial P/\partial t)_k$

$$\frac{i}{\hbar}\frac{\hbar^2}{2m}\int\left[-\frac{\partial^2\psi^*(q+y)}{\partial q^2}\psi(q-y)+\psi^*(q+y)\frac{\partial^2\psi(q-y)}{\partial q^2}\right]e^{2ipy/\hbar}\,dy.\qquad(5)$$

The second derivatives with respect to $q$ can be replaced by second derivatives with respect to $y$ and a partial integration then be carried out. The two terms in which the products of both first derivatives appear then cancel and the terms in which the exponential is differentiated gives

$$\frac{i\hbar}{2m}\frac{2ip}{\hbar}\int\left[\frac{\partial\psi^*(q+y)}{\partial y}\phi(q-y)-\psi^*(q+y)\frac{\partial\psi(q-y)}{\partial y}\right]e^{2ipy/\hbar}dy.\qquad(5a)$$

The differentiations with respect to $y$ can be replaced by differentiations with respect to $q$ – changing the sign of the second term. The result then is the same expression which appears in classical theory for the kinetic energy part $(\partial P/\partial t)_k$ of $\partial P/\partial t$ – if written in detail it is

$$\left(\frac{\partial P}{\partial t}\right)_k=-\sum_\kappa\frac{1}{m}p_\kappa\frac{\partial P}{\partial q_\kappa}.\qquad(5b)$$

The potential part of $\partial P/\partial t$ can be expressed in two ways. One can expand the potential energy expression in

$$(\pi\hbar)^n\left(\frac{\partial P}{\partial t}\right)_p=\frac{i}{\hbar}\int\psi^*(q+y)[-V(q+y)+V(q-y)]\psi(q-y)e^{2ipy/\hbar}dy\qquad(6)$$

either into a power series of $y$, or represent it as a Fourier transform. The second possibility shows again that $p$ and $q$ play similar roles in the theory of distributions. But the expansion of $V(q-y)-V(q+y)$ as a power series of $y$ gives

$$-V(q+y)+V(q-y)=-\sum_{\lambda_1\cdots\lambda_n}2\left(\frac{\partial^{\lambda_1+\cdots+\lambda_n}}{\partial q_1^{\lambda_1}\cdots\partial q_n^{\lambda_n}}V(q)\right)\left(\frac{y_1^{\lambda_1}y_2^{\lambda_2}\cdots y_n^{\lambda_n}}{\lambda_1!\cdots\lambda_n!}\right),\qquad(6a)$$

the summation to be extended to all non-negative (integer) $\lambda$ the sum of which is odd. This gives for the potential caused part of the time derivative

$$\left(\frac{\partial P}{\partial t}\right)_p = \sum_{\lambda_1 \cdots \lambda_n} \left(\frac{(\hbar/2i)^{\lambda_1+\cdots+\lambda_n-1}}{\lambda_1!\lambda_2!\cdots\lambda_n!}\right) \left(\frac{\partial^{\lambda_1+\cdots+\lambda_n}}{\partial q_1^{\lambda_1}\cdots\partial q_n^{\lambda_n}}V\right)\left(\frac{\partial^{\lambda_1+\cdots+\lambda_n}}{\partial p_{-1}^{\lambda_1}\cdots\partial p_n^{\lambda_n}}P\right) \tag{6b}$$

in which, however, all the $\lambda$ are non-negative and their sum odd. The first term of the series, in which one $\lambda$ is 1, all others 0, gives the classical expression for $(\partial P/\partial t)_p$. The lowest order corrections contain the second power of $\hbar$. And they constitute for

$$\frac{\partial P}{\partial t} = \left(\frac{\partial P}{\partial t} =\right)_k + \left(\frac{\partial P}{\partial t}\right)_p \tag{6c}$$

the lowest order corrections. None of the preceding considerations is new, neither is the last point of this section.

7. The oldest use of the quantum mechanical distribution function was based on the calculation of the quantum effects on the equations of states of gases. If Bose or Fermi statistics of these is disregarded, (this was treated later, in 1984 by O'Connell and Wigner) the distribution function of these is the normalized form of $e^{-H/kT}$. Setting $1/kT = \beta$, this can be written as $e^{-\beta H}$ and the equation which replaces the equation for $\partial P/\partial t$ becomes

$$\frac{\partial P}{\partial \beta} = -HP. \tag{7}$$

The expansion of $P$ in terms of $\beta$ has been discussed when the expression (1) or (2a) was first proposed and is reviewed also in the article by Hillery, O'Conell, Scully and Wigner mentioned several times before. There is no point repeating the calculation which replaces the calculation of the distribution function for $e^{tH/i\hbar}$ by the calculation of the distribution function for $e^{-\beta H}$. Perhaps I mention that the first application of the quantum mechanical distribution function concerned the equation of state of the He gas. At very low temperatures the experimental results deviated considerably from that given by classical theory, that is by the classical distribution function. The correction introduced by the quantum corrections to this discussed here were in the right direction but accounted only for about 2/3 of the deviations from the experimental measurements. It is possible that the reason for this was that the potential energy function was not known well enough. It would therefore be worthwhile to repeat that calculation. Its desirability was actually the stimulant for the introduction of our $P$.

## The Spin Variable

The preceding discussion largely disregards the spin variable – which is natural in the case of the He gas, since the He atoms have no spin. A possible way to add the description of the spin state to that of the other variables was discussed before (1983) for systems with spin 1/2 but that is easily generalized for larger spin.

For every particule of spin $s$ the density matrix has $(2s+1)^2$ components. The problem is only to find such linear combinations of these components which have relatively simple properties. We can specify the $(2s+1)^2$ components with two index symbols: $\mu$ and $\mu'$ – the first giving the row index of the density matrix, and $\mu'$ the column index. Both run for each particle of spin $s$ from $-s$ to $s$ in integer steps. We can then form, for each particle, another description of the spin state by combining the row and column components to have simple transformation properties. They will have transformation properties which correspond to the direct product of two representations $D^{(s)}$. It is possible then to produce linear combinations of the components characterized by $\mu$ and $\mu'$ which transform under rotations by the representations $D^{(0)}, D^{(1)}, \cdots, D^{(2s)}$. In the case of $s = 1/2$, which was considered before, there is a scalar and a vector component – the former giving the total probability, the others being formed by the components of $D^{(1)}$.

Let us denote the density matrix by $M(\xi, \mu; \xi', \mu')$, $\mu$ giving the row index of the spin variable of the particle in question, $\xi$ denoting all other variables of the row, $\mu'$ and $\xi'$ the same interpretation for the columns. The distribution function proposed would replace the $\mu$ and $\mu'$ by the indices $S$ and $m$:

$$M'(\xi, \xi'; S, m) = \sum_{\mu\mu'}(S, m; s, \mu, s', \mu')M(\xi, \mu; \xi', \mu') \, . \qquad (8)$$

the first factor after the summation sign being the coefficient which transforms the representation of the direct product $D^{(s)^*} \times D^{(s)}$ into $D^{(S)}$, and $m$, $\mu$ and $\mu'$ are the row indices of the representations $S$, and $D^{(s)^*}$ and $D^{(s)}$. It would not be reasonable to produce here these coefficients in general but it may be worth noting that

$$M'(\xi, \xi; 0, 0) = \frac{1}{2s+1}\sum_{\mu} M(\xi, \mu; \xi, \mu) \, . \qquad (8a)$$

For the case of $s = 1/2$, the $S$ assumes only two values: 0 and 1. The coefficients for 0 are given in (8a), those for $S = 1$ the transferred $M$, that is the $M'$, were given as the expectation values of the $x, y$, and $z$ components of the spin operator, that is of $s_x$, $s_y$, $s_z$. In many cases the effects of the higher $S$ components of $M'$ are insignificant and in those cases the same transformation of the spin coordinates can be recommended.

The total transformation to the quantum mechanical distribution function $P$ obeys then the same equation as in the absence of spin (4) and the $\mu$ and $\mu'$ for every particle are replaced, in terms of (8), by $S$ and $m$.

This is a somewhat superficial description of the transformation of the spin variables for what I call the quantum mechanical distribution function, but I hope that it gives the proposed transformation clearly enough.

## Problems of the
## Proposed Quantum Mechanical Distribution Function

The quantum mechanical distribution theory here described has two weaknesses. One of these was mentioned before: given an arbitrary real function $P(p,q)$, it is not clear whether it is a possible distribution function. If it is, it can be written in the form (2a) in terms of an acceptable density matrix $M$ but the acceptability of a density matrix is also not easily verified. In particular, it must be positive definite, or semidefinite – that is no expectation value of the transition to any state, that is no $(\phi, M\phi)$ can be negative. This applies also to our $P$: no integral of the product of two quantum mechanical distribution functions can be negative.

Just as in the usual theory, it is sufficient to demonstrate that the product is non-negative with any distribution function representing a single state, that is having the form (1), but even this is an infinite task – just as it is in ordinary quantum mechanics dealing with density matrices.

The other difficulty well worth mentioning is one also shared, at least to some degree, with the usual formulation of quantum mechanics: the postulate of the coherence with relativity theory. This causes difficulties also in the usual theory – it is necessary to introduce a field, that is an infinitely more complex definition of the state than is used in Schrödinger's old fashioned theory. In addition, the equations often lead to infinities and these must be eliminated by "renormalization". In summary, even the usual theory has weaknesses – I would say that its beauty is not absolute.

But the weakness of the theory here discussed is much more fundamental – at least it is so at present. It assumes that the interaction of the particles is instantaneous – that it depends only on their same-time positions. This is acceptable, and in fact generally accepted, in non-relativistic theory but is in conflict with the theory of relativity in which simultaneity is not independent of the state of motion of the coordinate system describing it. This renders, quite generally, the description of the states of systems by phase space functions unattractive – in phase space the interaction is assumed to depend on the simultaneous position of the particles and is, therefore, not relativistically invariant. It is possible to make it invariant, for instance by postulating that it depends on the distance in the coordinate system at rest with the temporary center of mass of the particles, or to depend on the two positions at the time when their relativistic distance is zero – when one is on the light cone of the other or conversely. It is even possible to assume a "force" depending on the integral of the distances between the light cones. But these possibilities have not been explored to my knowledge and the present theories assume interactions of fields – i. e. only interactions at points of the same positions and times. These gave many apparently correct results but needed the introduction of "fields", in particular electromagnetic potentials, and are not in harmony with the phase-space theories. Perhaps this could be amended, but I do not know of serious attempts in that direction – not even by myself.

# Covariant Phase-Space Representation and Overlapping Distribution Functions

Y. S. Kim and E. P. Wigner

Physical Review A *39*, 2829–2834 (1989)

(Received 1 September 1988)

The Lorentz deformation property of the phase-space distribution function is studied for harmonic oscillators. The overlap effect of two distribution functions is discussed in detail. It is shown that the Lorentz deformation of the phase-space distribution is responsible for the polynomial cutoff behavior of the proton form factor in the harmonic-oscillator quark model.

## I. INTRODUCTION

The phase-space representation of quantum mechanics[1] is of current interest.[2-4] We have shown in our previous papers[5,6] that the light-cone coordinate system is the natural language for the Lorentz-covariant phase space representation of quantum mechanics. The Lorentz-deformed phase-space distribution was discussed in detail for localized light waves[5] and harmonic oscillators.[6] It was shown in Ref. 6 that the covariant harmonic oscillator can illustrate the Lorentz-deformed phase-space distribution for a relativistic free particle with a space-time extension.

The purpose of the present paper is to study the overlapping phase-space distributions functions. We are of course interested in Lorentz-deformed distributions, and in how their effect manifests itself in the real world. For this purpose, we shall study the electromagnetic structure of nucleons. Because the radius of the proton is $10^5$ times smaller than that of the hydrogen atom, the proton had long been thought to be a point particle. However, Hofstadter's discovery in 1955 clearly demonstrated that the proton has a spread-out charge distribution.[7]

Although there had been many attempts to understand the structure of nucleon since 1955, the first comprehensive approach to the probability distribution of hadronic matter was the quark model, in which the nucleon is a bound state of three quarks.[8] Among the mathematical models for this bound state, the harmonic-oscillator model gives a simple explanation for a wide range of hadronic phenomena observed in high-energy physics laboratories.[9-11] While the major strength of the oscillator model is its mathematical simplicity, its most useful property for our present purpose is that the oscillator model constitutes a representation of the Poincaré group.[12-16] The harmonic-oscillator model can be made covariant.

Another important property of the oscillator model is that it is the natural language for phase space.[2,6,17] Therefore the harmonic-oscillator model is an effective scientific language for the covariant description of phase space. We shall study in this paper overlapping distribution functions in the covariant phase-space formulation of harmonic oscillators.

In Sec. II we summarize the earlier works on overlapping phase-space distribution functions and their physical interpretations. Section III explains how the form factor is defined and why a Lorentz-covariant formalism is needed for studying the form factor. Section IV deals with the covariant phase-space distribution functions which are needed for calculating the form factors. In Sec. V we study how the overlap of two Lorentz-deformed phase-space distribution functions lead to the correct form-factor behavior in the harmonic-oscillator model for hadrons.

## II. OVERLAPPING PHASE-SPACE DISTRIBUTION FUNCTIONS

If $\psi(x)$ is a solution of the time-independent Schrödinger equation, its phase-space distribution function is

$$W_\psi(x,p) = \frac{1}{\pi} \int \psi^*(x+y)\psi(x-y)e^{2ipy}dy . \quad (1)$$

For simplicity, we shall use the term "PSD function" for phase-space distribution function. This is a function of $x$ and $p$ which are $c$ numbers. This function is real but is not necessarily positive everywhere in the two-dimensional phase space of $x$ and $p$. We can, however, recover the positive distribution functions in the position and momentum coordinates by integrating the PSD function over $p$ and $x$, respectively,[1,2]

$$\rho(x) = \int W_\psi(x,p)dp, \quad \sigma(p) = \int W_\psi(x,p)dx . \quad (2)$$

In this paper, we are interested in two overlapping PSD functions. Indeed, the overlap integral becomes the absolute square of the inner product of the two wave functions in the Schrödinger picture. If $W_\psi(x,p)$ and $W_\phi(x,p)$ are the PSD functions for $\psi(x)$ and $\phi(x)$, respectively, then[1,2]

$$\int W_\psi(x,p)W_\phi(x,p)dx\,dp = (1/2\pi)|(\phi(x),\psi(x))|^2 . \quad (3)$$

This expression is non-negative, but can be zero if the two functions are orthogonal, indicating that the PSD functions are not always positive everywhere in phase space.

In studying the interaction of a photon with an atomic system, we often encounter the matrix element of the form

$$M_{fi} = (\phi, e^{ikx}\psi) . \tag{4}$$

This is the inner product of the wave functions $\phi$ and $\psi'$ with $\psi' = e^{ikx}\psi$. Thus we have to construct the PSD function for $\psi'$,

$$W_{\psi'}(x,p) = \frac{1}{\pi} \int \psi^*(x+y)\psi(x-y)e^{2i(p-k)y}dy . \tag{5}$$

This leads to

$$W_{\psi'}(x,p) = W_\psi(x,(p-k)) . \tag{6}$$

Therefore

$$|(\phi, e^{ikx}\psi)|^2 = 2\pi \int W_\phi(x,p)W_\psi(x,(p-k))dx\,dp . \tag{7}$$

We have thus far carried out the formalism for one-dimensional space. The generalization to the three-dimensional space is straightforward and has been discussed in the literature.[1,2]

## III. FORM FACTORS

If electrons are scattered by a charged point particle, the scattering amplitude in the Born approximation is $f(\theta) = 2me^2/(\mathbf{k}_f - \mathbf{k}_i)^2$. On the other hand, if the electron is scattered by a spread-out charge due to quantum probability distribution, the scattering amplitude is

$$f(\theta) = (2me^2/K^2)F(K^2) . \tag{8}$$

where $\mathbf{K} = \mathbf{k}_f - \mathbf{k}_i$, and $K^2 = (\mathbf{k}_f - \mathbf{k}_i)^2$. $F(K^2)$ is called the form factor and takes the form

$$F(K^2) = (\psi_f, e^{-i\mathbf{K}\cdot\mathbf{x}}\psi_i) = \int [\psi_f(\mathbf{x})]^\dagger \psi_i(\mathbf{x})e^{-i\mathbf{K}\cdot\mathbf{x}}d^3x , \tag{9}$$

with $F(0) = 1$, if the initial- and final-state wave functions are the same. If $\{[\psi_f(\mathbf{x})]^\dagger\psi_i(\mathbf{x})\}$ describes a point-charge distribution with $\delta(\mathbf{x})$, $F(K^2) = 1$ for all values of $K^2$. According to Eq. (3), the form factor should take the form

$$|F(K^2)|^2 = \int W_f(\mathbf{x},\mathbf{p})W_i(\mathbf{x},(\mathbf{p}-\mathbf{K}))d^3x\,d^3p . \tag{10}$$

This is a generalization of Eq. (7) to the three-dimensional space.

As the energy of incoming electrons becomes higher for the fixed nucleon target, $K^2$ becomes very large, and the problem becomes relativistic. For the electromagnetic interaction of point particles, we have to use quantum electrodynamics, where the scattering amplitude is expanded in a power series of the fine-structure constant $\alpha = e^2/4\pi$ in the Lorentz-Heaviside unit. The lowest nontrivial term in this expansion is essentially a relativistic version of the Born approximation.

In lowest order in $\alpha$, we can describe the scattering of an electron by a proton using the diagram given in Fig. 1. The corresponding matrix element is given in many textbooks on elementary particle physics.[18] It is proportional

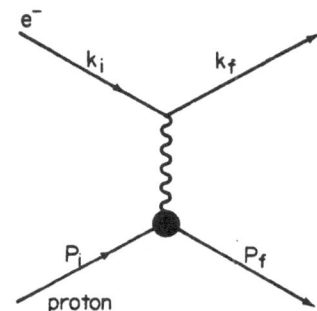

FIG. 1. Elastic electron-proton scattering. The electron behaves like a point charge. However, the proton has its hadronic structure, and has a spread-out charge distribution.

to

$$e^2[\overline{U}(P_f)\Gamma_\mu(P_f,P_i)U(P_i)](1/K^2)[\overline{U}(k_f)\gamma^\mu U(k_i)] , \tag{11}$$

where $P_i$, $P_f$, $k_i$, and $k_f$ are the initial and final four-momenta of the proton and electron, respectively. $U(P_i)$ is the Dirac spinor for the initial proton. We use here the four-vector convention $x^\mu = (x,y,z,t)$. $K^2$ is the four-momentum transfer squared given by

$$K^2 = K^2 - K_0^2 = (P_f - P_i)^2 = (k_i - k_f)^2 . \tag{12}$$

The $1/K^2$ factor in Eq. (11) comes from the virtual photon being exchanged between the electron and the proton. In the metric we use, the quantity is positive for physical values of the four-momenta for the particles involved in the scattering process.

The function $\Gamma_\mu(P_f, P_i)$ in Eq. (11) represents the closed circle in Fig. 1 and carries the effect of the nucleon structure. If the proton were a point charge, we would have $\Gamma_\mu = \gamma_\mu$. If the proton has an extended charge structure, we will be inclined to write it as $\Gamma_\mu = \gamma_\mu F(K^2)$. However, the proton and neutron have anomalous magnetic moments whose values are 2.79 and $-1.91$ in units of $e/2M$ for the proton and neutron, respectively, where $M$ is the nucleon mass. If we include these observed anomalous magnetic moments, $\Gamma_\mu$ should be written as

$$\Gamma_\mu = \gamma_\mu F_1(K^2) + i(\sigma_{\mu\nu}K^\nu/2M)F_2(K^2) . \tag{13}$$

The form factors are scalar functions in the Lorentz-invariant variable $K^2$. When we compare $F_1(K^2)$ and $F(K^2)$ with experimental data, it is more convenient to use the following linear combinations:

$$\begin{aligned} G_M(K^2) &= F_1(K^2) + F_2(K^2) , \\ G_E(K^2) &= F_1(K^2) + (K^2/4M^2)F_2(K^2) . \end{aligned} \tag{14}$$

These form factors should be written for the proton and the neutron separately. We may use the superscripts $p$ and $n$ to distinguish them. When $K^2 = 0$,

$$G_M^p(0) = \mu_p = 2.79, \quad G_E^p(0) = 1 \text{ (proton) ;}$$

$$G_M^n(0) = \mu_n = -1.91, \quad G_E^n(0) = 0 \text{ (neutron) .}$$

(15)

These numbers are the magnetic moments and electric charges of the proton and neutron, respectively.

Among the many attempts to understand the form factors, the quark model appears to be the most promising approach.[8,10,19] In this model, the nucleon consists of three nonstrange quarks. There are two nonstrange quarks called u (up) and d (down) which have electric charges $\frac{2}{3}$ and $-\frac{1}{3}$ respectively. The proton consists of two u quarks and one d quark, and the neutron is made up of one u quark and two d quarks. The neutron charge is therefore zero, while the proton charge is 1.

Indeed, one of the early successes of the quark model was the calculation of the magnetic moment ratio $\mu_n/\mu_p = -\frac{2}{3}$.[20] However, it is even more challenging to calculate the form factors for increasing values of $K^2$.[10,19] At present, we can make the following experimental observation. For the four form factors in the nucleonic system given in Eq. (14), the neutron charge form factor is zero at $K^2 = 0$, and remains small (not zero) for all values of $K^2$.[21] The three remaining form factors decrease like $1/(K^2)^2$ as $K^2$ increases beyond the value of the nucleon mass squared.[22] This behavior is usually called the dipole fit. We are interested in the question of whether each of the form factors can be written in terms of the single form factor $G(K^2)$, multiplied by a constant, where $G(K^2)$ is normalized as $G(0) = 1$, and is proportional to

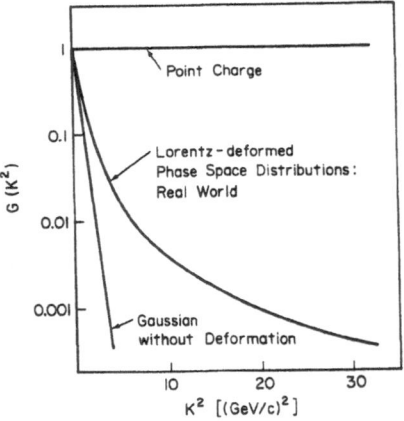

FIG. 2. Form-factor behaviors for increasing values of $K^2$. If the proton is a point charge, the form factor should be independent of $K^2$, as is illustrated by the horizontal line. If the charge distribution is Gaussian, the nonrelativistic calculation leads to an exponential cutoff in $K^2$. The relativistic calculation gives a reasonably accurate description of the real world. At present, the experimental data are available for $G_M^p(K^2)$, $G_E^p(K^2)$, and $G_M^n(K^2)$ from $K^2 = 0$ to 25, 15, and 7 $(\text{GeV/c})^2$, respectively. They are all consistent with the relativistic calculation with Lorentz deformation.

$1/(K^2)^2$ for large values $K^2$, as is illustrated in Fig. 2. In the case of the neutron charge form factor, we have to multiply $G(K^2)$ by zero within the framework of the model in which the spin, unitary spin, and spatial wave functions are factorized.[21] The question is whether it is possible to calculate the above-mentioned dipole behavior of $G(K^2)$ using the wave functions obtained from the quark model.

Another important aspect of the quark model is that the forces between the quarks are like harmonic oscillators, and the hadronic mass spectrum is consistent with the equal mass-squared spacing predicted by the oscillator model.[9,16] In addition, the parton distribution shows a Gaussian shape in the region where the structure function can be measured accurately.[11]

It is therefore a reasonable approach to calculate the nucleon form factor assuming that the nucleon is in the ground state of a harmonic-oscillator system. However, the Gaussian distribution gives an exponential cutoff of the type $\exp(-K^2/4\Omega)$, where $\Omega$ is the spring constant of the oscillator system. This contradicts the experimental observation, as is indicated in Fig. 2. It is therefore interesting to see whether the effect of Lorentz deformation in phase space could transform this exponential decrease into a polynomial cutoff.

## IV. COVARIANT PHASE-SPACE DISTRIBUTION FUNCTIONS

The covariant phase-space representation for harmonic oscillators has been discussed in Ref. 6. As in Ref. 6, we start with two quarks bound together by a harmonic-oscillator potential. Then the convenient coordinate variables are[15]

$$X = (x_a + x_b)/2, \quad x = (x_a - x_b)/2\sqrt{2} .$$

(16)

With these variables, the hadronic wave function takes the form

$$\varphi(X,x) = e^{\pm iPX}\psi(x) ,$$

(17)

where $P$ is the hadronic four-momentum. The wave function $\psi(x)$ describes the internal motion of the two-quark system. The preceding form as a representation space of the Poincaré group[12] for relativistic extended hadrons has been discussed in the literature.[16]

As for the four-momenta of the quarks $p_a$ and $p_b$, we can combine them into the total four-momentum and momentum-energy separation between the quarks,[15]

$$P = p_a + p_b, \quad q = \sqrt{2}(p_a - p_b) ,$$

(18)

where $P$ is the hadronic four-momentum conjugate to $X$. The internal momentum-energy separation $q$ is conjugate to $x$ provided that there exist wave functions which can be Fourier transformed.

If the hadron moves along the z direction with velocity parameter $\beta$, the hadronic rest frame is important. In this frame, the coordinate variables are

$$x' = x, \quad y' = y ,$$

$$z' = (z - \beta t)/(1-\beta^2)^{1/2} ,$$

$$t' = (t - \beta z)/(1-\beta^2)^{1/2} .$$

(19)

Since the x and y variables are not affected by boosts

along the $z$ direction, and since the harmonic-oscillator system is separable in the Cartesian coordinate system, we can drop these variables from the wave function. It is important to note that $t$ and $t'$ in Eq. (19) are the time-separation variables between the quarks. It was shown that the ground-state harmonic-oscillator function for the moving hadron takes the form

$$\psi_\beta^0(z,t) = \left[\frac{1}{\pi}\right]^{1/2} \exp[-\Omega(z'^2 + t'^2)/2] . \tag{20}$$

$\Omega$ is the spring constant for the oscillator system. For simplicity, we can use the unit where $\Omega = 1$, and restore this factor when we are ready to compare our calculation with experimental data.

This wave function can be written in the light-cone coordinate system, where the coordinate variables are

$$u = (t+z)/\sqrt{2}, \quad v = (t-z)/\sqrt{2} , \tag{21}$$

and

$$q_u = (q_z - q_0)/\sqrt{2}, \quad q_v = (q_z + q_0)/\sqrt{2} . \tag{22}$$

In this coordinate system, the Lorentz boost of Eq. (19) takes a form

$$u' = \left[\frac{1-\beta}{1+\beta}\right]^{1/2} u, \quad v' = \left[\frac{1+\beta}{1-\beta}\right]^{1/2} v ,$$

$$q_u' = \left[\frac{1+\beta}{1-\beta}\right]^{1/2} q_u, \quad q_v' = \left[\frac{1-\beta}{1+\beta}\right]^{1/2} q_v . \tag{23}$$

The wave function of Eq. (20) then becomes

$$\psi_\beta(z,t) = \left[\frac{1}{\pi}\right]^{1/2} \exp[-(u'^2 + v'^2)/2]$$

$$= \left[\frac{1}{\pi}\right]^{1/2} \exp\left[-\frac{1}{2}\left[\frac{1-\beta}{1+\beta}u^2 + \frac{1+\beta}{1-\beta}v^2\right]\right] . \tag{24}$$

The Lorentz-deformation property of this wave function has been discussed in the literature.[16]

For the ground state, the PSD function can now be defined as[6]

$$W_\beta(u,q_u;v,q_v) = \left[\frac{1}{\pi}\right]^2 \int [\psi_\beta(u+x,v+y)]^*$$

$$\times \psi_\beta(u-x,v-y)$$

$$\times \exp[2i(q_u x + q_v y)]$$

$$\times dx\, dy . \tag{25}$$

After the evaluation of this integral, the PSD function becomes

$$W_\beta(u,q_u;v,q_v) = J_\beta(u,q_u) J_{-\beta}(v,q_v) , \tag{26}$$

with

$$J_\beta(u,q_u) = \frac{1}{\pi}\exp\left[-\left[\frac{1-\beta}{1+\beta}u^2 + \frac{1+\beta}{1-\beta}q_u^2\right]\right] .$$

The PSD function $W_\beta(u,q_u;v,q_v)$ can be separated into two phase spaces consisting of $(u,q_u)$ and $(v,q_v)$, respectively. When the hadron is at rest with $\beta=0$, this PSD function is localized in the regions

$$(u^2 + q_u^2) < 1, \quad (v^2 + q_v^2) < 1 . \tag{27}$$

When the hadron moves, these regions undergo elliptic deformations.[6]

It is straightforward to generalized the preceding calculation to the three-quark system in the harmonic-oscillator regime.[15,16] If we let $x_a$, $x_b$, and $x_c$ be the space-time coordinates of the quarks, it is more convenient to use the variables

$$X = \tfrac{1}{3}(x_a + x_b + x_c) ,$$

$$r = \tfrac{1}{6}(x_a + x_b - 2x_c), \quad s = \tfrac{1}{2}(x_b - x_a) , \tag{28}$$

and their conjugate variables,

$$P = p_a + p_b + p_c ,$$

$$q = p_a + p_b - 2p_c, \quad k = \sqrt{3}(p_b - p_a) . \tag{29}$$

In terms of these variables, the covariant harmonic-oscillator wave function for the three-particle bound system takes the form

$$\psi_\beta(r,s) = (1/\pi)\exp[-(\Omega/2)(r_z'^2 + r_0'^2 + s_z'^2 + s_0'^2)] , \tag{30}$$

where, as in Eq. (20), the transverse components have been ignored. The primed coordinate variables are those in the hadronic rest frame. This function can also be written in terms of the light-cone coordinate variables, just as in the case of Eq. (24). It is then straightforward to construct the covariant phase-space distribution function.

## V. CALCULATION OF THE FORM FACTOR

Let us now go back to the expression for the form factor in Eq. (10). Since the harmonic-oscillator model gives a reasonable description for the mass spectrum of non-strange baryons,[9] and since the shape of the proton structure function shows a Gaussian behavior where the experimental data are accurate,[11] we are compelled to calculate the nucleon form factors with the ground-state harmonic-oscillator wave function.

For a two-body bound state, the nonrelativistic calculation without the Lorentz-deformation effect gives the form factor of the form

$$g(K^2) = \exp(-K^2/4\Omega) . \tag{31}$$

We use $g(K^2)$, instead of $G(K^2)$, for the two-body bound state. This expression does not lead to a polynomial cutoff for large values of $K^2$, and therefore is not consistent with the real world as is described in Fig. 2. The story is the same for $G(K^2)$ for the three-body bound state.

We are interested in the question of whether the

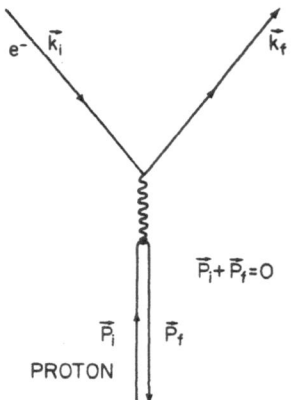

FIG. 3. Breit frame for electron-nucleon scattering. The momentum of the outgoing nucleon is equal in magnitude but opposite in direction to that of the incoming nucleon.

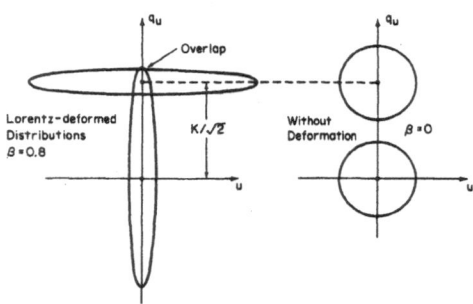

FIG. 4. Lorentz-deformed phase-space distribution functions and their overlaps. According to Eq. (27), the $J_\beta$ function of Eq. (26) is localized within a circular or elliptic region. As the momentum transfer increases, the PSD functions become separated. Without Lorentz deformation, the PSD functions become completely separated in the overlap integral of Eq. (35). This lack of overlap is the cause of an unacceptable exponential cutoff in $K^2$. However, the Lorentz-deformed PSD functions maintain a small overlapping region as $K^2$ increases. This leads to a polynomial decrease of the form factor.

Lorentz effect on the Gaussian distribution will lead to a dipole fit. For this purpose, let us go to the Lorentz frame in which the momenta of the incoming and outgoing nucleons have an equal magnitude but opposite signs, as is described in Fig. 3. Then

$$P_f + P_i = 0 . \tag{32}$$

The Lorentz frame in which this condition holds is usually called the Breit frame.[23] We assume that the proton comes in along the $z$ direction and goes out along the negative $z$ direction after the scattering process. In this frame, the four-vector $\mathbf{K} = (\mathbf{k}_f - \mathbf{k}_i) = (\mathbf{P}_i - \mathbf{P}_f)$ has no timelike component. Thus the exponential factor $\exp(-i\mathbf{K}\cdot\mathbf{x})$ can be replaced by the Lorentz-invariant form $\exp(-i\mathbf{K}\cdot\mathbf{x})$.

We can use the covariant harmonic-oscillator wave function discussed in Sec. IV for the proton. If we assume for simplicity that the proton is a bound state of two quarks, and the form factor should take the form

$$g(K^2) = \int \psi_f(x)\psi_i(x)e^{-i\mathbf{K}\cdot\mathbf{x}}d^4x . \tag{33}$$

Then the only difference between this form and the nonrelativistic cases is that the integral of Eq. (33) requires an integration over the timelike variable. This time-separation variable has been thoroughly discussed in the literature.[6,14,16] If $\beta$ is the velocity parameter for the incoming proton, $\psi_i$ and $\psi_f$ in Eq. (33) should be replaced by $\psi_\beta$ and $\psi_{-\beta}$ of Eq. (24), respectively. The form-factor integral in the Breit frame takes the form

$$g(K^2) = \int \psi^*_{-\beta}(z,t)\psi_\beta(z,t)e^{-iKz}dz\,dt . \tag{34}$$

In terms of the light-cone variables,

$$g(K^2) = \int \psi^*_{-\beta}(u,v)\psi_\beta(u,v)e^{-iK(u+v)/\sqrt{2}}du\,dv , \tag{35}$$

where $K$ is the magnitude of the vector $\mathbf{K}$. The form fac-

tor can then be computed from the overlap integral of two Lorentz-deformed PSD functions,

$$|g(K^2)|^2$$
$$= (2\pi)^2 \int W_{-\beta}(u,q_u;v,q_v)$$
$$\times W_\beta(u,(q_u - K\sqrt{2};v,(q_v - K/\sqrt{2}))$$
$$\times du\,dq_u\,dv\,dq_v . \tag{36}$$

Then, in terms of the $J_\beta$ function defined in Eq. (26), $g(K^2)$ takes the simpler form

$$g(K^2) = 2\pi \int J_{-\beta}(u,q_u)J_\beta(u,(q_u - K/\sqrt{2}))du\,dq_u . \tag{37}$$

This overlap integral of the PSD function is illustrated in Fig. 4. The evaluation of this integral is straightforward, and the result is

$$g(K^2) = [2M^2/(2M^2 + K^2)]$$
$$\times \exp\{-M^2K^2/[2\Omega(2M^2 + K^2)]\} . \tag{38}$$

This expression becomes the nonrelativistic form of Eq. (31) for small values of $K^2$ and becomes 1 for $K^2 = 0$. It decreases like $1/K^2$ as $K^2$ becomes large, but does not decrease like $1/(K^2)^2$. How are we going to get an extra $1/K^2$ factor?

Since there are three quarks inside the nucleon, there are two oscillator modes. We can therefore expect that each mode will contribute a $1/(K^2)$ factor to give the net decrease of $1/(K^2)^2$ as $K^2$ becomes very large. The generalization of the form factor calculation to this three-quark system is straightforward,[10,15,19] for the harmonic-oscillator wave functions. Since the oscillator wave functions are separable, the construction of the PSD function

2834    Y. S. KIM AND E. P. WIGNER    39

is also straightforward. The result of the calculation is

$$G(K^2)=[2M^2/(2M^2+K^2)]^2$$
$$\times \exp\{-M^2K^2/[\Omega(2M^2+K^2)]\} , \qquad (39)$$

which is 1 at $K^2=0$, and decreases as $1/(K^2)^2$. The behavior of this function is illustrated in Fig. 2.

The experimental curves are nicely summarized in Ref. 22. For protons, the data for the magnetic and electric form factors are available from $K^2=0$ to 25 and 15 $(GeV/c)^2$, respectively. For neutrons, the data for the magnetic form factor is available from $K^2=0$ to 7 $(GeV/c)^2$. All these data are consistent with the form given in Eq. (39) and illustrated in Fig. 2.

As for the charge form factor of neutrons, the coefficient to be multiplied to $G(K^2)$ of Eq. (39) is zero in

the harmonic-oscillator model in which only the ground state is taken into account. However, the observed neutron charge form factor is not zero for nonzero values of $K^2$. This is a clear indication that excited oscillator states should also be taken into account. This point has been discussed by Hussar and Haberman in their recent paper in the conventional harmonic-oscillator formalism.[24]

Throughout this paper, we ignored the effect of spins and assumed that the nucleon form factor can be decomposed into the form of Eq. (13) in the quark model. Indeed, there are models in which the quark spins can be combined for the nucleon to give the form of Eq. (13).[24,25] On the other hand, we still do not know how to treat the spins in the covariant phase-space formalism. This is a challenging future problem.

[1]E. P. Wigner, Phys. Rev. 40, 749 (1932); E. P. Wigner, in *Perspectives in Quantum Theory*, edited by W. Yourgrau and A. van der Merwe (MIT Press, Cambridge, MA, 1971).

[2]For a review article, see M. Hillery, R. F. O'Connell, M. O. Scully, and E. P. Wigner, Phys. Rev. 106, 121 (1984).

[3]R. G. Littlejohn, Phys. Rep. 138, 193 (1986); D. Han, Y. S. Kim, and M. E. Noz, Phys. Rev. A 37, 807 (1988); C. T. Lee, *ibid.* 38, 1230 (1988).

[4]For a discussion of overlapping distribution functions, see W. Schleich, D. F. Walls, and J. A. Wheeler, Phys. Rev. A 38, 1177 (1988). See also F. J. Narcowich, J. Math. Phys. 29, 2036 (1988).

[5]Y. S. Kim and E. P. Wigner, Phys. Rev. A 36, 1293 (1987).

[6]Y. S. Kim and E. P. Wigner, Phys. Rev. A 38, 1159 (1988).

[7]R. Hofstadter and R. W. McAllister, Phys. Rev. 98, 217 (1955); R. Hofstadter, Rev. Mod. Phys. 28, 214 (1956). See also R. Hofstadter, *Nuclear and Nucleon Structure* (Benjamin, New York, 1963).

[8]M. Gell-Mann, California Institute of Technology Report No. CTSL-20, 1961 (unpublished); Phys. Rev. 125, 1067 (1962); Phys. Lett. 13, 598 (1964).

[9]O. W. Greenberg, Phys. Rev. Lett. 133, 598 (1964); P. E. Hussar, Y. S. Kim, and M. E. Noz, Am. J. Phys. 48, 1043 (1980), and references contained therein.

[10]K. Fujimara, T. Kobayashi, and M. Namiki, Prog. Theor. Phys. 43, 73 (1970).

[11]P. E. Hussar, Phys. Rev. D 23, 2781 (1981).

[12]E. P. Wigner, Ann. Math. 40, 149 (1939); V. Bargmann and E. P. Wigner, Proc. Natl. Acad. Sci. U.S.A. 34, 211 (1948).

[13]P. A. M. Dirac, Proc. R. Soc. London, Ser. A 183, 284 (1945).

[14]H. Yukawa, Phys. Rev. 91, 415 (1953); 91, 416 (1953); M. Markov, Suppl. Nuovo Cimento 3, 760 (1956).

[15]R. P. Feynman, M. Kislinger, and F. Ravndal, Phys. Rev. D 3, 2706 (1971).

[16]Y. S. Kim and M. E. Noz, *Theory and Applications of the Poincaré Group* (Reidel, Dordrecht, 1986).

[17]S. Shlomo, J. Phys. A 16, 3463 (1983); H. J. Groeneworld, Physica 12, 405 (1946); J. E. Moyal, Proc. Cambridge Philos.

Soc. 45, 99 (1949); T. Takabayasi, Prog. Theor. Phys. 11, 341 (1954); J. P. Dahl, Phys. Scr. 25, 499 (1982).

[18]W. Frazer, *Elementary Particle Physics* (Prentice-Hall, Englewood Cliffs, NJ, 1966).

[19]K. Fujimara, T. Kobayashi, and M. Namiki, Prog. Theor. Phys. 44, 193 (1970); S. Ishida, *ibid.* 46, 1570 (1971); 46, 1905 (1971); R. G. Lipes, Phys. Rev. D 5, 2849 (1972); Y. S. Kim and M. E. Noz, *ibid.* 8, 3521 (1973); M. L. Haberman, *ibid.* 29, 1412 (1984); M. A. Maize and Y. E. Kim, Phys. Rev. C 35, 1060 (1987).

[20]M. A. B. Beg, B. W. Lee, and A. Pais, Phys. Rev. Lett. 13, 514 (1964).

[21]The charge of the neutron vanishes. However, the neutron form factor for nonvanishing $K^2$ does not vanish. This is believed to be due to the departure from the exact SU(6) symmetry scheme for the quark model. Although this is a very interesting current research problem, particularly in relation to the detailed study of the proton-neutron structure function ratio, we shall not go into the subject in the present paper. For a discussion of this problem, see A. Le Yaouanc, L. Oliver, O. Pene, and J. C. Raynal, Phys. Rev. D 18, 1733 (1978).

[22]D. P. Stanley and D. Robson, Phys. Rev. D 26, 223 (1982).

[23]When we do relativistic physics, we usually perform calculations in a fixed Lorentz frame, such as the center-of-mass frame. The condition for the Breit frame given in Eq. (32) is different from that for the center-of-mass system. What enables us to put an arbitrary condition on the momentum? The answer to this question is very simple. Lorentz boosts can be made in three different directions. Therefore it is possible to impose three independent conditions without loss of generality. Equation (32) indeed contains three independent conditions.

[24]M. L. Haberman and P. E. Hussar, Z. Phys. C 40, 153 (1988).

[25]A. B. Henriques, B. H. Kellet, and R. G. Moorhouse, Ann. Phys. (N.Y.) 93, 125 (1975); S. Ishida, A. Matsuda, and M. Namiki, Prog. Theor. Phys. 57, 210 (1977).

# Bibliography of Papers
# on Physical Chemistry

## Papers Reprinted in Volume IV

### 1923

(With A. Szegvari) Über elektrische Erscheinungen bei Stäbchensolen. Kolloid-Zeitschrift *33*, 218–222 (1923)

### 1924

(With H. Mark) Die Gitterstruktur des rhombischen Schwefels. Zeitschrift für physikalische Chemie *111*, 398–414 (1924)

### 1925

(With M. Polanyi) Bildung und Zerfall von Molekülen. Zeitschrift für Physik *33*, 429–434 (1925)

### 1928

(With M. Polanyi) Über die Interferenz von Eigenschwingungen als Ursache von Energieschwankungen und chemischer Umsetzungen. Zeitschrift für physikalische Chemie A (Haber-Band) *43*, 439–452 (1928)

### 1929

The Statistics of Composite Systems According to the New Quantum Mechanics. (Original in Hungarian with German Abstract). Összetett Rendszerek Statisztikája az Új Quantum-Mechanika Szerint. Magyar Tudományos Akadémia Matematikai és Természettudományi Értesitöje *45*, 576–582 (1929). Translated from Hungarian by Akos Sebestyén

### 1932

(With H. Pelzer) Über die Geschwindigkeitskonstante von Austauschreaktionen. Zeitschrift für physikalische Chemie B *15*, 445–471 (1932)

Über das Überschreiten von Potentialschwellen bei chemischen Reaktionen. Zeitschrift für physikalische Chemie B *19*, 203–216 (1932)

On the Quantum Correction for Thermodynamic Equilibrium. Physical Review *40*, 749–759 (1932)

Contributions to the Theory of the Neutron. (Original in Hungarian with a German Abstract). Adalékok a Neutron Elméletéhez. Magyar Tudományos Akadémia Matematikai és Természettudományi Értesitöje *49*, 142–146 (1932). Translated from Hungarian by Akos Sebestyén. Editor's Note: The neutron under discussion here is the "light" neutron, now called neutrino.

## 1933

Über die paramagnetische Umwandlung von Para-Orthowasserstoff. III. Zeitschrift für physikalische Chemie B *23*, 28–32 (1933)

## 1935

On a Modification of the Rayleigh-Schrödinger Perturbation Theory. Magyar Tudományos Akadémia Matematikai és Természettudományi Értesitöje *53*, 477–482 (1935). (Pages 475 and 476 are in Hungarian and have been omitted.)

## 1936

(With L. Farkas) Calculation of the Rate of Elementary Reactions of Light and Heavy Hydrogen. Transactions of the Faraday Society *32*, 708–723 (1936)

## 1937

(With H. Eyring) On the Rate of Chemical Reactions. Scientific Monthly *44*, 564–567 (1937)

Calculation of the Rate of Elementary Association Reactions. Journal of Chemical Physics *5*, 720–725 (1937)

## 1938

The Transition State Method. Transactions of the Faraday Society *34*, 29–41 (1938)

## 1939

(With J. O. Hirschfelder) Some Quantum-Mechanical Considerations in the Theory of Reactions Involving an Activation Energy. Journal of Chemical Physics *7*, 616–628 (1939)

Some Remarks on the Theory of Reaction Rates. Journal of Chemical Physics *7*, 646–652 (1939)

## 1942

(With C.W. Ufford) On the Calculation of the Distribution Function. Physical Review *61*, 524–527 (1942)

## 1953

(With T. Teichmann) Electromagnetic Field Expansions in Loss-Free Cavities Excited Through Holes. Journal of Applied Physics *24*, 262–267 (1953)

## 1954

The Problem of Multiple Scattering. Physical Review *94*, 17–25 (1954)
Derivations of Onsager's Reciprocal Relations. Journal of Chemical Physics *22*, 1912–1915 (1954)

## 1955

Lower Limit for the Energy Derivative of the Scattering Phase Shift. Physical Review *98*, 145–147 (1955)

## 1963

Review of Collision Theory. In: Transfert d'energie dans les gaz. Douzième conseil de chimie 1962 (R. Stoops., ed.). Interscience Publishers, New York 1963, pp. 211–239, 303–306, 308, 453–457, 515–516. Translated into Hungarian: Áttekintés az ütközések elméletérol. Fizikai Szemle *14*, 35–44 (1964)

## 1971

Quantum-Mechanical Distribution Functions Revisited. In: Perspectives in Quantum Theory; Essays in Honor of Alfred Landé (W. Yourgrau, A. van der Merwe, eds.). M.I.T. Press, Cambridge, MA, 1971, pp. 25–36

## 1981

(With R. F. O'Connell) Quantum-Mechanical Distribution Functions: Conditions for Uniqueness. Physics Letters *83A*, 145–148 (1981)
(With R. F. O'Connell) Some Properties of a Non-negative Quantum-Mechanical Distribution Function. Physics Letters *85A*, 121–126 (1981)

## 1984

(With M. Hillery, R. F. O'Connell and M. O. Scully) Distribution Functions in Physics: Fundamentals. Physics Reports *106*, 121–167 (1984)
(With R. F. O'Connell) Manifestations of Bose and Fermi Statistics on the Quantum Distribution Functions for Systems of Spin-0 and Spin-$\frac{1}{2}$ Particles. Physical Review A *30*, 2613–2618 (1984)

## 1987

The General Properties of the Distribution Function and Remarks on Its Weakness. In: The Physics of Phase Space (Y. S. Kim, W. W. Zachary, eds.). Lecture Notes in Physics, vol. 278. Springer, New York 1987, pp. 162–170

**1988**

(With Y. S. Kim) Covariant Phase-Space Representation for Harmonic Oscillators. Physical Review A *38*, 1159–1167 (1988)

**1989**

(With Y. S. Kim) Covariant Phase-Space Representation and Overlapping Distribution Functions. Physical Review A *39*, 2829–2834 (1989)

## Papers Related But Reprinted in Other Volumes of The Collected Works

(With Y. S. Kim) Cylindrical Group and Massless Particles. Journal of Mathematical Physics *28*, 1175–1179 (1987); Vol. III, Part I
(With Y. S. Kim) Covariant Phase-Space Representation for Localized Light Waves. Physical Review A *36*, 1293–1297 (1987); Vol. III, Part I

## Papers Related But Not Reprinted in The Collected Works

New Theory of Chemical Binding. In Hungarian: A kémiai kötés újabb lemétete. Chemische Rundschau für Mitteleuropa und den Balkan *VI*, 24–27 (1929)

# PART II

## Solid State Physics

# Wigner on Solid State Physics

Annotation by Walter Kohn

## I. Introduction

Although the published work of Eugene Wigner on the theory of solids essentially extends over only 6 years, from 1933 to 1938 (not including three overview papers with F. Seitz between 1953 and 1956), and consists of only 12 papers out of a total of over 500, Wigner is justly regarded as one of the fathers of the modern theory of the electronic structure of metals.

Important insights had preceded his work on the electronic structure of metals: The realization that the Pauli exclusion principle is essential for an understanding of metals (Pauli and Sommerfeld); the origin of the strong magnetic coupling in ferromagnets (Heisenberg); the importance of energy band structure (Bloch). What Wigner and his students brought to this subject was, primarily, quantitative quantum mechanical theories, needed to calculate *total electronic energies*. From these cohesive energies, structures and elastic coefficients could be calculated. Wigner's interest in energies, and derivative quantities, brings to mind his background in chemistry, where energies play a dominant role. Much of the work [1–4] has a distinctly practical character. One senses that, as a former engineer, he was after a good, at least semiquantitative answer, and was ready to use any tools which would lead there. The problem was formidable because of the combined major effects of strong ionic potentials, the Pauli exclusion principle and strong electron-electron interaction effects. There is no small parameter which would permit systematic perturbation expansions, so in many places Wigner had to rely on brilliant intuitive insights (see especially [2]). In the course of his studies of the alkali metals, Wigner was led to investigate the idealized model (later known as the jellium model) in which electrons move in a large volume filled by a static, uniform charge background. This model had been previously considered by Pauli, Sommerfeld and others under neglect of the dynamical effects of the electron-electron interaction (the so-called free electron model). Wigner realized that for estimates of the cohesive energies these dynamical effects, called correlation energies, could not be neglected. Partly together with his student F. Seitz, he succeeded, by hook and by crook, in successively improved calculations, to obtain this correlation energy $E_c$, as function of the electronic density ([2–4]). This work, which has withstood the test of time astonishingly well, may be considered the beginning of a later very important field of physics – *many body theory*, – the theory of very many strongly interacting particles (atoms in classical and quantum liquids, electrons in solids, nucleons in nuclei, etc.). Wigner's correlation energy (or recent slight improvements) also enters the local density approximation of density functional theory for atoms, molecules and solids.

A significant by-product of Wigner's studies of interacting electrons in the presence of a uniform positive charge background was his realization that, while at high densities the electrons are in a gas/liquid phase, at sufficiently low densities and low temperatures the phase of lowest free energy is a crystalline phase (now called the Wigner crystal) [4], in which the electrons localize themselves near the lattice points of a closepacked lattice. After many decades of search such Wigner – (or closely related) crystals have been discovered for electrons on He-surfaces and for ions in non-neutral plasmas in strong magnetic fields. Apart from its intrinsic interest, the Wigner crystal which owes its very existence to the Coulomb interactions between electrons, highlighted the importance of these interactions.

In 1953 – 15 years after his last research paper in solid state physics – Wigner, together with Seitz, published a review entitled *Qualitative Theory of Cohesion of Metals*, an attempt to apply the lessons learned in their classic work on the alkalis, to metals in general [5]. As the word "qualitative" in the title suggests, this was a rather sketchy overview which did not introduce significant new concepts and has had little impact on the field.

Another outgrowth of Wigner's work on the electronic structure of the alkali metals was his work, with J. Bardeen, on the electronic structure of their surfaces. Because translational symmetry is lost in the presence of a surface, this problem is even more difficult than that of the bulk metals. Nevertheless Wigner and Bardeen correctly recognized the essential role of the exchange energy and were able to calculate, with semiquantitative success, the electronic work functions [6,7]. This constituted the first serious theoretical insight into the electronic structure of metal surfaces, which was of enormous technical interest in that age of the vacuum tube. The first significant further advance in this area did not occur until more than three decades later.

An important contribution to solid state theory of a very different character was the elucidation of the symmetry properties of the wave functions in crystals (i.e. Bloch waves) by Wigner and collaborators [8]. This work, on the one hand, was a rather straightforward "exercise" in group theory, for someone with Wigner's profound command of the subject; on the other hand it established the principles for all subsequent work dealing with the rich variety of different topologies of energy bands in crystals.

During World War II Wigner played a leadership role in the physics and engineering of chain reactions (see Vol. V). A critical issue was the structural damage caused to reactor components (at the time called "Wigner's disease") by neutrons and other fission products. In unpublished work Wigner estimated that, in a wide class of materials, an atom would be permanently displaced from its lattice site if an energy $\sim 25\,\mathrm{eV}$ was imparted to it. This "Wigner threshold" provided an important design guideline. Following the war, the field of "radiation damage", was extremely active for one or two decades, with much of the leadership coming from former members of Wigner's wartime group, especially F. Seitz (see [9]).

In conclusion, I would like to speak of the great influence of Wigner on the subsequent development of solid state theory in the US and elsewhere. In the US the field was initiated at approximately the same time, the early 1930's, by three towering figures – J. C. Slater, J. H. Van Vleck and E. P. Wigner – in close and cordial contact with each other. Of these probably Slater was the broadest. Van Vleck's main contribution was his brilliant quantum mechanical elucidation of magnetism in solids, for which – about 50 years later – he received the Nobel prize. In my view Wigner may have been the deepest, the first to really understand, physically and mathematically, the dynamics of electrons under the combined influence of the periodic crystal potential, the Pauli principle and their mutual Coulomb repulsion.

All three had some excellent students. Three of Wigner's students became highly influential, during the 1940's and 50's and beyond: John Bardeen, Conyers Herring and Fred Seitz. Bardeen and Herring who joined the Bell Telephone Laboratories in the middle 40's played an essential role (together with W. Shockley, P.W. Anderson and others) in creating there what was, by common consensus, regarded as the word's leading solid state group. Bardeen, Shockley and the experimentalist W. Brattain developed the transistor in 1948, for which they received the Nobel prize. Seitz in 1940 published *Modern Theory of Solids*, which many regard as the defining publication for this field. At the University of Pennsylvania, the Carnegie Institute of Technology and, for many years, at the University of Illinois he was the key person to establish three of the strongest early University groups in solid state physics. In the 50's Bardeen moved from Bell Labs to the University of Illinois where, in 1957, he, L. N. Cooper and R. Schrieffer published the now famous BCS theory of superconductivity. In due course, this earned them the Nobel prize, the second for Bardeen. The writer of these lines (who, through a misunderstanding, missed becoming Wigner's Ph.D. student) can attest from personal experience the inspiration which Wigner instilled in students and younger colleagues, by sharing with them his deep physical and mathematical insights into the nature of solids.

# II. Commentaries

• (with F. Seitz), *On the Constitution of Metallic Sodium*, 1933 [1]
• (with F. Seitz), *On the Constitution of Metallic Sodium*, 1934 [2]
• *On the Interaction of Electrons in Metals*, 1934 [3]

These three papers represent the beginning of the many-electron theory of metals. The problem was formidable. As Wigner knew well from his acquaintance with the quantum theory of small atoms and molecules, any serious estimates of the energy as function of the nuclear positions – which is required for the calculation of lattice parameters, cohesive energies and elastic constants – had to go beyond mean field theories (Sommerfeld, Hartree, Hartree-Fock) and include the effects of dynamical correlations. The Rayleigh-Ritz variational method had resulted in extremely accurate calculations for the He-atom (Hylleraas) and the hydrogen molecule (James and Coolidge), but the effort increased very rapidly with the number of atoms involved, and the method was totally inapplicable to the case of an "infinite" metal.

In the absence of a systematic quantitative procedure, the authors proceeded largely heuristically [1]. They pointed out that in a metal each electron is surrounded by a neutralizing hole of total charge $+e$ in the charge distribution of the other electrons. They calculated this hole in the Hartree-Fock (HF) approximation for the case of a uniform electron gas. (There is no such hole in the Hartree approximation). They noted, however, that in the HF approximation the hole was due to *statistical* correlations of electrons of parallel spin and that the important *dynamical* correlations due to the electron-electron repulsion, which affected electrons of both parallel and antiparallel spin were ignored. In the event they adopted the heuristic viewpoint that any electron, when located in a particular atomic cell, say $j$, excluded all other electrons from that cell, while the ions of the other cells were perfectly screened by the charges of the other electrons. This led them to an effective electronic potential

$$V(r) = V_k(r), r \in \Omega_k$$

where $k$ denotes the cell, $\Omega_k$ is the volume surrounding ion $k$, and $V_k(r)$ is an effective potential for the valence electron in *atomic* sodium. They pointed out that the lowest energy wavefunction of the $3s$-band in metallic sodium differed from the atomic $3s$-function by the boundary condition

$$n \cdot \bigtriangledown \psi = 0 \text{ on all } S_k$$

where $S_k$ is the boundary of cell $k$, and $n$ is the normal to that boundry. Because of the periodicity of the lattice, the Schrödinger equation needed to be solved in only one cell, say $k = 0$, and to do this the authors replaced the actual, nearly spherical, cell $\Omega_0$ by a equivolumic sphere of radius $r_s$ (later known as the Wigner-Seitz sphere and radius). To allow for the exclusion principle they added to the energy $E_0$, so calculated, the mean Fermi-energy, $E_F$, of a uniform electron gas of density $n \equiv (\frac{4\pi}{3}r_s^3)^{-1}$. The total energy per ion, $E_{\text{tot}}(r_s)$ was taken to be the sum of $E_0$ and $E_F$ and minimized with respect to $r_s$. This yielded a lattice parameter, cohesive energy and compressibility of 4.2 Å, 25.6 kilocalories/mole and $1.6 \times 10^{-11}$ c.g.s. units compared to the experimental values (at the time) of 4.23, 26.9 and $1 \times 10^{-11}$; the agreement was embarrassingly good.

In retrospect, we can say that this first paper identifies all major issues and deals with them in a rough yet very judicious manner. The excellent agreement with experiment, which the authors themselves attributed to compensating errors, nevertheless showed that all major aspects of the problem had been reasonably treated. Paper [2] with the same title as [1], goes over the same ground but more systematically. It examines at length the question of the Fermi energy, a relatively minor issue leading to a small correction of the free electron result obtained in [1]. The main contribution of this paper is the first serious attack on the problem of the *correlation energy*, the change in total energy due to electronic correlations resulting from their mutual repulsion. The authors realized that this energy was due mostly to the fact that electrons of anti-parallel spin would be kept apart, since, even without repulsion, those with parallel spin are kept apart by the Pauli exclusion principle. They made the inspired Ansatz Eq. (22), [2] for the many-electron wave-function, notwithstanding its violation of strict anti-symmetry. Further inspiration was needed to deal with the functions $\psi_\nu(y_1, \cdots y_n; x_1)$ occurring in this Ansatz. So the authors assumed that, in dealing with the spin-up electrons of coordinate $x_1$, an appropriate "mean configuration" for the spin down electrons $y_m$, was a close-packed lattice occupied by pairs of the latter! With this brilliantly outrageous assumption, the correlation energy for a uniform electron gas was approximately calculated as a function of $r_s$. When this was added to the appropriate Hartree energy and exchange energy, and minimized with respect to $r_s$, a lattice parameter of 4.75 Å and cohesive energy of 26.9 kcal were obtained. (The lattice parameter is now in much poorer agreement with experiment than in [1].)

In [3], Wigner returns once again to the crucial issue of the correlation energy. This time, in considering the function $\psi_1(y_1, \cdots y_n; x_1)$, instead of assuming that the $y_k$ constitute a close-packed lattice of electron pairs, he makes the more rational Ansatz

$$\psi_1(y_1, \cdots y_n; x_1) = \psi_\nu(x_1) \left( 1 + \sum_{m=1}^n f_\nu(y_m - x_1) \right),$$

$\psi_\nu(x_1)$ is a plane wave and the functions $f_\nu$, expected to be small, short range and negative, are to describe the effect of the repulsion. (This is closely related to what much later became known as the Jastrow Ansatz.) This time the distribution of the $y_m$ is taken to be that in a Slater-determinant of plane waves. $f_\nu$ is calculated by second order perturbation theory on the assumption that it is small.

Wigner who draws attention to the remaining uncertainties of the calculation, reassures himself, and the reader, by examining the *known* correlation energy of small atoms and the low density limit of the metallic correlation energy. He points out that in the limit ($r_s \to \infty$) the kinetic energy of the electrons ($\alpha r_s^{-2}$) becomes negligible compared to their potential energy ($\propto r_s^{-1}$) and that to minimize the latter the electrons will occupy the points of a close-packed lattice, later known as the Wigner lattice. The correlation energy in this limit could be simply, reliably and accurately estimated as $E_c = 0.292 e^2/r_s$ per electron.

The theoretical estimates for lattice parameter and cohesive energy resulting from this paper's correlation energy were 4.62 Å and 26.1 kcals/mole. The agreement with experiment was similar to [2].

Nearly 60 years after [3], Wigner's calculation of the correlation energy has been improved by only a few percent. Current theoretical values for the lattice parameter and cohesive energy are $4.07 \, \mathring{A}$ and $25.8 \, \mathrm{kcals/mole}$. The best experimental values are $4.22 \, \mathring{A}$ and $26.1 \, \mathrm{kcals}$. The agreement with experiment of the lattice parameter is significantly better than it was in [3] (today theoretical lattice parameters are generally accurate); the agreement of the cohesive energy is similar to what it was then. (In view of the considerable error of the lattice parameter in, [3] the *exact* agreement of the cohesive energy ($26.1 \, \mathrm{kcal/mole}$) must be regarded as coincidental.)

- *On the Structure of Solid Bodies*, 1936 [10].

This is a very elementary, qualitative overview of the principal classes of solids, molecular, metallic, valence and ionic. It was written shortly after Wigner and his collaborators had clearly understood the nature of metallic cohesion. The cohesive energies of ionic and molecular solids (like NaCl and He) had been previously understood, and those of valence solids (like Si) had just been clarified by the work of Pauling and others on the chemical bond.

In the introduction Wigner presents his view of solid state physics, in relation to physics as a whole. According to Wigner physics research proceeds on two fronts. He sees one front, in the main the physics of nuclear phenomena, as leading to "important changes in our fundamental concepts". The efforts on the other front, including the physics of solids, are "...directed toward the deepening and broadening of our knowledge of phenomena which, we believe, *can* be understood on the basis of existing concepts and theories. Doubtless the second front is of less importance. It rarely leads to fundamental discoveries in physics proper..." Would Wigner still maintain this view after the discovery of electron-pairing superconductivity, the Aharonov-Bohm effect, universality classes in critical phenomena, the quantum Hall effect, etc.?

- (with J. Bardeen), *Theory of the Work Function of Monovalent Metals*, 1935 [6]

This paper, [6] and its sequel, [7], by J. Bardeen on the surface dipole barrier, are the foundation of the theory of electronic structure of metallic surfaces. The modest introduction, "It is perhaps not quite superfluous to have an approximate treatment which merely shows how the quantity in question is determined···", is followed by a clarification of the issues involved. In current terminology, the work function $\varphi$ can be written as $\varphi = eD - \mu$, where $eD$ is the surface dipole barrier and $\mu$ the internal chemical potential. Paper [6] identifies the two parts and estimates the quantity $\mu$; a semiquantitative estimate of $D$ is provided in [7].

The calculation of $\mu$ takes advantage of the previous calculations by Wigner and Seitz [1–3] of the cohesive energies of the alkali metals. For the present purpose the energy of an alkali metal is regarded as a function of the number of ions, $n_i$, and electrons, $n_e$: $E = E(n_i, n_e)$. The total energy, of a neutral alkali crystal, the subject of the earlier papers is $E(n_i, n_i)$ and the workfunction is the energy required to remove an electron, was taken to be $\varphi = E(n_i, n_{i-1}) - E(n_i, n_i) = [\partial E(n_i, n_e)/\partial n_e]_{n_e=n_i}$. The agreement with existing experimental data was, somewhat deceptively, excellent when $D$ was neglected, but it did establish that for these metals $D$ was small and that all main contributions to $\varphi$ (kinetic energy, exchange energy,...) had been properly included.

• (with H.B. Huntington), *On the Possibility of a Metallic Modification of Hydrogen* 1935 [11]

For Wigner, the chemist, it was natural – after the work on the alkali metals – to consider hydrogen, which likewise lies in group I of the periodic table. (Note that this paper was published in the Journal of Chemical Physics.) At the same time the different character of solid hydrogen was obvious: At standard temperature and pressure the alkalis crystallize in Bravais lattices, one atom per unit cell, and are metallic, while hydrogen forms an insulating molecular crystal, bound by Van der Waals forces between the hydrogen molecules. Following an idea of J. D. Bernal, the authors explored the question if, under sufficiently high pressure, hydrogen would undergo a transition into a metallic phase.

As the authors state, "The calculation itself was a rather mechanical application of the principles outlined elsewhere," i.e. earlier papers by Wigner and Seitz (and one by J.C. Slater). But the questions raised and the, mostly tentative, answers offered by the authors – specific to hydrogen – were, in retrospect, the most pertinent. What the calculation showed beyond doubt was that, at zero pressure, an assumed metallic phase is *much* less stable (by about 40 kcal) and many times denser than the known molecular phase. The interesting question whether under pressure – and, if so, at what pressure – the metallic phase becomes stable is raised but not answered. Unavailable, at the time, was the energy as a function of volume of the *molecular* phase, which of course, is also needed for this purpose. The authors do, however, quote a lower estimate of 250 kbar. Today the metallic phase has still not been conclusively identified in the laboratory. But both theory and experiment suggest very strongly that it should be stable at pressures of about 2–4 Mbar. Finally the authors advance chemical reasoning to suggest the possible existence of other allotropic forms, specifically layer structures. Later theoretical research has confirmed that the energy differences between several "exotic" allotropic structures are quite small. Finally, there is today good reason to believe that the hydrogen on the planet Jupiter is indeed metallic.

• (with L.B. Bouckaert and R. Smoluchowski), *Theory of Brillouin Zones and Symmetry Properties of Wave Functions in Crystals* 1936 [8]

This paper presents the group theoretic implications of crystal space groups for the symmetry properties of energy bands and eigenfunctions, and for their topological properties as function of $k$ [12]. A representation of the space group is labeled by $k^2$, and by the representation of the small point group, which leaves $k$ invariant. A useful concept is the "star of $k$", the totality of all distinct $k$-vectors in the principal Brillouin zone arising from the original $k$ by application of the elements of the full point group. Energy bands are characterized by the representations of the small groups associated with $k$ values at symmetry points in the Brillouin zone. The topology of the energy bands *near* such symmetry points is elucidated. Compatibility rules are established between possible representations at points of high symmetry in $k$-space (e.g. $(0,0,0)$) and on connected manifolds of lower symmetry (e.g. $(k,0,0)$). From this the question of "sticking together" of bands (e.g. the $p_x, p_y, p_z$ – bands of a composite $p$-band at $k = (0,0,0)$) is clarified. Any physicist who has seen the tousled "spaghettis" $(E_n(k))$ of typical band structure calculations (e.g. hybridized $s$-$d$ bands) appreciates the value of the group theoretic ordering principles so clearly presented in this paper.

• *On the Constant A in Richardson's Equation*, 1936 [13].

The standard expression for the thermionic emission current from a metal was (and is) Richardson's Equation

$$i = AT^2 e^{-W/RT} ,$$

where $W$ is the work function of the metal and – in simple models – $A$ has the *universal* value of 120 amp $cm^{-2} K^{-2}$. In practive the *form* of this expression is well confirmed by experiment but fitting experimental data gives values of $A$ that differ from the canonical "120" by factors ranging typically from $10^{-1}$ to $10^{+1}$. The simplest extensions of Richardson's model are reflection effects, replacing $A$ by $A(1 - r)$, obviously unable to account for *high* values of $A$; and temperature dependence of the work-function, $W = W_0 + \gamma(T - T_0)$, resulting in $A \rightarrow Ae^{-\gamma/R}$. In this paper Wigner estimates the temperature dependence of $W$ due to thermal expansion and, while unable to make quantitative estimates obtains changes which are of the right order of magnitude to account for the experimental data. However, since several other important effects are ignored (e.g. temperature dependent spacing between surface layers) little can be concluded except, as the author states, "that it is purely accidental if the (experimental) constant $A$ in Richardson's equation… has the value 120 amp $cm^{-2} K^{-2}$".

• *Effects of the Electron Interaction on the Energy Levels of Electrons in Metals*, 1938 [4]

Here Wigner continues his considerations of interaction effects of metallic electrons in a positive uniform charge background begun in earlier papers, especially [3].

He first addresses two issues. Does a uniform interacting electron gas become ferromagnetic? He concludes that, while without exchange and correlation effects the answer is no, for all electron densities; when exchange is included the ferromagnetic state becomes stable at low densities ($r_s \geq 5.47$); and when correlation effects are also included it appears likely that the non-magnetic state is again stable for all densities.

Secondly he addresses the question of the density of low-lying excited states, which determines the low-temperature electronic specific heat. In simple Hartree/Sommerfeld theory, which ignores exchange and correlation, the density of states at $T = 0$ is energy independent in accord with the $T^1$ behavior of the electronic specific heat. However, when exchange is included the level density tends to zero with the excitation energy, and – as had been noted by J. Bardeen – the agreement with the experimentally linear specific heat is lost. Wigner discusses how correlation effects might save the day but does not actually solve the problem.

Once again [3] he is led to consider electrons at low densities, where correlation effects become large, and, in particular, the electron *lattice* (today called the Wigner lattice) which eventually becomes the stable phase. Because of his concern with specific heats he now considers also the vibrations of the lattice. But the problem of the wrong $T$-dependence of the specific heat remains. Later many-body theories showed that the electron-electron interaction in the gas phase (relevant to physical metals) is effectively screened at large distances, which restores the correct linear $T$-dependence of the specific heat.

• *Qualitative Theory of the Cohesion in Metals*, 1953 [5]

This paper, presented 15 years after Wigner's last published research paper on metals [4], offers qualitative comparative ideas about the cohesion of various classes of metals (alkalis, alkaline earths, transition metals,···). Metallic cohesion is, as much as possible, related to the atomic energy levels of the constituent element. However, Wigner did no further work in this area to firm up these ideas. The important work of J. Friedel and co-workers in the 1960's on transition metals, using tight binding models, is in a similar spirit.

• (with F. Seitz), *The Effect of Radiation on Solids*, 1956 [9]

Radiation effects – usually damage – in solids became a practically extremely important issue in connection with nuclear reactors, some of whose components are subject to intense radiation by neutrons and other particles. This paper explains in simple terms how radiation can permanently displace crystal atoms from their equilibrium positions. It concludes by speaking of the "promise to provide us with useful information about the solid state in general". In the writer's view this promise has since been fulfilled on a modest scale.

# References and Footnotes

[1] (with F. Seitz) On the constitution of metallic sodium. Phys. Rev. *42*, 804–810 (1933); Vol. IV, Part II

[2] (with F. Seitz) On the constitution of metalic sodium II. Phys. Rev. *46*, 509–524 (1934); Vol. IV, Part II

[3] On the interaction of electrons in metals. Phys. Rev. *46*, 1002–1011 (1934); Vol. IV, Part II

[4] Effects of the electron interaction on the energy levels of electrons in metals. Transactions of the Faraday Society *34*, 678–685 (1938); Vol. IV, Part II

[5] Qualitative theory of the cohesion in metals, Proceedings, International Conference of Theoretical Physics, Kyoto and Tokyo, 649–663 (1953); Vol. IV, Part II. (Reprinted in slightly expanded form with F. Seitz, Solid State Physics *1*, 97–126 (1955), Academic Press)

[6] (with J. Bardeen) Theory of the work function of monovalent metals. Phys. Rev. *48*, 84–87 (1935); Vol. IV, Part II

[7] J. Bardeen: Theory of the work function II. The surface double layer. Phys. Rev. 653–663 (1936)

[8] (with L.B. Bouckaert, R. Smoluchowski) Theory of Brillouin zones and symmetry properties of wave functions in crystals. Phys. Rev. *50*, 58–67 (1936); Vol. IV, Part II

[9] (with F. Seitz) The effect of radiation on solids. Scientific American *195*, no. 2, 76–84 (1975); Vol. IV, Part II

[10] On the structure of solid bodies. Scientific Monthly *39*, 415–419 (1936)

[11] (with H.B. Huntington) On the possibility of a metallic modification of hydrogen. Journal of Chemical Physics *3*, 764–770 (1935); Vol. IV, Part II

[12] The paper restricts itself to crystals without glide planes and scalar axes. For such crystals the space group is the direct product of a translation group and a point group, more precisely, the group generated by the $\exp(ik\Gamma_j)$, where $\Gamma_j$ are fundamental translation vectors

[13] On the constant $A$ in Richardson's equation. Phys. Rev. *49*, 696–700 (1936); Vol. IV, Part II

# On the Constitution of Metallic Sodium

E. P. Wigner and F. Seitz

Physical Review *43*, 804–810 (1933)

(Received March 18, 1933)

Previous developments in the theory of metals may be divided clearly into two parts: that based principally upon the hypothesis of free electrons and dealing with conductivity properties, and that based upon calculations of valence forces and dealing with the chemical properties. In the present article an intermediate point of view is adopted and the free-electron picture is employed in an investigation of chemical properties of metallic sodium. The assumption is made that in the metal the $K$ and $L$ shells of an atom are not altered from their form in the free atom. The properties of the wave functions of the electrons are discussed qualitatively, first of all, and it is concluded that the binding energy will be positive even when the Pauli principle is taken account of. This is followed by a quantitative investigation of the energy to be associated with the lowest state. First of all it is shown to what extent the present picture takes account of the interactions of electrons with both parallel and antiparallel spins, and to what extent remaining effects may be neglected. Next a Schroedinger equation is solved in order to determine the lowest energy level for various values of the lattice constant. To this a correction is made to account for the Pauli principle and from the result the lattice constant, binding energy and compressibility are calculated with favorable results.

## I.

THE investigations which have been carried out so far on the constitution of metals by quantum mechanics may be divided into two classes, the work on conductivity and related phenomena, carried out chiefly by Bloch, Peierls, Nordheim and Brillouin[1] are mainly based on the hypothesis of free electrons[2] and are concerned with the interaction between the electronic motion and the vibrations of the lattice, which is responsible for the electric resistance. The works of the other class[3] are mainly concerned with the chemical properties and crystal structure of the metals and are based on calculations of valence forces. They encounter great mathematical difficulties because the application of the usual methods to calculate valence forces becomes more and more difficult as the number of atoms increases.

The present work intends to take an intermediate point of view by applying the free electron picture but aiming at a calculation of chemical properties of metallic sodium such as lattice constant, heat of vaporization, compressibility, etc. The method of calculation is the same one as that proposed by Hund for molecules[4] and more recently applied by Lenz and Jensen[5] to ionic lattices, and by Lennard-Jones and H. J. Woods[6] to two dimensional metallic lattices. The electrons are assumed to move freely in a potential field and their interaction is supposed to be contained to a large extent in this field, much as in Hartree's method of the self-consistent field which is actually the field adopted in the calculations of Lenz (not in ours). The initial assumptions which one makes about the statistical connections of positions of different electrons are necessarily rather rough in this picture and should be improved afterwards.

## II.

We assume first that the electrons in the $K$ and $L$ shells are not affected by the metallic bond and their wave functions the same as in the

[1] Cf. the comprehensive treatment by L. Brillouin, Die Quantenstatistik. Berlin, 1932.

[2] Cf. W. Pauli, Zeits. f. Physik **41**, 81 (1927); A. Sommerfeld, Zeits. f. Physik **47**, 1 (1928).

[3] J. C. Slater, Phys. Rev. **35**, 509 (1930); E. A. Hylleraas, Zeits. f. Physik **63**, 771 (1930); and especially H. S. Taylor, H. Eyring, A. Sherman, J. Chem. Phys. **1**, 68 (1933).

[4] F. Hund, Zeits. f. Physik **40**, 742 (1927) and applications of this point of view to crystals, Zeits. f. Physik **74**, 1 (1932).

[5] W. Lenz, Zeits. f. Physik **77**, 713 (1932); H. Jensen, Zeits. f. Physik **77**, 722 (1932).

[6] J. E. Lennard-Jones and H. J. Woods, Proc. Roy. Soc. **A120**, 727 (1928).

free state. This is justified since the corresponding wave functions practically vanish in half the interatomic distance. For the valence electron, however, such an assumption is quite out of the way, since the maximum of the corresponding wave function is (quite necessarily, as we shall see) just about half way between two atoms. Contrary to the conditions which exist in the free state, however, the wave function must not drop to zero after the maximum but can continue periodically through the whole crystal. It will therefore be much smoother than the wave function of the free atom, and the kinetic energy of the corresponding state will consequently be much smaller than that of the electron in the free atom. The potential energy, on the other hand, will be negatively larger in the lattice than in the free state because outside of the above-mentioned maximum the wave will not be under the influence of the nucleus considered originally, but under that of the next nucleus of the lattice, which is nearer. The electron with the wave function just described will have a larger negative energy than that in the free atom and we consider this to be the essence of the metallic state.

Of course, the wave function which one obtains by continuing the atomic wave function periodically in the lattice is not the real wave function of the free electron in the lattice, but the energy of the latter will be even smaller than that of the former. We shall try to find an approximation to the real wave function by actually solving a differential equation.

It must be added that not all the free electrons can be in the state given above, because of the Pauli principle. This reduces the magnitude of the metallic bond, because the electrons must have an additional kinetic energy, which is known as the zero-point energy of a Fermi gas. One sees easily, however, that this additional energy is smaller than the reduction of the kinetic energy which was obtained by continuing the wave function periodically through the whole lattice, so that there certainly remains a positive amount for the metallic bond.

### III.

First, we shall calculate the energy of the free electron in the lowest state. We shall do this by numerically solving a Schroedinger equation. It will not be necessary to solve it for the entire lattice, because it will have the same symmetry as the crystal and hence will merely repeat itself a great number of times. Because of this symmetry, the derivative of the wave function at every crystallographic symmetry plane will be zero perpendicular to this plane. This will be used as a boundary condition. The crystallographic symmetry planes which we shall use in this way bisect perpendicularly the lines connecting the second nearest atoms. If we draw lines connecting the nearest atoms and consider the planes bisecting these perpendicularly, we have every atom surrounded by a truncated octahedron (Fig. 1).

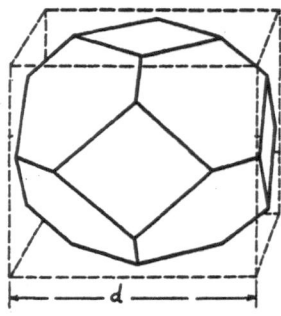

FIG. 1.

The middle points of the planes of the latter possess such symmetry ($S_6$) that the derivative of the wave function must vanish at these points in every direction. It will be quite a good approximation to replace the polyhedron of Fig. 1 by a sphere of equal volume, and to take as boundary conditions that the derivative of the wave function vanishes at the boundary of this sphere.

The determination of the potential function to be used inside this sphere is more difficult. It would be quite out of the way to use Hartree's method as the density of electrons in the greatest part of the domain is very small. This fundamental difference between metallic and ionic lattices was already pointed out by Lenz.[5] If we assume that two electrons are never around the same ion, every ion may be supposed to be surrounded by a spherical electron cloud which will

E. WIGNER AND F. SEITZ

exactly[7] cancel its potential outside of the sphere. Hence it seems to be the best simple assumption to take the potential as that of the ion inside the sphere mentioned above. The knowledge of this potential will allow us to set down the differential equation for the free electron which will be solved with the boundary condition that the derivative vanish at the boundaries of a sphere. The solution will be obviously spherically symmetric about each atom, which is of course not true for the actual wave function of the free electron but is a consequence of our approximations. It is not very far from the truth, however, since the wave function will actually have the highest crystallographic symmetry $(O^h)$ which is not very far from spherical symmetry, for to every direction there are not less than 47 other equivalent directions.

The justification of the assumption that two electrons are never around the same ion arises from two sources Consider first the statistical connections between the positions of two electrons in an ideal Fermi gas. The complete wave function will be a determinant

$$\Psi(1, 2, \cdots n) = \frac{1}{(n!)^{\frac{1}{2}}} \begin{vmatrix} \psi_1(1) & \psi_2(1) & \cdots & \psi_n(1) \\ \psi_1(2) & \psi_1(2) & \cdots & \psi_n(2) \\ \cdot & \cdot & \cdot & \cdot \\ \cdot & \cdot & \cdot & \cdot \\ \cdot & \cdot & \cdot & \cdot \\ \psi_1(n) & \psi_2(n) & \cdots & \psi_n(n) \end{vmatrix} \quad (1)$$

where the numbers in parentheses represent the Cartesian and spin coordinates of the corresponding electrons and the $\psi$ are the wave functions of the different states.[8] In order to obtain the statistical relations between the positions of two electrons we have to square (1) and integrate it over the coordinates of all electrons except those considered, which will be taken as 1 and 2. Because of the orthogonality relations of the $\psi_\kappa$ the result will be, apart from a constant,

$$\sum_{\kappa=1}^{n} \sum_{\lambda=1}^{n} [|\psi_\kappa(1)|^2 |\psi_\lambda(2)|^2 - \psi_\kappa(1)\psi_\lambda(1)^* \psi_\kappa(2)^* \psi_\lambda(2)]. \quad (2)$$

This still contains the spin coordinates $s_1$ and $s_2$ of 1 and 2 and reads more explicitly, if the edge of the cubic-shaped crystal is $L$,

$$\sum_{\nu_1\nu_2\nu_3} \sum_{\mu_1\mu_2\mu_3} [|e^{2\pi i(\nu_1 x_1 + \nu_2 y_1 + \nu_3 z_1)/L}|^2 |e^{2\pi i(\mu_1 x_2 + \mu_2 y_2 + \mu_3 z_2)/L}|^2 \delta_{s_1\sigma_1}\delta_{s_2\sigma_2}$$
$$- e^{2\pi i[(\nu_1-\mu_1)(x_1-x_2)+(\nu_2-\mu_2)(y_1-y_2)+(\nu_3-\mu_3)(z_1-z_2)]/L}\delta_{s_1\sigma_1}\delta_{s_1\sigma_2}\delta_{s_2\sigma_1}\delta_{s_2\sigma_2}]. \quad (2a)$$

There are really two questions to discuss: the statistical connection between electrons with antiparallel spin $(\sigma_1 = -\sigma_2)$ and between those with parallel spin $(\sigma_1 = \sigma_2)$. For a pair of the first kind the second term of (2a) vanishes so that they are statistically independent. For two electrons with parallel spin, on the other hand, we have to evaluate the sum of (2a) after having omitted the spin factors. We shall denote the distance of the two electrons by $r$, and may assume that the line joining them lies in the $X$ direction, since the

probability will not depend on the direction. With these conditions, (2a) becomes

$$\sum_{\nu_1\nu_2\nu_3} \sum_{\mu_1\mu_2\mu_3} (1 - e^{2\pi i(\nu_1-\mu_1)r/L}). \quad (3)$$

Here the summation over $\nu_2$, $\nu_3$, $\mu_2$, $\mu_3$ can be carried out at once. The limitation on the $\nu_1$, $\nu_2$, $\nu_3$ and $\mu_1$, $\mu_2$, $\mu_3$ being

$$\nu_1^2 + \nu_2^2 + \nu_3^2 \le \nu^2; \quad \mu_1^2 + \mu_2^2 + \mu_3^2 \le \nu^2$$
$$\nu = (3n/8\pi)^{\frac{1}{3}} \quad (4)$$

it gives

$$P(r) = \sum_{\nu_1=-\nu}^{\nu} \sum_{\mu_1=-\nu}^{\nu} \pi^2(\nu^2 - \nu_1^2)(\nu^2 - \mu_1^2)(1 - e^{2\pi i(\nu_1-\mu_1)r/L}). \quad (5)$$

Now the summation over $\nu_1$ and $\mu_1$, after dividing by $\pi^2\nu^2(4\nu^2-1)^2/9$ gives for the probability of the electrons with parallel spin being a distance $r$ apart

---

[7] There is nothing like exchange forces in our picture.

[8] This consideration is contained implicitly in the work of Uhlenbeck and Gropper for the case of only slightly degenerated gases, Phys. Rev. 40, 1029 (1932). We are interested in the case of complete degeneracy, however.

$$4\pi r^2 P(r) = 4\pi r^2\left\{1-\left(\frac{3}{2}\,\frac{\cos{(\pi r/L)}\sin{(2\pi\nu r/L)}-2\nu\sin{(\pi r/L)}\cos{(2\pi\nu r/L)}}{\nu(4\nu^2-1)\sin^3{(\pi r/L)}}\right)^2\right\}$$

$$=4\pi^2 r^2\left\{1-\left(3\,\frac{\sin{(r/d')}-(r/d')\cos{(r/d')}}{(r/d')^3}\right)^2\right\}. \tag{6}$$

Here $d'=v_0^{1/3}/3^{1/3}\pi^{1/3}$ and $v_0=L^3/n$ is the atomic volume. The function $P(r)$ is sketched in Fig. 2, it vanishes for $r=0$ and approaches 1 as $r$ becomes large compared with the lattice constant. It attains its half-value for $r=1.79\,d'$ or $0.460\,d$ for a body-centered lattice with a cube edge $d$. The radius of the sphere described above is about the same, namely $(\frac{3}{8}\pi)^{1/3}d=0.492\,d$. So we see that two electrons with parallel spin will be very rarely at the same ion, simply in consequence of the exclusion principle. This will be also true for a Fermi gas subject to a periodic potential, as the potential does not materially alter this argument.

It remains to investigate the case of antiparallel spins somewhat more closely. As it is not possible that three electrons have antiparallel spins, the probability of three electrons being at the same ion will be very small anyway. For two electrons with antiparallel spins and without interaction, however, there is no statistical connection of the positions. If one should take the interaction into account, it would turn out, however, that there is a connection of such a kind that they are but rarely in the neighborhood of each other. This is already indicated in the well-known solution of Hylleraas[9] for He and in the similar solution of Bethe[10] for the negative hydrogen ion. These solutions also show that the connection is of such an order of magnitude that the choice of such a potential function, which corresponds

to our rather rough picture of the metallic bond, is justified to some extent. It must be admitted, however, that the lower limit of the energies of the free electrons, which we calculate in this way, will certainly give too large a binding energy for the lattice, as all of the electrons will not be at different ions with certainty, and also because the terms of Hylleraas and Bethe just discussed will increase the kinetic energy above the value which we obtain by our boundary conditions. We shall not take up this question in more detail this time as it is deeply connected with the interaction problem of the electron and the justification of the notions of the free and bound electrons and we hope to return to it at another time.

## IV.

The calculation of the wave function inside the proper spheres of the ions is very simple in principle. The potential function of Prokofjew[11] was used for the purpose. This was obtained by Prokofjew following a method of Kramers[12] in which one employs experimental values of the terms of Na. The differential equation for the radial function $R=r\psi(r)$ is

$$-(h^2/8\pi^2 m)(\partial^2 R/\partial r^2)+V(r)R=ER(r) \tag{7}$$

and in units of the Bohr radius of $H$, the quantity $Q(\rho)=-a_0\rho^2 V/e^2$, is approximated for various intervals by parabolas, as follows:

| $\rho=0.00$ to $0.01$ | | $Q=11\rho$ |
|---|---|---|
| 0.01 | 0.15 | $=-26.4\ \rho^2+11.53\ \rho-0.00264$ |
| 0.15 | 1.00 | $=-2.84\ \rho^2+4.46\ \rho+0.5275$ |
| 1.00 | 1.55 | $=+1.508\rho^2-4.236\ \rho+4.876$ |
| 1.55 | 3.30 | $=0.1196\rho^2+0.2072\rho+1.319$ |
| 3.30 | 6.74 | $=0.0005\rho^2+0.9933\rho+0.0222$ |
| 6.74 | $\infty$ | $=\rho$ |

The boundary condition $\partial\psi/\partial r=0$ requires that at the boundary $R$ should satisfy

$$\partial R/\partial r=R/r. \tag{8}$$

[9] E. A. Hylleraas, Zeits. f. Physik 48, 469 (1928).
[10] H. Bethe, Zeits. f. Physik 57, 815 (1929).
[11] W. Prokofjew, Zeits. f. Physik 58, 255 (1929). (Note: Prokofjew's table of $Q(\rho)$ (p. 258) contains obvious errors in two places, one in decimal point and one in sign. These were easily detected by the continuity conditions. The form given here is corrected.)
[12] H. A. Kramers, Zeits. f. Physik 39, 828 (1926).

808    E. WIGNER AND F. SEITZ

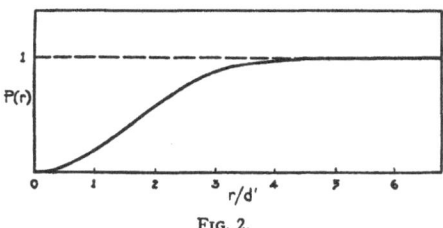

FIG. 2.

Instead of finding the energy value $E$ for different radii of the sphere, the radii corresponding to different energy values were determined. Thus an arbitrary energy value $E$ was taken and the corresponding wave function was obtained from (7) using the method of finite differences employed by Prokofjew. The calculation was started at $r=0$, however, so that every calculated wave function could be used. For $r=0$ the radial function $R$ vanishes and the solution up to $r=0.025$ was calculated by means of a power series in $r$. After this the method of finite differences was employed, first with differences of 0.005 and then with larger ones when allowable. The largest difference employed was 0.32 ($r>4.6$). The wave function had practically no dependence on $E$ in the neighborhood of the origin so that it was not always necessary to repeat this part of the calculation. As a check, the energy of the electron in the free atom was also determined, the calculated value lies between 0.3820 and 0.3800 Rydberg

units, while the experimental value is 0.3778. In Fig. 3, the wave function of the free atom and the wave function for $E=0.500$ are plotted. The numerical tables will be published at another time.

After having the wave function, it was easy to determine the radius of the sphere, for which the boundary condition (8) is satisfied by drawing the tangents to $R$ from the origin. The figure shows, that the boundary conditions are satisfied for two different radii, so that every numerical integration yields two points of the $E(r)$ curve, which gives the energy of the most strongly bound free electron as a function of the lattice constant $d=(8\pi/3)^{\frac{1}{3}}r$. In Fig. 4 the $E(r)$ curve is given (lower line), the unit of energy being the ionization energy of $H$. For very large $r$ it approaches the ionization energy of atomic Na, possesses a minimum around $r=3$, and rises again for smaller values of $r$. This latter behavior is due to the fact that a further compression of the lattice would push the valence electron inside the closed $L$ shell, which of course requires energy. The lattice, unlike a similar $H$ lattice,[13] would be stable, therefore, even without taking into account Fermi statistics.

The calculation of a wave function took about two afternoons, and five wave functions were calculated on the whole, giving the ten points of

---

[13] Cf. E. A. Hylleraas, reference 3.

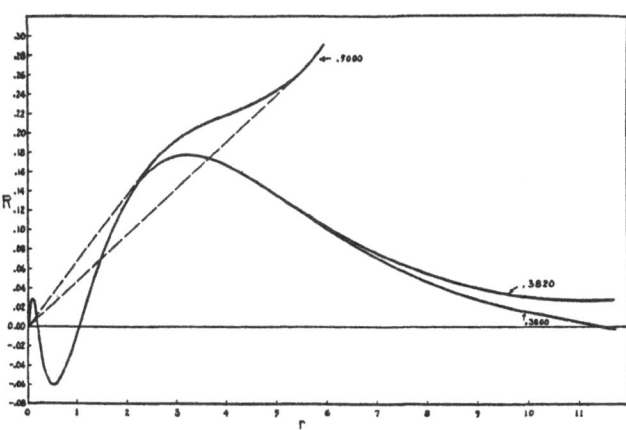

FIG. 3.

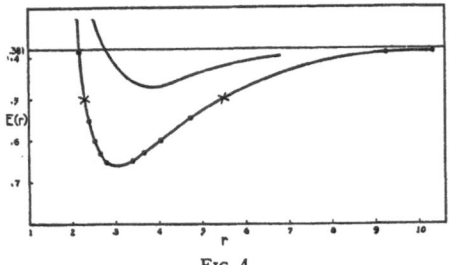

FIG. 4.

the figure. The points of the wave function of Fig. 4 are marked by a cross.

Another point which should be mentioned is that concerning the change in energy of the inner shells. The change is not due chiefly to the change of the boundary conditions as discussed above for the valence electron. This, of course, does increase the binding energy of these electrons, but only by a very small amount. A greater change arises from the increase in the probability of the valence electron being inside the $L$-electrons, because of the material change of the normalization factor. This decreases the binding energy of the inner electrons and hence lowers the heat of vaporization. A calculation of this effect has been made and shows that the decrease in binding energy is 0.008 Rydberg units per atom, or 2500 small calories per mole. This was obtained by evaluating the change of the potential of the inner electrons in the field due to the valence electron for the free and bound atom. The inner charge distribution was taken to be that given by Hartree.*

## V.

The last question we have to investigate is concerned with the additional energy of the other free electrons due to the Fermi distribution. This energy was calculated by the simple Fermi formula and it gives a mean additional energy for every electron

$$\frac{3h^2}{10m}\left(\frac{3}{8\pi}\right)^{\frac{2}{3}}\frac{1}{V_0^{\frac{2}{3}}}=\frac{9h^2}{80\pi m}\left(\frac{3}{2\pi}\right)^{\frac{2}{3}}\frac{1}{r^2} \qquad (9)$$

or $(9\pi/10)(3/2\pi)^{\frac{2}{3}}r^{-2}$ if the energy is measured in Rydberg units and $r$ in Bohr units. As a matter of fact, this formula is valid only for free electrons

* Hartree, Proc. Camb. Phil. Soc. 24, 111 (1927).

and it certainly gives too large a value for bound electrons. The fact that the energy differences for bound electrons are smaller than for free electrons was shown first by Bloch.[14] It also follows from the following argument. Let the wave function of the electron with the lowest energy be $\psi_{000}$ $(x, y, z)$, which is invariant with respect to an addition of an identity period to the coordinates. This invariance is not possessed by the wave functions of the other free electrons, and that with the quantum numbers $\nu_1$, $\nu_2$, $\nu_3$, will be multiplied by $e^{2\pi i\nu_1 d/L}$, $e^{2\pi i\nu_2 d/L}$ and by $e^{2\pi i\nu_3 d/L}$ if $x$, $y$, or $z$ are increased by $d$, respectively. Now $\psi_{000}$ gives the lowest possible energy of all wave functions, which are orthogonal to the wave functions of the $L$ and $K$ shells. For the wave function with the quantum numbers $\nu_1$, $\nu_2$, $\nu_3$ this is true if we compare it only with functions which have the same symmetry character, i.e., are multiplied by the same factors if one replaces $x$, $y$ or $z$ by $x+d$, $y+d$ or $z+d$, respectively. There is, however, the function

$$\psi_{\nu_1\nu_2\nu_3}=e^{2\pi i(\nu_1 x+\nu_2 y+\nu_3 z)/L}\psi_{000}(x, y, z), \quad (10)$$

which has all the required properties and the energy of which differs from that of $\psi_{000}$ only by $(h^2/2mL^2)(\nu_1^2+\nu_2^2+\nu_3^2)$, the Fermi energy for free electrons. This is easily seen upon calculating $(\psi_{\nu_1\nu_2\nu_3}, H\psi_{\nu_1\nu_2\nu_3})$ for (10) and remembering that $\psi_{000}$ may be assumed to be real. The energy of the real wave function with the quantum numbers $\nu_1$, $\nu_2$, $\nu_3$ is certainly less than that of (10) and so the average additional energy for the free electrons in higher states is also certainly less than (9). Nevertheless (9) was adopted in the subsequent calculation and the corresponding expression added to the energy of the lowest electron $E(r)$. The result is given in Fig. 4 (upper line). It is probably true that the fact that (9) gives a too high value largely compensates the error which was made by the assumption of the free electrons, as discussed in Section III.

## VI.

The upper curve in Fig. 4 gives at once all quantities we desire to calculate. The position of the minimum gives the radius of the sphere for which the energy is the smallest and

[14] F. Bloch, Zeits. f. Physik.

810                    E. WIGNER AND F. SEITZ

when multiplied by $(8\pi/3)^{\frac{1}{3}}$ it should give the lattice constant for the absolute zero point. Similarly the depth of the minimum below the line of the energy for the free atom should give, after subtracting the correction for the energy gained by the inner shells (2.5 kilo cal.) the energy difference between the gas and the solid state (i.e., the heat of vaporization per atom) in Rydberg units at the absolute zero point. Finally the curvature at the minimum $r_m$ is in a simple connection with the compressibility: the energy change for a linear compression in the ratio $\alpha$ per volume $v_0$ is $v_0\kappa(3\alpha)^2/2$, where $\kappa$ is the compressibility, and, on the other hand, it is $\frac{1}{2}r_m^2\alpha^2 d^2E(r_m)/dr^2$ when calculated from the figure. This gives

$$\kappa = (1/9)(r_m^2/V_0)(d^2E(r_m)/dr^2). \qquad (11)$$

The quantity $d^2E/dr^2$ was calculated as if the lower curve were linear at $r_m$ and all the curvature arose from (9), which is approximately true according to the figure. The final values obtained for the three quantities $d$, $\lambda$ and $\kappa$ are $d=4.2$A, $\lambda=25.6$ kilo cal./mol, $\kappa=1.6\times10^{-11}$ c.g.s. The depth of the minimum of the lower curve below the energy of the free atom is 88.5 kilo cal.; and the depth at the point where the upper curve has its minimum is 70.4. The Fermi correction at this point is 42.3 kilo cal. In order to have a fair comparison with experiment, the experimental values for these quantities must be extrapolated to the absolute zero point. It must be remembered, however, that we did not treat the motion of the nuclei by quantum me-chanics and in consequence, the extrapolation should be done in such a way as to neglect the quantum effects. This was done by taking the values for room temperature and correcting them linearly. The three values for room temperatures are[15]: $d=4.30$A; $\lambda=26.00$ kilo cal./mol; $\kappa=1.67$ $\times10^{-11}$ c.g.s. The coefficient of thermal expansion is $62\times10^{-6}$;[16] the corrected value of $\lambda$ is determined by adding the difference in heat content of solid and gas $((6-3)$ cal./deg. $\times300$ deg. $=900$ cal.) to the room temperature value of $\lambda$; and the value of $\kappa$ at $0°$K was obtained by extrapolating values given along with the above. The final values are: $d=4.23$A, $\lambda=26.9$ kilo cal./mol; $\kappa\sim1.0\times10^{-11}$ c.g.s. The theoretical values compare favorably with these, partly, without doubt, as a consequence of compensating errors.

The work on sodium is being extended with particular reference to a more exact determination of the distribution of energy levels in the neighborhood of the lowest one. Moreover, the corresponding calculations on Li, K, Rb, by using Hartree's and Hargreaves' fields,[17] are being undertaken by one of us.

---

[15] $d$: P. P. Ewald, Hand. d. Phys. XXIV, 331.

λ: J. Sherman, Chem. Rev. 11, 93 (1932).

κ: Landolt Bornstein, Erster Ergaenzungsband 5 auf., 25. Int. Crit. Tables III, 47.

[16] Int. Crit. Tables II, 461.

[17] We wish to offer our thanks, at this time, to Professor J. C. Slater and hence to Dr. Hartree, for the use of the unpublished tables of K+.

# On the Constitution of Metallic Sodium. II

E. P. Wigner and F. Seitz

Physical Review *46*, 509–524 (1934)

(Received June 18, 1934)

The present work represents an extension to a previous development by the same authors, on the theory of metallic sodium. In the first part of this paper a completely self-consistent solution of Fock's equations for the sodium lattice is carried through indirectly, this being the approximation in which one-electron functions are employed. The question of the correlations between electrons with parallel spin is investigated quantitatively and the Fermi "zero-point energy" is calculated using the proper effective field. The results show that the electrons behave almost exactly as if they were entirely free, the binding energy being 9 kg cal and the lattice constant 4.86A, as compared with the observed values 26.9 kg cal and 4.23A. To complete the picture, the correlations between electrons with anti-parallel spins are investigated in the latter part, since these are not included in the Fock picture. A general discussion of this question is presented and a quantitative treatment of its effect is made which yields a new binding energy of 23.2 kg cal and a lattice constant of 4.75A. The source of the remaining discrepancy is discussed.

## I. THE POTENTIAL INSIDE THE LATTICE

### 1.

In a previous paper by the same authors a method of calculating the binding properties of metals was developed[1] and applied to sodium. The procedure employed was essentially one of solving the Fock system of equations[2] for the valence electrons (i.e., the system of equations to which the Schrödinger equation reduces when one electron functions are assumed). This solution did not proceed from a formal investigation of Fock's differential equations, but was developed indirectly under the guiding principles of the picture afforded by the free electron theory.

To begin with, the lattice was subdivided into polyhedrons of equal size and form, which we shall call s-polyhedrons, each of which surrounds one ion lying in its center, and is bounded by the planes which bisect, perpendicularly, the lines connecting the corresponding ion with its 14 neighbors (the alkali metals form body centered lattices). Since these polyhedrons closely resemble spheres, they may be replaced by spheres of equal volume for many purposes and these we shall designate as s-spheres, their radius being $r_s = (3v_0/4\pi)^{\frac{1}{3}}$, where $v_0$ is the atomic volume.

Concerning the nature of the electronic states of the lattice, we know that there will be bands of allowed levels, no more than two electrons occupying each level because of the restrictions imposed by the Pauli principle, and that the lowest state in the lowest band will possess the symmetry of the lattice. From this it follows that its normal derivative will vanish on the boundaries of the s-polyhedrons and to obtain its wave function, it is only necessary to solve the Schrödinger equation within one polyhedron, by using a suitable effective field and this boundary condition. Approximate wave functions of other electrons may be obtained from this by multiplying it with factors of the form

$$e^{2\pi i(\nu_1 x + \nu_2 y + \nu_3 z)/L}, \tag{1}$$

where $\nu_1$, $\nu_2$, $\nu_3$ are positive or negative integers and $L$ is the length of the crystal-edge, corresponding to free electrons with nonvanishing eigenvalues of momentum.

In order to find the effective potential field inside of an s-polyhedron, we first replace them by s-spheres, correcting later for the small error arising from this. For an electron at a given point the field consists of three parts: first, the potential arising from the ion at the center of the s-sphere, second, the potential of the other ions; and third, the potential arising from the other free electrons. Instead of making a direct calculation of each of these in the order given it is found far more advantageous to begin with an investigation of the third for it happens that simplifying assumptions concerning the nature of this interaction may be fully justified at a later stage and allow a simple treatment of the first two.

[1] E. Wigner and F. Seitz, Phys. Rev. **43**, 804 (1933).
[2] J. C. Slater, Phys. Rev. **35**, 210 (1930); V. Fock, Zeits. f. Physik **61**, 126 (1930).

The potential of the electron under consideration arising from the other electrons is equal to that arising from a continuous electron fluid of density which is $e$ times the probability of having an electron at the corresponding point of the lattice. If we assume that all wave functions arise from one by multiplication with a factor of the form (1), the probability distribution will be the same for each. Moreover, since it will be spherically symmetric within each sphere the effective field of each ion will be neutralized outside of the sphere in which it is contained, and the only field which remains to be taken into account for our chosen electron is that arising from the charge within the $s$-sphere which contains it. Since this will give rise to a spherically symmetric field and a spherically-symmetric wave function, the assumption of spherical symmetry is completely self-consistent. The error introduced by replacing the $s$-polyhedrons by spheres is calculated in appendices 1 and 2 and turns out to be negligibly small.

In the previous work only the field arising from the ion in each $s$-sphere was taken into account and it was assumed that correlations were such that other electrons did not penetrate the given sphere. Under this assumption, the wave function turned out to be practically constant for more than 90 percent of the volume of the sphere (cf. Fig. 1), a fact which allows us to take the other electrons into account very simply and correct for the inaccuracies involved. For if we now introduce the assumption that all of the electrons have the distribution given by $\psi_{000}$ of

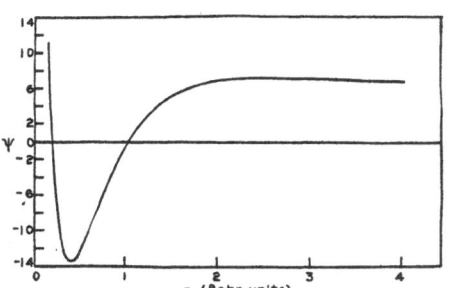

FIG. 1. Metallic wave function for lowest electronic state. The corresponding energy is $-0.6$ Rydberg units and the proper boundary condition is satisfied at 4.04 Bohr units.

Fig. 1 (with factors (1)), it is found that this set of one-electron functions forms a practically self-consistent solution of Fock's equations for the lattice.

The potential $V(x, y, z)$ which enters into the wave equation of an electron in Fock's system is not the ordinary average of the potentials of the other particles for all configurations, but the average of the potentials with each configuration taken with the weight it has under the assumption that the considered electron is in the point $x, y, z$. We have seen Eq. (6) that there is a hole in the otherwise uniform electron fluid around every electron because the probability of two electrons having parallel spin being very near is very small. We shall call this the Fermi hole. Its effect is such as to make the potential more negative than it would otherwise be. Since its shape is practically independent of the position of the electron, the change of the potential arising from it is constant and it alters the energy but not the wave function. The previous work merely gave an estimate of this change and we shall give a more accurate calculation here.

Because the wave functions of electrons of higher energy are given by multiplying $\psi_{000}$ by (1) only in the case of free electrons (i.e., when $\psi_{000}$ is constant), we have investigated this very important point in more detail (cf. Section II), and have found that the energy differences between the different states is the same as that obtained when the electrons are free. The wave functions are not simply $\psi_{000}$ multiplied by (1), however, but each is multiplied in addition by a factor which is nearly constant within a sphere one Bohr unit in radius but becomes exponential outside.

In all of the work sketched in the previous paragraphs we have been able to restrict ourselves to the one electron picture, that is, the wave function for all of the electrons may be assumed to be a determinant of single electron functions. This leads to correlations between electrons of parallel spin of the type expressed by the Fermi holes and to none whatever between electrons of antiparallel spin. There are such correlations, however, arising from the mutual repulsion terms, but they lie beyond the scope of Fock's equations. We shall call these holes "correlation holes" and the corresponding energy "correlation energy," and shall attempt to handle them in Section III.

**2.**

Under the assumption of a uniform charge distribution the potential within the $s$-sphere at the distance $r$ from the ion is

$$V(r) + (3e^2/2r_s) - (e^2r^2/2r_s{}^3), \qquad (2)$$

where the first term arises from the ion, and the other two from the electron fluid inside the $s$-polyhedron. In reference 1 the wave equation for the potential $V(r)$ alone—which was taken from Prokofjew's work[3]—was solved and the corresponding characteristic value determined. The correction for the fact that this is not exactly the energy of the corresponding wave function is taken care of in Appendix 1. The second term in (2) is independent of $r$, leaving the wave function unchanged, and gives a positive contribution

$$E_{11} = 3e^2/4r_s \qquad (3)$$

to the energy. A factor $\frac{1}{2}$ enters, since the interactions between electrons have to be counted only once.

In order to take into account the last term, the Schrödinger perturbation theory may be applied and this yields

$$E_{12} = \frac{1}{2}\left( -\frac{e^2}{2r_s{}^3} \int \psi^2 r^2 dv \right.$$
$$\left. + \frac{e^4}{4r_s{}^6} \sum_{\kappa}' \frac{(\int \psi\psi_\kappa r^2 dv)^2}{E - E_\kappa} \right) \qquad (4)$$

up to the second approximation where $\psi$ and $E$ are the solutions of the wave equation with the potential $V(r)$ alone, which we know from reference 1. The factor $\frac{1}{2}$ arises for the same reason given before. To determine the first part of (4) we have numerically evaluated the integral over $\psi^2 r^2$ for three different values of the lattice constant, i.e., $r_s$ and the corresponding energy. The results are tabulated in Table I, the energy

[3] W. Prokofjew, Zeits. f. Physik **58**, 255 (1929).

being given in Rydberg units, $r_s$ in Bohr units, and the integral in squares of Bohr units. For the actual lattice constant $r_s = 4$, the integral has the same value that it would have if $\psi$ were a constant. The last column gives the ratio $f$ of the numerically calculated integral to that calculated with a constant $\psi$ for the other cases.

In the second part of (4) the denominator may be replaced by a mean value $E - E_k$ and the summation in the numerator gives $\int \psi^2 r^4 dv - (\int \psi^2 r^2 dv)^2$ in the usual way.[4] By taking $\psi$ to be constant for this small term one obtains

$$-\frac{1}{E_k - E} \cdot (e^4/r_s{}^2) \cdot \left( \frac{3}{56} - \frac{9}{200} \right) = -0.005e^2/r_s. \qquad (5)$$

The mean value of $E_k$ has been estimated to be $+0.3$ Rydberg units. The effect of the perturbing terms in (2) on the wave functions can safely be said to be extremely small because of the smallness of (5), and their total contribution to the energy is

$$E_{12} = (-0.15f - 0.005)e^2/r_s. \qquad (6)$$

The fact that we must deal with the $s$-polyhedron instead of $s$-sphere is taken into account in Appendix 2.

**3.**

In the preceding section we have calculated using the ordinary average potential. We know, however, that in order to calculate the potential of an electron at a certain point we should average over the configuration of the other electrons with the weights which they would have if the electron under consideration were at the given point. It was shown in reference 1 that if there is an electron at a given point the probability of another electron being at a distance $r$ from it is

$$\frac{1}{2}\frac{e}{v_0}\left[ 1 - 9\left( \frac{\sin (r/d') - (r/d') \cos (r/d')}{(r/d')^3} \right)^2 \right]. \qquad (7)$$

The factor $\frac{1}{2}$ expresses the fact that only the distribution of charge with parallel spin is affected, $v_0$ is the atomic volume and $d' = (v_0/3\pi^2)^{\frac{1}{3}}$. (7) was derived under the assumption that $\psi_{000}$ is constant and that the higher ones are obtained from it by multiplication with (1). Both assumptions will be shown to be practically correct

[4] A. Unsöld, Zeits. f. Physik **43**, 563 (1927).

TABLE I.

| $E$ | $r_s$ | $\int \psi^2 r^2 dv$ | $f$ |
|---|---|---|---|
| −0.60 | 4.05 | 9.90 | 0.994 |
| −0.55 | 4.74 | 12.95 | 0.962 |
| −0.50 | 5.48 | 16.50 | 0.916 |

insofar as energies are concerned. The total amount of charge removed because of the hole expressed by (7) is exactly $e$ and may be viewed as being concentrated in the electron at the center of the hole. To this approximation it has

been assumed that all of this hole is in the parallel charge distribution—an assumption not exactly correct, however.

The decrease in the potential energy of the electron at the center of the hole is

$$\frac{9}{2}\frac{e^2}{v_0}\int_0^\infty \frac{1}{r}\left(\frac{\sin (r/d') - (r/d') \cos (r/d')}{(r/d')^3}\right)^2 4\pi r^2 dr = \frac{3^{4/3}e^2}{2\pi^{1/3}v_0^{1/3}} = \frac{3^{5/3}}{2^{5/3}\pi^{2/3}}\frac{e^2}{r_s} = 0.916\frac{e^2}{r_s}. \tag{8}$$

Since this energy[5] does not depend on the position of the electron, it is to be carried forward at once as an energy with a factor $\frac{1}{2}$, in order that the interactions between pairs of electrons is taken into account only once. We have, therefore,

$$E_2 = -0.458e^2/r_s. \tag{9}$$

If the wave function were not entirely flat in the lowest state, (7) should be more rigorously multiplied by $v_0\psi_{000}$. The integral (8) will hardly be affected by this, however, because the regions in which $\psi_{000}$ changes appreciably are small compared to the distance in which (7) has appreciable variation and the mean value of (8) for different positions of the electron will differ from the calculated value by even a smaller percentage.

The sum $E_{11}+E_{12}+E_2$ from the viewpoint of Fock's equation, gives the effect arising from the fact that the potential at any point is not exactly $V(r)$ (where $r$ is the distance from the nearest ion), as it would be if the hole in the electron fluid of the other electrons extended exactly over the $s$-sphere. It really surrounds (cf. Fig. 2) the electron under consideration spherically in such

a way that part $(A)$ of the $s$-sphere is left uncovered. On the other hand, it extends over regions $(C)$ outside the $s$-sphere. Now $E_2$ represents the effect of the entire hole $(B+C)$, while $E_{11}+E_{12}$ represents the effect of the lack of hole in the regions $A$ and $B$. The fact that we made all calculations under the assumption that the electron density is constant, will not greatly affect our final result, since in the largest region, $B$, it does not matter which density we use, as long as we use the same density in both cases.

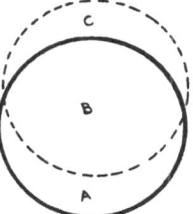

FIG. 2. Schematic diagram of prominent regions in the unit cell.

An error will be introduced only by using an improper density in $A$ and $C$—but here $\psi_{000}$ is very nearly constant.

## II. THE FERMI ENERGY

4.

Our next task is the calculation of the zero-point energy of the free electrons. First of all, an error in reference 1 should be corrected, for it was stated there that the zero-point energy is actually smaller than it would be for free electrons. It was observed that $\psi^0_{\nu_1\nu_2\nu_3} = \psi_{000} \exp [2\pi i(\nu_1 x + \nu_2 y + \nu_3 z)/L]$ possessed the proper transformation properties for a state with the quantum numbers $\nu_1$, $\nu_2$, $\nu_3$ (the "components of momentum" being[6] $2\pi\hbar\nu_1/L$, $2\pi\hbar\nu_2/L$, $2\pi\hbar\nu_3/L$). Moreover its mean energy was higher than that of $\psi_{000}$ by the amount $2\pi^2\hbar^2(\nu_1^2 + \nu_2^2 + \nu_3^2)$ which is just the Fermi cor-

[5] This formula was first found by F. Bloch, Zeits. f. Physik **57**, 545 (1929).

[6] $\hbar$ in this paper is Dirac's $\hbar$, namely, Planck's constant divided by $2\pi$. We would like to correct another error of reference 1 here. The subtraction of 2500 cal. from the binding energy at the end of Section IV is incorrect. The fact that the binding energy of the inner shells decreases with increasing binding of the valence electrons does not appear in the Prokofjew picture because the potential energy of the valence electron is taken to be $\rho V$ instead of $\frac{1}{2}\rho V$, so that it is not necessary to consider the change of the potential acting on the other particles.

rection for an entirely free electron. As the real $\psi_{\nu_1\nu_2\nu_3}$ is the solution of a minimum problem, it was concluded that its energy is even lower than that of $\psi^0{}_{\nu_1\nu_2\nu_3}$.

This argument is not correct, however, since $\psi_{\nu_1\nu_2\nu_3}$ has the lowest energy of only those states which are orthogonal to the states of the $K$ and $L$ shells. This is correct for $\psi_{000}$ but not for $\psi^0{}_{\nu_1\nu_2\nu_3}$ and the real $\psi_{\nu_1\nu_2\nu_3}$ should be calculated just as $\psi_{000}$ was—by solving a differential equation.

At any rate, it can be assumed from the transformation property of $\psi_{\nu_1\nu_2\nu_3}$ that it has the form

$$\psi_{\nu_1\nu_2\nu_3} = \chi_{\nu_1\nu_2\nu_3} \exp\left[2\pi i(\nu_1 x + \nu_2 y + \nu_3 z)/L\right], \tag{10}$$

where $\chi_{\nu_1\nu_2\nu_3}$ has the translational periods of the lattice. By inserting (10) into the Schrödinger equation, one obtains

$$H\chi_{\nu_1\nu_2\nu_3} + \frac{2\pi i\hbar^2}{mL}\left(\nu_1\frac{\partial}{\partial x} + \nu_2\frac{\partial}{\partial y} + \nu_3\frac{\partial}{\partial z}\right)\chi_{\nu_1\nu_2\nu_3} = F_{\nu_1\nu_2\nu_3}\chi_{\nu_1\nu_2\nu_3},$$

$$E_{\nu_1\nu_2\nu_3} = F_{\nu_1\nu_2\nu_3} + (2\pi^2\hbar^2/mL^2)(\nu_1^2 + \nu_2^2 + \nu_3^2). \tag{11}$$

If we neglect the second term on the left side to begin with, the equation is identical with the equation for $\psi_{000}$, which we shall denote by $\psi_0$ in this section, the other solutions of (11) with $\nu_1 = \nu_2 = \nu_3 = 0$ being designated by $\psi_\epsilon$. To this approximation $E_{\nu_1\nu_2\nu_3} = E_{000} + 2\pi^2\hbar^2/mL^2(\nu_1^2 + \nu_2^2 + \nu_3^2)$ which is just the result of reference 1.

We shall take into account the second term of (11) by the Rayleigh-Schrödinger perturbation method. The first approximation gives zero, since $\int\psi_0\partial\psi_0/\partial x$ vanishes, and the second approximation will be proportional to the $\nu$'s. From considerations of the $s$-polyhedron, the boundary conditions were found to be such that the value and derivative of the wave functions should be equal in the two points at which a line perpendicular to the boundary plane cuts the boundary. We know furthermore from group theory that all solutions $\psi_\epsilon$ of the unperturbed problem are either even or odd functions of $x$, $y$, $z$. Now if we calculate the second approximation for the energy, we need integrals of the type $\int\psi_0\partial\psi_\epsilon/\partial x$, and since $\psi_0$ is an even function of the coordinates, $\psi_\epsilon$ must be even in $y$, $z$ and odd in $x$, or the integral vanishes. Therefore, all three perturbing terms in (11) will give nonvanishing matrix elements only with different $\psi_\epsilon$. The second approximation for the energy is, therefore

$$F_{\nu_1\nu_2\nu_3} = \frac{4\pi^2\hbar^4}{m^2L^2}(\nu_1^2 + \nu_2^2 + \nu_3^2)\sum_\epsilon{}' \frac{|\int\psi_\epsilon\,\partial\psi_0/\partial x|^2}{E_0 - E_\epsilon}, \tag{12}$$

since the integral with the $y$ and $z$-derivatives can be replaced by integrals with $x$ derivatives, $E_\epsilon$ is the energy of the state $\psi_\epsilon$. The sum in (12) contains one positive term (arising from the $2p$ level) and all other terms are negative. Since the value of the positive term is very sensitive to the actual shape of the wave function $\psi_0$, we tried to transform (12) in a form more suitable for the calculation in the following way.

If $\psi_0$ and $\psi_\epsilon$ were the characteristic functions of an atom, (12) could be evaluated by the Thomas-Kuhn sum rule and would give just the negative Fermi correction, so that the sum of the first and second approximation would be zero. In the present case of periodic boundary conditions, however, the partial integrations cannot be performed in the same way. We obtain from the Schrödinger equation for $\psi_\epsilon$ by multiplication with $x\psi_0$ and subtracting the corresponding equation for $\psi_0$ multiplied by $x\psi_\epsilon$

$$(E_\epsilon - E_0)\int x\psi_0\psi_\epsilon = -\frac{\hbar^2}{2m}\int\left[x\psi_0\Delta\psi_\epsilon - \psi_\epsilon\Delta(x\psi_0)\right] - \frac{\hbar^2}{m}\int\psi_\epsilon\frac{\partial\psi_0}{\partial x}. \tag{13}$$

The first term can be transformed by Green's theorem into a surface integral and we have

$$\int\psi_\epsilon\frac{\partial\psi_0}{\partial x}dv = \frac{m(E_0 - E_\epsilon)}{\hbar^2}\int x\psi_\epsilon\psi_0 dv - \tfrac{1}{2}\int\left(x\psi_0\frac{\partial\psi_\epsilon}{\partial n} - \psi_\epsilon\frac{\partial(x\psi_0)}{\partial n}\right)df, \tag{14}$$

where $\partial/\partial n$ denotes the normal derivative. For the usual boundary conditions, the last term in (14) would vanish. After replacing *one* of the integrals in (12) by the expression (14), the summation over $\kappa$ can be carried out in the first term by the completeness-relation[7] and yields

$$(4\pi^2\hbar^2/mL^2)(\nu_1^2+\nu_2^2+\nu_3^2)\int x\psi_0(\partial\psi_0/\partial x)dv. \tag{15}$$

Since $\psi_0$ is a function of $r$ alone, we have

$$\int x\psi_0\frac{\partial\psi_0}{\partial x}dv = \frac{2\pi}{3}\int_0^{r_e}r^3\frac{\partial\psi_0^2}{\partial r}dv = \tfrac{1}{2}v_0\psi_0(r_e)^2-\tfrac{1}{2}.$$

The $-\tfrac{1}{2}$ gives the negative Fermi correction and cancels the first approximation. The first term, however, gives an expression very similar to the Fermi correction, namely

$$\frac{2\pi^2\hbar^2}{mL^2}(\nu_1^2+\nu_2^2+\nu_3^2)v_0\psi_0(r_e)^2, \tag{16}$$

the difference being that the value of $\psi_{000}$ at the boundary enters instead of the mean value.

The second term of (14), inserted into (12), gives

$$-\frac{2\pi^2\hbar^4}{m^2L^2}(\nu_1^2+\nu_2^2+\nu_3^2)\sum_{\kappa}'\frac{\int(\psi_\kappa\,\partial\psi_0/\partial x)dv}{E_0-E_\kappa}\int x\psi_0\frac{\partial\psi_\kappa}{\partial n}df, \tag{17}$$

since all odd $\psi_\kappa$ vanish at the boundary. To this sum the $2p$ state does not contribute anything more, because its wave function is practically zero outside $r=3$. Similarly we found upon direct computation, that the integral $\int\psi_\kappa\partial\psi_0/\partial x d\vartheta$ vanishes "accidentally" for $r_e=4$ for the next higher odd state. The energy of the next is of the order of magnitude $+5$ Rydberg units and we have replaced $E_\kappa$ in the denominators of (17) by a mean value $\bar{E}_\kappa$ of this magnitude. Now $\int\psi_\kappa\partial\psi_0/\partial x d\vartheta$ is the expansion coefficient of $\psi_\kappa$ in the series for $\partial\psi_0/\partial x$ so that the summation can be carried out in the numerator and gives $\int x\psi_0(\partial/\partial n)(\partial\psi_0/\partial x)df$. Since $\psi_0$ is a function of $r$ alone, (17) can further be transformed into

$$-\frac{2\pi^2\hbar^4}{m^2L^2}(\nu_1^2+\nu_2^2+\nu_3^2)\frac{4\pi}{3}r_e^3\psi_0(r_e)\frac{\partial^2\psi_0(r_e)}{\partial r^2}. \tag{18}$$

Here $\partial^2\psi_0/\partial r^2$ can be calculated from Schrödinger's equation and we obtain for the entire zero-point energy the sum of (16) and (18)

$$E_{\nu_1\nu_2\nu_3}-E_{000}=(2\pi^2\hbar^2/mL^2)(\nu_1^2+\nu_2^2+\nu_3^2)v_0\psi_0(r_e)^2[1+2(E_0-V(r_e)/(E_0-\bar{E}_k)]. \tag{19}$$

This equation shows that the ratio of the actual zero-point energy of the electrons to the zero point energy of free electrons is

$$v_0\psi_0(r_e)^2[1+2(E_0-V(r_e)/(E_0-\bar{E}_k)] \tag{19a}$$

can be larger than 1. It approaches zero for very large $r_e$ but increases greatly with decreasing $r_e$ when the pressing of the valence electrons into the inner shells becomes appreciable. The

[7] This is exactly London's reasoning (Zeits. f. Physik 39, 322 (1926)) with the only difference that the partial integration must be carried through more carefully than in his case.

second term of (19) is negligible, and several values of the first are given in the table below:

| $r_e$ | 3.67 | 4.05 | 4.74 | 5.48 |
|---|---|---|---|---|
| $v_0\psi_0(r_e)^2$ | 1.08 | 0.99 | 0.89 | 0.76 |

Once again, it is found that $\psi_{000}$ behaves at $r=4$ as if it were constant. Moreover it is remarkable that the second term in (19a) increases the Fermi energy over the classical value by a very small amount, if the potential is taken as arising entirely from the ion, and it decreases it somewhat below that value if $\psi_0$ is supposed to be the solution of a Schrödinger equation with the potential (2).

**5.**

If one plots the Fermi energy against $(\nu_1^2 + \nu_2^2 + \nu_3^2)^{\frac{1}{2}}$ Eq. (19) should give the curvature of the plot at the zero point. It would not be right, however, to assume that it is correct for all $\nu$. As a matter of fact we know from the work of F. Bloch[8] that for higher $\nu$ it may depend upon the ratios $\nu_1 : \nu_2 : \nu_3$ and from the work of P. M. Morse, R. Peierls[9] and especially L. Brillouin[10] that it is not a continuous function of the $\nu$'s everywhere but has discontinuities corresponding to wave-lengths for which the Bragg-Laue conditions of x-ray reflections are fulfilled.

According to Brillouin this discontinuity in the space of the $\nu$'s is on the rhombdodecahedron
$$\nu_1 \pm \nu_2 = \pm L/d, \quad \nu_1 \pm \nu_3 = \pm L/d, \quad \nu_2 \pm \nu_3 = \pm L/d$$
where $d$ is the lattice constant in the space of the $\nu$'s. In order to obtain an idea of the size of these discontinuities, we have taken three points of this rhombdodecahedron, corresponding to the three directions (100), (110), (111) and tried to obtain the two energy values for each of them, the transformation properties of the corresponding waves being determined for the symmetry elements of the space group.

It was an easy matter to find a function which had the proper transformation properties within the $s$-polyhedron but it seemed difficult for us to satisfy the continuity conditions at the boundary. The procedure finally adopted possesses no rigorous justification, but it was decided most reasonable to select wave functions which were continuous only at the midpoint of the fourteen boundary planes, with linear combinations of $s$, $p$, $d$, $f$ and $g$ functions. For the point $\nu_1 = \nu_2 = L/2d$, $\nu_3 = 0$ ((110) direction) it was sufficient to take a linear combination of the $s$, $d$ and $g$ functions for the lowest function at the discontinuity and a $p$ function for the upper. For the (111) direction the lower was represented by a linear combination of $p$ and $d$ functions and the upper by a combination of $s$, $f$ and $g$ while for the (100) direction the lower part was a linear com-

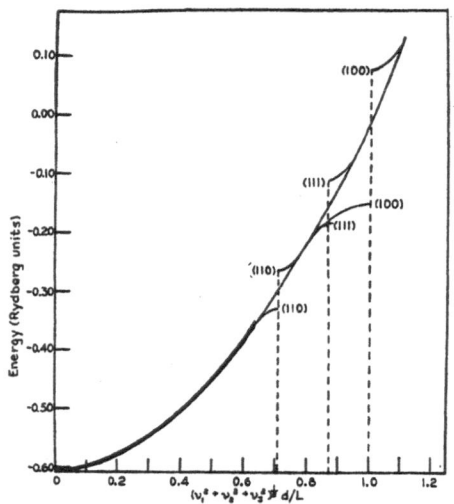

FIG. 3. Graph showing the energy as a function of $(\nu_1^2 + \nu_2^2 + \nu_3^2)^{\frac{1}{2}}d/L$ for wave functions of electrons 'moving' in three prominent crystallographic directions.

bination of $p$ and $d$ functions and the upper was a $d$ function. The results for the observed lattice constants are shown in Fig. 3 in which the abscissa is $(\nu_1^2 + \nu_2^2 + \nu_3^2)^{\frac{1}{2}}d/L$ and the energy is in Rydberg units. The parabola corresponds to free electrons; the heavy section representing the extent to which electrons would occupy the levels, and the other three curves correspond to the · designated direction $\nu_1 : \nu_2 : \nu_3$. In accord with the remarks in previous paragraphs that the electrons behave almost as if free, the discontinuities are not very pronounced. The calculations leading to Fig. 3 do not justify an accuracy greater than 0.05 Rydberg units for absolute values of energy, although the energy differences at the discontinuities are probably more accurate.

In a recent paper which has appeared while this manuscript was in preparation, Slater[11] has investigated this phase of the problem from a viewpoint similar to that presented here. He has made a formal solution satisfying the proper boundary conditions at the centers of the eight hexagonal faces of the $s$-polyhedron, which

[8] F. Bloch, Zeits. f. Physik **52**, 555 (1928).
[9] P. M. Morse, Phys. Rev. **35**, 1310 (1930); R. Peierls, Ann. d. Physik **4**, 121 (1930).
[10] L. Brillouin, Die Quantenstatistik, Berlin 1931, p. 281–316.

[11] J. C. Slater, Phys. Rev. **45**, 794 (1934). Cf. also H. Jones, N. F. Mott, H. W. B. Skinner, Phys. Rev. **45**, 379 (1934).

requires a general function with eight arbitrary parameters that he chose to be a linear combination of one $s$, three $p$, three $d$ and one $f$ function each with arbitrary coefficients. This yields a secular equation of the eighth order which he solves in a degenerate case, namely, when the wave is traveling in a direction orthogonal to one of the coordinate axes. For the case of the (110) direction, which he considers in detail, the results agree, essentially, with those we have obtained and bear out our conclusions completely regarding the accuracy with which the free electron picture represents facts.

6.

The unsatisfactory feature of the previous paragraph is that the wave functions employed are not continuous so that they do not possess an energy-value in a strict sense. It is evident, however, as Slater points out, that a continuous wave function would be obtained by superposing an infinite number of wave functions with different angular factors, which is also possible in the case of a constant potential, though it would be a rather awkward procedure, since we know the wave functions accurately. This would correspond to a development of $e^{2\pi i \nu z/L}$ into a series

$$e^{2\pi i \nu z/L} = f_0(r) + P_1(\theta, \varphi)f_1(r) + P_2(\theta, \varphi)f_2(r), \quad (20)$$

where the $P_l$ are the spherical harmonics and the $f_l$ depend only on the distance from the center of the $s$-polyhedron and are essentially Bessel-functions. If we determine (20) in each $s$-polyhedron, the resulting functions will join each other continuously, since they all represent the same function $e^{2\pi i \nu z/L}$. For the energy value of 0.15 Rydberg units which in the case of free electrons corresponds to $\nu d/L = \frac{1}{2}$, the quantities $rf_l'/f_l$ are given at the surface of the $s$-sphere in the first line of the table below:

| $l$ | $s$ 0 | $p$ 1 | $d$ 2 | $f$ 3 |
|---|---|---|---|---|
| Free electron | $-0.9666$ | 0.049 | 1.647 | 2.7 |
| Electron in lattice | $-1.00$ | 0.044 | 1.674 | 2.8 |

The last line gives the same quantities for the Prokofjew-field of Na for the distance $r_s = 4$ and the energy value $-0.45$ Rydberg units, which lies 0.15 above the energy value $-0.60$ of the electron with the lowest energy. The first two numbers were obtained by graphical integration,

and the others are those for a simple Coulomb field, since the field of the Na ion does not give different values from those of a simple Coulomb field for $d, f$ and higher terms. It is seen that the deviation of the two sets of numbers in the table are very small. Hence, if one adds the $s$, $p$, $d$, etc., solutions of the Prokofjew field with such coefficients that the value of each function at $r = r_s$ is equal to the value of the corresponding function $f_l$ in (20) for the same radius, one obtains a function within the $s$-polyhedron which will join similar functions in the other $s$-polyhedra continuously and even with practically continuous first derivatives. This shows that the energy of the waves for $(\nu_1^2 + \nu_2^2 + \nu_3^2)^{\frac{1}{2}} d/L = \frac{1}{2}$ is higher than that of the wave $\nu_1 = \nu_2 = \nu_3 = 0$ by the same amount (0.15 Rydberg units) in both cases. The same is true for all $\nu$'s which characterize states occupied by electrons.

If the numbers of the first line were smaller than those of the second, the Fermi energy would be smaller for free electrons than in the actual lattice. A more detailed numerical consideration shows that this is actually the case, though only to a very small extent. The situation seems to be opposite, however, if one takes into account the perturbation given in (3). The difference does not amount to more than 1 or 2 percent in either case, however, and we shall employ the numbers corresponding to free electrons in the following.

Of course the coincidence of the two sets of numbers arises simply from the fact that the electrons in our picture behave very nearly like free electrons. We know that the optical properties of metals can also be explained very well on this assumption,[12] so it possesses additional backing.

When all the energy terms calculated in section 1 are added, one obtains the lower curve of Fig. 4 for the energy of the bottom of the Fermi distribution as a function of the lattice constant, which is represented by the radius of the $s$-sphere. The upper curve contains the Fermi energy and gives the total energy of the electrons in the Fock picture. It yields a lattice constant of 4.76A and a binding energy of 9.0 cal.

---

[12] C. Zener, Nature **132**, 968 (1933); R. de L. Kronig, Nature **133**, 211 (1934).

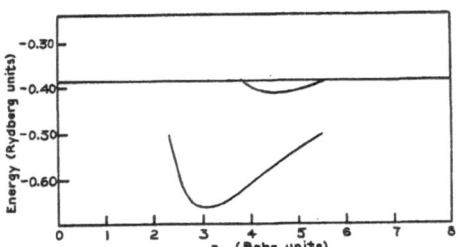

FIG. 4. Energies as functions of $r_s$. The lower curve is the energy of the lowest electronic state while the upper curve represents the energy of the entire lattice computed on the basis of the one-electron picture.

### III. CORRELATIONS BETWEEN ELECTRONS WITH ANTIPARALLEL SPIN

#### 7.

In proceeding with a generalization of the foregoing sections to include electron correlations more general than those allowed by the one electron picture, we no longer have at hand guiding principles that are as definite as those which we had there. Since the energy is to be minimized, the extensions to be made rest principally upon the possibility of forming a hole around every electron in the charge distribution of the other electrons. This possibility is, of course, given by the multi-dimensionality of the wave function, which allows a different probability distribution of the other electrons for different positions of the one considered. Since such a hole is already present in the charge distribution of the electrons with parallel spin, we have principally to consider the possibility of a similar hole in the distribution of the electrons with antiparallel spin. These holes are not included in the one electron picture (cf. reference 3), but we know them to exist in atomic eigenfunctions from the works of Bethe[13] and Hylleraas.[14] They seem to play an even more important rôle in metals than in free atoms.[15]

The wave function employed in the previous sections for $2n$ electrons, may be written under neglect of the spin part, as[16]

$$\Psi(x_1 \cdots x_n, y_1 \cdots y_n) = (1/n!) \begin{vmatrix} \psi_1(x_1) & \cdots & \psi_1(x_n) \\ \cdot & & \cdot \\ \cdot & & \cdot \\ \psi_n(x_1) & \cdots & \psi_n(x_n) \end{vmatrix} \begin{vmatrix} \psi_1(y_1) & \cdots & \psi_1(y_n) \\ \cdot & & \cdot \\ \cdot & & \cdot \\ \psi_n(y_1) & \cdots & \psi_n(y_n) \end{vmatrix}. \tag{21}$$

Here $x_k$ represents the three Cartesian coordinates of the $k$th electron with upward spin and the $y$ refer in a similar way to the electrons with downward spin. The functions $\psi_1, \cdots, \psi_n$, previously designated by $\psi_{r_1, r_2, r_3}$ are the $n$ lowest energy wave functions for an electron in the field of all the ions and the other electrons, regarded as a charge distribution. The task of the previous section, which we believe to be solved to a sufficient degree of accuracy, was to calculate the functions $\psi$ and the energy of the total $\Psi$ of (21) as functions of the lattice constant.

A natural suggestion for the generalization of (21) is to include the coordinates of all electrons of antiparallel spin in $\psi_r(x_k)$ as parameters. This function will then be a different function of $x_k$ for different values of $y_1, \cdots, y_n$, and will have a minimum around every value of the set. We shall consider the effect on the function $\psi_r$ and the energies if we replace the continuous charge distribution of the electrons by point charges at $y_1 \cdots y_n$. This replacement will alter the field acting on the electron $x$ and we shall use a perturbation method to calculate its consequences.

We shall set

$$\Psi(x_1, \cdots, x_n; y_1, \cdots, y_n) = (1/n!) \begin{vmatrix} \psi_1(y_1, \cdots, y_n; x_1) & \cdots & \psi_1(y_1, \cdots, y_n; x_n) \\ \cdot & & \cdot \\ \cdot & & \cdot \\ \psi_n(y_1, \cdots, y_n; x_1) & \cdots & \psi_n(y_1, \cdots, y_n; x_n) \end{vmatrix} \begin{vmatrix} \psi_1(y_1) & \cdots & \psi_1(y_n) \\ \cdot & & \cdot \\ \cdot & & \cdot \\ \psi_n(y_1) & \cdots & \psi_n(y_n) \end{vmatrix} \tag{22}$$

[13] H. Bethe, Zeits. f. Physik 57, 815 (1929).
[14] E. A. Hylleraas, Zeits. f. Physik 48, 469 (1928).
[15] We are much indebted to Professor Slater for the remark that possibly the disagreement between experiment and his theory for the order and distance of atomic terms

due to the same configuration, also arises from the fact that the correlation energy is greater for singlets than for higher multiplets.
[16] M. Delbrück, Proc. Roy. Soc. (London) A129, 686 (1930).

and denote by

$$\Psi_0(x_1, \cdots, x_n; y_1, \cdots, y_n) = \psi_1(y_1, \cdots, y_n; x_1) \cdots \psi_n(y_1, \cdots, y_n; x_n) \cdot |y|, \qquad (22a)$$

the part of (22) which is not yet antisymmetrized in the $x$'s, where $|y|$ is the second determinant in (22). When no ambiguity arises, we shall designate $\psi_\nu(y_1 \cdots y_n, x)$ by $\psi_\nu(x)$, in which it is to be understood that both $x$ and $\nu$ correspond to sets of three numbers. In addition, it will be assumed that both $\psi_\nu(y_1 \cdots y_n, x)$ and its complex conjugate appear in the first determinant of (22) when this function is not real, that the same is true regarding $\psi_\nu(y)$ in $|y|$, and that the $\psi_\nu(y_1 \cdots y_n, x)$ as functions of $x$ are solutions of a characteristic value problem so that

$$\int \psi_\nu(y_1, \cdots, y_n; x)^* \psi_{\nu'}(y_1, \cdots, y_n; x)dx$$
$$= \delta_{\nu\nu'} \quad (23)$$

for all values of $y$.

It is well known that the total energy of $\Psi$ can be written in the form [17]

$$\int\int \Psi^*(K_x + K_y + V_0 + V)\Psi_0 dxdy,$$

where $K_x$ and $K_y$ are the operators for the kinetic energy of the electrons of upward and downward spin respectively, $V_0$ is the old potential function calculated from the viewpoint in which the electrons are regarded as smeared out over the entire crystal, and $V$ is the potential function required to account for the change incurred by viewing the $y$-electrons to be point charges with regard to their interaction with the $x$-electrons. The first three terms are those dealt with in the previous work and yield the wave function (21), while the last is a perturbing term and will yield a perturbation energy. If we retain the old wave functions in the neighborhood of the ions and replace them by a more general solution of (22) in regions where they behave like $e^{2\pi i\nu x/L}$, the expectation value of $V_0$ will be left unchanged since the electron density is changed nowhere, and in addition, it may be seen that (7) remains essentially true. The sum of the energy correction $E$ and the Fermi energy

$F$ will be given, therefore, by the integral

$$E + F = \int\int \Psi^*(K_x + K_y + V)\Psi_0 \qquad (24)$$

taken over the part of the configuration space in which the wave function is changed and which we shall assume to be the entire space.[18] It is to be noted that for an entirely flat wave function the expectation value of $V$, namely

$$\int\int V|y|^2 dxdy \qquad (25)$$

is zero. We shall proceed with a calculation of the energy correction $E$.

**8.**

To begin with, we shall transform the three terms of (24) to a more suitable form. In calculating the expectation value of $\hbar^2/2m \cdot \Delta_x$, we can use (23) to carry out the integration over all of the remaining $x$'s and obtain

$$\int\int \Psi^* K_x \Psi_0 dxdy = -\frac{\hbar^2}{2m}\frac{1}{n!}\Sigma_\nu \int\int \psi^*$$
$$\times \Delta_x \psi_\nu |y|^2 dxdy. \qquad (26)$$

If the same is done for $-\hbar^2/2m \cdot \Delta_y$, the terms arising from the second derivative of the determinant are

$$(4\pi^2\hbar^2/2m)\sum \nu^2. \qquad (27a)$$

Those terms containing the derivatives of the $\psi_\nu(y_1 \cdots y_n, x_\nu)$ and of the determinant may be integrated over all of the $x$'s except $x_\nu$, and lead to expressions of the form

$$\int\int \psi_\nu(x_\nu)^* |y| \frac{\partial \psi_\nu(x_\nu)}{\partial y_\kappa}\frac{\partial}{\partial y_\kappa}|y| dx_\nu dy.$$

If this is added to the similar term containing the derivative of $\psi_\nu^*$ it follows from (23) that the sum will vanish upon integration over $x_\nu$. Finally, those terms which contain the second derivative of $\Psi_0$ are of two types, namely,

---

[17] Cf. e.g. E. Wigner, Gruppentheorie und ihre Anwendung. Braunschweig 1931. p. 323.

---

[18] The difference of the two volumes is negligible.

$$- (\hbar^2/2m)(1/n!)\sum_\nu \int\int \psi_\nu(x_\nu)^* \Delta y_k \psi_\nu(x_\nu) |y|^2 dx_\nu dy \qquad (27b)$$

and

$$- (\hbar^2/2m)(1/n!) \sum_{\nu \neq \mu} \int\int [\psi_\nu(x_\nu)^* \psi_\mu(x_\mu)^* - \psi_\nu(x_\mu)^* \psi_\mu(x_\nu)^*]\cdot \text{grad}_{\nu_k} \psi_\nu(x_\nu)\, \text{grad}_{\nu_k}\psi_\mu(x_\mu)|y|^2 dx_\nu dx_\mu dy.$$

If $\psi_\nu \neq \psi_\mu^*$ we can add to the latter the term in which $\psi_\nu^*$ replaces $\psi_\nu$ and the sum will vanish upon integration over $x$. Hence the only terms resulting from the first term in the bracket are

$$- h(^2/2m)(1/n!)\sum_\nu \int \left| \int \psi_\nu(x_\nu)^*\, \text{grad}_{\nu_k}\, \psi_\nu(x_\nu)dx_\nu \right|^2 |y|^2 dy \qquad (27c)$$

and those from the second are such as to yield

$$- (\hbar^2/2m)(1/n!) \sum_{\mu \neq \nu} \int \left| \int \psi_\mu(x)^*\, \text{grad}_{\nu_k}\, \psi_\nu(x)dx \right|^2 |y|^2 dy. \qquad (27d)$$

If $\mu = \nu$ in (27d), we obtain (27c), so that the sum of (26), (27a, b, c, d) and

$$(1/n!)\sum_\nu \int\int |\psi_\nu(y_1, \cdots, y_n; x)|^2 V(y_1, \cdots, y_n; x)dx |y|^2 dy$$

leads to

$$\frac{1}{n!}\sum_\nu \int\int \psi_\nu(x)^* \left[ V(y_1, \cdots, y_n; x) - \frac{\hbar^2}{2m}\left(\Delta_x + \Delta_{y_1} + \cdots + \Delta_{y_n} + \frac{4\pi^2\nu^2}{L^2}\right)\right]\cdot \psi_\nu(x)dx|y|^2 dy.$$

$$- (\hbar^2/2m)(1/n!)\sum_k \sum_{\mu, \nu} \int \left| \int \psi_\nu(x)\, \text{grad}_{\nu_k}\, \psi_\nu{}^*(x)dx \right|^2 |y|^2 dx = E \qquad (28)$$

for the change in energy since

$$F = (\hbar^2/2m)2\sum_\nu (4\pi^2\nu^2/L^2)$$

is the Fermi energy. The summations over $\mu$ and $\nu$ are to be extended over all occupied states. In the following it will appear from the form given $\psi_\nu(y_1\cdots y_n, x)$ (i.e., see (29)) that the last expression in (28), which could only lower the energy, vanishes.

We shall proceed under the assumption that $\psi_\nu(y_1\cdots y_n, x)$ has the form

$$\psi_\nu(y_1, \cdots, y_n; x) = e^{2\pi i \nu \cdot x/L}\chi_\nu(x - y_1, \cdots, x - y_n), \qquad (29)$$

in which $\chi_\nu$ to the approximation considered here, is a constant plus a sum of functions, each of which depends on one argument $x - y_k$. When substituted in (28) this yields

$$E = \sum_\nu \int\int \chi_\nu{}^*[V(y_1, \cdots, y_n; x) - (\hbar^2/2m)(2\Delta_x + (4\pi i/L)\nu \cdot \text{grad}_x)]\psi_\nu \cdot |y|^2 dx dy. \qquad (30)$$

In order to minimize this $\chi_\nu$ must satisfy the equation

$$- (\hbar^2/2m)\Delta\chi_\nu - (2\pi i\hbar^2/mL)\nu \cdot \text{grad}\, \chi_\nu + V(y_1, \cdots, y_n; x)\chi_\nu = \epsilon_\nu\chi_0(y_1, \cdots y_n; x). \qquad (31)$$

Eq. (31) is much the same as (11), the principal difference being that the relative mass $\tfrac{1}{2}m$ appears instead of $m$.

**9.**

The characteristic values $\epsilon_\nu$ of (31) will depend on the position $y_1 \cdots y_n$ of the electrons with downward spin, and we must average this characteristic value using $|y|^2$ as a weight-function to obtain a final value. It is possible to perform this general calculation to a sufficient degree of accuracy using a modification of the usual Rayleigh-Schrödinger perturbation process, since it does not converge in its usual form, but this involves a rather lengthy computation, so we shall content ourselves here with an estimate of the solution to (31) and present the more general solution at a later time.

It is evident that if the $y$'s are to be most favorable for a large negative value of the correction energy. we shall want the regions in which no $y$ occurs to be as large as possible. In these regions $\chi_\nu(y_1 \cdots y_n, x)$ will have large maxima and the electron $x$ will have the greatest probability of being as far as possible from the electrons of opposite spin. A secondary effect resulting from this will be an alteration of the original Fermi distribution $|y|^2$, and, as a consequence, (8), but we shall not take this into account in the present calculation.[19]

The most disadvantageous configurations for the $y$'s, that is those giving the smallest $-\epsilon_\nu(y \cdots y_n)$, are the close-packed lattices, while the most advantageous is one in which all of the $y$ electrons are at one corner-point of the lattice. This latter is practically impossible for the Fermi distribution, however, while the former possess the greatest probability. A graphical integration of (8), however, shows that the mean distance of the nearest electron with parallel spin to a given electron is $1.1r_s$ in contrast with the nearest distance of $2.3r_s$ for close-packed configurations. For this reason it seems reasonable to us, as far as energy is concerned, to make use of a "mean configuration" of the electrons with downward spin in which pairs of electrons form a body-centered lattice. In this the nearest electron will be much closer than it should be but there are

no clusters of more than two electrons present, which compensates the fact that the pairs are at the most disadvantageous position.

The energy $\epsilon_\nu(y_1 \cdots y_n)$ of this mean configuration can be calculated very simply by the method employed in reference 1; the radius of the sphere surrounding every electron is $(4)^{\frac{1}{3}}r_s$ and the potential energy within the sphere is

$$V = 2e^2/r + e^2r^2/r_1{}^3 - 3.6e^2/r_1, \qquad (32)$$

in which the last term has been introduced in order to satisfy (25). We shall designate $\chi_\nu(y_1, \cdots, y_n, x)$ by $\chi_\nu(x, y, z)$ when the configuration of the $y_1, \cdots, y_n$ introduced here is implied.

In order to determine the $\chi_0$ associated with the lowest energy, we shall regard $V$ in (31) as a perturbation, for which the unperturbed problem is

$$-(\hbar^2/m)\Delta\xi(x, y, z) = \eta\xi(x, y, z). \qquad (33)$$

Each of the solutions of this possess an "azimuthal quantum number" $l$, these for $l=0$ and $l=1$ being of the form $r\xi_{0\kappa} = \int m\omega_\kappa r$, and $r\xi_{1\kappa} = \partial/\partial x \sin \omega_\kappa r$, etc., where the $\omega_\kappa$ are to be determined from the relation $\tan \omega_\kappa r_1 = \omega_\kappa r_1$ arising from the boundary conditions. The normalized solutions for $l=0$, are, to a sufficient degree of accuracy

$$\xi_{00} = (3/4\pi r_1{}^3)^{\frac{1}{2}}; \quad \xi_{01} = \frac{1.024}{(2\pi r_1)^{\frac{1}{2}}}\frac{\sin 1.43\pi r/r_1}{r} \qquad (34)$$

$$\xi_{0\kappa} = (1/2\pi r_1)^{\frac{1}{2}}\frac{\sin (2\kappa+1)\pi r/2r_1}{r} = (1/2\pi r_1)^{\frac{1}{2}}\frac{\sin \omega_1 r}{r}$$

for which the corresponding unperturbed energies are

$$\eta_{00} = 0, \qquad n_{0\kappa} = \hbar^2\omega_\kappa{}^2/m. \qquad (34a)$$

The matrix elements $V_{0\kappa}$ of $V$ may be obtained as follows. We have

$$V_{0\kappa} = \int \xi_{00}V\xi_{0\kappa}dV$$

$$= -(3/4\pi r_1{}^3)^{\frac{1}{2}}(1/\omega_1{}^2)\int V\Delta\xi_{0\kappa}dV, \qquad (35)$$

in which the Laplacian may be made to operate on the $V$ by use of partial integration. From Poisson's equation $\Delta V$ is $-8\pi e^2(\delta(r) - \frac{3}{4}\pi r_1{}^3)$ in which the last term gives rise to a vanishing contribution since it, like $\xi_{00}$, is constant and the

---

[19] It cannot be taken into account without also replacing the effect of electrons with parallel spin on each other by point charges in a similar way to that carried through here for the electrons with antiparallel spin. If this is not done, the two kinds of electrons will push each other into different parts of the crystal, thus invalidating the assumption under which we have calculated their mean effect.

integral over it vanishes in (35). The first term yields

$$V_{0\kappa} = (24)^{\frac{1}{2}} e^2 / \omega_\kappa r_1^2 \qquad (36)$$

so that the solution of (31) for $\nu = 0$ is

$$\chi_{00} = (3/4\pi r_1^3)^{\frac{1}{2}} - \sum_\kappa{}' \frac{V_{0\kappa}}{\hbar^2 \omega_\kappa^2 / m + V_{\kappa\kappa}} \xi_{0\kappa} \qquad (37)$$

and the corresponding energy is[20]

$$1/(16.2 + 3.07 r_s) + 0.005. \qquad (38)$$

Although this is given in atomic units, the correction is true for only half of the electrons and must be taken in Rydberg units. The solution (37) has also been obtained by use of numerical integration and is illustrated in Fig. 5.

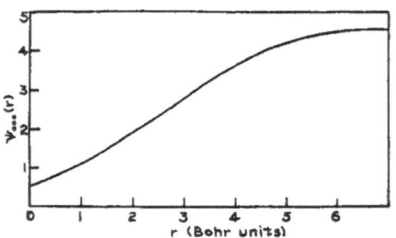

FIG. 5. Wave function corresponding to a solution of (30) with $\nu = 0$ and $r_s = 4$.

The wave function of an electron has only about 30 percent of its mean value in the region where two other electrons are situated and only 50 percent in the region where there is an electron with antiparallel spin.

For $\nu$ other than 0, we shall have greater energy gains because the Fermi distribution will be narrower as a result of the presence of the field arising from the $y_1 \cdots y_n$ electrons than it would be without this field. In this case we can take the second term of (35) into account as a perturbation, the calculation being very similar to that employed in Section 4. The first approximation is zero, while the second, namely

$$(4\pi^3\hbar^4 / m^2 L^2)\nu^2 \sum_\kappa |\int(\partial\Psi_{00}/\partial x)\chi_\kappa|^2 / - E_\kappa \qquad (39)$$

is evidently negative, and would vanish if $\chi_{00}$ were entirely flat. Because of the selection rules,

it is sufficient to take wave functions with $l = 1$ for $\chi_\kappa$, which are simply derivatives of the functions (34) to a first approximation, while $\chi_{00}$ must be taken from (37). The sum (39) may be conveniently compared with the Fermi energy of free electrons and a simple calculation shows that it is 4 percent of this quantity.

If this result and (38) are subtracted from the upper curve in Fig. 4, one obtains Fig. 6 which

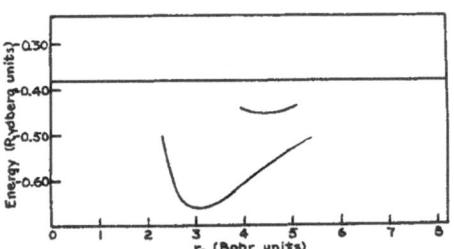

FIG. 6. The upper curve replaces that of Fig. 4 and contains the results arising from the correction to the one-electron picture.

yields $r_s = 4.40$ or a lattice constant of 4.75A, and a heat of sublimation of 23.2 kg. cal. The experimental values are 4.23A and 26.9 kg. cal., respectively. For the observed lattice constant the calculated energy is 18.6 cal. It may be mentioned at this point that the preceding calculation of the correlation energy is valid only in the neighborhood of $r_s = 4$.

It is to be noted that the actual $\chi_\nu$ will not have, in general, forms such as those we have calculated here. For irregular configurations of the electrons with downward spin they will be rather complicated linear combinations of these functions, possessing energies that are nearly the same, however. Each function $\psi_\nu$ going with an energy $\epsilon_\nu$ will have a maximum in that region in which the potential has a trough allowing such a characteristic value and will be small everywhere else as a result of interference of the superposed waves.[21] This does not affect the actual wave function (22), however, since this is a determinant of the $\psi_\nu$ and does not change upon taking such a linear combination.

---

[20] The $V_{\kappa\kappa}$ must be calculated graphically. We obtained $V_{11} = 2.4\ e^2 / r_1$.

[21] Cf. the discussion of a very similar problem by F. Hund, Zeits. f. Physik 40, 742 (1927).

522                          E. WIGNER AND F. SEITZ

**10.**

It is hardly necessary to mention that the calculation of the last section must be regarded only as an attempt to find the correct wave function for the electrons in the metal, and we are well aware that we could guess its form only roughly. We should like to mention that the deviation of the wave function from the determinantal form (21) using plane de Broglie waves is much more pronounced here than in the one electron picture (cf. Fig. 5) in which the plane waves were a surprisingly good representation. They still form a reasonably good approximation to the true state of affairs, however, and for this reason we believe that the fact that (22) is not symmetric under interchange of $x$ and $y$ is not of extremely great importance. The function (22) still possesses this property if we consider the differences of the $\psi$, from the de Broglie waves as small quantities and neglect the second order terms.

The remaining discrepancy of 3.7 kg. cal. we believe to arise from two sources. First it is not evident that the Prokofjew field is completely suitable for calculations of this kind and, in fact, it is surprising that the atomic levels are as good as we have found them to be. Rigorously the exchange terms between the valence electrons and the closed shells cannot be transformed into the form of potentials because they are represented by more general operators, and there is a possibility that the effect of these is not described accurately enough for the states occurring in the crystal by an effective potential field. This important point is being more accurately investigated in connection with a treatment of Li to appear shortly and it is hoped that more light may be thrown upon it.

On the other hand, it is also possible that the actual wave function is not represented to a sufficient degree of accuracy by a wave function of the form (2). It does not seem to be easy to use wave functions of a greater generality, however.

### Appendix 1

It is possible to continue the wave function with the same spherically symmetric potential over the whole volume of the $s$-polyhedron. If one assumes this $\psi$ to be valid in all the $s$-polyhedrons of the lattice, the resulting wave function will be continuous everywhere, but the derivatives will be discontinuous at the boundary planes. The energy, therefore, is not the characteristic value of the differential equation, but must be calculated as the sum of kinetic and potential energy. The correction for the potential energy will be given in Appendix 2, while the correction for the kinetic energy is

$$(\hbar^2/2m)\left\{\int_{V_1} (\text{grad } \psi)^2 d\vartheta - \int_{V_2} (\text{grad } \psi)^2 d\vartheta\right\}, \tag{1}$$

where $V_1$ is the region of the $s$-sphere outside of the $s$-polyhedron and $V_2$ is the equally large volume of the $s$-polyhedron outside of the $s$-sphere. In the neighborhood of the $s$-sphere, we can set

$$\frac{\hbar^2}{2m} \text{ grad } \psi = \frac{\hbar^2}{2m} \frac{d^2\psi(r_s)}{dr^2}(r - r_s) = (V(r_s) - E)\psi(r_s)(r - r_s), \tag{2}$$

so that (1) becomes

$$(2m/\hbar^2)(V(r_s) - E)^2 \psi(r_s)^2 \left\{\int_{V_1} (r - r_s)^2 dv - \int_{V_2} (r - r_s)^2 dv\right\}. \tag{3}$$

Now the mean value of $r - r_0$ is certainly smaller than $r_0/20$, $V_1$ is smaller than $v_0/10$, and $\psi(r_s)^2 v_0 \sim 1$, so that the first part of (3) is

$$(2mr_s^2/\hbar^2)(V(r_s) - E)^2/4000 = 0.0004 \text{ Rydberg units.}$$

In addition to this, the second term of (3) essentially cancels the first, so that the total correction really is negligible.

## APPENDIX 2

It is necessary to calculate the potential energy of a body-centered lattice of positive point charges in a fluid of negative charge of density $\rho = \psi^2$ and then to compare this energy with the electrostatic energy of the $s$-sphere.

To begin with, we can draw $s'$-spheres around each positive charge which are smaller than the $s$-spheres and just touch one another. Outside of these $s'$-spheres we may assume the density to be uniform and of magnitude $\rho$. Relative to the regions of space outside of the lattice, the energy of the $s'$-spheres is as large as if the negative charge had a constant density within the sphere. This is because the potential outside a sphere arising from a spherically symmetric charge distribution is independent of the radial distribution inside of the sphere.

The total electrostatic energy of the lattice may now be divided into two parts: first, the energy $E_1$ of a body-centered lattice of positive charges $v_s\rho = e$ with a uniform negative charge-density and second the difference, $E_2$, of the inner energy of an $s'$-sphere with the actual $s'$-sphere on the one hand and an $s'$-sphere with uniform charge distribution on the other. Since the density is actually $\rho$ outside of the $s'$-sphere, the second energy is equal to the difference in energy of the actual $s$-sphere and one with constant negative distribution $\rho$. This second part is

$$E_2 = E_t - (3^{5/3}\pi^{1/3}/5)(\rho^2 v_0^2/d) = E_t - 3.6557\,\rho^2 v_0^2/2d,$$

where $E_t$ is the energy of the actual $s$-sphere, that is, the energy employed in the text, and $-3.6557\,\rho^2 v_0^2/2d$ is the energy of an $s$-sphere with uniform charge distribution.

If $E_1$ were $-3.6557\,\rho^2 v_0^2/4d$ our calculation would be exactly correct. Since it is actually $-3.6391\,\rho^2 v_0^2/4d$, as will be shown at once, the total energy is smaller than that used in the text by the amount

$$0.0166\,\rho^2 v_0^2/2d = 0.004 e^2/r_s,$$

which for $r_s = 4$ is only 0.62 kg. cal.

The calculation of $E_1$ is a problem of electrostatics, we have used essentially the procedure of Appell-Madelung. Since the proper energy of the positive point charge, which must be obtained, is infinite, we must assume positive charges of finite size concentrated into small cubes with edges $2\delta$.

The Fourier expansion of the density is, then,

$$\sum_{\kappa,\,\lambda,\,\mu = -\infty}^{\infty} a_{\kappa\lambda\mu} \exp\left[2\pi i(\kappa x + \lambda y + \mu z)/d\right]$$

with

$$a_{\kappa\lambda\mu} \begin{cases} 2\rho v_0 \sin\alpha\kappa \sin\alpha\lambda \sin\alpha\mu/d^3\alpha^3\kappa\lambda\mu, \\ 0 \end{cases}$$

where $\alpha = 2\pi\delta/d$ and the first line is valid for $\kappa + \lambda + \mu$ even and the second for this odd in addition to the case $\kappa = \lambda = \mu = 0$.

The total energy per atom of the lattice is

$$E_{11} = \sum d^5 |a_{\kappa\lambda\mu}|^2/4\pi(\kappa^2 + \lambda^2 + \mu^2).$$

The density distribution of the positively charged cube may be written in the form of the Fourier integral

$$\int\int\int_{-\infty}^{\infty} a(\kappa\lambda\mu) \exp\left[2\pi i(\kappa x + \lambda y + \mu z)/d\right] d\kappa d\lambda d\mu$$

with

$$a(\kappa\lambda\mu) = \rho v_s \sin\alpha\kappa \sin\alpha\lambda \sin\alpha\mu/d^3\alpha^2\kappa\lambda\mu$$

E. WIGNER AND F. SEITZ

and the corresponding energy is

$$E_{12} = (d^5/2\pi) \int\int\int_{-\infty}^{\infty} |a(\kappa\lambda\mu)|^2/(\kappa^2+\lambda^2+\mu^2)d\kappa d\lambda d\mu.$$

The energy $E_1$ is now simply $E_{11}-E_{12}$. If we put $\delta=0$ both $E_{11}$ and $E_{12}$ diverge, but if we take the difference before passing to the limit and then set $\delta=0$, we obtain

$$E_1 = \lim_{N\to\infty}\lim_{\delta\to 0} \sum_{\kappa,\lambda,\mu=-N}^{N} \frac{d^5}{4\pi} \frac{|a_{\kappa\lambda}|^2}{\kappa^2+\lambda^2+\mu^2} - \frac{d^5}{2\pi}\int\int\int_{(-N-\frac{1}{2})}^{(N+\frac{1}{2})} \frac{|a(\kappa\lambda\mu)|^2}{\kappa^2+\lambda^2+\mu^2}d\kappa d\lambda d\mu$$

$$= \lim_{N\to\infty} \frac{\rho^2 v_0^2}{\pi d} \left\{ \sum_{-N}^{N}{}' \frac{1}{\kappa^2+\lambda^2+\mu^2} - \frac{1}{2}\int\int\int_{(-N-\frac{1}{2})}^{(N+\frac{1}{2})} \frac{d\kappa d\lambda d\mu}{\kappa^2+\lambda^2+\mu^2} \right\}.$$

In $\sum'$, the values for which $\kappa+\lambda+\mu$ is odd and the case $\kappa=\lambda=\mu=0$ are to be omitted.

The summation and integration over $\mu$ were carried out directly and yield

$$E_1 = \frac{\rho^2 v_0^2}{2d} \left\{ \sum_{\kappa+\lambda \text{ odd}} \left(\frac{1}{\kappa^2+\lambda^2}\right)^{\frac{1}{2}} \text{tgh} \frac{\pi}{2}(\kappa^2+\lambda^2)^{\frac{1}{2}} + \sum_{\kappa+\lambda \text{ even}} \left(\frac{1}{\kappa^2+\lambda^2}\right)^{\frac{1}{2}} \text{ctgh} \frac{\pi}{2}(\kappa^2+\lambda^2)^{\frac{1}{2}} - \int\int \frac{d\kappa d\lambda}{(\kappa^2+\lambda^2)^{\frac{1}{2}}} \right\}.$$

which was then calculated directly. The tgh and ctgh may be replaced by 1 when either $\kappa$ or $\lambda$ is greater than 3. For $\kappa$ or $\lambda$ greater than 9, the bracketed difference is $1/24(\kappa^2+\lambda^2)^{\frac{3}{2}}$ with sufficient accuracy, and the integration over $\kappa$ and $\lambda$ was carried out in this region. The terms for which $\kappa\leqq 3$, $\lambda\leqq 3$ yielded $-3.57221$, while the second sum gave

$$E_1 = -3.6391\rho^2 v_0^2/2d.$$

The van der Waals attraction of the ions is $-0.12$ cal. per mol.

*Note added in proof.* In Fig. 3 the upper point of the (110) discontinuity, given by a pure $p$ function, lies at $-0.33$ Rydberg units in place of the value shown. The lower is at $-0.34$. Although the first lies below the parabola, this is undoubtedly of no real significance.

# On the Interaction of Electrons in Metals

E. P. Wigner

Physical Review *46*, 1002–1011 (1934)

(Received October 15, 1934)

The energy of interaction between free electrons in an electron gas is considered. The interaction energy of electrons with parallel spin is known to be that of the space charges plus the exchange integrals, and these terms modify the shape of the wave functions but slightly. The interaction of the electrons with antiparallel spin, contains, in addition to the interaction of uniformly distributed space charges, another term. This term is due to the fact that the electrons repel each other and try to keep as far apart as possible. The total energy of the system will be decreased through the corresponding modification of the wave function. In the present paper it is attempted to calculate this "correlation energy" by an approximation method which is, essentially, a development of the energy by means of the Rayleigh-Schrödinger perturbation theory in a power series of $e^2$.

## 1.

THE attempt has been made in previous work[1] to give a more general expression for the wave function of free electrons in metals than that provided by Hartree's method of the self-consistent field[2, 3] or Fock's equations. The form of the wave function assumed in Fock's equations for a system of $2n$ electrons, occupying $n$ doubly-degenerate states is

$$\frac{1}{n!} \begin{vmatrix} \psi_1(x_1) & \cdots & \psi_1(x_n) \\ \cdot & & \cdot \\ \cdot & & \cdot \\ \psi_n(x_1) & \cdots & \psi_n(x_n) \end{vmatrix} \begin{vmatrix} \psi_1(y_1) & \cdots & \psi_1(y_n) \\ \cdot & & \cdot \\ \cdot & & \cdot \\ \psi_n(y_1) & \cdots & \psi_n(y_n) \end{vmatrix}, \quad (1)$$

where $x$ stands for three Cartesian coordinates of electrons with upward spin, and $y$ for those of electrons with downward spin. The $\psi_\nu$ are the solutions of a Schrödinger equation in which the potential of the charge distribution of the other electrons enters as well as the potential arising from the ions.

In a metal the charge distribution of all electrons is practically unaltered by removing one so that the second quantity may be replaced by the former and the potential for a given electron at the point $u$ is given by adding to the Coulomb field of the ions the fields of all electrons with parallel and with antiparallel spin. The former distribution may be obtained by inserting $u$ for $x_n$ in (1) and integrating over all coordinates except $x_1$ and $u$, while the latter is obtained by a similar operation with the exception that the integration should be carried out over all coordinates except $y_1$ and $u$.

Actually, it had been shown in[1, 4] that the wave functions $\psi_\nu$ of the free electrons in a Na-lattice are very nearly plane waves $e^{2\pi i \nu \cdot x/L}$ where $L$ is the cube edge of the crystal and $\nu$ stands for a set of three integers, $\nu \cdot x$ denotes the scalar product of $\nu$ and $x$. Hence the charge distribution of the electrons with opposite spin is practically uniform, that of the electrons with parallel spin uniform with a "hole" around $u$.[5]

In no wave function of the type (1) is there a statistical correlation between the positions of electrons with antiparallel spin. The purpose of the aforementioned generalization of (1) is to allow for such correlations. This will lead to an improvement of the wave function and, therefore, to a lowering of the energy value. This energy gain will be called "correlation energy."

## 2.

The new form of the wave function, assumed in[1] was

$$\frac{1}{n!} \begin{vmatrix} \psi_1(y_1 \cdots y_n; x_1) & \cdots & \psi_1(y_1 \cdots y_n; x_n) \\ \cdot & & \cdot \\ \cdot & & \cdot \\ \psi_n(y_1 \cdots y_n; x_1) & \cdots & \psi_n(y_1 \cdots y_n; x_n) \end{vmatrix} \begin{vmatrix} \psi_1(y_1) & \cdots & \psi_1(y_n) \\ \cdot & & \cdot \\ \cdot & & \cdot \\ \psi_n(y_1) & \cdots & \psi_n(y_n) \end{vmatrix}, \quad (2)$$

[1] E. Wigner and F. Seitz, Phys. Rev. 46, 509 (1934).
[2] D. R. Hartree, Proc. Camb. Phil. Soc. 24, 89 (1928).
[3] J. C. Slater, Phys. Rev. 35, 210, 1930; V. Fock, Zeits. f. Physik 61, 126 (1930).
[4] J. C. Slater, Phys. Rev. 45, 794 (1934); A. Sommerfeld and H. Bethe, Geiger-Scheel's Handbuch der Physik, Vol. 24, 2nd part, 2nd edition, p. 406.
[5] E. Wigner and F. Seitz, Phys. Rev. 43, 804 (1933).

which contains the functions $\psi_\nu(y_1\cdots y_n;x)$ which are different functions of $x$ for different configurations $y_1,\cdots,y_n$ of the electrons with opposite spin, instead of the $\psi_\nu(x)$. It is proposed to find the best wave function, i.e., that with the lowest total energy of the form[6] (2). The $y_1,\cdots,y_n$ in $\psi_\nu(y_1\cdots y_n;x)$ are to be viewed merely as parameters. (Cf. Eq. (23) ref. 1.) The total energy of the wave function (2) was previously calculated and relative to the solution of Fock's equations yielded the following energy

$$E_1=\frac{1}{n!}\int dy\,|y|^2\left\{\sum_\nu \epsilon_\nu-\frac{h^2}{2m}\sum_{\nu\nu'}\sum_\kappa\left|\int\psi_{\nu'}(y;x_1)^*\frac{\partial\psi_\nu(y;x_1)}{\partial y_\kappa}dx_1\right|^2\right\},\tag{3}$$

where $|y|$ denotes the second determinant of (1), the summation $\kappa$ runs over all $3n$ coordinates $y$ and that over $\nu$ and $\nu'$ over all occupied states, i.e., over all indices occurring in the wave functions in (1) or (2); $\psi_\nu(y;x)$ stands for $\psi_\nu(y_1\cdots y_n;x)$ and $dy$ for $dy_1\cdots dy_n$. The quantities $\epsilon_\nu$ are the integrals

$$\epsilon_\nu(y_1\cdots y_n)=\int\psi_\nu(y;x_1)^*\{V-(h^2/2m)(\Delta_{x_1}+\Delta_{y_1}+\cdots+\Delta_{y_n})\}\psi_\nu(y;x_1)dx_1$$

$$+\int\psi_\nu(x_1)^*(h^2/2m)\Delta_{x_1}\psi_\nu(x_1)dx_1,\tag{3a}$$

where again $V(y_1\cdots y_n;x)$ is the difference between the potentials at the point $x$ of a charge distribution corresponding to $\psi_1(y),\psi_2(y),\cdots,\psi_n(y)$, on the one hand, and point charges at $y_1,\cdots,y_n$, on the other.

It is necessary now to assume for $\psi_\nu(y_1,\cdots,y_n;x)$ the form·

$$\psi_\nu(y;x)=\psi_\nu(x)\{1+f_\nu(y_1-x)+f_\nu(y_2-x)+\cdots+f_\nu(y_n-x)\}\tag{4}$$

and that if $\nu$ and $\nu'$ are both occupied states

$$\int\psi_{\nu'}(y;x_1)^*(\partial\psi_\nu(y;x_1)/\partial y_\kappa)dx_1=0\tag{4a}$$

so that the second term in (3) vanishes. Both (4) and (4a) will turn out to be correct in the approximation to be used. By means of (4) it is possible to transform (3a) so as to get rid of all derivatives with respect to the $y$, after which one may minimize (3) by minimizing $\epsilon_\nu(y_1,\cdots,y_n)$ for every combination of the $y$. This would lead to a differential equation for the $\psi_\nu(y_1\cdots y_n;x)$ in which the $y$ would be merely parameters. (The solutions $\psi_\nu(x)$ of Fock's equations which also enter into this equation are supposed to be known.) The result of the transformation is especially simple if one uses for $\psi_\nu(x)=\epsilon^{2\pi i\nu\cdot x/L}$, namely,

$$\epsilon_\nu(y_1\cdots y_n)=\int\psi_\nu(y;x)^*\{V-(h^2/m)(\Delta_x-(2\pi i\nu/L)\,\mathrm{grad}_x)\}\psi_\nu(y;x)dx.\tag{3b}$$

In addition to the energy contribution (3) which is negative and was calculated in[1] there is a further one which is generally positive. This arises from the fact that the probabilities of the relative distances of electrons with upward spin are changed by the transition from the $\psi_\nu(x)$ to the $\psi_\nu(y_1,\cdots,y_n;x)$. Since the latter will be large for $x$'s, which lie in regions comparatively free from $y$'s, the distribution of the $x$'s will not be uniform throughout space and they will be nearer together than they were under the previous assumption.

---

[6] The form (2) of the wave function is certainly not the correct one. It does not belong even to one single multiplicity but is a linear combination of functions of different "multiplicities" (belonging to different representations of the symmetric group). If, however, the functions $\psi_\nu(y_1\cdots y_n;x)$ are not too different from the functions $\psi_\nu(x)$, they all belong to very low multiplicities. This is the only case, anyway, in which the present approximation is good and it can be expected that the real wave function is in this case near to (2).

In order to evaluate this energy change, one may first calculate the probability of two electrons with parallel spin being at the points $x_1$ and $x_2$, respectively, if the complete wave function is given by (1). Under the approximate assumption that $\psi_\nu(x) = \psi_0(x)e^{2\pi i \nu \cdot x/L}$ one obtains for this in the way given in reference 5,

$$|\psi_0(x_1)\psi_0(x_2)|^2 g(|x_1 - x_2|),\tag{5}$$

where

$$g(r) = 1 - 9\left(\frac{\sin(r/d) - (r/d)\cos(r/d)}{(r/d)^3}\right)^2\tag{5a}$$

is the probability of the distance $|x_1 - x_2| = r$ for free electrons with parallel spin,

$$d = (v_0/3\pi^2)^{\frac{1}{3}} = (4/9\pi)^{\frac{1}{3}}r_s = 0.521 r_s\tag{6}$$

is $1/(2\pi)$ times the wave-length of the fastest electron, $v_0$ is the atomic volume and $r_s$ the radius of the sphere with this volume. If we make the assumption $\psi_\mu(y, x) = \psi_\nu(y, x)e^{2\pi i(\mu-\nu)\cdot x/L}$, an expression of the form (5) is valid for the wave function (2) as well as for (1) and the change of mutual potential energy of the electrons with upward spin arising from the transition from (1) to (2) is

$$E_2 = \frac{1}{n!}\int dy\,|y|^2 \sum_\nu \epsilon_\nu'(y_1\cdots y_n),\tag{7}$$

where

$$\epsilon_\nu'(y_1\cdots y_n) = \int dx_1 dx_2 (ne^2/2r)\{|\psi_\nu(y;x_1)\psi_\nu(y;x_2)|^2 - |\psi_\nu(x_1)\psi_\nu(x_2)|^2\}g(r).\tag{7a}$$

The $\frac{1}{2}$ enters once again because the interaction of a pair of electrons should be counted once only. In order to be able to evaluate the integral (7a), $g(r)$ has been replaced by

$$1 - e^{-1.6r/d}(1 + 1.6r/d + 1.2(r/d)^2),\tag{8}$$

which, as is shown in Fig. 1, runs rather near to $g(r)$.

### 3.

The task of Section 4 will be to calculate the wave functions $\psi_\nu(y_1\cdots y_n; x)$ which minimize the sum of expressions (3b) and (7a) and to calculate $E_1 + E_2$ corresponding to these wave functions. Before doing this, however, an estimate of the order of magnitude of the effect to be expected should be given. This can be taken from calculations of atomic spectra by the method of Fock's equation or Hartree's field, and their comparison with experimental results. The best result in this connection seems to be that on the normal state of He, where Fock's equation is identical with Hartree's. The discrepancy here is[7] $0.077 Ry$ (Rydberg units) for both electrons, or 12 Cal. per electron. This must be the amount of correlation energy in He.

The situation is somewhat more complicated in the calculation of the terms of $O^{++}$, $O^+$, and O by Hartree and Black,[8] since Hartree's method has been used instead of Fock's, and also because in the latter cases more than two electrons play important rôles. For $O^{++}$ the differences between observed values (in brackets) and theory are

$$^3P(4.050)0.074; \quad {}^1D(3.868)0.090;$$
$$^1S(3.658)0.176.$$

If we denote the radial wave function for the electrons by $P(r)$, these terms correspond to the linear combinations $P(r)P(r')$ multiplied by

$$xy' - yx'; \quad (x+iy)(x'+iy'); \quad xx'+yy'+zz'.$$

In the first case the correlation energy is very small, since the electrons are probably far away

[7] Cf. D. R. Hartree and A. L. Ingman, Mem. of the Manchester Lit. and Phil. Soc. 77, 69, 87 (1933).

[8] D. R. Hartree and M. M. Black, Proc. Roy. Soc. A139, 311 (1933).

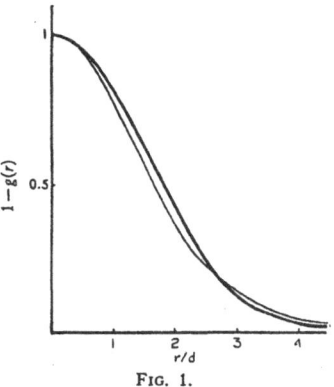

FIG. 1.

from each other anyway, because their spins are parallel. In such cases in the theory of metals we do not take into account any correlation energy at all. The discrepancy in this case may, to a considerable extent, be due to the use of Hartree's instead of Fock's method, and certainly would be further diminished by the use of the latter. On the other hand, in the case of singlet's, Hartree's equations correspond much more closely to Fock's equation and the increased discrepancy of about $0 \cdot 050 Ry$ is probably due to the neglect of the correlation energy. It is not quite clear, however, why it is so much greater in the $^1S$ than in the $^1D$ term.

A comparison of experimental and theoretical values in O⁺ and O points in a similar direction, the correlation energy is smaller though in these cases by a factor of the order 2. It is evident that it must diminish eventually if one goes over to more and more loosely bound electrons, since because of the lower electron densities the total interaction energy diminishes and the correlation energy is only the non-appearance of part of this.

If one goes over to a metal like Na, on first sight the effect could be expected (because of the low electron density) to be much smaller than in He, about as great as in O. This will not be quite so, however, because the fluctuations in the potential of the electrons with downward spin will be greatly increased by their great number. The effect to be calculated in the next section will be about equal, therefore, to the effect per electron in He.

4.

Although the actual wave functions $\psi_\nu(x)$ in the Na lattice are actually different from plane waves $e^{2\pi i \nu \cdot x/L}$ in the Hartree-Fock approximation, we shall use plane waves for $\psi_\nu(x)$ in the subsequent calculation. Since only integrals over the unperturbed functions occur in the perturbation calculation, this will not introduce a great error, because the $\psi_\nu(x)$ are extremely near to plane waves in much the greatest part of the volume.

The whole following calculation will be performed in the approximation which corresponds to the second approximation in the Rayleigh-Schrödinger perturbation theory. The wave functions will be developed into a Fourier series

$$\psi_\nu(y;x) = L^{-\frac{1}{2}}(e^{2\pi i \nu \cdot x/L} + \sum_\mu \alpha_{\nu\mu} e^{2\pi i \mu \cdot x/L}), \quad (9)$$

$e^{2\pi i \nu \cdot x/L}$ being taken as unperturbed function and $V(y_1, \cdots, y_n; x)$ the perturbation. The $\alpha_{\nu\mu}$, which are functions of the $y$, are supposed to be small, so that third order terms of $\alpha$ and $V$ will be neglected for the energy, and second order terms for the wave function.

For the actual calculation it can be seen, first of all, that one can replace a set of $\psi_\nu(y;x)$ by any orthogonal linear combination of them without affecting the final result. The orthogonality condition between $\psi_\nu(y;x)$ and $\psi_{\nu'}(y;x)$ gives

$$\alpha_{\nu\nu'} + \alpha_{\nu'\nu}{}^* + \sum_\mu \alpha_{\nu\mu} \alpha_{\nu'\mu}{}^* = 0. \quad (10)$$

By Schmidt's method one can build a set of orthogonal $\psi_\nu(y;x)$ such that $\alpha_{\nu\nu'} = 0$ for $\nu < \nu'$

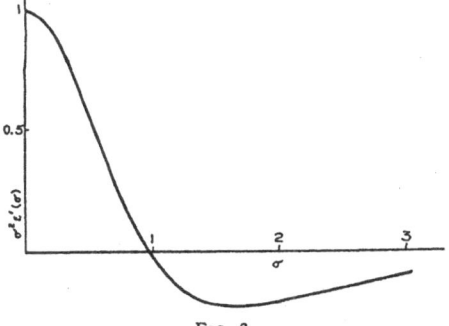

FIG. 2.

if $\nu'$ is occupied. It then follows under omission of the last term in (10) that $\alpha_{\nu\nu'}=0$ if $\nu$ and $\nu'$ are both occupied states, whence the summation over $\mu$ in (9) must be extended only over the unoccupied states. It is then seen at once that (4a) is satisfied in our approximation.

In order to calculate $\epsilon_\nu(y_1\cdots y_n)$ one must first calculate the matrix elements of $V(y_1,\cdots,y_n;x)$. The matrix elements $V_{\nu\nu}(y_1,\cdots,y_n;x)=V_{00}$ will

not be zero for all values of the $y$, but they evidently will not change the wave functions and merely give a contribution to the energy. The average value over all configurations of this contribution is zero, however, since the mean value of the potential of different configurations of charges is equal to the potential of the mean charges. Therefore, we shall set $V_{\nu\nu}(y_1,\cdots,y_n;x)=0$. For $\mu\neq\nu$ we have

$$V_{\nu\mu}(y_1\cdots y_n;x)=L^{-3}\int e^{2\pi i(\mu-\nu)\cdot x/L}V(y;x)dx$$

$$=-(1/4\pi^2(\mu-\nu)^2 L)\int e^{2\pi i(\mu-\nu)\cdot x/L}\Delta V(y;x)dx \tag{11}$$

$$=(e^2/\pi(\mu-\nu)^2 L)\sum_{\kappa=1}^{n}e^{2\pi i(\mu-\nu)\cdot y_\kappa/L}$$

because of Poisson's equation.

For $\epsilon_\nu$, (3b) yields

$$\epsilon_\nu(y_1\cdots y_n)=\sum_\mu(\alpha_{\nu\mu}V_{\nu\mu}+\alpha_{\nu\mu}{}^*V_{\nu\mu}{}^*)+\sum_{\mu\mu'}\alpha_{\nu\mu}{}^*\alpha_{\nu\mu'}V_{\mu\mu'}+\sum_\mu\frac{4\pi^2 h^2}{mL^2}(\mu^2-\mu\cdot\nu)|\alpha_{\nu\mu}|^2, \tag{12}$$

In order to calculate the $\epsilon_\nu'(y_1\cdots y_n)$ by means of (7a), the charge distribution for $\psi_\nu(y;x)$ will first be found:

$$|\psi_\nu(y;x)|^2=L^{-3}+L^{-3}\sum_\mu(\alpha_{\nu\mu}e^{2\pi i(\mu-\nu)\cdot x/L}+\alpha_{\nu\mu}{}^*e^{2\pi i(\nu-\mu)\cdot x/L})+L^{-3}\sum_{\mu\mu'}\alpha_{\nu\mu}{}^*\alpha_{\nu\mu'}e^{2\pi i(\mu-\mu')\cdot x/L}. \tag{13}$$

Now $g(r)$ is 1 minus the function of the "hole," which decreases rapidly with increasing $r$. The 1, inserted into (7a), simply gives the energy difference of a uniform charge distribution and that corresponding to $\psi_\nu(y;x)$, namely,

$$4\pi ne^2 L^{-3}\left\{\sum_\mu\frac{|\alpha_{\nu\mu}|^2}{4\pi^2(\mu-\nu)^2/L^2}+R\sum_{\substack{2\nu-\mu\\ \text{unocc.}}}\frac{\alpha_{\nu\mu}\alpha_{\nu 2\nu-\mu}}{4\pi^2(\mu-\nu)^2/L^2}\right\}, \tag{14a}$$

in which the terms higher than the second order in $\alpha$ are omitted, and $R$ means that the real part of the following expression is to be taken. The second sum must be taken only over those $\mu$'s for which $2\nu-\mu$ is unoccupied, while $\mu$ itself is an unoccupied state in all summations. For the calculation of the other part of (7a), arising from the function of the hole, one sets $x_1+r$ for $x_2$ in (7a), introduces (13), and performs the integration over $x_1$. The result

$$2L^{-3}R\{\sum_\mu|\alpha_{\nu\mu}|^2+\sum_{\substack{2\nu-\mu\\ \text{unocc.}}}\alpha_{\nu\mu}\alpha_{\nu 2\nu-\mu}\}e^{2\pi i(\nu-\mu)\cdot r/L}$$

must be multiplied with $ne^2/2r$ and the function of the hole $g(r)-1$, which must be taken from (8), and integrated over $r$. Setting $\sigma=2\pi(\mu-\nu)d/L$, this yields

$$-\frac{4\pi d^2 ne^2}{2.56L^3}R\left(\sum_\mu|\alpha_{\nu\mu}|^2+\sum_{\substack{2\nu-\mu\\ \text{unocc.}}}\alpha_{\nu\mu}\alpha_{\nu 2\nu-\mu}\right)\cdot\left(\frac{1}{1+\sigma^2/2.56}+\frac{1.063}{(1+\sigma^2/2.56)^2}+\frac{3.75}{(1+\sigma^2/2.56)^3}\right) \tag{14b}$$

Added to (14a) this gives with help of the relation $4\pi r_s{}^3/3=v_0$

$$\epsilon_{\nu}{}'(y_1 \cdots y_n) = \frac{2^{\frac{1}{3}}}{3^{\frac{1}{3}}\pi^{\frac{1}{3}}} \frac{e^2}{r_s} R\left(\sum_\mu |\alpha_{\nu\mu}|^2 + \sum_{\substack{2\nu-\mu \\ \text{unocc.}}} \alpha_{\nu\mu}\alpha_{\nu 2\nu-\mu}\right)\epsilon'(\sigma),$$

$$\epsilon'(\sigma) = \frac{1}{\sigma^2(1+\sigma^2/2.56)} - \frac{0.415}{(1+\sigma^2/2.56)^2} - \frac{1.465}{(1+\sigma^2/2.56)^3}. \tag{14}$$

$\sigma^2\epsilon'(\sigma)$ is given graphically in Fig. 2.

This formula for the increase of potential energy between electrons with upward spin is not exact and may be viewed as containing two parts: First (14a), the increase of the potential of the space charges, due to the less even charge distribution for $\psi_\nu(y; x)$ than for $\psi_\nu(x)$. This increase is lowered by the second part (14b), caused by the greater efficiency of the Fermi hole in a non-uniform charge distribution. The first neglection was made in setting $\psi_\mu(y; x) = \epsilon^{2\pi i(\mu-\nu)\cdot x/L}\psi_\nu(y; x)$ when calculating $\epsilon_\nu'$. This tends to increase the $\epsilon_\nu'$ especially for those $\nu$, for which it is large anyway, because it overemphasizes the unevenness in the charge distribution. The second neglection was to keep no terms higher than the second order terms in $\alpha$. This certainly decreases $\epsilon_\nu'$, because part of the uneveness in the charge density is due to the higher terms, especially to those of the fourth order. Finally, the normalization constant, which is smaller than 1, enters in the second power in (14) and only in the first in (12). Its omission again increases (14). In the whole these errors will about compensate.

The final quantity to be minimized is, after omission of the higher order terms,

$$\epsilon_\nu + \epsilon_\nu' = R\sum_\mu \{2\alpha_{\nu\mu}V_{\nu\mu} + (t_{\nu\mu} + t_{\nu\mu}')|\alpha_{\nu\mu}|^2 + t_{\nu\mu}'\alpha_{\nu\mu}\alpha_{\nu 2\nu-\mu}\}, \tag{15}$$

where

$$t_{\nu\mu} = \frac{4\pi^2 h^2}{mL^2}(\mu^2 - \mu\cdot\nu), \qquad t_{\nu\mu}' = \frac{2^{\frac{1}{3}}}{3^{\frac{1}{3}}\pi^{\frac{1}{3}}} \frac{e^2}{r_s}\epsilon'\left(\frac{2\pi(\mu-\nu)d}{L}\right). \tag{15a}$$

$V_{\nu\mu}$ is given in (11), $\epsilon'(\sigma)$ in (14), $\nu$ is an occupied state, the last term in (15) should be taken only if $2\nu-\mu$ is an unoccupied state. By setting the derivative of (15) with respect to $\alpha_{\nu\mu}{}^*$ equal to zero, one obtains

$$V_{\nu\mu}{}^* + (t_{\nu\mu} + t_{\nu\mu}')\alpha_{\nu\mu} = 0 \tag{16a}$$

if $2\nu-\mu$ is occupied. If it is unoccupied

$$V_{\nu\mu}{}^* + (t_{\nu\mu} + t_{\nu\mu}')\alpha_{\nu\mu} + t_{\nu\mu}'\alpha_{\nu 2\nu-\mu} = 0, \qquad V_{\nu\mu}{}^* + (t_{\nu 2\nu-\mu} + t_{\nu\mu}')\alpha_{\nu 2\nu-\mu}{}^* + t_{\nu\mu}'\alpha_{\nu\mu} = 0. \tag{16b}$$

The last equation is obtained by differentiating (15) with respect to $\alpha_{\nu 2\nu-\mu}$ and considering that $V_{\nu 2\nu-\mu} = V_{\nu\mu}{}^*$ and $t_{\nu 2\nu-\mu}' = t_{\nu\mu}'$. Solving (16), one finds if $2\nu-\mu$ is occupied

$$\alpha_{\nu\mu} = -V_{\nu\mu}{}^*/(t_{\nu\mu} + t_{\nu\mu}') \tag{17a}$$

and

$$\alpha_{\nu\mu} = -V_{\nu\mu}{}^* t_{\nu 2\nu-\mu}/[(t_{\nu\mu} + t_{\nu\mu}')t_{\nu 2\nu-\mu} + t_{\nu\mu}t_{\nu\mu}'] \tag{17b}$$

if $2\nu-\mu$ is unoccupied, while $\mu$ is, of course, always unoccupied. These formulas show that the $\psi_\nu(y_1, \cdots, y_n; x)$ do have the form (4) with

$$f_\nu(y-x) = -\frac{e^2}{\pi L}\left(\sum_{\substack{2\nu-\mu \\ \text{occ.}}} \frac{e^{2\pi i(\nu-\mu)\cdot(y-x)/L}}{(\mu-\nu)^2(t_{\nu\mu}+t_{\nu\mu}')} + \sum_{\substack{2\nu-\mu \\ \text{unocc.}}} \frac{e^{2\pi i(\nu-\mu)\cdot(y-x)/L}}{(\mu-\nu)^2(t_{\nu\mu}+t_{\nu\mu}'+t_{\nu\mu}t_{\nu\mu}'/t_{\nu 2\nu-\mu})}\right). \tag{18}$$

Inserting (17) into (15) one obtains for the total energy

$$\epsilon_\nu + \epsilon_\nu' = -\sum_{\substack{2\nu-\mu \\ \text{occ.}}} \frac{|V_{\nu\mu}|^2}{t_{\nu\mu} + t_{\nu\mu}'} - \sum_{\substack{2\nu-\mu \\ \text{unocc.}}} \frac{|V_{\nu\mu}|^2 t_{\nu 2\nu-\mu}}{(t_{\nu\mu}+t_{\nu\mu}')t_{\nu 2\nu-\mu} + t_{\nu\mu}t_{\nu\mu}'}. \tag{19}$$

Instead of the second term one could write half of the sum of the terms for $\mu$ and $2\nu-\mu$, which makes it somewhat more symmetric.

### 5.

It would be rather difficult to perform the summation over $\mu$ in (19) for an arbitrary set of $y_1$, $\cdots$, $y_n$. Fortunately only the mean value of (19) with the weight $|y|^2/n!$ is needed and this can be computed quite easily. Since the $t$ do not depend on the $y$, one finds

$$\frac{1}{n!}\int |V_{\nu\mu}|^2|y|^2dy=\frac{e^4}{\pi^2(\mu-\nu)^4L^2}\sum_{\kappa,\lambda=1}^{n}\int e^{2\pi i(\mu-\nu)\cdot(\nu_\kappa-\nu)/L}|y|e^{2\pi i(\nu_1 y_1+\cdots+\nu_n y_n)/L}dy \tag{20}$$

$$=\frac{e^4}{\pi^2(\mu-\nu)^4L^2}\left(\sum_{\kappa=1}^{n}1-\sum_{\kappa\neq\lambda}\delta(\nu-\mu+\nu_\kappa-\nu_\lambda)\right).$$

The first term comes from $\kappa=\lambda$, the second from $\kappa\neq\lambda$. In the second, the summation over $\lambda$ can be carried out, and it yields 1 if $\nu-\mu+\nu_\kappa$ is an occupied state, zero otherwise. The whole bracket is, therefore, equal to the number of occupied states $\nu'$, for which $\nu+\nu'-\mu$ is not occupied. Fig. 3 shows a cross section of the $\nu'$-space through the origin and the point $\nu-\mu$. The radius of the circles is $(3n/4\pi)^{\frac{1}{3}}=L/2\pi d$. The sphere through the weak circle contains the points $\nu-\mu+\nu'$ and the hatched part is unoccupied. As a consequence, it is

$$\frac{1}{n!}\int |V_{\nu\mu}|^2|y|^2dy=\frac{e^4n}{\pi^2(\mu-\nu)L^2}\eta(2\pi(\mu-\nu)d/L),\quad \eta(\sigma)=\begin{cases}1 & \text{for } |\sigma|>2 \\ \dfrac{3|\sigma|}{4}-\dfrac{|\sigma|^3}{16} & \text{for } |\sigma|<2\end{cases} \tag{21}$$

and hence

$$F_\nu=\frac{1}{n!}\int (\epsilon_\nu+\epsilon_\nu')|y|^2dy=-\frac{2Ry}{3\pi^3}\int_{\substack{|\sigma+\rho|>1 \\ |\sigma-\rho|<1}}\frac{\eta(\sigma)d\sigma}{\sigma^4[\sigma^2+\sigma\cdot\rho+c\epsilon'(\sigma)]}-\frac{2Ry}{3\pi^3}\int_{\substack{|\sigma+\rho|>1 \\ |\sigma-\rho|>1}}\frac{\eta(\sigma)d\sigma}{\sigma^2[\sigma^4-(\sigma\cdot\rho)^2+2c\sigma^2\epsilon'(\sigma)]}. \tag{22}$$

Here $\rho=2\pi\nu d/L$, $\sigma=2\pi(\mu-\nu)d/L$ and the summation over $\mu$ has been replaced by an integration over $\sigma$. The total energy is expressed in Rydberg units. The constant $c$ is

$$c=(2^{\frac{1}{3}}/3^{\frac{1}{3}}\pi^{\frac{1}{3}})(e^2md^2/h^2r_s)=0.1106r_s/a_0 \tag{22a}$$

when expressed as function of the radius of the "$s$-sphere" in Bohr units $a_0$.

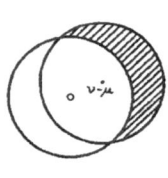

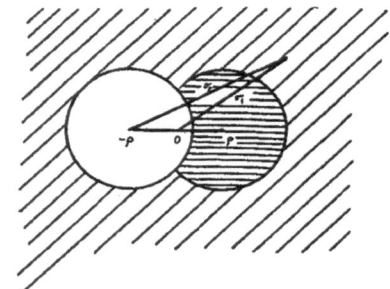

FIG. 3. Cross section of $\nu'$ space.        FIG. 4. Section of $\sigma$ space.

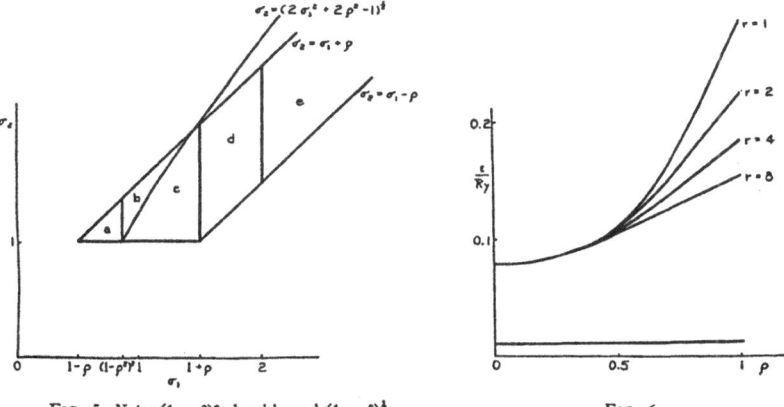

FIG. 5. *Note:* $(1-\rho^2)^2$ should read $(1-\rho^2)^{\frac{1}{2}}$.                    FIG. 6.

In order to perform the integration of (22), one may first introduce elliptic coordinates $\sigma_1 = |\sigma|$, $\sigma_2 = |\sigma+\rho| = |2\pi\mu d/L|$ in the $\sigma$-space, using as centers the origin and the point $-\rho$ (cf. Fig. 4). The first integral is to be extended over the horizontally, the second over the obliquely hatched region. The first will be a sum of two integrals (a) and (b) in Fig. 5, in both of which $\eta(\sigma) = 3\sigma_1/4 -\sigma_1^3/16$. The second integral is a sum of three integrals (c), (d) and (e), and in the first two of them the same expression is valid for $\eta(\sigma)$ while $\eta(\sigma) = 1$ in the last. For $d\sigma$ one writes $2\pi\sigma_1 d\sigma_1\sigma_2 d\sigma_2/\rho$, $\sigma_1^2$ for $\sigma^2$, and $\sigma\cdot\rho = \frac{1}{2}(\sigma_2^2 - \sigma_1^2 - \rho^2)$. The integration over $\sigma_2$ can be carried out simply and gives the five integrals

$$-F_\rho = \frac{4Ry}{3\pi^2\rho} \int_{1-\rho}^{(1-\rho^2)^{\frac{1}{2}}} \ln \frac{2c\epsilon'(\sigma_1)+2\sigma_1^2+2\sigma_1\rho}{2c\epsilon'(\sigma_1)+\sigma_1^2+1-\rho^2} \left(\frac{3\sigma_1}{4} - \frac{\sigma_1^3}{16}\right)\frac{d\sigma_1}{\sigma_1^3}$$

$$+\frac{4Ry}{3\pi^2\rho} \int_{(1-\rho^2)^{\frac{1}{2}}}^{1+\rho} \ln \frac{2c\epsilon'(\sigma_1)+2\sigma_1^2+2\sigma_1\rho}{2c\epsilon'(\sigma_1)+3\sigma_1^2+\rho^2-1} \left(\frac{3\sigma_1}{4} - \frac{\sigma_1^3}{16}\right)\frac{d\sigma_1}{\sigma_1^3}$$

$$+\frac{4Ry}{3\pi^2\rho} \int_{(1-\rho^2)^{\frac{1}{2}}}^{1+\rho} \ln \frac{2\sigma_1 v+\sigma_1^2+\rho^2-1}{2\sigma_1 v-\sigma_1^2-\rho^2+1} \left(\frac{3\sigma_1}{4} - \frac{\sigma_1^3}{16}\right)\frac{d\sigma_1}{\sigma_1^2 v}$$

$$+\frac{4Ry}{3\pi^2\rho} \int_{1+\rho}^{2} \ln \frac{v+\rho}{v-\rho} \left(\frac{3\sigma_1}{4} - \frac{\sigma_1^3}{16}\right)\frac{d\sigma_1}{\sigma_1^2 v} + \frac{4Ry}{3\pi^2\rho} \int_{2}^{\infty} \ln \frac{v+\rho}{v-\rho} \frac{d\sigma_1}{\sigma_1^2 v},$$

where $v$ stands for $(2c\epsilon'(\sigma_1)+\sigma_1^2)^{\frac{1}{2}}$. A calculation of this quantity for $r_s = 4$ shows that it is practically equal $\sigma_1$ if $\sigma_1 > 2$ and the last integral can be evaluated accordingly. It yields

$$\frac{4Ry}{3\pi^2\rho}\left\{\left(\frac{1}{8} - \frac{1}{2\rho^2}\right) \ln \frac{2+\rho}{2-\rho} + \frac{1}{2\rho}\right\} \approx \frac{Ry}{9\pi^2}\left(1+\frac{\rho^2}{20}\right).$$

The other integrals were evaluated numerically and the results are plotted against $\rho$. Fig. 6 shows the plots for $r_s = 1, 2, 4, 8$. From this, the mean value of the correlation energy which is the mean value of $-\frac{1}{2}F_\rho$ with the weight $\rho^2$, was calculated, and plotted against $r_s$. The $\frac{1}{2}$ enters, because the whole energy correction is present only for half of the electrons, that is, those with upward spin. In Fig. 7 the upper curve represents the values calculated in this way. The energy is given in multiples of $e^2/r_s$.

### 6.

One must remember, of course, that the preceding calculation is only an approximate one. Even if one confines oneself to wave functions of the form (2), the upper curve of Fig. 7 gives the correlation energy in first approximation only. The neglections are due to three causes: first to the use of an unnormalized wave function, second to the neglection of the terms with higher than the second power of $\alpha$ in $\epsilon_\nu$, when going over from (12) to (14a) and third to the non-complete orthogonality of the wave functions employed. Our approximation is good if the $\alpha$'s are small. One gets an idea about the accuracy of the approximation by calculating $\sum_\mu \int |\alpha_{\nu\mu}|^2 |y|^2 dy/n!$ (which is the reciprocal square of the normalization constant minus one), though an idea only, since $\sum_\mu |\alpha_{\nu\mu}|^2$ should really be small for all configurations of the $y$, not only its mean value. A calculation of the former quantity shows[9] that it stays well below one, except for large $r_s$ and for $\rho$ which are very near to 1. It is in these cases that our approximation must be expected to break down.

The real value of the correlation energy will be smaller in these cases than the calculated one. The correction with the normalization constant could easily be taken into account, as has been shown at another place.[10] It always decreases the correlation energy, not very much, however, as the magnitude of the normalization constant or the formulas in[10] show. The second neglection is probably more dangerous and also more laborious to correct. It has been done for one point ($r_s = 4$) only and for this one very roughly. The second neglection also increased the calculated value of the correlation energy, since it amounts to taking $1 - 2f$ instead of $(1 - f)^2$ for the probability of the electron being at a certain point, and the minima of $1 - 2f$ are much deeper than those of $(1 - 2f)^2$ (the maxima are lower but less important).

If the electrons had no kinetic energy, they would settle in configurations which correspond to the absolute minima of the potential energy. These are closed-packed lattice configurations, with energies very near to that of the body-centered lattice. Here, every electron is very nearly surrounded with a spherical hole of radius $r_s$ and the potential energy is smaller than in the random configuration by the amount $0.75 = e^2/r_s$. This would be the sum of the correlation energy and that due to the Fermi hole. Since the latter one is,[11, 1] $0.458e^2/r_s$, the maximum amount of the correlation energy is $0.292e^2/r_s$. This value will be attained only if the kinetic energy can be neglected, i.e., for $r_s = \infty$, and represents the asymptote to the real correlation energy curve, which is attempted to be drawn into Fig. 7. It appears to run much higher than one would have thought without calculation. I believe it to be in error everywhere by less than 20 percent.

The dotted line at $0.142e^2/r_s$ corresponds to a correlation energy as great as assumed in the first calculation[5] giving $0.6e^2/r_s$ together with the energy of the Fermi hole.

The calculated constants of the Na lattice with the correlation energy of Fig. 7 and the other quantities as in reference 1 are as follows: lattice constant 4.62A as compared with the observed value of 4.23A. The binding energy associated with this is 26.1 Calories, to be compared with the observed value of 26.9. The calculated value of the binding energy for the observed lattice

---

[9] The greatest part of the numerical work has been done by Dr. M. Vermes of Budapest. A table of the calculated values is given here:

| $\rho$ | $r = 1$ | 4 | 8 |
|---|---|---|---|
| 0 | 0·006 | 0·10 | 0·45 |
| 0·4 | 0·009 | 0·14 | 0·54 |
| 1 | 0·04 | 0·30 | 0·94 |

[10] To appear shortly in the Bull. of the Hung. Acad.

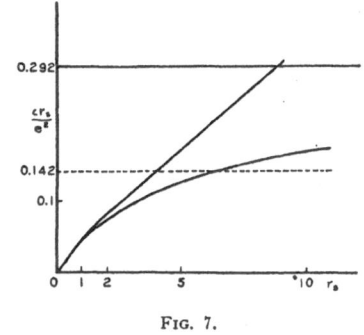

FIG. 7.

[11] F. Bloch, Zeits. f. Physik 57, 545 (1929).

constant is 22.3 Cal. As far as the lattice constant goes, one must remember, however, that both the correlation energy and that due to the Fermi hole are calculated for a flat wave function and the wave functions are flat only for $r_s = 4$ and its neighborhood.

The magnitude of the correlation energy is important for questions of paramagnetism and ferromagnetism as well as for questions of lattice energy. It modifies Bloch's original theory on the ferromagnetism of free electrons[11] in such a way that it yields ferromagnetism in fewer cases than in its original form.[12] I hope to return to this question at another time.

I wish to express my gratitude to Dr. F. Seitz for his kind help in connection with the preparation of this manuscript.

[12] A paper of S. Schubin and S. Wonsowsky, Proc. Roy. Soc. A145, 159 (1934) which appeared recently, points in the same direction.

# Theory of the Work Functions
# of Monovalent Metals

E. P. Wigner and J. Bardeen

Physical Review *48*, 84–87 (1935)

(Received April 30, 1935)

The factors which determine the work function of a metal are described in a qualitative way. The work function is defined as the difference in energy between a lattice with an equal number of ions and electrons, and the lattice with the same number of ions, but with one electron removed. The work function is then found by first calculating the energy of a lattice with $n_i$ ions and $n_e$ electrons. The final formula gives the work functions of monovalent metals in terms of the heats of sublimation. This formula is approximate, and can claim validity only in a qualitative way, as one of the important factors, the electric double layer on the surface, is omitted entirely, and it is assumed that the Fermi energy is as great as if the electrons were entirely free. The values obtained from this formula check very closely with the experimental values for the alkalis, so that it can be concluded that the double layer is probably small for these metals. Finally, the deviations to be expected for other than monovalent metals are considered. A more exact calculation of the work function of one substance (Na) will be given by one of us in an ensuing paper.

## (1)

IT is perhaps not quite superfluous to have, in addition to a more exact calculation of a physical quantity, an approximate treatment which merely shows how the quantity in question is determined, and the lines along which a more exact calculation could be carried out. Such a treatment often leads to a simple formula by means of which the magnitude of the quantity may be readily determined. A treatment of this nature has been given by O. K. Rice[1] for the binding energies of the alkalis.

We intend to give here an approximate calculation, on somewhat different lines, of the work functions of monovalent metals, the main result being a formula which relates the work function with the heat of sublimation. We do this for two reasons; first, because a more exact treatment seems to involve a great deal of computational labor, and second, because the connection between the work function and other properties is empirically very pronounced. In fact, Sommerfeld,[2] in his original paper on the electron theory of metals, has already noticed that if one orders the metals according to their work functions (or according to their Volta potentials), the series is the same as if one orders them according to the Fermi energy of the electrons, calculated on the basis of the free electron theory. The greater the Fermi energy, the greater the work function. This fact was rather puzzling on the basis of the naïve free electron theories of that time, but an explanation was given later by Frenkel.[3] Although Sommerfeld and Bethe[4] have shown that the details of his calculations cannot be maintained, we feel that his basic idea is correct, namely that the binding energy of the electrons (i.e., work function) increases with the kinetic energy. The connection between the kinetic and potential energy, as given by the virial theorem in a Coulomb field, has been utilized by Tamm and Blochinzev[5] on the basis of the Fermi-Thomas model. It seems to us, however, that it is dangerous to use the Fermi-Thomas model in this connection, because this model does not yield the metallic bond.[6]

## (2)

The work function may be defined as the difference in energy between a lattice with an equal number of ions and electrons, and the lattice with the same number of ions, but with one electron removed. It is assumed in both cases that the lowest electronic states are completely filled, so that the electron is removed from the highest energy state of the neutral metal. It is necessary to specify the position of the electron after its removal from the lattice, because, in general, the work function is different

[1] O. K. Rice, Phys. Rev. **44**, 318 (1933).
[2] A. Sommerfeld, Naturwiss. **15**, 825 (1927); **16**, 374 (1928).
[3] J. Frenkel, Zeits. f. Physik **49**, 31 (1928).
[4] See article by A. Sommerfeld and H. Bethe in the *Handbuch der Physik*, Vol. 24, Berlin, 1933, p. 424.
[5] J. Tamm and D. Blochinzev, Zeits. f. Physik **77**, 774 (1932).
[6] For a discussion of the Fermi-Thomas model as applied to the metallic state, see J. C. Slater and H. M. Krutter, Phys. Rev. **47**, 559 (1935).

for different crystallographic planes. We shall suppose that the electron is at a point in the neighborhood of a surface plane of the crystal, the distance from this plane being small compared with the dimensions of the plane, but large in comparison with atomic dimensions. We shall calculate this energy difference by calculating the energy $E(n_i, n_e)$ of a lattice with $n_i$ ions and $n_e$ electrons where we may suppose that $|n_i - n_e| \ll n_i$ so that effects connected with the finite capacity of the sample may be neglected. For the calculation of $E(n_i, n_e)$ we shall use the method of orbital wave functions. This method requires hardly any modification of the scheme[7] used for $n_i = n_e$, i.e., for uncharged metals.

There is only one important difference, that due to the electric double layer which the electron cloud and the ions may form on the surface of the metal.[8] These double layers are due to the fact that the electron distribution will not be symmetric around the surface ions, as it is around the inner ones. It will extend partly outside of the limits of the spheres surrounding the ions[*] of the surface, and in the inside of the spheres it will have partly greater, partly smaller densities than in the inner $s$ spheres. This alteration of the charge density may result in double layers on the surface of the metal. The double layers will be such that they generate a constant potential inside the metal. The potential outside the metal, however, will not be constant, but will vary in such a way that the potential differences between the outer neighborhoods of differently oriented crystal surfaces will be equal to the differences in the moments of the double layers of the corresponding surfaces. These differences are revealed experimentally as the differences in the work functions of the different surface planes of the same crystal. This shows, conversely, that the double layer is of the same order of magnitude as these differences, i.e., around $\frac{1}{2}$ to 1 electron volt.[9]

The total energy of the lattice with $n_i$ ions and $n_e$ electrons can be considered as containing

three parts: (1) the energy which the lattice would have if the surface ions were surrounded with the same symmetric charge distribution which prevails in the interior; (2) the energy change of the inner ions and electrons due to the double layer on the surface; (3) the energy of this double layer. We shall assume that the electron comes from the interior of the metal, i.e., that the double layer remains unchanged, and that the density of the electrons changes in the interior only. This is, strictly speaking, incorrect, as one can easily see that the density will be practically unchanged in the interior, and that the electron will come from the surface. The result for the work function cannot be incorrect in spite of this, because the energy required to move an electron from the surface to the interior of a metal is of a much smaller order of magnitude than the energy required to remove an electron from the metal (or, the work function). If the energy necessary to move an electron from the surface to the interior were appreciable, the electrons would have already rearranged themselves in the neutral metal. We shall omit in the final result the change in energy of the second and third part of $E(n_i, n_e)$. This is not allowable, of course, but this part of the work function is just equal to the moment of the double layer, and cannot be taken into account without an explicit calculation of the latter, which is reserved for a later paper by one of us.

### (3)

We can now use for the calculation of $E(n_i, n_e)$ the scheme of reference 7. The kinetic energy of the ions will be neglected, so that the total energy contains three parts, $\frac{1}{2}\Sigma e_i V_i$ for the ions, $\frac{1}{2}\Sigma e_i V_i$ for the electrons, and the kinetic energy of the latter. The first part can be calculated as follows: First one assumes that the density of electrons everywhere, except in the $s$ sphere of the ion under consideration, is as great as it would be if $n_e = n_i$. Then, the charges outside this $s$ sphere will have no effect on our ion, because the potential of the electrons in every other $s$ sphere will be nullified by the potential of the ion inside.[10] Only the double

[7] E. Wigner and F. Seitz, Phys. Rev. **43**, 804 (1933); **46**, 509 (1934) (I and II); E. Wigner, Phys. Rev. **46**, 1002 (1934) (III); F. Seitz, Phys. Rev. **47**, 400 (1935) (IV).
[8] J. Frenkel, Zeits. f. Physik **51**, 232 (1928).
[*] "s spheres," reference 7, II.
[9] H. E. Farnsworth and B. A. Rose, Nat. Acad. Sci. **19**, 777 (1935); B. A. Rose, Phys. Rev. **44**, 585 (1933).

[10] We neglect the small correction of Appendix 2, reference 7, II.

layer of the surface will be effective, and we obtain as the first contribution to the energy of the crystal:

$$n_i[(n_e/2n_i)\int|\psi|^2Vdv+\tfrac{1}{2}eD],\qquad(1)$$

where $V$ is the potential of the ion core, $D$ is the moment of the double layer, and the integration must be extended over the $s$ sphere of the ion. The factor $n_i$ comes in, because we have $n_i$ ions, the factor $n_e/n_i$ because there are only that many electrons in the mean in the $s$ sphere; (1) should still be averaged over the wave functions of the different free electrons. The fact that there are $((n_i-1)/n_i)(n_i-n_e)$ electrons missing outside the $s$ sphere introduces a further correction:

$$\tfrac{1}{2}\Sigma e_j\Phi_j,\qquad(1a)$$

where $\Phi_j$ is the potential of the missing electrons at the $j$th ion, and is of the order of magnitude $(n_i-n_e)e/R$ where $R$ is a lateral dimension of the crystal.

The potential energy of the electrons is more complicated. First, again, we can supplement the missing electrons outside the $s$ sphere in which our electron happens to be. The potential energy due to the nearest ion is $\tfrac{1}{2}V(r)$. To this the electron interaction energy must be added, which, if the charge distribution is assumed to be uniform, amounts to $(n_e/n_i)(3e^2/2r_s-e^2r^2/2r_s^3)$ where $r_s$ is the radius of the $s$ sphere. These two terms give, together with the contribution from the double layer, if averaged over all positions of the electron under consideration in the $s$ sphere:

$$n_e[\tfrac{1}{2}\int|\psi|^2Vdv+0.6e^2n_e/n_ir_s-\tfrac{1}{2}eD]\qquad(2)$$

with definitions similar to those used in (1). The electrons missing outside the $s$ sphere, yield the following correction to the energy:

$$-(n_e/2n_i)\Sigma e\Phi_j.\qquad(2a)$$

Next we must take into account the effects of the different holes. The Fermi hole (exchange energy) gives $-0.458e^2/r_s$ if $n_e=n_i$. One easily sees that $r_s$ must be replaced by $(n_i/n_e)^{\frac{1}{3}}r_s$ if $n_e\neq n_i$. Similarly, if we denote by $r_sf(r_s)$ the correlation function,[11] the correlation energy for

[11] Represented by the lower curve, Fig. 7, reference 7, III.

one electron is $-e^2f((n_i/n_e)^{\frac{1}{3}}r_s)$. The total contribution from these sources is:

$$-n_e[(0.458e^2/r_s)(n_e/n_i)^{\frac{1}{3}}+e^2f((n_i/n_e)^{\frac{1}{3}}r_s)].\qquad(3)$$

Finally, the kinetic energy of the electrons gives:

$$-(\hbar^2n_e/2m)\int\psi^*\Delta\psi dv,\qquad(4)$$

which again should be averaged over the different free electron wave functions.

The energy of the double layer due to the electrons and ions inside the crystal is simply:

$$-\tfrac{1}{2}eD(n_e-n_i).$$

This energy will change, even though the double layer itself is unchanged.

If we add all these quantities together, (4) together with the first terms of (1) and (2) gives, because of the Schrödinger equation $-(\hbar^2/2m)\Delta\psi+V\psi=E\psi$ just $n_eE$, where still the mean value of $E$ over all occupied states of the electrons must be taken. This gives the energy $E_0$ of the lowest free electron level, augmented by the mean value of the Fermi energy, $F$, i.e.,

$$n_e(E_0+F(n_e/n_i)^{\frac{2}{3}});$$
$$F=(9\pi/10)(3/2\pi)^{\frac{1}{3}}(R_y/r_s^2).\qquad(5)$$

Thus the total energy is:

$$E(n_i,n_e)=n_e(E_0+F(n_e/n_i)^{\frac{2}{3}})$$
$$+(n_e^2/n_i)(0.6e^2/r_s)+(1/2n_i)(n_i-n_e)\Sigma\Phi_j$$
$$-n_e[(0.458e^2/r_s)(n_e/n_i)^{\frac{1}{3}}+e^2f((n_i/n_e)^{\frac{1}{3}}r_s)]$$
$$-eD(n_e-n_i).\qquad(6)$$

$$(4)$$

The derivative of $E(n_i,n_e)$ at $n_i=n_e$ is the work function $\varphi$ with the negative sign.

$$-\varphi=E_0+5F/3+1.2e^2/r_s-(4/3)(0.458e^2/r_s)$$
$$-e^2f(r_s)+e^2r_sf'(r_s)/3-eD.\qquad(7)$$

The derivative of the sum in (6) tends to zero with increasing size of the crystal, and has therefore been omitted. If we omit the double layer term, and replace the Fermi energy by its value for free electrons, the only unknown quantity in (7) is $E_0$. One can express $E_0$ in terms of the ionization energy $I$, and the heat of

TABLE I. *Work functions of the alkalis.*

| Metal | $r_s$ | $I+H$ | $F$ | $\varphi_{calc.}$ | $\varphi_{exp.}$ |
|-------|-------|-------|-----|-------------------|------------------|
| Li | 3.28 | 7.04 | 2.07 | 2.19 | 2.28[1] |
| Na | 4.00 | 6.25 | 1.89 | 2.15 | 2.25[2] |
|    |      |      |      |      | 2.46[1] |
| K | 4.97 | 5.27 | 1.22 | 2.20 | 2.24[1] |
|   |      |      |      |      | 2.24[3] |
|   |      |      |      |      | 2.17[2] |
| Rb | 5.32 | 5.03 | 1.07 | 2.20 | 2.19[3] |
|    |      |      |      |      | 2.16[1] |
| Cs | 5.73 | 4.70 | .92 | 2.15 | 1.96[3] |
|    |      |      |      |      | 1.87[1] |
|    |      |      |      |      | 1.81[4] |

[1] A. R. Olpin (see reference 13).
[2] Z. Berkes, Math. Phys. Lapok **41**, 131 (1934).
[3] J. J. Brady, Phys. Rev. **41**, 613 (1932).
[4] K. H. Kingdon, Phys. Rev. **25**, 892 (1925).

sublimation, $H$,

$$I+H = -(\partial/\partial n_i)E(n_i, n_e) - (\partial/\partial n_e)E(n_i, n_e)$$
$$= -E_0 - F - 0.6e^2/r_s + 0.458e^2/r_s + e^2 f(r_s), \quad (8)$$

which, of course, is the result of reference 7. We have, from (7) and (8)

$$\varphi = I + H - \tfrac{2}{3}F - 0.6e^2/r_s + 0.458e^2/3r_s$$
$$- e^2 r_s f'(r_s)/3 + eD. \quad (9)$$

Table I gives the work functions of the alkalis as calculated from (9) under the assumption that $D=0$. The Fermi energy for Li was taken from the work of F. Seitz,[12] and the free electron values were used for the remaining metals. The check with the experimental values[13] is extremely good, considering the approximate nature of the theory, and shows that the double layer is probably small for these metals. The double layer may, however, give an important contribution to the work functions of other metals.

The solid curve of Fig. 1 gives the theoretical values of $I+H-\varphi$ in electron volts plotted against $r_s$ in Bohr units. It is assumed that the Fermi energy has its classical free electron value, and that $D=0$. The light broken line connects the experimental values for the monovalent metals. Most of the discrepancy for Li ($r_s=3.28$) is accounted for by the deviation of the Fermi

[12] See reference 7, IV.
[13] For the experimental values of the work function, see A. L. Hughes and L. A. DuBridge, *Photoelectric Phenomena*, McGraw-Hill, 1932, p. 75. For the heats of sublimation, see J. Sherman, Chem. Rev. **11**, 93 (1932).

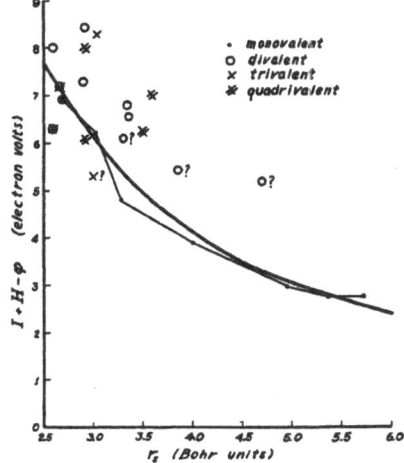

FIG. 1. Theoretical curve of $I+H-\varphi$ for monovalent metals, assuming no surface double layer. Broken line connects experimental values for these metals.

energy from its free electron value.[12] The experimental values of $I+H-\varphi$ for other than monovalent metals are also shown in Fig. 1. A calculation similar to the preceding one could be made for these metals by treating all valence electrons as free. The work function could then be obtained as the difference of quantities which are even greater than those for monovalent metals. One can obtain an orientation, however, by considering only one electron as free, the rest belonging to the "ion" core. There will then be two main modifications of the foregoing considerations. First, the Fermi energy will be even further from its classical value. It will in general be smaller than this if the levels of the last electron all belong to the same Brillouin zone; it may be greater or smaller if they extend over both sides of a gap between two zones. Second, a large part of the cohesion will be due to the interaction of the "ions." Thus in (9), only that part of the heat of sublimation should be inserted which results from the outermost electron. This correction is, indeed, in the right direction to bring the experimental points for these metals closer to the theoretical curve.

# On the Possibility of a Metallic Modification of Hydrogen

E. P. Wigner and H. B. Huntington

Journal of Chemical Physics *3*, 764–770 (1935)

(Received October 14, 1935)

Any lattice in which the hydrogen atoms would be translationally identical (Bravais lattice) would have metallic properties. In the present paper the energy of a body-centered lattice of hydrogen is calculated as a function of the lattice constant. This energy is shown to assume its minimum value for a lattice constant which corresponds to a density many times higher than that of the ordinary, molecular lattice of solid hydrogen. This minimum—though negative—is much higher than that of the molecular form. The body-centered modification of hydrogen cannot be obtained with the present pressures, nor can the other simple metallic lattices. The chances are better, perhaps, for intermediate, layer-like lattices.

## (1) INTRODUCTION

THE present calculation of the constants of metallic hydrogen has a threefold purpose. First, we wanted to explore the possibility of preparing metallic hydrogen under high pressures.[1] The hope for this, however, appears to be small, as the pressures under which metallic hydrogen would be stable seem to be unattainable with the present techniques. Second, it seemed to us important to check the free electron method of calculating wave functions and heats of vaporization[2] in a case where experiments give an upper limit to the energy. The obvious instability of metallic hydrogen under ordinary pressure shows that its heat of formation is smaller than that of the molecular hydrogen. This condition proved to be only too well fulfilled. Finally, we hoped to obtain some information concerning the conditions for the metallic state and the line separating metals from metalloids in the periodic table.[3] This purpose has been achieved by the present work to the extent expected, but this point, although perhaps apparent to the reader, will not be stressed in the present paper.

## (2) CALCULATION

The calculation itself was a rather mechanical application of the principles outlined elsewhere.[2] First, the energy $E(r_s)$ was found for which the solution of the Schrödinger equation with the ordinary Coulombic potential $-e^2/r$ had zero derivative at distance $r_s$. This was done by a numerical method and, in addition, checked by an analytic approximation,[4] which gave a direct

expression for the wave function needed for subsequent work. The plot (lower curve of Fig. 1) shows the results of the numerical method which gave the more accurate results. Its values lie about 0.02 Ry (Rydberg units) below those of the analytic formula. In this case $E(r_s)$ has no minimum, as in case of Na, but instead continues to fall to $-\infty$ for $r_s \rightarrow 0$. In Na the rising of $E(r_s)$ for small $r_s$ results from the energy necessary to press the valence electron into the inner shells, whereas no such shell exists for hydrogen.

The upper curve of Fig. 1 is obtained by adding to $E(r_s)$ the Fermi energy for free electrons,

$$(2.2099/r_s^2)\text{Ry},$$

where $r_s$ is expressed in Bohr units $\hbar^2/e^2m$. This corresponds to the calculation adopted in ref. 2 I and gives a minimum of $E = -1.156$ for $r_s = 1.50$. This would give a density of 0.80 (as compared to 0.087 in the solid molecular form!) and a heat of vaporization of 48.7 cal. into atomic hydrogen, as compared with 52.5 cal., the heat of formation of the molecular form. We see that the molecular form is the more stable, even according to this very approximate calculation, but if this were all, it would probably still be possible to transform it under pressure to a metallic form.

Unfortunately, a great number of corrections must be applied to the simple picture given in ref. 2 I which decreases the stability of the metallic form in this instance. These corrections are given for the flat wave functions in refs. 2 II and 2 III. They are:

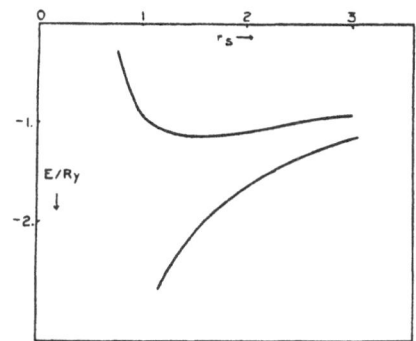

FIG. 1. Lower curve—the energy of the electrons without momentum, the upper curve—total energy as a function of lattice constant. Both according to a naïve free electron theory.

[1] It was J. D. Bernal who first put forward the view that all substances go over under very high pressures into metallic or valence lattices.
[2] (I) E. Wigner and F. Seitz, Phys. Rev. 43, 804 (1933); (II) 46, 509, 1934; (III) E. Wigner, ibid. 46, 1002 (1934); (IV) F. Seitz, ibid. 47, 400 (1935); (V) J. C. Slater, ibid. 45, 794 (1934). E. L. Hill remarks in a "Note on the Statistics of Electron Interaction" (Physik. Zeits. Sowjetunion 7, 447 (1935)) that the expressions "parallel spin" and "antiparallel spin" should be replaced by "parallel spin Z components" and "antiparallel spin Z components," respectively. The former expressions are not quite equivalent with the latter ones according to the general formalism of the statistical interpretation of quantum mechanics and can be used only as an abbreviation for them. (This is done in the present paper also.) We believe that Dr. Hill's further remarks concerning the papers of this reference are based on a misunderstanding.
[3] Interesting quantum-mechanical considerations on this point have been put forward by F. Hund (International Conference on Physics, October 1934, London).
[4] To be given in Section 3.

(a) The difference between the energy of the electron clouds (§2 in II) and the exchange energy (§3 in II)

$$0.284/r_s - 0.017/r_s^2(E_k - E), \quad \text{(in Rydberg units)}, \quad (1)$$

where $E_K$ is a mean value of the energy of the higher $s$ terms and has been estimated[4] to be about 20 Ry/$r_s^3$. The second term in (1) is, therefore, almost negligible.

(b) The correlation energy, calculated in ref. 2 III, which can be represented for $r_s > 1$ rather closely by

$$-0.584/(r_s + 5.1). \quad (2)$$

(c) A correction for the Madelung number will be added in the next section.

The energy after the corrections (a) and (b) have been added is plotted on a magnified scale on the lowest curve in Fig. 3. One sees that a great part of the binding energy is lost.

### (3)

It is very probable that the wave functions have nearly the form $e^{ip \cdot x/h}$, not only in Na, but also in the other alkalies excepting Li.[5] In H this is not the case, however, and there will be further corrections introduced into the energy curve similar to those calculated for Li by F. Seitz.[2 IV] Still the situation for H is rather different from that for Li. In H the wave function of the lowest energy electron $(p=0)$ is strongly modulated but the modulation does not increase much for higher $p$ functions. In Li the wave function for $p=0$ was quite flat and the modulation increased with increasing $p$.

For the calculation of these effects an approximate method has been used. The equation for the wave function of the electron with the momentum $p$, multiplied by $e^{ip \cdot x/h}$, is[6]

$$H_\mu \chi_\mu = -\Delta \chi_\mu - i\mu \cdot \text{grad } \chi_\mu - 2\chi_\mu/r = \epsilon_\mu \chi_\mu,$$
$$\mu = 2\hbar p/e^2 m. \quad (3)$$

Here the $\epsilon_\mu$ does not include the kinetic energy of the flat wave functions, $p^2/2m = \mu^2/4$ Ry. The $\epsilon_\mu - \epsilon_0$ will give us the change in the Fermi energy, and the wave functions $\chi_\mu$ the change in charge distribution, which are both caused by the fact that the electrons are not free in the lattice. The charge distribution is used in a

[4] Cf. F. Mott and C. Zener, Proc. Camb. Phil. Soc. 30, 249 (1934); also J. Bardeen and E. Wigner, Phys. Rev. 48, 84 (1935); for Li see reference 2 (IV) and J. Millman, ibid. 47, 286 (1935).
[5] Cf. Eq. (11), reference 2 (II). The energy in (3) is in Rydberg units Ry and the length in Bohr units.

recalculation of the interaction of the electron clouds and the Fermi hole (exchange energy). We shall see that $\chi_\mu$ can be assumed to be the sum of a function of $r$ alone ($s$ part) and a function of $r$ multiplied by $(\mu \cdot r)$ ($p$ part). The boundary condition for the $\chi_\mu$ is periodicity in the lattice, which means zero derivative for the $s$ part, and vanishing of the $p$ part on the surface of the $s$ sphere.

Eq. (3) has been solved by a perturbation method. We shall need only those solutions of the unperturbed equation

$$-\Delta \xi_{nl} = \eta_{nl}\xi_{nl}, \quad (4)$$

which are functions of $r$ alone ($l=0$, or the $s$ part of $\chi$), and those which are functions of $r$ multiplied by $(\mu \cdot r)$, ($l=1$ or the $p$ part of $\chi$). The boundary conditions for these $\xi$ functions are the same as for the corresponding part of $\chi$. We have[2 II]

$$\xi_{00} = (3/4\pi r_s^3)^{\frac{1}{2}}, \quad \xi_{n0} = c_n(\sin \omega_n r)/r,$$
$$\xi_{n1} = (3^{\frac{1}{2}}c_n/\mu\omega_n)(\mu \text{ grad})(\sin \omega_n r)/r, \quad (5)$$
$$\eta_{00} = 0, \quad \eta_{n0} = \eta_{n1} = \omega_n^2,$$

with

$$\omega_1 = \frac{1.430\pi}{r_s}, \quad c_1 = \frac{1.024}{(2\pi r_s)^{\frac{1}{2}}};$$

$$\omega_n = (n+\tfrac{1}{2})\frac{\pi}{r_s}, \quad c_n = \frac{1}{(2\pi r_s)^{\frac{1}{2}}}$$

when the last two expressions hold well for $n > 1$ only.

Next we must calculate the matrix elements of $H_\mu$. All terms give nonvanishing elements between $\xi_{nl}$ and $\xi_{n'l}$ only, except for $i\mu \cdot$ grad which gives elements between $\xi_{nl}$ and $\xi_{n'l+1}$. We have

$$(H_\mu)_{00:\,00} = -3/r_s;$$
$$(H_\mu)_{10:\,10} = 20.2/r_s^2 - 5.65/r_s;$$
$$(H_\mu)_{n0:\,n0} \approx (H_\mu)_{n1:\,n1} \approx (n+\tfrac{1}{2})^2\pi^2/r_s^2;$$
$$(H_\mu)_{11:\,11} = 20.2/r_s^2 - 3.66/r_s; \quad (6)$$
$$(H_\mu)_{n'0:\,n1} = -(i\mu\omega_n/3^{\frac{1}{2}})\delta_{n'n};$$
$$(H_\mu)_{00:\,10} = 1.36/r_s;$$
$$(H_\mu)_{00:\,n0} = 2(6)^{\frac{1}{2}}/(n+\tfrac{1}{2})\pi r_s.$$

For $\mu = 0$, this gives, by Rayleigh-Schrödinger formulas,

$$\epsilon_0 = \frac{3}{r_s} - \frac{(1.36)^2}{20.2 - 2.65 r_s} - \sum_{n=2}^{\infty} \frac{24}{(n+\frac{1}{2})^4 \pi^4},$$

$$\chi_0 = \xi_{00} + \frac{1.36 r_s}{20.2 - 2.65 r_s}\xi_{10} + \sum_{n=2}^{\infty} \frac{2(6)^{\frac{1}{2}} r_s}{(n+\frac{1}{2})^3 \pi^3}\xi_{n0}.$$

(7)

These are the formulae to which we referred in Section 2. For $\mu \neq 0$, we diagonalized the three rows and columns of $H_\mu$ referring to $\xi_{00}$, $\xi_{10}$ and $\xi_{11}$ and used only for the remaining ones the Rayleigh-Schrödinger method. This was advisable, because $(H_\mu)_{10;\,10}$ and $(H_\mu)_{11;\,11}$ were close together. This gave

$$\epsilon_\mu = \frac{3}{r_s} - \frac{1.85}{20.2 - 2.65 r_s - 6.72\mu^2(20.2 - 0.6 r_s)^{-1}}$$
$$- 0.0093. \quad (8)$$

The mean value of $\epsilon_\mu - \epsilon_0$ with the weight $\mu^2$, from $\mu = 0$ to $\mu = 3.84/r_s$ is the correction to the Fermi energy; it is plotted in the lowest curve of Fig. 2. It is negative, just as in Li, because there is no $p$ level below the $s$ level of the valence electron. The energy $\epsilon_\mu - \epsilon_0$ is no longer very accurately proportional to $\mu^2$, because the effect of the next Brillouin zone is appreciable.

(4)

Next the interaction of the electron cloud and the more accurate exchange energy must be derived. The greater density near the nucleus makes it evident that the first will be greater than the free electron value assumed in the previous section. This is true also for the exchange energy, because the Fermi hole is more effective in consequence of the uneven charge distribution. The two effects work against each other, but in contrast to Li, the former is greater here and decreases the binding.

The calculation of these quantities is a little laborious and contains nothing of principal interest. It will be sketched only.

The square of the calculated wave function, i.e., the electric density $\rho$, was approximated by an analytical expression[7] containing a constant

[7] The reason for this is that only for this form were we able to perform the integrations.

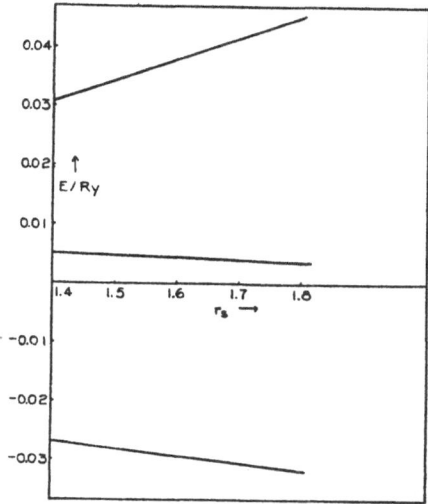

FIG. 2. Bottom line, correction to the Fermi energy; middle line, correction for the Madelung number; top line, correction for the interaction energy.

and an exponentially decreasing charge distribution around each ion:

$$\rho = \rho_1 + \rho_2, \quad \rho_1 = e(1-c)/(4\pi r_s^3/3). \quad (9)$$

$\rho_2$ itself contains two parts: first the exponential charge distribution around the ion, in the $s$ sphere in which we are, and second the exponential charge distribution around the other ions. The latter was averaged over all directions so that $\rho_2$ also became spherically symmetric in each $s$ sphere

$$\rho_2 = \frac{ec\beta^3}{8\pi}\left[e^{-\beta r} + 4k\frac{\text{Sh}(\beta r)}{\beta r}\right],$$
$$k = e^{-2\beta r_s}(1 + \beta r_s + \beta^2 r_s^2/2)(\beta r_s - 1)^{-1}. \quad (9a)$$

The two constants $c$ and $\beta$ were adjusted so as to fit the square of the calculated wave function (7) best.

The potential corresponding to this charge distribution contains also two parts: $\Phi_1$, corresponding to the constant part $\rho_1$ and $\Phi_2$ corresponding to $\rho_2$ which could be easily calculated, and hence also the interaction of the electron clouds.

For the calculation of the exchange energy the wave functions of the electrons with momentum $p$ were assumed to be in the form $e^{i(p \cdot z)/\hbar}\chi(r)$ where $\chi(r)$ is the wave function of the lowest state. Under these conditions, the exchange energy is[2 II, 2 III]

$$-\frac{e^2}{2}\int \rho(r)\rho(r')g(|r-r'|)\frac{1}{|r-r'|}drdr', \quad (10)$$

where

$$g(r) = 9\left[\frac{\sin(r/d) - (r/d)\cos(r/d)}{(r/d)^3}\right]^2$$

$$\approx e^{-1.6r/d}\left[1 + 1.6\left(\frac{r}{d}\right) + 1.2\left(\frac{r}{d}\right)^2\right] \quad (10a)$$

with $d = 0.521r_s$. All the integrations could be carried out in elliptic coordinates. The result is evidently a quadratic function in $c$. The constant term is simply the exchange energy for flat wave functions ($c=0$), i.e., $0.916\text{Ry}/r_s$. The linear terms vanish as may be seen from the following. The integral of the first term in $\rho_1(r)$ times $g(|r-r'|)/|r-r'|$ over $r$ is independent of $r'$. The product of this quantity and the terms containing $c$ in $\rho(r')$, integrated over $r'$, must vanish, since the integral of $\rho(r')$ is independent of $c$. By analogy the other linear term vanishes also and the sole remaining correction to the exchange energy of $0.916\text{Ry}/r_s$ goes with $c^2$. Hence this is a second order effect[8] of the deviation from flatness of the wave function (9), and smaller than the increase of the interaction of the charge clouds which is a first order effect.

Next the change of the correlation energy, due to the uneven charge distribution, has to be considered. In the case of even distribution, the correlation energy is certainly smaller than $0.292e^2/r_s$, even for a very great lattice constant. For the actual uneven distribution in a greatly expanded lattice, however, this energy is as follows. Hartree's equations, solved for many electrons in the field of many ions, give as wave functions of the electrons a sum of spherically symmetric maxima around every ion. If all ions are sufficiently far removed from each other,

these waves—let their form be $u(r)$—no longer overlap. The total interaction energy per electron is

$$\tfrac{1}{2}\int \Phi(r)|u(r)|^2 4\pi r^2 dr \quad (11)$$

where $\Phi(r)$ is the potential of the charge distribution $|u(r)|^2$. Since no such interactions actually exist, Hartree's equation gives an incorrect result by this amount, which will be compensated by the exchange and correlation energy.

If Fock's picture is used instead, the interaction energy between electrons with parallel spins vanishes in consequence of the enormously increased Fermi hole. There remains still the interaction between electrons with antiparallel spin, by which amount Fock's picture is still in error. This remaining half of (11) must be nullified by the correlation energy. It follows that this energy must greatly exceed $0.292e^2/r_s$, the largest value possible for flat functions. We see that the correlation energy is increased by the modulation of the waves by a greater factor than the exchange energy, and it becomes in the limiting case of an infinite lattice just one-half of (11), approximately $\tfrac{1}{4}\text{Ry}$ per electron.[9]

We have increased the correlation energy for flat waves, given in,[2 III] by the same factor by which the exchange energy appeared to be increased. The total change of the electron interaction energy was positive, and plotted in Fig. 2 also.

Finally, the correction for the Madelung number, as given in appendix 2, Reference[2 II] is given in the same plot. The sum of all three corrections has been added to the curve of Fig. 3 and the upper full line of this figure obtained. This curve assumes its minimum of $-1.05\text{Ry}$ (heat of vaporization about 16 Cal.) for $r_s = 1.63$ Bohr units, corresponding to a density of 0.62. The possible error in these figures is quite large, because the correction quantities are so great.

(5)

The last quantity which had to be calculated for the thermal constants of metallic H was the zero-point energy corresponding to the half-

---

[8] This differs from the case for Li.

[9] We emphasize this point since doubts have been raised whether the correlation energy actually exists.

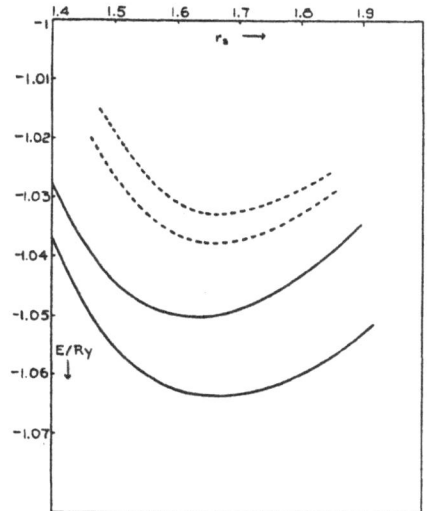

FIG. 3. Energy of the lattice as a function of lattice constant. Lowest curve, for flat wave functions; second curve with all corrections, except zero-point energy of nuclei; the dotted lines contain the zero-point energy, the lower for heavy, the upper for ordinary H.

quanta vibration of the nuclei. If the elastic constants of the medium were known, the elastic spectrum and the zero-point energy could be calculated following Blackman[10] or a paper of Barnes, Brattain, and Seitz.[11] Since, however, only the compressibility can be derived from Fig. 3 (and this only very inaccurately, as it may be in error by as great a factor as two), it was performed by the older method of Debye.[12]

From the curvature of the upper curve in Fig. 3 one obtains for the reciprocal compressibility of a lattice of very heavy hydrogen,

$$\tfrac{1}{3}(c_{11}+2c_{12}) = 2.9 \times 10^{12} \text{ dyne/cm}^2 \text{ for } r_s = 1.5$$
$$= 1.5 \times 10^{12} \text{ dyne/cm}^2 \text{ for } r_s = 1.6$$
$$= 1.1 \times 10^{12} \text{ dyne/cm}^2 \text{ for } r_s = 1.7$$
$$= 0.8 \times 10^{12} \text{ dyne/cm}^2 \text{ for } r_s = 1.8.$$

By analogy with other metals $c_{11} = 2c_{12} = 4c_{44}$ was assumed. The zero-point energy is then about

$0.0244(E'')^{\frac{1}{2}}$Ry, where $E''$ is the second derivative of the energy in Ry with regard to $r_s$ in Bohr units. For the isotropic hydrogen with mass two the value is $0.0173(E'')^{\frac{1}{2}}$Ry. The dotted lines of Fig. 3 give the final results for the lattice energy, the upper one is for light, the lower for heavy hydrogen. The density of the former is 0.59, its heat of formation out of atomic hydrogen 10 Cal. We think that this figure is somewhat too low because the correlation energy is greater than we assumed it to be.

## (6) CONCLUSION

The great density of the hypothetical metallic form of H is not surprising. It seems to be the general rule that the density of the metallic and valence lattice of a substance be greater than that of the layer lattice, and this in turn has greater density than the molecular lattice. Table I illustrates the point.

This table makes it natural to assume that all atoms are bound together by shared electron forces in metallic and valence lattices, and that this is the case also for the atoms within a layer in layer lattices. The different layers are bound together, however, more weakly, perhaps by van der Waals forces. This is well in accord with the general geometric considerations of A. Reis and K. Weissenberg.[13] It seems sensible, therefore, to speak about valence-like forces between two atoms in a lattice, if the density of the outermost electrons of both overlaps considerably at this distance.[14]

The heat of formation of the ordinary, molecular $H_2$ lattice is 52.5 Cal., practically one-half of the heat of dissociation. Although the figures of the preceding section may be in error by quite a few calories, it is evident that the molecular form is so much more stable (by about 40 Cal.) than the metallic, that the chances of obtaining the latter are extremely small.

In the ordinary lattice, there are two "nearest distances." The distance between two atoms of the same molecule is $0.75A = 1.45a_0$, while the nearest distance between two H atoms of

[10] M. Blackman, Proc. Roy. Soc. London A148, 365, 384, (1935); 149, 117, 126 (1935).
[11] Barnes, Brattain and Seitz, Phys. Rev. 48, 582 (1935).
[12] Cf. e.g., E. Schrödinger, Geiger-Scheel Handbuch der Physik, Vol. X.

[13] A. Reis, Zeits. f. Physik 1, 204 (1920); 2, 57 (1920); K. Weissenberg, Zeits. f. Krist. 62, 12, 52 (1925).
[14] This corresponds also to the ideas developed by L. Pauling, J. Am. Chem. Soc. 53, 1367 (1931); J. C. Slater, Phys. Rev. 38, 1109 (1931).

TABLE I.

| SUBSTANCE | DENSITY | | DENSITY |
|---|---|---|---|
| As, metallic | 5.72 | yellow | 2.03 |
| Diamond | 3.51 | graphite | 2.24 |
| Black phosphorus | 2.70 | yellow | 1.83 |
| Se, metallic | 4.82 | red | 4.47 |
| Sn, white | 7.28 | grey | 5.76 |

different $H_2$ molecules is about $3.3A = 6.3a_0$.[15] If one compresses the ordinary lattice, the latter distance will decrease, the former increase, as follows already from the calculations o F. London.[16] Thus, the increase of the pressure brings the intermolecular distances nearer to the intramolecular distances, and the lattice will become more similar to a simple lattice than it is at low pressures. It is possible that this goes on continually, and that one reaches a simple close packed lattice with metallic properties only in the limiting case of infinite pressures.

During the increase of pressure, the total energy of the substance will increase by the work done (we assume very low temperatures). It is possible, thus, that the energy of the continuously compressed lattice becomes at a certain volume, assumed at a very high pressure, greater than that of another modification. In this case there will be a *discontinuous change* to the latter.[17] Whether or not such a transition will occur into the metallic modification, depends on the magnitude of the compressibility of the molecular form at high pressures. The smaller this is, the greater will be the pressure necessary for a certain change in volume, and also the energy increase at this compression. If the extremely high

compressibility at ordinary pressures,[18] which is $2.10^{-9}$ cm$^2$/dyne, would hold throughout, the molecular form would be stable for all volumes. This is not possible, however, and the compressibility certainly decreases with increasing pressure, but one cannot tell, at present, whether it decreases sufficiently to make the metallic form stable at *any* pressure. One calculates easily, that even under the assumption of the most advantageous compressibility at high pressures, the pressure necessary for the transformation is 250,000 atmos., which is outside the scope of the present technique.

The objection comes up naturally that we have calculated the energy of a body-centered metallic lattice only, and that another metallic lattice may be much more stable. We feel that this objection is justified. Of course it is not to be expected that another simple lattice, like the face-centered one, have a much lower energy,— the energy differences between these forms are always very small. It is possible, however, that a layer-like lattice has a much greater heat of formation, and is obtainable under high pressure. This is suggested by the fact that in most cases of Table I of allotropic modifications, one of the lattices is layer-like,[19] the other either metallic, valence, or molecular. The difficulties for such an experiment will be greatly increased by the necessity of the formation of a nucleus for the new lattice.

---

[15] The calculated nearest distance for the metallic form is 1.5A.

[16] F. London, Zeits. f. Elektrochemie 35, 552 (1929); H. Eyring, J. Am. Chem. Soc. 53, 2537 (1931).

[17] Such polymorphic transitions induced by pressure in solids are described by P. W. Bridgman, Rev. Mod. Phys. 7, 1 (1935).

[18] Measured by A. Eucken for the liquid form, Ber. deutschen Phys. Ges. 4 (1916).

[19] Diamond is a valence lattice, but graphite is a layer lattice (A. W. Hull, Phys. Rev. 10, 661 (1917)). Yellow arsenic and phosphorus are evidently molecular lattices, black phosphorus a layer lattice (R. Hultgren and B. E. Warren, Phys. Rev. 47, 808 (1935)), and metallic As is also only approximately a simple lattice. The red, monoclinic selenium can be said to form a molecular lattice (F. Halla, F. Bosch, E. Mehl, Zeits. f. physik. Chemie B11, 455 (1930)), while the metallic modification is a thread lattice (A. J. Bradley, Phil. Mag. 48, 477 (1924)). The situation seems to be most complicated with tin. Grey tin forms a diamond lattice, but shows otherwise no similarity to the valence lattice of diamond, while the metallic lattice has a rather complicated structure (H. Mark and M. Polanyi, Zeits. f. Physik 18, 75 (1923)).

# On the Structure of Solid Bodies

E. P. Wigner

Scientific Monthly *42*, 40–46 (1936)

(1) PHYSICS always develops in two directions. One front pushes forward towards phenomena which do not yet fit into the general picture, and the victories on this front are marked by important changes in our fundamental concepts. On this front to-day the main struggle is for a better understanding of nuclear phenomena by the application of both theory and experimentation. But, apart from this search for new concepts, there is a constant effort directed toward the deepening and broadening of our knowledge of phenomena which, we believe, *can* be understood on the basis of existing concepts and theories. Doubtless this second front is of less importance. It rarely leads to fundamental discoveries in physics proper but supports rather the studies on the borderline of this science, such as physical chemistry and the applied sciences. Spectroscopy suddenly changed, about six years ago, from the first to the second category, and not much later it became apparent that the study of the solid body belongs also to this second class. In spite of this, it remains one of the most attractive of all fields, since it deals in a scientific way with those subjects with which we must deal in our everyday experience. For example, we are never afraid when dropping a key that it will fly to pieces, as glass would, nor do we fear that a gold coin will dissolve in water nor evaporate if left for awhile in the open air.

X-ray studies have revealed that most of the solid bodies in our surroundings are crystalline. This does not necessarily mean that they are formed by one single crystal—although even this is true for bodies of such enormous size as icebergs. More commonly, they are poly-crystalline, like the metal parts of ordinary tools, *i.e.*, a conglomerate of microscopic crystals of various sizes. Crystalline in this connection does not mean a regularly shaped body of the kind we see in our crystallographic collections, but only that the grains have a regular *inner structure* arising from the arrangement of the atoms in surprisingly regular *lattices*. Samples of such lattices are shown in Fig. 1. (The circles represent the centers of atoms; the lines have no physical significance and are drawn only in order to facilitate space-vision.) The region over which the regular arrangement has a certain orientation is called a microcrystal and may have a size anywhere from .00001 mm to 1 mm or even more. These microcrystals, generally possessing irregular boundaries, are heaped together in an apparently random manner to form the polycrystalline body. (Very little is known about how the microcrystals with their different orientations fit and stick together. Some assume a separate very thin non-crystalline phase which "pastes" them together, but there is no definitive evidence for this.)

The crystalline and polycrystalline substances constitute by far the greater part of all solid bodies found in nature. Practically all rocks are conglomerates of crystals, ice is crystalline, and so are all metals. The grains of sand are minute crystals and loam also is crystalline. Apart from the glasses and substances of organic origin, like wood, there are very few non-crystalline solids.

(2) A distinction not necessary in the case of gases or liquids must be made between different kinds of properties of solids.

## ON THE STRUCTURE OF SOLID BODIES    41

Evidently, the consideration of a regular lattice is much simpler than that of an irregularly spaced heap of atoms. It is important, therefore, that many of the properties of a polycrystal are the same as those possessed by a perfect single crystal. These properties are connected with phenomena which affect the bulk of the material, like vaporization, fusion, specific gravity and compressibility. Our understanding of these "insensitive properties" is naturally the farthest advanced, and we shall devote most of our attention to them.

Unfortunately, a great many very important properties belong in a second "sensitive" class. The breaking strength, for instance, is determined by the very weakest part of the crystal; one single imperfection of certain types may suffice to cause rupture under a very low stress, a tenth or even a hundredth of that which a perfect crystal could support. Fig. 2 gives a rough picture of how this can happen: the stress, characterized by the stress lines, concentrates in the neighborhood of the imperfection and attains values which are many times those in the bulk of the material. This highly concentrated stress can widen the notch and finally break the whole body. Thus, the parts of a solid which lie above and below a crack not only do not increase the strength of the material but very definitely weaken it. One can say that the strength of a solid is much smaller than that of its weakest part.

The situation for the electric breakdown of insulators parallels that for the elastic limit (the smallest stress which causes a permanent deformation), and the study of these sensitive properties of crystals involves besides a knowledge of the crystal in bulk, its criminology, *i.e.*, a knowledge of the most usual faults and imperfections.

In addition to these extremes, there are, of course, a number of borderline properties. These are partly connected

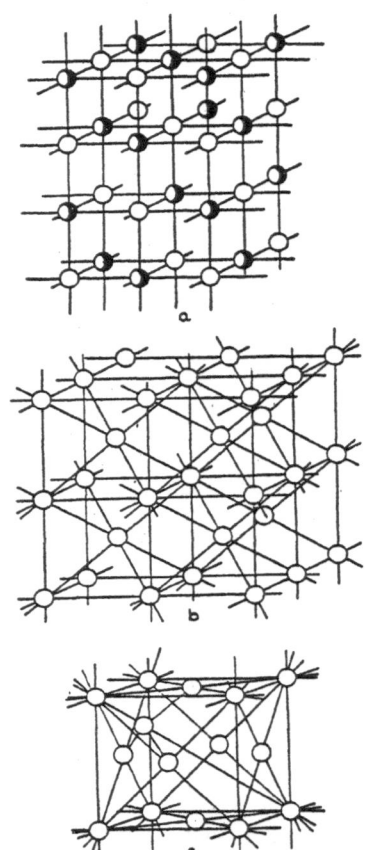

FIG. 1a.  PART OF A KCl LATTICE.  THE SHADOWED SPHERES DENOTE THE POSITIONS OF THE K IONS, THE EMPTY ONES THE POSITIONS OF THE Cl IONS.  THE DISTANCE BETWEEN NEAREST NEIGHBORS IS .00000031 mm.  ORDINARY ROCKSALT HAS THE SAME LATTICE WITH SOMEWHAT SMALLER DIMENSIONS.

FIG. 1b.  PART OF THE LATTICE OF AN ALKALI METAL.  THE SPHERES REPRESENT THE CENTERS OF MASS OF THE ATOMS.  THE DISTANCE BETWEEN NEAREST NEIGHBORS IS .000000372 mm IN SODIUM.

FIG. 1c.  UNIT CELL OF THE DIAMOND LATTICE. THE DISTANCE BETWEEN NEXT NEIGHBORS IS .000000154 mm.  Si HAS A SIMILAR LATTICE WITH A DISTANCE OF .000000234 mm BETWEEN NEAREST NEIGHBORS.

42                    THE SCIENTIFIC MONTHLY

with the external surface, as, for example, the thermionic emission of electrons, or with the internal boundaries of crystallites, exemplified by the electric conductivity of compressed salts. All these properties are influenced to some extent by small contaminations. With extreme care and sufficient experimental skill reproducible results are sometimes obtainable for these phenomena, and they are then frequently as amenable to theoretical interpretation as the insensitive properties.

(3) Let us return now to the insensitive properties. Even with regard to these, the variety found in solids is much greater than that in gases. From the empirical point of view, four main classes, with many transitions between them, can be distinguished. This classification, which in its essentials goes back to Grimm, contains:

(a) *Molecular lattices.* Inert gases or saturated compounds like He, Ne, A,

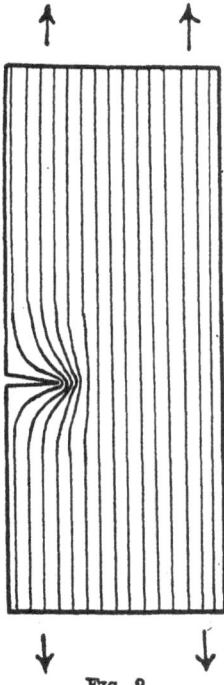

FIG. 2.

etc., $H_2$, $N_2$, $O_2$, etc., $CH_4$, $C_2H_6$, $H_2O$, $H_2S$, etc., and all organic compounds form such crystals. They all have low heats of vaporization and condense only at comparatively low temperatures. They are soft and moderately brittle, are good insulators and are transparent, except in spectral regions in which the building molecules themselves show absorption.

(b) Metals have in many respects properties opposite to those of class a. The binding forces between the atoms are much higher and the heat of vaporization greater, and they have an increased hardness. Their most remarkable property is, of course, that they are good conductors for electricity and heat. They are opaque and owe many of their important applications in industry to their plasticity; that is to say, they break only after great deformations.[1] Their solubility in each other is considerable (alloys), but they never dissolve in solids of other classes.

(c) *Valence lattices* (diamond, quartz, carborundum) and

(d) *Ionic lattices* (salts) are rather similar types. They both have high heats of vaporization, strong cohesive forces, are transparent like molecular lattices, are good insulators and are hard and brittle. The main difference between them is that while the former are formed from neutral atoms, the building stones of the salts are electrically charged ions, held together by the electrical attraction between opposite charges. They dissolve, therefore, in liquids with high dielectric constants like water, which diminish the electrical attraction of the ions down to a small fraction of its original value.

[1] This is why they do not break if dropped. The sudden stopping on the ground causes great stresses. In consequence of this, the metal will suffer a plastic deformation which will not cause rupture, however. In consequence of the plastic deformation the metal will act as its own shock absorber by allowing more time for the stopping of the bulk of the material.

## ON THE STRUCTURE OF SOLID BODIES

This characterization of the four groups of solids should be understood in the same sense as should a similar characterization of a class of plants in botany. It does not give ironclad rules, but rather ideals from which the real cases often deviate; especially is this true for the more complicated compounds. Also various kinds of transitions occur between the four groups. Sometimes inside individual layers we have a lattice of one kind, while the forces *between* the layers are characteristic of another of our classes. There are also cases which are really transitional in all their behavior between two (or even three) groups, especially between valence and ionic lattices.

These exceptions are rare, however. The importance of the four groups becomes most evident, perhaps, if we realize that instinctively we classify into one of these groups all solid bodies of inorganic origin, which happen to fall into our hands. The above characterization of the four groups is the scientific description of what all of us would expect with regard to vaporization, hardness, electric conductivity and brittleness after some inspection and handling of such substances as condensed $CO_2$, rhodium, carborundum and Glauber's salts, even if we had never seen them before. On the other hand, we wouldn't quite know what to expect from transition lattices such as carbide or even graphite.

(4) The enormous differences between the physical properties of different kinds of lattices make it evident that the forces holding the atoms or molecules together are very different in the four cases. In order to understand the origin and nature of these forces, we must first recall the structure of isolated atoms and molecules. This is probably well known to the readers of the SCIENTIFIC MONTHLY. It is only recently that Professor Eyring gave an excellent review of this subject

in these pages.[2] According to Rutherford, the atom contains, first of all, a heavy nucleus, containing all the positive charge and (except for about one part in two thousand) all the mass of the atom. The center of gravity of the atom practically coincides with the nucleus, so that in Fig. 1 the circles may be regarded alternatively as the positions of atoms or nuclei. This nucleus, though small, is full of mysteries, which fortunately are of no importance in understanding the solid state. The negative charges, which exactly compensate the positive charge of the nucleus of a neutral atom, are carried by light particles, the electrons. These electrons surround the nucleus like an enormous cloud with dimensions a hundred thousand times that of the nucleus, although the cloud is itself only about .0000001 mm thick. Quantum mechanics, created by Heisenberg, Schrodinger and Dirac, unravelled for us about eight years ago the exact laws of motion of this electron cloud. It is now possible to calculate the density of this cloud at different distances from the nucleus, and from this one would naturally expect to obtain important information concerning the structure of solids, by comparing the density distribution of the electrons for different distances in the lattice. The outermost or valence electrons are responsible for the entire chemical behavior of the atoms. In Fig. 3, the full line represents the density of the valence electrons as a function of the distance from the nucleus. In addition to this, the position of the nearest neighbor is marked on the abscissa and the density distribution of the valence electron of this neighbor is plotted *in the direction of the first atom* as the dotted line. The first plot is for He, the most characteristic representative of a molecular lattice; the second is for sodium, a typical metal; the third for

[2] SCIENTIFIC MONTHLY, 39: 415–419, November, 1934.

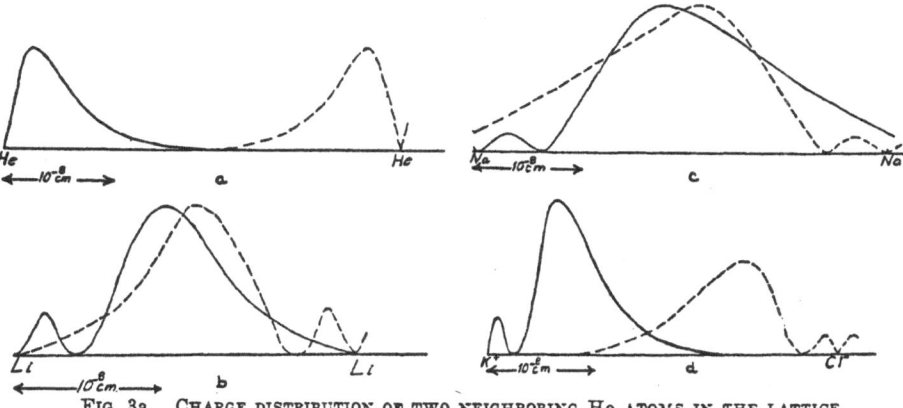

FIG. 3a.   CHARGE DISTRIBUTION OF TWO NEIGHBORING He ATOMS IN THE LATTICE.

FIG. 3b.   CHARGE DISTRIBUTION OF EXTERNAL ELECTRONS IN FREE Si ATOM (FULL LINE). THE DOTTED LINE IS THE CHARGE DISTRIBUTION OF THE VALENCE ELECTRONS OF ANOTHER Si ATOM, PLACED AT THE SAME DISTANCE FROM THE FIRST AS IN THE LATTICE.

FIG. 3c.   CHARGE DISTRIBUTION OF THE VALENCE ELECTRON IN FREE SODIUM ATOM (FULL LINE). THE DOTTED LINE IS THE CHARGE DISTRIBUTION OF THE VALENCE ELECTRON OF ANOTHER SODIUM ATOM, PLACED AT THE SAME DISTANCE FROM THE FIRST AS THE NEAREST NEIGHBOR IN THE LATTICE.

FIG. 3d.   CHARGE DISTRIBUTION OF EXTERNAL ELECTRONS IN K ION (FULL LINE) AND Cl ION (DOTTED LINE). THE DISTANCE OF THE ZEROS OF THE TWO PLOTS IS EQUAL TO THE DISTANCE OF THE NEAREST IONS IN THE KCl LATTICE.

the valence lattice of silicon and the last one is for KCl, which closely resembles ordinary rocksalt.

We realize at once an important difference between the molecular and ionic lattices (first and last pictures) on the one hand, and the metallic and valence lattices on the other. For the former, the overlapping of the electron clouds is small, in the latter ones it is so great that it is impossible to tell to which atom a certain valence electron belongs. In the former cases the constituent atoms or ions, although attracted by their neighbors, have their charge distribution but slightly affected. This is not so for the metals and valence lattices. There is no region between the atoms with a small charge density and consequently no forbidden region for the electrons. The electrons are able to pass from one atom to the next. Thus the valence electrons move freely and are common to the whole lattice. This is of decisive importance for the properties of these substances.

In molecular and ionic lattices, it is possible to consider the constituents as different entities. Born's classical theory of mechanical electric and thermal properties, which treats the atoms and ions of the lattice as individuals, attained its great successes for these lattices.

The great differences in the behavior

 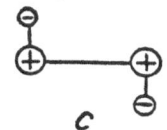 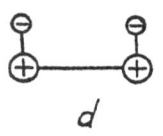

FIG. 4.

## ON THE STRUCTURE OF SOLID BODIES

of the two classes are due to the different character of the constituents. These are neutral atoms in the first case, in the ionic lattices they are charged particles. The electric forces between ions are very strong, and this makes the cohesive forces, the heat of vaporization and hardness great. The distances between neighboring ions are given by the charge distribution, as illustrated in Fig. 3. The calculations of ionic radii were carried out by L. Pauling in California and show remarkable agreement with the values derived by Goldschmidt from observations. These lattices are always so constructed that the positive ions are surrounded by negative ions, the negative ions by positive ones. (*Cf.* the NaCl lattice in Fig. 1.) Since opposite charges attract each other, there are considerable forces holding these lattices together.

The nature of the forces in molecular lattices is not so simple. Van der Waals was the first to assume that condensation is caused by the same forces which are responsible for the deviation in the behavior of real gases from the ideal gas laws. This proved to be true in the case of molecular lattices, and the laws of these forces have been recognized by London and Wang on the basis of quantum mechanics and called van der Waals forces.

Of course, there is no attraction due to electric charges in molecular lattices, since the constituents are uncharged. And, indeed, the attraction can not be understood as long as we consider the electrons as charge clouds. But if we remember their corpuscular nature, we realize that they can form *dipoles* with the nucleus. The direction of this dipole will vary quickly because of the quick motion of the electrons. There will be no force, in the mean, on a dipole of constant orientation, since the attraction for one dipole orientation is as great as the repulsion for the opposite orientation—

and all dipole directions are equally probable. But if two variable dipoles face each other, it will be possible that the two attractive configurations *a* and *c* of Fig. 4 will occur more often than the repulsive configurations *b* and *d*, although all the orientations for the *single* dipoles are equally probable. London and Wang have shown that this is actually the situation, and thus laid the foundation not only for a satisfactory theory of molecular lattices but also for a theory of the behavior of real gases.

Naturally, the van der Waals forces are much smaller than the Coulombic forces between ions. Thus, the cohesion in molecular lattices is small, the vaporization easily giving volatile substances. Also it is evident that these very small forces will be important only if no other stronger forces are present. Molecular lattices will be formed by saturated compounds and inert gases.

(5) Fig. 3 shows us that the metallic and valence lattices form the more compact modification of matter, as contrasted with molecular lattices and salts. This gives important information concerning the question of the behavior of solids under extremely high pressures: according to Bernal, who first emphasized this point, they will go over into metals or valence lattices. A convincing piece of evidence for this point of view, which is quite independent of calculations of charge distribution, is furnished by the phenomenon of *allotropy*. This is the name given the phenomenon of the appearance of the same chemical element in different "modifications" with widely different physical and chemical properties. The ordinary, yellow (white) form of phosphorus forms a somewhat complicated molecular lattice. It is a good insulator, soft, dissolves in organic solvents and has a density of 1.83. Bridgman at Harvard subjected this element to very high pressures, and the lattice "collapsed." It transformed into *black*

*phosphorus,* which has a density of 2.70, is a fairly good conductor of electricity and insoluble in organic liquids. And this is the general rule: Whenever an element has two allotropic modifications, the *metallic or valence form has the higher density.* The following table illustrates this point:

| | | | |
|---|---|---|---|
| As, metallic | 5.72 | yellow | 2.03 |
| diamond | 3.51 | graphite | 2.24 |
| black phosph. | 2.70 | yellow | 1.83 |
| Se, metallic | 4.82 | red | 4.47 |
| Sn, white | 7.28 | grey | 5.76 |

Calculations made in our laboratory by H. B. Huntington show that metallic hydrogen should also exist, though only under extremely high pressures, and that it should have a density many times higher than that of the usual molecular form.

I shall not go into detail with regard to the next question which naturally arises—the cause of the fundamental difference between valence and metallic lattices. Although both form in the compact modification of matter, apart from the high heat of vaporization and boiling point, they have nothing in common. The reason for this is deeply rooted in the principles of quantum mechanics and has been brought out but

lately by works of Peierls and Brillouin. According to their investigations, it is essential for a valence lattice that the number of valence electrons be *even,* and this rule holds without exceptions. We owe much valuable information concerning the structure and crystal form of valence lattices to Pauling and Slater, but a review of their work would greatly exceed the scope of this report.

I hope that I have succeeded in imparting to the reader the impression that the foundations for the understanding of the nature of the solid state are laid. Still, it will require much thorough work, perseverance and many new ideas before we will be able to add the theory of solids as a finished story to the building of physics and before we will be able to apply with success our knowledge in industry.

The progress in the explanation of the properties of solid bodies is due on the theoretical side to the newly developed quantum mechanics, and experimentally mainly to the study of crystal structure by x-rays. Without these tools we would face these problems as helplessly as we still face the problem of liquids where x-ray studies have proved less efficient so far.

# Theory of the Brillouin Zones and Symmetry Properties of Wave Functions in Crystals

L. P. Bouckaert*, R. Smoluchowski, and E. P. Wigner

Physical Review 50, 58–67 (1936). Reprinted in: Group Theory and Solid State Physics, Part I (P. H. Meijer, ed.). Gordon and Breach, New York 1964, pp. 58–67

(Received April 13, 1936)

It is well known that if the interaction between electrons in a metal is neglected, the energy spectrum has a zonal structure. The problem of these "Brillouin zones" is treated here from the point of view of group theory. In this theory, a representation of the symmetry group of the underlying problem is associated with every energy value. The symmetry, in the present case, is the space group, and the main difference as compared with ordinary problems is that while in the latter the representations form a discrete manifold and can be characterized by integers (as e.g., the azimuthal quantum number), the representations of a space group form a continuous manifold, and must be characterized by continuously varying parameters. It can be shown that in the neighborhood of an energy value with a certain representation, there will be energy values with all the representations the parameters of which are close to the parameters of the original representation. This leads to the well-known result that the energy is a continuous function of the reduced wave vector (the components of which are parameters of the above-mentioned kind), but allows in addition to this a systematic treatment of the "sticking" together of Brillouin zones. The treatment is carried out for the simple cubic and the body-centered and face-centered cubic lattices, showing the different possible types of zones.

## I.

INVESTIGATIONS of the electronic structure of crystal lattices in particular in metals, made on the basis of Bloch's theory, led to the conception of the so-called Brillouin zones.[1] In spite of these investigations, which cover a large part of the field, it seems desirable to develop the theory from a unique point of view. It appears that taking into account special symmetry properties of different lattices brings out interesting features of the constitution of the B-Z which are not evident from the existing general theory. These features can be dealt with

---

* C. R. B. Fellow.
[1] The existence of these zones was first noticed by I. J. O. Strutt, Ann. d. Physik 85, 129 (1928); 86, 319 (1929); and then, independently, by F. Bloch, Zeits. f. Physik 52, 555 (1928); cf. also P. M. Morse, Phys. Rev. 5, 1310 (1930). From another point of view, they were discussed by R. Peierls, Ann. d. Physik 4, 121 (1930). Their connection with x-ray reflection was first pointed out by L. Brillouin (cf. e.g., Die Quantenstatistik (Berlin, 1931)). Important physical applications were given by H. Jones, Proc. Roy. Soc. A144, 225 (1934); 147, 396 (1934); H. Jones, N. F. Mott and H. W. B. Skinner, Phys. Rev. 45, 379 (1934); J. C. Slater, Phys. Rev. 45, 794 (1934); Rev. Mod. Phys. 6, 209 (1934); F. Hund and B. Mrowka, Ber. Sachs. Akad. D. Wiss. 87, 185, 325 (1935). Compare

also F. Hund, Zeits. f. tech. Physik 16, 331, 494 (1935); Zeits. f. Physik 99, 119 (1936). Hund's work deals with those properties of the Brillouin zones which are common to all zones of the same lattice (as matter of fact he does not discriminate between different types of zones at all). We consider here the different types of zone separately. The differences between the different types are of the same kind as e.g. the difference between even and odd terms in atomic spectra. It is surprising that there are at all common properties of all zones but Hund has shown that this is the case for the more complicated crystal structures.

uniformly by the methods of group theory,[2] and we propose to take up the subject here from this point of view. The first start in this direction has been made by F. Seitz,[3] and we shall use his results extensively, though a knowledge of his work should not be necessary for the understanding of this paper.

In the theory of Bloch, every electron has a separate wave function. This assumption is identical with the Hartree-Fock approximation method and amounts to neglecting the statistical correlations between electrons. If we neglect these correlations, every electron obeys a separate Schrödinger equation of the type

$$-\frac{\hbar^2}{2m}\left(\frac{\partial^2}{\partial x^2}+\frac{\partial^2}{\partial y^2}+\frac{\partial^2}{\partial z^2}\right)\psi+V\psi=E\psi \qquad (1)$$

in which $V$ contains the ordinary and exchange potentials of the ions and electrons.[4] The potential $V$ has the whole symmetry of the lattice, that is, the group of our Schrödinger equation (1) is the space group of the lattice.

It is clear from the ordinary group theory[2] that every characteristic value of (1) belongs to a certain representation of the space group and the dimension of the representation is equal to the number of characteristic functions belonging to this characteristic value.[5] Thus far the group theory of the B-Z is not different from the group theory of any other system. But while in atoms, molecules, etc., the characteristic values of (1) are well separated, the characteristic values of (1) for a crystal form a continuous manifold. There will be several characteristic values in the neighborhood of any one $E$ and the representations of these characteristic values will be said to form the neighborhood of the representation of $E$ for this B-Z. Thus a certain topology for the representations must exist and it will be shown that part of this topology is independent

[2] Cf. e.g., E. Wigner, *Die Gruppentheorie und ihre Anwendungen* (Braunschweig, 1931). The first application of group theory to crystal lattices has been given by H. Bethe, Ann. d. Physik 3, 133 (1929).
[3] F. Seitz, Ann. of Math. 37, 17 (1936).
[4] L. Brillouin, *Actualités Scientifiques et Industrielles* (Paris, 1933).
[5] To the symmetry operations of the space group, the "reversal of time" (cf. E. Wigner, Gott. Nachr. 546 (1932)) should be added. It has been remarked by F. Hund (reference 1) that this will often be of great importance. It can be omitted, however, in the case of the cubic lattices investigated here.

of the special B-Z. Even if $E$, $E'$, $\cdots$ be in different B-Z but have the same representation, there will be energy values neighboring $E$ (with a few exceptions) with the same representations as those of energy values neighboring $E'$, etc. The investigation of the "topology" of representations will be essentially the subject of this paper, from the mathematical point of view.

## II.

We must review next, the theory of representations of space groups. F. Seitz[3] has shown that all space groups are soluble groups and their representations can be obtained according to the general theory for these.[3, 6] Seitz first considers the invariant subgroup formed by the translations. Since these commute, the corresponding matrices in the representation can be assumed to have the diagonal form. This means that we shall consider such linear combinations $\psi_\mu$ ($\mu = 1, 2, \cdots n$, where $n$ is the dimension of the representation) of the wave functions, which are merely multiplied by constant factors ("multipliers") $\omega_{\mu 1}$, $\omega_{\mu 2}$, $\omega_{\mu 3}$ if a displacement by the three elementary identity periods is made. In other words, the matrix corresponding to the displacement by the first elementary identity period is a diagonal matrix with the diagonal elements $\omega_{11}$, $\omega_{21}$, $\cdots$, $\omega_{n1}$, with similar matrices for the representatives of the other displacements. Since all matrices must be unitary, $|\omega_{\mu 1}| = |\omega_{\mu 2}| = |\omega_{\mu 3}| = 1$; and if one writes

$$\omega_{\mu 1} = e^{i(k_x x_1 + k_y y_1 + k_z z_1)}$$
$$\omega_{\mu 2} = e^{i(k_x x_2 + k_y y_2 + k_z z_2)} \qquad (2)$$
$$\omega_{\mu 3} = e^{i(k_x x_3 + k_y y_3 + k_z z_3)}$$

with $x_1$, $y_1$, $z_1$, $x_2$, $y_2$, $z_2$, $x_3$, $y_3$, $z_3$, the $x$, $y$, $z$ components of the first, second and third identity periods, the vector $\mathbf{k}$ is called[7] "the reduced wave number vector." Of course, $\mathbf{k}$ will be, in general, different for the different wave functions $\psi_1$, $\psi_2$, $\cdots$, $\psi_n$. It must be remembered, how-

[6] G. Frobenius, Berl. Ber. 337 (1893); I. Schur, Berl. Ber. 164 (1906).
[7] Cf. A. Sommerfeld and H. Bethe's article in *Handbuch der Physik*, Vol. 24 (Berlin, 1933), chapter 3. Also J. C. Slater, Rev. Mod. Phys. 6, 209 (1934). For a simple cubic lattice $x_1 = y_2 = z_3 = d$; $y_1 = z_1 = x_2 = z_2 = x_3 = y_3 = 0$. For a face centered lattice $y_1 = z_1 = x_2 = z_2 = x_3 = y_3 = d/2$; $x_1 = y_2 = z_3 = 0$, etc.

ever, that the reduced wave vector **k** is defined by (2) only up to an integer multiple of a vector **r** of the reciprocal lattice, i.e., a vector **r**, for which

$$r_x x_1 + r_y y_1 + r_z z_1 = 2\pi n_1$$
$$r_x x_2 + r_y y_2 + r_z z_2 = 2\pi n_2 \qquad (2a)$$
$$r_x x_3 + r_y y_3 + r_z z_3 = 2\pi n_3$$

always can be added to **k**, without changing its meaning. The space of **k** is periodic with all the periods **r**, satisfying (2a); two reduced wave vectors differing by such an **r** are considered identical. If there are no essential gliding planes and screw axes in the space group,[8] one needs to consider, in addition to the above translations, rotations and reflections only. If such a transformation is applied to $\psi_\mu$, it will be transformed into a wave function, say $\psi_\lambda$, the reduced wave vector of which arises from that of $\psi_\mu$ by just the rotation or reflection considered. Thus the reduced wave vectors of the wave functions of one representation all arise from one another by the pure rotations and reflections of the group, i.e., the elements of the crystal class. If the reduced wave vector of one $\psi_\mu$ is transformed by every element of the crystal class into a different vector, this will be true for all of them, and we shall have as many wave functions $\psi_1, \cdots, \psi_n$ as the crystal class has elements. The matrices of the representation corresponding to rotations and reflections will merely interchange the different $\psi_\mu$. If there are symmetry elements which leave a wave vector invariant, they form a group which we shall call *the group of the wave vector*. So, for example, if the wave vector lies in the $x$ direction, its group will contain all rotations around $x$ and all reflections in planes through $x$.

A wave function $\psi_\mu$ with a wave vector **k** either is left invariant under the transformations of the group of **k**, or else transformed into a new $\psi_\lambda$ with the same wave vector, **k**, however. In the first case there will be only one wave function with the wave vector **k**. In the second case there will be several of them and they will transform

under the transformations of the group of **k** by an irreducible representation of this group, which we shall call the *small representation*. These are the results of Seitz.

Hence the representations of the space group must be characterized by two symbols. The first gives the reduced wave vectors (or set of $\omega$) which occur in the representation; the figure of all these wave vectors forms a "star" with all the rotational and reflection symmetries of the lattice. Three such stars are given in Fig. 1 for

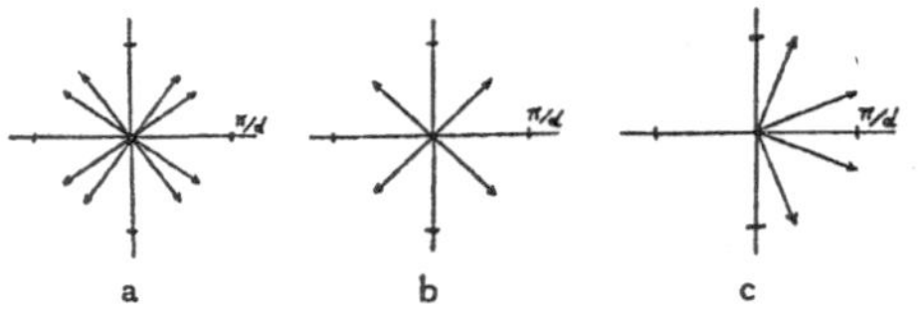

FIG. 1.

a two-dimensional quadratic lattice. The second symbol characterizes the small representation, which is an irreducible representation of the group of one wave vector (the groups of all wave vectors of a star are holomorphic). If the wave vectors lie in general positions (Fig. 1a) their group will contain the unit element only. In this case the second symbol may be omitted. It may be emphasized again that two wave vectors must be considered identical, if the corresponding set of $\omega$'s is the same. Thus for example, if the three $k_x x_i + k_y y_i + k_z z_i$ are all integer multiples of $2\pi$ (not necessarily of $2\pi$) the wave vector $k_x$, $k_y$, $k_z$ is identical with the wave vector $-k_x$, $-k_y$, $-k_z$ and the inversion ($x \rightarrow -x$, $y \rightarrow -y$, $z \rightarrow -z$) always belongs to the group of the wave vector.[9]

## III.

We now consider an energy value $E$ with a certain representation $D$ and the wave functions $\psi_1, \cdots, \psi_n$. If we multiply one of these by $e^{i(\kappa_x x + \kappa_y y + \kappa_z z)}$ where $\kappa_x$, $\kappa_y$, $\kappa_z$, are the components

---

[8] We mean by this that all symmetry elements can be considered as products of two symmetry elements, the one of which is a pure translation, the other a pure rotation or reflection. This is the case in the most important space groups.

[9] It is in this connection that the time reversal is important (cf. F. Hund, reference 1). If the crystal class does not contain the inversion, $k_x$, $k_y$, $k_z$ will still be carried over into $-k_x$, $-k_y$, $-k_z$ by the "time reversal." Since, as we shall see, the above consideration determines the surface of the B-Z, this will be fundamentally affected by the operation of time reversal.

of a very small vector, it will have the wave vector $\mathbf{k}+\mathbf{\kappa}$ and belong to a new representation $D'$. The set of new representations obtained in such a way will be called the neighborhood of $D$. It is clear that there will be near $E$, an $E'$ with a representation $D'$. For if $\psi_1$ satisfies (1), $\psi_1 e^{i(\kappa_x x + \kappa_y y + \kappa_z z)} = \psi_1'$ satisfies

$$\left(-\frac{\hbar^2}{2m}\Delta + V\right)\psi_1' + \frac{\hbar^2 i}{m}\left(\kappa_x\frac{\partial}{\partial x} + \kappa_y\frac{\partial}{\partial y} + \kappa_z\frac{\partial}{\partial z}\right)\psi_1'$$
$$= \left(E + \frac{\hbar^2}{2m}(\kappa_x{}^2 + \kappa_y{}^2 + \kappa_z{}^2)\right)\psi_1'. \quad (3)$$

In this equation the second term is small, and its negative value may be treated as a perturbation. Performing the perturbation calculation, we shall obtain a characteristic value $E'$ of (1) which is near $E$ and the wave function of which will have the same translational symmetry as $\psi_1'$, since both the original operator in (3), and the perturbation

$$-\frac{\hbar^2 i}{m}\left(\kappa_x\frac{\partial}{\partial x} + \kappa_y\frac{\partial}{\partial y} + \kappa_z\frac{\partial}{\partial z}\right) \quad (3a)$$

have the whole translational symmetry of the lattice.

This is all the general theory we need. If $E$ had a star of the general type, the star of $E'$ also will be of the general type and our result merely states the well-known fact, that the energy is a continuous (and even differentiable) function of the components of the wave vector. The set of all energies and wave functions which may be obtained from one single energy level continuously by this operation, never touching a point in which the star degenerates, is properly defined as one Brillouin zone. The restriction to such representations, the stars of which are of the general type, is necessary for the definition of a Brillouin zone, since, as we shall see, two or more Brillouin zones may stick together for degenerated stars (as those in Figs. 1b and 1c).

If we consider an energy value, the wave vectors of which are left invariant by some of the rotation or reflection operations, the situation still will be left essentially unchanged, if no two wave functions have the same reduced wave vector (the same multipliers). If, however, two or more (say $s$) wave functions have the same

wave vector, and we choose $\kappa_x$, $\kappa_y$, $\kappa_z$ in such a way that the new wave vector $(\mathbf{k}+\mathbf{\kappa})$ has the general position, there will be $s$ orthogonal wave functions, with the wave vector $\mathbf{k}+\mathbf{\kappa}$ and with energies near $E$. Since for general wave vectors it never happens that two wave functions with the same wave vector belong to the same energy value, we must conclude that they all belong to different B-Z which are very close for small $\kappa$ and that for the original energy value $E$ these $s$ B-Z "stick together." The sticking together will, therefore, always occur for such wave vectors which are left invariant by some symmetry operations.[10]

We must investigate two more cases. First let $\kappa$ be such a vector that $\mathbf{k}+\mathbf{\kappa}$ still has the group of $\mathbf{k}$. In this case, the small representation of $E$ is equivalent to the small representation of $E'$. Otherwise the wave functions would have to change abruptly even for a small change of $\mathbf{k}$. The sticking together will be the same along symmetry elements.

In the second case, the group of $\mathbf{k}+\mathbf{\kappa}$ is only a subgroup of $\mathbf{k}$, but still contains more than the identity. This case occurs, for instance, if we pass from a symmetry axis to a symmetry plane through this axis, or from the vector $\mathbf{k}=0$ to a symmetry axis. The small representations of $E'$ will be irreducible representations of the subgroup, and if the small representation of $E$ is not irreducible as representation of the group of $\mathbf{k}+\mathbf{\kappa}$, the B-Z which stuck together for $\mathbf{k}$ will be partly separated for $\mathbf{k}+\mathbf{\kappa}$. The small representations in these B-Z will be, for $\mathbf{k}+\mathbf{\kappa}$, the irreducible parts of the small representation of the group of $\mathbf{k}$.

The proposed characterization of a B-Z is given, hence, by the small representations of the groups of all wave vectors, which have a group greater than unity. For wave vectors lying in equivalent symmetry elements, the small representations are equivalent, and for a symmetry element which is a subgroup of another, the small representation must be contained in the small representation of the latter. Wherever the small representation is $s$ dimensional, we have a sticking together of $s$ B-Z, all of them having this same small representation for the

---

[10] Including the time reversal.

symmetry element under consideration. Again, it is important to remember that the group of the wave vector for which

$$k_x x_i + k_y y_i + k_z z_i = n_i \pi \quad \text{(for } i = i_1) \quad (4)$$

holds for one $i$, say $i = i_1$, contains all elements which transform $\mathbf{k}$ in such a way, that $(k_x x_i + k_y y_i + k_z z_i)/\pi$ remains an integer for $i = i_1$ and is unchanged for the two other $i$, since the corresponding wave vectors are all the same.

The argument which shows that the small representation will be the same all along a symmetry element, breaks down for such points in which two B-Z touch each other, if no such touching is required by symmetry considerations. In the case that the energy for a certain value of $\mathbf{k}$ is the same in two B-Z, without this being the result of the symmetry, we speak of an accidental degeneracy.[10a] In points of accidental degeneracy, the small representations of the two B-Z may be interchanged, but the case of such an accidental degeneracy is explicitly excluded from the following considerations. One can see that it does not occur for very large lattice constants, though it may occur for the actual ones.

In the following sections, these results will be applied to the three most important cubic lattices, the simple, the face-centered, and the body-centered cubic lattices. Since, for instance all small representations of wave vectors in the fourfold axes are the same, this small representation will be called "the representation along the fourfold axis" and a similar notation will be used for the other symmetry elements.

It has been pointed out by J. C. Slater[1] that the energy as function of $\mathbf{k}$ should be considered as a periodic, multivalued function, the periods being the vectors of the reciprocal lattice. The "discontinuities" then arise from considering for some $\mathbf{k}$ one, for other $\mathbf{k}$ other branches of this multivalued function. In our way of talking, the periodicity is expressed by the fact that two wave vectors differing by a vector $\mathbf{r}$ of the reciprocal lattice, are considered identical. It is convenient to single out from all sets of "identical" vectors one (generally the shortest), and not to consider the rest at all. The manifold

of these "reduced wave vectors" forms the inner of the B-Z, their boundary in the $k_x k_y k_z$ space (where the discontinuities are assumed ordinarily) forms the surface of the B-Z.

The energy as function of $\mathbf{k}$ has, furthermore, all the symmetry of the (reciprocal) lattice. This is clear, since wave functions with all the $\mathbf{k}$ of a star belong to the same representation, and have the same energy, hence.

### IV.

We want to consider the effect of the time reversal, first. This transforms $\mathbf{k}$ into $-\mathbf{k}$. Thus $-\mathbf{k}$ is always in the star of $\mathbf{k}$, even if there is no inversion center present: the energy as function of $\mathbf{k}$ is always equal for $\mathbf{k}$ and $-\mathbf{k}$. Just as for x-ray reflection, the inversion is always added to the symmetry of the problem.[11]

For a triclinic lattice, for instance, this means that the derivative of energy with respect to $\mathbf{k}$ is zero in the middle of the faces, edges and at the corner points of the B-Z, i.e., for

$$k_x x_i + k_y y_i + k_z z_i = n_i \pi \quad (i = 1, 2, 3). \quad (5)$$

One can see directly also, that the group of these $\mathbf{k}$ contains the time reversal and the wave functions are real, hence. Thus the average value of the perturbation operator (3a) vanishes for these wave functions and the energy change goes with $\kappa^2$.

This cannot be claimed, however, for all the surface of the B-Z, i.e., for points for which only one of Eq. (5) is satisfied. The derivative of energy with respect to $\mathbf{k}$ will not vanish in these points and they will not really form the surface of the Brillouin zone.[12]

According to the program of section III, we shall determine now the small representations and their connections in the different types of B-Z for the simple cubic, body-centered, and face-centered cubic lattice. We shall begin with the simple cubic lattice, although no metal with this structure is known.

---

[10a] The case of an accidental degeneracy will be treated in a paper by C. Herring, to appear shortly. We wish to thank Mr. Herring for interesting discussions on this subject.

[11] Cf. G. Friedel, Comptes rendus 157, 1533 (1913). For a more critical discussion of Friedel's rule, cf., however, e.g., P. P. Ewald's article in *Handbuch der Physik*, Vol. 23/2 (Berlin, 1933).

[12] It is not always true, thus, that the $k$ for which $\partial E/\partial k = 0$, are those for which the Bragg conditions are satisfied.

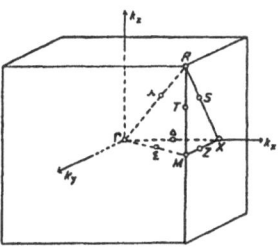

FIG. 2.

TABLE I. *Characters of small representations of* $\Gamma$, $R$, $H$.

| $\Gamma$, $R$, $H$ | $E$ | $3C_4{}^2$ | $6C_4$ | $6C_2$ | $8C_3$ | $J$ | $3JC_4{}^2$ | $6JC_4$ | $6JC_2$ | $8JC_3$ |
|---|---|---|---|---|---|---|---|---|---|---|
| $\Gamma_1$ | 1 | 1 | 1 | 1 | 1 | 1 | 1 | 1 | 1 | 1 |
| $\Gamma_2$ | 1 | 1 | $-1$ | $-1$ | 1 | 1 | 1 | $-1$ | $-1$ | 1 |
| $\Gamma_{12}$ | 2 | 2 | 0 | 0 | $-1$ | 2 | 2 | 0 | 0 | $-1$ |
| $\Gamma_{15}'$ | 3 | $-1$ | 1 | $-1$ | 0 | 3 | $-1$ | 1 | $-1$ | 0 |
| $\Gamma_{25}'$ | 3 | $-1$ | $-1$ | 1 | 0 | 3 | $-1$ | $-1$ | 1 | 0 |
| $\Gamma_1'$ | 1 | 1 | 1 | 1 | 1 | $-1$ | $-1$ | $-1$ | $-1$ | $-1$ |
| $\Gamma_2'$ | 1 | 1 | $-1$ | $-1$ | 1 | $-1$ | $-1$ | 1 | 1 | $-1$ |
| $\Gamma_{12}'$ | 2 | 2 | 0 | 0 | $-1$ | $-2$ | $-2$ | 0 | 0 | 1 |
| $\Gamma_{15}$ | 3 | $-1$ | 1 | $-1$ | 0 | $-3$ | 1 | $-1$ | 1 | 0 |
| $\Gamma_{25}$ | 3 | $-1$ | $-1$ | 1 | 0 | $-3$ | 1 | 1 | $-1$ | 0 |

## V.

*Simple cubic lattice.* Here the surface of the B-Z is a cube as represented in Fig. 2, with the cube edge $2\pi/d$. The inner symmetry elements are: the center $\Gamma$, the threefold axis $\Lambda$, the fourfold axis $\Delta$, the twofold axis $\Sigma$, the symmetry planes $\Delta\Sigma$, $\Sigma\Lambda$ and $\Lambda\Delta$. The simplest way to obtain the group of a wave vector ending on the surface is to draw in all equally long wave vectors which are "identical" with it. The group of the figure constructed in this way is the group of the wave vector. For the arbitrary vector of the surface $\pi/d$, $k_y$, $k_z$, for instance, the figure contains the vector $-\pi/d$, $k_y$, $k_z$, and the group of the wave vector is, hence, the symmetry plane $k_y k_z$. Similarly, for the point $T$, there are four vectors $\pm\pi/d$, $\pm\pi/d$, $k_z$, and the group contains the fourfold axis $k_z$ and all the symmetry planes through it. It is holomorphic with the group of the wave vector ending at $\Delta$ which contains the fourfold axis $k_z$ and the symmetry planes through this. The group of $S$ is holomorphic with that of $\Sigma$; that of $Z$ contains the symmetry planes $k_z k_y$ and $k_y k_z$ and the rotation by $\pi$ about $k_y$. $R$ has the full cubic group like $\Gamma$; $M$ has the group of $T$ and, in addition, the symmetry plane $k_z k_y$. $X$ has the same symmetry.

The tables[13] give the characters of the irreducible representations for the groups of the wave vectors designated in the upper left corner. The corresponding representations will be the "small representations" characterizing the B-Z. The upper right corner contains the group elements. $E$ is the identity, its character will be

TABLE II. *Characters for the small representations of* $\Delta$, $T$.

| $\Delta$, $T$ | $E$ | $C_4{}^2$ | $2C_4$ | $2JC_4{}^2$ | $2JC_2$ |
|---|---|---|---|---|---|
| $\Delta_1$ | 1 | 1 | 1 | 1 | 1 |
| $\Delta_2$ | 1 | 1 | $-1$ | 1 | $-1$ |
| $\Delta_2'$ | 1 | 1 | $-1$ | $-1$ | 1 |
| $\Delta_1'$ | 1 | 1 | 1 | $-1$ | $-1$ |
| $\Delta_5$ | 2 | $-2$ | 0 | 0 | 0 |

the dimension of the representation. $C_3$ is the threefold axis; $C_4$, the rotation by $\pm\pi/2$ about the fourfold axis; $C_4{}^2$, the rotation by $\pi$ about the same axis; and $C_2$ is the rotation about the twofold axis; $J$ is the inversion. $JC_4$ is the product of $J$ and $C_4$, etc. $JC_2$ and $JC_4{}^2$ are the reflections in the symmetry planes perpendicular to the twofold and fourfold axes, respectively. The figures before the symbols of group elements denote how many group elements of that kind are present in the group. The lower left corner gives the notation to be used to designate the small representation in question; it is always given for one of the wave vectors only, as, for instance, for $\Gamma$ in Table I. The small representation of the wave vector $R$ which has the same character as $\Gamma_{12}'$ will be designated by $R_{12}'$, etc. The lower right corner contains the character of the group element above it, for the representation to the left.

In order to save space we have included in the

TABLE III. *Characters for the small representations of* $\Lambda$, $F$.

| $\Lambda$, $F$ | $E$ | $2C_3$ | $3JC_2$ |
|---|---|---|---|
| $\Lambda_1$ | 1 | 1 | 1 |
| $\Lambda_2$ | 1 | 1 | $-1$ |
| $\Lambda_3$ | 2 | $-1$ | 0 |

[13] The representations of most crystallographic groups were given already by H. Bethe, loc. cit., reference 2. All of them are given in E. Wigner, Gött. Nachr. (1930), p. 133.

TABLE IV. *Characters for the small representations of* $\Sigma$, $S$.

| $\chi$, S | E | $C_2$ | $JC_4^2$ | $JC_2$ |
|---|---|---|---|---|
| $\Sigma_1$ | 1 | 1 | 1 | 1 |
| $\Sigma_2$ | 1 | 1 | $-1$ | $-1$ |
| $\Sigma_3$ | 1 | $-1$ | $-1$ | 1 |
| $\Sigma_4$ | 1 | $-1$ | 1 | $-1$ |

tables some wave vectors ($H$ and $F$) important for the body-centered lattice only.

For the points dealt with so far, it was sufficient to denote the group elements by the symbols $C_2$, $C_3$, etc., all the rotations about twofold axes being in the same class and having the same character in all representations. But for the point $M$, the rotation by $\pi$ about the fourfold axes $k_x$, $k_y$ is not equivalent to the rotation about the fourfold axis, $k_z$, which is perpendicular to the wave vector. The latter will be denoted by $C_4^2\perp$. Although the groups of the wave vectors ending at $M$ and at $X$ are holomorphic, the element in the second group which corresponds to $C_4^2\perp$ of the first is the rotation by $\pi$ about $k_z$, which is the axis parallel to the wave vector $\Gamma X$. It will be denoted by $C_4^2\|$.

TABLE V. *Characters of small representations of* $M$, $X$.

| $M$<br>$X$ | E<br>E | $2C_4^2$<br>$2C_4^2\perp$ | $C_4^2\perp$<br>$C_4^2\|$ | $2C_4\perp$<br>$2C_4\|$ | $2C_2$<br>$2C_2$ | $J$<br>$J$ | $2JC_4^2$<br>$2JC_4^2\perp$ | $JC_4^2\perp$<br>$JC_4^2\|$ | $2JC_4\perp$<br>$2JC_4\|$ | $2JC_2$<br>$2JC_2$ |
|---|---|---|---|---|---|---|---|---|---|---|
| $M_1$ | 1 | 1 | 1 | 1 | 1 | 1 | 1 | 1 | 1 | 1 |
| $M_2$ | 1 | 1 | 1 | $-1$ | $-1$ | 1 | 1 | 1 | $-1$ | $-1$ |
| $M_3$ | 1 | $-1$ | 1 | $-1$ | 1 | 1 | $-1$ | 1 | $-1$ | 1 |
| $M_4$ | 1 | $-1$ | 1 | 1 | $-1$ | 1 | $-1$ | 1 | 1 | $-1$ |
| $M_1'$ | 1 | 1 | 1 | 1 | 1 | $-1$ | $-1$ | $-1$ | $-1$ | $-1$ |
| $M_2'$ | 1 | 1 | 1 | $-1$ | $-1$ | $-1$ | $-1$ | $-1$ | 1 | 1 |
| $M_3'$ | 1 | $-1$ | 1 | $-1$ | 1 | $-1$ | 1 | $-1$ | 1 | $-1$ |
| $M_4'$ | 1 | $-1$ | 1 | 1 | $-1$ | $-1$ | 1 | $-1$ | $-1$ | 1 |
| $M_5$ | 2 | 0 | $-2$ | 0 | 0 | 2 | 0 | $-2$ | 0 | 0 |
| $M_5'$ | 2 | 0 | $-2$ | 0 | 0 | $-2$ | 0 | 2 | 0 | 0 |

TABLE VI. *Characters of small representations of* $Z$, $G$, $K$, $U$, $D$.

| $Z$<br>$G$, $K$, $U$<br>$D$ | $E$<br>$E$<br>$E$ | $C_4^2$<br>$C_4^2$<br>$C_2$ | $JC_4^2$<br>$JC_4^2$<br>$JC_2$ | $JC_4^2\perp$<br>$JC_2$<br>$JC_2\perp$ |
|---|---|---|---|---|
| $Z_1$ | 1 | 1 | 1 | 1 |
| $Z_2$ | 1 | 1 | $-1$ | $-1$ |
| $Z_3$ | 1 | $-1$ | $-1$ | 1 |
| $Z_4$ | 1 | $-1$ | 1 | $-1$ |

This finishes the investigation of the symmetry axes in Fig. 2, and there remain only the symmetry planes. A somewhat closer inspection will show, however, that the small representations

TABLE VII. *Compatibility relations between* $\Gamma$ *and* $\Delta$, $\Lambda$, $\Sigma$.

| $\Gamma_1$ | $\Gamma_2$ | $\Gamma_{12}$ | $\Gamma_{15}'$ | $\Gamma_{25}'$ |
|---|---|---|---|---|
| $\Delta_1$ | $\Delta_2$ | $\Delta_1\Delta_2$ | $\Delta_1'\Delta_5$ | $\Delta_2'\Delta_5$ |
| $\Lambda_1$ | $\Lambda_2$ | $\Lambda_3$ | $\Lambda_2\Lambda_3$ | $\Lambda_1\Lambda_3$ |
| $\Sigma_1$ | $\Sigma_4$ | $\Sigma_1\Sigma_4$ | $\Sigma_2\Sigma_3\Sigma_4$ | $\Sigma_1\Sigma_2\Sigma_3$ |

| $\Gamma_1'$ | $\Gamma_2'$ | $\Gamma_{12}'$ | $\Gamma_{15}$ | $\Gamma_{25}$ |
|---|---|---|---|---|
| $\Delta_1'$ | $\Delta_2'$ | $\Delta_1'\Delta_2'$ | $\Delta_1\Delta_5$ | $\Delta_2\Delta_5$ |
| $\Lambda_2$ | $\Lambda_1$ | $\Lambda_3$ | $\Lambda_1\Lambda_3$ | $\Lambda_2\Lambda_3$ |
| $\Sigma_2$ | $\Sigma_3$ | $\Sigma_2\Sigma_3$ | $\Sigma_1\Sigma_2\Sigma_4$ | $\Sigma_1\Sigma_2\Sigma_4$ |

TABLE VIII. *Compatibility relations between* $M$ *and* $\Sigma$, $Z$, $T$.

| $M_1$ | $M_2$ | $M_3$ | $M_4$ | $M_1'$ | $M_2'$ | $M_3'$ | $M_4'$ | $M_5$ | $M_5'$ |
|---|---|---|---|---|---|---|---|---|---|
| $\Sigma_1$ | $\Sigma_4$ | $\Sigma_1$ | $\Sigma_4$ | $\Sigma_2$ | $\Sigma_3$ | $\Sigma_2$ | $\Sigma_3$ | $\Sigma_2\Sigma_3$ | $\Sigma_1\Sigma_4$ |
| $Z_1$ | $Z_1$ | $Z_3$ | $Z_3$ | $Z_2$ | $Z_2$ | $Z_4$ | $Z_4$ | $Z_2Z_4$ | $Z_1Z_3$ |
| $T_1$ | $T_2$ | $T_2'$ | $T_1'$ | $T_1'$ | $T_2'$ | $T_2$ | $T_1$ | $T_4$ | $T_5$ |

TABLE IX. *Compatibility relations between* $X$ *and* $\Delta$, $Z$, $S$.

| $X_1$ | $X_2$ | $X_3$ | $X_4$ | $X_1'$ | $X_2'$ | $X_3'$ | $X_4'$ | $X_5$ | $X_5'$ |
|---|---|---|---|---|---|---|---|---|---|
| $\Delta_1$ | $\Delta_2$ | $\Delta_2'$ | $\Delta_1'$ | $\Delta_1'$ | $\Delta_2'$ | $\Delta_2$ | $\Delta_1$ | $\Delta_5$ | $\Delta_5$ |
| $Z_1$ | $Z_1$ | $Z_4$ | $Z_4$ | $Z_2$ | $Z_2$ | $Z_3$ | $Z_3$ | $Z_2Z_3$ | $Z_1Z_4$ |
| $S_1$ | $S_4$ | $S_1$ | $S_4$ | $S_2$ | $S_3$ | $S_2$ | $S_3$ | $S_2S_3$ | $S_1S_4$ |

prevailing on the symmetry axes already determine the representations for the symmetry planes, i.e., they determine whether the wave function will remain unchanged or assume the negative value, if reflected in one of the symmetry planes.

A B-Z must be characterized by one each of the following 10 symbols: $\Gamma$, $\Delta$, $\Lambda$, $\Sigma$, $R$, $T$, $M$, $S$, $X$ and $Z$. If there is an accidental degeneracy, however, the representation may change on an axis, etc. Not all the combinations of symbols correspond to possible B-Z. The small representation $\Delta$ on the fourfold axis must be contained in the representation $\Gamma$ of the center, if this is considered as a representation of the group of $\Delta$, and similar conditions exist between all pairs of adjoining symmetry elements. Table VII shows with which $\Delta$, $\Lambda$, $\Sigma$, a certain $\Gamma$ can be combined in the symbol of a possible B-Z. The compatibility relations between $R$ and $T$, $\Lambda$, $S$, are the same as those between $\Gamma$ and $\Delta$, $\Lambda$, $\Sigma$. (Table VII.) These compatibility relations reduce considerably the number of possible types of B-Z. In addition to these compatibility relations, there are others originating from the four sets of wave vectors characterized by $k_z=0$; $k_x=k_y$; $k_y=k_z$; $k_z=\pi/d$. Every wave vector satisfying one of these equations, has a group consisting of a symmetry plane, and the corresponding wave function will belong either to the symmetric, or to the antisymmetric representation of this group. This representation must be contained in the small representations of the axes lying in this plane. Table X gives, under $+$,

TABLE X. *Compatibility relations on symmetry planes.*

| SYMMETRY PLANE | $+$ | $-$ |
|---|---|---|
| $k_z = 0$ | $\Sigma_1\Sigma_4$ $\Delta_1\Delta_2\Delta_5$ $Z_1Z_3$ | $\Sigma_2\Sigma_3$ $\Delta_1'\Delta_2'\Delta_5$ $Z_2Z_4$ |
| $k_x = k_y > k_z$ | $\Sigma_1\Sigma_3$ $\Lambda_1\Lambda_3$ $T_1T_2'T_5$ | $\Sigma_2\Sigma_4$ $\Lambda_2\Lambda_3$ $T_2T_1'T_5$ |
| $k_y = k_z < k_x$ | $\Lambda_1\Lambda_3$ $S_1S_3$ $\Delta_1\Delta_2'\Delta_5$ | $\Lambda_2\Lambda_3$ $S_2S_4$ $\Delta_2\Delta_1'\Delta_5$ |
| $k_x = \pi/d$ | $S_1S_4$ $T_1T_2T_5$ $Z_1Z_4$ | $S_2S_3$ $T_1'T_2'T_5$ $Z_2Z_3$ |

those representations along the axes, which are compatible with the symmetric representation in the plane, and under $-$ those which are compatible with the antisymmetric representation. It shows that, for instance, $\Sigma_1$ is incompatible with $\Delta_1'$, $\Delta_2'$, $Z_2$, $Z_4$, $\Lambda_2$, $T_2$, $T_1'$.

As an example, we may consider the three B-Z which stick together at $k_x = k_y = k_z = 0$ having for this wave vector the representation $\Gamma_{25}$. These three B-Z will be separated along the twofold axis, having there the small representations $\Sigma_1$, $\Sigma_2$, $\Sigma_4$, respectively (Table VII). We may consider the one with $\Sigma_2$. This necessarily goes with $\Delta_5$ along the fourfold axis (Tables VII and X), and sticks together with one of the other zones there. Along the threefold axis it may have one of the two representations $\Lambda_2$ or $\Lambda_3$. We shall assume that it has $\Lambda_2$. For $R$, we still have the choice of $R_2$, $R_{15}'$, $R_1'$ or $R_{25}$. We shall choose $R_2$. This requires, then, $S_4$ and $T_2$, and hence $Z_4$. According to Table VIII, it will have $M_3'$ and according to Table IX, $X_5'$. Its whole symbol will be $\Gamma_{25}\Sigma_2\Delta_5\Lambda_2R_{15}T_2S_4Z_4M_3'X_5'$ and we see that most small representations were uniquely given by the compatibility tables and the previous choices.

We believe that the above description of B-Z for the simple cubic lattice is complete from the

point of view of symmetry. We are well aware, of course, that many of the types which are possible geometrically will not be important physically, since they have, for example, too high energies. It appeared to us, however, that for the sake of clarity a complete geometric discussion should be given once for a simple case.

The construction of the compatibility tables is very easy. If one is interested, e.g., in the compatibilities between $\Sigma$ and $M$, one considers for $M$ the characters corresponding to elements which are contained in $\Sigma$. These elements are $E$, $C_2$, $JC_4^2\bot$, $JC_2$ (one must take $JC_4^2\bot$, not $JC_4^2$, since the latter are the symmetry planes $k_xk_z$, $k_yk_z$ which do not occur in the group of $\Sigma$). The corresponding characters in $M_5$, for instance, are 2, 0, $-2$, 0. One sees that this is the sum of the characters of $\Sigma_2$ and $\Sigma_3$ and these are, consequently, compatible with $M_5$. Thus it will not be necessary to give the compatibility relations for the other lattices explicitly.

## VI. BODY-CENTERED CUBIC LATTICE

The shape of the surface of the B-Z is self-evident in the simple cubic lattice but not in the body-centered lattice. The identity periods can be taken as three space diagonals, with coordinates $1/2d$, $\pm 1/2d$, $\pm 1/2d$. The shortest vectors of the reciprocal lattice are the face diagonals, with coordinates 0, $\pm 2\pi/d$, $\pm 2\pi/d$; $\pm 2\pi/d$, 0, $\pm 2\pi/d$; $\pm 2\pi/d$, $\pm 2\pi/d$, 0. Since the inner of the B-Z should contain only different vectors $k$, the addition of a vector of the reciprocal lattice to a $k$ lying inside the B-Z must lead to a vector in the outside. This is most simply accomplished by choosing the rhombododecahedron of Fig. 3 as the surface, in which

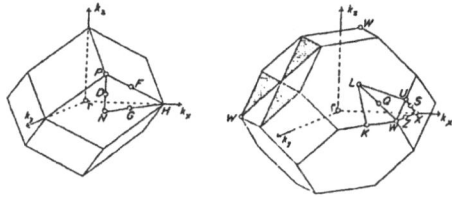

FIG. 3.                    FIG. 4.

opposite faces just differ by a vector of the reciprocal lattice. The distance $\Gamma H$ is $2\pi/d$.[14]

The symmetry elements in the inside of the B-Z are the same as in the simple cubic lattice, $\Gamma$, $\Delta$, $\Lambda$, $\Sigma$, and the compatibility relations between these are also maintained. The point $H$ has, however, the full cubic symmetry, since the vectors of the reciprocal lattice transfer it to all the end points of the coordinate axes. The point $P$ is identical with three similar vertices, forming a tetrahedron.

TABLE XI. *Characters for the small representations of P.*

| $P$ | $E$ | $3C_4{}^2$ | $8C_3$ | $6JC_4$ | $6JC_2$ |
|---|---|---|---|---|---|
| $P_1$ | 1 | 1 | 1 | 1 | 1 |
| $P_2$ | 1 | 1 | 1 | $-1$ | $-1$ |
| $P_3$ | 2 | 2 | $-1$ | 0 | 0 |
| $P_4$ | 3 | $-1$ | 0 | $-1$ | 1 |
| $P_5$ | 3 | $-1$ | 0 | 1 | $-1$ |

TABLE XII. *Characters for the small representations of N.*

| $N$ | $E$ | $C_4{}^2$ | $C_2\|\|$ | $C_2\perp$ | $J$ | $JC_4{}^2$ | $JC_2\perp$ | $JC_2\|\|$ |
|---|---|---|---|---|---|---|---|---|
| $N_1$ | 1 | 1 | 1 | 1 | 1 | 1 | 1 | 1 |
| $N_2$ | 1 | $-1$ | 1 | $-1$ | 1 | $-1$ | $-1$ | 1 |
| $N_3$ | 1 | $-1$ | $-1$ | 1 | 1 | $-1$ | 1 | $-1$ |
| $N_4$ | 1 | 1 | $-1$ | $-1$ | 1 | 1 | $-1$ | $-1$ |
| $N_1'$ | 1 | $-1$ | 1 | $-1$ | $-1$ | 1 | 1 | $-1$ |
| $N_2'$ | 1 | 1 | 1 | 1 | $-1$ | $-1$ | $-1$ | $-1$ |
| $N_3'$ | 1 | 1 | $-1$ | $-1$ | $-1$ | $-1$ | 1 | 1 |
| $N_4'$ | 1 | $-1$ | $-1$ | 1 | $-1$ | 1 | $-1$ | 1 |

The small representations for the other points were already given in the previous tables. We shall not give the compatibility relations between axes and points, since they are easily obtained by the method outlined in the previous section. It may be mentioned that the group of the vectors ending in a general point of the surface is the symmetry plane $JC_2$. The following relations are analogous to those of Table X.

TABLE XIII. *Compatibility relations for symmetry planes.*

| SYMMETRY PLANE | + | − |
|---|---|---|
| $k_z = 0$ | $\Sigma_1\Sigma_4, \Delta_1\Delta_2\Delta_5, G_1G_4$ | $\Sigma_1\Sigma_2, \Delta_1'\Delta_2'\Delta_5, G_2G_3$ |
| $k_x = k_y > k_z$ | $\Sigma_1\Sigma_3, \Lambda_1\Lambda_3, D_1D_3$ | $\Sigma_2\Sigma_4, \Lambda_2\Lambda_3, D_2D_4$ |
| $k_y = k_z < k_x$ | $\Lambda_1\Lambda_3, \Delta_1\Delta_2'\Delta_5, F_1F_3$ | $\Lambda_2\Lambda_3, \Delta_2\Delta_1'\Delta_5, F_2F_3$ |
| $k_x + k_y = 2\pi$ | $D_1D_4, F_1F_2, G_1G_3$ | $D_2D_3, F_1F_3, G_2G_4$ |

[14] The vectors $\mathbf{k}$ in the inside of the $B$-$Z$ are transformed under this choice again into vectors in the inside by every symmetry element.

Since the surface of the B-Z is a symmetry plane, the derivative of the energy perpendicular to this plane is zero on the surface.

## VII. FACE-CENTERED CUBIC LATTICE

The B-Z of the face-centered cubic lattice have a rather complicated structure. The reciprocal lattice is the body-centered lattice, the shortest vectors of which are the space diagonals with components $\pm 2\pi/d, \pm 2\pi/d, \pm 2\pi/d$. If we assume the inner of the B-Z to be bounded by the octahedron with the 8 planes $\pm x \pm y \pm z = 3\pi/d$, then no wave vectors of the inside will differ by one such vector. Nevertheless, some of them will be equivalent, differing by the sum of two shortest vectors of the reciprocal lattice, $\pm 4\pi/d, 0, 0; 0, \pm 4\pi/d, 0; 0, 0, \pm 4\pi/d$. In order to exclude these, one must cut off the corners of the octahedron by planes parallel to the coordinate planes at the distance $\pm 2\pi/d$ from these. The resulting figure is the well-known truncated octahedron of Fig. 4. With this choice of the surface of the B-Z, every wave vector of the inside will go over into a wave vector of the inside by all the symmetry operations. This requirement, however, which determines the whole shape of the surface for the simple cubic and body centered cases, fixes the surface here only at the truncating planes, but not at the octahedral planes. One could, for instance, bulge out in all octahedral planes the part which is shaded on one of the planes in Fig. 4 and bulge in by an equal amount the unshaded regions. The resulting surface would still satisfy all requirements. The truncating planes, on the other hand, cannot be deformed. If we pushed out a point on the $k_z = 2\pi/d$ plane, we would have to push in the corresponding point on the $k_z = -2\pi/d$ plane. After this, however, the reflection on the $k_y k_z$ plane would carry over wave vectors of the inside to the outside of the surface of the B-Z.

The B-Z is always uniquely determined if there is a symmetry plane perpendicular to the vector $\mathbf{r}$ of the reciprocal lattice,[15] which generates that part of the surface. In this case the surface lies at the distance $r/2$ on both sides of the symmetry plane. This was true for the vectors parallel to the coordinate axes which generated the surface

[15] Cf. Eq. (2a).

for the simple cubic lattice, it was true for the vectors parallel to the face diagonals in the body-centered structure and it is true for the vectors generating the truncating planes in Fig. 4. The situation for the octahedral plane of Fig. 4, however, is similar to that for the triclinic lattice and will be shown to have similar consequences.

Although the surface of the B-Z is thus left undetermined by general requirements, it is certainly allowable to assume it to have the shape of Fig. 4.

In the inside of the B-Z we have again the same situation as for the simple cubic lattice with the same compatibility relations holding between the small representations of $\Gamma$ and the two-, three- and fourfold axes. This also applies to the points $X$, $S$ and $Z$ on the cubic plane. The point $W$ is identical with three other points of the surface, two of which are shown on the figure, while one at the bottom is hidden. The small

TABLE XIV. *Characters of small representations of W.*

| $W$ | $E$ | $C_4{}^2$ | $2C_2$ | $2JC_4$ | $2JC_4{}^3$ |
|---|---|---|---|---|---|
| $W_1$ | 1 | 1 | 1 | 1 | 1 |
| $W_1'$ | 1 | 1 | 1 | $-1$ | $-1$ |
| $W_2$ | 1 | 1 | $-1$ | 1 | $-1$ |
| $W_2'$ | 1 | 1 | $-1$ | $-1$ | 1 |
| $W_3$ | 2 | $-2$ | 0 | 0 | 0 |

TABLE XV. *Characters of small representations of L.*

| $L$ | $E$ | $2C_3$ | $3C_2$ | $J$ | $2JC_3$ | $3JC_2$ |
|---|---|---|---|---|---|---|
| $L_1$ | 1 | 1 | 1 | 1 | 1 | 1 |
| $L_2$ | 1 | 1 | $-1$ | 1 | 1 | $-1$ |
| $L_3$ | 2 | $-1$ | 0 | 2 | $-1$ | 0 |
| $L_1'$ | 1 | 1 | 1 | $-1$ | $-1$ | $-1$ |
| $L_2'$ | 1 | 1 | $-1$ | $-1$ | $-1$ | 1 |
| $L_3'$ | 2 | $-1$ | 0 | $-2$ | 1 | 0 |

representations for the points $K$ and $U$ were given in Table VI. $L$ is identical with its antipode. The points $Q$ on the line $LW$ cannot be moved in or out. They belong to the surface, since they

are carried over into themselves (i.e., into the "identical" point on the opposite face) by the twofold axis bisecting the $Z$ and $-X$ axes. The wave function of the wave vector ending at $Q$ will be either symmetric or antisymmetric with respect to this rotation. In the former case it is compatible with $L_1$, $L_1'$, $L_3$, $L_3'$ on one side and with $W_1W_1'W_3$ on the other. If it is antisymmetric, it is compatible with $L_2$, $L_2'$, $L_3$, $L_3'$ and $W_2$, $W_2'$ and $W_3$.

The group of the points on the lines $LK$, $KW$, $LU$, $UW$ contains only the symmetry plane on which they lie, they have no additional symmetry owing to their position on the surface. This is natural, since the surface can be shifted away from them. The Compatibility Table X holds for $k_x = k_y > k_z$ between $\Sigma$ and $\Lambda$, but there is nothing to replace $T$, and $T$ must be omitted also from the last section of ($k_z = \pi/d$) of this table. The rest of the table remains valid, however, and should be supplemented by the compatibilities just given, owing to the symmetry of the point $Q$.

The surface of the B-Z at the octahedral planes cannot be chosen in such a way that the "identical" point $k_x - 2\pi/d$, $k_y - 2\pi/d$, $k_z - 2\pi/d$ to every point of the surface could be reached by a symmetry operation also. This has the consequence that the derivative of the energy perpendicular to the somewhat arbitrarily chosen plane octahedral face will only vanish on the diagonals ($LW$) corresponding to the separating lines between shaded and unshaded regions. On the other hand, it will have the consequence also that the energy for $k_x = \pi/d + \mu$, $k_y = \pi/d - \mu - v$, $k_z = \pi/d + v$ will be equal to the energy for $-\pi/d + \mu$, $-\pi/d - \mu - v$, $-\pi/d + v$ and, because of the twofold axis, also equal to the energy for $\pi/d - v$, $\pi/d + \mu + v$, $\pi/d - \mu$. The energy as function of $\mathbf{k}$ will be symmetric with respect to the line $LW$ on the surface and, hence, will have on the octahedral *surface*, a sixfold rotational symmetry.

# On the Constant $A$ in Richardson's Equation

## E. P. Wigner

Physical Review *49*, 696–700 (1936)

(Received March 16, 1936)

According to Herzfeld, the "chemical constant" of the electron gas is different in the high temperature region, where all actual measurements are carried out, from its value at very low temperatures. Hence the value of the constant $A$ in Richardson's equation must be different at high temperatures from the well-known low temperature value of 120 amp./cm² °K². The present paper gives a calculation of the high temperature value in terms of thermodynamic quantities and a numerical estimation of these.

### 1

IT is well known that the constant $A$ in Richardson's formula[1] for the thermionic current $i$

$$i = A T^2 e^{-W/RT} \qquad (1)$$

($W$ is the measured work function) should have under ideal conditions the value 120 amp./cm² °K². Under "ideal conditions" we mean that:

(a) The reflection coefficient $r$ for electrons should be zero, i.e., all electrons striking the surface should be absorbed by the same.

(b) The work function $W$ should be constant. The work function is defined as the energy necessary to remove one electron from the top of the Fermi distribution and bring it to a point near the surface, but still to a distance from it which is large as compared with the lattice constant. The "top of the Fermi distribution" again means the energy of that electron which has the highest energy if all electrons are in the lowest possible state. This is the work function which is measured experimentally by the photo-electric effect if Fowler's method[2] is used for the evaluation.

(c) That the electron gas in free space should behave like an ideal gas.

(d) Finally, the surface must be uniform and thus free from patches and contaminations.

This last condition and its effects have been carefully analyzed by J. A. Becker.[3] It seems that it can account in the case of the most careful experiments, and for low accelerating potentials for small deviations of $A$ from 120

only, and generally only for a lowering of this value. Under the conditions (a)–(d), (1) is a consequence of the principle of detailed balance and general quantum statistical laws.[4]

Condition (c) is fulfilled for not too high temperatures[5] and will be assumed further on. If the reflection coefficient is different from zero, we should only expect an additional coefficient $1-r$ in (1). Since $1-r$ is certainly in the order of magnitude of 1 and smaller than this, the reflection can explain only a deviation of $A$ from 120 by such a factor.[6] In spite of this, the experimental values[7] are in most cases rather far from 120. It must be emphasized, however, that these experimental values are still subject to comparatively large variations and in many cases hardly reproducible. They are most trustworthy if the work function is measured photoelectrically also. It is only too common an experience that a new set of experiments changes the measured values of both $A$ and $W$ considerably, generally both in the same direction, which indicates that not so much the measurement of $i$, as that of its temperature coefficient, is crucial. In spite of this, it is generally accepted[8]

[1] For the history of Richardson's discovery cf., e.g., W. Schottky, *Handbuch der Experimentalphysik*, Vol. XIII (Leipzig, 1928), or A. L. Reimann, *Thermionic Emission* (London, 1934).
[2] R. H. Fowler, Phys. Rev. **38**, 45 (1931).
[3] J. A. Becker, Rev. Mod. Phys. **7**, 95 (1935).
[4] Cf. reference 1 and K. F. Herzfeld, Phys. Rev. **35**, 248 (1930), also J. H. Becker and W. H. Brattain, Phys. Rev. **45**, 694 (1934).
[5] Cf. M. v. Laue, *Jahrbuch der Radioaktivität und Elektronik*, Vol. 15, 205, etc. (1918) and W. Schottky, reference 1.
[6] L. Nordheim, Zeits. f. Physik **46**, 833 (1928); Proc. Roy. Soc. A121, 626 (1928); R. H. Fowler and L. Nordheim, Proc. Roy. Soc. A119, 173 (1928) show that the reflection coefficient is very small in general, around 0.07. See also Langmuir and Jones, Phys. Rev. **31**, 401 (1928).
[7] Cf. A. L. Reimann, reference 1 and A. L. Hughes and L. A. DuBridge, *Photoelectric Phenomena* (New York, 1932). For Cb, H. B. Wahlin and L. O. Sordahl, Phys. Rev. **45**, 886 (1934), for Rh unpublished data of H. B. Wahlin.
[8] I am much indebted to Professor L. A. DuBridge for a discussion on this point. Cf. also his monograph, *Actualités Scientifiques et Industrielles*, No. 268, Paris, 1935.

TABLE I. *Measured values of the constants in Richardson's equation.*

| | $A$ | $W$ | | $A$ | $W$ |
|---|---|---|---|---|---|
| Cs | 162 | 1.81 | Ta | 60 | 4.12 |
| Ba | 60 | 2.11 | Hf | 14.5 | 3.53 |
| Ni | 1,380 | 5.02 | Th | 70 | 3.38 |
| Pd | 60 | 4.98 | Re | 200 | 5.1 |
| Pt | 17,000 | 6.27 | Cu? | 65 | 4.33 |
| Mo | 55 | 4.15 | Ag? | 0.76 | 3.56 |
| W | 60 | 4.53 | Au? | 40 | 4.32 |
| Zr | 330 | 4.12 | Cb | 57 | 3.96 |
| | | | Rh | 35 | 4.81 |

that the deviations of $A$ from 120 are real and must be explained by the breakdown of assumption (b). The present paper will be devoted to a discussion of (b).

**2**

The consequences of the temperature variation of the work function have often been analyzed,[9] especially by Bridgman and Herzfeld. I shall follow the latter's treatment quite closely. The reason this subject is taken up again, is that our somewhat improved knowledge of the metallic structure allows a somewhat closer analysis which shows that in addition to the effect, analyzed by Herzfeld (which is reproduced), there is another effect which tends to compensate for it in many cases. This is the reason, perhaps, that the measured $A$'s are not all higher than 120, as one may expect from Herzfeld's paper.

It has been emphasized by Becker and Brattain[10] that the quantities measured in case of variable work function (and thus contained in Table I) are $W^*$ and $A^*$, defined by

$$W^* = RT^2(d/dT) \ln i/T^2,$$

$$\ln A^* = \ln i/T^2 + T(d/dT) \ln i/T^2 \qquad (2)$$
$$= \ln i/T^2 + W^*/RT,$$

where $i$ is the thermionic current.

The procedure to obtain theoretical expressions for $W^*$ and $A^*$ is to calculate first $i$ and obtain then $W^*$ and $A^*$ by (2).

**3**

In order to calculate the current, one may first calculate the vapor pressure $P$ of the electron gas at a point with zero electrostatic

---

[9] P. W. Bridgman, Phys. Rev. **31**, 90 (1928); K. F. Herzfeld, reference 4.
[10] Reference 4.

potential by thermodynamics. If $P$ is given, the number of electrons striking the surface can be calculated by elementary kinetic theory ($m$ is the electronic mass) to be $P/(2\pi mkT)^{\frac{1}{2}}$. If $r$ is the reflection coefficient, this gives a current $i$ per cm$^2$ of the metal.

$$i = Pe(1-r)/(2\pi mkT)^{\frac{1}{2}}. \qquad (3)$$

The current from the metal is equal to this in equilibrium and is supposed to be equal also under the conditions of thermionic emission. Thus the problem reduces to the calculation of the vapor pressure $P$.

Both the vapor pressure and the work function[11] depend, in addition to the temperature $T$, on the (mechanical) pressure $p$, under which the metal is kept. We shall assume that the metal carries always only an infinitesimally small charge only. If we allow $\epsilon$ electrons per mole of metal to escape and keep the volume of the metal constant, the mechanical pressure will increase by $\epsilon p_e(v, T)$. The energy necessary for the removal of $\epsilon$ electrons, when the metal is kept at constant volume, is $\epsilon W_v(v, T)$. It seems to be simplest to apply the Clausius Clapeyron cycle by keeping the metal at constant volume. We then have[12]

$$\partial \ln P(v, T)/\partial T = W_v(v, T)/RT^2 + 5/2T, \qquad (4)$$

which gives

$$\ln P(v, T) = -\frac{W_v(v, 0)}{RT} + (5/2) \ln T + j$$

$$+ \int_0^T [W_v(v, \tau) - W_v(v, o)] \frac{d\tau}{R\tau^2}$$
$$\qquad (5)$$

$$j = (3/2) \ln (2\pi mk/h^2) + \ln 2.$$

The well-known Sackur-Tetrode value of the chemical constant must be increased by $\ln 2$ because of the spin. In order to determine $p_e$, we can apply the usual cycle to a piece of metal

---

[11] For the definition of the work function see the preceding section. The vapor pressure $P$ is a function of the mechanical pressure $p$ for ordinary vaporization also, but this effect is rarely considered.
[12] In this section, temperature $T$ and volume $v$ will be the independent variables. Thus temperature derivatives are always to be taken at constant volume, volume derivatives at constant temperature.

which is deprived of $\epsilon$ electrons permole. The pressure is $p+\epsilon p_e$, the energy $E+\epsilon W_v$, so that we have

$$\frac{\partial E}{\partial v}+\epsilon\frac{\partial W_v}{\partial v}+p+\epsilon p_e=T\frac{\partial}{\partial T}(p+\epsilon p_e), \qquad (6)$$

which gives in addition to the usual

$$\partial E/\partial v+p=T\partial p/\partial T \qquad (7)$$

valid for any substance, the required

$$\partial W_v/\partial v+p_e=T\partial p_e/\partial T \qquad (8)$$

since (6) holds for every $\epsilon$. From (8) follows

$$p_e=-\frac{\partial W_v(v,\,0)}{\partial v}$$

$$+T\int_0^T\left[\frac{\partial W_v(v,\,\tau)}{\partial v}-\frac{\partial W_v(v,\,0)}{\partial v}\right]\frac{d\tau}{\tau^2}. \qquad (9)$$

We shall now calculate $W^*$ and $A^*$ by (5), (3) and (2). We have to remember, for this, that when the thermionic current is measured at different temperatures, not the volume, but the mechanical pressure of the metal is kept at constant value, namely zero. The temperature derivatives in (2) are meant to be taken at constant pressure, therefore. Hence, if $\alpha$ is the volume expansion coefficient, $v_0$ the molal volume, we have

$$W^*=RT^2\left(\frac{\partial}{\partial T}+v_0\alpha\frac{\partial}{\partial v}\right)\ln\frac{i}{T^2}$$

$$=RT^2\left(\frac{\partial}{\partial T}+v_0\alpha\frac{\partial}{\partial v}\right)\ln\frac{P}{T^{5/2}} \qquad (10)$$

or $\qquad W^*=W_v(v,\,T)+v_0\alpha Tp_e \qquad (11)$

as the comparison with (9) shows. For $A^*$, we have

$$\ln(A^*/A(1-r))=-W_v(v,\,0)/RT+W_v(v,\,T)/RT$$

$$+\int_0^T[W_v(v,\,\tau)-W_v(v,\,0)](d\tau/R\tau^2)+v_0\alpha p_e/R. \qquad (12)$$

Here $A$ is the constant for the ideal case, 120 amp. cm$^{-2}$ °K$^{-2}$. The last term in (12) is that of Herzfeld, the other ones have been omitted by him, since the work function chiefly depends on the volume. Its contribution to $A^*$ is not negligible, however. In general, $p_e$ is negative, the pressure decreases if an electron is removed.

### 4

For the evaluation of (12) we shall make assumptions very similar to those of Herzfeld. The total energy of the metal contains two parts. The first, $V(v)$ arises from the motion of the electrons, depending, therefore, on the volume only.[13] The second arises from the motion of the nuclei. The first part is the most important for the actual value of the work function, its change by the removal of electrons from the metal gives practically all of this quantity. For the second part, an expression $RTD(\Theta/T)$ will be used. In this, $\Theta$ is a function of the volume and will be changed also by the removal of electrons by the amount $\epsilon\Theta_e$ if $\epsilon$ electrons per mole are removed.

[13] We shall neglect the small specific heat of free electrons (cf. A. L. Reimann, reference 1).

The total energy of a mole of the metal, out of which $\epsilon$ electrons have been removed and brought to a point near the surface but still sufficiently distant from it in order to make the image force negligible, is thus

$$E(v,\,T)+\epsilon W_v(v,\,T)=V(v)+\epsilon W_v(v,\,0)$$

$$+RTD(\Theta/T)+\epsilon R\Theta_e D'(\Theta/T). \qquad (13)$$

These assumptions can be justified on the basis of the usual assumption of fast electronic and slow nuclear motion,[14] but I shall not enlarge upon this subject. From (13) one readily obtains

$$W_v(v,\,T)=W_v(v,\,0)+R\Theta_e D'(\Theta/T), \qquad (14)$$

which gives with (12)

$$\ln\frac{A^*}{A(1-r)}=\frac{\Theta_e}{T}D'\left(\frac{\Theta}{T}\right)+\Theta_e\int_0^T D'\left(\frac{\Theta}{\tau}\right)\frac{d\tau}{\tau^2}+\frac{v_0\alpha p_e}{R}$$

$$=(\Theta_e/T)D'(\Theta/T)-(\Theta_e/\Theta)D(\Theta/T)+v_0\alpha p_e/R,$$

$$\ln(A^*/A(1-r))=-C_v\Theta_e/R\Theta+v_0\alpha p_e/R$$

$$=-3\Theta_e/\Theta+v_0\alpha p_e/R, \qquad (15)$$

[14] Cf., e.g., H. Pelzer and E. Wigner, Zeits. f. physik. Chemie B15, 445 (1932); F. London, Zeits. f. Physik 74, 143 (1932).

where $C_v \approx 3R$ is the specific heat of the metal at constant volume, it has assumed its classical value at the temperatures under consideration.

The thermionic emission can be considered as a generalized dissociation reaction, and calculated, hence, according to the transition state method for calculating reaction rates.[15] The result is identical with the one obtained here, but I preferred this method because it is the more usual one in these problems. As has been remarked by Herzfeld, (15) corresponds to the expression for the chemical constant of the electron gas at high temperatures, while the familiar 120 amp./°K² cm² is the low temperature expression. This is quite similar to the situation for $H_2$, e.g.: at low temperatures the chemical constant is $(3/2) \ln (2\pi m k^{5/3}/h^2)$ while at high temperatures there is an additional $\ln 8\pi^2 kJ/h^2$. In contrast to $H_2$, the change of the "chemical constant" in our case is due to a change in the condensed state. Also the temperature dependence does change in $H_2$ (from $T^{5/2}$ to $T^{7/2}$) while there is no such change in our case.

**5**

Next we shall try to estimate the quantities $\Theta_e$ and $p_e$ occurring in (15). It must be remarked at the outset that these two quantities are of a rather different nature: $\Theta_e$ is a function of the volume only, but $p_e$ depends, according to (9), on the temperature and can be reduced to the more basic quantities, $\partial W_v(v, 0)/\partial v$, $\Theta_e$ and $d\Theta_e/dv$ by (14). One obtains thus

$$\ln (A^*/A(1-r)) = -C_v\Theta_e/R\Theta$$
$$-(v_0\alpha/R)\partial W_v(v, 0)/\partial v$$
$$-v_0\alpha T(\partial/\partial v)(C_v\Theta_e/R\Theta) \quad (16)$$

since the integral in (9) is just the $v$ derivative of the expression we have calculated in the previous section.

[15] For the history of this method cf. E. Evans and M. Polanyi, Trans. Faraday Soc. **31**, 875 (1935). The formula

$$i = AT^2e^{-w/RT} \quad (*)$$

used by J. A. Becker and W. H. Brattain (reference 4) has been derived in the papers to which they take reference only for the case when the work function is independent of temperature. Of course, it is always possible to represent the thermocurrent by a formula (*) with variable $w$, but this $w$ will not be equal to the work function. It follows furthermore from the third law that $w$ can have no term linear in $T$ at low temperatures (cf. reference 9).

It would be rather difficult to estimate the last expression in (16) since it contains $d\Theta_e/dv$. If one assumes however, that $v(\partial/\partial v)(C_v\Theta_e/R\Theta)$ is of the order of magnitude of $C_v\Theta_e/R\Theta$ itself—which seems reasonable, the last term can be neglected altogether, because $\alpha T = 10^{-2}$. Since $W_v(v, 0)$ is the energy necessary at the absolute zero, to remove one electron, $\partial W_v(v, 0)/\partial v$ is the negative increase in pressure, due to the removal of one electron. Since the electrons do the binding in the metal, the removal of one will loosen the binding and the actual pressure decrease will not be quite as great as calculated on the free electron hypothesis. For alkalis, one can obtain an estimate of this magnitude from the formulas, derived for the calculation of the work function.[16] Our $W_v(v, 0)$ is in the present approximation what is denoted there by $\varphi$. Eq. (7) reference 16 reads in our notation:

$$-W_v(v, 0) = 13.5(E_0 + 5 \cdot 2.21/3r_s^2 + 1.18/r_s$$
$$-0.58/(r_s+5.1) - 0.19r_s/(r_s+5.1)^2 - eD). \quad (17)$$

The work function is expressed here in volts, $r_s$ in Bohr's units ($0.528 \cdot 10^{-8}$ cm) and is defined by $4\pi r_s^3/3 = v_0/L$ = atomic volume. $D$ is the momentum of the double layer on the surface and will be neglected hereafter. From (17) we obtain

$$v_0\partial W_v/\partial v = (r_s/3)\partial W_v/\partial r_s =$$
$$-4.5(r_s dE_0/dr_s - 7.4/r_s^2 - 1.18/r_s$$
$$+0.58r_s/(r_s+5.1)^2 + 0.38r_s^2/(r_s+5.1)^3). \quad (18)$$

In order to eliminate $dE_0/dr_s$, we set the derivative of (8) reference 16 equal to zero. This expresses that the total energy is a minimum for $r_s$.

$$-r_s dE_0/dr_s + 4.42/r_s^2 + 0.28/r_s$$
$$-0.58r_s/(r_s+5.1)^2 = 0. \quad (19)$$

This gives

$$v_0\partial W_v(v, 0)/\partial v$$
$$= 4.5(3/r_s^2 + 0.9/r_s - 0.38r_s^2/(r_s+5.1)^3), \quad (20)$$

which is for $r_s = 4$ about 1.5 volts. The second term in (16) thus becomes with $\alpha = 21.10^{-5}$ about $-3.6$.

[16] J. Bardeen and E. Wigner, Phys. Rev. **48**, 84 (1935).

The quantity $\Theta_e/\Theta$ is the percent change of the vibration frequencies for one percent change in the number of electrons. It is, therefore, half the percent change of the restoring force $F$, if one percent of the electrons is removed.

In order to obtain an estimate for this, we may consider first the whole restoring force if only one ion is displaced. This is the force which acts on the displaced ion and originates from the electron cloud which is itself distorted by the displacement of the ion.

It is difficult, in general, to calculate the distortion of the electron clouds. Naturally, the inner electrons can be considered to be rigidly attached to the ion. We may assume that the valence electrons have an even distribution throughout the lattice and show wriggles in the neighborhood of the ion only. Then, the total charge distribution of the valence electron can be considered to contain two parts: $\rho_1$, the variations of which are appreciable only in distances comparable with the lattice constant, and $\rho_2$ which contains the small wriggles only. It seems reasonable to assume that the ion carries the wriggles with itself but as long as only one ion is displaced, the slowly varying part of the charge will not be much affected. If the ion was displaced by $x$ in the $X$ direction, it will be acted upon by a force

$$Fx = xZe\partial E_x/\partial x, \qquad (21)$$

where $E$ is the field due to the slowly varying part $\rho_1$ of the electronic charge and $Ze$ the ionic charge. We have, by Poisson's equation div $E$ $= 4\pi\rho_1$ and for a cubic lattice, this gives because of $\partial E_x/\partial x = \partial E_y/\partial y = \partial E_z/\partial z$

$$F = 4\pi\rho_1 Ze/3. \qquad (22)$$

The percent change of this,[17] for a one percent decrease of the number of electrons is simply $-\rho_{1t}/\rho_{1m}$ where $\rho_{1t}$ is the slowly varying part of the density of the highest energy electrons, at the point where the ion is; $\rho_{1m}$ is the mean value of the same quantity for all valence electrons. The first term of (16) is therefore, approximately

$$(3/2)(\rho_{1t}/\rho_{1m}). \qquad (23)$$

This will be comparatively large ($\sim1.5$), if the last valence electron is an $s$ electron, small otherwise. If the highest energy electron happens to be such an electron in the lattice, for which $\rho_{1t}$ is zero, the whole first term will be negative also, because of the approximations we made.

## 6

It is hardly necessary to mention that while the considerations of sections 2 and 3 are, (12) inclusive, strictly based on thermodynamics and those of 4 also should be correct within a very small error, the considerations of section 5 are very crude and will not give, in general, more than the sign of the effects and their order of magnitude. What can be claimed safely, is only

(a) that it is purely accidental if the constant $A^*$ in Richardson's equation (defined by (2)) has the value 120 amp./$°K^2$ cm$^2$ sec.

(b) that the deviation in the ln of $A^*$ should be of the order of magnitude 1–3, rather negative than positive. It should be negative always when the last valence electron is not essentially an $s$ electron in the lattice.

It should be emphasized once more that rather large deviations in the constant $A^*$ can be caused by a patchy character of the surface, as considered by Becker and Rojansky.[3] These patches can consist in surface contamination or also in the polycrystalline character of the surface, since different crystal planes will have different work functions. It would be desirable, from this point of view, to measure the work function and $A^*$ on definite crystal planes, i.e., to use single crystals for the experiments.[18]

In some cases, zero energy outside of the metal may correspond to a forbidden region inside. This would have as consequence an unusually large reflection coefficient.[19] The effect of a reflection coefficient has also been omitted in this paper.

It is a pleasure to express my gratitude to Professor Herzfeld for his valuable discussion and criticism of this paper.

---

[17] One can calculate, from (22), the characteristic temperature of the metal, and it comes out in the right order of magnitude. A more satisfactory calculation of the vibrational frequencies of a metal has been given lately by K. Fuchs, Proc. Roy. Soc. **A153**, 622 (1936).

[18] Professor H. B. Wahlin has pointed out to me that part of the effect of the continued heating and flashing of the samples is perhaps due to recrystallization. Cf. H. B. Wahlin and J. A. Reynolds, Phys. Rev. **48**, 751 (1935).

[19] P. M. Morse, Phys. Rev. **35**, 1310 (1930).

# Effects of the Electron Interaction
# on the Energy Levels of Electrons in Metals

## E. P. Wigner

Transactions of the Faraday Society *34*, 678–685 (1938)

*Received* 11*th February*, 1938.

## 1.

The very simplest form of the theory of the energy bands in metals [1]
gave for many problems such accurate explanations of often very in-
tricate properties of metals and alloys that it may well appear superfluous to consider extensions of the simple form of the theory.   Nevertheless I believe that if the theory of metals is to broaden so as to include
an even greater variety of phenomena, however satisfying the simple
picture is for some purposes, the development of a more rigorous theory
would not be superfluous.

The above mentioned " simple " theory considers the metallic electrons as moving (without interacting among themselves) in a pre-existing
field with the symmetry given by the space group of the crystal.   This
picture fails in two respects :

(1) The field is not preformed, but owes its origin partly to just those
electrons whose motion we are going to consider.   To meet this objection,
one must base the considerations on the Hartree-Fock equations. [2]
This will have the advantage of allowing a determination of the field in
which the electrons move, whereas the properties of the pre-existing
field of the simple theory had to be taken entirely from experiment.
Surprisingly, enough, in many cases, the field has such a form that plane
waves form a reasonably accurate solution for the outer electrons.   In
these cases, the only major difference between this stage of the theory
and the simplest form lies in the energy values which the two pictures
attribute to the same plane waves.   This difference of energy values
originates in the exchange energy, characteristic for Fock's equations.
This Hartree-Fock stage of the theory has been worked out for a great
number of phenomena, [3] and in many cases the results differ significantly
from those of the simple theory.

(2) The simple theory assumes that the electrons move independently
of each other. [4]   This is the approximation characteristic for the Hartree-

---

[1] *Cf. e.g.*, N. F. Mott and H. Jones, *The Theory of the Properties of Metals and Alloys*. Oxford, 1936. Chapters II, III, VI, VII and especially Chapter V. Also several chapters in H. Fröhlich, *Elektronentheorie der Metalle*, Berlin, 1936.

[2] The term Hartree-Fock picture will be used for the approximation in which the wave function can be written as a single determinant of single particle wave functions containing the spin. Such a wave function has, in general, neither any definite spacial symmetry, nor any single multiplicity.

[3] F. Bloch, *Z. Physik*, 1929, **57**, 545, made the first calculation on this basis. *Cf.* also L. Brillouin, *J. Physique*, 1932, **3**, 565. It was developed subsequently for many problems by J. Bardeen, *Physic. Rev.*, 1936, **49**, 653 ; 1936, **50**, 1098 ; 1937, **52**, 688. Perhaps the most important effect of the exchange energy which has not yet been treated thoroughly is the increase of the energy gap between occupied and unoccupied Brillouin zones.

[4] One should say, more accurately, that in every stationary state the electrons show only those statistical relations which are required by the antisymmetry of the whole wave function. However, if one compares different stationary states, the energy values for one electron will be different for different states of the others.

Fock model. Unfortunately, it is difficult to evade this approximation and but few attempts have been made in this direction. The old Heitler-London scheme does not use this assumption; it makes approximations, however, which in the case of the conduction electrons in metals, are probably even more objectionable.[5]

The first question which immediately arises is: Why is it that the very simplest form of the theory gives such satisfactory results for many—indeed most—questions? I shall put forward the view that although the two above-mentioned improvements both increase the absolute value of the binding energy (" correlation energy "), their effects on the energy differences between different states of the metal tend to compensate each other. I shall try to show this in two cases.

## 2.

In the simplest form of the theory, the average energy of the electrons if half of them have spins in one direction, the other half in the other direction (non-ferromagnetic state) is :—

$$\epsilon_n = E_0 + \frac{3^{2/3} \pi^{4/3} h^2}{10m} \frac{n^{2/3}}{V^{2/3}} = E_0 + \frac{2 \cdot 21 Ry}{r_s^2}. \qquad . \qquad . \quad (1a)$$

Here $E_0$ is the average potential in the lattice, the second term is the average zero point energy, $V$ and $n$ total volume and total number of electrons, $r_s$ is defined by $(4\pi/3) (r_s a_0)^3 = V/n$ with $a_0$ the radius of Bohr's first orbit in H; $Ry$ denotes the ionisation energy of H, and $h$ Planck's constant divided by $2\pi$. The energy for the state in which all spins are parallel is

$$\epsilon_m = E_0 + 2^{2/3} \frac{2 \cdot 21 Ry}{r_s^2} = E_0 + \frac{3 \cdot 51 Ry}{r_s^2}. \qquad . \qquad . \quad (2a)$$

According to these well-known formulæ, the more stable state would be always non-magnetic; the ferromagnetism could not be due to free electrons. The energy difference between the non-magnetic and magnetic state is given in Fig. 1 by the line marked $(a)$.

The Hartree-Fock picture changes this situation radically. Bloch[6] has made the calculation for the case where the modulations of potential do not affect the energy differences between electrons with different values of the momentum. This would certainly be the case if one could replace the positive ions by a uniformly distributed space charge. It is probably very nearly true,[7] however, in the alkalis also. The only change from the simple picture occurs then through the appearance of the exchange energy, as the ordinary interaction is supposed to be included in $E_0$. One has

$$\epsilon_n = E_0 + \frac{2 \cdot 21 Ry}{r_s^2} - \frac{0 \cdot 916 Ry}{r_s} \qquad . \qquad . \quad (1b)$$

for the non-magnetic and

$$\epsilon_m = E_0 + 2^{2/3} \frac{2 \cdot 21 Ry}{r_s^2} - 2^{1/3} \frac{0 \cdot 916 Ry}{r_s} \qquad . \qquad . \quad (2b)$$

for the magnetic state. The energy difference between the non-magnetic and the magnetic state is plotted in Fig. 1 (line $(b)$). It becomes negative

[5] Cf. e.g., A. Sommerfeld and H. Bethe, *Handbuch der Physik*, Vol. 24/2. Berlin, 1933.
[6] Cf. ref. 3.    [7] Cf. Mott and Jones,[1] p. 315, and references given there.

680    ENERGY LEVELS OF ELECTRONS IN METALS

for $r_s > 5.47$ and elements with $r_s > 5.47$ (*i.e.*, Cs) should be ferromagnetic, even if the effect of the modulations of the potential on the electronic energy can be neglected.

As pointed out before, it is somewhat difficult to go beyond the Hartree-Fock picture. However, with the above assumption, a calculation has been performed [8] which adds a term to (2a)

$$\epsilon_n = E_0 + \frac{2.21\,Ry}{r_s^2} - \frac{0.916\,Ry}{r_s} + \frac{0.58Ry}{r_s + 5.1} \quad . \quad . \quad (1c)$$

while it leaves (2b) practically unchanged. Fig. 1 shows that the difference between (1c) and (2b) practically coincides with the difference between (1a) and (2a), again leading to the result that free electrons (*i.e.*, electrons for the motion of which the lattice potential is of no material importance) do not

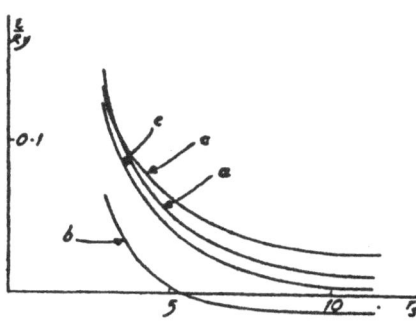

FIG. 1.—Energy difference between the ferromagnetic and the non-magnetic states. Line (*a*) for the simple picture, line (*b*) for the Hartree-Fock picture, line (*c*) taking into account the correlation energy for the non-magnetic state, line (*e*) estimated correct.

lead to ferromagnetism. This accords well with recent ideas on ferromagnetism [9] according to which ferromagnetism is due to the *d*-electrons which are rather tightly bound to the lattice. It must be admitted, however, that Fig. 1 gives an exaggerated picture of the agreement between the simple form and the accurate theory. The additional term in (1c) is only approximately correct and for higher $r_s$ a "correlation term" should be added to (2b) also. An estimate of the real energy difference between the non-magnetic state and the ferro-magnetic state is given in Fig. 1 (curve *e*). We shall see later that for very large $r_s$, $\epsilon_m - \epsilon_n$ goes to zero exponentially. Thus the inclusion of the correlation energy will cancel only a part (though probably the major part) of the effect of the exchange energy. It must further be emphasised that one should consider also states with intermediate momenta.[10]

**3.**

The next question is the density of energy levels. The magnitude of this determines a great many effects, such as paramagnetism,[11] specific heat at low temperatures;[12] moreover, conductivity and optical properties are crucially dependent on it.[13] If the effect of the zones is negligible, the difference between the energies of the states with the wave numbers $k$ and $k_m$ is given by

$$\epsilon(k) = \frac{h^2}{2m}\,(k^2 - k_m^2) \quad . \quad . \quad . \quad . \quad (3)$$

[8] E. Wigner, *Physic. Rev.*, 1934, 46, 1002.
[9] *Cf.* J. H. Van Vleck, *ibid.*, 1937, 52, 1178, and further literature given there.
[10] S. Schubin and S. Wonsowsky, *Proc. Roy. Soc.*, A, 1934, 145, 159.
[11] H. Fröhlich,[1] § 27.    E. C. Stoner, *Proc. Roy. Soc.*, A, 1936, 154; 656.
[12] *Cf. e.g.*, Mott and Jones,[1] p. 178.
[13] *Cf.* Mott and Jones,[1] Chapters VII and III.

where $k_m$ will denote the wave number of the highest state occupied at the absolute zero temperature. From (3) the specific heat and hence the logarithm of the level density of the whole metal can be calculated by the methods of Bethe and Bardeen.[14] The latter becomes

$$\ln \rho\,(\epsilon) \approx 1 \cdot 62 n r_s\,(\epsilon/nRy)^{\frac{1}{2}} \qquad . \qquad . \qquad . \quad (4)$$

Taking into account the exchange energy, Dirac[15] obtains instead of (3)

$$\epsilon(k) = \frac{h^2}{2m}\,(k^2 - k_m{}^2) + e^2\frac{k^2 - k_m{}^2}{2\pi k}\ln\left|\frac{k + k_m}{k - k_m}\right|. \qquad . \quad (5)$$

This equation gives (for not too high $k$) a stronger dependence of $\epsilon$ on $k$, than (3), and hence a much lower level density. For $k = k_m$, the derivative of $\epsilon$ with respect to $k$ even becomes infinite. It must be remembered, however, that it was assumed in the derivation of (5) that the electrons fill up all the levels with smaller wave numbers than $k_m$, except for the electron under consideration, the wave number of which is $k$. Since at the temperature $T$, the top of the Fermi distribution is fuzzed out to a distance of about $kT$ in the energy scale, (5) will be expected to fail if $\epsilon$ is not large compared with $kT$.

For $\epsilon \sim kT$, one must take into account the fact that some levels with a larger wave number than $k_m$ are occupied, while others with a smaller wave number are unoccupied. Denoting by $\nu(k)$ the probability that the level $k$ is occupied, we have at low temperatures ($\beta = 1/kT$).

$$\nu(k) = \frac{1}{e^{\beta\epsilon(k)} + 1}. \qquad . \qquad . \qquad . \quad (6)$$

On the other hand, the energy $\epsilon(k)$ is no longer given by (5) but by

$$\epsilon(k) + \epsilon(k_m) = \frac{h^2}{2m}k^2 - \frac{1}{V}\sum_{k'}\nu(k')\int\frac{e^2}{r}\,e^{i(k-k')\cdot r}dv \qquad . \quad (6a)$$

where the last term contains the exchange integrals of the wave function with wave number $k$ with all occupied states. Equations (6) and (6a) give a " self-consistent " occupation probability $\nu(k)$ and energy $\epsilon(k)$. Both $\nu(k)$ and $\epsilon(k)$ depend on the temperature, the latter because the electron $k$ is under the influence of the most probable distribution of the electrons, which depends on the temperature. Bardeen,[16] to whom this consideration is due, has solved (6) and (6a) approximately and obtained also the level density for $k = k_m$. This decreased logarithmically with decreasing temperature, i.e. with decreasing total energy of the system. The temperature dependent part of the total energy is affected also: it is smaller than Sommerfeld's value by the factor

$$1 + \frac{e^2 m}{\pi h^2 k_m}\ln\frac{2e^2 k_m}{\pi kT} = 1 + 2 \cdot 15 r_s - (r_s/6)\ln Tr_s \;. \qquad . \quad (7)$$

Equation (7) is not exact but is sufficiently accurate for temperatures between 1 and 100 degrees absolute. It bears on the comparison of observed and calculated electronic specific heats, given by Jones and Mott.[17] The disagreement between calculated and observed values

[14] H. A. Bethe, *Physic. Rev.*, 1936, **50**, 332 ; J. Bardeen, *ibid.*, 1937, **51**, 799 ; H. A. Bethe, *Rev. Mod. Physics*, 1937, 9, 69, § 53. There is a misprint in equ. (284), the exponent of $U^a{}^{+1}/\alpha$ should be $1/(2n)$.
[15] P. A. M. Dirac, *Proc. Camb. Phil. Soc.*, 1930, **26**, 376 ; L. Brillouin, ref. 3, also J. Bardeen, *Physic. Rev.*, 1936, 49, 653.
[16] J. Bardeen, *ibid.*, 1936, **50**, 1098, and unpublished work.
[17] H. Jones and N. F. Mott, *Proc. Roy. Soc., A*, 1937, 162, 49.

becomes several times greater and the excellent agreement for Cu seems to be lost. However, as mentioned in the first section, the correlations neglected in Hartree's picture tend to increase the level density and the specific heat again and thus restore the agreement between calculated and observed values. They will be considered next.

### 4.

We shall now investigate the way in which an improvement of the wave functions beyond the Hartree-Fock scheme will affect the density of energy levels. I can see but two ways, both involving very rough approximations, of obtaining information on this question.

(1) Apply the method used previously for the lowest state of the metal [8] to excited states. The wave function obtained in this way is certainly only very approximate. In particular, even for the ground state (for which the solution of the Hartree-Fock equation belongs to a single multiplet system (*viz.*, the singlet system) the improved wave function is a linear combination of wave functions for several multiplet systems (all of which are low, however). This is true also for the excited states. On the other hand, the wave function behaves qualitatively in the expected way, *i.e.*, the electrons with antiparallel spins try to keep apart also. This method of calculation is quite laborious, however, and I shall not go into it here.

(2) The improvement of the wave function is made in two steps : (*a*) Build wave functions out of the Hartree-Fock solutions,[18] which have the proper symmetry, *i.e.* belong to a definite multiplet system. (*b*) Combine many wave functions with the same symmetry to obtain a better wave function. Clearly the first step is much easier than the second. It may be remarked that the first step is unnecessary for the lowest state ; here the wave function has already a singlet character. Nevertheless, the second step gives a very great improvement in the energy.

Let us consider, first, a state [19] in which $2s$ orbits, all close to the top of the Fermi distribution, are singly occupied. We have, on the whole, $2^{2s}$ wave functions. From these, at least all with an equal number of spins in both directions (total $Z$ component of spin zero) were given the same energy in the previous calculation. Of these $\binom{2s}{s}$ functions

$$\binom{2s}{s-S} - \binom{2s}{s-S+1} \sim \frac{2^{2s}}{\sqrt{\pi}} \frac{2S+1}{s^{3/2}} e^{-S^2/s} . \qquad . \quad (8)$$

have the multiplicity $S$. Those with a higher multiplicity will be pulled in general lower, since their energy must be equal to the energy of a state in which $\frac{1}{2}n + S$ spins are directed one way, $\frac{1}{2}n - S$ the other way. For these partially magnetic states, a correction such as that in (2*b*) must be introduced, which amounts to $- 0.916. 8S^2/(9nr_s)$. Thus the $\binom{2s}{s}$ states which had the same energy in the Hartree-Fock approximation spread out. This will doubtless increase the density of levels everywhere, since the Hartree-Fock density is an extremely rapidly

[18] *I.e.*, out of Slater determinants of plane waves.
[19] Similar considerations have been carried out for nuclei by J. Bardeen and E. Feenberg. I am much indebted to these authors for a discussion of their results.

increasing function and every energy value will gain more from its higher energy neighbour than it loses.  However, the average spread is only

$$\frac{8}{9} \frac{0\cdot916}{nr_s} \frac{\int S^2 (2S+1) e^{-S^2/\sigma} \, dS}{\int (2S+1) e^{-S^2/\sigma} \, dS} = \frac{8}{9} \frac{0\cdot916}{r_s} \frac{s}{n} \quad . \quad . \quad (9)$$

which is essentially independent of the number of particles.  Thus, for macroscopic bodies, the first step in the improvement of the wave function does not increase the level density sufficiently to change the specific heat.[20]  This is because we have assumed a spreading only as *required* by the symmetry of electron exchange, which materially affects states with large $S$ only.  These however are scarce.

If we improve our wave functions further by forming linear combinations of wave functions with the same symmetry, we shall expect a spreading of levels of all multiplicities.  Especially, the interaction between levels of different configurations now becomes important—the more important, the larger the crystal.  At present, there is no well-developed method for taking all these interactions into account.  Their effect can, however, be readily visualised in the case of a very strongly expanded lattice, and may be expected to be in the same direction in the case of ordinary lattice constants.

Of course, if one could really expand a lattice, the band structure of the electronic states would become more and more pronounced and our original assumption become quite unjustified.  For an orientation about the problem in hand it is equally useful, however, to replace the positive ions by a uniform space charge first (which, in many cases, does not alter matters appreciably) and then to expand this space charge.  Thus our expanded lattice will not be a lattice at all and will in many ways have very little similarity to an actual metal.  On the other hand, it can readily be treated mathematically and it may also be expected that the effect of the above-mentioned improvement of the wave functions will have a similar effect on the level density in this model as in actual crystals.

For a very large lattice constant, the kinetic energy of the electrons will play only a secondary rôle.  As a first approximation, the electrons will assume definite positions such that their potential energy will be a minimum.  This is the case for a regular lattice arrangement, probably the body-centered lattice.[21]  As a second approximation, the electrons will move a little around their equilibrium positions.

We are thus led to consider a model corresponding to an " inverted alkali metal."  The positive charges are uniformly distributed (because we know that the effect of the ions can be replaced in alkalis by such a charge distribution), while the negative electrons are localised because of their repulsive action.  For the potential on one electron, only the effect of the positive space charge of the $s$-sphere surrounding this electron has to be taken into account, the effect of the rest of positive space charges being very nearly cancelled by the other electrons.[22]  The electron thus moves in the field

$$(- 3/r_s + r^2/r_s^3) \, Ry \quad . \quad . \quad . \quad (10)$$

($r$ denotes the distance from the centre of the sphere in Bohr units), *i.e.*,

[20] This calculation assumes $S < s \ll n$, *i.e.*, that we have no ferromagnetism. The situation is the same, however, for the ferromagnetic state also.
[21] K. Fuchs, *Proc. Roy. Soc.*, *A*, 1935, 151, 585.
[22] E. Wigner and F. Seitz, *Physic. Rev.*, 1934, 46, 509 : K. Fuchs.[21]

684    ENERGY LEVELS OF ELECTRONS IN METALS

it forms an oscillator with the frequency $Ry/\pi h r_s^{3/2}$. The energy in the lowest state is

$$(- 3/r_s + 3/r_s^{3/2})Ry \qquad . \qquad . \qquad . \qquad . \qquad (11)$$

while the wave function becomes

$$\psi = (\alpha/\pi)^{3/4} e^{-\alpha r^2/2}; \quad \alpha = 1/r_s^{3/2}. . \qquad . \qquad . \qquad (12)$$

Of course, this picture is only consistent if the wave function (12) has practically dropped to zero at the surface of the $s$-sphere, *i.e.*, if $\alpha r_s^2 = r_s^{1/2} > 2$. A somewhat more detailed investigation of the conditions of validity of our picture shows that it must be quite accurate for $r_s \sim 10$ and that even for $r_s \sim 6$ it is probably more nearly true than we are used to think. This is partly borne out by the magnitude of the correlation energy, and means that the electron gas under this density is already rather far from the ideal condition and more nearly a liquid with high compressibility than a gas.[23]

In the treatment just given we have neglected the effect of the displacement of one electron on the field acting in another $s$-sphere. This treatment corresponds to Einstein's treatment of the solid body, while the cross terms between displacements are taken into account only in Debye's, Born-Karman's, and Blackman's theories. This would yield a somewhat lower [24] value for the zero-point energy, and a somewhat different expression for the wave function, since in (12) the normal co-ordinates would appear in the exponent, with coefficients $\alpha$ proportional to the corresponding frequencies. However, the frequency spectrum of our model is rather different from the frequency spectrum of ordinary solids.

It may be interesting to note that for $r_s \sim 6$ (11) gives a value for the correlation energy which is one-half of that of the previous calculation [25] although the wave function (12) is certainly still very inaccurate for this radius.

Of course, the foregoing considerations are valid only for very large $r_s$, especially only if the $\psi$ of (12) is already practically zero for $r = r_s$. However, the (positive) energy change necessary to accomplish this boundary correction goes exponentially to zero with $r_s^{1/2}$. Apart from such a correction we have in (12) in fact $n$ l solutions of Schrödinger's equation, as every electron can be allocated to every lattice point. On account of Pauli's principle, this means that we have, introducing the spin, $2^n$ wave functions with approximately the same energy, giving

$$\binom{n}{\tfrac{1}{2}n - S} - \binom{n}{\tfrac{1}{2}n - S + 1}$$

terms of multiplicity $S$. This shows that for large $r_s$, the energy difference between the ferromagnetic and non-magnetic states decreases

[23] The electron gas differs rather curiously from an ordinary gas in so far as in the latter the potential between the atoms can be neglected if the density is small. In the electron gas, the greater the density, the more perfect the degeneracy and the behaviour becomes ideal for high densities.

[24] One can take this into account by a method due to F. London, *Z. Physik*, 1930, 63, 245, which gives 2·7 instead of the second 3 in (11).

[25] The correlation energy is obtained from (11) by subtracting it from the energy, calculated for the same density, by the Hartree-Fock model. The latter contains three parts : the electrostatic energy $(- 1\cdot2/r_s)$, the zero point energy $2\cdot21/r_s^2$, and the exchange energy $(- 0\cdot916/r_s)$. On account of the somewhat different picture used here, the low density limit of the correlation energy is somewhat different from that obtained in ref. 8. For $r_s = 6$ and $r_s = 8$, (11) gives $0\cdot15/r_s$ and $0\cdot21/r_s$, while the estimates of ref. 8 are $0\cdot32/r_s$ and $0\cdot36/r_s$. For these radii, equation (11) cannot yet be expected to be valid.

E. WIGNER                          685

much more rapidly than appears from $(2b)$, $(1c)$ demonstrating the existence of a substantial correlation energy for the ferromagnetic state also.

One can calculate from the frequency of the electron vibrations, using Debye's theory, their specific heat, and from this the density of energy levels, following Bethe's and Bardeen's procedure.[14]  One obtains for the logarithm of the density

$$\ln \rho(\epsilon) \approx n(\ln 2 + 2 \cdot 2 r_s^{3/4} (\epsilon/n \; Ry)^{3/4}) \qquad . \qquad . \quad (13)$$

where $n$ is the total number of electrons, and the term $n \ln 2$ arises from the above-mentioned $2^n$ fold approximate degeneracy.  Comparing (13)

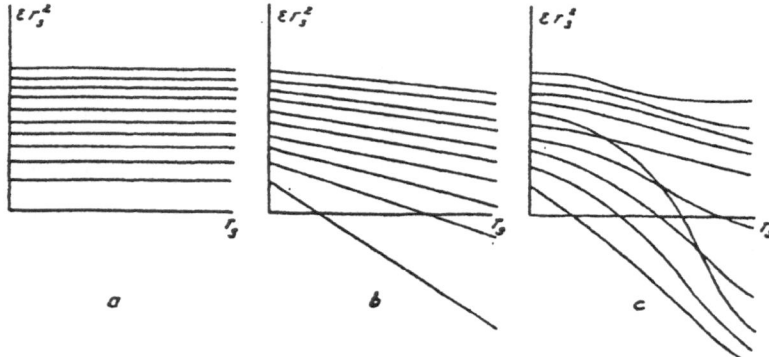

FIG. 2.—Schematic picture of the behaviour of energy levels for changing $r_s$. The density of the lines of the figure corresponds to the logarithm of the energy level density of the whole metal or to the density of one particle levels.  Figure (a) for the simple picture, (b) for the Hartree-Fock picture, (c) for the picture with correlations.

even with (4) shows clearly that, at any rate for large lattice constants, the inclusion of correlations increases the density of levels.  The same can be expected for the real lattice constants also, and the effect of improving the wave functions beyond the Hartree-Fock scheme partly counterbalances the effect of the exchange energy.

Fig. 2 shows diagrammatically the behaviour of $r_s^3$ times the energy, as a function of $r_s$, for the three models.  Although, of course, the irregularities indicated in the third picture come in only at those values of $r_s$ for which there is no practical example, it would not be surprising to find some anomalies in the electronic specific heat of alkalis at low temperatures.  However, the effect of the exchange must still be expected to predominate in these cases.

I am very much indebted to the Wisconsin Alumni Research Foundation for the support which they have given my research.

# Qualitative Theory of the Cohesion in Metals

## E. P. Wigner

Proceedings of the International Conference of Theoretical Physics, Kyoto and Tokyo, September, 1953. Published by the Organizing Committee International Conference of Theoretical Physics Science Council of Japan, Ueno Park, Tokyo 1954, pp. 649–663. Reprinted in expanded form in collaboration with F. Seitz in: Solid State Physics I. Academic Press 1955, pp. 97–126

## 1. Introduction.

The subject on which I wish to speak has occupied me on and off for many years. It consists only of qualitative considerations and much of what follows will not be new to those who are actively working in the field. The occasion to present these ideas was chosen not only because of the kind invitation extended by the Science Council of Japan but also because some recent quantitative work has given some strength to the rather crude ideas which will be described below. I am referring particularly to Wohlfarth's and Fletcher's work[1] on the structure of the Brillouin zones of the $3d$ shell and to the doctoral dissertation of F. Stern in

Princeton.

Before starting, let me make a few comments on the purpose of this and similar work. If I had a great calculating machine, I would perhaps apply it to the Schrödinger equation of each metal and obtain its cohesive energy, its lattice constant, etc.. It is not clear, however, that I would gain a great deal by this. Presumably, all the results would agree with the experimental values and not much would be learned from the calculation. What would be preferable, instead, would be a vivid picture of the behavior of the wave function, a simple description of the essence of metallic cohesion and an understanding of the causes of its variation from element to element. Hence the task which is before us is not a purely scientific one; it is partly pedagogic. Nor can its solution be unique: the same wave function can be depicted in a variety of ways (just as a cubic close-packed lattice can), the same energy can be decomposed in a variety of ways into different basic constituents. Hence the value of any contribution to the problem will depend on the taste of the reader. In fact, from the point of view of the present article, the principal purpose of accurate calculations is to assure us that nothing truly significant has been overlooked.

Recent work, in particular Korringa's[2] and the one which was just reported[3], gave the theory of the cohesion of alkalis a simplicity and roundedness which will be difficult to surpass. More important, the latter one derives the properties of the alkali atoms, which are used in the calculation of the cohesive energy, directly from the observed spectra. One is tempted to ask, therefore, what inferences one can draw from the spectra of more complicated atoms on their cohesive properties. This, in a sense, is the thesis of my talk, but before broaching my subject proper, I wish to recall a few figures from the theory of the alkali metals. Let us take therefore a glance at the calculations.

The cohesive energy is repesented as the sum of three quantities. The first I shall call boundary correction. It gives the difference between the binding energy of the valence electron of the free atom and the bottom of the Brillouin zone in the crystal. It is the direct result of the circumstance that the radial derivative of the crystal wave function vanishes at the cell boundary while it must be, at the same distance, negative in the atom. Hence its name. Its magnitude is about 72 calories/mol in the case of sodium. The second part of the cohesive energy is the kinetic energy of the free electrons (Fermi energy). This amounts to about 43 calories in Na. The last part are the contributions from the various "holes": the circumstance that the atomic electron is always in the field of a charged ion, i.e. in a field of positive potential, has its counterpart in the metal in the "holes" of electron density which the electrons carry around themselves and which result from statistical correlation between the positions of the electrons. The net total of this last contribution to the cohesive energy is very small and it became customary to neglect it. Thus the cohesive energy of Na is $72-43 = 29$ calories/mole. The 1st Figure illustrates this situation. It shows, in particular, that the energy of the electron at the top of the Fermi distribution is just about equal to the energy of the electron in the free atom. Furthermore, if one assumes that the energy is a quadratic function of the momentum throughout the whole Brillouin zone, the kinetic energy of the electrons would just compensate the boundary correction and no binding would result. These results, obtained in the first place for sodium, can be considerably generalised and form the basis of the second of the following observations.

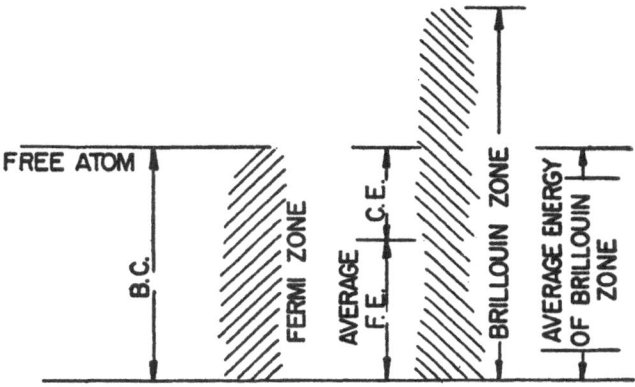

Fig. 1. Comparison of the various quantities contributing to the cohesive energy (C.E.) of Na. The boundary correction (B.C.) determines the bottom of the Fermi zone. The electrons at the top of the Fermi zone (the occupied part of the Brillouin zone) have just about the same energy as the electron in the atom. The average energy of the whole Brillouin zone would also be very nearly equal to the energy of the atomic electrons. Hence, no binding would result if the whole zone were full.

## 2. Preliminary Observations.

1. The multitude of allotropic transformations with very low energy of transition shows that the cohesive energy is reasonably independent of the crystal structure. The following considerations could be based, therefore, on any one of the common crystal structures, and the body centered one was chosen because the Brillouin zones show the greatest symmetry for this structure. In most cases, I have convinced myself that the results would be quite similar for a face-centered lattice.

2. If one builds up the wave functions of a Brillouin zone by properly superposing atomic wave functions centered at the different lattice points

$$\varphi_{\vec{k}} = \sum_r e^{i\vec{k}\cdot\vec{a}_r}\psi(\vec{r}-\vec{a}_r) \tag{1}$$

($a_r$ are the different sites of atoms, $\psi$ the atomic wave function) the average energy of all the electrons of the Brillouin zone is greater than the energy of the electron in the free atom. This is what the right side of Figure 1 illustrates. It is also contained in Heitler and London's valence theory's statement that closed shells repel each other. It means, in our language, that the boundary correction is smaller in the approximation of (1) than the average Fermi energy of the zone and

it is not difficult to see this also directly.     Actually, of course, the implication of
this observation is well borne out in the case of the noble gases; the origin of
the cohesion of these elements cannot be described in the individual particle picture. It is not borne out at all in the earth alkali elements all of which have in
fact larger cohesive energy than the alkalis — a behavior opposite to the one
which would be expected from Figure 1. It will be necessary, therefore, to modify
the underlying picture rather drastically.

This modification shall consist, from a mathematical point of view, in considering
boundary correction and Fermi energy together as a boundary correction. This
has been done, of course, before, in particular lately by W. Kohn[4]. The principal
point which I shall stress is, from the physical point of view, the effect of the
excited states of the atom on the band structure of the metal.

3. The third observation relates to the extent to which the effect of the ion's
charge on the atomic electron is compensated by the effect of the various " holes "
in the metal. We have seen that this compensation is nearly perfect in one electron
systems. To be sure, the absolute value of the potential, where it is particularly
large, is larger in the atom. On the other hand, the electron carries along its
particular potential hole wherever it travels in the metal and this hole, though
more diffuse than the positive charge of the ion, is right there where it is needed.

The compensation of the holes cannot be expected to prevail in the case of
elements the structure of which is more complicated, particularly if the normal
state's configuration gives rise to levels of different multiplicities (such as $p^2$). In
these cases there is a " hole " around each electron already in the atom which
follows it around. The correlation between the positions of electrons is weak if
there are only two valence electrons but becomes stronger as the electron number
increases*. It manifests itself also in the difference between the normal state of

---

* There is no correlation, in the Hartree-Fock picture, between the distances of the various
electrons from the nucleus. As long as the number of electrons does not exceed $2l+1$, the
average probability, for all states of highest multiplicity, for an angle $\vartheta$ between the directions
of any two electrons is proportional to $[(2l+1)/2l][1-P_l(\cos\vartheta)^2]$. In this case, all the electron
spins are parallel. If the number $n$ of the electrons exceeds $2l+1$, their spins cannot be parallel
any more. In this case, the probability for an angle $\vartheta$, averaged over all states of highest
multiplicity, becomes the average of the above expression with the weight $\frac{1}{2}(2l+1)(2l+$
$\frac{1}{2}(n-2l-1)(n-2l-2)$ and of 1 with the weight $(2l+1)(n-2l-1)$. If the position of one electron
is fixed, the distribution of the other electrons has a quadrupole moment in the direction fixed
by the first electron. This does not correspond to a very close correlation.

In particular, in the first case mentioned above, the probability vanishes for two electrons
lying on the same straight line through the nucleus no matter whether they lie on the same
or on opposite sides thereof. Correlations which go beyond those given by the Hartree-Fock
picture will surely improve on this situation and furnish, in addition to the quadrupole moment
mentioned above, a dipole moment in the distribution of the electrons around a given electron.
However, the total effect of correlations in the atom will remain lower than it is in the metal
because the correlations in the metal give not a quadrupole or dipole moment, but a net positive charge around any given electron. This positive charge is being balanced, in the present
method of bookkeeping, against the equally large net positive charge of the ion plus the effect
of the quadrupole and other moments described above. (Note added December 1953)

the atom, and the energy calculated in the orthodox Hartree fashion; it is given by the Slater integral. This results in a more effective hole in the charge distribution around the atomic electron than around the metal electron. It has two effects. First, the binding due to the $p$ and $d$ electrons decreases as their number increases, on this account as well as on account of their increasing kinetic energy. Second, the picture based on the independent particle model with traveling (Bloch) waves becomes inaccurate toward the middle of the $d$ and even the $p$ shell. It is indeed very unlikely that, in the case of half filled shells, the electrons in the metal should not be able to correlate their positions better than the free electron picture would indicate and that they could not approximate the correlations present in the atom. I shall, indeed, show wave functions which do this.

There is, of course, no effect of the type here considered in the $s$ shell and the effect virtually disappears also for closed shells. The third observation is summarized in Table 1.

Table 1.  Hole Energies  (cf. Appendix)

| $s$ | $-0.6e^2/r_s$ | $+0.458e^2/r_s+C'$ | $= -0.142e^2/r_s+C'$ |
|---|---|---|---|
| $s^2$ | $-2\times0.6e^2/r_s-C$ | $+2\times0.458\times2^{1/6}e^2/r_s+2C'$ | $=  0.045e^2/r_s+2C'-C$ |
| $\begin{matrix}p^n\\(n\leq3)\end{matrix}$ | $-0.6ne^2/r_s-Jn(n-1)/2+0.458n\ n'^{1/6}e^2/r_s+J'n^2/6+nC'$ | | |
| $p$ | $-0.28e^2/r_s+J/6+C'$ | | |
| $p^2$ | $-0.39e^2/r_s-J/3+2C'$ | | |
| $p^3$ | $-0.42e^2/r_s-3J/2+3C'$ | | |
| $p^4$ | $-0.40e^2/r_s-J/3$ | | |
| $p^5$ | $-0.25e^2/r_s+J/6$ | | |
| $p^6$ | $-0.046e^2/r_s$ | | |

4. The last observation only records the fact that in very heavy atoms the magnetic interaction becomes comparable with the cohesive energy. This makes it desirable to restrict our considerations to the cases in which the magnetic interaction is not too large.

### 3. The Structure and Width of $p$ and $d$ bands.

The second of the above observations shows that, as long as the wave function can be assumed to have the form (1), the bands are centered at the value of the atomic level. The binding will be, in this approximation, largest for the first electron of each shell, will be about zero for the electron in the middle of the shell and will be negative for the electrons in the second half of the shell. This means that the cohesive energy should be largest at the middle of the shell and decrease toward both sides of it.

Figure 2 gives the cohesive energies of all solids as function of the charge number $Z$. One sees that the behavior as described above obtains at least approximately in the $3p$ and $4p$ shells — the behavior of the cohesive energy as function of $Z$ is indeed very similar in these shells. The situation seems somewhat more involved in the $3d$ and $4d$ shells which are also rather similar to each other. They differ from the above expectations principally in giving a non negligible binding for the closed shells — Zn and Cd.   The cohesion in solid carbon is too large to conform with the above picture and this is one of the facts for which I shall be unable to tender any explanation.   It was mentioned before that the binding energy of the earth alkalis is larger than that of the alkalis which directly contradicts the above picture.   Nevertheless, the degree of agreement makes it desirable to arrive at some more detailed picture for bands of electrons with wave functions of the form (1).

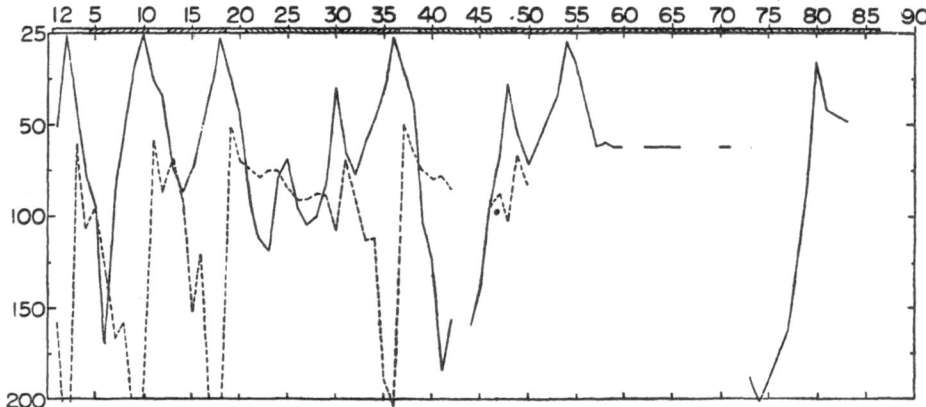

Fig. 2.   The full line gives the cohesive energy of the condensed state in kcal as function of $Z$.   The electronic shells are marked on the horizontal axis.   The broken line gives one half of the first ionization potential for comparison.

This can be done, of course, on the basis of detailed calculations and such calculations have been carried out repeatedly, most recently for the $d$ shell by Wohlfarth and Fletcher[1].   Let me consider, therefore, the $p$ band of a body centered lattice.   The three wave functions have the same energy for $k = 0$, i.e. at the center of the Brillouin zone and Figure 3a shows the sign of the wave function at the boundaries of the cells in the (110) plane.   One sees that on most boundaries the wave function originally has opposite signs on the two sides.   This means that the wave function will actually vanish at these boundaries, i.e. the boundary correction is positive in this case and the $k=(0,0,0)$ point lies at the top rather than the bottom of the Brillouin zone.   This illustrates, by the way, why it was desirable to consider Fermi energy and Boundary correction together: they have, at least in this case, little meaning separately.

Let us consider next a point at the surface of the Brillouin zone and choose, first, the point at which the three branches again coincide (Figure 3b).   One sees that the fit is, in this case, by no means perfect (as it is for $s$ electrons) but nevertheless sufficiently good to give a very substantial negative boundary correction. This point therefore lies near the bottom of the band.

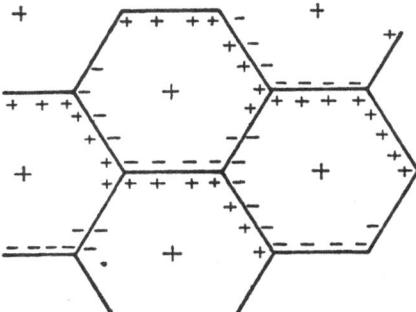

Fig. 3a. Section through a body centered lattice parallel to the (110) plane. The crosses at the centers of the hexagons give the positions of the nuclei. The hexagons are the intersections with the polyhedra surrounding the atoms. The + and − signs at the hexagons give the sign of a $p$ wave function of an atom which is supposed to be at the center of the hexagon. In this Figure, the $p$ wave functions of all the atoms have the same phase corresponding to a wave vector $k = (0, 0, 0)$. Over most of the boundary, the sign of the atomic wave functions is opposite on the two sides of the boundary. This indicates poor fit and high energy of the $k = (0, 0, 0)$ state. The Figure gives only one $p$ wave function. However, the situation is the same with the other two.

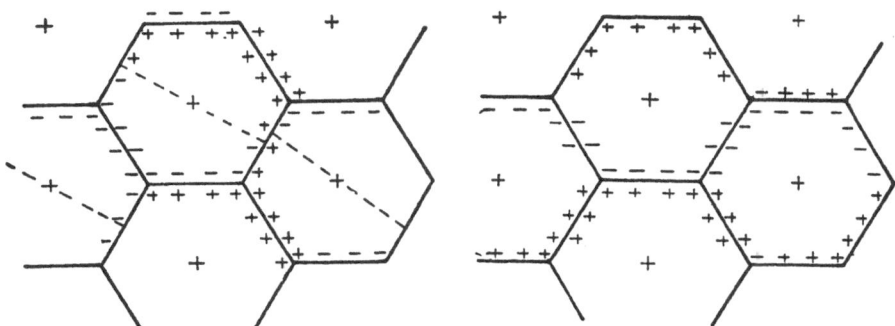

Figs. 3b, 3b′. The arrangement is the same as in Figure 3a, except that the wave vector is $k = (0, 0, 2\pi/a)$ where $a$ is the lattice constant. It corresponds to a surface point of the Brillouin zone. For this $k$, the atomic wave functions of nearest neighbors have opposite sign at corresponding points. The three $p$ functions give two energy values. Figure 3b gives the sign of the atomic wave functions at the boundary of the polyhedron for the wave function of one of these energy values, 3b′ for one of the wave functions of the other. The fit is rather good in both cases, indicating low energy values.

Figure 3c shows another point at which three zones coincide — its boundary correction is very small so that it must lie near the center of the band. Figure 3d, finally, illustrates the sign of the wave functions for a wave vector for which the three bands are separate. The fit is very poor for two of the wave functions in this case but is the best of all for one of them. Hence, the three energy surfaces

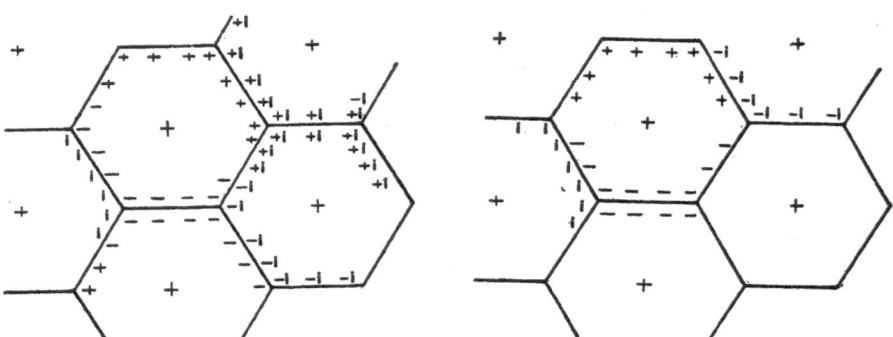

Figs. 3c, 3c′.  Same as Figures 3b, 3b′ for the wave vector $k = (\pi/a, \pi/a, \pi/a)$, another point on the surface of the Brillouin zone.  The two atomic wave functions differ, over much of the boundary, by a factor $i$.  Hence the boundary correction is very small, the energy is close to the middle of the Brillouin zone.

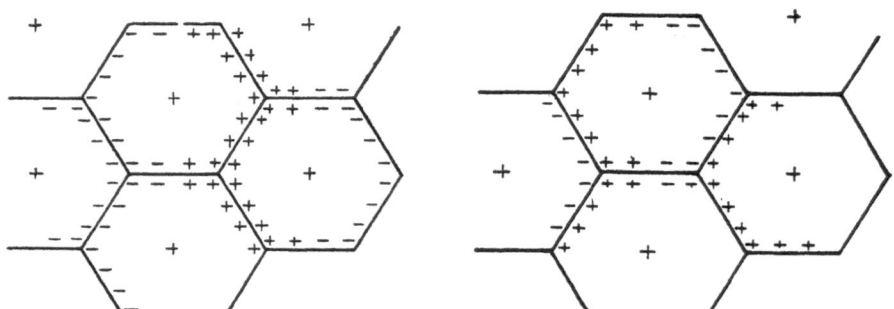

Figs. 3d, 3d′  Same as other Figures 3, for the wave vector $k = (\pi/a, \pi/a, 0)$, also a point on the surface of the Brillouin zone.  The atomic wave functions fit together at the boundaries of the polyhedra very well in Fig. 3d.  The energy which corresponds to the wave function represented is presumably the lowest in the zone.

are widely separated for this wave vector ($k = \pi/a$, $\pi/a$, 0), one of them forming, presumably, the bottom of the zone.

The picture which we obtain for a $p$ band which is well separated from all other bands and has wave functions of the form (1) is, therefore, as follows.  There is no wave function with entirely perfect fit at the boundaries and the bottom of the band is, therefore, depressed less than in the $s$ zone.  The total width of the band is slightly smaller than that of an $s$ band of similar energy and it contains three times more states than the latter.  The density of states seems particularly large near the bottom of the band because the energy varies rather little between $k = (0, 0, 2\pi/a)$ and $k = (\pi/a, \pi/a, 0)$ on one of the energy surfaces.

The Brillouin zone due to $d$ electrons — actually, there are two of them — can be discussed in the same way and the results are similar to those of Fletcher with-

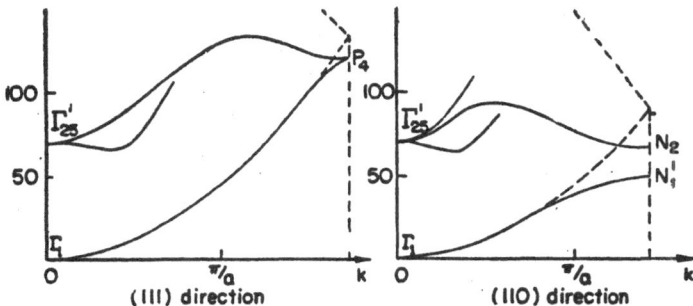

Fig. 4. Schematic diagram giving the energy (in kcal/mole) as function of the wave number $k$ in the (111) and (110) directions. The broken line gives the energy of a free electron, the full lines indicate how the existence of low lying excited states of the atom might influence the energy. The point $\Gamma_1$ represents an $s$ orbit, the points $P_4$ and $N_1'$ $p$ orbits, the points $\Gamma_{25}'$ and $N_3$ represent $d$ orbits. The wave functions for $P_4$ and $N_1'$ are illustrated in Figures 3c, 3c', and 3d. The $p$ and $d$ states increase the number of available states below about 100 Cal and thus decrease the Fermi energy. The line connecting $\Gamma_{25}'$ and $P_4$ has a twofold degeneracy. The symbols $\Gamma$, $P$, etc. refer to the symmetry symbols given by Bouckaert, Smoluchowski and Wigner (Phys. Rev. **50** (1938), 58).

out giving, of course, the quantitative results he obtained. The fit of the wave function which gives the lowest state is very good in this case. The total width of the two zones is, under the assumption of well separated zones, very nearly the same as in the case of an $s$ zone. Naturally, the width of the zone will be much diminished if the lattice constant is kept larger than would correspond to the minimum of the energy of the $d$ electrons alone. The electrons of the $s$ shell which preceeds the $d$ shell exert such an influence.

Much of the qualitative behavior of the energy surfaces can be seen in the following way. Let us insert for the atomic wave function $\psi$ in (1) the product of $xy+xz+yz$ and a radial function. The $\psi$ obtained in this way will be much larger in the (111) direction, i.e. in the direction of two of the nearest neighbors, than in the direction of any other neighbor. The bulk of the boundary correction will arise, therefore, from the part of the surface which lies between the original atom and these two neighbors. The boundary correction will have the largest negative value if the $e^{i\vec{k}\cdot\vec{a}_r}$ of (1) corresponding to the $a_r$ of these two neighbors, is equal to 1, i.e. if the wave vector $k$ is perpendicular to the $(1,1,1)$ direction. However, the correction will depend rather little on the component of $k$ perpendicular to this direction so that one has a rather flat portion of the energy surface (the minimum apparently does not lie at $k=(0,0,0)$ as one might first suspect).

One can further augment this lack of sensitivity of the energy on the components of $k$ perpendicular to the direction of the maximum of $k$, by skipping, in (1), some of the lattice sites $a_r$. The resulting wave will not be a Bloch wave any more but, around certain $k$, the energy dependence on the perpendicular component of $k$ will be very slight indeed. One will be tempted, therefore,

to attribute the same spin direction to all these states thus recovering some
of the correlations between electronic positions which lower the atomic energy
and which are responsible for the $J$'s of Table 1. There will be, of course, other
wave functions of the form (1) which contain those $a_r$ which were skipped in the
wave function considered before. If the spin of the corresponding states have
opposite directions, no non-orthogonality corrections between the two types of wave
functions will apply. It is conceivable that the state of the whole system based
on such non-Bloch type wave functions is lower than the one based on Bloch type
wave functions and one is also reminded faintly, of Zener's theory of ferromagnetism[5]. It seems to me, however, that it would be premature to follow these
ideas further at this time particularly since the variety of the non-Bloch type wave
functions is so large.

Returning to our principal line of discussion, it should be evident from what
we know about the widths of Brillouin zones, in particular from Slater's explanation
of ferromagnetism, that the picture of well separated $p$ and $d$ zones is far from
accurate. Rather, the $p$ and $d$ zones strongly overlap at least with the preceding
$s$ zones. This causes them — particularly the $d$ zones — to be much narrower because
$s$ electrons of energy equal to that of $d$ electrons, have a much longer tail than
the $d$ electrons. The structure of the region between the $s$ and the $p$ and $d$ zones
might be similar to that given by Figure 4. This gives the energies of an $s-p$
and $s-d$ pair of zones with wave vectors along the (111) and (110) directions.

It will be attempted now to interpret the cohesive energies, i.e. the contents of
Figure 2, in terms of the above observations.

### 4. Interpretation of cohesive energies.

No attempt will be made here to reproduce the interpretation of the cohesion
of alkalis beyond the remark that the decrease of the cohesive energy, in the Li-
Na-K-Rb-Cs series, is principally due to the decreased boundary correction

$$\frac{\hbar^2}{2m}\int_s \psi^* \mathrm{grad}_n \psi \, ds \, . \tag{2}$$

This in turn is due to the increasing size and decreasing ionization energy along
the above series.

The first real problem arises in connection with the cohesive energy of the
earth alkalis which, as mentioned before, would vanish, were the $s$ zone well
separated from the following zones. Instead, the cohesion is invariably larger in
the earth alkalis than in the preceding alkali. One can understand this in the
following way[6].

The effective charge and hence the ionization energy is larger in the earth
alkalis than in the alkalis. This increases the boundary correction (2). This, in
itself, does not explain the cohesion because the Fermi energy should increase
simultaneously. However, the Fermi energy increased in this way becomes much
larger than the excitation energy of the next $p$ electron (or $d$ electron) even if

this is augmented by the difference of the boundary corrections of the $p$ (or $d$) and $s$ electrons. Because of the higher density of states in the $p$ (and $d$) zones, as compared with the $s$ zones, this puts an effective stop to the further increase of the Fermi energy and causes the boundary correction to prevail.

In agreement with the above picture is the interpretation, given by Mott and Skinner, to Skinner's experiments on the absorption of soft X rays. They find that the top of the $s$ zone has $s$ character in alkalis but $p$ character in earth alkalis. It is also interesting to follow, from this point of view, the ratio of the cohesive energies of alkalis and earth alkalis. This is greatest in the Be-Li pair in which the ratio of $p$ excitation energy to boundary correction is smallest. In the Mg-Na pair the difference between $p$ and $s$ levels is relatively larger and, hence, the ratio of cohesive energies is much smaller. The cohesive energy ratio is again large for the Ca-K pair either because the $p$ level is very close in this case or because the $d$ level, with its relatively large boundary correction, is also quite close. The same is true for the Sr-Rb pair although, in this case, the $d$ zone is more likely to be involved than the $p$ zone.

The weakness of the above description of the cohesion in the earth alkalis is that it presents no general reason for this cohesion exceeding that of the preceding alkali. On the other hand, it is well to remember that the hole correction (Table 1) is positive in this and only this case.

Let us go over now to the consideration of the $p$ shell. Figure 4 already indicates that if the $p$ shell follows an $s$ shell, part of its states are filled already when the $s^2$ configuration is reached. The same is true, presumably, if the $s^2$ shell is followed by a $d$ shell. Furthermore, Table 1 shows that the hole correction becomes strongly negative when the $p^3$ configuration is reached. One will expect, therefore, that the third $p$ electron will already decrease the cohesive energy — a phenomenon which the naive theory would expect to happen at the fourth $p$ electron. In fact, in every case excepting that of C, maximum binding is reached with the $p^2$ configuration. As was mentioned before, no explanation is being offered for the exceptionally large cohesive energy of C.

Considerations quite similar to the above ones apply to the $d$ shell. Maximum cohesion is reached in both $d$ shells at the $d^5$ configuration. However, in contrast to the $p$ shell, addition of further $d$ electrons leaves the solid metallic. As a result, one would expect, on the basis of Table 1, a certain recovery of the cohesion just beyond the half shell, i.e. at $d^6$. In fact, a secondary maximum appears in both $d$ shells at this point which may be an indication herefor.

Because of the low $d$-$p$ energy difference in the atom, the $p$ state may play some role already while the $d$ shell is being filled. If so, one sees little evidence therefor : the maxima and minima come at the points where one would expect them without the interference of the $p$ shell. Only the non negligible, though small cohesive energy of the $d^{10}$ configurations (Zn and Cd) gives evidence of an interaction between $p$ and $d$ states.

5. **Summary.**

There are four points which emerge from the above analysis and which can be stated very simply,

1. Configurations of the atom which have a low excitation energy increase the cohesion in the solid.

2. The various zones, $s, p, d$ are rarely separate. Usually, they overlap, hang together at boundary points of the Brillouin zone, etc.

3. As a result of this circumstance, the Fermi energy is relatively smaller and less important in transition metals and even in the earth alkalis than in alkalis.

4. On the contrary, the hole corrections cannot be expected to cancel in general as closely as they do in alkalis and they are responsible at least for some of the fluctuations of the cohesive energy.

### Appendix (added December 1953)

### CALCULATION OF THE HOLE ENERGIES FOR TABLE 1.

The calculation leading to Table 1 closely follows that presented by F. Seitz and the present writer in an earlier article[7]. In first approximation, the potential in each cell is assumed to be the same as if the atom, which is at the center of the cell, were the only atom present. The wave functions are supposed to satisfy, on the surface of the cell, the proper metallic boundary conditions. The energy is then calculated for the wave functions so obtained taking into account all interactions, such as the effect of the atoms outside of the cell, the interaction of all the electrons, etc. The term "hole correction" means the difference between the energy evaluated in this way and the energy obtained under the assumptions of the first approximation, i.e. atomic field with metallic boundary conditions. This latter energy includes the zero point energy of the the free electrons so that this quantity does not appear in Table 1. The enumeration of the terms which follow applies to the third line of the Table, leading to the corrections for $n \leq 3 p$ electrons. The other entries in the Table can be obtained in the same way. Further details and a more complete justification of the procedure can be found in the aforementioned article[7].

The electrons move, in the free atom, in the field of an ion. The charge of this ion is neutralized in the metal by the free electrons. Putting it differently: removal of a single electron does not change the potential in the metal and this remains the same as in a neutral atom. The difference between the two potentials is approximated by that of a uniform charge distribution, giving an average potential of $1.2\,e^2/r_s$ and an average potential energy of $1.2\,e^2/r_s$ per electron. The total energy is one half of the product of charge and average potential. This gives the term $-0.6\,ne^2/r_s$.

The atomic energy includes, in the Hartree-Fock approximation, not only the interaction of the charges but also certain exchange integrals which cause the difference between the various terms arising from the same configuration[8]. The origin of these terms are the correlations which are described in footnote on page 652. Strictly speaking, one should compare the energy of the metal with that of the lowest state of the atom. Instead, the average energy of all terms of highest multiplicity is considered. This does not differ too much from the energy of the lowest term (it is, actually, the same in the $p$ shell) and is much easier to calculate. One easily finds that it is equal, per electron pair, to the average value of

$$J_{\mu\nu} = \int \psi_\mu(1)^* \psi_\nu(2)^* \frac{e^2}{r_{12}} \psi_\nu(1) \psi_\mu(2) \, d1 d2 , \qquad (A1)$$

where $\psi_\mu$ and $\psi_\nu$ are two wave functions of the $p$ shell with magnetic quantum numbers $\mu$ and $\nu$. The average is to be taken over all $\mu, \nu$ combination:

$$J = \frac{1}{(2l+1)l} \sum_{\mu<\nu} J_{\mu\nu} . \qquad (A2)$$

This $J$ must be multiplied with the number of pairs of electrons, $n(n-1)/2$, and given the negative sign because it appears in the energy of the atom, rather than the metal. No term similar to (A2) appears in the calculation of the cohesion of alkali metals[7] because $n=1$ in that case.

The above contributions, which decrease the binding energy, are compensated at least partially by the result of the correlations in the metal. We remain again within the Hartree-Fock picture, i.e. calculate only the exchange terms. When calculating these, we have to remember that $p$ electrons give rise to three energy surfaces, i.e. three zones, in momentum space. These energy surfaces coincide at certain points and even lines but are distinct for a general wave vector. It appears convenient, therefore, to distinguish between exchange integrals involving wave functions of the same zone and exchange integrals involving different zones. The first one can be particularly easily calculated if one assumes that one has to deal with plane waves and becomes

$$\frac{0.458 \, e^2}{r_s} n'^{1/3} \qquad (A3)$$

per electron if there are altogether $n'$ electrons per atom in the zone[9]. This $n'$ was 1 for alkalis because there was only one zone and one electron per atom; in the present case $n'=n/3$ because the $n$ electrons are distributed to three zones. (A3) gives the contribution per electron, hence the factor $n$ in the term $0.458 \, nn'^{1/3} e^2/r_s$ of Table 1.

The exchange integral between electrons of different zones will be denoted by $J'$. If the wave functions are given by expressions of the form (1), $J'$ can be calculated along the lines given in reference[7] and is then approximately equal to $J$. It appears only between electrons with parallel spin and for these only if they are in different zones. Half of the electrons have their spin parallel to a given electron and $2/3$ of these are not in the zone of the given electron. This would give a factor $\frac{1}{2}(\frac{2}{3})n$ but since one counts every interaction twice in this way, the factor of $J'$ is $n/6$ per electron or $n^2/6$ altogether. No expression similar to $J'n^2/6$ applies in the case of alkalis because there is only one zone in this case. Hence, this term has no counterpart in earlier calculations[7].

The quantities $C$ and $C'$ stand for the energy due to correlations which go beyond the Hartree-Fock picture and which will not be described in further detail. $C$ refers to the atom, $C'$ for the metal; the latter is probably much greater than the former.

The fourth and successive lines of the Table give the values of the third line for $n=1,2,3$ and of a similar expression for $n=4,5,6$ which applies when the $p$ shell is more than half full. To simplify the final expression, $J=J'$ was set in the last six lines and $C$ and $C'$ omitted in the last three. Even though the above consideration must be regarded as extremely crude and sketchy, it is believed that a large part of the errors committed would compensate.

The Table shows clearly how the "hole correction" increases up to the middle of the shell and then decreases again. At the middle of the shell, it contributes a substantial correction which decreases the cohesive energy calculated in the first approximation. On the other hand, the hole correction practically vanishes at the end of the shell.

## REFERENCES

1) G. C. Fletcher and E. P. Wohlfarth: Phil. Mag. 42 (1951), 106. G. C. Fletcher: Proc. Phys. Soc. London A 65 (1952), 192.
2) J. Korringa: Physica 13 (1947), 392.
3) Cf. also T. S. Kuhn and J. H. Van Vleck: Phys. Rev. 79 (1950), 382.
4) W. Kohn: Phys. Rev. 87 (1952), 472.
5) A similar proposal was made independently by J. C. Slater in a personal communication to P. Löwdin.
6) A. quantitative discussion in the case of Be was given by C. C. Herring and A. G. Hill: Phys. Rev. 58 (1940), 132. For further references concerning applications of the cellular method, cf. J. C. Slater, Technical Report 4, Solid State and Molecular Theory Group, Mass. Inst. of Techn. July 1953.
7) E. P. Wigner and F. Seitz: Phys. Rev. 46 (1934), 509. Also F. Seitz: Modern Theory of Solids, McGraw-Hill Book Co., New York 1940, Chapters IX and X.
8) J. C. Slater: Phys. Rev. 34 (1929), 1293, also E. U. Condon and G. Shortley: The Theory of Atomic Spectra, University Press, Cambridge 1953, Chapter VII.
9) This expression is contained already in F. Bloch's article Zeits. f. Physik 57 (1929), 545. Cf. also F. Seitz, reference 7, p. 240 ff where the connection between correlations and exchange energy is brought out particularly clearly.

## DISCUSSION

**F. Seitz:**  What is the broken line?

**Wigner:**  The broken line in Figure 2 represents the ionization energy. It is given in the Figure because it illustrates an interesting relation: the ionization energy, as function of the atomic member $Z$, is usually convex where the cohesive energy is concave, and vice versa. Such a behavior could be expected from the picture on which I was reporting.

**N. F. Mott:**  What is $J$ which you just mentioned?

**Wigner:**  $J$ is the exchange energy, the Slater exchange integral between two $p$ functions of the same atom. It is principally responsible for the energy difference between the triplet and singlet states of the two electron system and, more generally, for the energy difference between states of higher and lower multiplicities in many electron systems. If we denote the three $p$ functions by $\psi_+, \psi_0, \psi_-$, it is given analytically by

$$J = \int \psi_+(1)^* \psi_0(2)^* \frac{e^2}{r_{12}} \psi_0(1) \psi_+(2)\, d1\, d2.$$

More generally and more accurately, I mean by $J$ the average value

$$J = \frac{2}{(2l+1)2l} \sum_{\mu < \nu} \int \psi_\mu^*(1) \psi_\nu^*(2) \frac{e^2}{r_{12}} \psi_\nu(1) \psi_\mu(2)\, d1\, d2,$$

where $\mu$ and $\nu$ run from $-l$ to $l$ and denote the magnetic quantum number, i.e. the projection of the orbital angular momentum into a fixed direction.

**J. H. Van Vleck:**  What's the significance of $J'$?

**Wigner:**  $J'$ is the same quantity in the metal that was denoted by $J$ in the atom. Its value may be different from $J$ because it is calculated with metallic rather than atomic wave functions. However, the difference may not be too great and the last part of the Table assumes that $J$ and $J'$ are equal.

**Van Vleck:**  What are $C$ and $C'$?

**Wigner:**  These are the correlation energies in the atom and in the metal, i.e. they stand for an estimate of that part of the energy which is neglected in the Hartree-Fock picture. It appears that $C'$ is considerably greater than $C$ and the last part of the Table omits $C$.

# Bibliography of Papers
# on Solid State Physics

## Papers Reprinted in Volume IV

### 1933

(With F. Seitz) On the Constitution of Metallic Sodium. Physical Review *43*, 804–810 (1933)

### 1934

(With F. Seitz) On the Constitution of Metallic Sodium. II. Physical Review *46*, 509–524 (1934)

On the Interaction of Electrons in Metals. Physical Review *46*, 1002–1011 (1934)

### 1935

(With J. Bardeen) Theory of the Work Functions of Monovalent Metals. Physical Review *48*, 84–87 (1935)

(With H. B. Huntington) On the Possibility of a Metallic Modification of Hydrogen. Journal of Chemical Physics *3*, 764–770 (1935)

### 1936

On the Structure of Solid Bodies. Scientific Monthly *42*, 40–46 (1936)

(With L. P. Bouckaert and R. Smoluchowski) Theory of the Brillouin Zones and Symmetry Properties of Wave Functions in Crystals. Physical Review *50*, 58–67 (1936). Reprinted in: Group Theory and Solid State Physics, Part I (P.H. Meijer, ed.). Gordon and Breach, New York 1964, pp. 1–10

On the Constant $A$ in Richardson's Equation. Physical Review *49*, 696–700 (1936)

### 1938

Effects of the Electron Interaction on the Energy Levels of Electrons in Metals. Transactions of the Faraday Society *34*, 678–685 (1938)

### 1954

Qualitative Theory of the Cohesion in Metals. Proceedings of the International Conference of Theoretical Physics, Kyoto and Tokyo, September, 1953. Published by the Organizing Committee International Conference of Theoretical Physics Science Council of Japan, Ueno Park, Tokyo 1954, pp. 649–663. Reprinted in expanded form in collaboration with F. Seitz in: Solid State Physics I, Academic Press, New York 1955, pp. 97–126

## Papers Related But Reprinted in Other Volumes of The Collected Works

(With F. Seitz) The Effect of Radiation on Solids. Scientific American *195*, no. 2, 76–84 (1956); Vol. V

## Papers Related But Not Reprinted in The Collected Works

Appendix C of M. L. Goldberger: Density of States in Periodic Systems. In: Problems of Theoretical Physics. A Memorial Volume to Igor E. Tomm. Nauka, Moscow 1972, pp. 396–416. This appendix describes the calculation of an infinite sum by a technique suggested to the author by Professor E. P. Wigner

# Springer
# and the
# environment

At Springer we firmly believe that an international science publisher has a special obligation to the environment, and our corporate policies consistently reflect this conviction.
We also expect our business partners – paper mills, printers, packaging manufacturers, etc. – to commit themselves to using materials and production processes that do not harm the environment. The paper in this book is made from low- or no-chlorine pulp and is acid free, in conformance with international standards for paper permanency.